Nuc n
Light Wa ctors
Severe Ac nt Phenomenology

Edited by

Bal Raj Sehgal

Support provided by the SARNET (Severe Accident Network)
in the Framework Programmes of Research
of the European Commission

ELSEVIER

AMSTERDAM • BOSTON • HEIDELBERG • LONDON
NEW YORK • OXFORD • PARIS • SAN DIEGO
SAN FRANCISCO • SINGAPORE • SYDNEY • TOKYO

Academic Press is an imprint of Elsevier

Academic Press is an imprint of Elsevier
225 Wyman Street, Waltham, MA 02451, USA
525 B Street, Suite 1900, San Diego, CA 92101-4495, USA
The Boulevard, Langford Lane, Kidlington, Oxford OX5 1GB, UK
Radarweg 29, PO Box 211, 1000 AE Amsterdam, The Netherlands

First edition 2012

Library of Congress Cataloging-in-Publication Data
Sehgal, Bal Raj.
Nuclear safety in light water reactors / Bal Raj Sehgal. – 1st ed.
p. cm.
Includes bibliographical references and index.
ISBN 978-0-12-388446-6 (alk. paper)
1. Light water reactors–Risk assessment. 2. Light water reactors–Safety measures.
I. Title.
TK9203.L45S44 2012
621.48′35–dc23
2011039984

British Library Cataloguing in Publication Data
A catalogue record for this book is available from the British Library

ISBN: 978-0-12-388446-6

Contents

Preface

The idea of documenting the advances achieved in the field of light water reactor severe accident safety research in the form of a book originated during the development of the research program for the Severe Accident Research Network of Excellence (SARNET), coordinated by IRSN France, that was started in March 2004, under the auspices of the Sixth Framework Research Program of the European Commission. After some discussion, it was decided that the book should be a textbook for students and young researchers in the field and not a handbook. It was also decided that the book should not be a compendium of all the research in the field, but rather should be written to impart understanding and knowledge about the complex physics of severe accidents. The physics of severe accidents involves several disciplines, including probability theory, neutron physics, thermal hydraulics, high-temperature material science, chemistry, and structural mechanics. Thus, imparting understanding and knowledge of severe accidents is not a simple task. It is nonetheless an important task in reactor safety, since a severe accident is the only source of risk to the public from an operating light water reactor power plant. Preventing and managing the consequences of a severe accident, which is the main goal of severe accident research, contributes greatly to reducing the public risk of nuclear power.

The importance of severe accident research was recognized by the EURATOM Part of the Framework Research Programs of the European Commission. The Framework Program No. 4 was totally focused on severe accident (SA) research. The European Commission is continuing its support of SA research and, through its support of SARNET, is encouraging the focus of national research efforts on an integrated European program of this research. This book, *Nuclear Safety in Light Water Reactors: Severe Accident Phenomenology,* describes the results obtained from these research programs, conducted over the last 15 years in Europe. The book also contains the results of SA research conducted over the years in the United States, Japan, Korea, Russia, and other countries. The research conducted in the United States, in particular, was the forerunner of the research that has been conducted in Europe. The knowledge gained in the U.S. research forms a very important base for SA research conducted throughout the world. The book, therefore, documents the data, phenomenology, and methodology developed for the description of severe accidents in all countries.

This book is a joint effort since it is a product of SARNET, a network. It was conceived as a pedagogical effort, however, written by acknowledged experts in the different areas of the SA field. It was coordinated, chapter by chapter, by different experts and finally compiled and edited by the undersigned.

Chapter 1 of the book provides a historical review of the whole field of reactor safety, with short introductions on the various severe accident phenomenological topics. It also attempts to provide an insight into the logic of advancements in rector safety since the birth of nuclear energy. The most recent tragic event at Fukushima is also briefly described in Chapter 1, based on the information gained as of the end of June, 10, 2011. This description may need corrections as more complete information about these severe accidents at Fukushima emerges in time. Chapter 1 also briefly deals with the advances in mitigating severe accidents achieved in the designs of some of the new (GEN III+) LWRs.

The remaining contents of the book follow the severe accident scenario, starting with the loss of cooling of the decay-heated core. The resulting core heat-up, core degradation, hydrogen production, core melting, accumulation of melt in lower head, failure of the lower head, hydrogen combustion, steam explosion, molten corium– concrete interactions (MCCI), fission product release, transport in the primary system, containment, and the like, are the subjects treated in the book.

A question that arose early in deliberations on the contents of the book concerned the maturity of the SA field. We believe that nearly all the knowledge gained through SA research, and described in the book, is mature enough and will stand the test of time. It should also be stated that severe accident is still an active field of research, and some issues remain open; notably, the knowledge base still has uncertainties, and more research (experimentation, modeling, validation, etc.) is needed, and indeed is being performed. Even for these issues, however, a sufficient knowledge base has already been acquired to document these areas of uncertainties.

The book is a product of the efforts and dedication of the contributors listed with the text in the book. Very substantial efforts were involved in editing the various contributions and organizing the book. We hope that it will be a worthwhile book for the education of nuclear engineering students and a reference text for the young researchers who want to work, or are already engaged, in the field of LWR severe accident safety.

Last, but not the least, we the contributors (authors) and the Editor wish to acknowledge with thanks, the steadfast and continous support of M. Michel Hugon, the EU Program Manager for the SARNET Network of Excellence. The Editor, also wishes to acknowledge with thanks, the able assistance of Dr. Van Dorsselaere in the “final-edit” of the book.

Bal Raj Sehgal

Contributors

Hans Alsmeyer, Karlsruher Institut für Technologie (KIT), retired, Technologies Institut für Kern- und Energietechnik (IKET), Hermann-von-Helmholtz-Platz 1, D-76344 Eggenstein-Leopoldshafen, Germany

Eberhardt Alstadt, Helmholtz-Zentrum Dresden-Rossendorf (HZDR), Institute of Safety Research, P.O.B. 51 01 19, D-01314 Dresden, Germany

Marc Barrachin, Institut de Radioprotection et de Sûreté Nucléaire (IRSN), Cadarache, BP3, 13115, Saint-Paul-lez-Durance, France

Ahmed Bentaib, IRSN, BP 17, F- 92262, Fontenay aux Roses, FRANCE

Jonathan Birchley, Paul Scherrer Institut (PSI), CH- 5232, Villigen PSI, SWITZERLAND

Manfred Burger, Universität Stuttgart (IKE), Pfaffenwaldring 31, D-70569 Stuttgart, Germany

Cataldo Caroli, IRSN, BP 17, F- 92262, Fontenay aux Roses, FRANCE

Michel Cranga, IRSN, Cadarache, BP 3, 13115, St Paul-lez-Durance, France

Truc Nam Dinh, Royal Institute of Technology (KTH), Alba Nova, 10691 Stockholm, Sweden and Idaho National Laboratory, Idaho falls, USA

Jean-Pierre Van Dorsselaere, SARNET coordinator, IRSN, Cadarache, BP3, 13115 Saint-Paul-lez-Durance, France

Gerard Ducros, Commissariat à l'Energie Atomique et aux Energies Alternatives (CEA), Cadarache Bât.315, F- 13108, St Paul-lez-durance, FRANCE

Peter Eisert, Gesellschaft fûr Anlagen Und Reaktorsicherheit mbH (GRS), Schwertnergasse 7, D- 50461, Köln, Germany

Florian Fichot, IRSN, Cadarache, BP 3, 13115 St-Paul-lez-Durance, France

Manfred Fischer, AREVA GmbH, D-91050, Erlangen, Germany

Jerzy J. Foit, KIT, Technologies Institut für Kern- und Energietechnik (IKET), Hermann-von-Helmholtz-Platz 1, D-76344 Eggenstein-Leopoldshafen, Germany

Salih Guentay, PSI, CH- 5232, Villigen PSI, Switzerland

Tim Haste, IRSN, Cadarache, BP 3, 13115 Saint-Paul-Lez-Durance Cedex, France

Luisen Herranz, Centro de Investigationes Energeticas Medio Ambientales Y Tecnologicas (CIEMAT), Avda. Complutense, 22, E- 28040, Madrid, SPAIN

Zoltan Hozer, KFKI Atomic Energy Research Institute (AEKI), Konkoly Thege ut 29-33, 49, H- 1525, BUDAPEST, HUNGARY

Christos Housiadas, "Demokritos" National Center for Scientific Research, PO Box 60228, 15310 Agia Paraskevi Attikis, Greece

Ivan Ivanov, Technical University of Sofia (TUS), 8, Kl. Ohridski Blvd., Block 12, Office 12440, 1797, Sofia, BULGARIA

Thomas Jordan, KIT, Technologies Institut für Kern- und Energietechnik (IKET), Hermann-von-Helmholtz-Platz 1, D-76344 Eggenstein-Leopoldshafen, Germany

Christophe Journeau, CEA, Cadarache, F- 13108, St Paul-lez-Durance, FRANCE

Martin Kissane, IRSN, Cadarache, BP 3, F- 13115, St-Paul-lez-durance, FRANCE

Ivo Kljenak, Jozef Stefan Institute (JSI), Reactor Engineering Division, Jamova 39, Ljubljana, Slovenia

Marco Koch, Ruhr-Universität Bochum, LEE, IB 4/127, Universitätsstr.150, D- 44801, Bochum, GERMANY

Jean-Sylvestre Lamy, Electricité de France (EDF), R&D, SINETICS, 1 avenue du Général de Gaulle, 92140 Clamart, France

Jean-Claude Latche, IRSN, Cadarache, BP3, F- 13115, St-Paul-lez-Durance, France

Terttaliisa Lind, PSI, CH- 5232, Villigen PSI, SWITZERLAND

Weimin Ma, Royal Institute of Technology (KTH), Nuclear Power Safety, Alba Nova, 10691 Stockholm, Sweden

Daniel Magallon, retired Scientist, Joint Research Centre (JRC), European Commission, Institute for Energy and Transport (IET), P.O. Box 2, NL-1755 ZG Petten, The Netherlands, seconded to CEA/Cadarache

Mani Mathews, Atomic Energy of Canada Limited (AECL), Chalk River Laboratories, Chalk River, Ontario, K0J 1J0, Canada

Leonhard Meyer, KIT, Technologies Institut für Kern- und Energietechnik (IKET), Hermann-von-Helmholtz-Platz 1, D-76344 Eggenstein-Leopoldshafen, Germany

Christoph Mueller, GRS mbH, retired, Forschunginstitute, 85748, Garching b. München, Germany

Pascal Piluso, CEA, Cadarache, 13108 Saint-Paul-lez Durance, France

Horst Schnadt, Tüv Rheinland Industrie Service GmbH, retired, Cologne, Germany

Andreas Schumm, EDF - R&D, SINETICS, 1 avenue du Général de Gaulle, 92140 Clamart, France

Bal Raj Sehgal, Emeritus Professor, Royal Institute of Technology (KTH), Nuclear Power Safety, Alba Nova, 10691 Stockholm, Sweden

Jean-Marie Seiler, CEA, 17, av. des Martyrs, F- 38054, Grenoble, France

Juergen Sievers, GRS mbH, Schwertnergasse 1, 50667 Köln, Germany

Claus Spengler, GRS, Schwertnergasse 7, D- 50461, Köln, Germany

Bertrand Spindler, CEA, 17 av. des Martyrs, F- 38054, Grenoble, France

Bruno Tourniaire, CEA, 17 av. des Martyrs, F- 38054, Grenoble, France

Klaus Trambauer, GRS mbH, retired, Forschunginstitute, 85748, Garching b. München, Germany

George Vayssier, Consultant, Nuclear Services Corporation (NSC), Kamperweg 1, 4417 PC Hansweert, The Netherlands

Jean-Michel Veteau, CEA, retired, 17 av. des Martyrs, F- 38054, Grenoble, France

Chapter 1

Light Water Reactor Safety: A Historical Review

Bal Raj Sehgal

Chapter Outline

Nuclear Safety in Light Water Reactors. DOI: 10.1016/B978-0-12-388446-6.00001-0

1.1. INTRODUCTION

The light water reactor (LWR) safety that we are concerned with in this book is basically about estimating the risks posed by an individual or a population of nuclear power plants (NPPs) to the public at large and the efforts to reduce these risks. The public of most concern is that which resides in the vicinity of a nuclear power plant but also at other locations, which could be affected by an accident in a NPP located anywhere.

The basic goal of safety is to ensure that a LWR will not contribute significantly to individual and societal health risks. This basic goal translates to the prevention of the release of radioactivity into the environment from the NPP. A complementary aim is to prevent damage to the plant and to protect the personnel at the plant from injury or death in an accident.

Since LWR safety aims to protect the public at large, it is heavily regulated. Each nuclear power country (and even some without NPPs) has regulatory commissions (bodies) that regulate every aspect of a NPP from design and construction to operation and any modifications. They require very extensive analyses, documentation, and quality control. The reactor safety design has to follow definite rules and regulations. Some of these requirements will be described in this chapter.

The reactor performance, on the other hand, is concerned with long-term steady-state operations, since most LWR plants are base-loaded and strive to operate at full power, without interruption, between scheduled outages for maintenance. Reactor performance is also concerned with efficiency, the capacity factor, fuel cycle costs, maintenance costs, and the radiation dose to the operating staff. Thus, it is not regulated. However, it has been found that a well-running LWR plant is, generally, a safer plant with a much lower frequency of incidents, which, generally, are the precursors to more serious events.

1.2. THE EARLY DAYS

The nuclear era started with the natural uranium-graphite pile built by Fermi and his associates at Stagg Field at the University of Chicago [1]. It did not involve light water as a coolant since only natural uranium was available and criticality could be achieved only with graphite or heavy water. The safety concepts developed there, however, were adopted by the LWR plants that developed several yeas later. Enrico Fermi and his associates recognized that:

- Nuclear fission reactions, which are the basis of nuclear power, emit high levels of radioactivity and thus could be a health hazard to any person exposed to it. This implied shielding, containment, and remote siting.
- The safe operation of the reactor (or pile) would require protective and control measures, as evidenced by the provision of a control rod in the pile that Fermi and his associates built.

Shielding and remote siting were required for the plants that were built for the production of plutonium in the United States and other countries during the years before and after World War II. Remote siting of these plants not only protected the public but also maintained the secrecy surrounding the production of nuclear weapons for a number of years.

The containment aspect of protecting the public from a nuclear accident was not considered or employed for the plants generating plutonium. Those were the years of above-ground nuclear weapons tests, which in any case were releasing considerable amounts of radioactive fission products in the atmosphere. Fortunately, there were no reported accidents of any great significance in the plutonium production plants in either the United States or other Western countries.

Leak-tight containment as a safety system for a civilian NPP was not long in coming. It was proposed in 1947 [2] for a sodium-cooled fast reactor that was the focus of the power reactor development by the U.S. Department of Energy at that time. Later, the LWR plant developers adopted leak-tight containment for their plants.

1.3. THE DEVELOPMENT OF CIVILIAN LWRs

The LWR development started as a military program in the United States and stemmed from the initiative of Admiral Hyman Rickover, who is considered the father of the U.S. nuclear navy [3]. His team conceived the pressurized water-cooled reactor (PWR) as the NPP for submarine propulsion, since a sodium-cooled fast reactor, the focus of the U.S. national program, was considered unsuitable for a nuclear submarine submerged in water. Admiral Rickover obtained the necessary funding and the considerable intellectual resources needed to generate the extraordinarily rapid development of the PWR plant for the U.S. submarine fleet.

President Dwight Eisenhower issued the call for Atoms for Peace in 1954 [3], which became the signal for adapting military developments for civilian purposes. The construction of the Shippingport PWR, Pennsylvania, USA. [3], which was completed in 1957, provided the prototype for NPPs, generating a reasonable amount of electrical power for public consumption. EBR-1, a fast reactor, was the first nuclear reactor in the United States to generate electrical power. However, the quantity generated was insufficient to transmit for public consumption.

The development of the other civilian water-cooled nuclear power reactor, that is, the boiling water reactor (BWR), was started almost in parallel with that of the PWR and the construction of the Shippingport PWR plant. The BWR development was spearheaded by the General Electric (GE) Company, a private enterprise, which, in fact, invested its own funds to develop the BWR as a commercial NPP. In this effort they were aided by national laboratories in the United States—for example, Argonne National Laboratory, which built a 5-MW BWR system [3], and the Idaho Laboratories, where experiments were performed [4] to demonstrate the stability and safety of the BWR system. The first prototype commercial BWR plant was designed and built, as a dual-cycle (i.e., it had a steam generator for the steam that went to the turbine) plant, already in 1960 by GE.

In the United States, the first truly commercial NPP was the Yankee-Rowe plant, a PWR, which was also built in 1960. This plant was conceived as a commercial venture and was specifically commissioned by a utility company that supplied electricity to the public. The Yankee-Rowe plant was constructed with a leak-tight containment, and it was approved for commercial operation by the regulatory authorities in the United States Atomic Energy Commission (AEC). The plant designers at that time did not realize that their decision to employ a leak-tight, pressure-bearing containment was their most important safety decision.

The civilian use of nuclear energy was very popular with the public during 1960s. Claims were being made that nuclear energy could provide unlimited and cheap electric power: too cheap to meter. Projections were being made of constructing hundreds (or even a thousand) power reactors in the United States alone. Some proposals involved the location of plants very close to the cities to provide generation sources near large consumption centers, in order to become more economic in the total cost of the electricity to the consumers. The 1970s saw a large number of orders placed by U.S. utility companies with U.S. vendors. The most prominent of these companies were: Westinghouse for the PWR plants, since it was the vendor for the naval PWRs; General Electric for the BWRs, since they were the developers of this reactor type; and Babcock

Wilcox for PWRs, since they had extensive experience in the construction of conventional power-generation equipment. Later Combustion engineering, another vendor of conventional power-generation equipment, joined their ranks and constructed PWRs. There was a quick scale-up of reactor power from 300 to 600 to 1000 MWe. LWR plant construction programs were also started in Germany, France, Japan, Sweden, the Soviet Union, among other countries. Great Britain chose to construct gas-cooled NPPs.

1.4. EARLY SAFETY ASSESMENTS

The dangers of a major accident in a NPP were first assessed in the early 1950s. The 1955 Geneva Conference, which was the first gathering of nuclear reactor scientists from both the East and the West, provided the first estimates of the possible hazards of an accident in a LWR. The paper presented by the U.S. investigators [5] estimated 200 to 500 fatalities and 3,000 to 5,000 high exposures to radioactivity. Even before these results were fully digested, the study WASH-740 [6] was published. This study was performed with the stated purpose of estimating the consequences of a "worst-case" nuclear accident, in order to provide data for NPP insurance legislation. The authors of WASH-740 assumed that 50% of the radioactive inventory of a 500-MWe reactor would be released in the atmosphere, with the most unfavorable weather conditions prevailing. They estimated that up to 34,000 fatalities, 43,000 injuries, and contamination of 240,000 square kilometres of land could occur. A probability estimate of ~10% was quoted pertaining to these consequences. The WASH-740 authors stated categorically that the estimates of deaths, injuries and land contamination were highly conservative because of the assumptions made in deriving these estimates.

1.5. THE SITING CRITERIA

The consequences and risks estimated in the WASH-740 study hastened the enactment of the site criteria by the Atomic Energy Commission (AEC). These criteria, recommended in the report TID-14844 [7] [8], published in 1959, are based on the recognition that the pressure-bearing containment provided on the projected LWRs would most probably survive in a hypothetical accident and that radiation would be released into the atmosphere only through the leakage of fission products deposited in the containment (the Source-Term). The study assumed that a certain fraction of the gaseous and solid fission products, contained in an irradiated core, would be deposited in the containment as it was done for the WASH-740 study. The difference between these two studies, of course, is that the WASH-740 study stated that the radiation would be released immediately to the atmosphere, while the authors of TID-14844 stated that it would be released only at the leak rate of the containment, that is, 0.1%/ per day.

The TID-14844 required the establishment of exclusion and of a low-population zone (LPZ) on whose boundaries the limits of exposure that could be suffered by the thyroid and the whole body of a person, situated there, were prescribed.

The TID-14844 recommendations were considered in the site criteria enacted by the AEC in 1962. Those criteria provided for the minimum distance that a NPP should be situated away from a low population center as a function of its thermal (or electric) power capacity. These criteria were part of the first regulatory action recognizing the potential of using a nuclear reactor to generate electricity if sited correctly.

1.5.1. Assumptions and Requirements of TID-14844 and 10 CFR 100

The authors of the TID 14844 and the code of federal regulations that resulted—10 CFR 100—made the following main assumptions about the source term, that is, the fission products released into the containment during the nuclear accident:

- 100% of the noble gas inventory in the core.
- 50% of the halogen inventory.
- 1% of the solid fission products.
- 50% of the released halogens to remain available for further release from the containment; spray, wash-down features, and filtering devices could provide additional reduction. (However, these were not credited.)
- Containment leak rate of 0.1% per day.

The radioactive fission product transport in the atmosphere was assumed to be under the following conditions:

- Atmospheric dispersion under inversion-type conditions; no shift in wind direction for the duration of the leakage.
- No ground deposition of particulates.

The dose limits to the population provided in these documents were as follows:

- For 2 hours' exposure at the boundary of the exclusion area, maximum whole body dose of 25 rem and thyroid dose of 300 rem.
- For 30 days, or infinite exposure, at the outer boundary of the LPZ, the maximum whole- body dose of 25 rem and thyroid dose of 300 rem.

The definitions of the areas around the site are also provided in 10 CFR 100, as follows (see Figure 1.1):

- The exclusion area is the fenced area around the plant where public is normally not allowed
- The low-population zone is the area around the exclusion area whose extent is determined by the dose rates established above

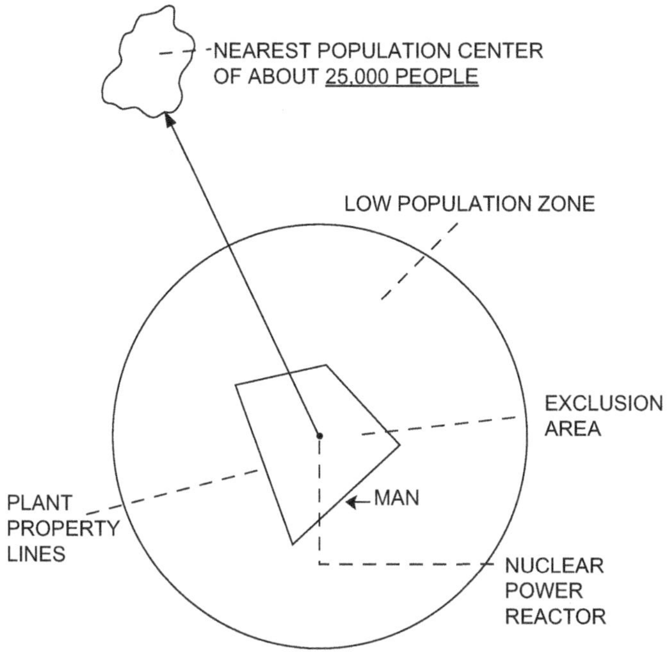

FIGURE 1.1 Part 100 CFR distance requirements (typical plant).

- The nearest population center containing about 25,000 residents should be away from the reactor by at least 1.33 times the distance from the reactor to the outer boundary of the LPZ.

The other requirements on siting involved the following:

- Regulations on land use for plant and transmission lines.
- Regulations for water use and the temperature of the water body to which the plant discharges water.
- The access and corridors for evacuation of the residents of at least the LPZ.
- Other environmental regulations.
- Attitudes of the local population and government bodies.

Figures 1.2 and 1.3, taken from TID-14844, show the distances needed for the exclusion area, the LPZ, and the population center as a function of the thermal power of the LWR to be sited at a particular location.

1.6. SAFETY PHILOSOPHY

Before a fleet of nuclear power reactors was to be constructed for the commercial market, it was important that a philosophy for the safety design and a safety design basis be developed [8]. This philosophy was developed as the

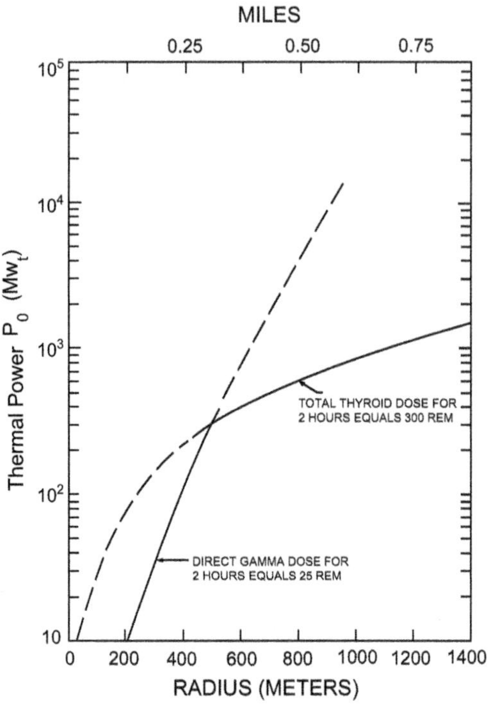

FIGURE 1.2 Determination of the exclusion radius.

orders for the reactors started coming into the United States and the AEC's regulatory role started to function in earnest.

1.6.1. The Defense-in-Depth Approach

The defense-in-depth approach to the safety design followed intuitively from the configuration of a LWR, which provides three important physical barriers to the release of the fission products to the environment—that is, the clad on the fuel element, where the fission products are generated; the reactor vessel, which contains all the fuel elements forming a reactor core; and the leak-tight containment, which is supposed to keep any fission products inside the containment from escaping to the environment. Assuring the integrity of each of these physical barriers in any accident scenario becomes the defense-in-depth approach against the release of radioactivity to the public environment.

In practical design aspects, the defense-in-depth approach for safety design was refined as the following set of preventive measures:

- Perform careful reactor design, reactor construction, and reactor operations so that malfunctions, which could lead to major accidents, will be highly improbable.

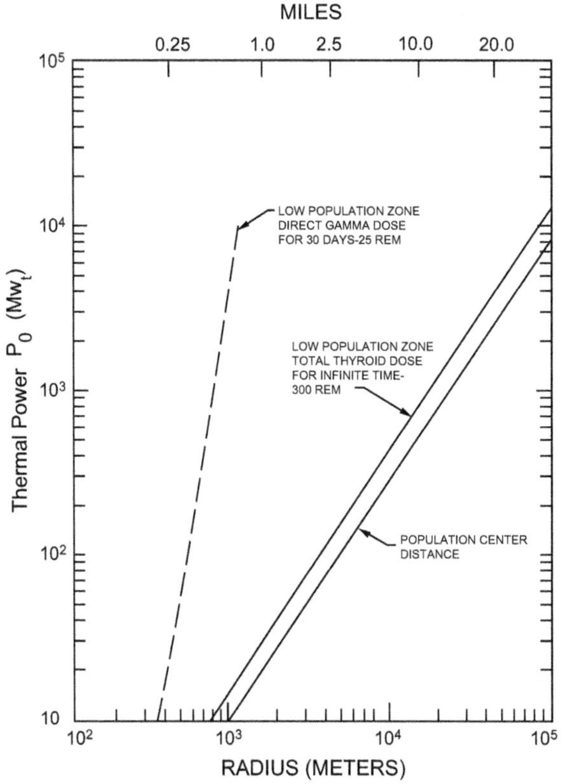

FIGURE 1.3 Determination of the population radii.

- Provide systems and equipment, which would prevent such malfunctions, as do occur, from turning into major accidents. Examples are: Scram systems to shut down the fission reactions in the core and leak-before-break detection equipment to anticipate serious loss of coolant from the reactor primary system.
- Provide systems to reduce and limit the consequences of the postulated major accidents, for example, the emergency core cooling systems (ECCS).

There are at least three echelons for the defense-in-depth approach. The first echelon provides accident prevention through sound design, which:

- can be built and operated with very stringent quality standards,
- provides a high degree of freedom from faults and errors,
- provides high tolerance for malfunctions should they occur,
- employs tested components and materials,
- employs considerable redundancy in instrumentation, control, and mitigation systems.

The second echelon of the defense-in-depth approach assumes that there will be human or equipment failure. It provides detection and protection systems to maintain safe operation or shut the NPP down safely when incidents occur due to human or equipment failure. Examples of the detection and protection systems are as follows:

- Sensitive detection systems to warn of incipient failure of fuel cladding or the coolant systems,
- Redundant sources of in-plant electricity,
- Systems for automatic shut down on nuclear fission reactions in the core (SCRAM) on signals from the monitoring systems. This is generally achieved through insertion of control rods in the core.

The third echelon of the defense-in-depth approach is to provide additional margins to protect the public should severe failures occur despite the first two echelons. Examples of systems and equipment that provide such additional margins are:

- The steel-lined concrete building containing the whole high-pressure primary system of a LWR. This containment should be constructed with a pressure-bearing and leakage-prevention rating.
- The ECCS to flood the core with water and to keep it covered if the high-pressure coolant of a LWR is lost through a break in the piping of the primary system; and residual heat removal systems.

This defense-in-depth approach is followed all over the world. It is quite comprehensive and has served the nuclear enterprise well over the plus 50 years that the commercial plants have been in operation. There has not been a single catastrophic break in the large pipes of the primary systems in the LWRs installed so far.

1.7. SAFETY DESIGN BASIS

A basis for the design of the safety systems had to be provided to the LWR designers. In the United States, this was provided by the AEC, the precursor to the U.S. Nuclear Regulatory Commission (NRC), which was formed a few years later (1974) and was established as an independent civilian agency charged specifically to regulate the fast developing nuclear power industry in the United States. The NRC has other functions besides the regulation of the nuclear power industry, ranging from radiation sources used in medical profession to waste management.

The safety design basis selected for the LWRs was, and remains, the large-break loss of coolant accident (LOCA). This is the two-sided, guillotine break of the largest pipe in the primary system; that is, the coolant discharge from such a break is supposed to occur from both sides of the break. The large LOCA is considered to be an enveloping accident that removes water from the primary system at the largest rate. The consequence of a large LOCA is the uncovering of the reactor core in a very short time (~30 seconds). This in turn requires a supply of water to the reactor at a commensurate rapid rate to fill the vessel and submerge the core in water, before the decay heat raises the temperature of the Zircalloy clad above the threshold temperature for the exothermic Zircalloy–steam oxidation reaction, which can lead to clad and fuel melting.

The PWR and BWR have a significantly lower probability of a power increase accident due to reactivity insertion. In these reactor designs, as the core heats up and increases the void fraction due to the boiling of the water coolant, there is a large negative reactivity and power feedback. This shuts down the fission reaction, even when the control rods are not inserted. Thus, the reactivity-induced accident (RIA) was not considered as the safety design basis accident.

Besides the large LOCA as the enveloping accident, other accident and/or transient events were specified. In addition, it was required that the specified events be analyzed and documented for review in Chapter 15 of the Safety Analysis Report (SAR) that each plant owner has to submit before it can be granted a construction or operating license.

Some examples of transients specified for safety design basis are as follows:

- increase in heat removal by the secondary system,
- decrease in heat removal by the secondary system,
- decrease in reactor coolant system flow rate,
- reactivity and power distribution anamolies,
- increase in reactor coolant inventory,
- decrease in reactor coolant inventory,
- radioactive release from a subsystem or component,
- anticipated transients without scram.

These transients were chosen because they affect the state of the reactor and can lead to additional complications in operations. For example, an increase in heat removal by the secondary system in a PWR would lead to low temperature for the primary coolant, which would add reactivity to the core and increase power. A decrease in heat removal in the secondary system would lead to higher pressure in the vessel of a PWR.

A decrease in the core water inventory may be through a small-break LOCA, which complicated the transient experienced by the TMI-2 (as described later) and lead to the accident. The small-break LOCA, in particular, can go on for a considerable time to become a complex transient. The operator response and actions can change the course of the transient to a benign or a more demanding state for the reactor.

Besides Chapter 15 of the SAR, the regulatory authorities require the submittal of comprehensive information on other related topics in the SAR. These include:

- site description,
- functional performance,

- description of all safety systems and the engineered safety features,
- conformance with the general design criteria for the design, construction, and operation of the plant,
- quality assurance program and preoperational testing,
- periodic testing requirements for operations,
- failure mode analysis,
- radiological monitoring and surveillance requirements,
- possible R&D needed to confirm the design chosen.

1.7.1. LOCA and the ECCS Controversies

With the successful operation of the PWR and BWR demonstration reactors at Shippingport and Dresden, respectively, the U.S. electric power utility industry wanted to construct plants with much greater power rating. Meanwhile, the large LOCA was approved as the design basis, which demanded the design and performance of a very robust emergency core cooling system (ECCS). The AEC appointed a task force to study the various ECCS designs submitted by the vendors for the NPPs that the utility companies wanted. The task force was also chartered to evaluate the consequences in the event the ECCS did not function sufficiently well. Their findings [9] that this could lead to core meltdown and possible containment failure posing a great hazard to the public created much uncertainty all around. The AEC responded to this uncertainty by requiring improvements in the new ECCS designs of the vendors, for example, providing greater capacity, redundancy, diversity, and assurance of electrical supply. The other reactors were also required to install or improve their ECCS.

A major recommendation of the ECCS task force to the AEC national research program was to (1) perform experiments to observe the thermal hydraulic processes of the ECCS, (2) obtain data, (3) develop models for predictive analyses, and (4) validate the models against the measured data. The AEC started an experimental program that built a small-scale thermal-hydraulic loop simulating a large-break LOCA with ECC injection. The first experimental results obtained led to uncertainty about the efficacy of the ECCS. It was observed that the injected water bypassed the core and did not reach the hot rods in the electrically heated core. The analysis models at that point in time had not recognized that the steam generated in the hot core would not let the water enter the core due to countercurrent flooding (CCF). It was later observed that the CCF breaks down during an extended ECCS injection, so that water could reach the hot rods. The AEC responded [10] [11] by demanding additional margins in the ECCS calculation models. Quite detailed criteria were issued for the assumptions to be made and the heat transfer correlations to be used in the models for predicting the plant's thermal hydraulic behavior during the large LOCA and the ECCS injection following the large LOCA. A limiting temperature for the Zircaloy cladding of the hot rods was proposed, which was kept below the temperature at which the exothermic Zircaloy-steam reaction accelerates.

The specification of these criteria did not satisfy the critics, and so public hearings on the ECCS performance in LWRs were organized in January 1972.

The ECCS hearings [11] lasted more than 18 months, and the conclusions reached pointed to the inadequacy of the knowledge and understanding of the phenomena defining thermal hydraulic behavior during the large LOCA, which is a very violent event. In addition, it was concluded that the calculational models available at that time could not be easily defended. Another study [12] of large LOCA and ECCS for LWRs was conducted by a group assembled by the American Physical Society. This study also concluded that there was an insufficient knowledge base to make reliable quantitative predictions of the plant behavior and consequences in reactor accidents. This group recommended an intensive research effort for 10 years or more, to acquire sufficient knowledge about the very complex phenomena that prevail during the large LOCA accident. They emphasized the development of validated models, which could be used for LOCA with ECCS for prototypic plants. They also pointed to the need to quantify the margins that may be available in mitigating a large LOCA by the ECCS.

The acrimonious debates during the ECCS hearings, the differences in the opinions of various experts, and the recommendations made by the various independent groups prompted the United States to start an ambitious research program on LOCA and ECCS. Simultaneously a code of Federal Regulation (10CFR 50) [13] was enacted, which had the force of federal law, providing the safety design basis and the general design criteria for the safe operation of a LWR plant. This design basis included a large LOCA and a set of operational transients for which results of analysis had to be submitted. The large LOCA analyses had to be performed on a very conservative basis, with prescribed assumptions and correlations for heat transfer. The clad temperature limit was specified to be 1200°C (2200°F), and the limit on clad oxidation was prescribed to be 17%. Several guideline documents were written, which, for example, provide categories of accidents, classes for various levels of quality control, and so on. As an example, the primary system had to be class 1, which required rigorous quality control and inspections of the materials as well as the manufacturing and welding processes employed. The cornerstone for LWR safety was established as (a) remote siting, (b) prevention of any radioactivity release in the design-basis accidents (DBAs), (c) defense-in-depth, (d) strong containment, and (e) deterministic safety analyses. These remain the basis for the safety design of LWRs; to this day.

Western Europe and Japan watched the developments on LWR safety in the United States. They chose to follow the U.S. rules and regulations for the design, construction, and safe operation of LWR plants. They may have added some more regulations, but they did not subtract any of the important criteria or regulations in 10CFR50 and 10CFR100. These countries also followed the U.S. ECCS Research Program; and they supplemented it by building several experimental facilities of their own.

The large LOCA and the ECCS research conducted in the United States and other countries was both comprehensive and expensive, since several large-scale integral-effect and separate-effect facilities were constructed. The largest of these was the loss of fluid test (LOFT) facility, which employed a nuclear core generating ~55MWth power. The scaling employed in all of these facilities was that the ratio of power/primary system volume was kept equal to the prototypic value from a 1,000 MWe LWR plant. This scaling was found to be appropriate for most of the thermal hydraulic processes that occur during the large-break LOCA and the ECCS injection. Hundreds of large- and small-scale, integral effect, and separate-effect experiments were performed in these facilities to understand the physics of the two-phase thermal hydraulic phenomena occurring and to obtain pertinent data for the validation of computational codes, for example, the RELAP series of codes and the TRAC code, which were developed later on. Many of the separate-effect experiments illuminated the details of the phenomena, which helped in formulating the computational models that were later employed in the integral codes. For example, the reflooding process, being so complex, was modeled with representative models for which insight and data were obtained from the separate-effect experiments.

Most of these experimental facilities were closed down in 1990s. There are, however, a few large-scale facilities left, for example, ROSA in Japan and PKL in Germany, where research on any new issue that may arise in LWR thermal hydraulics and safety would be performed. Presently, it is believed that the codes RELAP-5 and TRACE (successor to TRAC) are able to generate reasonable predictions of the thermal hydraulic behavior of PWRs and BWRs in the large LOCA accident with ECCS injection. These codes without the large LOCA heat transfer assumptions provide best-estimate analysis results for the large LOCA. The operational transients can also be analyzed, since, recently, these codes have incorporated the control systems with their time lags, the secondary systems of PWRs and the actions of the safety and the relief valves.

After the TMI-2 accident in 1979 (described later in this chapter), the integral- and separate-effect facilities built for research on large LOCA were employed for research on small-break LOCA, which posed its own unique thermal hydraulic phenomena, for example, phase separation (since more time is available), natural circulation, and so on. Again hundreds of separate-effect and integral-effect experiments were performed to delineate the physics of the new phenomena, and models were developed for incorporation in the codes. Later, the LOFT facility was also employed for a few tests in which severe accident conditions were simulated, and indeed clad and fuel damage occurred and fission products were released. These were the terminal tests for the LOFT facility; data obtained in those tests have been employed for validation of core-degradation models in the LWR severe accident codes.

In all of the experiments conducted, over many years, on the integral- and separate-effect facilities for LOCA and ECCS research, at no time has the clad on the heater rods or on nuclear fuel rods (in LOFT) experienced temperatures exceeding 1200°C. It has been re-assuring to the reactor safety community that the ECCS, as designed for the PWRs and BWRs, will be able to protect the core (with perhaps some minimal damage) and prevent any significant release of radioactivity to the containment or to the environment. It should be added that containments are designed for the large LOCA thermal and pressure loadings and their integrity should not be in question for the large LOCA accident.

1.8. PUBLIC RISK OF NUCLEAR POWER (WASH-1400)

The late 1960s and the early 1970s were the glory years for nuclear power in the United States and the world generally. The promise of cheap nuclear power was still in full bloom, and there were firm orders and many orders in the wings for NPPs in the United States. The power ratings were increasing, and more and more companies were becoming NPP vendors. The prospect of a large number of NPPs dotting the landscape of the United States and other countries in a relatively few years made some persons quite apprehensive, and questions arose about the risk that accidents in NPPs posed to the general public. Since there was no quantitative measure of public risk in the 1960s, F. R. Farmer [14] of the UK proposed such a measure through a curve of probability versus consequences, with the risk defined as probability multiplied by consequences. The proposed curve was basically intuitive and recognized that, as consequences increase, the probabilities of occurrence for such consequences should decrease. The risk of a certain enterprise would be acceptable to the public if the probability of a certain consequence remained below the proposed curve. In contrast, the probability values above the curve, for specific consequences, would not be acceptable to public.

Farmer also recognized that while the public may well accept accidents that have a low frequency of occurrence, it might not accept accidents with very high consequences at an equal risk level. Thus, the high consequence accidents should pose a low overall societal risk.

Farmer proposed the curve (A) shown in Figure 1.4, with the accident consequences represented by the release of curies of ^{131}I on the abscissa and the probability of occurrence on the ordinate. The risk level of 1 is chosen for the consequence level of 10^3 curies of radioactive ^{131}I released with a probability of 10^{-3}. The curve is flattened at the top so that the highest probability of some (10 curies) radioactive ^{131}I release is 10^{-2}. The curve can be given a slope of -1 for an equal risk for high consequence accidents, but more likely public acceptance would be for the line with a slope of -1.5, so that the very high-consequence events occur with a relatively lower public risk. For example, a release of 10^6 curies of ^{131}I would be acceptable only with an occurrence probability of 10^{-8}, that is, with a risk level of 10^{-2}.

Farmer's approach did not specify any risk values for accidents in NPPs, but it clarified societal acceptance of risk for a new technology and it also provided a base for quantifying the risk of nuclear power.

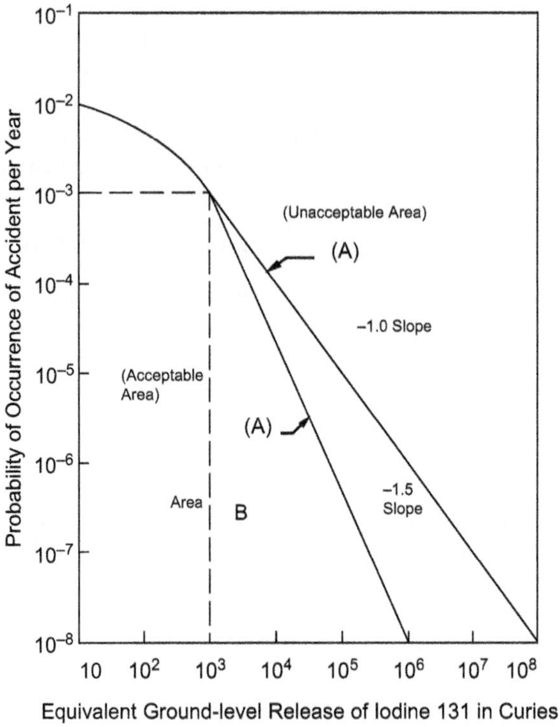

FIGURE 1.4 Farmer's curve.

1.8.1. The Reactor Safety Study

The United States Nuclear Regulatory Commission was established in 1974. One of the NRC's early initiatives was to sponsor a study of the public risk of nuclear power under the leadership of Professor Norman Rasmussen of MIT with the extremely able assistance of Saul Levine of NRC. This study, named the Reactor Safety Study (RSS), published as the report WASH-1400 [15], provided the first structured assessment of the public risks of accidents in U.S. LWRs.

The RSS employed a comprehensive and detailed fault and event tree methodology to obtain the probabilities of faults and to predict accident scenarios that could release radioactivity in the environment damaging the health of the public living in the vicinity of the plant and also contaminate the land around the site of the plant. A typical PWR and a typical BWR were chosen for the Level 1, 2, and 3 probabilistic safety analyses (PSAs). State-of-the art methodology was employed for the Level 2 and Level 3 consequence estimations. Clearly, the severe accident progression and consequence models employed were not as detailed and sophisticated as they would become later.

α	β	γ	δ	ε
CONTAINMENT RUPTURE DUE TO VESSEL STEAM EXPLOSION	CONTAINMENT LEAKAGE	CONTAINMENT RUPTURE DUE TO HYDROGEN BURNING	CONTAINMENT RUPTURE DUE TO OVERPRESSURE	CONTAINMENT RUPTURE DUE TO MELT-THROUGH

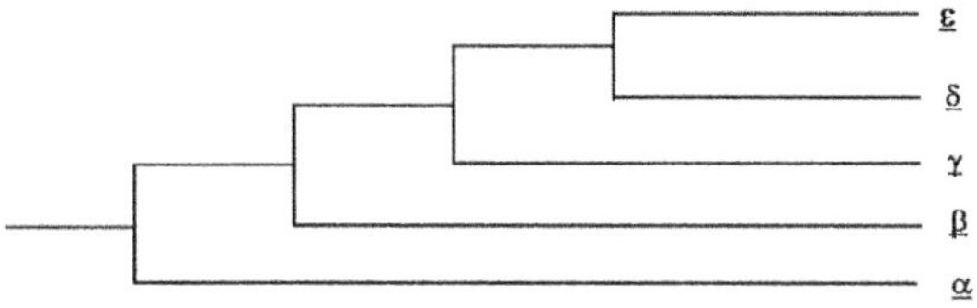

FIGURE 1.5 PWR containment event tree.

Looking back, however, it is remarkable that the estimates made for the consequences in many of the beyond the design basis (BDBA) scenarios were reasonably good. This speaks for the good engineering judgment capability of the U.S. researchers working on the RSS. This study was published in 1975.

In the following paragraphs we will provide some snapshots of the methodology employed by the RSS researchers and then describe the principal results they obtained, as reported in WASH-1400 [15].

The RSS researchers recognized quite early that the integrity of the containment, which is the last barrier to the release of fission products to the environment, is the key to determining the consequences of the severe accident. In this context the containment failure was categorized as shown in Figure 1.5. The α and γ modes of containment failure (rupture) were considered as catastrophic due to a fast-acting loading generated either by an in-vessel steam explosion or a hydrogen detonation in the containment. The ε mode of failure applied primarily to the Mark-1 BWRs in the United States. The δ mode of failure was ascribed to the excessive pressure created in the containment due to the steam released from the primary system during a break, but more significantly due to the molten corium concrete interaction that occurs when the molten core is discharged on the containment basement in the event of vessel failure. The energetic modes of containment failure would not provide any retention of the containment aerosol source term, but the δ and β (containment leakage) modes of failure could be credited with retention due to (a) the natural processes of aerosol deposition on walls, floors, and so on, and (b) operator-action or automatic remedial actions (e.g., spray actuation).

Figure 1.6 shows the event tree for the BDBA large-break, LOCA scenario. An event tree is inductive, and it looks forward. Its logic is very similar to that of a decision tree, as employed for decision making in business, economics, and the like. An event tree is generally drawn from left to right and begins with an initiator. This initiator is an event that could lead to shutdown or failure of a system or a component. In the event tree, the initiators are connected to other

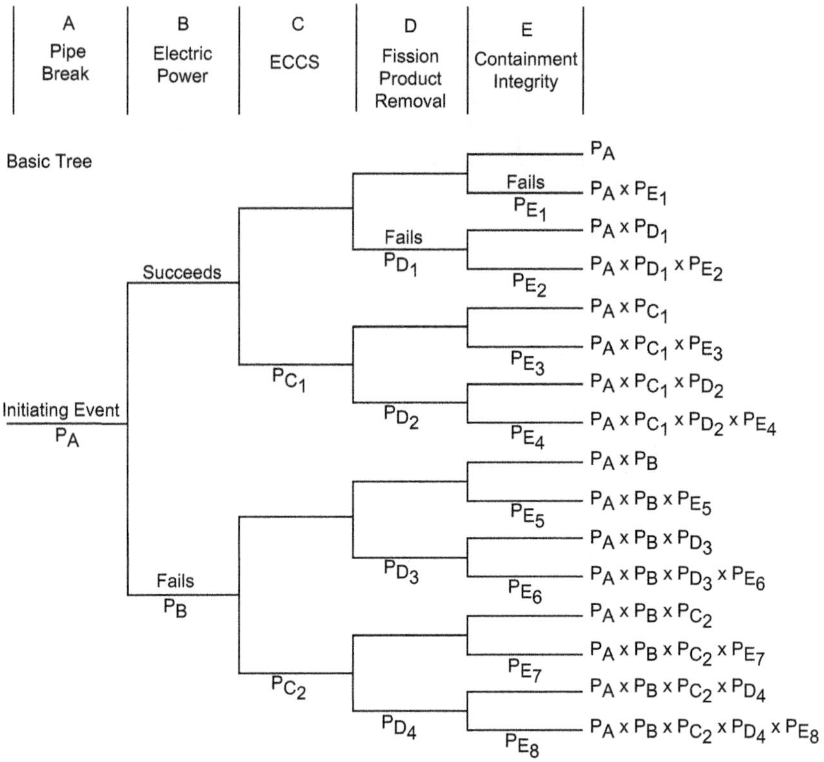

FIGURE 1.6 Simplified event tree for a large LOCA.

possible events by branches; a scenario is a path of these branches. The tree in Figure 1.6 shows the probabilities of success and failure for each of the systems or processes listed at the top. These, in turn, determine the final probability of the consequences represented by each branch of the tree on the right.

The event tree shown in Figure 1.6 can be reduced to that shown in Figure 1.7 to concentrate on the major consequences and their probabilities. This tree shows, for example, that with the nonavailability of electric power both the ECCS and fission product removal systems are unable to function. The containment failure probability is the same as the probability of the failure of electric power to function. The tree in Figure 1.6 or Figure 1.7 could be combined with the containment failure mode tree in Figure 1.5.

The fault tree (which is not illustrated here) is a construction to determine the probability of the initiating fault or failure. Thus for each of the tree branches shown in Figure 1.6, the probability shown (e.g., P_B for the failure of the electric power to function or P_{C_1} the failure of ECCS to function), in spite of the success of having electric power functioning, is determined by the fault tree

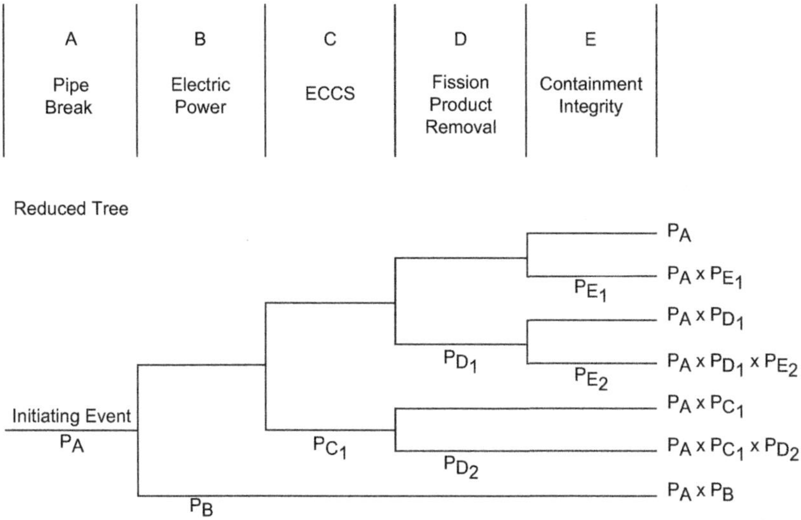

FIGURE 1.7 Reduced event tree for a large LOCA.

analysis in which the plant electrical and mechanical systems and components are examined for the probability of their failure, which leads to the probability P_B or P_{C_1} described above. The fault trees can be huge with many branches since many components, including electric relays, switches, and pumps, may be involved in the functioning of a safety system.

The fault trees and event trees provide the probability of the events occurring; however, the consequences of the events have to be determined by developing models for the physical process that occurs in the BDBA. The researchers of the RSS developed a code for estimating the source term (fission products resident in the containment as a function of time) and, then, the release and transport in the environment for the various modes of failure of the containment and the various metreologies assumed for the areas around the location of the PWR and the BWR that were considered in the study. The fission products released, being, so many were combined into certain number of categories and their biological damage, in terms of early fatalities, or exposures that would lead to early illnesses, was calculated. In addition, estimates were made of the total property damage that would occur and the land area that could be subject to contamination due to the release. In this respect it was found that a higher-consequence accident scenario would have a lower probability of occurrence. For example, catastrophic containment rupture that would release fission products, without retention, in the containment, as could occur with an in-vessel steam explosion or with a hydrogen explosion in the containment, is of much lower probability since these energetic events have much lower probabilities of occurrence.

Tables 1.1 and 1.2, extracted from the report WASH-1400, provide the main results of the RSS. Table 1.1 shows the consequences of early fatalities, early illness, total property damage in billions of dollars, decontamination area in square miles and the relocation area in square miles as a function of probabilities that vary from 5×10^{-5} (1 in 20,000) to 1×10^{-9} (1 in a billion). The highest probability selected was 5×10^{-5}, which was the calculated probability of core melt occurring in a LWR/year. Table 1.2 presents the long-term consequences of latent cancer fatalities, thyroid nodule cancers, and genetic effects in the population close to the PWR and the BWR plants considered in the RSS.

What is immediately clear is that the threshold for the public hazard is the occurrence of a core-melt accident in a LWR. There is no hazard to the public if such an accident does not occur, since no fission products are released until there is a core heat up and clad failure. The probability of core melt occurring is quite small. Certainly, an individual reactor could be decommissioned long before, and that is the hope and prayer of each owner of the LWR plants. Another conclusion from these two tables is that the early consequences are indeed very small. The early fatalities and illness values can be called significant only when the probability values are very small (10^{-7} to 10^{-9}). The longer term effects appear to be significant at the levels of probability equal to 10^{-6}. However, here the latent cancer fatalities due to the postulated core meltdown accident in a LWR are competing with the cancer fatalities caused by cigarette smoking and the environmental hazards that a public regularly, and by their own volition, accepts. In fact, the latent cancer incidence of 170 is less than the statistical uncertainty in the normal cancer fatalities/year.

The thyroid nodules are generally not associated with the other environmental hazards. Thyroid cancer is caused by the deposition of the radioactive iodine in a person's thyroid released as a fission product in the core-melt accident. Children are more susceptible to the thyroid malignancy. This is why iodine tablets are distributed to the population around a NPP, so that the thyroid is already saturated with nonradioactive iodine. The genetic effects supposedly are caused by mutations in the cells in the body. Again, the numbers are too small to be statistically significant since such effects are also caused by other environmental substances (e.g., chemicals or even airplane rides). This is illustrated in Table 1.3, also extracted from WASH-1400, in which a comparison is made for the latent (or long-term) health effects caused by the core-melt accident in one reactor against those that occur normally in the population that was exposed to the core-melt accident. It is seen that even for the very low probability of 10^{-6}—that is, a rather severe accident in which containment failure did take place and a large fission product release occurred—still the latent health effects are of the order of one-tenth those of normal incidence.

The consequence-probability estimates derived by the authors of WASH-1400 for one reactor can be extrapolated to a population of reactors in any country. Table 1.4 shows such an extrapolation for 100 reactors, which is

TABLE 1.1 Early Consequences of Reactor Accidents for Various Probabilities for One Reactor

Probability per Reactor-Year	Early Fatalities	Early Illness	Total Property Damage $\$10^9$	Decontamination Area ~Square Miles	Relocation Area (Square Miles)
One in 20,000 *	<1.0	<1.0	<0.1	<0.1	<0.1
One in 1,000,000	<1.0	300	0.9	2,000	130
One in 10,000,000	110	3,000	3	3,200	250
One in 100,000,000	900	14,000	8	-	290
One in 1,000,000,000	3300	45,000	14	-	-

**This is the predicted probability (chance) of core melt per reactor-year.*

TABLE 1.2 Late Consequences of Reactor Accidents for Various Probabilities for One Reactor

Probability per Reactor-Year	Latent Cancer** Fatalities (per year)	Thyroid Nodules** (per year)	Genetic Effects*** (per year)
One in 20,000*	<1.0	<1.0	<1.0
One in 1,000,000	170	1,400	25
One in 10,000,000	460	3,500	60
One in 100,000,000	860	6,000	110
One in 1,000,000,000	1500	8,000	170
Normal Incidence	17,000	8,000	8,000

**This is the predicted probability of core melt per reactor-year.*
***This rate would occur approximately in the 10- to 40-year period following a potential accident.*
****This rate would apply to the first generation born after a potential accident. Subsequent generations would experience effects at a lower rate.*

TABLE 1.3 Incidence per Year of Latent Health Effects Following a Potential Reactor Accident

	Probability per Reactor per year		
Health Effect (per year)	One in 20,000[(a)]	One in 1,000,000[(a)]	Normal[(b)] Incidence Rate in Exposed Population (per year)
Latent Cancers	<1	170	17,000
Thyroid Illness	<1	1400	8,000
Genetic Effects	<1	25	8,000

[(a)] *The rates due to reactor accidents are temporary and would decrease with time. The bulk of the cancers and thyroid modules would occur over a few decades, and the genetic effects would be significantly reduced in five generations.*
[(b)] *This is the normal incidence that would be expected for a population of 10 million people who might receive some exposure in a very large accident over the time period that the potential reactor accident effects might occur.*

TABLE 1.4 Consequences of Reactor Accidents for Various Probabilities for 100 Reactors

Chance per Reactor-Year	Latent Cancer[(b)] Fatalities (per year)	Thyroid Nodules[(b)] (per year)	Genetic Effects[(c)] (per year)
One in 200[(a)]	<1.0	<1.0	<1.0
One in 10,000	170	1,400	25
One in 100,000	460	3,500	60
One in 1,000,000	860	6000	110
One in 10,000,000	1,500	8,000	170
Normal Incidence	17,000	8,000	8,000

[(a)] *This is the predicted chance per year of core melt for 100 reactors.*
[(b)] *This rate would occur approximately in the 10- to 40-year period after a potential accident.*
[(c)] *This rate would apply to the first generation born after the accident. Subsequent generations would experience effects at a decreasing rate.*

approximately the current population of the LWR plants in the United States. The calculated incidence of one core-melt accident in such a population is 1 in 200 years or 5×10^{-3}/year. The consequences remain the same, that is, insignificant to statistically insignificant.

Interesting data are shown in Table 1.5, also obtained from WASH-1400; these data show the average risk of fatality by various man-caused and nature-caused events per year. These statistics are for the United States in the late 1960s and early 1970s. The highest number of fatalities is self-caused by the operation of motor vehicles, followed by falls, fires, drowning, firearms, air travel, and the like. It is seen that the riskiest enterprise that we engage in is that of operating motor vehicles. The numbers have improved since the late 1960s and early 1970s because of the many improvements made in modern cars such as airbags and seatbelts, but the traffic has worsened and speeds have increased.

The events caused by nature are also listed in Table 1.5; these include lightning, tornadoes, and hurricanes. The fatalities caused by hurricanes and tornadoes shown in this table may be lower than what they have been in more recent years. The probabilities for an individual suffering a fatal nature-caused accident may also have increased recently as these events have been of greater

TABLE 1.5 Average Risk of Fatality by Various Causes

Accident Type	Total Number	Individual Chance per Year
Motor vehicle	55,791	1 in 4,000
Falls	17,827	1 in 10,000
Fires and hot substances	7,451	1 in 25,000
Drowning	6,181	1 in 30,000
Firearms	2,309	1 in 100,000
Air travel	1,778	1 in 100,000
Falling objects	1,271	1 in 160,000
Electrocution	1,148	1 in 160,000
Lightning	160	1 in 2,000,000
Tornadoes	91	1 in 2,500,000
Hurricanes	93	1 in 2,500,000
All accidents	111,992	1 in 1,600
Nuclear reactor accidents (100 plants)	-	1 in 5,000,000,000

strength lately. The probability of an individual dying in a nuclear accident has been estimated to be 1 in 5×10^9 or 2×10^{-10}/year, which is highly insignificant.

Another comparison of human-caused, nature-caused, and one reactor severe accident in a population of 100 reactors is shown in Table 1.6, which shows probabilities for large- consequence accidents, that is, those accidents that may result in fatalities of 100 or 1,000 persons. It is seen that the most frequent cause is the airplane crash for a 100 fatality accident and a hurricane for a 1,000 fatality accident. The tsunamis that were spawned by the severe earthquake that occurred in Indonesia on Dec. 26, 2004, in which 230,000 persons died; or the earthquake in the Kashmir region of Pakistan-India on Oct. 8, 2005, in which more than 75,000 persons died; or the very recent earthquake in Japan on March 11, 2011 in which perhaps 26,000 persons perished, are, of course, very high-consequence unique events. The probability of 100 fatalities occurring in a nuclear accident for a country with 100 NPPs was estimated by WASH-1400 to be 1 in 100,000 years; for 1,000 fatalities, the probability was estimated to be 1 in a million years.

The most famous and most quoted results from WASH-1400 are shown in the Figures 1.8, 1.9, and 1.10; these compare, in turn, the fatalities and the

TABLE 1.6 Average Probability of Major Human-Caused and Natural Events

Type of Event	Probability of 100 or More Fatalities	Probability of 1,000 or More Fatalities
Human-Caused		
Airplane Crash	1 in 2 years	1 in 2,000 years
Fire	1 in 7 years	1 in 200 years
Explosion	1 in 16 years	1 in 120 years
Toxic Gas	1 in 100 years	1 in 1,000 years
Natural		
Tornado	1 in 5 years	Very small
Hurricane	1 in 5 years	1 in 25 years
Earthquake	1 in 20 years	1 in 50 years
Meteorite Impact	1 in 100,000 years	1 in 1,000,000 years
Reactors		
100 plants	1 in 100,000 years	1 in 1,000,000 years

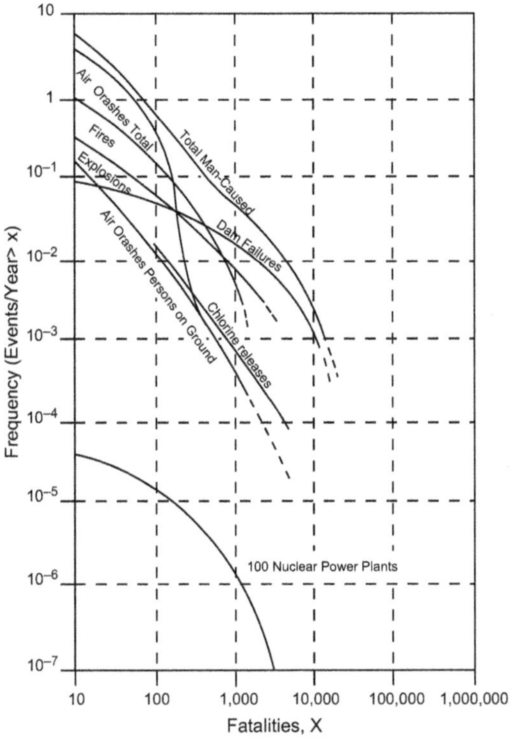

FIGURE 1.8 Frequency of fatalities due to man-caused events.

property damage caused by a nuclear accident in the United States, with its 100 NPPs, against the human-caused and nature-caused events. Clearly, the probabilities at any consequence level for the 100 NPPs are many orders of magnitude smaller than those for the other human-caused or for the nature-caused events. The close comparison of the public risk from the 100 NPPs to the nature-caused event of a meteorite hitting the earth is apt, but it was ridiculed by some critics of WASH-1400.

The other significant results from WASH-1400 were a comparison of the probabilities for the various consequences for a PWR versus those for a BWR. It was found that the risks were quite the same for those two types of LWRs. An example is shown in Figure 1.11 for the consequence of early fatalities/year from a severe accident in either reactor.

A startling finding of the RSS was that the operator errors could be a significant contribution to the probability of a core-melt accident occurring. Although there was no quantification of this contribution; it was clear that in complex events, operator actions could aggravate the situation, which could progress into causing damage to the core. A case in particular was that of the small-break LOCA, which may continue for 1 to 2 hours during which mistakes by the operator could result in a core-melt accident. This is exactly what

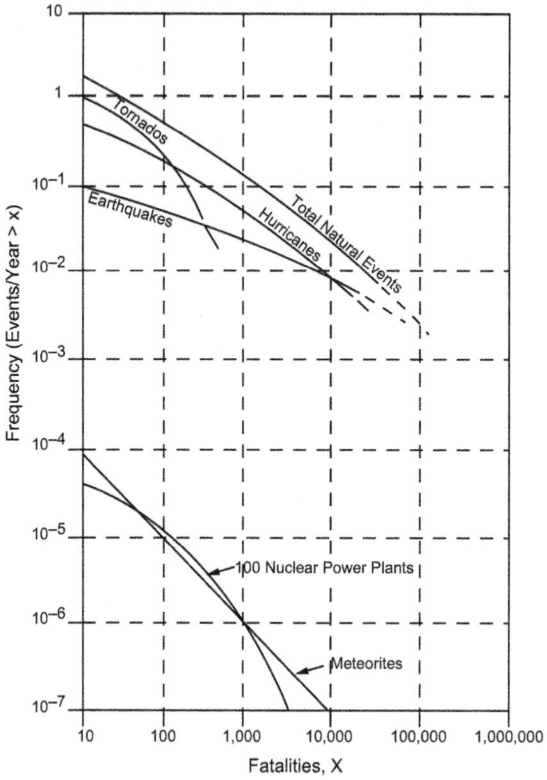

FIGURE 1.9 Frequency of fatalities due to natural events.

happened during the TMI-2 accident. It must be noted here that large-scale two-phase phenomena were not as well known at that time, and there were surprises, which were later understood and recognized.

WASH-1400 received an exhaustive review from a diverse group of scientists, including a panel set up by the NRC under the leadership of Professor H. W. Lewis [16]. The reviewers liked the methodology employed but questioned the estimation of the uncertainties and the final values for the core-damage frequency. No major discrepancies, however, were discovered. Germany followed the same procedures as in WASH-1400 for their plants and published the German Risk study in 1980 [17].

The WASH-1400 prediction of the risk to the public from a population of say 100 NPPs to be so much less than any other risk that the public faces was a great vindication for supporters of nuclear power. The order stream for new NPPs continued to grow and actually overwhelmed the capacity of the vendors. These were the years of optimism. It was still not clear whether the cost of nuclear electricity would be less than that from coal-fired plants, since the design criteria, the strict quality control and its documentation, and the

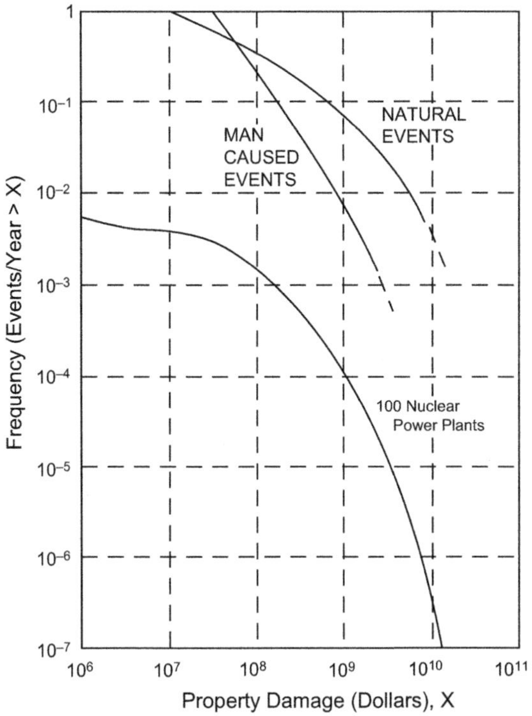

FIGURE 1.10 Frequency of property damage due to natural and human-caused events.

submission of safety analysis report and other documents were adding tremendously to the capital costs of LWRs. In addition, court challenges by interveners were delaying the construction of several NPPs in the United States, which were also adding significant sums to their capital costs.

1.9. THE TMI-2 Accident

1.9.1. Description of the Accident

The core-melt accident of the Three Mile Island – 2 (TMI-2) reactor near Harrisburg Pennsylvania, occurred on March 28, 1979—that is, less than four years after the publication of the RSS (WASH-1400). This accident was entirely unexpected, and it was a shock to nuclear establishments all over the world. The detailed results of WASH-1400 were not known to a large part of the nuclear community, and suddenly there was a general realization that we missed something vital in our perceptions. That a core can melt and melt so fast was never in anyone's imaginings. For a number of years, it was thought that, perhaps, only a small part of the core was damaged. Only after the removal of the upper internals did it become clear that, at least, half of the core had melted.

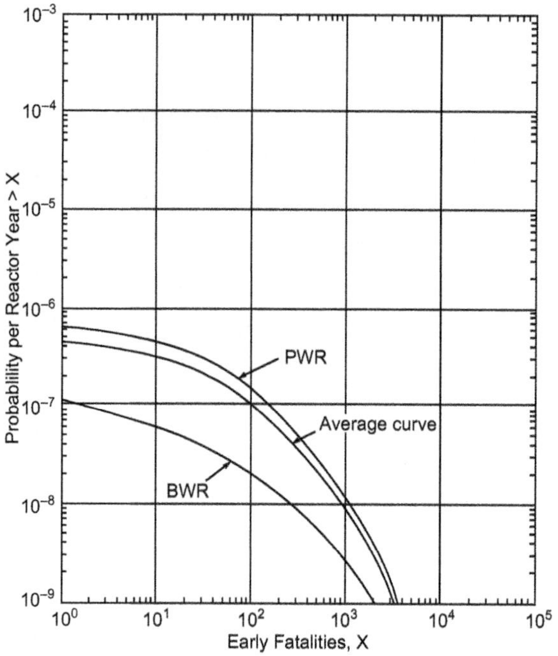

FIGURE 1.11 Probability distribution for early fatalities per reactor-year.

Later, it was found that some (20 tons) of the melted core had reached the lower head. If the operators had not filled the TMI-2 vessel with water, or if a much larger quantity of melt had dropped into the lower head, it is not clear whether the lower head would have survived and that all the melt would be retained in the vessel. Release of melt to the containment, and the possible melt-concrete interaction, would have created much greater uncertainty and untold additional issues in 1979.

The accident started with the loss of feed water to the steam generators, which resulted in the dry-out of the secondary side of the steam generators within 10–15 minutes. The dry-out of the secondary side stopped the heat removal from the core, and the reactor pressure started increasing. Meanwhile, this fault automatically tripped the turbine and scrammed the reactor. The increase in reactor vessel pressure opened the pilot-operated relief valve (PORV), as it should have, to decrease the pressure in the vessel. As the pressure in the vessel decreased, the PORV should have closed; it did not, however, and so the coolant kept discharging from the vessel and the vessel pressure started decreasing. This led to the start of the high-pressure ECCS, as it should have, and some water started to be added to the vessel. Meanwhile, the pressure decrease in the vessel created much steam, a phase separation, and the formation of a steam bubble. The water in the pressurizer, which could have come to the vessel, was blocked due to the steam bubble or by the flow of steam [Counter Current Flooding Limit (CCFL) phenomenon] and the pressurizer indicated full. The operators reacted too slowly to the fact that PORV was open, which they closed only after a considerable amount of water had been lost from the vessel. Another error made by the operators was that they closed the ECCS injection to the core, following their instructions in case of a full pressurizer. They also stopped the pumps since they had started cavitating due to the passage of steam along with water in the primary system.

This equipment failure, coupled with faulty operator actions, resulted in the loss of much water from the vessel, no water addition to the vessel, boil-off of water in the vessel, and finally the uncovering of the core approximately 130 minutes after the first malfunction. No primary feed water was being added, since the pumps were stopped and the ECCS was shut off; the continuing boil-off resulted in almost complete uncovering of the core. The decay heat and the absence of heat removal raised the clad temperature to values where the exothermic steam-Zircaloy reaction started. From then on, there was no turning back from a core-melt scenario! About 50% of the core was melted and flowed down to form blockages near the bottom (see Figures 1.12 and 1.13).

Luckily, the operators restarted the cavitating pumps and filled the vessel with water. This action stopped further melting of the fuel elements, but it fragmented much of the already molten fuel in the core region to form a debris

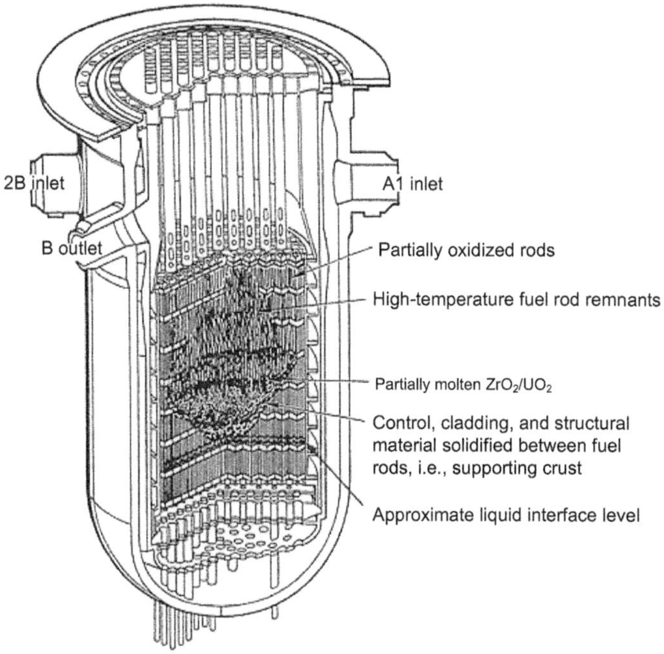

FIGURE 1.12 Hypothesized core damage configuration at 173 minutes.

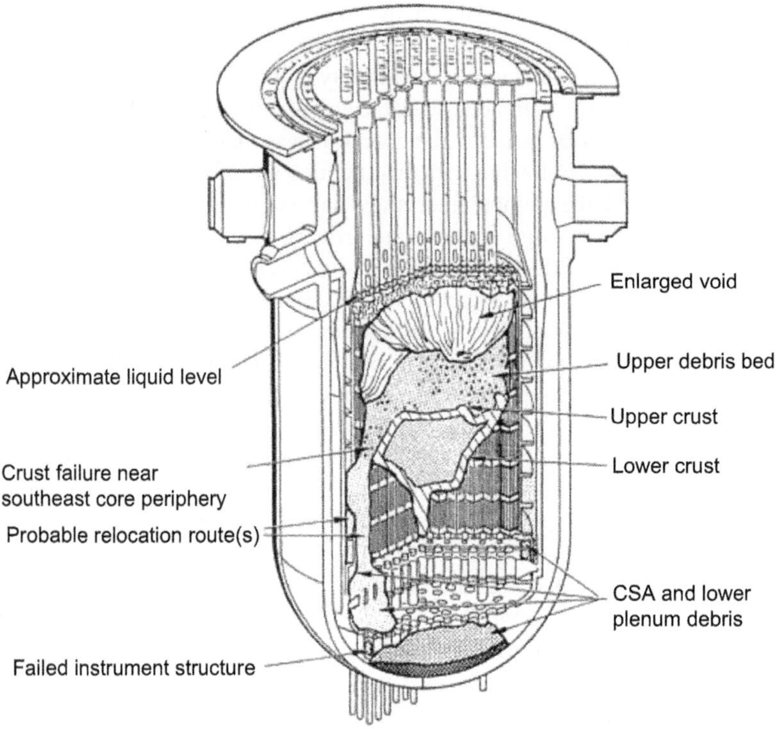

FIGURE 1.13 Hypothesized core damage configuration at 226 minutes.

bed, which could not be fully cooled, even when submerged in water. The fragmented debris in the core region reheated and about 20 tons of the molten material broke through the crust at the side and flowed down to the lower head, while ablating the core cylinder. The water present in the lower head and in other parts of the vessel finally quenched the melt, although it took a few hours. The vessel survived and the next day, the pumps were stopped. The natural circulation flow between the vessel and the steam generators was sufficient to remove the decay heat from the core. More details on the core degradation process and the vessel behavior can be found in the Chapter 2.

The Zircaloy-steam exothermic reaction produced hydrogen, which was released to the containment through the open PORV. It accumulated in the containment and several hours after the start of the accident burned, producing a pressure spike of 2 bars in the containment. The containment, designed for the pressure rating of 5 bars, had no problem with the hydrogen burn and did not sustain any damage.

The volatile fission products released during the core heat up, melting of cladding, and melting of the fuel also accumulated in the containment, and radioactivity was detected in the containment. At that time, a door was open from the containment to the auxiliary building, and some fission products were released to the auxiliary building, before the door was closed by the operators. Despite the fact that the auxiliary building was not built like a leak-tight containment, only ~0.01 % of fission products escaped from it to the environment. In total, less that 10^{-5} % of ^{131}I inventory of the core was released to the air. During the first 16 hours after the accident, only about 10 Ci was released to the environment, and ~70 curies of I was released over the next 30 days (see Figure 1.14). Radioactive material found within the exclusion area surrounding the reactor included ~0.5 Ci of ^{137}Cs and ~0.1 Ci of ^{90}Sr. In this context, it is perhaps, instructive to know that the inventories of ^{131}I, ^{137}Cs and ^{90}Sr in a prototypic LWR core, near the end of a cycle, are ~91, 5 and 4 million curies. The TMI-2 core was only 90 days old, and its inventory of fission products would have been somewhat less than these values.

The TMI-2 accident did not cause any injuries or deaths or property damage. It also did not release sufficient fission products to contaminate the

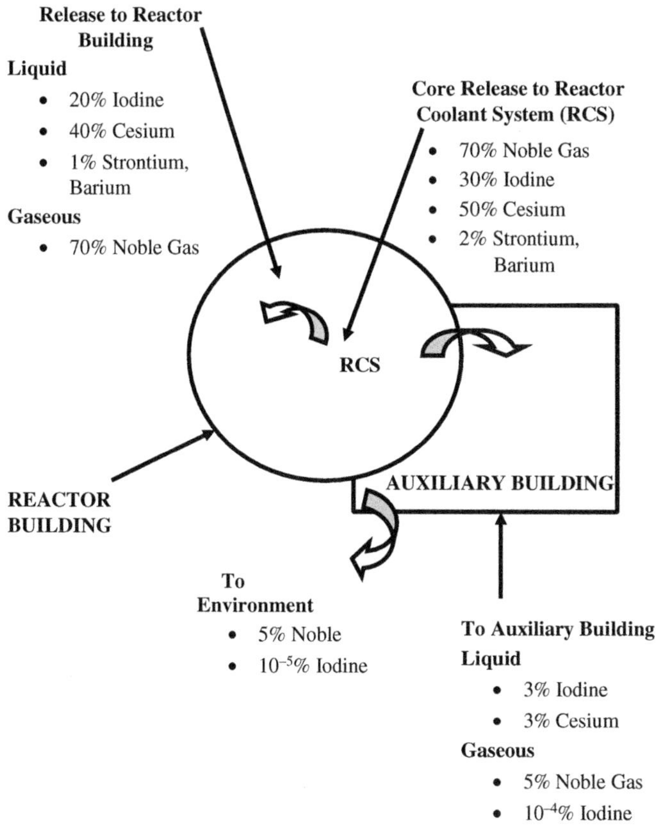

FIGURE 1.14 Release fractions at TMI-2.

surrounding soil, except perhaps the exclusion area slightly. It, however, caused serious psychological harm in the United States, especially to the population of Harrisburg and the state of Pennsylvania. The major cause was the duration of the perceived threat: for almost one week, there was news that a hydrogen bubble might be inside the vessel, which could explode, causing the vessel and the containment to fail, resulting in a devastating release of radioactivity at any moment. This was faulty science inasmuch as a hydrogen explosion cannot occur for lack of oxygen. But the NRC did not vigorously contradict this thought. The result was a panic-driven evacuation from a large area around the plant, even though the authorities never called for an evacuation. This publicity, the interviews with so many so-called experts, and the many "what- if:" projections soured the public on nuclear power. It is only very recently that the public view of nuclear power may have changed.

1.9.2. The Aftermath of TMI-2 Accident

The occurrence of the TMI-2 accident, in spite of the very small physical damage, was taken very seriously by the authorities and the nuclear industry. The U.S. President appointed a special commission to inquire into the causes and circumstances of the accident. The commission identified 18 faults and errors [18]: 5 in design, 2 in regulation and 11 in operation. An error identified for the NRC was the failure to inform TMI-2 plant personnel of a similar event that happened earlier at the Davis-Bessie plant (designed and built by the same vendor), which was successfully terminated. The commission also faulted the NRC for not admitting that it had made an error about the possibility of a hydrogen explosion and informing the public forcefully about it.

The TMI-2 accident was a wake-up call for the whole nuclear enterprise in the United States. It was realized that this accident was unlike the design-base accidents but more like the accidents postulated in WASH-1400. The equipment errors (valve malfunctioning) and operator errors that initiated it, together with the circumstances of the accident, called for much better equipment testing, control room instrumentation, operating procedures, lines of authority, lines of communication, technical support to the operator, emergency planning, public evacuation, and so on. It emphasized that nontechnical aspects (e.g., operator training, emergency procedures, organization, management) are as important as technical aspects (e.g., equipment design, construction, equipment qualification, and safety analysis). The industry responded to these new challenges immediately by forming a new organization called INPO, and the industry research arm EPRI started the Industry Degraded Core (IDCOR) research program. The focus of safety research was put on the beyond-the-design-base accidents. Further research was initiated both by the NRC and EPRI. The NRC worked with the utility industry to require TMI-2 back fits for the plants. These back fits included hydrogen control measures, since the TMI-2 containment was subjected to a hydrogen burn generating a 2-bar pressure spike during the accident and containments of several plants had either lower design pressure or smaller volume than the TMI-2 containment. Hydrogen combustion research was initiated to test igniter systems for control of hydrogen in the containment.

The TMI-2 accident also had several good results. The leak-tight containment, for example, did not suffer any damage when hydrogen burned. The heat sink was established with natural circulation flow between the core and the steam generators, and a safe stable state was reached without failure of either the vessel or the containment.

Another good result from the TMI-2 accident was the absence of significant airborne iodine and cesium radioactivity in the TMI-2 containment. Most of the radioactivity was found to be in the sump water. Analysis activity [19] initiated soon after the accident pointed out that almost all of the iodine and cesium fission products released in the accident, converted to the highly water-soluble compounds, CsI and CS OH, which were removed from the containment atmosphere either by the water spray activated or by the aerosol agglomeration and deposition on the floors and the walls of the containment and eventually transported to the sump. This reduced, by some orders of magnitude, release that could occur through the containment leakage to the environment.

Beginning in 1980, the LWR safety research became known as the LWR severe accident safety research; even though the LOCA experiments and analyses development research did not terminate. But the focus of the LWR safety research shifted to the beyond-the-design-base accidents. We shall address this topic after we describe the other major accident that affected the history of nuclear power safety.

1.10. THE CHERNOBYL ACCIDENT

1.10.1. Background and RBMK Specifics

On April 26, 1986, a core-melt accident occurred in one of the four RBMK reactors situated in a complex called the Chernobyl NPP in Ukraine, Soviet Union. The RBMK are water- cooled channel type reactors in which the water boils as in a BWR; however, they are moderated by graphite. The core configuration is that of a large graphite block in which about 2,000 channels are drilled, each of which contains a pressure tube, and a large fraction of them contains a fuel bundle through which the cooling water flows from the core bottom to the top. The channels are connected at the bottom through several headers, to water inlet. The channels at the top are connected to a multitude of pipes that bring the steam (two-phase mixture) to the steam drum, from which the separated steam is taken to the turbine and the water flows back to the inlet piping at the bottom of the channels. Fresh feed water is admitted to the steam drum. A picture of the RBMK configuration is shown in Figures 1.15 and 1.16. The RBMK core is physically much larger than that of a LWR since it is moderated by graphite, which has a much larger diffusion and slowing down

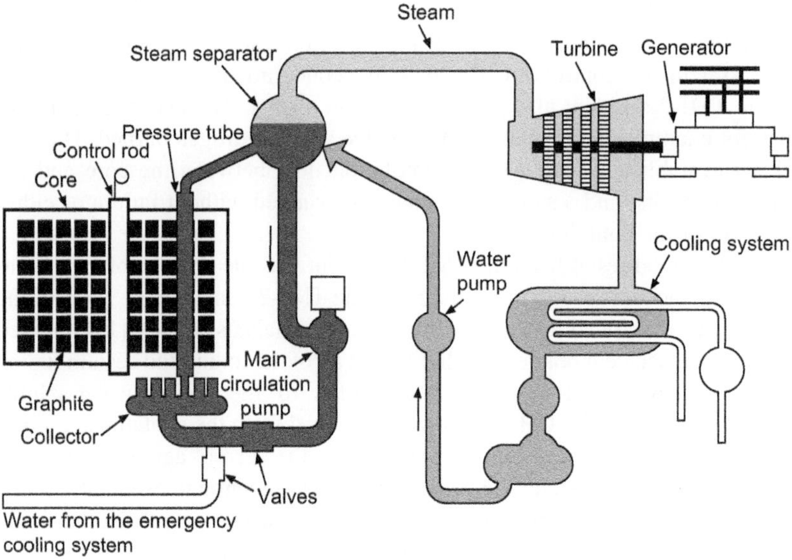

FIGURE 1.15 Schematic diagram of the RBMK-1000.

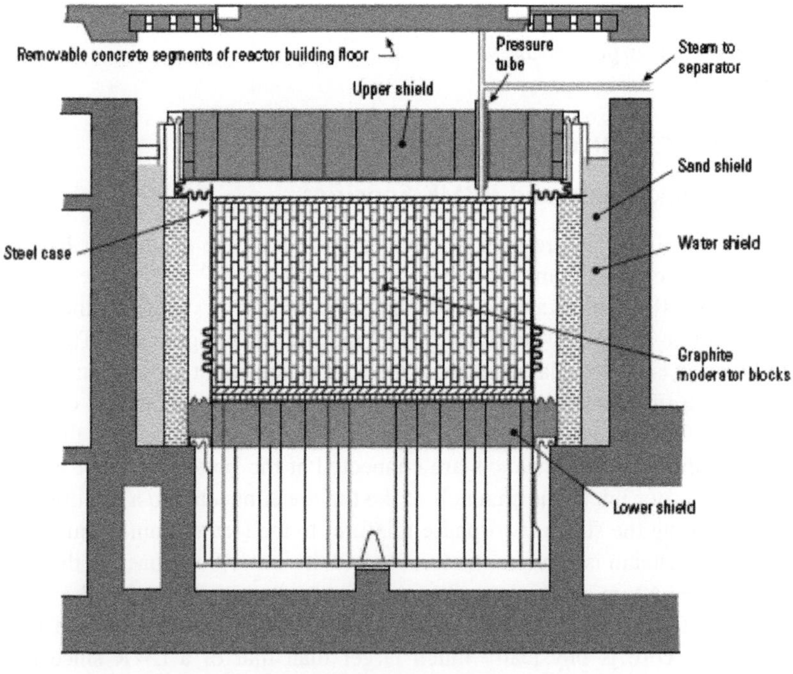

FIGURE 1.16 Cross-sectional view of RBMK-1000 reactor vault.

length for neutrons. The core is fueled, at power, by a fuel machine resting on the shielding above the core. The fuel machine takes a bundle out and replaces it, in general, with a fresh fuel bundle. It does this almost every day.

Besides the differences of moderator between the RBMK and a PWR or a BWR and that of multiple-pressure tubes (as in a CANDU reactor), instead of a pressure vessel, there is a major difference (deficiency) in the RBMK. The RBMK has a containment only on the bottom part of the core and on the piping underneath the core (see Figure 1.17). The outlet piping and the top of the shielded core are enclosed in a confinement building that is not pressure-bearing or leak tight. This building is accessible to the plant personnel, even when the reactor is operating. A water pool is situated below the containment at the bottom of the core to serve as a condenser pool for any steam release from a break in the inlet piping under the reactor core.

The reactor is controlled by the control rods that enter the core from the top. These control rods are different from those in a LWR in that they are follower-type rods in which a graphite region is inserted in the core followed by an absorber. This, as we shall see, failed badly during the accident.

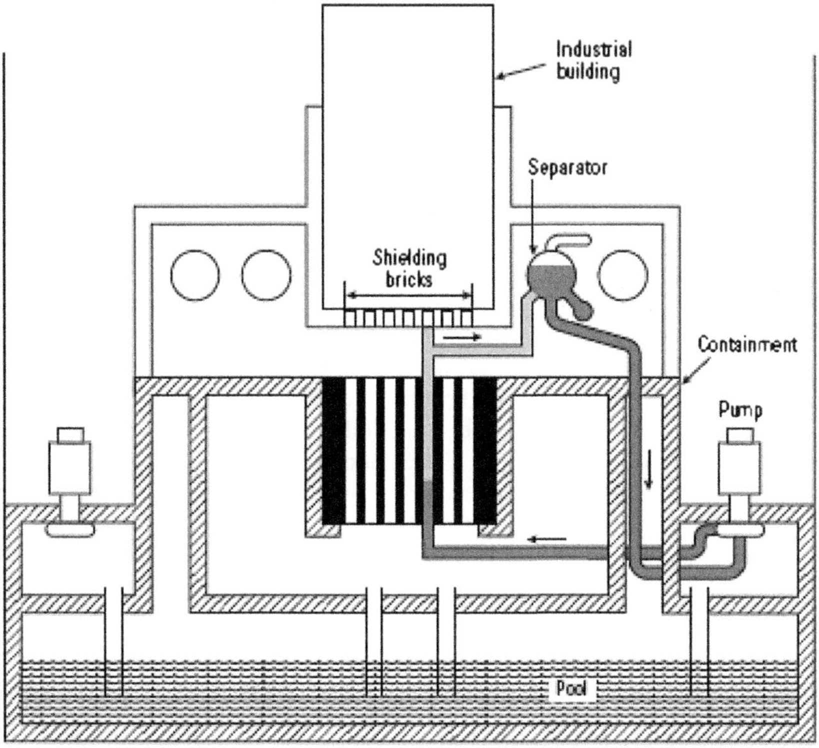

FIGURE 1.17 Chernobyl containment.

The RBMK is also different from a LWR in two other very important aspects: its reactivity feedback characteristics and its stability behavior at low-power level(s). The LWR being optimally moderated by the water loses reactivity as the water density decreases, for example, by boiling of the water coolant. Thus a BWR, for example, will experience a large negative reactivity, if for some reason (e.g., a rise in power, depressurization) the void fraction (quality) increases. This will decrease the power level and shut the fission reaction down even if the control rods are not inserted. The RBMK, on the other hand, being optimally moderated by graphite, will have a positive reactivity feedback, if the water density decreases from that in regular operation, since that reduces the absorption in the coolant. This behavior: positive void coefficient and the rather unstable operation at low power required the reactor operation to be limited by certain constraints. The principal restraints were that:

- The reactor should be operated only when a certain number of control rods are in the core.
- The operation of power levels below 20% of full power should be avoided.

The former is required, since many control rods do not have to travel as much to insert the additional absorber in the core. The latter avoids the large instabilities at low-power levels which are hard to control and adjust. The operation of a RBMK without the presence of the required number of control rods in the core was strictly forbidden.

It should be noted from the outset that the Chernobyl accident was a reactivity insertion accident (RIA) rather than a heat removal degradation accident as was the TMI-2. The accident happened on the night of April 26, 1986.

Most of the information (including the figures) in the following paragraphs is extracted from a Canadian report [20] that was published in September 1987 following the postaccident briefing that the Soviet scientists conducted at the International Atomic Energy Agency (IAEA) offices in Vienna. We believe this is the most cogent description of how and why the Chernobyl accident happened.

The four reactors of Chernobyl are located near the small town of Chernobyl about 105 kilometers (km) north of Kiev in Ukraine. The nearest city is Kiev, and the plant personnel lived in the specially constructed town of Pripyat, 3 km from the NPP with a population of 45,000. The Pripyat River flows next to the town of Pripyat on its way to the reservoir near Kiev. Figure 1.18 shows the geography of the area around the NPP, whereas Figure 1.19 shows the location of Chernobyl with respect to the neighboring countries. Normally, such figures are not required for describing a nuclear accident, since all studies for the atmospheric transport of radioactive releases from a LWR containment had predicted that fission products would not be found beyond a 20- to 30-km zone around the NPP. In the case of the Chernobyl accident, however, radioactivity was first detected at the Forsmark plant on the east coast of Sweden at a distance of more than a 1,000 km from Chernobyl. Forsmark plant personnel initially thought that something had gone wrong with

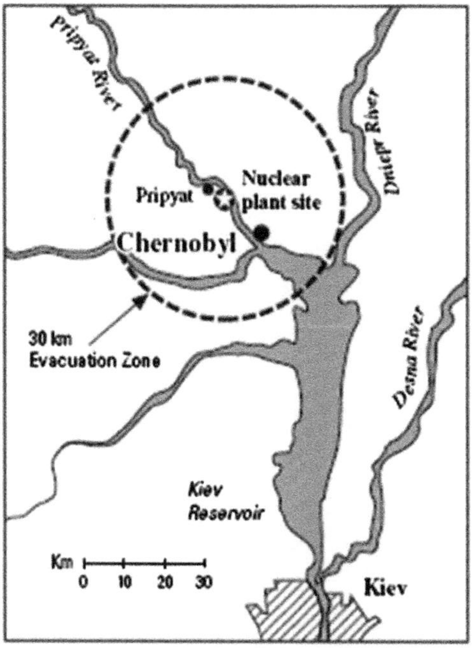

FIGURE 1.18 Area nearby the Chernobyl reactor site.

their plant but soon confirmed that the radioactivity found came from somewhere in the East. The source was confirmed only after 2–3 days when the Soviet authorities announced the accident in the Chernobyl reactor N°4.

The following description of the accident is taken verbatim from the Canadian Report (20).

1.10.2. How and Why Chernobyl Happened

A Test for Safety Sets It Off

It is one of history's ironies that the worst nuclear accident in the world began as a test to improve safety. The events of April 26 started as an experiment to see how long a spinning turbine could provide electrical power to certain systems in the plant. What was the reason for the test? Well, the Soviets, in common with most of the rest of the world, design their reactors not only to withstand an accident, but also to cope simultaneously with a loss of electric power. This may seem a little strange—to run out of power at a generating station—but in an accident the reactor is shut down right away, so cannot generate its own power directly. It would normally get power from the electrical supply to the station or from the other reactors at the same site. To ensure an extra layer of defence, it is considered that there is a possibility that these sources have also failed. The normal backup is to provide diesel engines at the site to drive emergency generators, just as hospitals do in case of a power failure. These diesels usually start up in 30 seconds, and in Western NPPs this is a short enough interruption to keep important

FIGURE 1.19 Chernobyl reactor location.

systems going. For the Chernobyl reactor, the Soviets felt this was not short enough, and they had to have almost an uninterrupted supply. Now even with the reactor shut down, the spinning turbine is so heavy, it takes a while to slow down, and the Soviets decided to tap the energy of the spinning turbine to generate electricity for the few seconds before their diesels started. The experiment was to see how long this electricity would power the main pumps which keep the cooling water flowing over the fuel.

The test had been done before on unit N°3, with no particular ill-effects on the reactor. However, the electrical voltage had fallen off too quickly, so that the test was to be redone on unit N°4 with improved electrical equipment. The idea was to reduce reactor power to less than half of its normal output, so all the steam could be put into one turbine; this remaining turbine was then to be disconnected, and its spinning energy used to run the main pumps for a short while. At the meeting in Vienna the Soviets were at some pains to point out that the atmosphere was not conducive to the operators performing a cautious test:

- The test was scheduled to be done just before a planned reactor shutdown for routine maintenance. If the test could not be done successfully this time, then the people would have to wait another year for the next shutdown. Thus, they felt under pressure to complete the test this time.

- Chernobyl unit N°4 was a model plant—of all the RBMK-1000 type plants, it ran the best. Its operators felt they were an elite crew and they had become overconfident.
- The test was perceived as en electrical test only, and had been done uneventfully before. Thus, the operators did not think carefully enough about the effects on the reactor. There is some suggestion that in fact the test was being supervised by representatives of the turbine manufacturer instead of the normal operators.

How the Trap Was Set

The accident really began 24 hours earlier, since the mistakes made then slowly set the scene that culminated in the explosion on April 26. The "Event Sequence" attached shows a summary of all the things the operators did and how the plant responded; here we describe the key events.

At 1 a.m. on April 25, the reactor was at full power, operating normally with steam going to both turbines. Permission was given to start reducing power for the test, and this was done slowly, with the reactor reaching 50% power twelve hours later at 1:05 in the afternoon. At this point only one of the two turbines was needed to take the steam from the reactor, and the second turbine was switched off.

Normally, the test would then have proceeded, with the next step being to reduce power still further to about 30%. However, the people in charge of the distribution of electricity in the USSR refused to allow this, as apparently the electricity was needed, so the reactor stayed at 50% power for another 9 hours. At 11:10 p.m. on April 25, the Chernobyl staff got permission to continue with the power reduction. Unfortunately, the operator made a mistake, and instead of holding power at about 30%, he forgot to reset a controller and the power fell to about 1%—the reactor was almost shut off. This was too low for the test. Now in all reactors, a sudden power reduction causes a quick build-up of a material called Xenon in the uranium fuel. Xenon is a radioactive gas, but more important it sucks up neutrons like a sponge, and tends to hasten the reactor down the slope to complete shutdown. As well, the core was at such a low power that the water in the pressure tubes was not boiling, as it normally does, but was liquid instead. Liquid water has the same absorbing effect as Xenon. To try to offset these two effects, the operator pulled out almost all the control rods, and managed to struggle back up to about 7% power—still well below the level he was supposed to test at, but as high as he could go because of xenon and water.

It was as if you were trying to drive a car with the accelerator floored and the brakes on: it is abnormal and unstable.

Indeed it is a very serious error in this reactor design to try to run with all the control rods out. The main reason is that some of these same rods are used for emergency shutdown,

Event Sequence

Time	Event	Comments
April 25		
01:00	Reactor at full power Power reduction began	As planned
13:05	Reactor power 50% All steam switched to one turbine	As planned
14:00	Reactor power stayed at 50% for 9 hours because of unexpected electrical demand	
April 26		
00:28	In continuing the power rundown, the operator made an error which caused the power to drop to 30 MW (th), almost shutting the reactor off.	This caused the core to fill with water & allowed Xenon (a neutron absorber) to build up, making it impossible to reach the planned test power
01:00–01:20	The operator managed to raise power to 200 MW (th). He attempted to control the reactor manually, causing fluctuations in flow and temperature.	The RBMK design is unstable with the core filled with water—i.e., small changes in flow or temperature can cause large power changes, and the capability of the emergency shutdown is badly weakened.
01:20	The operator blocked automatic reactor shutdown first on low water level, then on the loss of both turbines.	He was afraid that a shutdown would abort the test. Repeat tests were planned, if necessary, and he wanted to keep the reactor running to perform these also.
01:23	The operator tripped the remaining turbine to start the test	
01:23:40	Power began to rise rapidly. The operator pushed the manual shutdown button.	The reduction in flow as the voltage dropped caused a large and fast increase in boiling leading to a fast power rise. Too late. The damage was done in the next four seconds. The emergency shutdown would have taken six seconds to be effective.
01:23:44	The reactor power reached about 100 times full power, fuel disintegrated, and excess steam pressure broke the pressure tubes.	The pressure in the reactor core blew the top shield off and broke all the remaining pressure tubes.

and if they are all pulled out well above the core, it takes too long for them to fall back into the high power part of the reactor in an emergency, and the shutdown is very slow. The Soviets said that their procedures were very emphatic on that point, and that "Not even the Premier of the Soviet Union is authorized to run with less than 30 rods!"

Nevertheless, at the time of the accident, there were probably only 6 to 8 rods in the core. At any rate, the operator had struggled up to 7 % power by 1 a.m. on April 26, by violating the procedure on the control rods. He had other problems as well—all stemming from the fact that the plant was never intended to operate at such a low power. He had to take over manual control of the flow of water returning from the turbine, as the automatic controllers were not operating well at the low power. This is a complex task to do manually, and he never did succeed in getting the flow correct. The reactor was so unstable that it was close to being shut down by the emergency rods. But since a shutdown would abort the test, the operator disabled a number of the emergency shutdown signals.

After about half an hour trying to stabilize the reactor, by 1:22 a.m. the operators felt that things were as steady as they were going to be, and decided to start the test. But first they disabled one more signal for automatic shutdown, and this sealed the reactor's fate. Normally the reactor would shut down automatically if the remaining turbine were disconnected, as would occur in the test, but, because the staff wanted the chance to repeat the test, they disabled this shutdown signal also. The remaining automatic shutdown signal would go off on abnormal power levels, but would not react immediately to the test. The staff was now in the worst possible situation for a rise in power which could not be caught in time by the shutdown systems. And this is what happened.

The Test Begins

At 1:23.04, the turbine was disconnected and its energy fed to 4 of the 8 main pumps. As it slowed down, so did the pumps, and the water in the core, now moving more slowly over the hot fuel, began to boil. Twenty seconds later the power started rising slowly, then faster, and at 1:23:40 an operator pushed the button to drive in the emergency rods and shut down the reactor. We do not know for sure why he did it—the individual was one of the early casualties —but likely he saw either the power begin to rise or the control rods start to move too slowly in to overcome the power rise. But it was too late. The shutdown rods were so far removed from the core they would have taken six seconds to begin to shut the reactor down. Actually, the insertion of the graphite region of the control rods into the core added more reactivity initially, since it displaced water in the channel. Within four seconds, the power had risen to perhaps 100 times full power and had destroyed the reactor.

The destruction of the whole reactor and the release of the radioactivity to the environment were exacerbated by a common-cause fault in the Chernobyl design: the concrete lid on top of the reactor is lifted by the pressure due to steam entry from the failure of a small number of pressure tubes. In this accident with the large power increase the fuel disintegrated in small particles, failed the clad, mixed with water to generate steam very rapidly (explosively) which failed a number of pressure tubes, which in turn lifted the concrete lid, breaking all 1660 exit pipes, making the steam, the fission products and fuel fragments available for release. The pressure and energy generated were sufficient to hurl the concrete lid, topple the fuel machine, blow the roof off the confinement building and form a high pressure plume which rose to heights where the winds transported the radioactive products to the neighboring countries. The deposition pattern was dictated by the weather patterns over the neighboring countries. Most of the radioactivity deposited in Ukraine, Belarus and Russia; however, radioactive particles were detected in many neighboring European countries and even in countries far away from Chernobyl, e.g., Japan.

The heavy land contamination, which persists today, occurred in Ukraine and Belarus. Much cleanup was performed by thousands of liquidators, who came from the Soviet Union's scientific, civilian, and military sectors. The number of immediate deaths was quite small (31) for such a disastrous accident. Possibly, however, those who survived the accident have suffered a shortened life for part of the liquidator population absorbed substantial doses during the cleanup process of the site and environs. A recent UN report has stated that the effects of radioactivity on the public health and the environment of Chernobyl are not as large as were predicted earlier.

The accident was terminated by covering the core region, via helicopter, with almost 5,000 tons of sand, clay, and boron-bearing material. It is not clear whether this was the best thing to do, since it acted as a heat shield: inside, the graphite burned for at least 7 days during which time all the remaining fuel bundles melted, releasing volatile fission products to the environment. The molten material spread through the space under the reactor and flowed to the basement (see Figures 1.20 to 1.23). The water in the pool had been emptied to prevent any steam explosion at the cost of the lives of the two volunteers from the NPP staff. A sarcophagus was built on top of the destroyed reactor with haste. This building is still standing but it is not in good shape.

The G-7 countries are currently funding the construction of another confinement (sarcaphogus) building on top of the present one in Chernobyl. The material inside is in the form of very fine particles (dust) that still contain much radioactivity. The collapse of the roof of the present sarcophagus could generate a dust cloud that could again contaminate the areas surrounding the NPP.

The Chernobyl accident was the worst blow to nuclear power. The confidence of the public in nuclear power, already shaken by the TMI-2 accident, was now substantially reduced. This confidence was further eroded by the many so-called scientists who greatly exaggerated the long-term health and genetic effects of radiation.

The Chernobyl accident cannot be considered an LWR accident, since the reactors are so different and the Chernobyl reactor, basically, had some design flaws and no containment (where needed). The public, however, does not know the difference between a RBMK and a LWR. To them any

FIGURE 1.20 Photo of Chernobyl 4 and 3, right after the explosion.

scientist's assertion that an NPP cannot blow up like a bomb was found to be wanting. The magnitude of fission products released during the Chernobyl accident exceeded that released in the Hiroshima bomb, and their transport to many neighboring countries and their effect on the lifestyle of several sets of populations was destructive to the safety case of nuclear power as expounded by the WASH-1400. The costs of cleanup in the evacuation area around Chernobyl is estimated to have been 7 billion rubles (at that time a ruble was more expensive than a U.S. dollar), and these costs were with the use of personnel from the Soviet Army and nuclear laboratories. The Chernobyl accident exacted enormous economic, social, psychological and political costs from the Soviet Union and from nuclear enterprises all over the world. It led to the end of many nuclear power programs.

Recently, a United Nations report stated that the effects of Chernobyl on its surrounding populations around were not as dire as originally stated. Many evacuees have returned to Pripyat, and many of the liquidators are in reasonably

FIGURE 1.21 Looking into the debris-fill (look for the inverted lid on the left).

good health nearly 25 years after the accident. Regardless of this benign news, the calamitous effects of Chernobyl cannot be underestimated.

1.11. THE DIFFICULT YEARS

The TMI-2 accident and later the Chernobyl accident created difficult years for the nuclear industry. The nuclear opposition in the United States, which was already quite active before 1979, gathered tremendous force and opposed the completion of the plants under construction and the ordering of any new plants. Their tactic of delaying the construction and completion through legal challenges increased the capital costs of the plants. The plants were also subjected to inspections by INPO and the post-TMI-2 requirements that the NRC placed on the plants. These requirements included valve function, hydrogen management, specific plant improvements, operator training, human-machine interface, instrumentation, and safety culture. Very soon, nuclear electricity, which had been hailed as too cheap to meter, became too expensive to generate and bring online. The plants, also being new, did not operate too well: their capacity factors were in the range of 50 to 70%, which was too low to break even. The utility companies lost money and did not have sufficient funds to invest in plant improvements. The vendors lost many orders, and a number of plants were left incomplete, with great losses to the utility companies. Nothing went well for the nuclear industry during those years, which extended at least till the year 2000.

FIGURE 1.22 The photo is made from the helicopter on May 3, 1986. The smoke may be from the graphite fire.

1.12. SEVERE ACCIDENT RESEARCH

The wake-up call from the TMI-2 accident proved to be a wake-up call to the nuclear industry and the regulatory authorities: In spite of many years of earnest efforts to prevent a core-melt accident, there was now no doubt that such an accident could occur. It was also realized that only a core-melt accident would provide the public hazard of LWRs. The Chernobyl accident provided the public with a vivid demonstration of the hazard of the core-melt accident, if the containment fails. It was clear that such accidents had to be prevented and mitigated; and toward that end it was vital that a knowledge base on those accidents be acquired.

The core-melt accidents were initially called degraded core accidents; today they are called severe accidents. Sometimes they are also called

FIGURE 1.23 The "Elephant's Foot." Once molten, fuel/debris mixture that dripped down through the floors of the damaged RBMK-1000 reactor at Chernobyl.

beyond-the-design-basis accidents (BDBA) or design extension accidents (DEAs). The most generally accepted designation these days, however, is severe accident (SA).

The knowledge base for SA was very poor in 1979. Except for the work performed by the WASH-1400 team, which itself was quite preliminary, there was no organized ongoing effort on nuclear accidents anywhere in the West or East. Immediately after the TMI-2 accident, a resolution was made to acquire knowledge about the progression and consequences of severe accidents. This was done not only in the United States, but also in European countries, and accordingly both experimental and analysis development research programs were initiated [21]. A research center, the NSAC (Nuclear Safety Analysis Center), was formed at the Electric Power Research Institute (EPRI), the research arm of the utility industry. The NRC laboratories geared up for experiments in which fuel elements would be subjected to the kind of heat transfer degradation scenario that TMI-2 experienced. The EPRI effort also included development of the MAAP code [22] for determining the consequences of various severe accident scenarios for PWRs and BWRs. The objectives of the research were to determine whether the present LWR plants were sufficiently safe or whether they required substantial backfits, both for the prevention of severe accidents and to mitigate their consequences if they do start. Simultaneous with the initiative of the research and development, the tools developed by WASH-1400, namely, the fault tree and the event tree analyses were formalized into the Probabilistic Safety (Risk) analysis—1 (PS(R)A-1)—and the NRC required the plants to perform the PRA-1 studies in the so-called individual plant examination (IPE) program to discover any vulnerabilities in plant equipment, instrumentation, procedures, and so on, which could lead to a severe accident.

The severe accident research effort in the United States lasted almost as long as that for ECCS, but it was not as extensive. Later, the programs started by the European Union, the European national governments Japan and Korea have supplemented the United States' SA programs admirably. In particular, the Phébus FP program pursued by IRSN in France with international collaboration has performed core-melt experiments on a prototypic rod bundle with fission products through a representative primary system and containment of a PWR.

Clearly, the phenomena involved in a severe accident are extremely complicated, since the main characteristics of SA scenarios are the interactions of the core melt with the reactor structures and water; and the release, transport, and deposition of the fission product carrying aerosols and vapors. The interactions of core melt may lead to (a) ablation of structures, (b) steam explosions, (c) vessel failure, (d) concrete melting and gas generation, and (e) spreading/dispersion of heat-generating core melt (debris). These phenomena involve the disciplines of thermal hydraulics, high-temperature chemistry, high-temperature material interactions, and aerosol physics, among others. Predictions of the consequences of a severe accident have to be based on experimentation and models whose accuracy may be limited by the scale at which the information about phenomenology is derived. Scaling considerations become very important because large-scale experiments with prototypic melts are very expensive and very difficult to perform.

The emphasis in SA research was placed on the integrity of the containment. That this is the correct choice follows from the consequences of the TMI-2 and the Chernobyl accidents. The TMI-2 containment was full of fission products released from the core during the core heat-up and melting processes, but the fission products were retained in the containment. As mentioned earlier, the containment bypass due to the open door to the auxiliary building and the leakage over time contributed to the very small release of iodine. More recently, emphasis has also been placed on the survivability of the reactor vessel and in retaining the melt inside the vessel by flooding the containment with water and cooling the outer surface of the vessel. This has been adopted for the AP-600 [23], AP-1000, and the Korean APR-1400.

The loadings that can fail the containment were identified, and they were classified into two groups: (1) those that could fail the containment early and (2) those that could fail the containment much later. This early versus late distinction arose as a result of the natural processes that control the concentration of the fission product aerosols in the containment atmosphere. It was found that almost all of the fission products aerosols agglomerate and deposit on the walls and floor of the containment in approximately 4 hours, and from there they are transported to the sump.

Thus, the fission products and aerosols are not available for release to the environment upon the failure of the containment. In this process, the more toxic fission products, that is, ^{131}I and ^{137}Cs, which had formed the highly water-soluble compounds CsI and CsOH, are removed from the containment as soon as a spray action is activated. In this context, it should be mentioned that the NRC requires maintenance of containment integrity in a severe accident of at least 24 hours, with a conditional (on the occurrence of a severe accident) probability of 10%.

Hydrogen combustion-detonation, steam explosion, direct containment heating (DCH), and melt attack on the BWR Mark-1 containment liner were identified as the energetic processes that could fail the containment early. The longer-term gas-producing molten corium concrete interaction (MCCI), which would pressurize the containment and the lack of melt coolability, were identified as the processes, which could fail the containment later. It should be noted that a release during the later (after 24 hours) failure of the containment may be 4 to 5 orders of magnitude smaller than that for the early failure of the containment.

The in-vessel accident progression determines the containment loadings of fission products (the source term), the hydrogen, and the mass and composition of the melt delivered to the containment. Thus, any meaningful evaluation of the energetic processes, mentioned above, and their loads on the containment requires a good description of the in-vessel accident progression and estimates of its products: fission products, hydrogen, and melt characteristics. Accurate description of the in-vessel accident progression is also essential for evaluating the success of the accident management strategy of retaining and cooling the melt within the vessel by cooling the external surface of the vessel. This, of course, requires the flooding of the containment prior to the arrival of the melt to the lower head. Thus, in the following paragraphs, we will provide a short review of the phenomenology of in-vessel progression for a PWR and a BWR.

1.12.1. In-vessel Accident Progression for a PWR

The TMI-2 accident provides a vivid example of the in-vessel progression for a PWR. This accident was stopped by the cooling of the 20 tons of melt in the lower head and of what remained in the original confines of the core. The morphology of the melt disposition in TMI-2 is what is expected in a PWR, except that in the hypothetical severe accident, the water is not supplied to the vessel. Thus, the melting process in the core will be more prolonged, and the outer rows of the core will also melt and contribute to the melt volume that would transfer to the lower head. The melting process will follow the route of candling, blockage at the lower edge of the core, melt pool formation in the core, and its break through either on the side or at the bottom to pour into the lower head. Sufficient research results have become available to describe the early part of the in-vessel accident progression quite accurately.

The late phase of the in-vessel accident progression begins with the transfer of core melt from the core region to the lower head. This most probably would occur as a jet, which may break up or impinge on the wall of the bottom head. The issues for PWR have been: (a) the possibility of a sufficiently energetic steam explosion to rupture the lower head or to rupture the bolts on the upper head of vessel, producing a missile, which would tear the containment and (b) the failure of the lower head due to the thermal loads imposed by the melt jet. Both of these issues have invoked much research, analysis, and evaluation.

The lower head was found to be strong enough to withstand a strong explosion [27], and the failure of the vessel upper head and its subsequent flight to the containment and impact-failure of the containment was found to be [28] of very low (10^{-3}) conditional probability. The jet thermal loads were also found to be insufficient to make a hole in the vessel or to fail a penetration.

In the absence of the immediate failure of the lower head, the melt interacts with the water in the lower head and most probably will form debris, which may not be coolable. Most probably the water will be evaporated by the sensible heat delivered from the melt, resulting in a dried out particulate debris bed, which is generating decay heat. This leads to formation of a circulating melt pool with a metal layer on top, which would fail the lower head, not at the bottom but on the side of the lower head.

Some issues are still outstanding on the late phase in-vessel accident progression of PWRs, in particular, on the effects of melt composition and chemistry on the melt pool stratification, which affects thermal loading on the vessel wall. More research work is anticipated. Besides that, we believe sufficient knowledge has been gained to make reliable predictions about the consequences of the late phase in-vessel progression of the accident in a PWR.

1.12.2. In-vessel Accident Progression for a BWR

The in-vessel accident progression for a BWR is not as well known as the progression for a PWR, simply because there are no data on a BWR like that obtained from TMI-2 for a PWR. It appears that the early melting of the cruciform control rods and accumulation of their melt on the lower core plate may lead to the failure of the lower core plate, particularly in the higher probability dry core scenario. The melt resident in the core may discharge through the bottom of the core into the lower head. The channel boxes on the fuel bundles in a BWR core do not promote core-wide blockages, and melt from individual bundles may dribble down to the lower head, obviating the issues of melt jet impingement and steam-explosion-induced failure of the lower head. The melt jets will break up and form debris.

The quenching of melt debris and its subsequent re-melting are part of the late-in-vessel scenario for a BWR. The lower head of a BWR contains hundreds of in-vessel control rod guide tubes (CRGTs), which could have a small water supply. These tubes may provide some capability for cooling the melt debris and certainly can prolong the late-phase in-vessel accident progression. The failure of a BWR lower head would most probably occur at one or more of the many penetrations.

The BWR in-vessel accident progression may lead to greater hydrogen concentrations than for a PWR due to the presence of much more zirconium coming from the channel boxes. The BWR also contains much steel. Thus, the composition of the melt pool and of the discharged melt to the containment will include much more metal (stainless steel and Zr). The late-phase in-vessel accident progression for a BWR could last longer than for a PWR, since the BWR lower head contains much more water than does the PWR lower head. However, the possibility of the early failure of a penetration could shorten this phase.

The BWR would also form a circulating melt pool with a metal layer on top. This metal layer may be quite thick due to the large mass of the metals in the core and in the lower head of a BWR. The BWR lower head could also be cooled from the outside and the melt retained in the lower head. Such a scheme has been proposed by Siemens/Areva for their BWR design.

1.12.3. Fission Product Release and Transport during the In-vessel Accident Progression

The core heat-up and melting release 60 to 80% of the volatile fission products, for example, iodine and cesium isotopes. Tellurium is sequestered by the unoxidized zirconium in the core and is released when the Zr is oxidized. The volatile fission products form compounds with each other and with the steam carrier. The predominant compounds are CsI and CsOH, which are highly soluble in water. Some small fraction of the Iodine released may be in gaseous form.

The vapor compounds form aerosols as they encounter the cooler regions, for example, the upper internals in a PWR or the separators and driers in a BWR. The aerosol transport process from the core to the containment can result in the deposition of a sizable fraction of the released fission products on the surfaces of the primary system piping. This source term is not immediately available in the containment but can become available as the decay heat contained in the fission products heats up the primary system to revaporize the deposited fission product compounds. This late production can be a significant source term in the containment available for release to the environment. Again, if the containment does not fail at least 4 hours after the revaporized fission products reach the containment, there is no significant increase in the environmental or public impact.

1.12.4. Ex-vessel Accident Progression

The study of the ex-vessel accident progression of a severe accident basically focuses on the processes that may fail the containment. Direct containment heating (DCH), hydrogen detonation, steam explosion, and melt attack on the BWR Mark-1 containment liner were identified as the loads, which could cause early containment failure. We shall provide very brief discussions on these topics in the following paragraphs. Some of these subjects are treated in detail later in this book.

Direct Containment Heating (DCH)

Failure of the containment by the rapid pressure increase caused by the heating of its atmosphere by the fragmented core melt, as it is released at high pressure on vessel failure, is the process. A focused program for experiments and analysis was performed in the United States in the 1980s and early 1990s to show that the pressure reached for the Westinghouse reactor containments will be below the containment failure pressure and the containment failure for these plants is highly unlikely. This good result for the Westinghouse containments was due to the configuration of the lower part, which directs most of the fragmented melt particles to a dead-end room where they are trapped. Only a small fraction of the particles from the discharged mass will reach to the main containment volume. This may not be so for the containments of other reactor types, for example, the German and French PWRs. These containments are currently under study.

The DCH is prevented through unintended depressurization due to the failure of the surge line caused by the natural circulation flow of very hot steam from the core to the steam generator. It is also prevented by the intentional depressurization that can be affected by the opening of the relief valves by the operator in case of a high-pressure severe accident scenario or by the cooling of the primary water by increasing the flow in the secondary side of the steam generator. The BWRs generally depressurize for most accident sequences through the automatic depressurization system (ADS).

A model has been constructed [26] for the DCH, which can be employed for calculating the containment pressure response.

Hydrogen Detonation

The hydrogen combustion loads were the first that the NRC addressed immediately after the TMI-2 accident, since there was a 2-bar pressure spike in the TMI-2 containment due to the combustion of the hydrogen produced by the Zircaloy oxidation during the core heat-up. The hydrogen rule required management of the hydrogen combustion loads. The small containments of BWRs have all been inerted by replacing most of the air with nitrogen. The PWR ice condenser and the BWR Mark-3 containments have installed igniters to burn the hydrogen as it is produced and not allow it to reach concentration above 10%. The PWRs in the United States have not installed any special devices for hydrogen control. However, the PWRs in Europe have installed passive systems for hydrogen recombination. It should be mentioned here that high concentration of steam (generally present) in the containment atmosphere effectively acts as an inert gas.

Hydrogen detonation can also occur on turbulence generation due to flow or other sources. The phenomenon of transition to detonation is under investigation. It has been found to occur only for quite high hydrogen concentration mixtures [30]. Thus, hydrogen detonation can be prevented through management of hydrogen concentration in the PWR containments.

Ex-Vessel Steam Explosion

Ex-vessel steam explosion can be postulated for (Westinghouse) PWRs that flood their cavities and for the Swedish BWRs that flood their dry walls. These actions are intended in both of these reactors as an accident management strategy for cooling the melt discharged from the vessel and preventing the MCCI to occur in their basemats.

Steam explosion has a long history of research that appears to be increasingly likely to exclude the occurrence of a strong steam explosion. However, the results are not conclusive. The available experimental evidence from the various tests conducted [31] so far is that the oxidic melt: (noneutectic UO_2 + ZrO_2) does not easily explode, and when it does it has a very low energy yield. The noneutectic corium melt mixture has a small separation between the solidus and the liquidus temperature lines. The cooling transient that a melt droplet suffers in water tends to make a mushy layer at the boundary, which prevents its fine fragmentation needed for a steam explosion. There are other limiting mechanisms, such as vapor formation in the vicinity of the droplet due to the thermal radiation from the droplet. A quantification of these effects is under development but is not completed.

Elaborate three-field analysis codes [32] have been developed, but their validation is questionable. Fundamental single-drop experiments are being performed to discern the role of physical properties in the explosivity of melt and the energy yield.

Melt Attack on BWR Mark-1 Containment Liner

This mode of early containment failure is particular to the Mark-1 BWR due to the short distance between the vessel and the containment liner. The contention was that the melt discharged from the vessel, after its failure, would traverse that distance readily and attack the liner to fail it. Experiments and analysis [30] showed that if the dry-well were to fill with water, the thermal load on the liner would be insufficient to fail it. Thus, the preventive action for the Mark-1 is to add water to the dry-well before vessel failure in order to prevent early containment failure. Such provisions have been made for the BWR Mark-1 plants.

Molten Corium Concrete Interaction (MCCI) and Basemat Failure

The MCCI occurs for the dry containments in which no water has been added or present. It also occurs in wet containments if the melt pool cannot be cooled. The MCCI leads to concrete melting and gas (H_2O, CO, CO_2) generation. This results in: the pressurization of the containment by the noncondensible gases CO and CO_2 and the gradual sinking of the melt pool into the basemat. The melt pool forms crusts at its top surface, thereby making the heat loss to the containment environment quite low. It is feared, though not proven, that the heat-generating melt could, in time, melt through the basemat and attack the ground underneath. It would, however, probably pressurize the containment sufficiently to fail it before the complete melt-through of the basemat. This, of course, depends on the thickness of the basemat provided in a particular plant.

Another concern of MCCI is the ratio of the radial/axial ablation. It appears from the most recent MCCI experiments [31] that the radial ablation may be a factor of two greater than the axial ablation for the siliceous concrete employed in Europe. It is almost the same ablation for the limestone-common sand concrete, employed in U.S. NPPs, in both radial and axial directions. The primary difference between the two concretes is that the limestone-common sand concrete generates much more gas than the siliceous concrete does.

The codes for the MCCI do not describe the recent experiments as well as needed. Thus, more research is needed on the MCCI process. The MCCI can be avoided by cooling the melt pool to temperatures below the concrete ablation temperature of ~1000°C. The coolability of ex-vessel corium particulate beds and melt pools is discussed next.

Melt Debris Stabilization and Coolability

Cooling of Ex-vessel Particulate Beds

Melt coolability is perhaps the most vexing issue impacting severe accident containment performance in the long term [32]. As mentioned earlier, melt coolability is essential to prevent both the basemat melt-through and the continued containment pressurization, thereby stabilizing and terminating the accident, without the fear of radioactivity release from the containment.

Provision of deep (or shallow) water pools under the vessel may not ensure long-term coolability (quenchability) of the melt discharged from the vessel. Interaction of the melt jet with water may lead to a particulate bed, which may be difficult to cool if it has low porosity and/or large depth. Incomplete fragmentation could lead to a melt layer on the concrete basemat under a particulate debris layer and a water layer.

The coolability of the ex-vessel particulate debris beds in the BWR dry-well and in the Westinghouse PWR vessel cavity is determined primarily by the dry-out heat flux, since the bed will be waterlogged. The bed will most probably be radially and axially stratified. It could also have very low porosity and a small mean particle size if a steam explosion occurs, which produces very small size particles. It has been observed in the Porous Media Coolability-facility at the Royal Institute of Technology [Kungliga Tekniska Högskolan (KTH)] that beds with porosity ~40% and particle size ~2 mm are coolable with top flooding. Lower porosity and smaller particle beds are not easily coolable, except when water is injected from the bottom [36].

We believe that the ex-vessel debris beds will be three dimensional and stratified. There should be transverse paths available for water to penetrate the bed and cool the regions where dry-out may occur. Most of the debris bed experiments performed so far have provided one-dimensional addition of water, either from top or bottom. We certainly expect that the 3-D cooling will be much more efficient than top flooding.

Cooling of Ex-vessel Melt Pools

The coolability of a melt pool interacting with a concrete basemat by a water overlayer was under intense investigation in the Melt Attack and Coolability Experiments (MACE) project of Organization for Economic Cooperation and Development / Nuclear Energy (OECD/NEA). Three experiments were performed successfully in which melt pools of 30 × 30 × 15 cm depth, 50 × 50 × 25 cm depth and 120 × 120 × 20 cm depth were generated on top of limestone common sand (LCS) concrete basemats and water added on top. The melt material contained uranium oxide, zirconium oxide, zirconium, and some concrete products. The decay heat generation in the melt was simulated through electrical heating. It was found that for these three tests, the effect of the sidewalls dominated the phenomena, since the insulating crust formed on top of the melt pool attached itself to the sidewalls. The crust prevented intimate melt-water contact, and the heat transfer rate slowly decreased from approximately 2 to 0.1 MW/m^2, which is less than the decay heat input to the melt.

Four modes of heat removal from the melt pool were identified (1) the bulk boiling during the initial melt-water contact; (2) water ingression into the crust and conduction; (3) melt eruptions into water, when the heat generated in the melt is greater than that removed by conduction or water ingression through the crust; and (4) local crust break-up or fractures leading to renewal of melt-water contact, which may form another crust underneath. In the large test (120 × 120 × 20 cm), it appears that significant water ingression occurred, and/or water entry through local holes, since after the test the crust (or cooled melt) was 10 cm thick—that is, about half the melt was cooled. Continued concrete ablation led to the separation of the melt pool from the suspended crust, and the conduction heat transfer decreased substantially.

The integral test program was modified to investigate the modes of heat transfer through separate-effect tests with the intent of developing validated models that could be employed for the evaluation of prototypic coolability configurations.

A new project named MCCI was completed recently under the sponsorship of OECD/NEA. The objective was to continue the separate-effect tests. Tests were performed to study the water ingression mechanism. These tests appear to find that the water ingression mechanism is melt material dependent. In particular, it was found that the addition of concrete products to the oxidic melt pool decreases the water ingression rate markedly. The strength of the crust formed during the water ingression tests was measured and was found to be rather small. This probably indicated that large span crusts will fracture under the water loading imposed during the flooding process. The other mechanism of heat removal from the melt pool: melt eruption into water depends on the gas generation rate from concrete ablation. This mechanism will not be as active in the ablation of the siliceous concrete found in Europe since its gas content is quite low. The ablation of the limestone–common sand concrete may be able to support melt eruptions due to the larger gas generation. However, it is not clear what fraction of the melt pool could be cooled with this mode of heat transfer.

Currently, it is not evident that coolability of a corium melt pool by a water overlayer can be certified. Perhaps, at plant scale, with spans of several meters, the top crust will be unstable, and there would be periodic contact between the melt and water to eventually cool and quench the melt. It is clear that some basemat ablation will occur during the coolability process. One benefit of the water overlayer should be mentioned: the water will scrub most of the fission products that are produced during the MCCI.

Since melt coolability with a water overlayer may be hard to achieve, alternative and innovative means have been explored to cool and quench the melt. Experiments have been performed at the COMET facility in Forschung Zentrum Karlsruhe [FZK, today Karlsruhe Institude of Technology (KIT)] in which water is introduced at the bottom of a melt pool with a slight overpressure, either through nozzles or through a porous concrete substrate. It has been found [33] that melt cooling and quenching is quite readily accomplished and that no steam explosion occurred even with the Al_2O_3 melt. The COMET design has been optimized through a series of experiments at different scales. This concept has merit since it uses the same principle as in the coolability of a particulate debris bed with water injection at the bottom. The co-current water and steam flow are much more efficient in cooling and quenching a melt pool than the countercurrent flow that occurs when the melt pool is flooded at the top. It is advisable to inject water into the melt bottom boundary before a large quantity of siliceous concrete mixes with the corium melt, and imparts greater viscosity and a glass-like structure to the melt.

The COMET concept can be accomplished in the current plants with modifications in their containments.

Another concept that is under study at KTH is that of downcomers built into the containment that channel the water from the top to the bottom of the melt pool, thereby utilizing the already proven high-cooling efficiency of the bottom water injection. A loop will be established in which water goes down in the downcomer and the rising steam (after the evaporation of water entering at the bottom through the debris bed or the melt pool) provides the buoyancy head. Thus, quenching the debris bed or melt pool with top-flooding, which has to fight the CCFL, is enhanced by the much more efficient co-current cooling process brought on by the availability of water at the bottom. We believe that such an innovative cooling system can be installed quite easily in existing PWRs and BWRs, without jeopardizing the regular functioning of the plant and without periodic shutdowns or inspections.

Experiments performed in the POMECO (POrous MEdia Coolability) facility [37] [38] have demonstrated the benefits of the downcomers through a several-fold enhancement of the dry-out heat flux and the quenching rate. A similar experiment performed in the COMECO (COrium MElt COolability) facility with a melt pool, around a downcomer, flooded from the top, indicated [39] the substantial benefit of the downcomer, since a quench front progressed from both the top and bottom of the melt pool.

Containment Bypass

Containment bypass, as the term clearly states, would effectively negate all the beneficial effects of the containment. In the bypass scenario, a path is found for the fission product source term to escape from the containment without its failure. The PWR has two possible such paths: (1) the path from the containment to the auxiliary building caused by an interfacing LOCA and (2) the steam generator tube rupture (SGTR) providing a path to the environment through the dump valves on the secondary side of the damaged steam generator.

We believe that the interfacing LOCA path was identified, and the PWR plants have closed that probability or have made it highly improbable to occur. The SGTR containment bypass path cannot be closed, since the secondary side of the steam generator is a pressure vessel, which normally requires a safety (dump) valve. Some new designs have proposed the discharge of the dump valves into the containment rather than into the environment. However, dump valves in all of the current PWRs communicate with the environment.

The scenario of concern is that of the high-pressure accident like that of the TMI-2. The high-priority accident management action for the high-pressure accident scenario is to depressurize the vessel to obviate the risk of DCH and containment failure. The vessel pressure is reduced too much below that of the secondary side of the steam generator, so that the flow is from the secondary side to the primary side if any tube rupturs. A second accident management action is to fill the secondary side of the damaged steam generator with water immediately so that fission products that may escape to the secondary side from the primary side get absorbed in the water. The secondary side, being a very tall vessel full of water, will have a very high-value decontamination factor.

There is ongoing research on the fission product transport in the empty secondary side of a steam generator. Decontamination factors (DFs) are being measured for aerosol flows with different aerosol size distributions and flow rates. It appears that water filling of the secondary side will be needed to obtain a comfortable margin for DFs.

Fission Product Release and Transport during Ex-vessel Accident Progression

The fission products and the core materials during the core heat-up process arrive in the containment as aerosols. Their transport in the containment is governed by aerosol physics, which determines the fission product concentration in the containment atmosphere as a function of time. As mentioned earlier, if there is steam atmosphere in the containment (as it should be for a severe accident), the fission product aerosol concentration in the containment atmosphere decreases exponentially with time, largely owing to the process of aerosol particle size growth (due to steam condensation), agglomeration, and deposition. Another active aerosol deposition process is that of Stefan flow carrying aerosols to the walls of the containment where the steam is condensing. As mentioned earlier, typically, fission product concentration in the containment atmosphere can decrease by a factor of 10^{-4} in about four hours.

The release of fission products during the ex-vessel accident progression can occur during the MCCI due to the gas sparging and the high temperatures in the melt. The releases of interest are those of the less volatile fission products (e.g., Ba, Sr, Ce, Ru, Mo), since the volatile fission products have already been released.

The Advanced Containment experiments [37] provided systematic data on the release of the above-mentioned fission products. The measured values for releases were less than 1% of the inventory for all of the less volatile fission products. These values were much smaller than what were previously calculated.

Management of the iodine concentration in the containment immediately after the accident and for the long term is essential in order to reduce the potential of harmful releases due to containment leakage or other events. In this respect, the processes of concern are: (1) the interaction of iodine with paints on containment surfaces to form organic iodine, which is difficult to remove; and (2) the radiolytic formation of iodine. Thus, iodine chemistry in the containment is important and the use of pH control to reduce the iodine concentration is needed for the long-term management of the iodine concentration.

1.13. SEVERE ACCIDENT MANAGEMENT

After the TMI-2 and the Chernobyl accidents, it was clear that the great environmental and human costs of Chernobyl would be wholly unacceptable. It was also clear that the consequences for the public of the TMI-2 accident were minor, in spite of the great turmoil caused, and that was due to the TMI-2 containment preserving its integrity. Severe accident research since 1980 has shown that indeed dynamic loads can be imposed on the containment, which could fail the containment early and cause a very large release of radioactivity to the public. Such a large release would pose a greater public risk than that prescribed in TID14844, or even that calculated by WASH-1400. It also became clear that the accidents that may impose large dynamic loads on the containment had to be either prevented or mitigated completely.

This marked the birth of severe accident management (SAM) described in greater detail in chapter 6 as an active tool for minimizing the public risk of a severe accident. SAM may be defined as follows: "SAM is the use of existing and alternative resources, systems and actions to arrest and mitigate accidents that exceed the design basis of nuclear power plants."

The earliest SAM action in the LWR plants was that of managing the hydrogen concentration in the containment. The NRC required this action in view of the hydrogen combustion event in the TMI-2 containment. Other actions followed, some requiring backfits in the plants and others mandating operator actions for which training schedules had to be devised. Severe accident management guidelines (SAMGs) were produced for Westinghouse PWRs and General Electric's BWRs, which were appropriately modified for each specific plant. All of the utilities in the United States have already implemented the SAMGs for their individual plants. Some of the European plants also have implemented their individual SAMGs. They have closely followed the guidelines produced by the appropriate owner groups in the United States. They have adapted the set points, curves, and computing aids that were produced by the U.S. owner groups for their specific plants. The French, German, and Swedish plants have a rather open approach for SAMGs, since no generic standard guidelines are employed. Each plant or each set of plants (as in France) performs work on their individual plants to implement SAMGs. In general, each plant uses some equipment backfits that are specially designed to deal with a severe accident. They have also made many procedural and operational changes.

In the following paragraphs, the structure produced in the OECD Report [41] is employed to group the SAM actions under four main functions: (1) cooling a degraded core, (2) managing combustible gases, (3) managing the containment temperature, pressure, and integrity, and (4) managing the release of radioactivity.

1.13.1. Cooling a Degraded Core

Adding water to the reactor pressure vessel (RPV) is an action that is very similarly implemented in many countries. There is a general agreement that the hazards posed by increased hydrogen generation, possible recriticality and increased steam production, do not outweigh the benefits of retaining the degraded core inside the vessel. The criteria generally followed for this action involve supplying water to the reactor vessel as soon as injection capability is available. The generic guidelines issued by Westinghouse Owners Group (WOG) contain warnings about the negative effects of increased hydrogen production, and their computational aids take into account, in a simplified way, the additional risk of hydrogen combustion in the containment. The issue of recriticality is generally considered to have a greater effect on the BWR, where borated water sources are less available and early control rod material meltdown and relocation is a possibility. BWROG generic SAM guidelines specify use of the liquid control system with a maximum water level in RPV in case of core-melt criticality, but no criteria are given on the water flooding rate.

Reactor Cooling System (RCS) depressurization is also a generic SAM action that can be accomplished in a variety of ways. The preferred way for PWR is the "feed and bleed" actions in the EOP domain, i.e., adding water to the steam generators and depressurizing the secondary sides, thereby cooling down the primary side and reducing its pressure. If this action is ineffective, depressurization can be accomplished by direct opening of pressurizer valves. Intentional depressurization has numerous benefits, namely, alternative means of cooling become available, and high-pressure melt ejection is avoided. There are also possible drawbacks, such as increased hydrogen production and higher probability of in-vessel energetic fuel-coolant interaction. All PWRs have pressurizer valves that can be used, although sometimes pressurizer spray is a possibility. All BWRs are designed to be easily depressurized in the event of ADS failure.

The action of containment initial flooding in order to delay vessel failure by means of cooling through the vessel wall is one in which there is considerable variation among countries; it is recognized that the action cannot by itself guarantee vessel integrity, especially for reactors with high power, but the action may delay vessel failure.

Containment flooding to several levels is recommended in the standard WOG Severe Accident Management Guidelines, although specific implementation will depend on the design of the reactor cavity. Generic GE standard guidelines recommend dry-well or primary containment flooding as an integral SAM action that could provide a means of core cooling through the vessel wall, and also a possibility of alternative vessel flooding through the relief valve tail-pipes. German plants do not consider cavity flooding and continue the concept of "dry cavity." Swedish and Finnish BWRs have also implemented the strategy whereby a water pool is created under the vessel as soon as the water level falls below the top of the core. However, the level of water does not reach the vessel and the vessel wall is not cooled. This action is for ex-vessel cooling of debris/melt that deposits into the water pool on the failure of the vessel. This action is not for cooling of the vessel from outside.

1.13.2. Management of Combustible Gases

Considerable variations of strategies are followed to reduce H_2 and CO inventory in the containment because of the differences in existing equipment and the status of implementations. Many countries have decided on the use of catalytic recombiners in PWR containments, which can reduce H_2 and CO concentrations while keeping containment pressure low. Some BWRs and some PWRs use igniters in order to produce intentional H_2 or CO burning. Venting of the containment is a strategy considered also for reducing combustible gas inventory.

Catalytic recombiners have demonstrated their capability of reducing H_2 concentration under steam-inerted atmospheres, very low H_2 concentrations, and presence of aerosols [41]. Recombiners have been installed in Belgium, Germany, France, and the Netherlands and in some eastern European countries as well. Finland has decided on the installation of a new H_2 management system using catalytic recombiners, although currently igniters are being used. GE BWRs with Mark I and II containments and KWU German BWRs of old design are inerted and do not use ignition devices. GE BWRs with the larger Mark III containments have ignition systems. The PWRs in the United States do not have any hydrogen management system. Their containment volume is supposed to be large enough to not create a high (above 10%) concentration hydrogen mixture. Most PWRs in Europe are being fitted with passive catalytic recombiners.

1.13.3. Management of Containment Temperature, Pressure, and Integrity

Automatic or manual initiation of containment sprays to condense the steam released exists in most BWRs and PWRs, although significant variation exists in the equipment dedicated to the implementation of this action. Sprays are also used, in the longer term, in conjunction with heat exchangers, which can extract heat from the containment to avoid pressurization. German plants have no spray systems. Swedish plants have an independent dedicated spray system. Loviisa in Finland, Zorita in Spain, and two Belgian plants have external spray systems for their steel containments.

Fan cooler systems in PWRs can extract heat and avoid late pressurization due to release of noncondensable gases during MCCI, but not all plants have fan coolers as qualified safety grade equipment. The initiation of fan coolers for SAM in PWR containments is considered in Belgian, Spanish, and UK plants, and it is included as a standard action in WOG SAMG.

Containment flooding is considered in both PWRs and BWRs. Also, a consensus is developing that initial containment flooding will improve the chances of ex-vessel melt coolability in case of vessel breach, in spite of the higher risk of energetic ex-vessel melt-water interactions, and will reduce ex-vessel radioactive releases. Here we have to distinguish between PWRs with their larger and relatively strong containment and BWRs with their small containments and perhaps vulnerable vessel support structures, whose integrity may be threatened by a highly energetic steam explosion. Containment flooding before the discharge of the melt is not practiced in the French and German PWRs and BWRs. There the water is added only after the melt has been deposited on the basemat. Coolability of the melt is not expected. There will be ablation of the concrete basemat and possibly a basemat melt-through.

Many European plants include the strategy of containment venting, to avoid late failure due to overpressurization. Scenarios such as complete loss of containment heat removal capability, or full power ATWS in BWR, are typical examples where containment venting becomes essential. This accident management action can avoid late failure due to pressurization by noncondensable gases released during MCCI, for which containment heat removal systems are ineffective. Venting can also be used also to ease containment flooding and to reduce the inventory of combustible gases. Considerable variation exists in the implementation of this SAM feature. The standard WOG SAMGs do not include containment venting as an ultimate SAM strategy. Venting of containment with specially designed filtered vents systems is implemented in all PWRs and BWRs in France, Germany, Sweden, the Netherlands, and Switzerland. The UK, Belgian, and Spanish PWRs do not have venting. Spanish BWR have a dedicated manually operated venting system that connects the suppression pool airspace to the off-gas stack, without filtering. All the Mark-I BWRs in USA have installed hard vents, but no filters.

1.13.4. Management of Radioactive Releases

Standard strategies for mitigating the rate of radioactivity release through opening in the containment boundary include reducing the containment pressure by means of available containment heat-removal systems and through the venting systems. At later times in a severe accident revolatalization releases from the deposited aerosols in the RCS become a concern. Mitigation of those releases will involve cooling of the RCS walls.

A common strategy, for reducing the inventory available for release in the containment is the initiation of containment sprays in PWR and BWR. Sprays were designed for early operation and steam condensation after LOCA, and not for long-term operation during severe accidents. However, sprays can produce effective aerosol deposition due to the interception of droplets. Also, sprays can remove some of the gaseous molecular iodine as long as they do not become saturated with iodine. The effectiveness of sprays will depend on the availability of AC power and the extent of the areas covered by the spray system. Iodine volatility in many PWRs is reduced by means of additives that are included in the design of containment sumps, or the containment spray system.

Engineered filtering systems are installed in most PWR and BWR, with High Efficiency Particulate Air (HEPA) filters generally designed for conditions of normal operation. Use of engineered filtering systems during SA environmental conditions is possible, but the efficiency of the filter may be reduced if additional technical features have not been provided (i.e., emergency filtering systems). Swedish plant containments have a venturi-filtered venting system specially designed to deal with severe accident situations, which have a very high value of decontamination factors. The French plants have a sand-based filtered venting system for severe accidents.

Removal of radioactive aerosol by means of scrubbing in BWR suppression pools is a beneficial side effect of the suppression pool functional design. The scrubbing efficiency however decreases considerably if the suppression pool becomes saturated. Aerosol scrubbing by means of a water pool overlying the core debris is also considered in standard WOG and GE standard SAMG, as a strategy to reduce ex-vessel releases to the containment.

Secondary side flooding is a standard strategy, included in WOG SAMG, for mitigation of releases to the environment due to SGTR accidents and for protection of SG tubes from creep ruptures.

1.14. THE FUKUSHIMA ACCIDENTS

1.14.1. Introduction and Plant Characteristics

The Fukushima accidents are the most recent significant chapter in the history of LWR safety. These accidents occurred on March 11, 2011, and they were the result of a "common-cause" generated by an extreme natural event: an exceptionally strong earthquake accompanied by a colossal tsunami, off the northeastern coast of Japan, where the Fukushima Prefecture is located. The Fukushima plant is almost 240 km northeast of Tokyo, and the population density of the Fukushima prefecture is not very high. A map of Japan shown in Figure 1.24 shows the location of nuclear power plants relative to the epicenter of the earthquake. Figure 1.25 shows the architecture of the site with four reactors (Units 1 to 4) located very close to each other; and units 5 and 6 located next to each other but some distance away from Units 1-4. The plant is right next to the Pacific Ocean.

The earthquake that occurred on the afternoon of March 11, 2011 measured 9.0 on the Richter scale, and it involved the movement of part of the tectonic plate upward in the floor of the earth below the Pacific Ocean. As seen in Figure 1.24, this segment of the tectonic plate was not too far away from the Onagawa and the Fukushima Daichi and Daini stations. This type of movement of the tectonic plate can generate a colossal tsunami, as it did. The seismic design basis for the Fukushima plant was for an earthquake of 8.2 intensity on the Richter scale. The plant shut down automatically and did not suffer any visible structural damage. However, the AC electric power lines in the area were damaged, and the plant was without its off-site power supply.

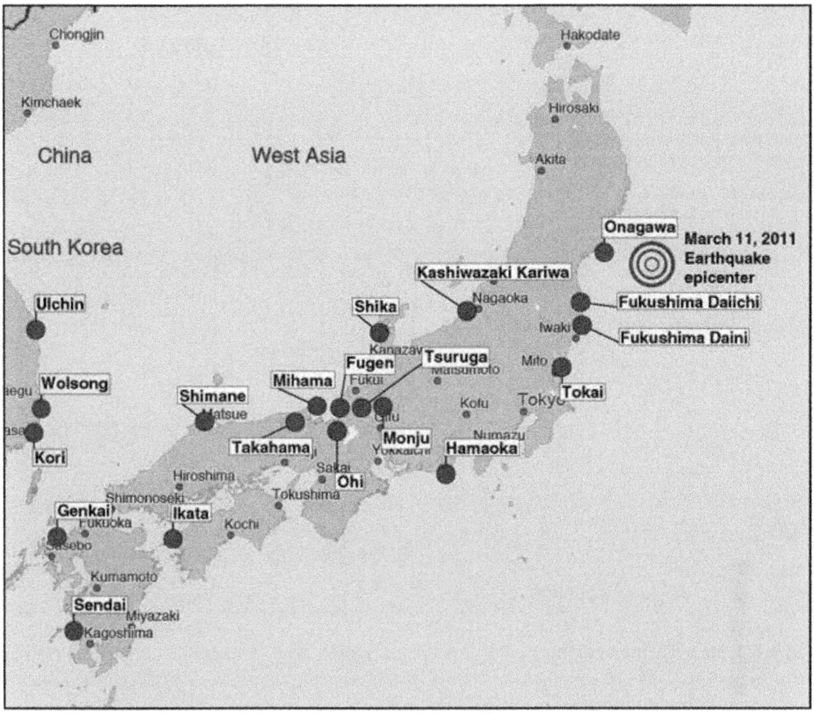

FIGURE 1.24 Location of earthquake and the Japanese nuclear power stations.

FIGURE 1.25 The Fukushima station.

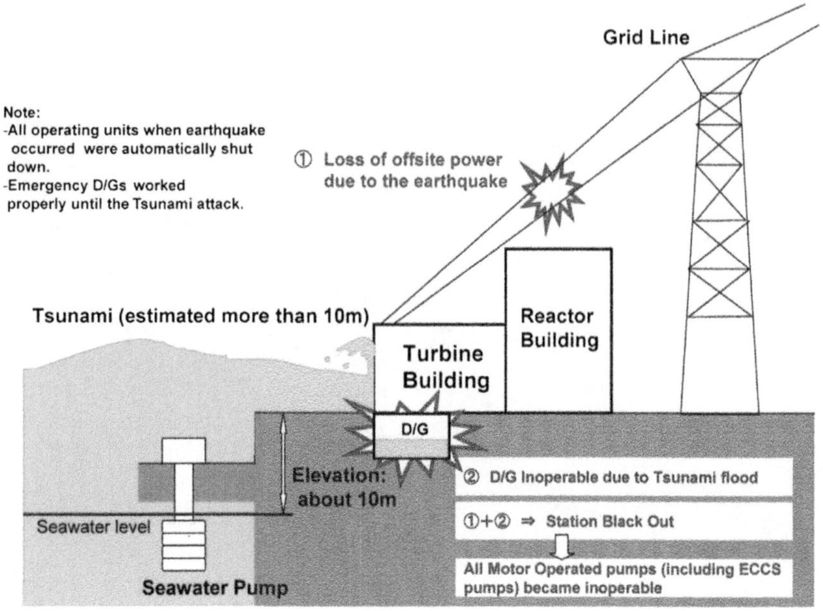

FIGURE 1.26 The topography of the Fukushima station with respect to the tsunami wave. *(From Report of NISA: Nuclear and Industrial Safety Agency, Japan and JNES: Japan Nuclear Energy Safety Organization, 4th April 2011)*

The tsunami that followed one hour later was video recorded and shown on TV news all over the world. The tsunami destroyed countless homes, commercial buildings, boats, cars, and property as seen on TV screens in "live action." It killed perhaps 26,000 people and left hundreds of thousands homeless, who are currently in many shelters some distance away from the Fukushima plant.

The tsunami wave near the Fukushima plant was colossal, probably due to the topography of the terrain. Its height was 14 meters as it reached the location of Units 1 to 6 of the Fukushima plant. The Fukushima plant designers had considered a tsunami in the design basis of the plant, though the maximum height of the wave was taken as ≈5.0 meters in the design basis. Figure 1.26 shows that the location of the turbine building was 10 meters above the sea level, and the two diesel engines provided for the emergency electrical power for the plant were located in the basement of the turbine building. They were operating and supplying power to the plant after the earthquake, but they stopped working as they got flooded. For the Fukushima Daiichi Station of six reactors, all (except one for Unit 6) diesel generators stopped functioning one hour after the off-site power supply was lost. This made the common-cause accident a long-duration-station blackout accident. Loss of off-site power has occurred in several nuclear plants over the years, but has generally been corrected within 30 minutes and the diesel generators are not needed after that time. There have been a few events of 24-hour loss of off-site power, as one happened in the eastern part of the United States a few years ago, during which the diesel generators provided the back-up power, operating all the time that the off-site power was lost. In general, station blackouts are rare events, let alone the long-duration blackout. However, they are of extreme concern, since without any power available, all instrumentation and safety systems are inoperable. Batteries have been provided in almost all the plants with length of operation of 4 to 8 hours. They are very helpful, providing illumination and power for instrumentation and for operation of some essential safety-related valves and pumps. The batteries by their very nature are not a solution for a very long-duration station blackout.

As mentioned earlier, the Fukushima plant has six reactors, five of which were Mark-I BWRs, the first BWR designed and manufactured by General Electric for commercial operation. Table 1.7 provides the specifications for these reactors. Unit 1 was the oldest, with a start of commercial operation in March 1971 and with the power level of ≈460 MWe. The next four units (N°2–5) followed with commercial operation in 1974 to 1978, and they had a power level of ≈800 MWe. Unit 6 was of Mark-II design with a power level of 1100 MWe—basically of the size that General Electric sold in the 1970s. It should be noted that all the Mark-I units had two emergency diesel generators, and the containment design pressure level was ≈4 bars.

The Mark-I and –II BWRs designed by General Electric Co. are distinguished by the small-size containment, since a body of water (suppression pool) is provided in the containment to condense the steam that is produced during the safety design base accident of the large LOCA. Condensation absorbs the heat energy discharged into the containment, and the containment volume can be reduced considerably, while keeping the design pressure level for the containment of three to five bars. Thus, the volume of the BWR Mark-I containment, shown in Figure 1.27, is only 12–15% of the volume of a PWR containment. The hydrogen generated during the severe accident in a BWR becomes an issue for the small-volume BWR Mark-I or Mark II containments since the BWR has much zirconium in the core and its oxidation, during a severe accident, generates perhaps more than a thousand kilograms of hydrogen. Its rapid accumulation in the Mark-I containment can pose a danger of detonation in the containment. That is why the BWR containments Mark-I and II are inerted with nitrogen to prevent any combustion in the containment volume itself. Since hydrogen is a noncondensable gas, its addition to the containment of Mark-I and Mark-II can raise the containment pressure sufficiently to reach the design base values shown in Table 1.7. For that eventuality, the Mark-I containments are provided with a vent line from containment to the stack, as shown in Figure 1.28. The vent line is connected to the gas volume in the suppression pool, since the passage of gases through the water would decontaminate the discharged gases by absorption of a considerable fraction

TABLE 1.7 The Specifications for the Reactors at the Fukushima Station

	Unit 1	Unit 2	Unit 3	Unit 4	Unit 5	Unit 6
PCV Model	**BWR-3** Mark-1	**BWR-4** Mark-1	**BWR-4** Mark-1	**BWR-4** Mark-1	**BWR-4** Mark-1	**BWR-5** Mark-2
Electric Output (MWe)	460	784	784	784	784	1100
Max. Pressure of RPV	8.24 MPa	8.24 MPa	8.24 MPa	8.24 MPa	8.62 MPa	8.62 MPa
Max. Temp of the RPV	300°C	300°C	300°C	300°C	302°C	302°C
Max. Pressure of the CV	0.43 MPa	0.38 MPa	0.38 MPa	0.38 MPa	0.38 MPa	0.28 MPa
Max. Temp of the CV	140°C	140°C	140°C	140°C	138°C	171°C(D/W) 105°C(S/C)
Commercial Operation	1971,3	1974,7	1976,3	1978,10	1978,4	1979,10
Emergency DG	2	2	2	2	2	3*
Electric Grid	275 kV × 4				500kV × 2	
Plant Status on Mar. 11	In Operation	In Operation	In Operation	Refueling Outage	Refueling Outage	Refueling Outage

**One Emergency DG is air-cooled.*

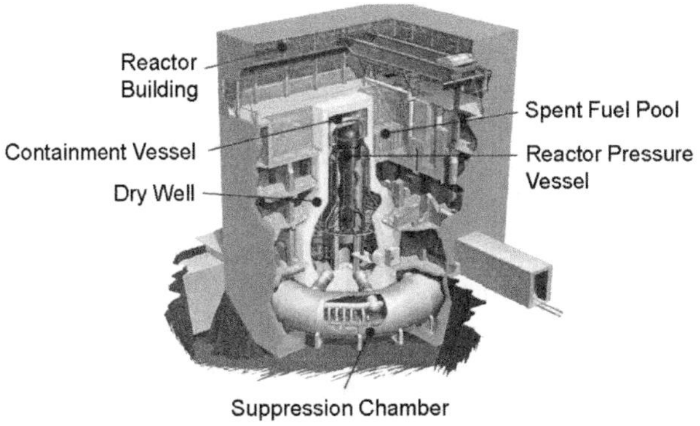

FIGURE 1.27 The configuration of the Mark-I BWR's vessel, containment and the reactor building.

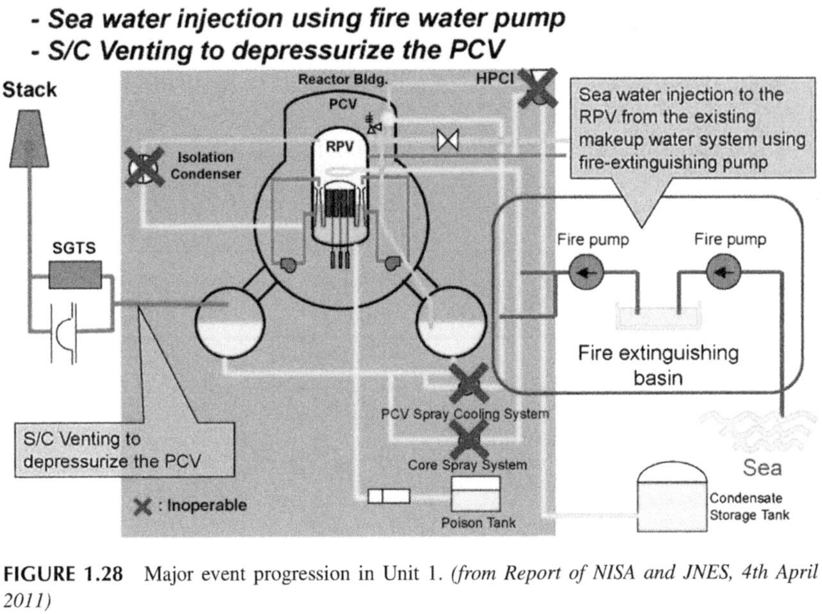

FIGURE 1.28 Major event progression in Unit 1. *(from Report of NISA and JNES, 4th April 2011)*

of the fission products in the water. Many European BWRs have installed filters in the containment vent to further decontaminate the discharge gases by filtering the radioactive fission products, since the decontamination achieved in the water of the suppression pool drops considerably when the water becomes saturated, due to absorption of steam in the suppression pool during the severe accident.

The initial conditions of the plant before the earthquake and the arrival of the tsunami were the following ones: Units 1, 2, and 3 were in operation, and

Unit 4 had been shut down recently and had its fuel discharged into the spent fuel pool on top of the containment. Units 5 and 6 had been shut down for a considerable period of time and were in the cold shutdown state.

1.14.2. Consequences of a Conservative Core-melt Scenario for Fukushima Reactors

Several computer codes can predict the consequences of severe accident scenarios for LWRs. These include the U.S. codes MAAP and MELCOR, the European code ASTEC, and the Russian code SOCRAT (see Chapter 8 for their detailed description). In general, predictions of severe accident consequences with these codes are quite similar within the uncertainties of the data and the complex processes they are modeling. In the following we will present one set of results [39] for the consequences calculated for a postulated most conservative (worst-case) severe accident scenario, which assumes that there is a station blackout following the failure of all active and passive emergency cooling systems. Table 1.8 shows the progression of events, and Table 1.9 provides the fission product radioactivity release.

Table 1.8 shows the importance of time after failure of all injections of the cooling water. It is seen that it takes only 2 hours and 20 minutes for the core water level to drop sufficiently to cause the fuel rods to heat up to the start-of-Zircaloy-oxidation temperatures. Ten minutes later the fuel melting starts and

TABLE 1.8 Progression of Events for the Conservative (Worst-Case) Scenario for the Fukushima Reactors

Time	Events
0	Station blackout with the failure of all emergency cooling systems including passive
2h20′	Start of steam—zirconium oxidation
2h 30′	Start of cladding melting
3h 30′	Depressurization of reactor, low-pressure scenario
5h	Start of fuel melting
7h	Melt release in the vessel, residual water evaporation, partial corium cooling
12 h	Pressure vessel dry-out, corium heating, start of start of vessel melting
13 h	Reactor vessel failure, melt release in the containment drywell, start of molten core concrete interaction
7 days	Concrete basement melt through (of about 6 m)

TABLE 1.9 Radioactive Releases for the Conservative (Worst-Case) Scenario for a Fukushima Reactor

	Isotope activity, Bq		
Unit No.	^{131}I	^{134}Cs	^{137}Cs
1	$6.1 \cdot 10^{17}$ (release after 31.2 hours)	$1.9 \cdot 10^{17}$ (release after 35.5 hours)	$1.3 \cdot 10^{17}$ (release after 35.5 hours)
2	$1.8 \cdot 10^{18}$ (release after 77.3 hours)	$8.3 \cdot 10^{17}$ (release after 84 hours)	$4.80 \cdot 10^{17}$ (release after 84 hours)
3	$1.0 \cdot 10^{18}$ (release after 60 hours)	$4.2 \cdot 10^{17}$ (release after 62.4 hours)	$2.4 \cdot 10^{17}$ (release after 62.4 hours)

the downward relocation of the melt causes the vessel failure about ten and a half hours later. Clearly, the time span available for action (addition of cooling water) is of the order of 2 hours, for the core to escape any damage, and of 4–5 hours to stop core-melt progression. The attack of the melt on the containment basemat, leading to basemat melt-through ("China syndrome") after 7 days is an event of considerable concern, since issues of long-range ground contamination and possible water- table contamination may arise. In general, the public would be deeply concerned about such an event.

The radioactive releases shown in Table 1.9 provide the release of the ^{131}I, ^{134}Cs and ^{137}Cs. Iodine-131 has a half life of $\approx$8 days, whereas ^{137}Cs has a half life of 30 years. The latter is the isotope important for the habitability of the nearby ($\leq$20–30 km) population in their own homes. These releases are substantial. The timing of these releases in Table 1.9 is based on the timings of the stop of all water injection into the individual Fukushima cores, as described in the next section.

1.14.3. The Actual Progression of the Fukushima Accidents

The material presented below is based on information collected through June 10, 2011. The information is perhaps not completely accurate and some what speculative in as much as the information on the accident progression in the three Fukushima reactors remains insufficient and the values quoted for various parameters may need to be revised in the future.

Long-Duration Total Station Blackout (SBO)

The flooding of the emergency diesel engines and the batteries at Fukushima led to the truly feared event of a long-duration-total-station-blackout, with no prospect of a corrective action of the 30-minute variety. The operators were operating in darkness, and most probably they were disoriented and worrying about the safety of their families. It is not known whether they had immediate communications with any technical support center or crisis center, where people more knowledgeable of the progression of the severe accident could advise them as to the course of appropriate action they should pursue. In Japan, any nuclear emergency has to be brought to the attention of the office of the Prime minister, and it appears that it was done, except that the timing of that report, in terms of the number of hours after the core cooling stopped, is not known to this author.

As mentioned earlier, Fukushima's Units 1 to 3 had shut down as soon as the first shock waves of the earthquake had arrived at the station; but the cores had to be cooled to extract the decay heat and heat had to be delivered to the ultimate heat sink (sea water in the case of Fukushima reactors) in order to keep the containment pressure within the design limits. The core damage (and melt) can be avoided if cold water can be supplied to the core to replenish water that is boiled off with the decay heat. The BWRs isolate the vessel with the operation of the main steam isolation valve (MSIV) from the turbine as soon as the reactor scrams in order to prevent the transport of any radioactivity to the turbine and other equipment. Steam formed in the vessel increases the vessel pressure that is relieved through the opening of automatic depressurization system (ADS) valves. The vessel steam is condensed in the suppression chamber, and a residual heat removal system is designed to cool the suppression chamber water to complete the decay heat removal process to the ultimate heat sink.

The lack of electrical power disabled both the delivery of cold water to the core in order to keep the core covered and the access to the ultimate heat sink. The boil-off in the cores produced steam that raised the vessel pressure, which was relieved by the opening of the ADS valves. This brought the steam to the suppression chamber, where the steam condensed until the suppression chamber water became saturated and the containment pressure started to rise.

Meanwhile, the water level in the reactor vessel dropped to a level about halfway below the top of the core, when the steam generation rate became too low to be able to cool the fuel rods in the dry portion of the core. The clad temperatures started escalating, and as the temperatures reached $\approx 1100°C$, Zircaloy oxidation by steam (producing hydrogen) set in and the exothermic reaction increased the clad and fuel temperatures exponentially. After that, there was no turning back and the core meltdown became a certainty. Fission products were released, and the combination of hydrogen, steam and fission products moved to the containment. The suppression pool was already (or soon) saturated, and the containment pressure started to increase rapidly.

What Else Went Wrong?

Something else went wrong! This was the venting system to relieve the pressure in the containment. When the containment pressure increased to 8.4 bars, that is, almost twice the design pressure in Unit 1, the operators were in panic and they had to vent the containment. The venting system was under the control of the operator, who attempted to vent the containment gases to the stack as shown in Figure 1.28. The valve system controlling the venting worked only partially, and it appears that the gases in the containment—hydrogen, nitrogen, and fission product vapors leaked perhaps through leak paths generated in the PCV when it was at high pressure into the building surrounding the light-bulb-shaped primary containment vessel (PCV). This building serves as a confinement building with regular atmosphere, and after the large quantity of hydrogen was released, it either detonated (had the necessary concentration) or had a strong deflagration burn. The roof of the building, which were lightly reinforced with steel, were blown out, opening the path to the environment for release of the fission products. The hydrogen explosions dispersed the volatile fission products (e.g., iodine, cesium, and tellurium) into the atmosphere. This happened in all three units. In a way it was reminiscent of Chernobyl, although at Chernobyl it was the explosion in the core itself and the release there was much larger. It should be recalled that the Chernobyl release was of all the core materials, and it was with orders of magnitude greater energy that carried the plume high up into the atmosphere and the radioactive fission products traveled, with the winds, to great distances. The Fukushima releases, without the hydrogen explosions, would have been of a much more local character; and even with hydrogen explosions, the dispersal of the radioactivity was primarily within a 50-km radius.

The hydrogen explosions were seen live by the world audience as they occurred, and they communicated the urgency of the situation, with respect to the injection of water in the vessel. In Unit 2, the hydrogen explosion damaged the suppression pool, and some water from this large body of water may have been released to the reactor building, from where it traveled to the turbine hall and from there to the other structures (tunnels, sumps etc.) next to the shores of the Pacific Ocean.

Some Details of the Actual Scenarios for Units 1, 2, and 3

The actual progression of the accidents in Units 1, 2, and 3 was different from the conservative scenario assumed in Table 1.8. Fukushima's Units 1, 2, and 3 contain core cooling equipment that does not need AC power. In Unit 1, the isolation condenser (IC) system had been installed; Unit 2 contained the reactor core isolation cooling (RCIC) system, and Unit 3 contained the RCIC and the high-pressure coolant injection (HPCI) ECCS systems. These systems operated for a certain period of time, and the operation of the plant without any water cooling or injection varied from reactor to reactor. The report from Japan to the IAEA [40] estimated that the stoppage of water injection to Unit 1 was for 14 hours and 9 minutes. This, of course, is a very long period of time; note that in Table 1.8 the grace time is 2–3 hours. The core of Unit 1 may have produced 800-1000 kg of hydrogen. This hydrogen and the fission products came to containment as the vessel valves opened and closed and from there were leaked to the reactor building. Unit 1's containment was vented at ≈14:00 on March 13 and a hydrogen explosion in the reactor building occurred at ≈15:36 on March 13, blowing out the roof and dispersing the fission products into the environment. It is quite probable that most (70–80%) of the core melted and moved to the lower head. It is also quite probable that the lower head was breached and that some of the core melt may be on the basemat in the containment dry-well. In Unit 1, water was added to the containment, which presumably has covered the core melt. However, the core melt may be reacting with the basemat concrete, if the melt depth is ≥20cm. It is well known from MACE experiments, conducted in the United States [25], that deep corium melt pools are difficult to cool by water flooding from the top alone.

The progression of the accident in Unit 2 involved core cooling by the RCIC system, which operated with a turbine-driven pump using whatever steam was generated in the reactor. This system worked from March 11 to 13:25 on March 14, when it stopped. The sea water injection started at 19:54 on March 14, thereby implying that the Unit 2 was without water injection for 6 hours and 29 minutes, which again is beyond the grace time for preventing a core-melt accident.

The core of Unit 2 also went through the process of hydrogen generation and meltdown just like that of Unit 1. Most of the fuel melted, and the core melt moved to the lower head, and the reactor pressure vessel (RPV) was breached about 5 hours after the fuel uncovery. The containment pressure-level changes recorded the event. The venting of the containment to the reactor building occurred at almost 3:00 on March 15 and at about 6:00 to 6:10 a hydrogen explosion occurred in the reactor building of Unit 2. This hydrogen explosion also damaged the suppression pool. The hydrogen accumulation into the reactor building may have resulted from leak paths generated in PCV by the earthquake.

The accident progression for Unit 3 also involved the cooling of the core water by the RCIC system. The RCIC stopped at 11:36 on March 12 for unknown reasons. The HPIC, which is also a water injection system, operated from 12:35 on March 12 to 2:42 on March 13, or a total of ≈ 14 hours. After the stoppage of HPIC, no water injection occurred for 6 hours and 43 minutes. The water level in the core was measured at 3:51 on March 14 to be 1.6 meter below the top of active fuel (TAF); that is, the water level in the core was a little bit above the middle level in the core, which implies that the Zircaloy oxidation and hydrogen generation was about to commence. In Unit 3 also, the scenario followed those of Units 1 and 2 and within 2 hours, the core melt had moved to the lower head and started attacking the wall of the RPV and damaged the RPV. Core melt may have come out into the dry-well. This increased the pressure in the containment to beyond its design pressure, and the containment had to be vented. The venting to the stack did not function since the rupture disc did not rupture, and the gases (hydrogen + nitrogen + fission products) leaked into the reactor building through leak paths in the PCV, which may have been generated during the earthquake. It is believed that some of this hydrogen traveled to the location of the spent fuel pool in Unit 4 through vent piping common to Units 3 and 4. The hydrogen explosion in Unit 3 occurred at 11:01 on March 14 and blew out the roof and part of the walls, allowing the fission products from the PCV (the source term) to be dispersed into the environment.

The Spent Fuel Pools Heat Up

The Fukushima reactors store spent fuel in the pools situated on top of the light-bulb-shaped primary containment vessel (PCV). These pools are not inside the PCV, but rather inside the reactor building around the PCV. The number of fuel assemblies for each of the Fukushima fuel pools was 292 in Unit 1, 587 in Unit 2, 514 in Unit 3, and 1,331 in Unit 4. Unit 4 contained the most since the core of that unit was recently unloaded into the fuel pool in order to perform major renovation work on that reactor. The Unit 4 pool thus had the largest amount of decay heat generation, which is estimated to be ≈3MWth. Water in the pools is cooled by a circuit in which the heat is transferred to sea water and the pool temperature is kept about 25°C.

After the tsunami, the circuits to cool the water pools in all of the Fukushima reactors became inoperative. The Unit 4 pool was of most concern due to its greater decay heat generation. The temperature of its water started to increase and was measured to be 84°C on March 14, 2011. A fire occurred on the third floor on March 15 and later on March 16, and the upper part of the building was severely damaged. Water was sprayed with hoses on the building and through holes in the building for 4 days. It was feared that the combustion observed came from the hydrogen produced in the overheated spent fuel rods, and it was presumed that the water level in the pool was below the top of the spent fuel rods leading to fuel overheating, Zr oxidation, and production of hydrogen, which burned. Later it was found that indeed hydrogen burned near the pool of Unit 4, but it came from that produced in Unit 3. These consequences were caused by the close proximity of the multi-units at one nuclear station.

The addition of water to the spent fuel pools of all the units 1–4, through the efforts of the self-defense forces, kept the fuel pools from drying out. The spent fuel pool in Unit 4 probably did not go through the heat up, Zircaloy oxidation, hydrogen generation, and fission product release sequence. There was a great scare of radioactive releases from the spent fuel in the pool of Unit 4, since the amount of fuel stored is very large and the releases would have been large. However, it may be that the fission products measured in Unit 4 were transported from Unit 3.

The Water Contamination Issue

As mentioned earlier, the Fukushima reactors did not follow the classic conservative scenario of complete nonavailability of water. The sea water became available and later the fresh water; in fact, 80,000 liters of fresh water, were added to the Unit 1 core even before the sea water became available.

However, it all was too late since the cores were not supplied with water for sufficient time to prevent them to melt, move the molten materials to the lower head, and in fact cause the pressure vessels to fail. Probably each vessel failed at some penetrations, and some melt may have been ejected onto the basemats in the respective dry-wells, which may have had some water in them. The sea water was supplied for a number of days and later fresh water was introduced, which is still being supplied. Some of this water boiled off, coming from the safety relief valves back into the containment, but the rest of the water supplied most probably just leaked from the bottom of the damaged vessels into the dry-wells of the respective PCVs. This water is highly contaminated, and the leakage rate for the three reactors is estimated to be 500 tons/day. From the dry-wells, this contaminated water leaks to the turbine buildings and from there to the connected tunnels and pits, some of which may have been filled with concrete recently. It is estimated that there were $\approx$100,000 tons of highly contaminated water, collected all around by the end of May 2011, which could find its way to the sea. TEPCO is currently trying to find ways to decontaminate it and dispose it in an environmentally responsible manner. TEPCO has also delivered tanks with a capacity of $\approx$30,000 tons of water to the Fukushima station.

The Radioactivity Released to the Environment and Its Spread

All the Fukushima accidents were initially classified as Accident Level 4 on the IAEA scale; but it was later raised to Level 5 (that of the TMI-2 accident). It was further raised to Level 7 after the release of iodine and cesium was measured. Upon raising the level of the accident to 5, it was stated that the Fukushima releases were of the order of 10% of those from the Chernobyl accident. The last upgrade of the accident level to 7 was accompanied by the statement that the releases from the four reactors of Fukushima were twice those reported earlier, that is, approximately 20% of those from Chernobyl or almost 10% of the volatile fission product inventory. These would be the noble gas, compounds of iodine, cesium, tellurium, strontium, barium, ruthenium, and so on. Almost all of these volatile compounds would be in the form of aerosols, and most of them are soluble in water. Their spread in the atmosphere would not be like the release of the compounds from Chernobyl (which also included the release of almost all of the nonvolatile isotopes due to the explosion in the core). As mentioned earlier, there are some similarities in the release patterns between the Fukushima and Chernobyl since both involved an explosion. However, the Chernobyl energy release was so large that a plume reaching high in the atmosphere, containing the volatile and the non-volatile fission products, was formed. This plume dispersed the Chernobyl fission products in a large number of European countries, although countries that were nearby, of course, received much greater deposition than the farther countries. The fact remains, however, that the dispersal area for the very large Chernobyl radioactivity release was also much larger than that for the Fukushima radioactivity release. The Fukushima radioactive depositions in the 20- to 30-km radius zone around the plant could be substantial, and decontamination measures would be needed. The best saving grace in the scenario of Fukushima accidents was that while the radioactivity was released to the atmosphere along with the hydrogen explosions, the winds were blowing toward the sea. For a calculated dispersal pattern of the Fukushima radioactivity releases, see chapter 7 of this book.

Early measurement of the ^{131}I and ^{137}Cs in soil samples approximately 40 km northwest (inland) from Fukushima station are shown in Table 1.10, and those of radioactive concentration in water at the site of Fukushima station are shown in Figure 1.29. Later measurements have shown much lower values. A slight amount of Strontium-90 was measured in the sea water near the station, in June 2011.

1.14.4. Concluding Remarks on the Fukushima Accidents

The Fukushima accidents represent a major landmark in the history of nuclear power safety. These accidents occurred in three reactors simultaneously, and they occurred in a technically advanced country. The lessons to be learned from the accidents are being enumerated by several individuals and organizations. There have been many adverse reactions to nuclear power; Germany, for example, immediately shut down seven of its older nuclear plants and announced plans to shut down the rest of their plants by 2022. Italy and Switzerland have also joined Germany in renouncing nuclear power for their future energy systems.

It should be stated again that the Fukushima accidents were caused by the "worst-case initiator" of a huge earthquake followed by a colossal tsunami that may have also disoriented the operators and limited their access to technical support or crisis centers for instructions from more knowledgeable persons. The current Severe Accident Management Guidelines (SAMG) in U.S plants assume that a technical support center will be available and be responsible for the management of a stricken plant through the mitigation of a severe accident.

Other major nuclear accidents have been caused by a confluence of equipment and human malfunctions. The TMI-2 accident progressed to a core-melt accident when the operators shut shown the ECCS; the Chernobyl accident resulted from the operators' deliberate disabling of the reactor shutdown system. The Fukushima accident would have been mitigated if the operators could have established a heat sink and had added water in the vessel within 2–4 hours after the loss of all injection, perhaps with water obtained with a fleet of helicopters (as in fire-fighting) to the station site. Proper venting could also have prevented the hydrogen leakage into the reactor building and its explosions, reduced the radioactive releases, and also provided a partial heat sink by releasing steam into atmosphere. Actual severe accidents do not follow a set scenario but rather a scenario altered by operator acts of omission or commission, which can increase or lessen the consequences.

The consequences of the Fukushima accidents for the public were mitigated by the Japanese authorities' timely actions in evacuating the population around

TABLE 1.10 Concentration of Radioactive Materials

Sampling Point	Address of Sampling Point	Sample	Sort of Region	Sampling Time and Date	Radioactivity Concentration (Bq/Kg)	
					^{131}I	^{137}Cs
[2-1] (About 40 km Northwest)	litate village	Land Soil	Soil	2011/3/19 11:40	300,000	28,100
	litate village	Land Soil	Soil	2011/3/20 12:40	1,170,000	163,000
	litate village	Land Soil	Soil	2011/3/21 12:32	207,000	39,900
	litate village	Land Soil	Soil	2011/3/22 12:00	256,000	57,400
	litate village	Land Soil	Soil	2011/3/23 12:25	135,000	32,200
	litate village	Land Soil	Soil	2011/3/24 13:05	45,500	1,870
	litate village	Land Soil	Soil	2011/3/25 13:05	265,000	27,900
	litate village	Land Soil	Soil	2011/3/26 12:00	564,000	227,000
	litate village	Land Soil	Soil	2011/3/26 15:20	82,000	28,000
	litate village	Land Soil	Soil	2011/3/27 11:40	169,000	29,100
	litate village	Land Soil	Soil	2011/3/27 12:00	69,800	20,800

	litate village	Land Soil	Soil	2011/3/28 11:50	14,000	2,040
	litate village	Land Soil	Soil	2011/3/28 12:10	23,100	860
	litate village	Land Soil	Soil	2011/3/29 11:50	53,700	5,650
	litate village	Land Soil	Soil	2011/3/29 12:10	58,400	25,100
	litate village	Land Soil	Soil	2011/3/30 12:25	89,000	32,300
	litate village	Land Soil	Soil	2011/3/30 12:45	11,900	408
	litate village	Land Soil	Soil	2011/3/31 11:30	149,000	27,600
	litate village	Land Soil	Soil	2011/3/31 11:45	60,800	26,500
	litate village	Land Soil	Soil	2011/4/1 11:30	146,000	43,700
	litate village	Land Soil	Soil	2011/4/1 12:05	21,400	1,410
	litate village	Land Soil	Soil	2011/4/2 11:24	55,500	8,140
	litate village	Land Soil	Soil	2011/4/2 11:48	61,900	30,800

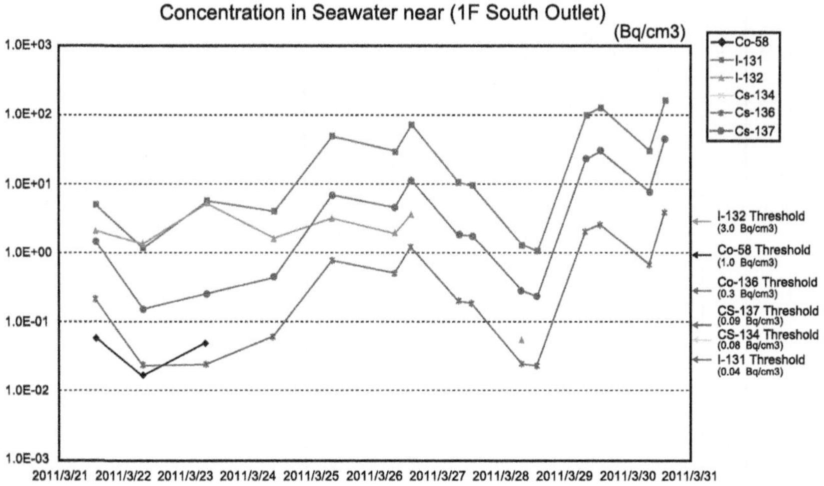

FIGURE 1.29 Radioactivity concentration of sea water samples near 1F South outlet (the concentration of radioactive iodine131 recorded on March 31 was approximately 4,385 times the limit set for water outside the environmental monitoring area).

the station. It should be noted that the Fukushima accidents did not cause any casualties and perhaps were responsible for only a few high exposures of workers at the station. It is also true, however, that the public has a great fear of radioactivity and that the public does not like to have to move away from their homes and perhaps not return for a long time; there by experiencing long-term disruptions of their lifestyle.

The Fukushima accidents have already affected the future of nuclear power in the world. The nuclear power renaissance which was about to spring forth in many countries has been buried back in the ground by the cold frost of these accidents. Currently, it is unclear whether it will be a permanent condition. The developing countries need electric power, and coal is not a good choice from environmental and climate change considerations. Finally, it will depend on what people will be willing to accept: an abundance of power versus the possible effects of a nuclear accident sometime in their collective future. Certainly, accepting exposure to low levels of radioactivity for a short time, which may not produce any ill health effects, could make a difference in people's attitudes toward nuclear power.

1.15. NEW LWR Plants

The presentlyinstalled LWR plants in Western countries have been addressing their safety performance from the day they were installed and operating. Prior to the TMI-2 accident, safety design-base issues, for example, the functioning of ECCS for various breaks, were of most concern. Plant concerns also revolved around the integrity of the primary system; for example, GE's BWRs plants needed replacement of some piping, the vessel weld material was of concern, and so on.

After the TMI-2 accident, concerns about safety centered on severe accident safety, that is, the prevention and mitigation of these accidents. This has been formalized into the programs of severe accident management (SAM) at most of the LWR plants. SA research results have led to backfits and accident management actions and procedures, which have enhanced plant safety, or provided the rationale for deliberate deciding not to require any backfits or SAM measures. A representative list of these actions and procedures follows:

- Hydrogen control with igniters and catalytic recombiners,
- Improved safety valves on PWRs,
- No inerting of MARK-3 BWRs,
- Water addition to the MARK-1 dry-well to prevent liner failure,
- Vessel depressurization for DCH protection,
- No backfits for protection against alpha mode failure,
- Use of BWR suppression (condensation) pools for fission-product removal,
- Hard vents for BWRs from the suppression pool,
- Flooding of the PWR vessel cavity for Westinghouse PWRs,
- Flooding of the dry-well for Swedish BWRs,
- Additional water delivery sources for accident termination,
- Reinforcement of containment penetrations,
- Realistic ex-vessel source term specification,
- Pressurized thermal shock prevention procedure,
- Filtered venting,
- Long-term management of iodine in the containment.

Clearly, not all the SA issues have been resolved for the presently installed plants. The most important of the unresolved issues is the coolability of the melt/debris produced during the postulated SA in order to stabilize and terminate the accident. This will ensure that the containment remains intact and that there is no significant radioactive release, precluding either the evacuation of the nearby population or their speedy return to their homes if any evacuation does occur. The issues of (a) ex-vessel steam explosion-induced containment failure, which is of concern for reactors that establish a deep water pool in their containments; (b) hydrogen detonation-induced containment failure, (c) DCH-induced containment failure, and (d) MCCI-induced basemat failure or containment pressurization failure can be addressed through operational or accident management actions, respectively, by (a) not establishing a water pool in the containment, (b) depressurizing the vessel in time and providing valves, which will bring the vessel pressure below 20 bars, (c) the availability of the hydrogen igniter and/or recombination systems, and (d) assuring the cooling of the melt/debris below the concrete ablation temperature of 1000°C.

One can conclude that if the melt/debris can be cooled and kept cool to stabilize and terminate the accident without having a preexisting pool in the containment, all the remaining concerns about the danger of severe accidents in the LWRs may be addressed adequately. Alternatively, if the melt can be cooled and retained in the vessel, thereby assuring containment integrity, the same conclusion may be reached. Recent concerns about the production of the fission product Ruthenium or the release of some small fraction of iodine as gaseous iodine are also addressed, since an intact and low-leakage containment will protect the public against the hazard of these releases to the containment.

We believe that the Generation 3+ LWRs or the near-future new LWRs have focussed on the issue of the long term coolability, stabilization, and termination of the severe accident as their goal. Two lines of design measures have been developed in these new LWRs: (1) in-vessel coolability and melt retention and (2) ex-vessel coolability and melt retention. We shall describe these very briefly in the following paragraphs.

1.15.1. The In-Vessel Melt Retention (IVMR) Strategy

The in-vessel coolability and retention strategy is based on the idea of flooding the PWR vessel cavity or the BWR dry-well with water either to submerge the vessel completely or at least to submerge the lower head. The PWR or BWR lower head containing the melt pool is cooled from outside, which keeps the outer surface of the vessel wall cool enough to prevent vessel failure. This concept is employed in the Loviisa VVER-440 in Finland, where it has been approved by the regulatory authority STUK. The concept is also employed in the PWR designs: AP-600, AP-1000, Korea's Advanced PWR-1400, and the 1000 MWe BWR design of AREVA.

The AP-600 design was analyzed [23] with a bounding accident assumption of the lower head full of convecting melt pool. It was found that the heat flux varied with angle, peaking near the equator. Fortunately, heat removal by the water outside also varied with the angle reaching highest value also near the equator. For a uniform corium pool for the 600 MWe AP-600 reactor, there was sufficient margin between the critical heat flux (CHF) on the water side and the incident heat flux from the corium pool. This margin of safety, however, may be reduced substantially in case there is a metal layer present on top of the oxidic corium pool. The metal layer results from the steel present in the PWR and the BWR lower heads that is melted by the corium pool, and since it is lighter it rises to the top of the corium pool (see Figure 1.30). The metal layer receives heat from the corium pool and performs Rayleigh-Benard convection transferring heat transversely to the vessel wall, which is then subject to a highly elevated heat flux. This heat flux focusing is most intense for a thin metal layer since the transverse area for heat transfer is smaller. It was found [42] that for metal layers of <30 cm depth the focused heat flux could overwhelm the critical heat flux near the equator. For the AP-600, it was found [26] that the

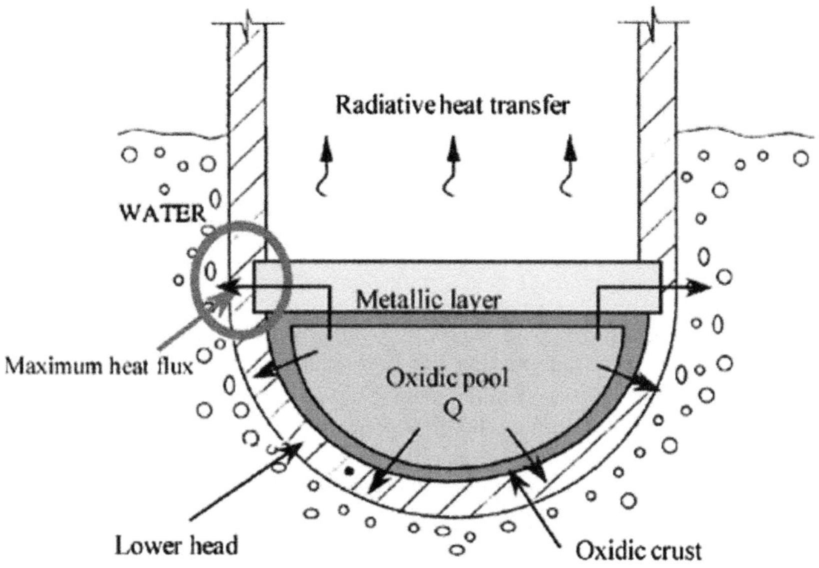

FIGURE 1.30 In-vessel melt retention.

metal layer would be thick and there was sufficient margin available between the focused heat flux and the CHF outside.

The power of the AP-1000 is 60% larger than that of AP-600 and 230% that of Korea Advanced PWR. For the 1400 MWe reactor, the focused heat flux would most probably be greater than the CHF on the water side. The strategy of the Korean plant is to simultaneously flood the metal layer with water inside the vessel, which, hopefully, could remove sufficient heat from the upper face of the metal layer to reduce the focused heat flux to values less than the CHF. A dedicated water system has been installed in the plant for water injection to reach the melt pool in the lower head at the appropriate time.

Further complications have been uncovered recently by the findings in the OECD-sponsored RASPLAV and the MASCA projects [43] [44] of chemical reactions between the melt constituents that may create different layer configurations in the melt pool. For example, the RASPLAV project revealed that the presence of even small amounts (<0.3%) of carbon in the system promotes the stratification of the melt pool by separating the oxides from the metals in the melt, thereby forming a light melt layer, rich in metals, and carbides residing on top of the oxide-rich melt pool. A finding from the MASCA project is that the combination of the steel components with uranium forms a metal compound, which being heavier than the oxidic pool sinks to the bottom of the oxide-rich melt pool. It is not clear whether all the steel will combine with the uranium metal. The initiator of this steel-uranium combination is the unoxidized Zr present in the melt. The worst situation would be one in which some of the steel is taken by uranium metal to the bottom of the pool, while some remains at the top to form a thin metal layer that can provide a strong, focused heat flux on the vessel wall. The melt pool composition and configuration situation is quite confused presently, since the more recent data obtained in the oxidizing atmosphere (steam) have shown that after Zr oxidation is completed the steel is released from uranium and rises back to the top of the pool. More research on the pool stratification issue is anticipated.

1.15.2. The Ex-Vessel Melt Retention Strategy

This strategy has been adopted by the European Pressurized Water Reactor (EPR) design currently in Finland and by the new Russian VVER-1000 designs for China and India. The EPR design [45] spreads the discharged corium, mixed with sacrificial concrete, on a flat steel surface coated with a high-temperature inert material, cools it from the bottom with water flowing in channels, and floods it with water from the top. The idea behind this design is that spreading will reduce the depth of the melt pool to the extent that it can be cooled by a water overlayer with some assistance from a cooling system at the bottom. Sacrificial concrete is mixed with corium discharged from the reactor pressure vessel (RPV) in a concrete-lined pit to reduce its temperature, and more so, its solidus temperature. Thus the mixture remains liquid over a much larger temperature range and, in fact, will spread more easily over a large floor area. Figure 1.31 shows the configuration of the EPR melt retention enclosure.

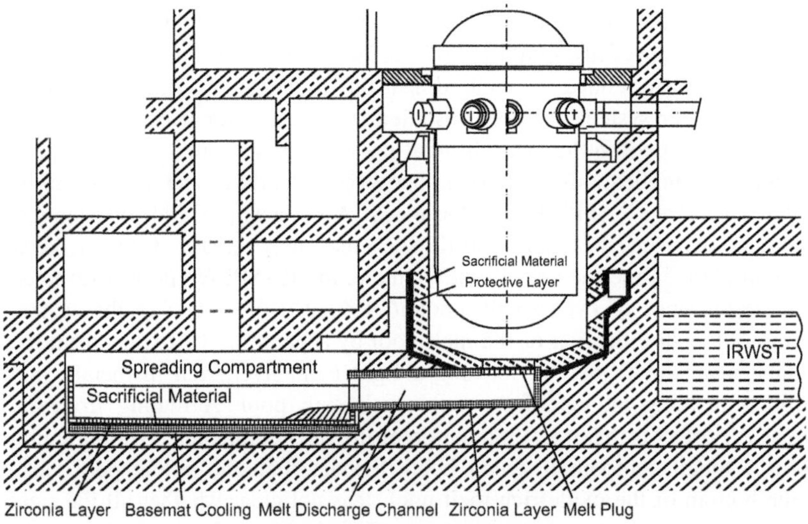

FIGURE 1.31 EPR core catcher.

Much research was performed on the efficiency of spreading the melt at various European laboratories, including that of the Nuclear Power Safety Division at KTH. We developed a very innovative scaling theory for spreading [46], which has been able to predict most of the spreading data obtained with simulant and prototypic melt materials. The EPR melt spreading analysis was also performed with this model, and it was found that even with conservative assumptions, uniform spreading of the discharged melt and concrete mixture can be obtained in the EPR design. Unfortunately, the depth of the melt (~40 cm) is greater than that which can assure melt coolability with water flooding alone. The cooling coils built in the base of the spreading chamber will be needed to cool the melt. It appears, however, that it will take considerable time before the center part of the spread melt pool will solidify.

The Russian VVER-1000 design employs a core catcher in the traditional sense. The core catcher shown in the Figure 1.32 is a separate vessel installed under the RPV, with an intake designed to cover almost the whole surface of the bottom head so that the melt discharged from the RPV is deposited in the core catcher, even if the RPV failure occurs at an angular position close to the equator (which it probably will). The core catcher is like a lower head but of much larger volume, and it is cooled from outside by a water pool as in the IVMR concept. The mixing process also reduces the mixture power density and the heat fluxes on the vessel walls. The core catcher is full of bricks made of oxidic material containing Fe_2O_3 and other oxides. The purpose is the same as in EPR: to reduce the temperature of the discharged corium and to keep it liquid over a larger range of temperature. The core catcher walls are steel, but they are lined with oxide bricks. The chemistry of the materials with the corium has been the subject of several experiments, and the chosen oxide composition is such that the uranium and the metals in the corium combine to form a dense metal layer that sinks to the bottom of the melt pool. There supposedly is no metal layer on the top of the oxidic pool. The melt pool is flooded with water to satisfy the argument that the probability of a stratified steam explosion is much reduced, since the metal is at the bottom under the oxidic material pool.

The melt pool in the Russian core catcher design may also remain molten for a considerable time and will perform natural convection. There is insufficient information in the literature to assess the long-term operation of the core catcher and the state of the melt pool inside.

General Electric has designed a Generation 3+ BWR, which also is equipped with an innovative core catcher below the vessel that is cooled by a set of steel pipes embedded in the floor and walls, lined with a non-ablating material. The steel pipes have natural-circulation water flow in them to remove the decay heat generated in the ex-vessel melt-debris pool. This core catcher design is currently under development, testing, and peer review.

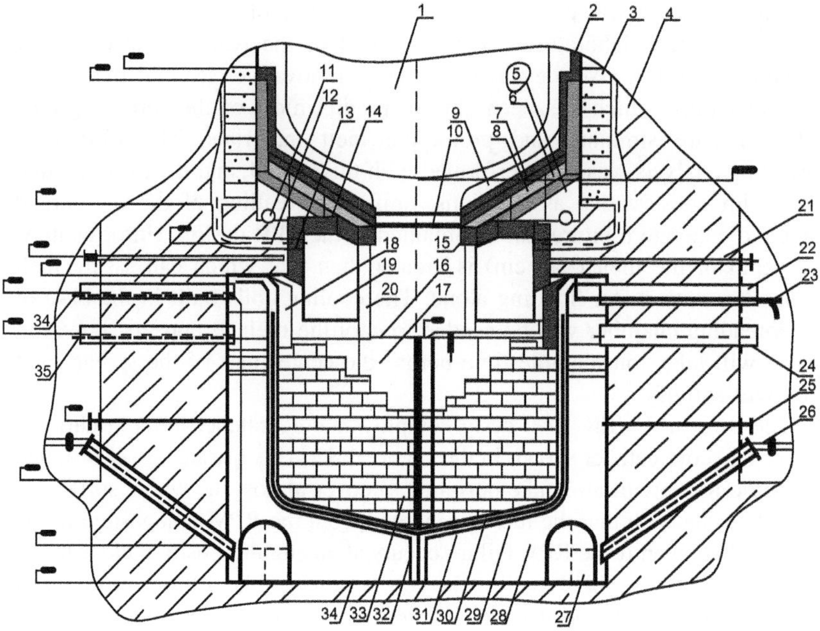

1 Reactor pressure vessel
2 Thermal insulation of dry protection
3 Dry protection
4 Concrete cavity
5 Lower plate
6 Channel of lower plate cooling
7 Corium-liquefying substratum
8 Heat-resistant concrete
9 Bearing slab
10 Protective shield
11 Damper
12 Channel of dry protection cooling
13 Thermal protection of concrete cantilever end
14 Thermal protection of lower plate
15 Deflector baffle
16 Maintenance floor
17 Support truss
18 Thermal protection of concrete cantilever base
19 Ventilation duct
20 Support column
21 Channel of cooling water supply to concrete cavity
22 Channel of steam removal from heat exchanger
23 Blowdown channels
24 Heat exchanger drain channel
25 Level indicator channel
26 Channel of cooling water supply to heat exchanger
27 Arch-shaped header
28 Heat exchanger
29 Thermal insulation of heat exchanger
30 Basket
31 Honeycomb-like thermal shock shield
32 Support truss slab
33 Filling compound
34 Channel for thermocouples location
35 Basket drain channel

FIGURE 1.32 Tian Wan core catcher.

CONCLUSIONS

The march of history with regard to LWR safety shows a definitly positive direction. The potential of nuclear power for public good, accompanied by its potential for serious public hazard, was recognized early on, and this is to the credit of the scientists and engineers who pioneered civilian nuclear power. Much credit also has to be given to the regulatory bodies of the various countries, which have been the guardians of public safety during the development and spread of nuclear power in the world. Most of the credit, however, has to go to the scientists and engineers, who have diligently raised every safety issue or question, performed research, and provided solutions. It has been a splendid history, and it deserves praise.

The challenges posed by the TMI-2 accident were met through patient hard work, severe accident research, and analyses developments. The presently installed LWR plants made improvements in components, systems, operator training, the man-machine interface, safety culture, and so on, thereby significantly reducing the probability of a severe accident. They also instituted severe accident management backfits, systems, and procedures, which are reducing the probability of an uncontrolled and large release of radioactivity in the event of a severe accident. However, some older plants may not have all the severe accident management systems operating when needed (as, for example, in the Fukushima reactors) to ensure the prevention of a large release of radioactivity. In addition, the presently installed plants cannot provide assurance of coolability of a melt pool/debris bed, which could be formed during a bounding severe accident. In that situation, the LWR owner cannot ensure the public that the accident has been stabilized and terminated and that there is no further danger of radioactivity release.

The new Generation 3+ LWR designs, exemplified by EPR, VVER-1000, AP-1000, APWR-1400, and ESBWR, which employ in-vessel, or ex-vessel, cooling and retention of the core melt/debris bed that would be produced in the postulated severe accident, are reaching the end state of development for public safety for LWRs. They not only provide systems, which have an extremely low probability for a severe (core-damage) accident, but also ensure that the core melt will be contained safely and cooled and that there will not be any large release of radioactivity to the environment. The public living near a NPP does not want to move from their homes. Although the designers (vendors) of these new LWRs have not explicitly provided such assurances, they may be able to do so. Convincing the public will not be easy, due to their fear of radioactivity.

Post-Fukushima, the challenges of LWR safety have not diminished for the current plants. The Fukushima accidents show that extraordinary events can occur (in spite of the extremely low probabilities calculated) and the consequences for a society may be unbearable. Focusing solely on probabilities and risk may have been overdone. Instead, the focus should shift to reducing the consequences of a severe accident by providing systems to mitigate the progression of any imaginable severe accident; and to eliminate the consequences for the public of a radioactivity release.

The human component of LWR safety needs to be improved. In particular, complacency has to be banned. The operators, staff, and management of the NPPs have to be more responsive and reliable than the plant components and systems. Plant aging is an issue that is becoming increasingly visible. Some of the LWR plants have reached, or are nearing the end of their estimated life spans. Components and systems will therefore become less and less reliable in the future. The need for vigilance on the part of plant staff and management has to increase. The utility companies have to recognize this fact and be prepared to spend the money to replace/renew old equipment, instrumentation, and systems.

It should always be remembered that the public has given civilian nuclear power a very short leash with respect to safety and in order to retain public confidence, the human and management components of LWR safety cannot afford to fail.

REFERENCES

[1] S. Glasstone, Sourcebook on Atomic Energy, third ed. D van Nostrand Company Inc, 1967.

[2] T.J. Thompson, J.G. Beckerley, The Technology of Nuclear Power Reactor Safety, vol. 1, The MIT Press, 1970.

[3] A.M. Weinberg, A Second Era, Prospects and Perspectives, Presented at the 40th Anniversary of the First Nuclear Chain Reaction, University of Chicago, 1982. December 1-2.

[4] J.R. Dietrich, Experimental Determination of the Self-Regulation and Safety of Operating Water-Moderated Reactors, Proceedings of the International Conference on the Peaceful Uses of Atomic Energy, United Nations, New York, 1956.

[5] H.M. Parker, J.W. Healy, Environmental Effects of a Major Reactor Disaster, Proceedings of the International Conference on the Peaceful Uses of Atomic Energy, United Nations, New York, 1956.

[6] U.S. Atomic Energy Commission, Theoretical Possibilities and Consequences of Major Accidents in Large Nuclear Power Plants, USAEC Report WASH-740 (March 1957).

[7] United States Atomic Energy Commission Report, Siting Criteria, TID14844 USAEC Report (1959).

[8] D. Okrent, Nuclear Reactor Safety, On the History of the Regulating Process, University of Wisconsin Press, 1981.

[9] U.S. Atomic Energy Commission, Emergency Core Cooling, Report of an Advisory Task Force, USAEC Report TID-24226 (January 1968).

[10] U.S. Atomic Energy Commission, The Safety of Nuclear Power Reactors (Light Water Cooled) and Related Facilities, USAEC Report WASH-1250 (July 1973).

[11] W.B. Cottrell, The ECCS Rule-Making Hearings, Nucl. Safety 15 (1) (1974).

[12] Report to the American Physical Society of the Study Group on Light-Water Reactor Safety, Rev. Mod. Phys. 47 (Suppl 1) (1975).

[13] U.S. Atomic Energy Commission, General Design Criteria for Nuclear Power Plants, Code of Federal Register, 10 CFR 50 Appendix A (1971).

[14] F.R. Farmer, Siting Criteria—A New Approach, Proceedings of a Symposium on Containment and Siting, International Atomic Agency, Vienna, 1967.
[15] U.S. Nuclear Regulatory Commission, Reactor Safety Study, An Assessment of Accident Risks in U.S. Commercial Nuclear Power Plants, USAEC Report, WASH-1400 (1975).
[16] H.W. Lewis, et al., Risk Assessment Review Group Report to the U.S. Nuclear Regulatory Commission, U.S. NRC Report NUREG/CR-0400, U.S. Nuclear Regulatory Commission, 1978.
[17] Federal Minister for Research and Technology, The German Risk Study: Nuclear Power Plants, Verlag TUV, Rheinland, 1980 (in German).
[18] Report of the President's Commission on the Accident on Three Mile Island, Washington, D.C., 1979.
[19] D. Cubicciotti, B.R. Sehgal, Vapor Transport of Fission Products in Postulated Severe Light Water Reactor Accidents, Nucl. Technology 65 (266) (May 1984).
[20] J.Q. Howieson, V.G. Snell, Chernobyl—A Canadian Technical Perspective, Nuclear Journal of Canada, No. 3 (September 1987). A technical executive summary of this is, in: V.G. Snell, J.Q. Howieson (Eds.), Chernobyl—A Canadian Technical Perspective— Executive summary, Atomic Energy of Canada Ltd. publication AECL-9334S, January 1987.
[21] B.R. Sehgal, Accomplishments and challenges of the severe accident research, Nucl. Eng. Design 210 (79) (2001).
[22] Electric Power Research Institute (EPRI), MAAP 4, Modular Accident Analysis Program User's Manual, EPRI Report prepared by Fauske & Associates Inc, 1994.
[23] T.G. Theofanous, et al., In-vessel coolability and retention of a core melt, DOE/ID-10460, U.S. Department of Energy Report (1995).
[24] T.G. Theofanous, W.W. Yuen, S. Angelini, J.J. Sienicki, K. Freeman, X. Chen, T. Salmassi, Lower head integrity under in-vessel steam explosion loads, DOE/ID-10541 U.S. Department of Energy Report (1996).
[25] SERG2, A reassessment of the potential for an Alpha-Mode containment failure and a review of the broader fuel coolant interaction (FCI) issues, NUREG-1529, U.S. NRC Report (1995).
[26] M.M. Pilch, et al., The probability of containment failure by direct containment heating in Zion, NUREG/CR-6075, U.S. NRC Report (1993).
[27] S. Dorofeev et al., Flame acceleration limits for nuclear safety applications, Proc. CSARP Meeting, Albuquerque, New Mexico, (1999).
[28] Huhtiniemi, D. Magallon, Insight into steam explosions with corium melts in KROTOS, Proceedings Nuclear Reactor Thermal Hydraulics-9 Meeting (1999).
[29] G. Berthoud, C. Brayer, First vapour explosion calculations performed with MC3D code, Proc. CSNI specialists meeting on FCIs, Tokai, Japan, 1997.
[30] T.G. Theofanous, et al., The probability of liner failure in a Mark-1 containment, NUREG/CR-5423 and NUREG/CR-5960, U.S. Nuclear Regulatory Commission Reports (1993).
[31] M.T. Farmer, S. Lomperski, S. Basu, The results of the CCI-3 reactor material experiments investigating 2-D core-concrete interaction and debris coolability with a siliceous concrete crucible, Proceedings of the ICAPP'06, Reno U.S. (June 4-8, 2006).
[32] B.R. Sehgal, Stabilization and termination of severe accidents in LWRs, Nucl. Eng. Design 236 (2006) 1941.
[33] H. Alsmeyer, B. Spencer, W. Tromm, The COMET concept for cooling of ex-vessel corium melts, Proceedings of ICONE-6 Conference, San Diego, USA, 1998.
[34] M.J. Konovalikhin and B.R. Sehgal, Investigation of volumetrically heated debris bed quenching, Proceedings of ICONE-9 Meeting, Nice, France (April 2001).
[35] A.K. Nayak, B.R. Sehgal, A.V. Stepanyan, An experimental study on quenching of a radially-stratified heated porous bed, Nucl. Eng. Design 236 (2006) 2189.

[36] M.J. Konovalikhin, A. Karbojian, B.R. Sehgal, Molten pool coolability. COMECO Experiments, Proceedings of ICONE-10 Conference, Arlington, VA, USA (April 14-18, 2002).
[37] J.K. Fink, D.H. Thompson, D.R. Armstrong, B.W. Spencer, B.R. Sehgal, Aerosol and melt chemistry in ACE molten corium concrete interaction experiments, High Temperature Science (1994).
[38] NEA/OECD Report, Implementing severe accident management in nuclear power plants, SESAM Group, NEA/OECD, 1996.
[39] V. Strizov, personal communication.
[40] Report of Japanese Government to IAEA Ministerial Meeting.
[41] M.T. Farmer, D.J. Kilsdonk, R.W. Aeschlimann, Corium Coolability under Ex-vessel Accident Conditions for LWRs, Nucl. Eng. Technol. 41 (5) (2009).
[42] J.M. Seiler, B.R. Sehgal, et al., European Group for analysis of corium concepts, Proceedings of the FISA Conference, European Commission, 2003.
[43] V.V. Asmolov, Last findings of the RASPLAV Project, Proceedings OECD/CSNI workshop on in-vessel core debris retention and coolability, Munich OECD/NEA Report (1999).
[44] V.V. Asmolov, et al., Zirconium and Uranium partitioning between oxide and metallic phases of molten corium, MPTR-9, RRC Kurchatov Institute Russia, 2003.
[45] M. Fischer, O. Herbst, H. Schmidt, Demonstration of the heat removing capabilities of the EPR core catcher, Nucl. Eng. Design 235 (2005) 1189.
[46] T.N. Dinh, M.J. Konovalikhin, B.R. Sehgal, Core melt spreading on a reactor containment floor, Progress in Nuclear Energy 36, No.4, Pergamon Press, 2000.

Chapter 2

In-Vessel Core Degradation

Chapter Outline

Nuclear Safety in Light Water Reactors. DOI: 10.1016/B978-0-12-388446-6.00002-2

2.1. INTRODUCTION

(F. Fichot, J.M. Seiler, K. Trambauer, and C. Mueller)

As a result of the research enhanced after the TMI-2 accident, there has been a significant increase in the ability to understand and model severe accident (SA) phenomena and to apply the resulting SA codes to the analysis of plants during SA conditions [1] [2]. Important trends have been identified through a wide range of experiments, examination of the TMI-2 core and vessel (Figure 2.1), as well as analysis of representative reactor designs. This chapter presents the SA phenomena related to core behavior, as well as other factors such as reactor design, initial and boundary conditions of the reactor core, the core degradation processes, and the fission product release. Relevant works and more comprehensive reviews are listed in the References; these give more detailed descriptions of SA trends and phenomena. Such reviews include the "In-Vessel Core Degradation Code Validation Matrix" [3] [4], status reports on VVER (the Russian abbreviation for *water-cooled water-moderated power reactor*) specific features [5], on molten material relocation [6], on core quench [7] [8], on molten fuel coolant interaction [9] [11], on the QUENCH experimental program [11], the Proceedings of the Workshop on In-Vessel Core Retention and Coolability [11], and the RASPLAV Application Report [12]. Detailed information can also be found in the "Catalogue of Generic Plant States leading to Core Melt in PWRs" [13].

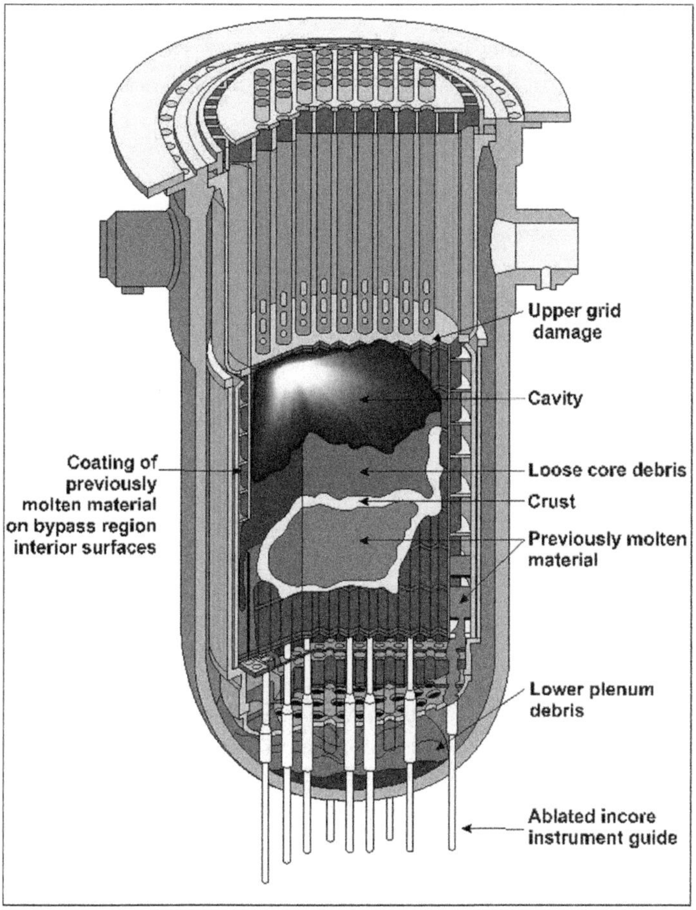

FIGURE 2.1 Schematic diagram of reactor core following TMI-2 accident.

The objective of this chapter is to define in a general way the initial and boundary conditions of the reactor core and the core degradation processes during the course of a severe accident. The scope is limited to the light water moderated reactors (LWRs). Detailed characterization of plant sequences, including accurate quantification of possible parameter ranges, is beyond the scope of this chapter.

The plant types considered in this chapter are those that are uranium dioxide (UO_2) fueled, and water moderated and cooled. This includes light water reactors (LWRs), that is, pressurized water reactors (PWRs and VVERs) and boiling water reactors (BWRs) of U.S. and European origin. Heavy water reactors (CANDU) are presented in Appendix 2. Advanced design plants are not explicitly discussed, although advanced light water reactors (ALWRs), including passive plants, are expected to have reactor cooling system (RCS) accident boundary conditions similar to low-pressure sequences for existing LWRs.

2.2. CORE DEGRADATION IN PWR

(F. Fichot, J.M. Seiler, K. Trambauer, and C. Mueller)

Despite the sometimes significantly different RCS designs or control systems, the arrangements of the fuel rods, spacer grids, control rods, and guide tubes are nearly identical. This means that local processes are similar for all Western PWRs discussed in this chapter.

The behavior of the core, particularly the heating of the core due to oxidation and the melting of core materials, is very strongly dependent on the composition and configuration of the core. Since UO_2 and zirconium-based cladding alloys are used in most commercial power reactors being operated around the world, the strong consequence of Zircaloy oxidation will be repeated with minor variations due to the actual alloying elements present. For example, Zr-Niobium alloy oxidation rates are somewhat lower than that for Zircaloy. As a result, the oxidation-driven heating rates at high temperatures may also be somewhat lower than the rate for an equivalent amount of Zircaloy. The melting points and chemical interactions between different core materials will also have a similar consequence for changes in core geometry. The general melting of core materials will depend on the melting temperatures of each core material but will be impacted by the formation of lower melting temperature alloys as different materials in the core chemically interact where they come in close contact. In most cases, the formation of the lower melting temperature alloys will result in the earlier and lower temperature failure of core structures, similarly to the impact of the chemical interactions between Zircaloy, stainless steel/Inconel, and control material structures on the early failure of Ag-In-Cd and B_4C control structures. The consequence of these chemical interactions can be determined using the appropriated phase diagrams and reaction kinetics rates. In the case of Zircaloy-stainless steel reactions, the reaction rates become very fast as temperatures approach 1500 K, typically resulting in the destruction of the structures at this temperature, even though lower melting point alloys are initially formed at much lower temperatures.

For other core materials, the oxidation and melting processes, though still important, may have significantly different consequences for the behavior of the core. For example, aluminum-based fuel elements, used in many research reactors, will respond very differently from Zr- UO_2 fuel rods due to the lower melting temperature of the aluminum alloys. In this case, although these alloys can react very strongly with steam, the early melting and relocation of the cladding effectively prevents the oxidation of these materials under most conditions. A more subtle difference, though of equal importance, is the consequence of the grid spacer materials and designs used in the core. Because of the strong interaction of Inconel and stainless steel with Zircaloy cladding, spacer grids made of these materials will be destroyed at relatively low temperatures (1500 to 1700 K) because of the formation of the lower melting point alloys. On the other hand, Zircaloy spacer grids will remain in place up to much higher temperatures (2000 to 2100 K, even higher if they are strongly oxidized). Moreover, the Zircaloy spacer grids will also oxidize much like the Zircaloy cladding materials. Since the intact spacer grids also act as debris catchers, the location of in-core blockages will depend on the spacer grid materials. The design of the core will largely determine the accident progression.

2.2.1. Thermal-hydraulics

During an SA, accurate prediction of the plant's overall thermal-hydraulic response is of fundamental importance for a successful analysis. However, since the plant's thermal-hydraulic response is also very sensitive to (a) the plant design, (b) the response of the plant systems and components to the initiating events, (c) operator actions, and (d) other external events, it is impossible in this chapter to define general trends for the thermal-hydraulic response of all plant types and conditions [13]. Nevertheless, once a specific reactor design and accident conditions are known, it should be possible to accurately predict the thermal-hydraulic response using currently available SA codes and models.

A wide range of phenomena can impact the thermal-hydraulic response of the plant. One of the most dominant factors for the management of SAs is the effect of the core and vessel reflooding, which is considered one of the few remaining outstanding technical issues on severe accident (SA). Some of the important trends associated with this process are discussed further in Section 2.2.4 dedicated to in-vessel reflooding.

For typical PWR designs, three general modes of natural circulation may have an impact on the response of the plant during an SA: in-vessel natural circulation, circulation within the hot leg and associated piping, and circulation through the primary loops.

For in-vessel natural circulation, experiments and detailed code calculations have shown that the natural circulation flow patterns are typically formed in the vessel as a direct result of the variation in temperature within the core and vessel. These flow patterns can be initially influenced by the ballooning and rupture of the fuel rod cladding, the formation of blockages over a long time period, and other damage inside the core. The primary impact of in-vessel natural circulation is to delay the overall heating of the core, due to the more effective heat removal from the hotter core regions to the colder core structures. As a result, radial temperature gradients in the core are reduced, and the resulting heating pattern becomes much more uniform. Once the peak core temperature approaches 1500 K, the effect of natural circulation is somewhat reduced due to the accelerating oxidation process in the hotter core region that is driven by the strong dependence of the reaction rate on temperature.

2.2.2. Oxidation of Core Materials

The oxidation of core materials is important in view of the SA progression, due to the generation of heat that may exceed the decay heat. Typical evolutions of decay heat are provided in [14] and [97]. Around 1500 K, the heat release rate associated to the transition of the metallic Zr into brittle ceramic ZrO_2 becomes comparable to the decay heat. It may even become ten times higher above 1800 K. It also leads to the production of hydrogen and possibly other combustible gases, which causes a risk of deflagration or even detonation in the containment, if those gases are mixed with air. Among all the oxidation reactions taking place in the core, the oxidation reaction of Zircaloy by steam is considered the most important with respect to hydrogen production and consequences on the degradation of the core.

The oxidation of other materials, particularly B_4C, can also be important in some cases. It should be noted that other Zr alloys behave differently from Zircaloy. When the oxidation kinetics of different alloys is considered, the quantitative trends may therefore be somewhat altered.

Zircaloy Oxidation in Steam

During an SA, the oxidation of the Zircaloy structures has a significant consequence for the overall behavior of the LWR core. Although the specific evolution might depend, to some extent, on the type of transient being analyzed, the exponential increase with temperature of the Zircaloy oxidation rate produces very sharp temperature increases in the core as soon as it exceeds 1500 K. This is a characteristic feature of any SA experiment or plant simulation. As shown in the LOFT-FP-2 experiment [20] [21], the initial core heat-up rate (less than 1 K/s due to decay heat) is rapidly increased by an order of magnitude, due to additional heat released by the Zircaloy oxidation. As a result, temperatures of the fuel rods and intact core structures increase rapidly above the melting point of Zircaloy (2100 K).

In this condition the subsequent thermal response of the core, particularly peak or average core temperatures, is strongly dependent on the sustained oxidation of the Zircaloy. While the total heat generated by the Zircaloy oxidation process is enough to drive the peak core temperature above 3000 K and up to the melting point of the fuel, the total amount of oxidation energy added to the core can be limited by several effects: (1) the availability of steam in the core, (2) the diffusion of steam through the hydrogen-rich boundary layer along the cladding (hydrogen blanketing), and (3) the diffusion rate of oxygen through the external ZrO_2 scale and underlying metallic Zircaloy (Figure 2.2). In general, the diffusion coefficient for Zircaloy is characterized by an

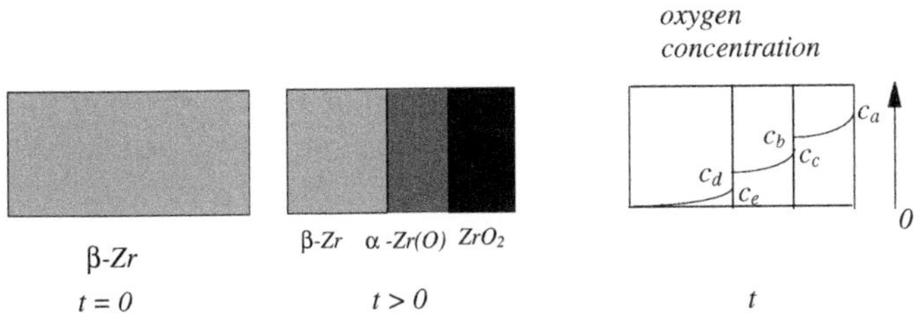

FIGURE 2.2 Sketch of the formation of the zirconia and α–Zr layers and view of the oxygen profiles in the layers (on the right).

exponential function of temperature. The diffusion-controlled reaction is governed by a parabolic kinetic law:

$\frac{dX^2}{dt} = K$, where X is the layer thickness or total oxygen mass gain in both $\alpha - Zr(O)$ and ZrO_2 layers.

In addition the temperature dependent kinetic constant verifies an Arrhenius formulation:

$K = Ae^{-B/RT}$ where T is the wall temperature and R is the perfect gas constant.

For typical transient sequences, the diffusion of oxygen into the Zircaloy tends to limit the oxidation process at lower temperatures. Once the peak temperature exceeds 1500 K, the positive feedback between core temperature and oxidation rate is limited by the growth of the external ZrO_2 scale. More specifically, the rate of oxidation decreases as the oxide thickness grows (at constant temperature, the rate is inversely proportional to the oxide layer thickness). However, the increase of the diffusion rate with temperature completely overwhelms the consequence of the protective effect of the ZrO_2 layer up to the complete oxidation of the cladding. As the total hydrogen production increases, the availability of steam and the diffusion of steam to the surface of the Zircaloy structures limit the oxidation rate, and the thermal-hydraulic conditions become more important than the temperature itself. In particular, the increase of hydrogen concentration in the upper core region and the decrease of steam generation rate due to the decreased water level become more effective in limiting the maximum oxidation rates, especially in the upper core region.

At a given location of the core, the total amount of oxidation that depends on the amount of Zircaloy present at that location can also be limited by the liquefaction and relocation of Zircaloy. For transient sequences characterized by relatively rapid initial heating rates (above 0.5 K/s), the build-up of a protective oxide layer on the outer surface of the Zircaloy cladding is limited, allowing, above the Zircaloy melting temperature, the cladding failure and relocation of molten Zircaloy to the lower core region. In this case, the oxidation process is terminated at the original location of the Zircaloy.

Although the relocating Zircaloy continues to oxidize, the enhanced cooling as the melt moves toward colder regions of the core tends to rapidly lower the temperature, in turn terminating the oxidation process as the material moves lower in the core. For transient sequences characterized by slower heating rates, typically lower than 0.3 K/s, the formation of a thicker protective oxide layer prevents the relocation of the molten Zircaloy and, as a result, the Zircaloy can be completely oxidized in place. For intermediate initial heating rates, a combination of relocating Zircaloy and complete consumption of the Zircaloy tends to control the oxidation process.

The total amount of hydrogen released to the RCS and containment building is mainly related to the oxidation of the Zircaloy. Although the oxidation of the core structures (stainless steel, B_4C) can contribute to the total amount of hydrogen generated in the vessel, the early melting of these structures tends to limit their contribution. The oxidation of the fuel can also contribute to the total hydrogen, but is limited by the exposure to steam and the rate of oxidation of UO_2.

The temperature response of the core is directly related to the Zircaloy oxidation process. Although the maximum core temperature is ultimately limited by the melting of the fuel, the peak core temperature is limited by the peak oxidation rate. At rapid heating rates, the peak core temperature occurring during rapid oxidation approaches the melting point of Zircaloy. At slow heating rates, the peak core temperature can be limited by the melting point of the ZrO_2 material. The bundle heating and melting experiments are normally terminated just after the rapid oxidation and melting occurs, and the peak core temperature measured in the experiments is directly related to the peak oxidation rate.

Oxidation of B_4C in Steam

A limited number of reflooding experiments with fuel rod bundles or B_4C control blades (see Section 2.2.4 on core reflooding) have shown that the oxidation of B_4C may contribute to the production of both hydrogen and other gases such as CO, CO_2, and large amounts of B-compound aerosols [15]. The latter gases are mainly produced during reflooding or the final rapid cooling phase. Nevertheless, the possible methane production that may react with the iodine released from the fuel to form volatile organic iodines (key factor of the radiologic risk) was found to be very low. More important are the potential effects of CO, CO_2, and B-componds on the chemistry of fission products in the primary circuit. Regarding the H_2 production, it should be noted that the amount of B_4C is relatively small compared to the amount of Zircaloy in typical NPPs so that the consequences of B_4C remain limited (maximum 15% of the total hydrogen production is due to B_4C). The B_4C was also found to favor the core liquefaction during the degradation phase of bundle experiments such as CORA and QUENCH tests and the Phébus FPT3 test. The degradation of fuel rods in the vicinity of B_4C rods occurs earlier. However, it is not clear whether this local effect has a significant impact at the reactor scale. The main impact of B_4C oxidation concerns the change of the chemistry in the RCS and in the containment as observed in the Phébus FPT3 test.

2.2.3. Loss of Core Geometry during a Severe Accident

The loss of the original core geometry can occur gradually over a period of minutes to hours, covering, for LWRs, a wide range of temperatures from 1500 K to 3000 K. The specific timing and temperatures of the core degradation processes are very strongly dependent on types of core materials, the initial uncovery and heating rates of the core, system pressure, and the overall thermal-hydraulic response of the plant. Therefore, the ranges quoted apply strictly to Western LWR designs with UO_2-Zircaloy fuel rods, and Ag-In-Cd or B_4C control rods or blades. The main causes of core degradation are chemical interaction between core materials leading to the formation of low melting point alloys. Other reactor designs may respond differently as a consequence of different formation of low melting point alloys. Since the core geometry is changed primarily with the local core temperature, many of these changes can occur simultaneously in different regions of the core. The typical transient sequences involve a general increase in the maximum and average core temperatures with time. The initial geometrical changes such as ballooning and rupture of the fuel rods occur at lower temperatures during the early phase of transient sequence. During the following phase, the core degradation increases with the temperature. A notable core geometrical change due to fragmentation of heavily oxidized materials can be associated with the core reflooding. This kind of geometrical change can occur at any time once the core structures have absorbed a sufficient amount of oxygen to become brittle.

Ballooning and Rupture of the Cladding

For low-pressure accident sequences, the Zircaloy cladding starts to balloon and rupture once the core temperature reaches 1000 to 1200 K. In that case, the timing and temperature of ballooning and rupture depend on the internal pressure of the fuel rods (including any fission gases that may be released into the gap), and the mechanical characteristics of the cladding material. For high-pressure accident sequences, due to the collapse of cladding onto the fuel at low temperature, the failure of Zircaloy cladding may be delayed until the core temperature reaches above 1500 K. Even though the cladding does not fail mechanically in that case, chemical interactions between the Zircaloy cladding and other core materials can cause local failures of the cladding due to the formation of low melting temperature alloys.

At this stage of core damage, the most significant consequences of cladding ballooning and rupture is the release of fission products, the exposure of the inner surface of the cladding to steam, and changes in the relocation of fuel rod materials later in the transient. Ballooning and rupture may also alter subsequent flow patterns in the core, particularly when the deformation is extensive.

Experiments (OECD Halden LOCA Project) [16] with high burn-up fuel rod segments have recently shown that the fuel fragments from higher elevations can fill in the lower ballooned region of the cladding, resulting locally in a higher linear power due to the increased fuel mass per length. This might impact the long-term coolability of the core even under Design Basis Accident conditions.

Liquefaction and Relocation of Control and Structural Materials

For typical LWR designs and at temperatures between 1500 and 1700 K, chemical interactions between Fe-Zr, B_4C-Fe, Ag-Zr, B_4C-Zr, and others can result in the early liquefaction and relocation of core structures such as grid spacers, Silver-Indium-Cadmium or B_4C control structures and the portions of the Zircaloy cladding material in direct contact with the other materials. For rapid transient sequences where the onset of chemical interactions is not delayed, the failure can occur at temperatures near 1500 K. For slower transients, the failure will be delayed until temperatures near 1700 K are reached. In the latter case, the formation of protective oxides on the Zircaloy tends to delay and restrict the kinetics of chemical interactions.

At this stage of core damage, the most significant consequence of the material interactions is that the control materials can become segregated from the fuel, increasing the potential for recriticality in case of core reflooding.

Liquefaction and Relocation of Zircaloy Cladding

The melting point of Zircaloy typically ranges between 2000 and 2250 K depending on the alloy and oxygen content. Just above 2000 K, the Zircaloy cladding can melt and, in some cases, drain into the lower core regions. Since the relocation of the molten Zircaloy cladding may be delayed or prevented by the formation of protective oxides on the outer surface of the cladding, the drainage of molten Zircaloy depends to a large extent on the early temperature history. For fast transients with heating rates in excess of 0.3 to 0.5 K/s or transients with the low core water level, the Zircaloy cladding will melt and drain with a short delay into the lower core regions. In this condition, the dissolution of UO_2 by molten Zircaloy is quite prevented. For much slower transients (typically with heat-up rates below 0.3 to 0.5 K/s), the relocation of molten Zircaloy is inhibited by the formation of a thick protective oxide layer (ZrO_2) on the outer surface of the cladding. The formation of this thick protective oxide layer can also be enhanced by transients characterized by multiple events of core heating and cooling with peak temperatures remaining below 2000 K, preventing relocation of molten Zircaloy. Examples might include the cycling of relief valves or periodic accumulator injection.

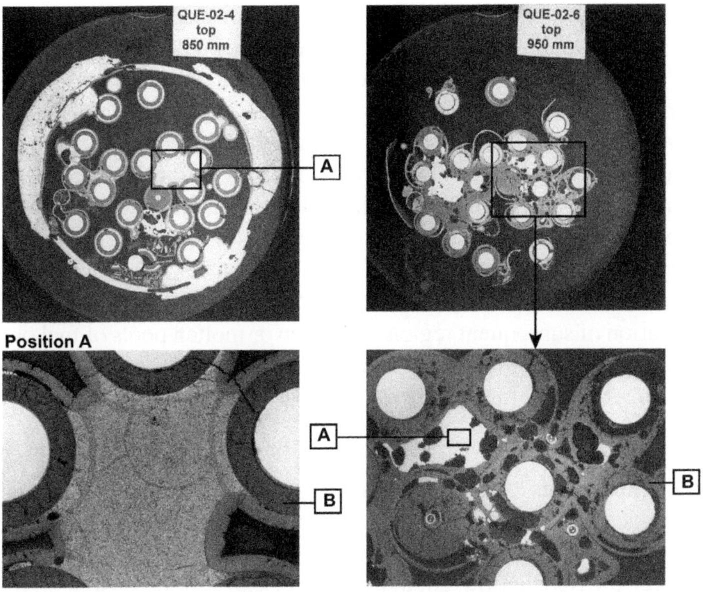

FIGURE 2.3 Photos of two cross sections from the QUENCH-02 test showing cladding failure and melt distribution in the assembly and impact on core geometry.

When maintained in place in between UO_2 pellets and the external ZrO_2 layer, the molten Zr can partly dissolve the UO_2 pellets and, at a lower extent, the external ZrO_2 layer of the cladding [89]. These chemical interactions induce the liquefaction of UO_2 and ZrO_2 at about 1000 K below their melting points and are responsible for the formation of U-O-Zr mixtures, which remain in place up to the failure of the ZrO_2 layer at temperatures greater than 2300 K. The greater the ZrO_2 thickness, the longer molten Zr can stay in contact with UO_2 and dissolve it. In this case the dissolution is limited by the concentration of Zr in the liquid phase (typically 20% by mass concentration). Figure 2.3 shows the two possible situations, in the same experimental bundle: liquid Zr remaining in contact with the UO_2 pellet or relocating along the external part of the rods.

After clad failure, the oxidation of the relocating mixtures can continue. Nevertheless, the most significant consequence of the drainage of molten Zr-rich mixtures is the reduction in the hydrogen production and of the large generation of heat due to the oxidation of the Zircaloy. Globally, the reduction in the core oxidation is due both to the relocation of the molten Zircaloy in colder zones of the core where its freezes on grid-spacers or near the water level and to the formation of blockages with reduced surface areas exposed to steam.

A secondary consequence of the fuel dissolution by the molten Zircaloy is the impact on fission product release. Enhanced release can occur due to the dissolution of the crystalline matrix and the migration of related gas bubbles through the bulk melt phase. However, as observed in some experiments carried out under some reducing conditions, the liquid phase could also act as a trap for large gas bubbles, producing a foam-like structure in which bubbles are stable.

After relocation, the subsequent heat-up of the core can be altered due to the fact that natural circulation flow patterns change as the coolant channels fill up with molten material. These flow blockage zones, typically located at the original grid-spacer elevations or near the water level in the lower portion of the core, are a preliminary step in the formation of subsequent blockages of molten fuel and remaining oxidized cladding materials. However, existing experiments indicate that the formation of such metallic Zr-rich blockages is not necessary to the formation of subsequent regions of frozen or molten pools of fuel or other oxidized materials. Ceramic crusts composed of oxidized materials can also form as a direct result of heat transfer in the core.

Liquefaction and Slumping of the Fuel

When the temperatures of the fuel or oxidized cladding material approach their melting points, the fuel and remaining oxidized cladding material start to slump lower in the core. In some cases, if the fuel has a sufficient level of burn-up and the RCS pressure is low enough, the fuel can swell, causing additional reductions in the flow area, as initial porosity of the fuel increases. Depending on the location of the slumping material, and the temperature gradients in the core, the ceramic fuel and oxidized cladding material will relocate to cooler regions of the core until it freezes, resulting in the formation of large blockages (Figure 2.4).

These blockages can then trap molten materials formed subsequently at higher levels in the core or upper plenum. Although the specific temperature

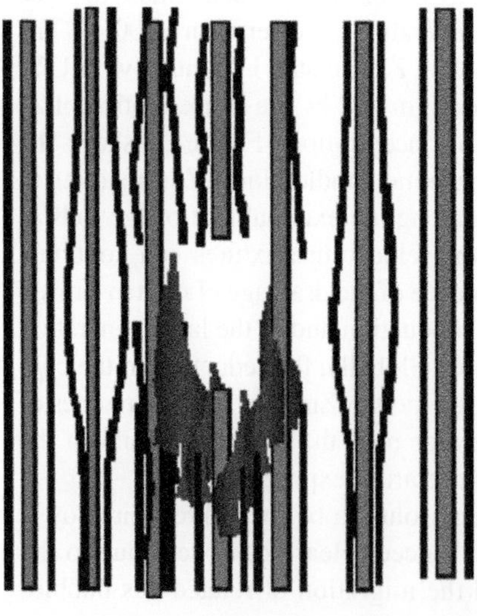

FIGURE 2.4 Scheme of first relocation of molten materials within the core.

range for fuel rod collapse depends on the composition of the fuel and oxidized cladding material, and on additional chemical interactions (UO_2 oxidation by steam, UO_2-ZrO_2 eutectic formation, etc.), the formation of ceramic melts will occur at temperatures well below the melting point of the ZrO_2 and UO_2 and even lower than the temperature of formation of the UO_2-ZrO_2 eutectic at ~2800 K. Bundle experiments with pre-irradiated fuel such as Phébus FP experiments carried out in oxidizing conditions showed fuel collapse temperatures around 2500–2600 K, quickly followed by a molten pool formation.

The consequence of the formation of large blockages characterized by molten pools of ceramic material can be significant. Because of the size of the blockages, typically extending over a large portion of the core, the flow patterns in the core can be greatly altered. However, the blockages may result in enhanced cooling of other unrestricted core regions, slowing the heat-up of cooler, typically lower powered core regions. Since the molten core pool may be surrounded by the peripheral frozen crust, fission products may be retained inside the fuel even though temperatures are above its melting point. When the size of the molten pool is relatively large, the resultant molten pool natural circulation can affect the heat transfer to the peripheral boundary through surrounding crust. The most notable impact of the molten pool natural circulation is that the heat transfer rate to its sides and top may be much greater than to the bottom. Thus, the frozen crust becomes thinner in the sides and top of the molten pool than the bottom, resulting at the upper level in the preferential failure of the corresponding crusts.

Relocation of Molten Pool Materials into the Lower Plenum

Along with the evolution of the scenario, the molten pool trapped in the core moves to lower elevations in the core or lower plenum. In the case of TMI-2, the melt moved through core bypass into the lower plenum (over 50 tons of corium in the core pool, more than 30 tons were relocated through the bypass). Although there are limited data for this process under protoypic conditions, the TMI-2 accident progression analysis indicates that the melt relocation is initiated by the penetration of the molten pool to the outer or lower periphery of the core and the failure of the frozen crust enclosing the melt. Prior to more dramatic relocation to the outside of core, the molten material can move radially and axially in the core region, but the movement appears to be a sporadic expansion of the frozen crust and molten pool due to the limited core flow areas. The melt relocation can occur even if the core is completely covered with water, although the additional cooling may slow or even prevent further movement of the melt.

In the case of TMI-2, the core was reflooded prior to the relocation of a significant amount of the melt into the lower plenum. In spite of the cooling water injection which filled the core region, the progression of the molten pool in the core could not be stabilized, the heat exchange through the surrounding crust being insufficient to extract the decay heat generated by trapped fission products in the pool. The radial progression was sufficient to reach the surrounding core's former wall. Although the exact details of the melt relocation in TMI-2 are not known, a small part of the melt was directly relocated axially through some fuel rod assemblies at the periphery of the core. The failure of the crust and relocation of a portion of the melt was accompanied by the partial collapse of loose debris and fuel rod fragments supported by the upper crust. Although confirmatory data are not available for prototypic core materials and scale, in the case of accidents where the core is not reflooded and the core remains uncovered, the molten pool will continue to grow axially and radially until it reaches the boundary of the core. The fraction of the core that is molten at this stage and the ultimate relocation path into the lower plenum will depend on the power distribution in the core, the design of the core and surrounding structures, and the thermal-hydraulic boundary conditions.

Because of the difference in axial peaking factors in the BWR core, with relatively flat flux distribution during some stages of the burn-up history, and the PWR core, with a more pronounced cosine shape distribution, it has been postulated that the molten pool is more likely to drain directly into the lower plenum in the BWR case or through the core bypass region in the PWR case. The drainage of the melt into the lower plenum is also made more likely in BWRs since many SA sequences in BWRs also result in the earlier depressurization of the system, resulting in a relatively low water level at the start of initial core heat-up.

The consequence of relocation of the melt into the lower plenum will depend strongly on the amount of water available in the lower plenum. In the unlikely case that the lower plenum is steam filled, the melt can directly contact the lower head structures and may melt-through the structures relatively quickly. In the more likely case where water is present in the lower plenum, the heating of the lower head structures will be delayed, but the system pressure may increase sharply because of the contact of the melt with the water. Although the likelihood of an energetic melt-water interaction within the vessel is considered to be low, any fragmentation of the melt due to the enhanced heat transfer between the melt and the water can alter the long-term coolability of the debris and vessel wall. The relative timing and nature of the relocation process also has an important effect on the stratification of molten core materials in the lower plenum of the vessel. For example, the early relocation of ceramic material from the upper core portion may result in the formation of multiple layers of ceramic and metallic materials in the lower plenum, whereas the late relocation may promote the mixing of core materials in the lower plenum, resulting in the reduction of the number of different material layers.

Fragmentation of Embrittled Core Materials During Quenching

In the case of addition of water to the core, the core geometry will change with the fragmentation of embrittled materials. At lower temperatures below 1500 K, the fragmentation of fuel rod materials has been relatively well characterized due to the research on cladding embrittlement under DBA

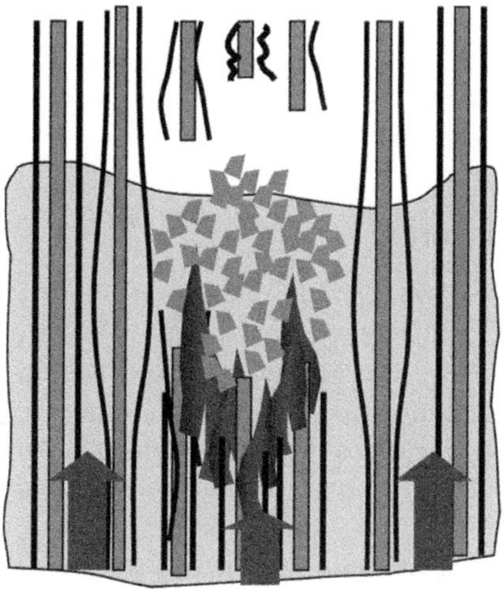

FIGURE 2.5 Fragmentation of core materials during quenching.

(Design Basis Accidents) conditions. In this case, experiments have shown that the cladding will fail and fuel pellets may fragment. At temperatures above 1500 K, the change in geometry is very much dependent on the geometry of the degraded core at the time of reflooding (Figure 2.5). For regions where either molten metallic or ceramic melts have refrozen, there may be some cracking of the refrozen material but there is little change in the geometry of the material. For regions where the fuel rods are relatively intact and peak temperatures remain below the melting point of Zircaloy, the fuel and cladding may fragment and partially collapse to form a solid debris bed. For regions where fuel rods are relatively intact but the peak temperatures have exceeded the melting point of the Zircaloy cladding, the fuel pellets will remain relatively unchanged even though much of the Zircaloy cladding has melted away.

The relative stability of the columns of fuel pellets, once Zircaloy melting temperatures have been reached, has been attributed to the dissolution of UO_2 by molten Zircaloy, resulting in penetration of molten Zr into cladding cracks and gaps between fuel pellet and cladding. This process effectively welds the fuel pellets together. Two notable exceptions have been noted. First, if the melting and draining of the molten Zircaloy occurs very quickly, the fuel pellets collapse during the addition of water. This has been attributed to the fact that there is insufficient time for the molten Zircaloy to penetrate into the pellet interfaces and cracks. Second, if the fuel is exposed to steam for an extended period of time, the oxidation of the UO_2 by steam tends to cause the fuel to break apart on grain boundaries during quenching.

The primary consequence of this change in geometry is associated with the break-up of the ZrO_2 layer and underlying brittle oxygen-rich Zr layer of the cladding. The break-up of fuel rods can result in a dramatic increase in fission product release, as the fuel is exposed to hot steam and the break-up of the fuel at grain boundaries is enhanced. A secondary consequence of the fuel rod break-up is that the resulting rubble debris bed may subsequently alter the flow patterns in the core and the heat transfer from the fuel.

2.2.4. Reflooding of Hot Damaged Cores

The reflooding of hot (temperatures above 1500 K), but relatively intact, fuel rods may result in a sharp increase of temperature of the fuel rods and surrounding core regions, large production of hydrogen and steam, enhanced fission product release and core material melting. This sounds counterintuitive, but the increase of temperature is caused by the oxidation of the regions of the core that have not yet been quenched but are exposed to large quantities of hot steam produced by the quenching process. Although this is still an area of active research, a large number of reflooding experimental data, as well as data from TMI-2, has demonstrated these characteristic trends. In detail (the capital letters below, after the name of the experiments, refer to the Figure 2.6), the

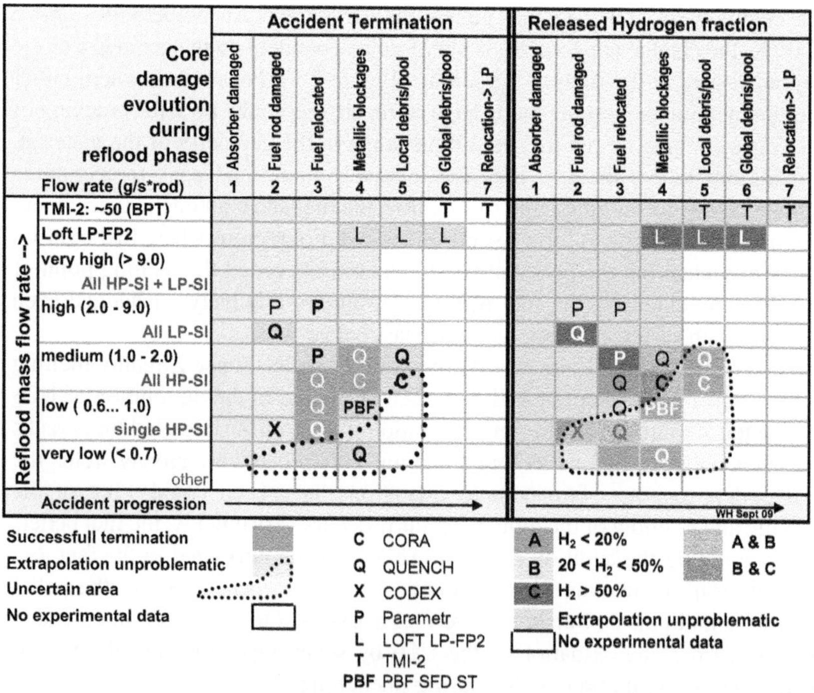

Accident Termination
Released Hydrogen fraction
Core damage evolution during reflood phase
Absorber damaged
Fuel rod damaged
Fuel relocated
Metallic blockages
Local debris/pool
Global debris/pool
Relocation-> LP
Flow rate (g/s*rod)
TMI-2: ~50 (BPT)
Loft LP-FP2
very high (> 9.0)
All HP-SI + LP-SI
high (2.0 - 9.0)
All LP-SI
medium (1.0 - 2.0)
All HP-SI
low (0.6... 1.0)
single HP-SI
very low (< 0.7)
other
Reflood mass flow rate -->
Accident progression
WH Sept 09
Successfull termination
Extrapolation unproblematic
Uncertain area
No experimental data
C CORA
Q QUENCH
X CODEX
P Parametr
L LOFT LP-FP2
T TMI-2
PBF PBF SFD ST
A H_2 < 20%
B 20 < H_2 < 50%
C H_2 > 50%
A & B
B & C
Extrapolation unproblematic
No experimental data

FIGURE 2.6 Reflooding map: Experiments aligned by available mass flow rate (vertical) and core damage state (horizontal) for damage progression (left) and additional hydrogen release during reflooding (right), from [96].

Hungarian CODEX (X) tests [17] are performed in 3 × 3 bundles; the rather recent Russian PARAMETER (P) tests in Podolsk [18], dedicated to top and bottom flooding, refer to 19 rods in VVER bundles; the reflooding experiments CORA (C) [19] and QUENCH (Q) [11] address bundle sizes between 21 and 31 rods; and the central fuel element in LOFT-LP-FP2 (L) [20] [21] amounts to 121 rods. Most of the experiments address PWR (P) conditions; there are only one BWR (B) and five VVER (V) reflooding experiments, one CODEX one QUENCH and three PARAMETR tests.

These data have shown that this behavior occurs as a direct consequence of the fast oxidation of the Zircaloy-rich structures. It turns out that phenomena like oxide spalling, crack formation, and hydrogen absorption by the remaining metal have only a minor effect on the integral hydrogen source term. Temperature escalation during reflooding was observed only in tests where quantities of zirconium-bearing melt were exposed to the flowing steam, as observed in recent small-scale experiments and in QUENCH tests [11]. As a result of the enhanced oxidation of the core during quenching, local hydrogen production rates may increase by an order of magnitude relative to the rates produced during core degradation. In turn, the local heat generation rates in the core also increase by an order of magnitude, completely overshadowing the local decay heat contributions by factors of 10 to 20. However, it must be reminded that such phenomena are only temporary and may be localized in some parts of the core only.

The rapid heating and cooling, and the associated hydrogen production rates, can also affect other processes within the reactor system. Fission product release rates can increase rapidly during quench due to microcracking of the fuel and the related release of fission products bubbles trapped on the grain boundaries. It should also be noted that the temperature escalation due to the enhanced oxidation accompanying the quenching can also cause enhanced fission product release. For example, the rapid increase in fission product release was observed in the only two experiments, SFD-ST and OECD LOFT-FP-2, where irradiated fuel rods, using trace-irradiated fuel, were quenched under SA conditions.

The pressure in the system can also increase because of the production of large amounts of steam and hydrogen during quenching. The rapid increase in pressure was observed in TMI-2 under similar conditions when one of the reactor pumps was turned on briefly after initial core melting had started. The amplitude of the pressure peak was about 30 bars (pressure before reflooding was around 60 bars).

The core design and the temperature history of the core before quenching can have a quantitative impact on the above-mentioned effects of quenching. Transients leading to an extensive oxidation of the core prior to reflooding will reduce the consequence of oxidation during reflooding since there will be less Zircaloy to oxidize. The presence of B_4C control rods (or blades) or Zircaloy fuel assembly shrouds in some types of BWR and VVER designs will have the opposite result because of the additional oxidation potential of these structures and associated mixtures.

When large blockages of molten fuel have been formed in the core, the consequences of core reflooding are not well known from a quantitative point of view. However, the TMI-2 accident analysis and other supporting calculations have shown that after peak core temperatures exceed 2800 K and large melts of UO_2 and ZrO_2 have been formed, reflooding the core is not effective in arresting sustained core heating and growth of the molten pool. The aforementioned result is mainly due to the reduced heat transfer surface area and the low thermal conductivity of the ceramic crust surrounding the melt. In the case of the TMI-2 accident, the molten pool located in the upper core region continued to grow, and ultimately a fraction of the melt relocated into the reactor lower plenum, even after the core had been totally covered with water.

The reflooding of the regions of the core with temperatures above the melting point of the Zircaloy cladding (~2250 K for Zircaloy saturated with oxygen) but below the melting point of the fuel (~3120 K) is expected to lead to a strong oxidation and an increase of the mass of molten materials, possibly initiating the growth of a molten pool. For temperatures higher than 2250 K, intact fuel columns may collapse due to the thermal shock associated with quenching. This is not likely to produce any significant quantities of hydrogen, however, since much of the remaining Zircaloy may either be completely oxidized or relocated in the low-temperature regions of the core. The collapse of the fuel may result in an increased release of the fission products trapped inside the fuel rods, due to the increase of the contact area between fuel fragments and the fluid. Additional hydrogen may be produced mainly from the core regions where molten Zircaloy may have accumulated in the form of metallic layers, but the production rate could be limited by the reduced surface area of those metallic blockages or the low temperature of those relocated materials.

Following the analysis made in [96] and restricting the further analysis only on two parameters, namely, the reflooding mass flow rate (RMFR) and the core damage state (CDS) prior to reflooding initiation, the results of the tests can be arranged as a reflooding map (Figure 2.6) focused on the damage progress (left) and the corrected ratio of hydrogen release during reflooding to total hydrogen mass (right). The range of the RMFR, together with typical emergency core cooling (ECC) systems is shown in the left column for both parts of Figure 2.6. The experiments are located depending on the CDS and indicated literally as mentioned above. The fields in the left part of Figure 2.6 indicate reflood without serious damage propagation, while the light color indicates significant bundle damage.

Calculations revealed that low mass flow rates can lead to unexpected adverse effects, if nearly all evaporated water is consumed by the Zircaloy in the core and nearly pure hydrogen is released into the containment. Such situations occur when ECC pumps cease or other systems, originally not designed for ECC purposes, was activated by SAM measures. From the results of QUENCH-11, the RMFR of 0.6 g/s per rod is sufficient to cool a 1.2-m-long rod bundle, but it is not sufficient to cool a full dry core without formation of a large in-core pool.

Both parts of the reflooding map—at left, the reflooding initiated core damage progression and at right the additional hydrogen production—show similar behavior, which indicates a border for successful core reflooding. Including TMI-2 as an extreme case, it may be concluded that with increasing core damage increased reflooding capability is necessary to maintain coolability. The unpleasant effect, however, is that hydrogen release increases with damage progression, and this affects the hydrogen countermeasures in the containment, if insufficient steam is fed into the containment to avoid hydrogen deflagration or explosion.

The consequence of reflooding of the lower plenum region and structures is not well known and, as is discussed in a later section of this chapter, is an active area of current research. The examination of the TMI-2 lower head, and similar experiments using simulant materials, have shown the potential for the enhanced cooling of the debris and vessel structures due to the combination of the formation of gaps between the debris and the vessel structures and the cracking of the debris itself.

2.2.5. Experimental Programs

The major experiments, including recent, ongoing, and future projects, are described briefly below. Most test programs have produced data used for code validation. All tests used for this purpose are catalogued in an OECD summary report [3], including:

- Separate effect tests on chemical interactions: Numerous tests, conducted by various teams (in particular, FZK being today KIT, Germany, and AECL, Canada) have helped determine the kinetics of zircaloy oxidation, UO_2 dissolution by molten zircaloy, B_4C oxidation, zircaloy dissolution by molten steel, and so on.
- Separate effects tests on cladding rupture mechanisms: These tests (e.g., EDGAR tests) helped determine creep laws for cladding based on cladding temperatures and oxidation states.
- LOFT-FP: This project, completed in 1985, was conducted by the Idaho National Laboratory (INL/INEL, USA) on an assembly of 121 UO_2 rods with nuclear heating (in-pile). It consisted of tests on rod degradation and fission product release, and it involved temperatures up to 2400 K (locally). Steam cooling was used, followed by water reflooding.
- PBF-SFD: This project, completed in 1985, was conducted by INEL (USA) on an assembly of 32 rods of nonirradiated UO_2 with nuclear heating (in-pile). It included tests on degradation and FP release, at temperatures up

FIGURE 2.7 QUENCH test bundle with bundle head, grid spacers, and thermocouples.

to 2600–3100 K (locally). Steam cooling was used, followed by water reflooding (for some tests).

- NRU-FLHT: This project, completed in 1987, was conducted by AECL on an assembly of 16 rods of nonirradiated UO_2 with nuclear heating (in-pile). It involved degradation tests that were unique because of their use of full-scale rods (3.7 meters in height).
- ACRR-MP: This project, completed in 1992, was conducted by SNL (USA). It involved in-pile tests of debris bed melting ($UO_2 + ZrO_2$) in a neutral atmosphere, with temperatures up to 3000–3200 K. The formation and growth of a molten pool were observed.
- CORA: This project, completed in 1993, was conducted by FZK on an assembly of 25 rods of nonirradiated UO_2 with electrical heating (out-of-pile). It involved degradation tests at temperatures up to 2200 K (locally). Steam cooling was used, followed by water reflooding (for some tests).
- QUENCH (Figure 2.7): This project, managed by KIT (it is not finished yet, and other tests are planned), involves an assembly of 25 rods of nonirradiated ZrO_2 and electrical heating (out-of-pile). It consists of degradation tests and involves temperatures above 2000 K (locally). Steam cooling is used, followed by water or steam reflooding.
- Phébus-SFD and -FP: These projects, completed in 2004, were conducted by IRSN (France) on a bundle of 21 UO_2 rods (irradiated in the case of Phébus FP) with nuclear heating (in-pile). They involved degradation and/or FP release tests with temperatures up to 2600–3100 K (locally). Steam cooling was used.
- ISTC 1648 (QUENCH): This is a project of the International Science and Technology Center (ISTC), overseen and funded by the European Commission, and is conducted by the NIIAR (in Russia). It aims to study reflooding under post-LOCA conditions, and it includes two experimental items: degradation and reflooding tests using irradiated VVER fuel

segments; reflooding tests using a new VVER assembly of 31 rods (without UO_2).

- PARAMETER: This ISTC project, initiated by LUCH (in Russia), involves degradation of prototypical nonirradiated VVER assemblies of 19 rods (similar to QUENCH, but with UO_2 pellets) and allows bottom and/or top-down reflooding, with temperatures up to 2300 K. Two tests are planned, one with slow degradation and top-down reflooding, and a second involving a mix of bottom and top-down reflooding.

As indicated, there are few experimental programs on late-phase effects, apart from Phébus FP and ACRR. The LOFT and PBF tests attained late-phase degradation, but did not involve detailed analysis of corium progression and fuel rod melting.

With regard to knowledge of the reactor core evolution after late-phase degradation, the TMI-2 accident remains the sole point of reference. Detailed analyses of the TMI-2 reactor have been conducted and are available in the literature. Figure 2.1 illustrates the state of the core following the accident. Of particular interest are the large molten pool within the core, the collapse of a large portion of the rods above the pool (forming a debris bed), and partial corium relocation toward the lower head. Some characteristics of the accident scenario are also worth noting, particularly the high pressures involved and the fact that corium relocation to the lower head occurred after core reflooding (the bottom part of the core was already quenched).

2.2.6. Status in the Modeling of Phenomena

The behavior of the plant during an SA results from a combination of the occurrence of several physical and chemical phenomena. These phenomena must be taken into account for proper modeling in the codes. The objective of this section is to give some details about the important physical phenomena that should be modeled in the codes for a SA analysis. Comments on the modeling are also provided to help users in their selection of a SA code and its modeling options.

Reflooding of Rod-like Geometry

For reflooding of a rod-like geometry, modeling approaches and correlations developed for DBA conditions are applicable with the following limits. First, these models are typically developed for the original geometry of the core (with the possible consideration of flow blockages due to ballooning and rupture). As noted in the previous section, as temperatures above 1500 K are reached, the reflooding of the core may result in the partial or total break-up of these structures as a result of the strong thermal shock associated with reflooding. Second, the reflood can result in large increases in the hydrogen production, so the flow conditions above the quench front may be altered by the presence of large concentrations of noncondensable gases, which may not have been considered in the correlations for DBA conditions. Third, physical changes in the rod surface, as break-up of the claddings or grids, occur. Thus, because of the strong effect of this process on the subsequent heating and melting of the core and the production of the hydrogen, it is necessary to model both the thermal-hydraulic and mechanical aspects of this process in detail.

Since this issue is still being addressed, the type of models available and the accuracy of these models vary widely from one code to another. Also, this process was the subject of an International Standard Problem Exercise (ISP-45 QUENCH-06, OECD/NEA); the reader should review the published results for that ISP to see how the different modeling approaches compare.

Reflooding of Debris Beds

Models and correlations developed for the original geometry of the core under DBA conditions are not generally applicable to the flow in debris beds, particularly during reflood conditions. However, a number of correlations have been developed specifically for porous media that are, in many cases, also applicable to flow in debris beds as long as the flow does not disrupt or levitate the debris. In many cases, these correlations were developed to address the coolability of debris beds. That is, these correlations were used to define under what conditions it was possible to quench and cool debris beds. In general terms, these correlations indicate that debris beds with fine particles or low porosity (or, more precisely, low permeability) are difficult or even impossible to cool (because of the internal heat generation) while beds with large particles or high porosity are easier to cool. In the case of reflooding of debris beds from above, the countercurrent flow of steam acts to further reduce the coolability of a debris bed because of the reduced ability of the water to penetrate into the bed.

However, many of the correlations that were developed for porous media or debris are either zero- or one-dimensional. The zero-dimensional correlations simply define whether or not the bed as a whole is coolable, while the 1-D models also attempt to describe the quenching elevation within the bed. However, the debris beds that could be formed under SA conditions tend to be somewhat nonuniform and may vary axially, radially, and circumferentially around the core. Thus 0-D or 1-D correlations that predict reflooding for uniform debris beds may be inaccurate under the conditions expected in actual conditions. In the case of reflooding from below, such correlations may not adequately predict the formation of hot spots that may ultimately melt and form local blockages.

In the case of reflooding from above, the correlations may be overly conservative in terms of the limiting effects of countercurrent steam flow since water may be able to penetrate from the side of the debris, even though the penetration of water from the top may be limited (see [95] for details about this).

Although the modeling of the reflooding of debris beds varies substantially from one code to another, reflooding models dependent entirely on coolability-based correlations may be increasingly inaccurate as the nonuniformity of the debris beds increase. Zero-D based correlational models are largely ineffective except in the limited situations where the debris beds are small and uniform. 1-D correlations are also limited but may be most accurate when predicting the reflooding of debris beds in the lower plenum where 2- and 3-D effects are more limited. In some cases (uniform debris bed), since such models may overestimate the effect of countercurrent steam flow, such models may provide a conservative bound on the reflooding of the debris in the lower plenum. In very general terms, the more detailed codes tend to rely less extensively on quenching or coolability correlations, while the integral codes are more reliant on such correlations. In all cases, however, the accuracy of the models used to predict the reflooding of debris beds is largely dependent on the ability of the codes to accurately predict the state of the core and debris beds at the time of reflooding. Thus, the ability of the codes to accurately predict the extent and timing of the break-up of the fuel and core structures, the formation of local blocked regions of previously molten material, and the characteristics of the debris bed (e.g., particle size, porosity, permeability) will ultimately determine how accurately the codes can predict the reflooding of those debris beds.

Reflooding of Molten Pool

The models for the reflooding of molten pools are unique to SA accident conditions and vary widely from code to code. However, in general terms, the models are varying from user input (e.g., switches that specify whether or not molten pools can exist, user-defined heat transfer coefficients in molten pool regions) to more detailed models that attempt to predict the convective and radiative heat transfer on the outside of the crust surrounding the molten pool. In general, the detailed models rely on the application of standard correlations for heat transfer on simple geometries such as plates, spheres, or cylinders in combination with a prediction of the general shape of the molten pool crust based on the history of the formation of the melt. One important constraint limits the accuracy of such models. That is, the difficulty in predicting the exact shape, orientation, and characteristics of the surface of the molten pool (actually the frozen crust surrounding the molten pool) results in large uncertainties in the selection and application of the most applicable correlations. In general, the detailed codes can predict the general configuration of the molten pool, but there are large uncertainties in the local geometry and surface characteristics of the crust; even the most detailed models include modeling parameters that are used to determine the consequence of large uncertainties in the calculated reflooding behavior.

In either extreme, user-defined behavior or detailed modeling of the reflooding process, the primary issue to be addressed by the models or the user is the mechanical stability of the molten pool during and following reflooding. If the molten pool crust is mechanically stable and does not fail, the behavior of the melt inside the crust is somewhat irrelevant as long as long-term cooling is sufficient to remove the resulting decay heat (otherwise, water will simply boil off again). On the other hand, as occurred in the case of TMI-2, if the crust is not stable and fails, the melt can continue to relocate downward even if the core is completely covered with water (and perhaps much of the crust surface adequately cooled). Thus the ultimate consequence of the reflooding of the molten pools will be determined either by models that can successfully predict the thermal and mechanical response of the crust surrounding the molten pool or by the choices by the user.

Flow Patterns

It is important that thermal-hydraulic models for SAs are able to calculate two specific complex flow paths: natural convection in the reactor pressure vessel (RPV) and in the RCS. This is essential to predict the temperature along the hot leg, up to the SG (Steam Generator) where weak points may break due to overheating.

The coolant flow paths go through the partially blocked core. After substantial degradation and relocation of material, there are regions of low porosity (blockages) and regions of high porosity (voids) in the core. The flow becomes at least 2-D, with steam being diverted from the center where the first blockages appear and then being attracted to the center again in the upper part where void regions exist.

The early models for predicting the thermal-hydraulic response of the core and vessel during SA accidents were 1-D models. As a result, these models could not accurately predict the flow patterns in the core or vessel and the resulting heat-up and melting of the core. Most, if not all, of the commonly used codes now use 2-D models, with varying degrees of sophistication. The integral codes generally use models that predict the flows within predefined patterns and so, can predict the general trends associated with the flow patterns in the vessel but have difficulty predicting the effects of changes in core geometry. The detailed codes generally use models that predict the flow patterns in the vessel using a 2-D or 3-D nodalization of the vessel and core. Some Computational Fluid Dynamics (CFD) codes (TRIO, PHOENICS, etc.) have been used to predict 3-D flow patterns under limited conditions (a few elements of the circuit, with imposed boundary conditions), primarily under single-phase steam flow conditions. Examples include the calculation of flow mixing in the lower plenum region of steam generators to help quantify the flow patterns during hot leg natural circulation and the heating of the upper core and upper plenum structures in the later stages of the TMI-2 accident.

Coolability of Debris Bed

As discussed in the previous section on the reflooding of debris beds, standard thermal-hydraulics correlations that have been developed for intact bundles cannot be directly applied to debris beds. As a result, either these correlations must be adapted for use in debris beds or correlations developed specifically for flow in debris beds must be utilized.

Hydrogen Transport

Hydrogen enhances the natural convection and the heat transfer but reduces steam condensation. One of the main results of in-vessel analysis of a SA is the prediction of the hydrogen flow rate and the hydrogen/steam ratio at the break(s). These key parameters are then used as boundary conditions by ex-vessel codes or models to estimate the hydrogen distribution in the containment. Hydrogen is essentially a threat to the containment because of the risks of combustion. To predict this risk, the hydrogen transport to the break(s) must be properly calculated. This is obtained by including the transport of noncondensable species in the thermal-hydraulics model (at least one additional conservation equation).

Radiation Heat Transfer

Radiation heat transfer becomes important at high temperatures (above 1000 K) and after collapse of materials, when some structures are in direct view with hot debris located below. Radiation is modeled in most of the codes, including absorption of heat by steam, but usually the models cannot deal with scattering media (water droplets for example) or large cavities with strong absorption by the gas. Such cases would require multidimensional models that require a lot of computation time. In any case, models for radiation heat transfer, with a relevant estimate of view factors, across rod assemblies or debris, and across large cavities should be available in the code. The lack of appropriate radiative heat transfer models will lead to an incorrect temperature distribution in the vessel.

Thermal Behavior of Debris Bed

Heat conduction in a debris bed is usually solved by using an effective conductivity characterizing the complex bed made of particles, liquid melt, and fluid. The type of models used varies from code to code, with the detailed codes tending to describe the temperature distributions within the debris beds, while the integral codes may use lumped parameter approaches. In some cases, a porous medium description is used to characterize the heat transfer. In this case, radiation heat transfer is included in the effective conductivity.

Zr Oxidation

Up to 1500 K, many experiments have been performed, and reliable correlations exist to reflect the kinetics of the growth of the oxide layer and the hydrogen production rate. These correlations may be more questionable for higher temperatures, which are in the range of interest for SA. Most codes use the same correlations (with options to be selected by the user), or more detailed models, including equations for diffusion of oxygen through the different layers (zirconia, α-Zr, β-Zr). Although the better accuracy of diffusion models as compared to correlations has been shown in separate effects tests, their positive effect is not so clear for large integral tests or for reactor calculations, and they require more computation time. Therefore the use of correlations is still a good choice even for mechanistic codes.

Steel Oxidation

Steel oxidation is modeled in the same way as Zircaloy, but it is less exothermic compared to the Zircaloy oxidation. The mass of steel in the core and the surface area is small. Thus, the steel oxidation does not contribute much to the amount of hydrogen produced. An exception may be Russian PWRs of type VVER that contain a larger amount of steel than the Western-type PWRs.

B_4C Oxidation

B_4C oxidation is very exothermic and may have a significant effect on the early phases of degradation, particularly during the reflooding. However, the complex chemistry of this reaction (production of boric acid, methane, and other gases) is still under investigation, and models have only recently been introduced into the codes. Boric acid and methane affect the fission product chemistry.

Oxidation by Air

The consequence of air ingression is starting to attract more attention because oxidation of Zircaloy by air is more exothermic than oxidation by steam. Nitrogen may also diffuse into the Zircaloy, particularly if the oxygen of the air is depleted. Fuel oxidation by air is resulting in hyperstoichiometric urania (slightly more than 2 atoms of O for one atom of U) and higher release rates for several fission products [98]. These effects are not typically modeled in SA codes.

Steam Starvation

When the water level has dropped significantly, the oxidation of the core may be limited by the availability of steam. This leads to steam starvation zones that cannot be oxidized, although they are at high temperature. Such zones are likely to produce Zr-rich melts that will be oxidized after relocation in another part of the core or during reflooding. Starvation may also occur before blockage formation, in the case of a low steam mass flow rate, downstream of the oxidizing region. This phenomenon is modeled by a limitation of the steam diffusion mass flux available at the cladding surface. Such modeling is not very controversial, and it is available in all codes.

Several oxidation processes were studied in the frame of the EC project OPSA [22].

Control Rods Degradation (Absorber Models)

Each of the chemical reactions between control rods materials ("nonfuel dissolution") is modeled in most codes by simple "eutectic temperature" models, or kinetic rates and/or equilibrium diagrams (B_4C oxidation and interactions with steel may not exist in all codes). Similar models are necessary for spacer grids. Models for the mechanical behavior of the claddings with internal overpressure calculate the deformation (ballooning), the possible fuel/cladding contact, and the creep failure (burst or flowering). At present, it appears that Ag-In-Cd rods degradation does not have a strong effect on the fuel rods' degradation and the subsequent evolution because Ag-In-Cd relocates at a lower elevation than the oxides.

As a result, simple interaction models may be sufficient (i.e., equilibrium models assuming extremely fast chemistry). For B_4C, the situation is more complex because of its very exothermic oxidation. Finite rate kinetics should be used for B_4C rod oxidation and degradation.

Fuel Rod Dissolution

Fuel dissolution and ZrO_2 dissolution by liquid Zr is modeled in most codes by kinetic rates (correlations) and/or equilibrium diagrams. Some recent models are able to deal with the two simultaneous reactions. They offer a strong potential for accurate description of complex situations such as the dissolution of zirconia layer in case of steam starvation. When available, the consequence of such models should be checked. It was observed experimentally that increasing the fuel burn-up increases the fuel dissolution by liquid Zircaloy. However, such irradiated fuel effects are not currently modeled in SA codes. Several experiments and models related to this subject were led during the EC projects CIT and COLOSS. The results are available in several reports, and a good overview may be found, respectively, in [23] and [15].

Cladding Failure

The modeling of oxide shell failure, which determines the start of relocation, is also very important. It is usually based on simple temperature and oxide thickness criteria, coming from experimental observations. After phase transition of the Zircaloy (first into alpha phase, then into zirconia), the mechanical resistance is strongly reduced. Therefore, it is usually assumed that the cladding integrity depends essentially on the thickness of unoxidised Zircaloy (alpha phase) and on the temperature (thermal stresses). Several experiments have shown that cladding failure occurs for temperatures in the range 2000 K–2500 K and oxide thickness in the range 0.2 mm–0.4 mm. In case of reflooding, cooling of the cladding may be faster than phase change, and therefore, more complex models taking into account finite-rate kinetics and the relaxation of stresses across the cladding are more accurate.

Kinetics of Materials Interactions

Oxidation and dissolution processes are controlled by diffusion of the species (especially oxygen). Diffusion coefficients are introduced in correlations or in oxygen diffusion models. They are usually described by an Arrhenius law. The growth rate of oxidized or dissolved layers is, in most cases, given by a parabolic law. The comparison between diffusion models and correlations has shown that parabolic laws are usually acceptable as long as the heat-up (or cool-down) rate is not too large. In general, this concerns only some short periods of the accident sequences.

Melt Progression

The modeling of corium relocation must take into account not only the state of the materials but also the geometry of the solid components along which it flows (reduced cross section, grids, debris, etc.) and may be either 1-D if one wants to consider only the candling along rods or at least 2-D for a more general description such as horizontal spreading above crusts or plates and relocation towards the lower plenum.

Several models exist to calculate the velocity of the melt: from a simple imposed velocity to more detailed calculation using porous media theory. Validation studies have shown that simple 1-D relocation models are acceptable for bundle calculations (integral tests) because of the fast melt relocation process. However, only 2-D models (with radial spreading) have been able so far to reproduce melt progression in experimental debris beds. Therefore, the user should be aware of the limitations of 1-D models. In particular, the relocation observed in TMI-2 showed a radial progression of the melt (molten pool), followed by an axial progression through the bypass, down to the lower plenum. Such relocation processes were studied in the frame of the EC project COBE [20], with an emphasis on the use of relevant material properties (density, viscosity) and advanced multidimensional flow models to predict melt relocation and progression.

Early Fuel Pellets Slumping

If the amount of fission gases inside the fuel is significant (high burn-up), the fuel pellets may be subject to swelling when the temperature reaches 2700 K. This corresponds to a large volumetric expansion of the fuel and a reduced mechanical resistance. This may be a cause of fuel slumping. Such a phenomenon is not yet modeled by the codes. Another possibility is the collapse of solid fuel fragments (relocation of non-molten structures), which is already modeled by some codes. Up to now, there has been some experimental evidence of irradiated fuel relocation at fairly low temperatures compared to the values predicted by phase diagrams. Such behavior could not be explained in a satisfactory way (especially because it was only observed in a few experiments) and therefore is not currently well modeled by codes.

Behaviour of Molten Pool

A natural circulation regime is established in the molten pool, and the heat fluxes at the boundaries of the pool depend on the internal Rayleigh number. This is usually modeled with standard heat transfer correlations applied at the external boundaries of the pool. It also requires the modeling of crust formation that significantly changes the heat transfers along the pool boundaries. Although this modeling may be sufficient for heat transfer, it is not sufficient for mass transfer aspects such as phase segregation between solid and liquid or separation of non miscible liquids. Models exist for this problem, but they require the complete resolution of Navier-Stokes equations with a very refined meshing that cannot be considered yet in SA codes. Dedicated codes may be used if necessary. A simple comparison between the internal heat generation and the maximum heat flux that may be removed (critical heat flux or CHF) at the external boundaries shows that there is a maximum size for the pool, above which it becomes impossible to cool it and stop its progression. If this size is reached, the pool expands and grows, and finally reaches the external baffle or the lower core plate. This leads to relocation of corium into the lower plenum, which is a significant step in the SA progression. Actually, for the majority of SA scenarios there will be no water pool within the core that is in contact with a molten pool. Rather the molten pool will be supported by non-molten structures. Consequently, heat from the pool can be only transferred to steam or other structures and CHF is not an issue.

Solidification

At the boundaries of the molten pool, the formation of crusts may have a strong impact on the heat transfer. It appeared clearly in experimental programs such as RASPLAV [12] that the solidification of corium mixtures may lead to material segregation, which may even result in a stratification of the pool (see the paragraph about the focusing effect below). Such a process is related to thermochemistry: the equilibrium compositions of the solid and liquid phases are different. Some models exist for calculating such complex phenomena, but they have been implemented in stand-alone codes and are not yet available in SA codes. However, they usually include crust formation models without phase segregation. Crust formation, especially on top of the pool, has a strong impact on heat transfers.

Relocation of Corium into the Lower Plenum

Three relocation scenarios in the lower plenum should be mentioned (Figure 2.8) [99]:

- Lateral crust failure and relocation via the core bypass
- Bottom crust failure and jet flow through the lower support plate (unlikely)
- Massive relocation due to failure of the lower support plate

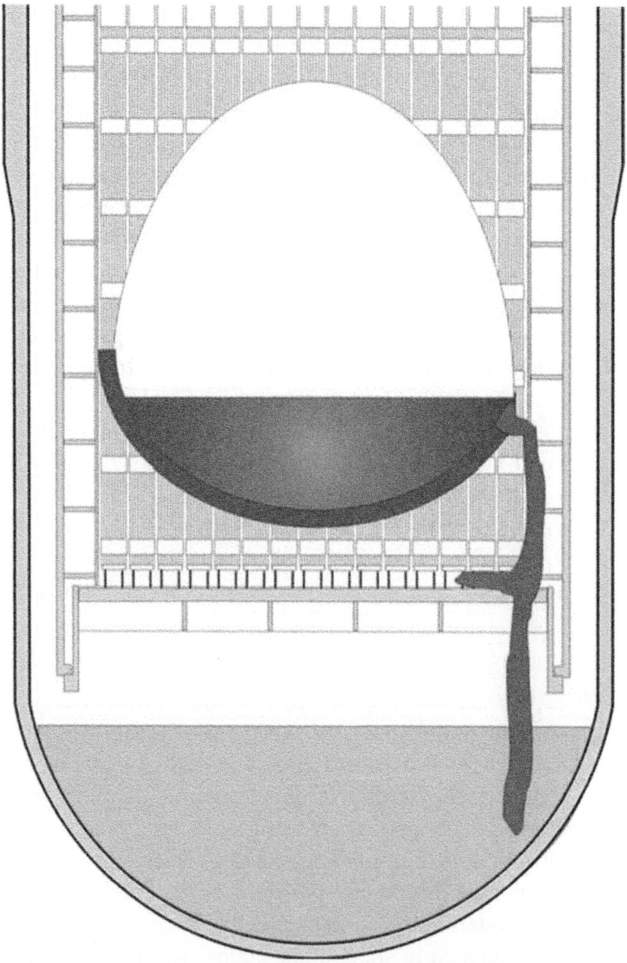

FIGURE 2.8 Schematic diagram of molten corium arriving in the lower plenum through the lateral baffle plates. The melt jet may fragment in the residual water.

2.3. ACCIDENT PROGRESSION IN THE LOWER PLENUM

(C. Mueller, J.M. Seiler, and F. Fichot)

This section deals with the part of core meltdown scenario that starts with the relocation of molten material from the reactor core to the lower head and ends with the vessel failure and relocation of molten material to the reactor cavity. The key research issue is the question of the heat flux from the molten material to the vessel wall and whether this heat flux can be extracted by accident management measures, while keeping the integrity of the vessel (secondary barrier).

It should be pointed out that, even without the vessel failure, there is a high probability that a substantial release of volatile fission products has taken place because either the core meltdown has been initiated by a leak in the primary circuit or the surge line has failed due to high temperatures and pressures. The main advantage of keeping the molten core inside the vessel is the fact that a direct attack on the containment basemat is avoided and the core materials are finally confined in the steel vessel.

The fundamental problem linked to the recovery of molten core materials in the lower head is heat removal. Decay heat is produced by the radioactive fission products that are still present in the melt. If this heat (also called residual power) is not removed in one way or another, the molten core material will inevitably melt-through the vessel wall.

Removing the decay heat is, potentially, an achievable task, provided there are no local high heat flux peaks that are beyond cooling capabilities. To give some numbers: in a PWR of 4500 MW (thermal) 2 h after shutdown the decay power is ~46 MW, which can be dissipated by evaporating ~23 l/s of water, which, at first glance, is not hard to provide. Obviously, it makes a lot of sense to look into the details of the behavior of the molten core in the lower head and to investigate measures to keep the melt safely inside the vessel.

The main problem is that the progression of molten corium in the core and in the lower head is a sequence of very complex phenomena that cannot be described in detail. Molten materials tend to form agglomerates and pools that are difficult to cool. Since the corium configurations are of fundamental importance for coolability investigation, a lot of work has been dedicated to the description of this progression. Different scenarios have been hypothezised, which take advantage of knowledge derived from the TMI-2 accident. In this reactor, the melt progression could be stopped inside the reactor vessel due to late water injection in the primary circuit under elevated pressure (approximately 100 bar), despite the relocation of ~20 tons of corium in the lower head and without external vessel cooling.

Many physical phenomena play an important role in the corium progression. This is illustrated in Figure 2.9, where some issues are shown.

In 2003 a concerted action of european experts in severe accidents was performed to establish a table on the severe accident issues where large

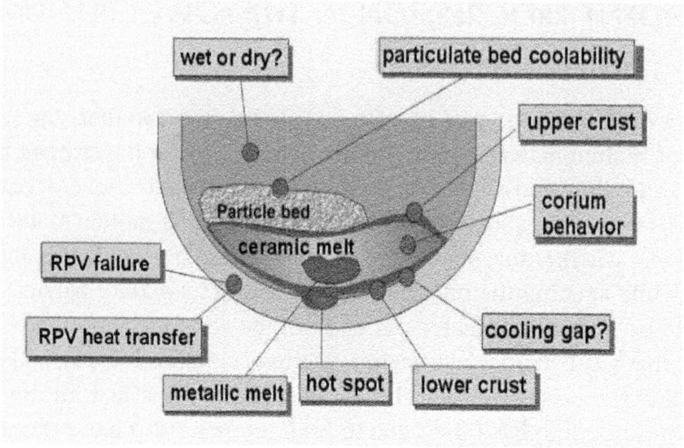

FIGURE 2.9 Corium configuration in the lower head.

uncertainties still subsist. All phenomena that have been investigated in connection with the evolution of a severe accident have been put in a Phenomena Identification and Ranking Table (PIRT). The outcome of the PIRT for late in-vessel issues is shown in the Table 2.1.

2.3.1. Main Physical Phenomena in the Lower Head, Melt Progression with or without Reflooding

In-vessel corium is formed of the core materials and some structure materials, after chemical and mechanical transformation linked to the core degradation and transfer into the lower head. Based on the TMI-2 core examinations [87], the following main forms of materials are expected:

- Dense layers of mainly oxidic material involving UO_2 and ZrO_2; this oxidic material may contain metallic species (U, O, Zr, Fe mixtures),
- Debris layers of the same oxidic materials,
- Metallic material coming from the core (steel, Zr) and from the structure steel. The metallic phases may also form debris and dense layers.

The whole process of material relocation from the core into the lower head is poorly understood and described. Thus, the relative amounts and locations of phases and layers after corium relocation into the lower head cannot be predicted in a deterministic way. Furthermore, the relocation process may take:

- only one to five minutes (if molten materials are considered to accumulate in the core and discharge into the lower head, through the bottom of the molten pool, after almost complete core melting)
- or may occur in several steps over a large time interval (up to several hours if molten materials are discharged into the lower head progressively and all

TABLE 2.1 Physical Phenomena Impacting Corium Transfer from the Core into the Lower Head

Physical situation	Issue	Phenomenon	Phenomena description other specific features
Relocation to lower corium head	Initial conditions	Molten pool failure modes	The various failure modes of the molten pool in the core region, as well as the failure location and the initial size of the crust break, are the initial conditions for the relocation to the lower head. The flow rate of corium leaving the pool will depend on the initial size of crust break, on hydrostatic head of molten pool, etc. The size of the break will increase due to corium heat transfer.
		Characteristics of corium arrival in lower plenum	The characteristics of corium arrival into the lower head are the chronology of successive slumps, the temperatures, masses and composition of corium flows, etc. *The timing and mode of the corium relocation process will modify the further behavior of corium in the lower head. It will also affect the risk of steam explosion.*
	Corium flow through the internals—wet lower head	Steam explosion	Vapor explosion in case of corium contact with water in the lower head. Vapor explosions can cause damage to the reactor structures.
	Oxidation and hydrogen production	Corium oxidation at arrival in lower head	The metallic components of the melt that slumps into the residual water pool in the lower head and breaks up could be oxidized by steam, which is intensively produced. *Melt-water interaction will only occur with residual water pool in lower plenum.*

(*Continued*)

TABLE 2.1 Physical Phenomena Impacting Corium Transfer from the Core into the Lower Head—cont'd

Physical situation	Issue	Phenomenon	Phenomena description other specific features
Lower head debris bed behavior	Heat transfer	Thermal-hydraulics within the debris bed	The heat-up or cooling of the debris depends on the external heat transfer and the debris porosity. If the debris bed is embedded in water and critical heat flux and porosity are not limiting, the debris does not heat up or will be quenched. If the convective heat transfer from the debris to the coolant is less than the heat generation, the debris will dry out and may melt. In this case, the debris porosity decreases. *Importance of nonuniform debris distribution.*
Lower head molten pool behavior	Pool configuration	Molten pool formation	The molten pool is formed by molten debris or by relocation of melt from the core region without significant fragmentation and its accumulation in the lower plenum. The relocation is either continuous or intermittent. Molten structure material might contribute to the melt pool. *A molten pool in the lower plenum behaves in principle similarly to that in the core region, but its size might be much larger due to the crucible-like pressure vessel wall and to supplementary material coming from internal structure melt-through. It might be below or/and above a debris bed.*
		Segregation and stratification of materials	Depending on the relative density of the different materials and their relative miscibility (existence of miscibility gaps), liquid phases (such as metallic and ceramic materials) may separate and form different layers. Metals may also come atop from melt-through of core structures. *It depends on physical and chemical properties as well as thermal and flow conditions and affects slightly the heat source distribution but significantly the heat flux distribution to the boundary in case that the heat conductivity varies a lot.*

	Heat transfer	Pool heat transfers to boundaries	The pool heat transfers include the phenomena: focusing effect, radiation upward heat transfer in case of dry lower head, heat transfer to a dry or wet particle bed, and heat transfer to possibly overlying water. It also includes the downward heat transfer by conduction.
		Vaporization of pool materials	In a large corium pool, heat-up due to decay heat could lead to a significant vaporization of metals and/or fuel. *Impact on fission product source term.*
Vessel external cooling	Wet cavity	Effect of lower head penetrations	Lower head penetrations as in TMI or BWRs may have a substantial impact on RPV wall external cooling by affecting the external convection flow and the steam formation. *Special case of BWR and of some PWR. Recently performed experiments indicate that penetrations do not significantly hinder the heater transfer to the water.*
Thermal and mechanical loadings and behavior of structures including the lower head	Lower head	Thermal and mechanical loadings	If different corium layers form by stratification in the lower head (oxide, metals, debris), this will induce axisymmetrical thermal loadings of the lower head with various distributions. In the absence of stratification, the 3-D distribution of the mixture of debris and molten corium in the lower plenum will induce local hot spots, and thus asymmetrical thermal loadings on the vessel. The mechanical loadings will be the primary pressure and the deadweight of vessel and corium. *Important for 3-D effects.*
		RPV mechanical failure	RPV modes of mechanical failure: plasticity, damage, creep.

along the core melting process) and remelting of debris in the lower head can begin before the core has been emptied of its fuel.

Jet Fragmentation and Wall Erosion by Jet Impact

Basically, two options concerning the water inventory in the lower head have to be considered in the analysis:

- The lower head is filled with water up to the core bottom plate. This is the most probable scenario in case of a LOCA with three loops remaining intact and natural circulation in these three loops. Condensed steam is flowing back from the steam generators to the lower head. The insulating crust of the in-core melt pool allows only poor downward heat transfer. One of the findings of the TMI-2 accident was that the bottom plate was always covered with water [86].
- The lower head is dry. This situation is purely hypothetical, and it cannot be reasonably expected in the course of an accident scenario. However, it is often considered in analyses because it is the situation where the possible failure of lower head occurs in the shortest time.

In case of a water-filled lower head, a bulk of hot molten metal or oxidic corium drops into water according to one of the three potential scenarios:

1. The hot melt sinks calmly down to the bottom.
2. The hot melt jet breaks up into debris and causes violent boiling.
3. A steam explosion occurs, spraying steam, boiling water, and the hot melt particulates in the upward direction.

In small-scale experiments with simulant materials, the jet break-up was observed. The medium-sized FARO experiments [10] [24] with 175 kg of corium showed jet fragmentation generating debris particulates with a size of 1–4 mm. Neither scenario 1 nor scenario 3 was observed in experiments. In real life, however, volcanic (ceramic) lava can produce spectacular steam explosions.

In case of a dry lower head, molten core material at ~2500°C impinges on the cold vessel steel, which exhibits a melting temperature of ~1500°C. It was hypothesized that the hot jet might erode the vessel wall and immediately might cause vessel failure. Since steel is a good heat conductor and oxidic corium only a poor heat conductor, and taking into account that the oxidic material freezes at elevated temperature and well above the melting point of steel, it has been argued that corium would form an insulating crust at first contact reducing drastically steel ablation. This behavior has been demonstrated in experiments with oxidic material impacting on steel plates.

Nevertheless, upon direct impact of superheated molten *metal* jets under dry conditions, the fast (up to ~1 cm/s ablation rate) melt-through of the vessel is possible. Jet instability and jet fragmentation under water will strongly reduce the direct attack of the melt on the vessel wall and the risk of immediate vessel failure.

Understanding Molten Pool Convection in the Bounding Situations

As stated previously, a deterministic description of corium progression in the vessel is presently not accessible. Nevertheless, for the evaluation of the in-vessel-retention (IVR) capability, knowledge of plausible heat loads is required. A way to solve the problem consists in evaluating heat loads associated with different potential corium configurations in the lower head. Doing this, the situations leading to the maximum heat loads are considered with peculiar attention. The objective is to demonstrate that the vessel is coolable (by reactor pit flooding also called "external" cooling), for these elevated heat loads. These situations are called "bounding situations" [90]. This approach has been developed for the AP 600 and AP 1000 safety demonstration [33].

The maximum heat load will be connected to corium compaction: the lower the surface to volume ratio, the higher the heat flux. Thus, bounding situations are connected to dense corium. Consideration of conduction heat transfer limitations in large masses of compacted corium leads straight to corium pools. Thus, the heat transfer in corium pools is of primary importance for analysis of the bounding situations.

The heat flux is usually considered as a result of temperature difference (K) multiplied by a heat transfer coefficient (W/m^{2}K). It might be uncommon that for molten pool with crust the same temperature difference results in substantially different heat fluxes along the pool boundary. This is the case, however, as illustrated in Figure 2.10.

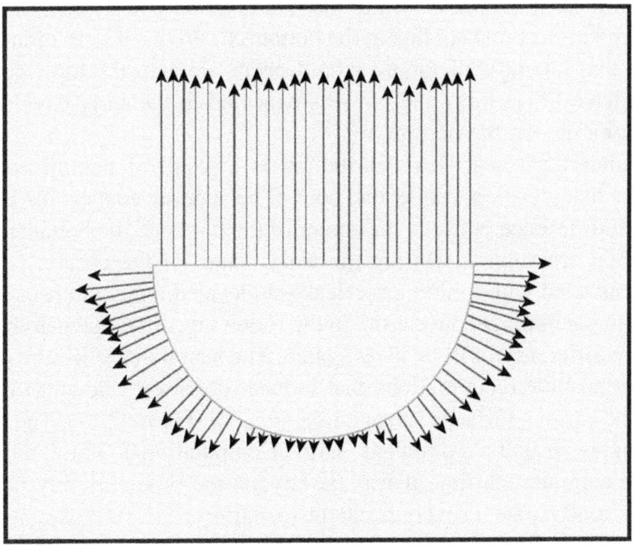

FIGURE 2.10 Heat flux distribution as measured in COPO test 7 [83] (molten pool thermalhydraulics).

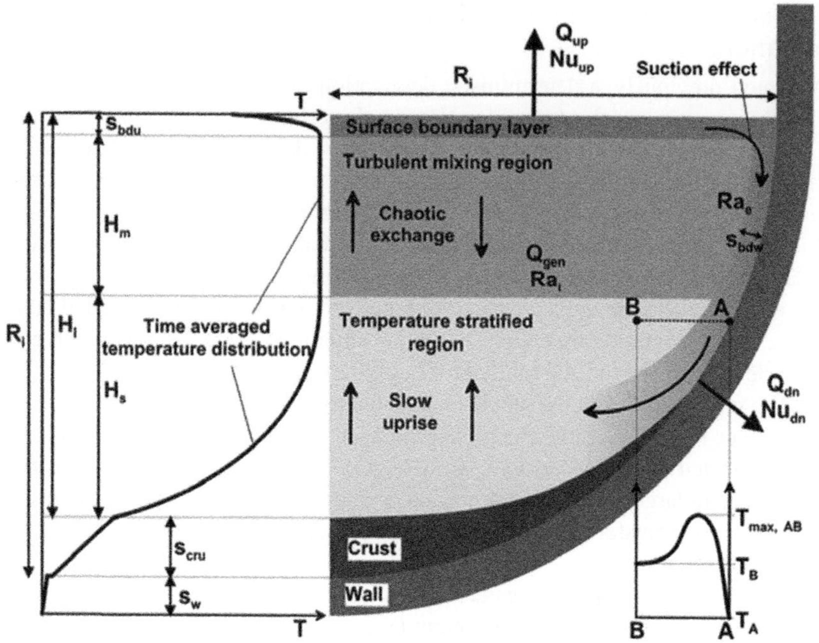

FIGURE 2.11 Illustration of natural convection flow in a homogeneous corium pool.

Figure 2.10 shows that about 50% of the power is going upward, with no variation in local distribution, while the lateral or downward heat flux varies strongly, with almost no heat flux at the bottom. From this distribution it can be concluded that lateral wall failure would occur close to the top (location of maximum lateral heat flux) and not—as expected in the early days of severe accident research—at the bottom.

This nonuniform heat flux distribution is a result of natural convection phenomena that develop inside the pool. The natural convection is due to temperature difference between the center of the pool and the boundary layers, where a solid crust exists. During the early stages of the corium pool, void fraction associated with fission gas release at elevated temperatures could also contribute to the internal convection. In the following, only effects linked to the temperature differences will be investigated. The heat transfer will be governed by the internal flow recirculations that induce (downward flowing) boundary layers on the various surfaces surrounding the pool (Figure 2.11). The heat flux distribution is usually computed with Computational Fluid Dynamics, requiring a computation time that is far beyond the time currently needed by most codes to simulate a severe accident scenario.

An engineer approach to the local heat flux consists in writing:

$$\varphi = h_{local}(T_{pool} - T_{boundary})$$

where:

φ is the local heat flux (W/m^{2}K)
h_{local} is the local heat transfer coefficient (W/m^{2}K)
T_{pool} is the temperature at the inner side of the boundary layer (K)
$T_{boundary}$ is the boundary temperature on the external interface (outer surface of the boundary layer)

The energy balance for the pool may be written simply as:

$$q^*V = h_{up}(T_{bulk} - T_{boundary})^* S_{up} + h_{down}(T_{bulk} - T_{boundary})^* S_{down}$$

with q: volumetric heat production (W/m^3), V pool volume (m^3), S surface (m^2) and h *average* heat transfer coefficients (W/m^2.K).

The average heat transfer coefficients can be deduced from a surface averaging of the local heat transfer coefficients.

In solving the problem, the engineer is faced with the following issues:

- First, concerning $T_{boundary}$: the melting temperature of steel is much smaller (~1000 K below) than the freezing temperature of oxidic corium. Thus, the presence of a solid crust is expected between the molten corium and the steel vessel. The interface temperature $T_{boundary}$ is then the "melting" temperature of the crust. But the corium is not a pure material: it is formed of mixtures of either oxidic materials (mixture of UO_2, ZrO_2, Zr dissolved in the oxidic phase, iron oxide, etc.) or metallic materials (mainly steel and zirconium). The different pure materials that are included in this mixture have very different melting temperatures (Fe_2O_3 ~1700 K, Zr ~2100 K, UO_2 ~3100 K). Thus, what does the "melting" of the crust really mean, and what precisely is the associated temperature? In the past decades, several authors considered the possibility that a mushy zone existed between the melt and the solid crust, in the zone where the temperature was between solidus (complete solidification) and liquidus (complete melting). It was considered that this mushy zone could have an influence on the flow in the boundary layer, and, as a consequence, also an effect on the heat transfer coefficient. This problem was solved during the last decade, as will be explained below.
- Second, concerning h_{local}: the local heat transfer coefficient has initially been derived from experiments that have been performed with simulant material (for instance water). Questions have been raised about the representativeness of simulant materials since these generally pure materials (like water) cannot reproduce potential effects related to the melting interval, as discussed previously. This problem has also found a solution (as explained later), and correlations derived from these experiments are used for reactor applications [91]. Generally, the local heat transfer coefficients derived from experiments are calculated on the basis of the local heat

flux and on the temperature difference between the maximum temperature in the pool and the local wall temperature. This will solve the third problem below. Note also that, in the later years, local heat transfer coefficients could be deduced from CFD calculations.

- Third, concerning T_{pool}: as the experimental heat transfer coefficients are generally based on the maximum pool temperature (as explained in the preceding paragraph), knowledge of local pool temperature inside the boundary layer is not required. Attention should, however, be paid to the fact that the temperatures in the bulk of the pool are not uniformly distributed.

The crust thickness is a result of the heat flux distribution; it is self-adjusting and, as a first approximation, inversely proportional to the heat flux.

2.3.2. Understanding In-vessel Corium

Definitions and Main Properties

In-vessel corium, with the abreviation "C," is not a new element, but rather a name for the mixture made up from molten core materials, which are composed of (U, Zr)O_2 + Zr + Fe + B_4C + fission products. The basic properties are:

- Solidus-liquidus temperature: 1500–2850°C (~1800–3100 K)
- Density 7 to 10 tons/m^3

Corium is characterized by, mainly, two parameters: the degree of oxidization of zirconium and the U/Zr ratio. The degree of oxidation is indicated by adding a number to C; thus, C-22 means corium with 22% oxidation of zirconium inventory.

Corium may be classified according to weight % or mol %. Table 2.2 explains how these calculations are performed and how the atomic ratio U/Zr and the degree of oxidation are defined.

Generally speaking, the term Zr(O) means a solution of O in Zr, (U, Zr)O_{2}-x a sub-stoichiometric solution of uranium and zirconium oxides. There are many forms of uranium oxide: UO_2, U_4O_9, U_5O_{13}, U_3O_8, UO_3.

The corresponding oxygen to metal molar ratios (O/M) are: 2, 2.25, 2.6, 2.67, 3.

In addition, there are many possible oxygen/metal ratios, as the oxygen atoms can be placed in the lattice in a nonstoichiometric fashion. Summing up, one typically uses the term *hyperstoichiometric* for UO_{2+x} and *hypostoichiometric* for UO_{2-x}.

In order to illustrate what in-vessel corium looks like, Figure 2.12 presents a SEM/EDX micrograph of in-vessel corium (SEM scanning electron microscopy, EDX electron probe microanalysis) of a probe from the MASCA 2/ MA-6 test.

Fission Products

At the end of a cycle, a core of initially ~120 tons of UO_2 contains about 4 tons of fission products and 1 ton of Pu. The fission products mass inventory at the

TABLE 2.2 Corium Calculations

Corium calculation
Weight %: m_{UO2}, m_{ZrO2}, m_{Zr}
Atomic weight: $A_{UO2} = 238 + 2 * 16 = 270$, $A_{ZrO2} = 91 + 2 * 16 = 123$, $A_{Zr} = 91$
at% or mol%: N_{UO2}, N_{ZrO2}, N_{Zr} number of species
Calculation of mol % $s = \frac{m_{UO2}}{A_{UO2}} + \frac{m_{ZrO2}}{A_{ZrO2}} + \frac{m_{Zr}}{A_{Zr}}$, $N_{UO2} = \frac{m_{UO2}}{s^* A_{UO2}}$, $N_{ZrO2} = \frac{m_{ZrO2}}{s^* A_{ZrO2}}$, $N_{Zr} = \frac{m_{Zr}}{s^* A_{Zr}}$
Atomic ratio k = U / Zr, $k = \frac{N_{UO2}}{N_{ZrO2} + N_{Zr}}$
Oxidation degree or oxidation %: $C_n = \frac{N_{ZrO2}}{N_{ZrO2} + N_{Zr}}$

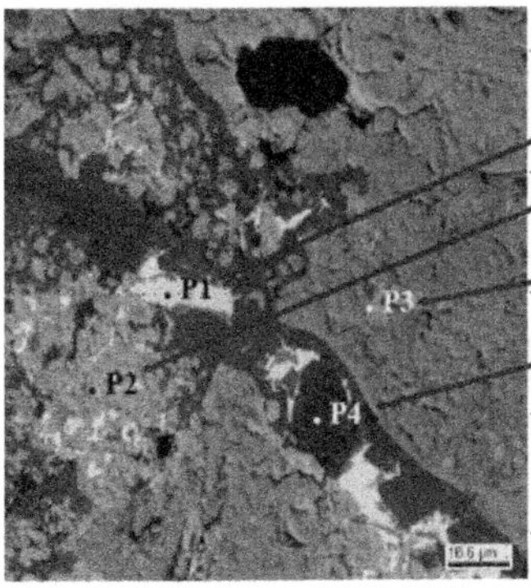

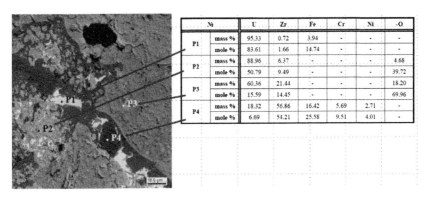

№		U	Zr	Fe	Cr	Ni	-O
P1	mass %	95.33	0.72	3.94	-	-	-
	mole %	83.61	1.66	14.74	-	-	-
P2	mass %	88.96	6.37	-	-	-	4.68
	mole %	50.79	9.49	-	-	-	39.72
P3	mass %	60.36	21.44	-	-	-	18.20
	mole %	15.59	14.45	-	-	-	69.96
P4	mass %	18.32	56.86	16.42	5.69	2.71	-
	mole %	6.69	54.21	25.58	9.51	4.01	-

FIGURE 2.12 Micrograph of in-vessel corium.

end of cycle, for 1 ton of U, is listed in Table 2.3 and presented under graphical form in Figure 2.13.

Nonmiscibility and Corium Stratification

Basically, molten steel phases (Fe, Ni, Cr) are not miscible with molten oxides (UO_2, ZrO_2), and metallic phases are lighter. Thus, in a molten pool situation, stratified metal layers are expected to float on top of molten oxidic material. This is the general representation of a molten pool in the bounding situations. This is, however, a simplified view of the real behavior.

TABLE 2.3 Typical Fission Products Inventory in the Core

Element	kg	Element	kg	Element	kg
Silver	0,072	Tellurium	0,465	Barium	1,480
Gadolinium	0,096	Yttrium	0,471	Ruthenium	2,280
Prometheum	0,100	Samarium	0,787	Cerium	2,670
Iodine	0,137	Technetium	0,847	Caesium	2,810
Europium	0,172	Strontium	0,903	Molybdenum	3,420
Rubidium	0,346	Praesodynium	1,160	Zirconium	3,620
Krypton	0,362	Lanthanum	1,250	Neodynium	3,890
Rhodium	0,446	Palladium	1,310	Xenon	5,380
Tellurium	0,465			rest	0,236

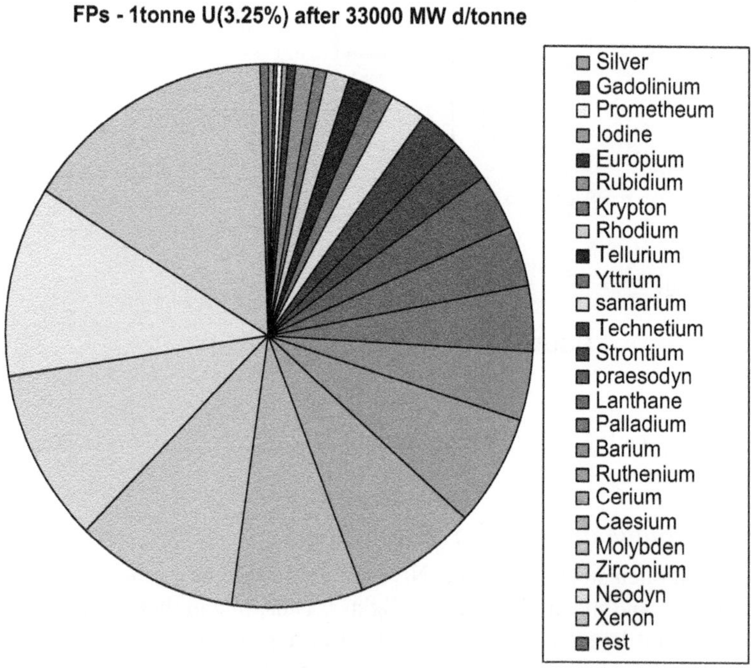

FIGURE 2.13 Fission product mass inventory at end of cycle.

The reader already knows that metallic zirconium is miscible in oxides (UO_2, ZrO_2). The phases formed between Zr, UO_2 and ZrO_2 can be described on the basis of thermodynamic calculations [25]. Figure 2.13 shows the microstructure of a real corium.

What the MASCA program has further demonstrated is that, in the presence of molten steel, Uranium may migrate from the oxidic melt into the molten metallic layer. The consequence is that the density of the metallic layer increases. Under some circumstances, the density of the metallic layer may even become greater than the density of the molten oxide. This results in a layer inversion: the metallic layer can relocate below the oxidic pool.

The RASPLAV and MASCA experiments contributed to the validation of thermodynamic databases such as NUCLEA [25] that can be used for calculating phase compositions (the phase diagram shows a miscibility gap). Complementary models [30] can then be used for the evaluation of phase densities. Table 2.4 presents calculation results of the maximum mass of steel that can undergo density inversion as a function of the U/Zr molar ratio and of the oxidation degree of zirconium.

The transfer of uranium into the molten metallic layer requires a direct contact between molten materials (i.e., no separation of metal and oxide by a solid crust) in order that the physico-chemical reactions can take place. This means that the metallic layer can come close to thermodynamic equilibrium with the oxidic layer. This aspect should not be forgotten when the reader searches for consistant boundary conditions ($T_{boundary}$) for the calculation of heat transfer in an inversely stratified pool, as will be discussed later.

The addition of small amounts of typical fission products in the experiments did not result in great variations in the miscibility of metallic and oxidic phases. But the unitentional addition of small impurities of carbon led to immiscibility

TABLE 2.4 Layer Inversion: Maximum Mass of Steel (in tons) That Stratifies Below the Oxidic Corium (the parameter *x*, in *Cx*, is the percentage of zirconium oxidation) [103]

	R1:U/Zr = 1.2 75 t UO_2	R2: U/Zr = 1.45 100 t UO_2	U/Zr = 1.2 100 t UO_2
C0	60	60	80
C10		48	
C20		35	
C25	32	32	43
C30	25	24	34
C40	14	10	18

in some RASPLAV experiments. The potential effect of other materials—for example, Pu that are present in the molten core material—has not been investigated to date.

2.3.3. The Crust-Melt Interface Conditions

For calculating the heat flux distribution at the melt-vessel interface, the interface condition (temperature) between the different layers of melt and the boundary must be known.

For oxidic melts, the liquidus temperature is generally far above (above 2000°C) the melting temperature of steel (1500°C). Thus, the interface is formed by a solid oxidic crust. The question is then: what is the interface temperature between the noneutectic melt (which can exhibit a large solidus to liquidus temperature difference) and this crust?

For metallic melts, the liquidus temperature can be below the melting temperature of the vessel steel (because of alloy formation with zirconium that can exhibit a melting temperature lower than 1000°C), and the solidus-liquidus temperature interval of the metallic melt may also be large. The questions in that case are: can the vessel steel be dissolved, and what is the resulting interface temperature?

General Considerations about Interface Temperature for a Noneutectic Melt under Thermalhydraulic Steady State

For the demonstration of long-term in-vessel retention, the heat load on the vessel is evaluated under steady-state conditions (the decrease of residual power takes months). This means that stable melt layers are considered in the vessel with constant power dissipation. As a consequence, stable crust and interface conditions are assumed. Considering, for instance, the crust that forms between the oxidic melt and the vessel steel, the steady-state situation corresponds to a time constant heat flux and crust thickness. In words of metallurgy, this means that the growth rate of the crust is zero. The consequence is that the interface temperature between the oxidic melt and its crust is the liquidus temperature of the oxidic melt [26].

Thus, in steady state:

$$T_{boundary} = T_{liquidus\ (actual\ melt\ composition)}$$

The delay time that is necessary to reach the liquidus temperature on the interface has been more precisely analyzed [31] [101] [102] for in-vessel corium and has been shown to be of the same order of magnitude than the delay time that is related to establishing the temperature field in the melt pool (establishment of thermal-hydraulic steady state).

The above formulation has several consequences:

- The liquidus temperature replaces the melting temperature of pure material that is used for thermal-hydraulic simulation of corium pools.

- When the liquidus temperature is reached on the interface, the mushy zone disappears and the interface becomes flat (in the sense of metallurgy). As a consequence, there is no impact related to a mushy zone on the heat transfer in the boundary layers in the final steady-state situation. Even for non-eutectic mixtures, the interface temperature is a well defined temperature. Thus, simulant materials, even pure, can be used for the investigation of natural convection-induced heat transfer in a corium pool.

Another consequence of uniform interface temperature was previously known. The interface temperature is uniform on the internal surface of a considered corium layer, provided that a stable solid crust completely surrounds this layer. As a result, under these conditions (stable solid crust), the heat flux distribution no longer depends on external conditions. Whatever happens outside the pool has no influence on the heat flux distribution, as long as the considered events do not lead to the destruction of the crust at the layer interfaces. As an example, the heat flux that goes from the oxidic pool to the molten metallic layer (stratified on top of the oxidic pool) does not depend on the thickness of this metallic layer.

The Crust-Melt Interface Temperature for a Homogeneous Melt

In steady state, the interface temperature between the melt and its crust is the liquidus temperature of the actual composition of the melt. This interface temperature has been experimentally validated by the following experimental programs: PHYTHER [27], SIMECO (RIT/KTH), RASPLAV [28] and more recently LIVE (KIT) [29].

The Crust-Melt Interface Condition for an Inversely stratified U-O-Zr-Fe Melt at Thermodynamic Equilibrium

This question rises for the situation of inverse stratification as depicted in Figure 2.14. This problem has been addressed in [31]. The outcome is that both layers, at thermodynamic equilibrium, are surrounded by a solid crust having the same composition on the interface with the same refractory oxides. The interface temperature with the crust is the transition temperature into the miscibility gap characterizing these mixtures (equivalent to the liquidus temperature for systems forming miscibility gaps).

The Crust-Melt Interface Temperature for a Stratified Melt out of Thermodynamic Equilibrium

This situation corresponds to the stratification of molten steel on top of the oxidic pool, such as is emphasized in the bounding case by Theofanous et al. [32] [33] (Figure 2.15). In the case that thermodynamic equilibrium between molten layers is not reached (they may be separated by a solid oxidic crust), the interface temperatures with the boundary are different for both layers.

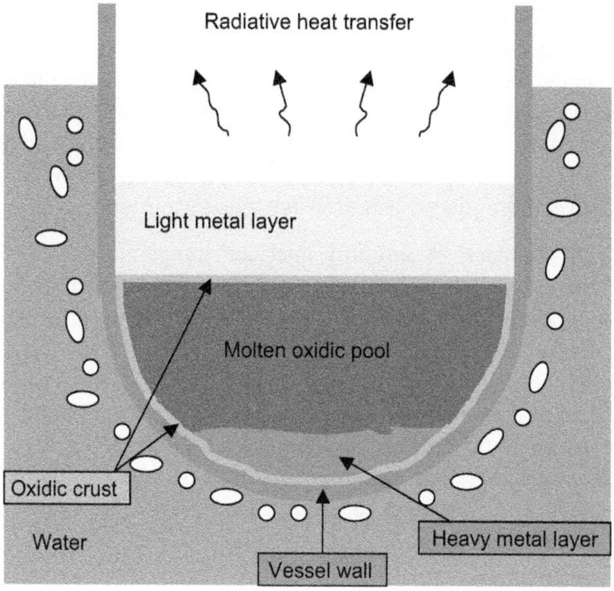

FIGURE 2.14 In-vessel retention configuration with inverted metal stratification.

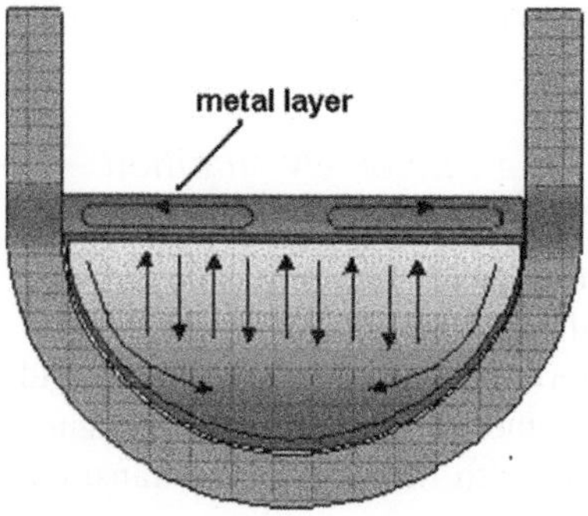

FIGURE 2.15 Illustration of the focusing effect [49].

Nevertheless, the relation $T_{interface} = T_{liquidus\ (actual\ melt\ composition)}$ still holds for both layers. But the liquidus is different for each layer.

For the metallic layer, the interface temperature with the vessel steel comes, in steady state, to the liquidus temperature corresponding to the composition of the metallic layer [31], which may be smaller than the melting temperature of steel (i.e., below 1500°C). Of course, the additional steel that is molten from the vessel due to a significant decrease of the interface temperature must be taken into account in calculating the composition of the metallic layer.

Interface Condition between Oxidic and Metallic Melt Layers

Oxidic and Metallic Layers Are in Thermodynamic Equilibrium

If oxidic and metallic layers come to thermodynamic equilibrium, there is no crust between these layers (which means direct liquid–liquid contact). This derives from the consideration that the interface temperature with the solid crust is the transition temperature into the miscibility gap, which is the same for both layers. This temperature corresponds to the first appearance of solid phase for both layers. Due to the power dissipation in the melt, the interface temperature between both liquid layers is then higher than their common interface temperature (liquidus temperature) with the crust. Therefore, no solid can appear at the interface between both layers.

Oxidic and Metallic Layers Are not in Thermodynamic Equilibrium

In this situation and considering that the liquidus temperature of the metallic layer is much smaller than the liquidus temperature of the oxidic phase, the interface temperature between the oxidic layer and the metallic layer drops below the liquidus temperature of the oxidic layer. As a consequence, a solid oxidic crust may appear at the interface between metallic and oxidic layers.

In that case, the interface temperature between the oxidic pool and this solid crust will be the liquidus temperature of the oxidic melt. The thickness of this crust adapts to the heat flux.

2.3.4. Heat Transfer in the Corium Pool

Average Heat Transfer

The decay heat (or residual power) will heat up the liquid corium, and the temperature difference between the bulk and the boundary is the motor of the natural convection in the pool (Figure 2.11). Heat transfer is then controlled by the flow near the boundaries, as discussed previously.

A number of experiments allowed measuring the heat flux distribution at the boundary of a volumetrically heated pool. From these experiments, heat transfer correlations have been derived. In this section, the reader will find heat transfer correlations that can be used for calculating the average heat transfer to the bottom (down) and to the top (up) of a pool contained in a hemisphere. "Bottom" designates the hemispherical interface with the steel vessel, and "Top" the flat upper surface (which is generally the interface with a stratified molten metal layer).

The heat transfer coefficient is related to the Nusselt number calculated with the pool height L:

$$Nu_{up} = \frac{h_{up} * L}{\lambda} \quad Nu_{down} = \frac{h_{down} * L}{\lambda}$$

The natural convection is characterized by a Rayleigh number ($Ra = Gr.Pr$). Usually, in practical problems, the temperature difference between the liquid and the wall is known. But this is not the case here inasmuch as the volumetric power dissipation is imposed and the temperature difference is unknown. Therefore, a specific Rayleigh number (called "internal Rayleigh number") is used:

$$Ra' = \frac{g\beta q L^5}{\alpha\lambda\nu}$$

Where g is gravitational acceleration (m/s^2), β thermal expansion coefficient of the liquid ($K-1$), q volumetric heat generation (W/m^3), L pool depth (length scale)(m), α thermal diffusity—bulk (m^2/s), ν kinematic viscosity—bulk (m^2/s)

According to preceding definitions, the following correlations have been derived. The validity range is indicated (experiments have been performed at different scales):

Upward heat transfer:

In the upward region, the top boundary is colder than the bulk of the pool. The density of the liquid near the boundary is higher than the density in the bulk. An instability due to this density difference develops. This instability, known as the Rayleigh-Bénard instability, controls the heat transfer to the top.

Kulacki-Emara [34]: $Nu_{up} = 0.34\ Ra'^{\,0.226}$ $2{*}10^4 < Ra' < 4.4{*}10^{12}$, Pr ~ 7
Steinberner-Reineke [35]: $Nu_{up} = 0.345 Ra'^{\,0.233}$ $10^7 < Ra' < 3{*}10^{13}$, Pr ~ 7
Mini-ACOPO [36]: $Nu_{up} = 0.345\ Ra'^{\,0.233}$
ACOPO [37]: $Nu_{up} = 2.4415\ Ra'^{\,0.1772}$
BALI [38]: $Nu_{up} = 0.381\ Ra'^{\,0.234}$ $10^{13} < Ra' < 10^{17}$

Downward heat transfer:

The heat transfer to the hemispherical surface is governed by a boundary layer flow that develops along the surface.

Mini-ACOPO: $Nu_{dn} = 0.02\ Ra'^{\,0.3}$ $10^{12} < Ra' < 7{*}10^{14}$, Pr ~ 7

Jahn-Reineke [39]: $Nu_{dn} = 0.54\ Ra'^{\,0.18}\left(\frac{L}{R}\right)^{0.26}$ $10^7 < Ra' < 5{*}10^{10}$, Pr ~ 7

Asfia-Dhir [40]: $Nu_{dn} = 0.54\ Ra'^{\,0.2}\left(\frac{L}{R}\right)^{0.25}$ $2{*}10^{10} < Ra' < 1{,}1{*}10^{14}$, Pr ~ 7

ACOPO: $Nu_{dn} = 0.1857\ Ra'^{\,0.2304}\left(\frac{L}{R}\right)^{0.25}$

BALI: $Nu_{dn} = 0.131\left(\frac{L}{R_V}\right)^{0.19} Ra'^{\,0.25}$ $10^{13} < Ra' < 10^{17}$ and

$$R_V = \left(\frac{3\ V_{pool}}{2\ \pi}\right)^{\frac{1}{3}}$$

As a rule of thumb, it can be assumed that $\frac{h_{up}}{h_{dn}} \sim 2$, which means that, for a hemispherical oxidic pool (i.e., $L = R$) with a crust surrounding the pool, the proportion of the decay power transmitted to the top is approximately 50%.

Local Heat Transfer

Due to the thickening of the boundary layer in the lower part of the pool, the downward heat flux is not uniformly distributed over the surface. The heat flux is maximum at the upper lateral part of the pool and minimum at the bottom. This appears clearly in Figure 2.10. Several authors provide relations for local heat transfer [36] [37] [40]. Nevertheless, it is difficult to provide a best-estimate relation for the local heat flux since it depends strongly on (i) the flow regime in the boundary layer (laminar [small scale] or turbulent [large scale]) and on (ii) the boundary conditions at the top (absence or presence of crust). For an order of magnitude evaluation, $\frac{\varphi_{dn,\ maxlocal}}{\varphi_{dn,\ average}} \approx 1{,}8$ at the top of the lateral surface of a hemispherical oxidic pool (i.e., $L = R$), with a crust surrounding the pool.

Heat Transfer between Oxidic Pool and Metal Layer Situated on Top of the Oxidic Pool

The residual power is mainly deposited in the oxidic phases. Heat is transfered from the oxidic pool to the metal layer when situated on top of the oxidic pool. Heat transfer on top of the oxidic pool and also at the bottom of the metallic layer is controlled by Rayleigh-Bénard instabilities. The correlations given above for the upward heat transfer are also valid for the calculation of heat transfer at the bottom of the metallic layer that receives heat from below.

Heat Transfer at the Lateral Surface of a Molten Metal Layer

The heat transfer at the lateral surface of a molten metal layer is governed by the development of a downward-flowing boundary layer. The correlations such as developed in [41] (and extended in [42]) fit well the experimental results.

- Laminar flow regime

$$Nu_x = 0.508\,\mathrm{Pr}^{\frac{1}{4}}\left(\frac{20}{21} + \mathrm{Pr}\right)^{-\frac{1}{4}} Ra_x^{\frac{1}{4}} \qquad Gr_x <\approx 10^9$$

- Turbulent flow regime

$$Nu_x = \frac{0.16}{\left[1 + \left(\frac{0.492}{\mathrm{Pr}}\right)^{\frac{6}{16}}\right]^{\frac{16}{27}}} Ra_x^{\frac{1}{3}} \qquad Gr_x >\approx 10^9$$

With $Nu_x = \frac{h_x x}{\lambda}$ and $Ra_x = \frac{g\,\beta\,\sin\theta\,\Delta T\,x^3}{\nu^2}$

x is the distance from the edge of the boundary layer (i.e., from the top of the molten metal layer).

To date, there is a substantial lack of information on whether multilayer pools form and on how the lateral heat flux may change in such a scenario. This information is of particular importance for scenarios in which it is assumed that the molten core material relocates stepwise from the core to the lower head and the composition of the materials in the lower head changes with time.

2.3.5. Gap Cooling in Case of Reflooding of the Primary Circuit

During the TMI-2 accident, the reactor vessel was filled with water at the moment when the molten corium began to flow down from the core. Around 10 tons of oxidic corium (~1 m^3) migrated to a compact mass in the lower head, and around 10 additional tons of debris were deposited in a layer around and above the compact corium mass. An analysis of samples taken from the reactor vessel has shown that the temperature of the inner surface of the vessel in contact with the corium mass rose to around 1100°C, while the external surface reached around ~800°C [85]. As mentioned previously, all the thermal simulations indicate that, assuming a perfect contact between the corium mass and the vessel, this temperature should have continued to rise until the vessel ruptured. The suggested explanation [93] is that a gap formed between the corium and the vessel. The formation of this gap would have been due to two phenomena:

- Water present in the porous steel boiled preventing contact between the corium and the steel.
- The solidifying corium was shrinking while the vessel was expanding due to heating.

The infiltration and recirculation of water within this gap would have cooled the vessel sufficiently to prevent it from failure in the TMI-2 conditions (~100 bar and a limited mass of ~20 tonnes of corium in the lower head).

A number of experimental studies have been carried out in order to investigate this gap cooling phenomenon:

- Small-scale tests of the flow of molten material into water at the base of a vessel in order to reproduce and analyze the mechanism experimentally were carried out by Fauske and associates [34] in Japan (JAERI) and in Korea (KAERI) [44]. All these tests were carried out using alumina thermite to simulate the corium.
- Tests to analyze the maximum power that could be evacuated by water in a gap where the limiting mechanism is the critical flux were carried out by IBRAE (Russia) [45], Siemens (Germany) [46] and JAERI (Japan) [47].

The conclusion at the present time is that the gap could only be formed by boiling water and that the differential expansion mechanism may plausibly not be a likely cause. One interpretation made by the CEA [48] shows that the possibility of power being evacuated by water in the gap is also unlikely at the low pressure used as the reference case for severe accidents in a PWR. For example, with a gap of 3 mm and a pressure of 1 bar, the critical heat flux is around 0.02 MW/m^2 compared with the 0.5 MW/m^2 that would have to be evacuated assuming that half the mass of the core relocates in the lower head.

The present conclusion is that there are too many intrinsic limitations (requirement for the permanent presence of water in the lower head, low critical heat flux, and closure of the gap by local melting of the vessel, etc.) to accept that the mechanism of cooling by the formation of a gap is credible in the vast majority of cases at low pressure.

In the light of current knowledge, it would appear difficult to demonstrate that the reflooding of the primary circuit alone would avoid a breach in the reactor vessel at the moment when a large amount of molten corium relocates down to the lower plenum, if the primary circuit is depressurized and in the absence of reflooding of the reactor pit (i.e., absence of external cooling of the vessel).

2.3.6. Analysis of the Bounding Configurations

Distribution of Heat Flux and Cooling for the Bounding Case Configurations

For a given configuration of the corium, and supposing that a crust surrounds the molten oxidic pool, the distribution of the heat flux from the melt depends on the geometry and on the internal pool convection only and is not affected by the boundary conditions outside the crust.

We will make use of the heat transfer correlations described previously and derived from tests with simulant materials (BALI, COPO, ACOPO, RASPLAV-Salt, etc.).

Orders of Magnitude of Heat Flux and Local Heat Flux Concentration (or Focusing Effect)

Given the configuration shown in Figure 2.15 and assuming that the entire oxidic mass of the core has relocated in the lower head, the residual power is approximately redistributed as follows:

- About 50% of the residual power deposited in the oxidic pool is transmitted to the reactor vessel that is in contact with the oxidic pool,
- The remaining 50% is transmitted from the oxidic pool to the above stratified liquid metal layer.

In the absence of water inside the vessel, the metal layer cannot be efficiently cooled on the top and will transmit most of the power coming from the oxidic pool and from its own deposition, that is, around 50% of the residual power to the steel wall of the reactor vessel that is in contact with the liquid metal. Depending on its thickness, the metal layer may concentrate the heat flux on the surface of the reactor vessel wall in contact with the metal layer (hence the term *focusing effect*). As a first-order approximation, the heat flux in that region is inversely proportional to the thickness of the metal layer. If we consider that the whole inventory of fuel is relocated in the oxidic pool, the thickness of the layer must be approximately greater than 50 cm (equivalent to 50 tons of molten steel) in order for the heat flux to fall below approximately 1.5 MW/m^2. Therefore, the integrity of the vessel can only be guaranteed if the heat flux transmitted to the vessel can be extracted by natural two-phase convection by means of water boiling outside the vessel. This naturally raises the question of the critical flux at the external surface of the reactor vessel wall.

Critical Flux in the External Natural Water Circulation

The limiting factor is therefore the critical heat flux (CHF) associated with the external cooling of the reactor vessel, particularly in the vicinity of the metal layer. It is for this reason that considerable work has been done to determine this critical flux and to attempt to increase it. A number of tests have been carried out using a variety of approaches. Among these, the most interesting large-scale tests are ULPU (UCSB) [33] [49] [50] [51] [52], SULTAN (CEA) [53], and KAIST (Korea) [54].

The first factor affecting the critical heat flux is the water recirculation system in the reactor pit in which the water is recirculated by natural convection. Simply reflooding the reactor pit is not sufficient. A water channel must be arranged to enhance recirculation and to maximize the flow of liquid along the walls of the reactor vessel. This implies the existence and geometric optimization of an ascending hot leg (along the reactor vessel) and a cold leg maximizing the difference in the weights of the water columns (two-phase on the hot side and liquid water on the cold side) in order to form a strong motor for recirculation, while minimizing pressure losses. The maximum critical flux is obtained when the steam quality remains negative in the heated zone—that is, when the flow is sufficient to restrict boiling near the wall surface in the heated zone (subcooled boiling). However, it is also necessary for the boiling to become saturated (positive quality) above the heated zone in order for the void effect to develop. There exists an optimum flow rate above which the aspiration effect of the void is insufficient and below which saturated boiling occurs in the heated zone and the critical heat flux decreases. The existence of this optimum flow limits the maximum critical flux to a value of around 1.5 MW/m^2 in the vertical region and for a smooth surface of a large vessel.

An analysis of the results of the tests cited above shows a wide variation in the values of the critical heat flux: ULPU shows values close to 2 MW/m^2, while the SULTAN and KAIST tests give values for the critical flux at a vertical wall between 1.2 and 1.5 MW/m^2.

A number of methods have been proposed for increasing the critical heat flux. Many of these methods are concerned with the effects associated with treating the external surface of the reactor vessel. According to Cheung [51], a deposit of porous metal on the external surface of the reactor vessel would result in a significant increase in the critical flux (by a factor of up to two times). Adding nanoparticles to the water coolant seems also to increase the critical heat flux [43]. There is, however, a lack of agreement as to the effectiveness of this kind of approach, and experimental verification at large scale would be necessary.

Limitation Associated with the Mechanical Strength of the Residual Thickness of the Vessel

For a heat flux of 1.5 MW/m^2, the residual thickness of the reactor vessel that has to withstand the mechanical stress (i.e., the thickness within which the temperature is less than 600°C) is no more than ~1 cm. This thickness is sufficient to withstand pressures of a few bars and to carry the weight of the lower head (filled with corium). An increase in the critical heat flux would lead to a reduction in this thickness (they are inversely proportional) and reduction in strength.

Limitation Associated with the Minimum Mass of Molten Steel Layer

In the light of the preceding discussion, one of the key points in the approach is the mass of steel available to form a metal pool on the surface of the oxidic corium. At values of critical heat flux between 1.3 and 1.5 MW/m^2, the minimum depth of molten steel needed to avoid the focusing effect would be around 50 to 60 cm for a 1000 MWe PWR. Given the geometry of this type of reactor, such a depth corresponds to a mass of molten steel of approximately 50 to 60 tons.

Studies on the AP600 and AP1000 have shown that a meltdown would release a sufficient quantity of steel from the following structures:

- Internal structures in the lower head,
- The reactor vessel walls,
- Structures in the lower section of the core.

Work carried out as part of the OECD MASCA program [29] has shown that the complex physical phenomena occurring could reduce the "available" mass of metal. These phenomena include:

- Trapping of some of the liquid metal (from the lower internal structures, for example) in the solid oxidic debris.
- Migration of part of the metal in the lower head as a result of physical and chemical effects (layer inversion). For a fixed metal inventory, this would

naturally lead to a reduction in the depth of the upper metal layer (Figure 2.14).

The chemical effects arise from the presence of unoxidized zirconium in the metal phase as it has been explained previously. The estimated densities of the phases resulting from these calculations can be used to estimate the mass of metal that would stratify below the oxide pool. Subtracting this from the original steel inventory gives the mass of metal present in the upper layer. This model has been developed by CEA and IRSN with the aim of quantifying the masses of metal needed to ensure that the heat flux transmitted from the metal layer to the vessel does not exceed the critical heat flux [55]. The method has been applied to a number of reactor types. The results of these studies show that a critical parameter is the fraction of unoxidized zirconium present in the pool. The greater is this fraction, the greater is the proportion of the total mass of metal that can migrate to the bottom of the lower head. The complexity of the demonstration of in-vessel corium retention therefore increases with the mass of available metallic zirconium. These studies also show that the results of the simulations are strongly affected by the choice of databases to be used in the thermodynamic calculations and the correlations used to estimate the density of the metallic phase, as well as the values of the critical flux at the outer surface of the reactor vessel.

These bounding cases were studied under steady-state conditions with a fixed pool configuration. The formation of the metal layers and oxide pool implies transients in the growth of the metal layer and an increase in the power from the oxide pool, both of which imply that the critical flux could be reached during these transients.

2.3.7. Main Objectives for Future Improvements

Many developments have been made in the past, including CFD calculations (not described here). The analysis of bounding situations has provided many detailed insights on critical phenomena and associated limitations that affect in-vessel retention. If conservative hypotheses are considered (for instance, shallow metal layers and transient effects), the analysis fails to demonstrate in-vessel retention. The question that arises is the question of how not to make too conservative hypotheses, a question that cannot be solved by assuming bounding situations.

The only way for significant improvement is to propose a realistic evolution of melt progression between the core and the lower head. This requires answering the questions raised in Table 2.1. As, despite further R&D efforts, the answer to these questions is not always straightforward and not deterministic, this effort must be accompanied by an evaluation of the effect of uncertainties on the final results. This approach has recently been developed, for instance, in the ASTEC or MELCOR integral codes (see Chapter 8).

2.3.8. Experimental Programs, Modeling, and Computer Codes

The major experiments, including recent or ongoing projects, are as follows.

- DEBRIS: This project, conducted by IKE (University of Stuttgart), aims to characterize two-phase flows in a debris bed with volumetric heating. The experimental system is one-dimensional and consists of steel balls heated by induction. The first phase involved characterizing pressure drops, which is necessary to predict the critical dry-out heat flux. The experimental system was recently modified to allow reflooding tests. As the preliminary tests were satisfactory, more quantitative tests are planned.
- SILFIDE: This project, conducted by EDF, was aimed at characterizing dry-out of a debris bed with volumetric heating [88]. The debris bed was large enough to observe 2-D flow effects, which made these tests a unique reference for validation of 2-D models. The debris bed consisted of steel balls heated by induction. Interesting results were obtained despite the challenge of establishing a homogeneous distribution of power. In particular, local flux was at times observed to be higher than the theoretical critical flux (for 3-mm particles, the maximal flux measured during SILFIDE was 1.7 MW/m^2 compared with the 1 MW/m^2 predicted by the Lipinski correlation). Researchers also observed temporary dry-out in localized zones, followed by rewetting. This experimental program was completed in 2000.
- RASPLAV: The Nuclear Energy Agency of the Organization for Economic Co-operation and Development (NEA-OECD) has, since 1994, sponsored the collaborative projects on severe accidents carried out in Russia. These projects, entitled the RASPLAV Project (from 1994 to 2000) and the MASCA1 (MAterial SCAling) Project (2000–2003) and MASCA2 Project (2003–2006), had the basic objective of providing data on the behavior of molten core materials under severe accident conditions. Studies of the impact of real materials in the frame of the OECD RASPLAV project demonstrated that the homogeneous corium melt behaved comparably to simulant materials in natural circulation. It was shown that such behavior could be expected if molten pool contained oxidic materials. Two of the four large-scale RASPLAV tests conducted with suboxidized corium showed that the melt pool, initially of a homogeneous composition and density, stratified into two layers of unequal density. RASPLAV project dealt with the corium of different oxidation degree. The RASPLAV Project objectives were realized through a combination of several major activities that included:
  - Development of technology for conduction of high-temperature corium tests, including design, fabrication, and assembly of a large-scale facility capable of heating up and sustaining molten corium at relevant temperatures.
  - Performance of confirmatory large-scale experiments to investigate the behavior of the melt of the core prototypical materials in the reactor

vessel lower head as well as to determine the principal differences in the behavior of the melt as compared with simulating liquids.

- Performance of supporting corium experiments and analysis to conduct the integral test and develop the methodology describing the phenomena of interest; among the objectives are measurements of material properties.
- Performance of salt tests to support the development of facility and facilitate the analysis and interpretation of corium tests.
- Development of computer tools for the pre- and post-test analysis of experiments to produce a consistent analysis, interpretation, and understanding of the results.

- MASCA: The MASCA projects were concentrated on the detailed investigation of the chemical behavior of corium in contact with structural materials, with emphasis mainly on steel. The major MASCA2 Project activities included:
  - Study of melt stratification and distribution of major species (U, Zr, O, Fe(SS)) between layers in inert atmosphere at large iron (SS) to corium ratio.
  - Study of control rod materials effects on interaction and distribution of major species (U, Zr, O, Fe(SS)) in inert atmosphere.
  - Study of melt stratification and distribution of major species (U, Zr, O, Fe(SS)) between layers in oxidizing atmosphere.
  - Investigation of molten metal alloy interactions with corium debris bed in the inert and oxidizing atmosphere.
  - Measurements of thermo-physical properties of the metallic phases such as density, surface tension, and thermal conductivity.
- SIMECO: This project, conducted by KTH (Royal Institute of Technology), aims to study the heat flux in stratified pools with different power density in each layer. In tests using simulant materials (salts and/or paraffins), three-layer pool configurations were reproduced (similar to those observed during MASCA tests, involving a heavy metal layer, an oxide layer, and a light metal layer). This allowed measurement of heat flux distribution along the vessel wall. The results require further analysis, but, according to the interpretation of KTH, they show a heat flux distribution different from what could be predicted by classical correlations.
- METCOR: The objective of this ISTC project, conducted by the NITI Institute (Saint Petersburg, Russia), is to study the corrosion of a steel sample (representing the vessel) by corium ($UO_2 + ZrO_2 + Zr$). The sample undergoes external cooling and is submitted to a heat flux representative of the conditions in a large molten pool (over 1000 K of variation across the sample). The project demonstrated the existence of a chemical interaction leading to the corrosion (i.e., melting) of steel at temperatures above 1100°C. A correlation was proposed to estimate the kinetics of corrosion as a function of the surface temperature of the steel element.

- LIVE: The main objective of this program at KIT is to study the core melt phenomena both experimentally in large-scale 3-D geometry and in supporting separate-effects tests, and analytically using CFD codes in order to provide a reasonable estimate of the remaining uncertainty band under the aspect of safety assessment. Several experiments have been performed with water and with noneutectic melts (mixture of KNO_3 and $NaNO_3$) as simulant fluids. The information obtained from the LIVE experiments includes heat flux distribution through the reactor pressure vessel wall in transient and steady-state conditions, crust growth velocity, and dependence of the crust formation on the heat flux distribution through the vessel wall. Supporting post-test analysis contributes to characterization of solidification processes of binary noneutectic melts. In complement to other international programs with real corium melts, the results of the LIVE activities provide data for a better understanding of in-core corium pool behavior. The experimental results are being used for development of mechanistic models to describe the in-core molten pool behavior and their implementation in SA codes such as ASTEC.
- INVECOR: The aim of this ISTC project, conducted by the IAE-NNC-RK (Kazakhstan) and completed in 2011, is to study corium interactions with reactor vessel steel by pouring premelted corium ($UO_2 + ZrO_2 + Zr$) into a hemispheric "lower head" (about 80 cm in diameter) and by using the appropriate setup to maintain power density (electrical discharges between carbon electrodes). The first tests, with 60 kg of corium poured into the vessel, have not produced any interesting results because of many experimental effects: corium solidification near cold points, uncontrolled pouring of corium, and impossibility of maintaining sufficient heating to keep the corium liquid. It was concluded that at least 200 kg of corium should be produced. An improved experimental device was proposed to reach that objective, but the tests have not been completed yet.

2.4. LOWER HEAD FAILURE

(E. Alstadt)

The lower head (LH) failure marks the final stage of the in-vessel accident progression. The initial situation is characterized by a molten corium pool in the lower head. The RPV is loaded by:

- the temperature field resulting from the heat flow from the corium pool into the vessel wall,
- the internal pressure, if the primary system pressure cannot be released,
- the deadweight of the vessel wall and the melt pool, and
- the thermo-chemical attack of the corium (corrosion).

For the evaluation of the further progression of the accident it is important to know whether or not the RPV will fail. If it fails, the location and the time of failure have to be evaluated. Scaled experiments and numerical simulations have been performed to investigate the lower head failure. The following section highlights the key phenomena important to evaluation of the lower head behavior under severe accident conditions.

2.4.1. Heat Flow and Temperature Field

The heat flux distribution is derived from considerations developed in the preceding sections. This distribution is input data for evaluating the mechanical behavior of the vessel. Figure 2.16 proposes an illustration of the temperature field in the corium and in the vessel wall (the vessel is externally cooled).

2.4.2. Mechanical Behavior of the Vessel

In an IVR scenario, the LH of the RPV is loaded by the deadweight of melt and the vessel, the internal pressure, and the temperature field. The internal pressure and the gravity loads cause primary stresses, which are not relieved by the deformation of the vessel wall, but are even increased due to wall thickness reduction. Unlike this, the temperature gradients cause secondary stresses, which is relieved by viscoplastic deformation. The main deformation mechanisms of the RPV wall are creep and plasticity. Creep is a time-dependent process, whereas plasticity occurs promptly, that is, simultaneously with the load. Creep is connected with elevated temperatures (above 600°C). With

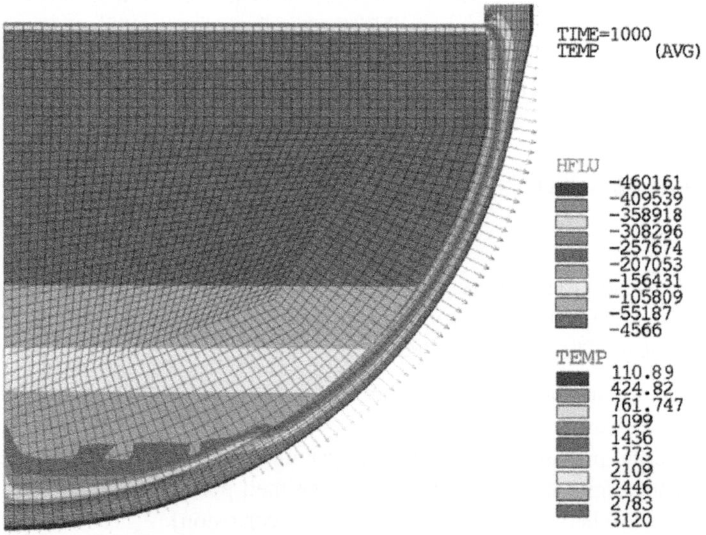

FIGURE 2.16 Temperature (°C) and heat flux (W/m^2) in the LH of a PWR (external RPV surface flooded with water), homogeneous oxidic melt pool.

regard to creep behavior, the most difficult part is the tertiary creep stage. The creep laws in FE (Finite Elements) algorithms must be based on true stress and strain. However, creep tests are usually performed with a constant force applied to the tension. Consequently, the true stress is not constant during the test because of the uniform reduction of the cross section in the early test stage and because of the necking later on. The increasing creep strain rate observed in the late test phase is a consequence of two effects:

- Geometrical creep acceleration due to reduction of cross section and necking,
- Decreasing material creep resistance due to microstructural changes (e.g., micro cracks, creep cavities).

In FE models, the geometrical creep acceleration is automatically considered if the large strain option is activated. An efficient way to describe the material tertiary creep is to use a creep strain rate that is coupled to the material damage:

$$\dot{\varepsilon}^{cr} = \frac{1}{1-D}\cdot f(\varepsilon^{cr};\sigma;T)$$

with D being the damage. The creep function f comprises the primary creep stage ($\ddot{\varepsilon}^{cr} < 0$) and the secondary creep stage ($\ddot{\varepsilon}^{cr} = 0$), whereas the decreasing creep resistance of the material in the tertiary stage ($\ddot{\varepsilon}^{cr} > 0$) is realized by the damage coupling, that is, by the factor $(1-D)^{-1}$. One possibility of quantifying the material damage increment is based on plastic and creep strain [56] [57]:

$$\Delta D = \left[\frac{\Delta\varepsilon^{cr}_{eqv}}{\varepsilon^{cr}_{frac}(\sigma,T)} + \frac{\Delta\varepsilon^{pl}_{eqv}}{\varepsilon^{pl}_{frac}(T)}\right]\cdot R_v$$

with $\varepsilon^{cr}_{frac}(\sigma,T)$ being the creep fracture strain of the uniaxial creep test at constant stress and temperature and $\varepsilon^{pl}_{frac}(T)$ the plastic fracture strain (true strain) of the tensile test. R_v is a function, which considers the damage behavior in dependence on the triaxiality of the stress tensor:

$$R_v = \frac{2}{3}\cdot(1+\nu) + 3\cdot(1-2\nu)\cdot\left(\frac{\sigma_h}{\sigma_{eqv}}\right)^2$$

where σ_h is the hydrostatic stress and σ_{eqv} is the von-Mises equivalent stress. The accumulated damage is:

$$D(t+\Delta t) = D(t) + \Delta D(t)$$

In the FE analysis, damage is calculated for each element. If the element damage reaches the value of $D = 1$, the element is deactivated, for example, by multiplying its stiffness matrix by a factor of 10^{-6}.

The experimental measurement of creep and the plastic properties of RPV steel is an expensive task, since the tests have to be performed at elevated temperatures (up to 1300°C). For French and American RPV steels this was done by CEA [58] [59]. It was found that creep behavior at temperatures above 700°C can considerably vary between different heats of the same steel. It was concluded that an increased sulphur (S) content could be the reason for a rather brittle failure (low creep fracture strain ε_{frac}^{cr}), whereas low S provides a rather ductile failure with large necking and high creep fracture strains [59]. This effect could also be observed in the FOREVER test series [60] [61].

2.4.3. Scaled Experiments on Vessel Failure

The FOREVER experiments were performed at KTH Stockholm between 1999 and 2002 [60]. The lower head of a RPV is simulated with a geometrical scale of 1:10. The vessel consists of a cylindrical part (material 15Mo3) that is welded to a hemisphere (material 16MND5 or SA–533B1). The internal radius of the vessel is 188 mm, and the wall thickness is approximately 15 mm (Figure 2.17). Instead of the prototypic melt, a simulant was used: an internally heated binary oxide mixture (70 wt.% CaO, 30 wt.% B_2O_3) with a melting point of approximately 1000°C was used. The main properties of the melt are: $\rho \approx 2500$ kg/m^3, Pr $\approx$ 70, $\lambda \approx 3$ W/(K·m), $c_p \approx 2200$ J/(kg·K). The electrical heater can provide a heating power of up to 45 kW. Using an Argon pressurization system, an internal vessel pressure of up to 40 bar can be created. The maximum melt temperature is around 1350°C, the maximum wall temperature is about 1000°C, and the corresponding heat flux reached about 140 kW/m^2 (Figure 2.18).

The maximum lateral heat flux occurs at the upper top boundary of the pool. It can be seen that the south pole remains cold and that the vessel failure will occur at 90°. Either a breach will open at this position and the molten material will be ejected horizontally, or the lower cap will drop down as whole still cold and stable holding some molten material like a cup.

Finite element models were developed for pre- and post-test calculations [56] [62]. The transient temperature field was calculated as well by CFD as by the ECM approach described later (see the Section 2.6.6). The viscoplastic model for the mechanical calculations was developed on the basis of the REVISA experiments [52]. The creep model is coupled to the material damage (see previous Section). There is also a mutual coupling between the thermal and the structural model to consider the following effects: (1) the decreasing melt level due to viscoplastic vessel wall deformation leading to a change of pool geometry; (2) the decrease of the wall thickness by viscoplastic deformation connected with a decrease of the heat resistance; (3) the increase of the heat radiation surface also caused by the vessel deformation; (4) thermal stresses and of course (5) the temperature-dependent viscoplastic properties. In this way a quite good agreement between experimental results and calculation could be achieved. Figure 2.19 shows the heating power, pressure, and measured and calculated displacement velocity of the vessel south pole in the experiment

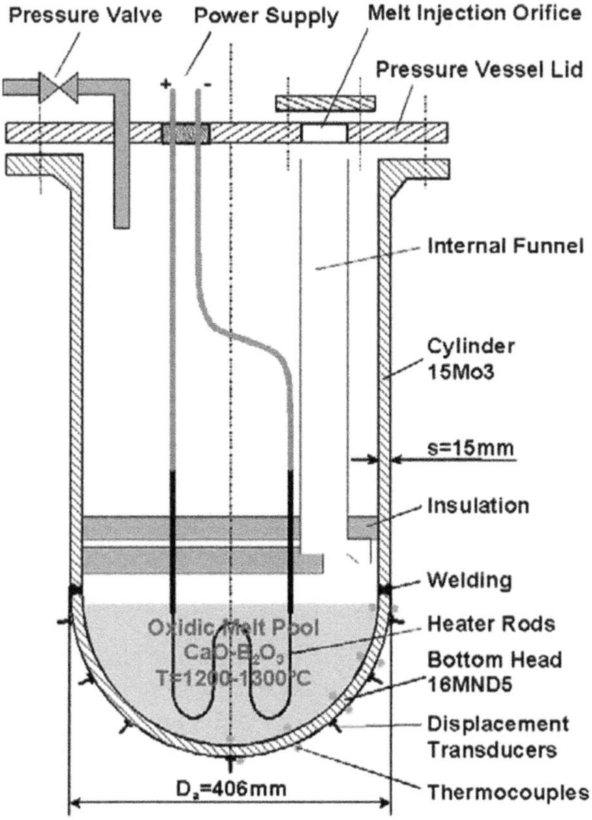

FIGURE 2.17 Principal scheme of the FOREVER experiments.

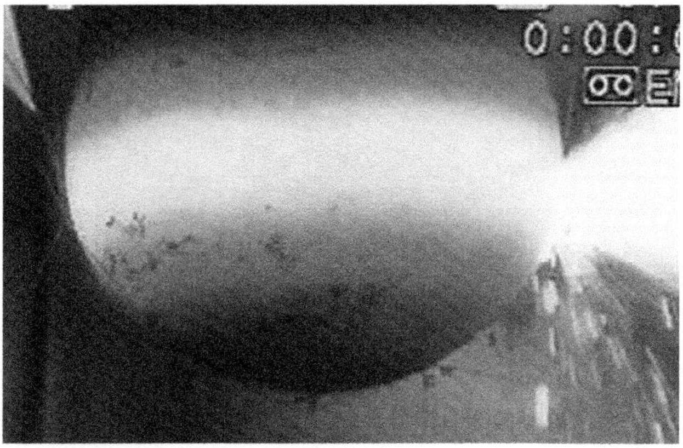

FIGURE 2.18 FOREVER vessel at failure time.

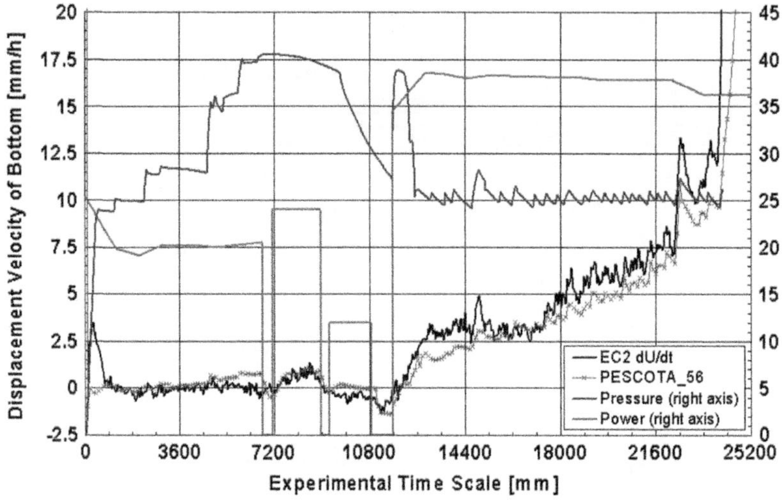

FIGURE 2.19 Heating power, pressure, and measured or calculated displacement velocity of the vessel south pole in the experiment FOREVER-EC2.

FOREVER-EC-2. Figure 2.20 shows the calculated temperature and vessel damage and comparison with metallographic post-test investigations.

The main results of the analyses of the FOREVER tests can be summarized as follows [62]:

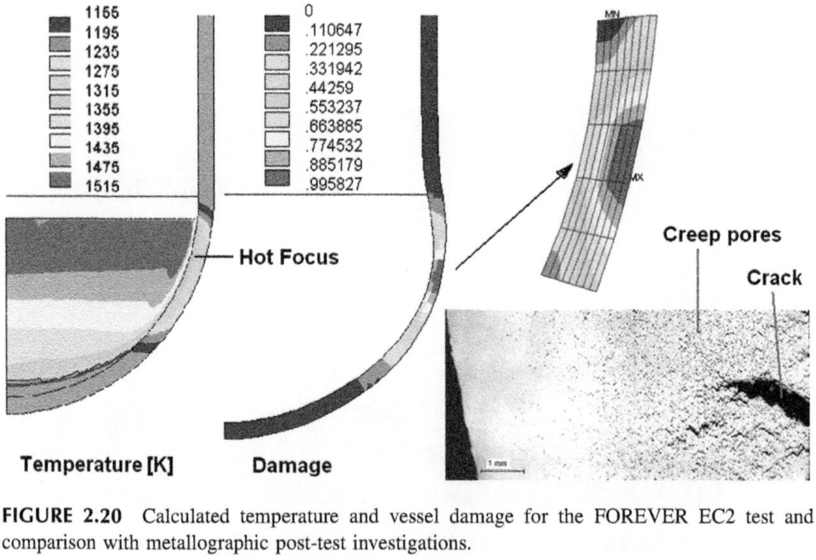

FIGURE 2.20 Calculated temperature and vessel damage for the FOREVER EC2 test and comparison with metallographic post-test investigations.

- The creep process is caused by the simultaneous presence of high temperature (above 600°C) and pressure (above 1 MPa). At low pressure and high temperatures, only a reversible expansion can be observed since the loads by the weight of vessel and melt are negligible.
- The hot focus region is the most endangered zone exhibiting the highest creep strain rates. The creep deformation leads to a wall thickness reduction, which accelerates the creep process.
- The failure position is in the zone of the highest temperatures. In addition, the height of the hot focus region has an influence on the failure time. Higher melt levels are more dangerous than lower ones, since the primary stresses in the cylinder are higher than those in the sphere.
- The levels of temperature and pressure have an influence on the vessel failure time but not on the failure position.
- The failure time can be predicted with an uncertainty of 20 to 25%. This uncertainty is caused by the large scatter and the high-temperature sensitivity of the viscoplastic properties of the RPV steel at higher temperatures.
- Contrary to the hot focus region, the lower center of the vessel head exhibits a higher strength because of the lower temperatures in this zone. The lower part moves down without significant deformation. Therefore it can be assumed that the vessel failure can be retarded or prevented by supporting this part.

The LHF/OLHF experiments were performed at Sandia National Laboratories, USA [63] [64]. The OLHF program (under the OECD umbrella) was the continuation of the LHF program. Models of a typical PWR lower head in the geometrical scale of 1:4.85 were used in the test (inner diameter 0.914 m, nominal thickness 70 mm). Prototypical steel for U.S. PWRs was used (SA 533B cl. 1). The vessel was internally heated with an induction-heated graphite-radiating cavity. During the tests temperatures, pressures, and displacements data were measured. The vessel surfaces were mapped both before and after the test to provide measurements of pretest thickness, post-test thickness, and cumulative vessel deformation.

The LHF series was focused on high internal pressures (10 MPa) associated with TMI2-like scenarios. In the OLHF series, large through wall temperature gradients were achieved by increasing (with respect to geometrical scaling) the wall thickness and leaving the wall un-insulated at lower RPV pressures (2 MPa … 5 MPa). The membrane stress was preserved by increasing the test pressure by a factor corresponding to the wall thickness distortion; that is, the test pressures ranged from 5 MPa to 12.3 MPa. The tensile and creep properties were measured up to ~1000°C.

Based on these properties, extensive numerical analyses were done for the experiments [64] [65] [94]. Different creep and damage models were applied to predict the vessel failure (ranging from a simple Norton-Bailey approach to the more sophisticated Lemaitre-Chaboche approach).

The results of benchmark calculations in the OLHF project can be summarized as follows [65]:

- Generally, the predicted failure times calculated by participants agree reasonably well with the test data. All the models, regardless of their complexity, gave reasonable prediction of the failure time (within 15 minutes). However, this does not really prove the quality of the analyses, since the creep failure time was governed by the course of the temperature in the test rather than by long term creep. The test temperature was increased with an approximately constant dT/dt.
- The maximum calculated vertical displacements at failure time are not in a good agreement with the experimental value. This is attributed to the fact that simple failure criteria like 1D strain values obtained from ultimate elongation in tensile tests or strain criteria for creep seem to underestimate the real failure. The multiaxial stress conditions have to be taken into account for the damage calculation.
- The predicted location of failure (only latitudinal position for 2-D axisymmetric calculations) is in quite good agreement with experimental results.
- Sensitivity studies with respect to material properties show that the failure time is extremely sensitive to creep and damage parameters (especially with respect to the time range where creep and plasticity occurs, that is, at elevated temperatures).

2.4.4. Scaling Considerations

From the geometrical, thermal, and mechanical parameters, statements can be derived about the scalability of the experiments in comparison with the prototypic scenario. However, it must be taken into account that thermal, mechanical, and geometrical dimensions cannot be scaled in the same way to a LWR scenario. Table 2.5 gives an overview of the scaling factors. The FOREVER tests have been designed in comparison to a smaller reactor than the KONVOI one that is considered here. In fact, the internal radius of the prototypic reactor, for which FOREVER was scaled, was 10 times the experimental radius. The KONVOI radius dimension is larger, while the wall thickness stays at 150 mm. Therefore the membrane stress scaling cannot be in line here, even without consideration of the ablation.

2.4.5. Corrosion (Thermo-chemical Corium–Steel Interaction)

In the METCOR experiments [66] it was found that the steel ablation at the interface between corium and vessel steel is not only a thermal phenomenon. Corrosion processes and the formation of eutectics lead to the erosion of the vessel steel at temperatures that are significantly lower than the melting temperature of steel. The METCOR experiments were conducted at the Alexandrov Institute of Technology (NITI) in Sosnovy Bor at the RASPLAV-3

TABLE 2.5 Scaling Factors of Different Physical Dimensions between the FOREVER-Experiments and a PWR of the German KONVOI Type

Parameter	FOREVER	KONVOI	Scaling
Geometry			
Internal radius [m]	0.19	2.5	1:12.5
Wall thickness of lower head [m]	0.015	0.15	1:10
Melt volume [m^3]	0.014	32.5	1:2300
Surface-volume ratio [m^{-1}]	24	1.8	1:0.08
Material properties			
Density of wall (steel) [kg/ m^3]	7850	7850	1:1
Density of melt (corium) [kg/ m^3]	2500	8000	1:3.2
Thermal boundary conditions			
Total heat generation in the pool [MW]	0.038	29.6	1:780
Volumetric heat generation density [MW/m^3]	2.7	0.91	1:0.33
Internal Rayleigh-number [-]	10^{10}	10^{17}	$1:10^7$
Wall surface heat flux at homogeneous distribution [kW/m^2]	112	500	1:4.5
Theoretical temperature difference over vessel wall without melting/ ablation [K]	56	2500	1:45
Temperature difference over vessel wall with melting/ablation [K]	56	1200	1:21
Mechanical loading			
Theoretical stress induced by the temperature difference with melting/ablation [MPa]	11	227	1:21
Weight of melt and vessel [Mg]	0.065	310	1:4800
Membrane stress by weight [MPa]	0.034	1.26	1:37
Internal pressure [MPa]	2.5	2.5	1:1
Membrane stress by pressure [MPa]	32	42	1:1.3

facility. These experiments have studied the interaction of a steel specimen with liquid corium. The liquid corium is inductively heated using the cold crucible technology and is positioned on the top of the steel specimen. The steel specimens were made from VVER 1000 base material (15 Kh2NMFA A). A large number of tests have been performed in which the melt composition, the thermal conditions, and the atmosphere above the melt have been varied.

The experimental correlation for the fast corrosion stage in suboxidized corium [66] is as follows:

$$\frac{dh}{dt} = \dot{h} = 0.46 \cdot 10^{-7} \frac{m}{s} \cdot \sqrt{\frac{T_{int} - T_{lim}}{1\,K}}$$

with: h – ablation depth, T_{int} – interface temperature, and T_{lim} – limit temperature at which the corrosion process starts. For a corium oxidation degree of 30%, the limit temperature is $T_{lim} \approx 1090°C$—that is, considerably lower than the melting temperature of steel. To demonstrate the effect of the corrosion within IVR scenarios, some results are shown from a VVER-1000 calculation [66] [61]. For the thermal simulation a total external RPV flooding is assumed, and for the mechanical analysis an internal pressure of 25 bar is assumed. The initial heat generation rate in the pool is 1.45 MW/m^3. Two simulations of this scenario are performed:

- Case A: no thermo-chemical interaction is considered, that is, melting of the wall takes place only if the melting point of the steel is exceeded.
- Case B: the thermo-chemical interaction is considered according to the preceding equation.

Figure 2.21 shows the ablation kinetics of both calculations. At the beginning both curves are identical ($0 < t < 4$ h); this is the pure wall ablation regime due to transgression of the steel melting point. Later on, the lower curve (case A) becomes constant. Contrary to this the upper curve for case B shows further progress in the wall ablation. The corrosion proceeds until the internal temperature is equal to the bounding temperature $T_{lim} = 1090°C$. At this point the corrosion is stopped.

Figure 2.22 shows the distribution of the equivalent stress in the lower part of the RPV within the final residual wall. It can be seen that the maximum values are nearly identical despite the fact that the residual wall thickness is lower by ~25% in case B. The reason is the strong temperature gradient over the vessel wall thickness. The internal part is hot, that is, close to the liquidus temperature of the steel for case A or close to the bounding temperature for case B. So especially in case A there is nearly no material strength to carry any relevant load at the wall inside. The load, consisting of deadweight and internal pressure, is carried almost completely by the external wall region because this region is cooled by the external flooding and therefore shows a much higher strength.

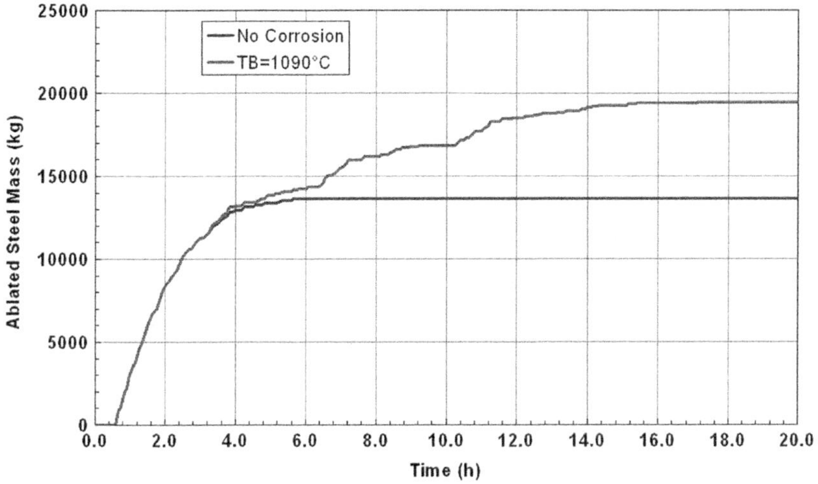

FIGURE 2.21 Ablation kinetics (steel mass eroded from the RPV wall over time); upper with corrosion damage consideration, lower without.

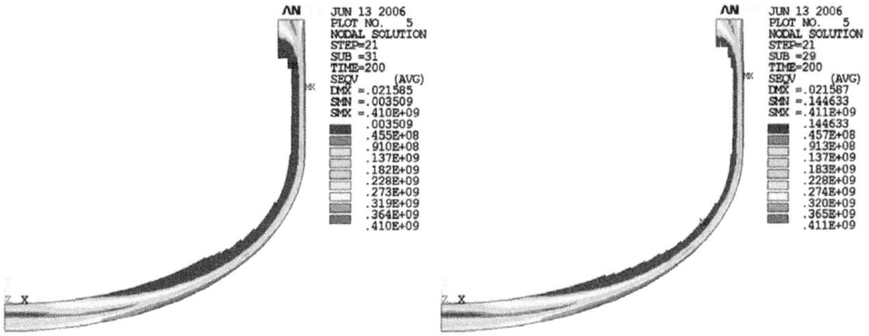

FIGURE 2.22 Equivalent stress (Pa) in the lower part of the RPV, left case A, right case B.

2.5. HIGH-PRESSURE ACCIDENTS IN PWR

(B.R. Sehgal)

2.5.1. Background

The high-pressure accident scenarios for the PWRs, resulting either from the station blackout or from the small-break-without-active emergency core cooling system (ECCS), or the loss of heat-sink events can be relatively large contributors to the nuclear power risk due to their large likelihood of occurrence. There is the likelihood of early containment failure, soon after vessel failure and melt release into the containment at high pressure, which can impose direct containment heating (DCH) loads on the containment. The TMI-2 accident was a high-pressure accident scenario, and fortunately the filling of the reactor vessel with water by activation of the main coolant pumps by the operators prevented vessel failure, thereby the DCH scenario and possible containment failure.

The current defense, practiced in practically all PWR plants, against the progression of a high-pressure accident is to depressurize the vessel through the deliberate opening of the relief valves by the operator(s) in the plant. This action is part of the severe accident management (SAM) strategy which has been adopted by almost all PWR plants.

Before the SAM strategy of vessel depressurization was considered and adopted for PWRs, it was discovered analytically [67] that the traditional "once-through" modeling of steam/hydrogen flows might not be valid during the core heat-up and degradation stages of severe accident sequences in PWRs. Owing to radial dependencies of steam temperatures, and of hydrogen generation within the core, as well as to unstable density stratification of primary coolant system (PCS) gas volumes, buoyancy-driven flows (natural convection) will develop between the core, upper plenum and the dome, and the upper plenum and adjacent gas volumes of the PCS. These flows will transfer, more effectively than the once-through flow, thermal energy from the overheating core to structural materials and resident gas in the upper parts of the RPV and the gas-filled portions of the PCS. For PWRs, therefore, these flows may strongly affect:

- the magnitude and the rate of hydrogen generation,
- the elapsed time before the onset of core melting,
- the transport of fission products within the PCS,
- the temperature levels of PCS piping/components,
- revaporization of fission products from the PCS surfaces,
- and lastly the ultimate course of the high-pressure accident scenarios.

Assessment of convective flow behavior during the core heat-up/degradation phase of high-pressure accident sequences in PWRs, and determination of the effects of buoyancy-driven flows on the parameters listed above, was the objective of a considerable research effort sponsored by the Electric Power Research Institute (EPRI). Of particular concern was the impact of natural convection on the temperature history of PCS piping and components, whereby a local failure in the PCS, prior to vessel breach, would radically alter the consequences of the high-pressure accident scenario in PWR plants. Such considerations formed the basis of the work reported in [67].

Given the relevance of natural convection to predictions of accident behavior, modeling in the EPRI-sponsored code CORMLT was enhanced [68] [69] to include it during the heat-up phase of high-pressure accidents. Later, as it became clear that the steam generators will also participate in the natural circulation flow loops, the modeling was extended to include natural convection flows between the PWR vessel and the PCS loops, which was required in order to predict the temperature history of PCS piping and components.

Subsequent to the early work on the CORMLT code, which showed the importance of the effects of natural convection flows on the characteristics of severe (core melt) accidents, it became clear that confirmation of the occurrence (and the assumed structure) of buoyancy-driven flow patterns should be obtained experimentally. A project at Westinghouse R&D Laboratories was initiated in which accident conditions were simulated in a one-seventh scale replica of a Westinghouse PWR core, upper plenum, piping, and steam generators. Experiments using low-pressure water and SF_6 gas were performed. The SF_6 gas at the pressure of 1 atmosphere was found to be an excellent simulant for high-pressure steam. A scaling analysis was performed to size the steam generators, the upper plenum in the vessel, and so on.

Concurrent with the experiments being performed at Westinghouse, an analysis project was initiated at Argonne National Laboratory in which the three-dimensional fluid flow code COMMIX-1A [70] was employed to analyze and predict the data measured in the Westinghouse tests [71].

2.5.2. Analysis of the High-pressure Scenarios

Flow Patterns

For the postulated high-pressure accident scenarios in PWRs (e.g., station blackout; small- break LOCA), core dry-out is essentially complete before major core damage takes place [67]. Thus, as core dry-out proceeds, spatially non-uniform heating of resident gas within the (intact) core, together with temporal lags in the heat-up of ex-core gas volumes, lead to systemwide variations in gas density that require consideration of buoyancy-driven flow patterns. Within the uncovered portion of the core, the combined effects of decay power at the center being approximately twice that at the periphery, and of heat being lost from peripheral fuel assemblies to surrounding structure, lead to radially increasing gas densities. The net result is an imbalance in hydrostatic head at any given elevation, which tends to promote radially inward flow. Above the core, heat-up of upper vessel gas lags that for core gas, and the RPV tends toward an unstably stratified configuration. Thus, if the momentum fluxes for the "once-through" flows (steaming, exhalation due to thermal expansion and depressurization) are sufficiently small, the in-vessel flow patterns will be buoyancy-driven and, likely, recirculating. A likely pattern is illustrated in Figure 2.23.

Between the upper plenum of the RPV and adjacent volumes of the PCS, other patterns may evolve. For designs involving U-tube steam generators

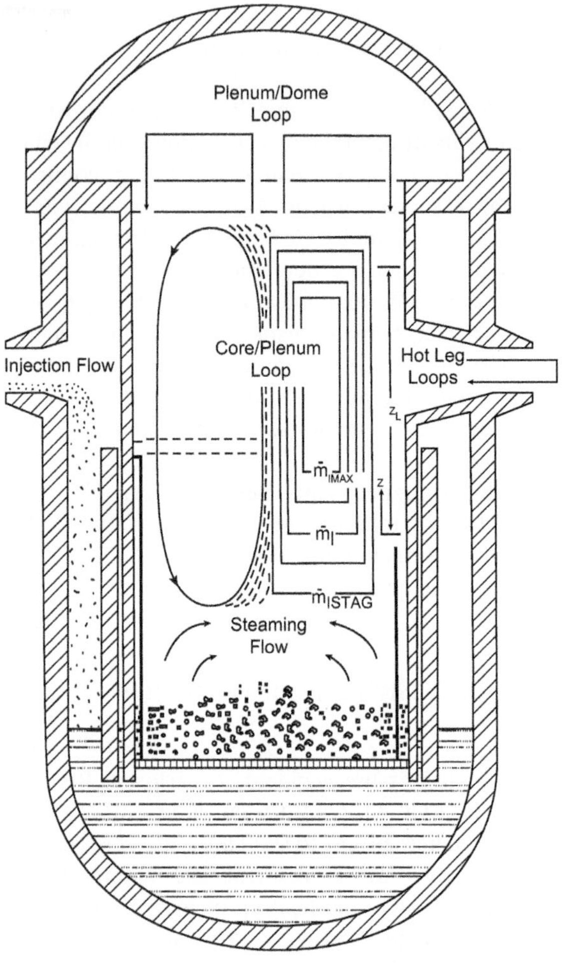

FIGURE 2.23 Schematic of core/plenum recirculation pattern.

(UTSGs), recirculating flow loops may develop between the upper plenum and the inlet plenum of the UTSG tube-ends by the outgoing "hot" flow from the RPV; a recirculating flow loop may be postulated via "outgoing" and "returning" subsets of the UTSG tubes (Figure 2.24). The proposed flow loop has hot gases from the upper plenum flowing out along the upper "half" of hot leg, with cold gases from the steam generator returning along the lower half. Part of the hot gas flow is presumed to loop back from the inlet plenum of the steam generator, while the rest circulates through the steam generator tubes as shown. Also depicted is the nodalization scheme chosen for the present

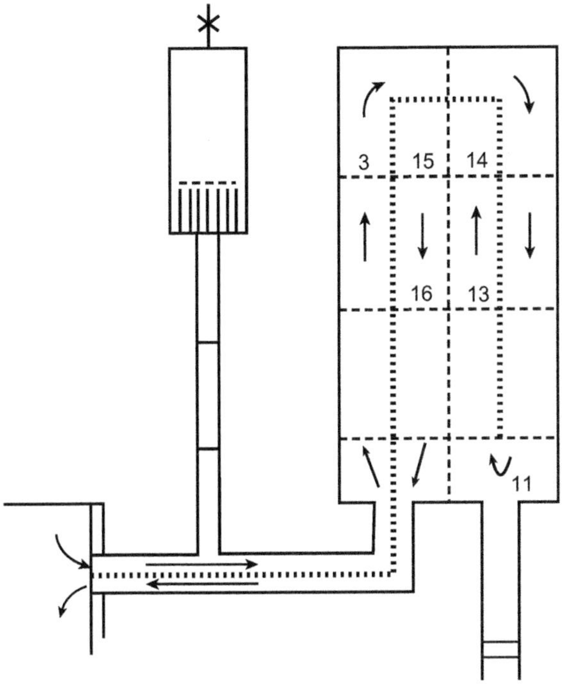

FIGURE 2.24 Schematic of ex-vessel flow patterns (Westinghouse plant).

calculation. It should be noted that the rest of the PCS loop is blocked by water seals in the cold leg and in the down comer.

Some Results from the Effects of Buoyancy-Driven Flows on Temperature Histories in the PWR Primary System

As a result of enhanced cooling of the core by recirculating steam from cooler regions, the onset of core damage would be delayed as compared with "once-through" predictions. This is further enhanced by the mass exchange that will occur between the upper plenum and UTSGS. A part of the core heat is transported to ex-core regions by the thermal convection flows. The resulting effects on core degradation and heat-up of ex-core regions of the reactor coolant system are illustrated in Figures 2.25 and 2.26. Two limits are presented—one for which only 10% of the estimated buoyancy-driven flow circulates through the steam generators (with 90% of the flow being bypassed at the inlet plenum) and a second for which only 50% of the flow bypasses the steam generator. As seen in Figure 2.25, significant delays in core degradation are obtained under conditions of extensive utilization of the steam generator heat sink. Shown in the Figure 2.26 are temperatures of

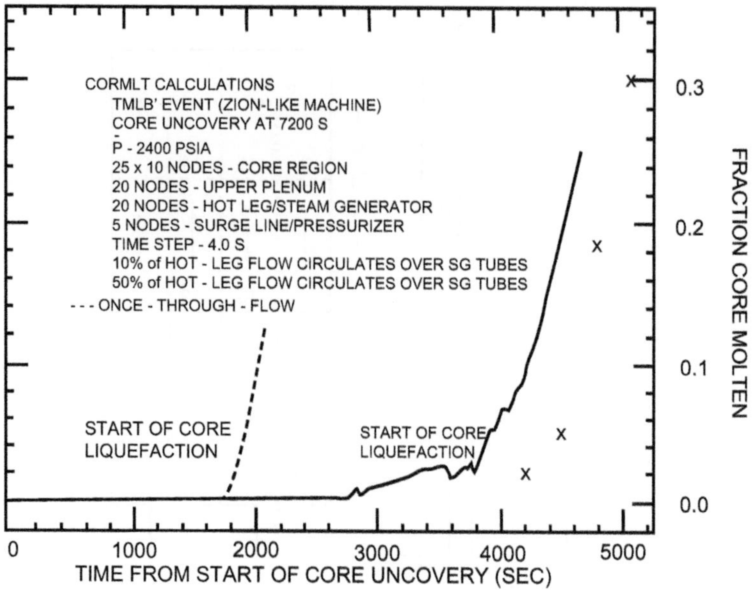

FIGURE 2.25 Effects of PCS flow options on onset of core degradation.

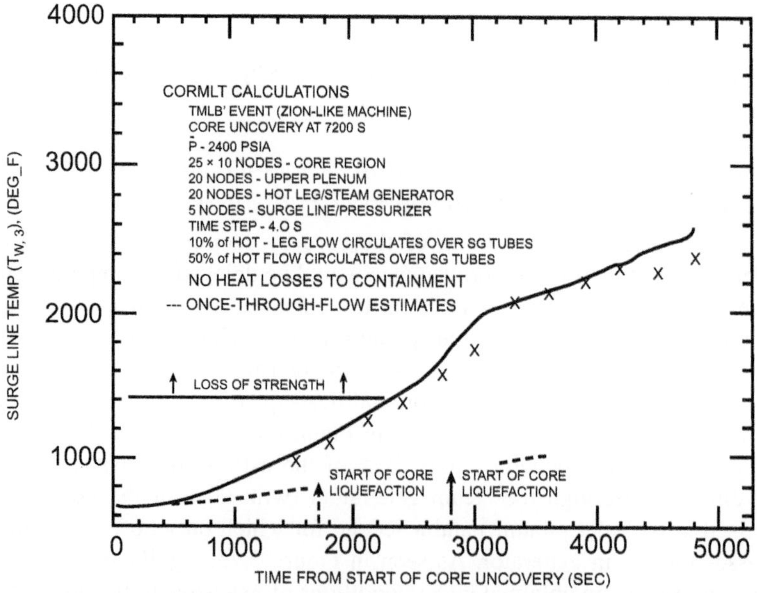

FIGURE 2.26 History of surge-line temperatures.

surge-line piping; which can exceed 2000°F (1366 K). If concern is restricted to times prior to core damage (say around 3000s), for which geometrical complications in flow modeling have yet to occur, it is seen that pipe-wall temperatures are still so high (approaching 1260 K) as to question mechanical survivability; At 1260 K, the yield stress of steel is less than 20% of its value at normal operating temperatures. It is expected that with the high pressure the surge line would fail mechanically. It is also expected that with the high gas temperatures the spring-loaded safety valves may not work.

2.5.3. Experiments on Natural Convective Flow Patterns in PWR Primary Systems

Tests have been conducted with a one-seventh linear-scale model of a PWR vessel and primary coolant system to investigate the natural circulation of superheated steam and hydrogen following core uncovery in a postulated accident. Water and sulphur hexafluoride (SF_6) at one atmosphere pressure were used as the analogue fluids. Water was used to enable visualization of flow patterns using dye injection in steady-state studies; while SF_6 was selected because its high density and high molecular weight partially offset reductions in scale and, further, simulate the smaller heat capacity of steam for transient cooling studies. The gravitational driving force depends on the Grashof number; by pressurizing SF_6 it is possible to offset fully the length scale reduction; experiments were performed at prototypic Grashof number and also at prototypic Reynolds numbers. Thus, complete thermal-hydraulic similarity between model and prototype was achieved for the postulated accident conditions.

A photograph of the reactor vessel model is presented in Figure 2.27. It is a scaled replica of one-half of a PWR reactor. Results of the measured velocity vectors for a typical steady-state experiment with water are shown in Figure 2.28. The "steam generators" are connected and remove 43% of the core heat input. There is no upflow from the lower plenum, nor is there any net flux of enthalpy out of the system via flow to the pressurizer.

Lower temperature fluid enters the core from the upper plenum only along those portions of the vessel periphery that lie beneath the hot leg openings. At the "back" of the vessel, flow actually rises from the core to the upper plenum. The front face flow pattern is, of course, directly displayed in Figure 2.28, where it is seen that the rising plume, which undergoes only modest accelerations within the core, exhibits rapid acceleration upon entering the upper plenum. The latter effect is caused by the abrupt decrease in flow resistance upon leaving the core. Of further interest is the spreading of the plume at the top of the plenum (Figure 2.28), resulting in a near-cup-mixing temperature of the above-nozzle fluid.

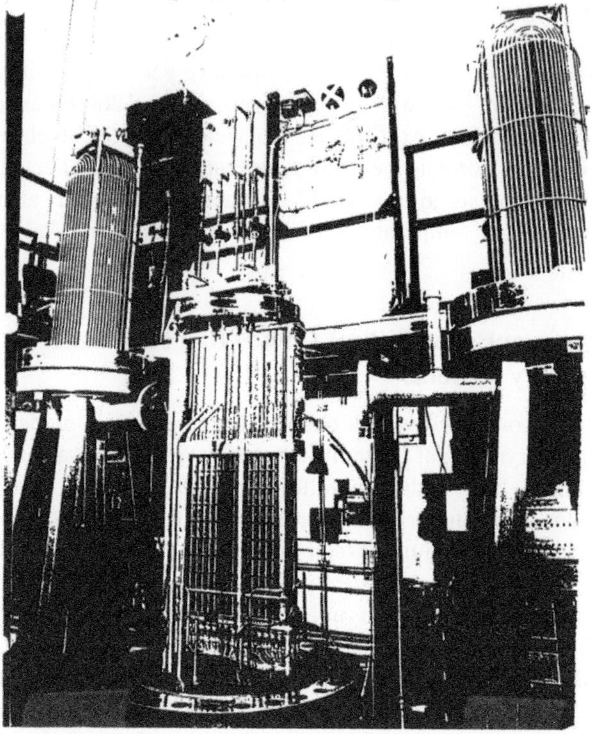

FIGURE 2.27 Picture of the experimental configuration.

2.5.4. Prediction of Westinghouse Test Results with the COMMIX Code

A three-dimensional COMMIX 1-A model was set up to simulate the one-seventh scale Westinghouse natural convection tests. The model included the upper plenum, the core, the lower plenum, two hot legs, and two steam generators. The pressurizer tee was modeled as an exit surface with a continuous-mass-flow boundary condition. A total of 1,464 computational cells and 81 surfaces were used for the COMMIX model. Three layers of vertical cells were used to model the hot leg, enabling reasonably accurate simulation of counter current flows.

In test 7A, nearly one-half of the total core power (28 kW) was removed by the simulant steam generators. As suggested earlier, the resulting strong countercurrent flow pattern in the hot legs tended to dominate the in-vessel thermal convection behaviour. This is illustrated in Figure 2.29, where COMMIX predicts maximum velocities in the hot-leg return flow and within the cascading flow at the plenum periphery. Other features of the test are also predicted by COMMIX including (a) rapid acceleration of the rising plume

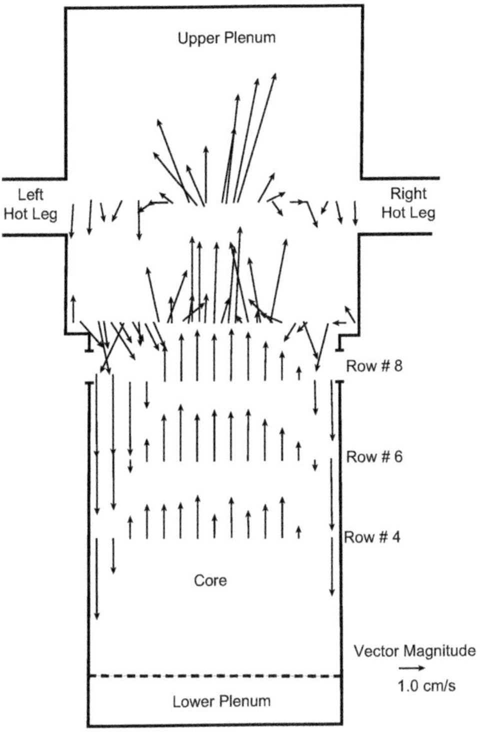

FIGURE 2.28 Velocity vectors from LDA fort natural circulation of water, 28 kW, with steam generators cooling hot leg flows.

upon entering the upper plenum, (b) plume-spreading above the nozzles, and (c) full penetration of the core by the reentrant flow. Of further interest is a comparison of velocity magnitudes. For example, the plume centerline velocities at the plenum floor are 0.036 m/s for the COMMIX prediction as compared to 0.031 m/s for the test.

2.5.5. Some Conclusions on High-pressure Scenarios

First of all, it is clear that natural convection flows will play an important role in the early part of the postulated high-pressure severe accidents in PWRs. The Westinghouse experimental project has obtained ample data to support this role. These tests, and simulant analyses, as performed with the COMMIX code, showed that flow patterns with water coolant and with gas flows are quite similar to those assumed in the CORMLT models.

Predictions by the CORMLT code of temperature levels within the primary coolant system for the postulated station blackout accident achieved high values quite early in the accident. A perusal of Figure 2.25 shows that at 3,000 seconds after core uncovery (before any substantial core melting has taken place: see Figure 2.24), the relief valve, which experiences the temperature of

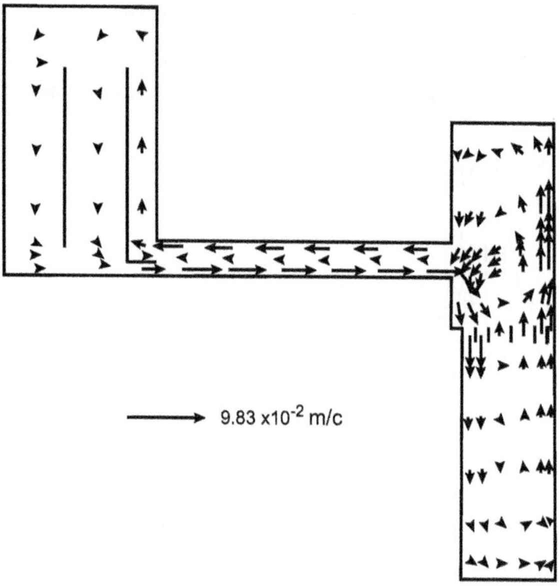

FIGURE 2.29 Flow patterns from COMMIX code calculations for test 7A.

the pressurizer exit gas, may reach temperature levels of 1200 K; and the surge-line temperature may be above 1260 K. At such high temperatures, it is probable that local failures in the primary circuit to the pressurizer may occur. As pointed out earlier, this will alter the course of a high-pressure accident; by (a) depressurizing the vessel and (b) adding water to the core from the accumulator tanks as the vessel pressure drops below the set pressure (approximately 30 bars in many plants). These occurrences will certainly reduce the consequences, in particular, those with respect to the effects of postulated direct heating on the duration of containment integrity.

Primary coolant system temperature histories are also important in assessing the nature of transport and retention of the fission products in the primary system. Natural convection flows will aid the movement of fission products to cooler regions (e.g., other loops and the steam generators). However the increased temperature levels will also imply earlier revolatilization of fission products and local failures in the primary system could discharge the fission products earlier into containment.

2.5.6. Additional Comments

The natural convection flow patterns discovered and later validated through the experiments performed in Westinghouse, as described above, became an important part of the risk evaluations for severe accidents. The reduction in the probability of failure of the vessel at high pressure decreased the risk of consequences of severe accidents. However, the discovery of high-temperature steam flows to the steam generator also raised issues with respect to the integrity of the steam generator tubes. Failure of the steam generator tubes would be a high-risk event, since the probability of containment bypass increases considerably. The steam generator tube failure was debated for a considerable period of time. The Westinghouse tests made available data on the temperatures of the SF_6 gas at the entry points of a number of tubes in the model steam generator. These data showed that there was a good mixing of the hot streams of the SF_6 entering the steam generator plenum from the vessel with the cold SF_6 returning from the cold tubes in the model steam generator. The temperatures measured in the steam generator tubes were much lower than the temperatures of the hot stream of SF_6 and of the SF_6 in the hot leg nozzle and the surge line.

The multidimensional natural convection flows could not be modeled naturally by the one-dimensional severe accident risk codes, for example, SCDAP-RELAP, MELCOR, MAAP and ASTEC (see the section on severe accident codes in Chapter 8 of this book). The code developers, however, were able to introduce a model for countercurrent flow in the hot leg and derive the flows of relatively hot and cold steam in the steam generator tubes. Thus, the PWR high-pressure accident can be reasonably well treated by the SA codes employed by the reactor safety community.

2.6. SPECIFIC FEATURES OF BWR

(T.N. Dinh and W. Ma)

2.6.1. Design

The boiling water reactor (BWR) is the second largest fleet of light water reactors. There are nearly a hundred BWR plants currently operating worldwide, with excellent track records in both safety and performance before the Fukushim accidents. The main characteristics of the modern (mainstream) BWRs [67] [72] [73] [74] are the use of (1) a direct steam cycle with an internal steam separator and dryer; (2) upward insertion of a control rod bank to the core through the lower head of the RPV, and (3) pressure suppression and inerted containment. Over its historical development, the BWR designs varied significantly, ranging from those using external recirculation loops to those using internal pumps (Figure 2.30). The containment design for BWR plants also evolved significantly, from "dry" to "Mark I," "Mark II," and "Mark III" types, and the more compact one (in ABWR).

The typical BWR reactor has a large core (e.g., 700 core assemblies, each with 60 to 100 fuel elements), generating power in the range from 3000 to 4500 MWt, producing steam at around 7 MPa. Due to both steam production and separation processes occurring inside the RPV, a large space is required in a BWR to incorporate large components (separator, dryer, internal pumps; see

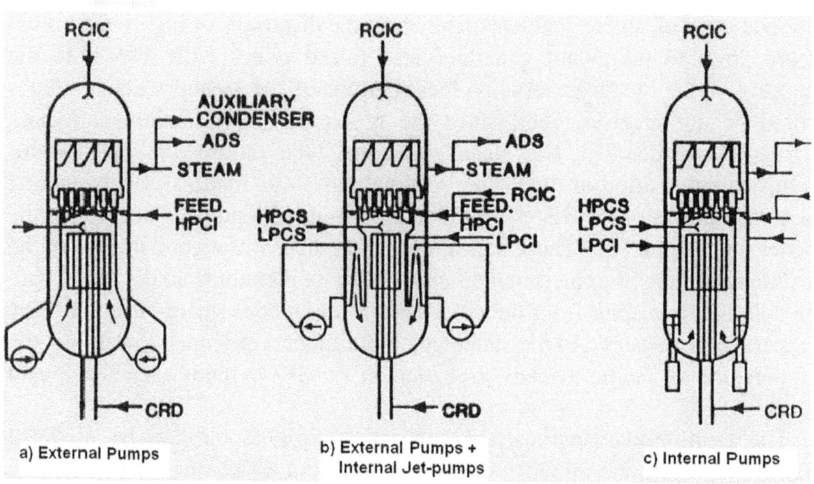

FIGURE 2.30 Three types of main recirculation system of BWRs.

Figure 2.31). The reactor pressure vessel (RPV) for BWRs is larger than that of PWR. The BWR RPV dimensions are typically 20 to 30 m high, 6–7 m in diameter, and 150 mm thick.

For the same power rating, the height of the BWR core is almost the same as that of the PWR core, and the numbers of fuel rods are also comparable. But the fuel rod diameter in BWR (e.g., 12.25 mm) is greater than that (9.5 mm) in PWR [72]. As a result, the mass inventory of the BWR core fuel and structural materials is much larger than that of PWR (cf. Table 2.6).

As a main difference of a BWR core with a PWR core, the fuel assembly is surrounded by a square fuel box (called a canister in GE BWRs), which also serves as a fuel channel (Figure 2.32). The core of a BWR is therefore composed of parallel channels for coolant flow. In each channel, the fuel rods are arranged in an array (8×8 to 10×10). A gap between the fuel assembly boxes is designed to accommodate cruciform control rods, neutron flux detectors, and other instrumentation sensors. Typically, four assemblies sit on a support piece of a control rod guide tube (CRGT), together forming a fuel module. The CRGTs are welded on the bottom head via control rod drive housing and a stub tube. A small water flow rate, about 60g/sec in each CRGT, is available during plant operation for purging and cooling purposes. The forest of control rods, whose number varies from 150 to over 200 in different designs, is an important feature to be considered in the assessment of severe accident progression.

The other design features of BWRs are summarized as follows (also see Table 2.7):

- The control rod guide tubes are the primary support of the core, and the core plate of BWR does not bear the weight of the fuel and its canisters. This is

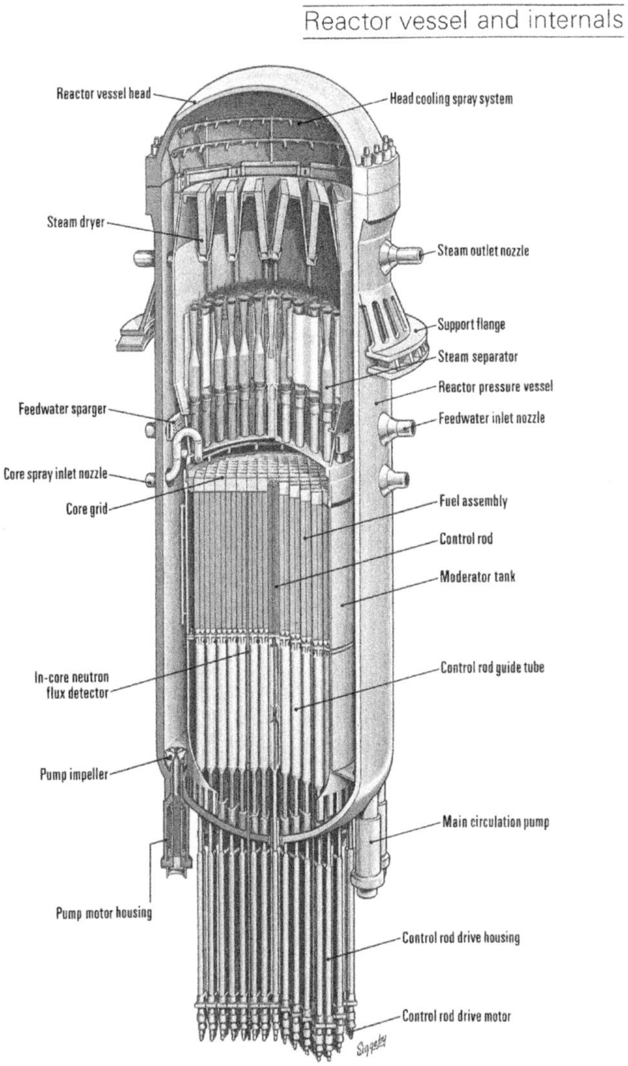

FIGURE 2.31 A boiling water reactor with internal recirculation pumps.

different from PWR where the core sits on the edge-supported core plate, that is, the core is supported by the core plate.

- The control rods of BWR are inserted from beneath the core. This means the control rods are supported by the understructure, unlike the PWR control rods that are held from above the core. This difference may affect the failure mode of control rods.

TABLE 2.6 PWR and BWR Masses of Metal and Fuel

Items	Evolutionary PWR	GE-BWR	ABB-Atom BWR
Thermal capacity, MWt	4000	3440	3300
Total metal (steel), tons	115	121	172
Total Zr, tons	30	62	53
Total UO_2, tons	110	172	140

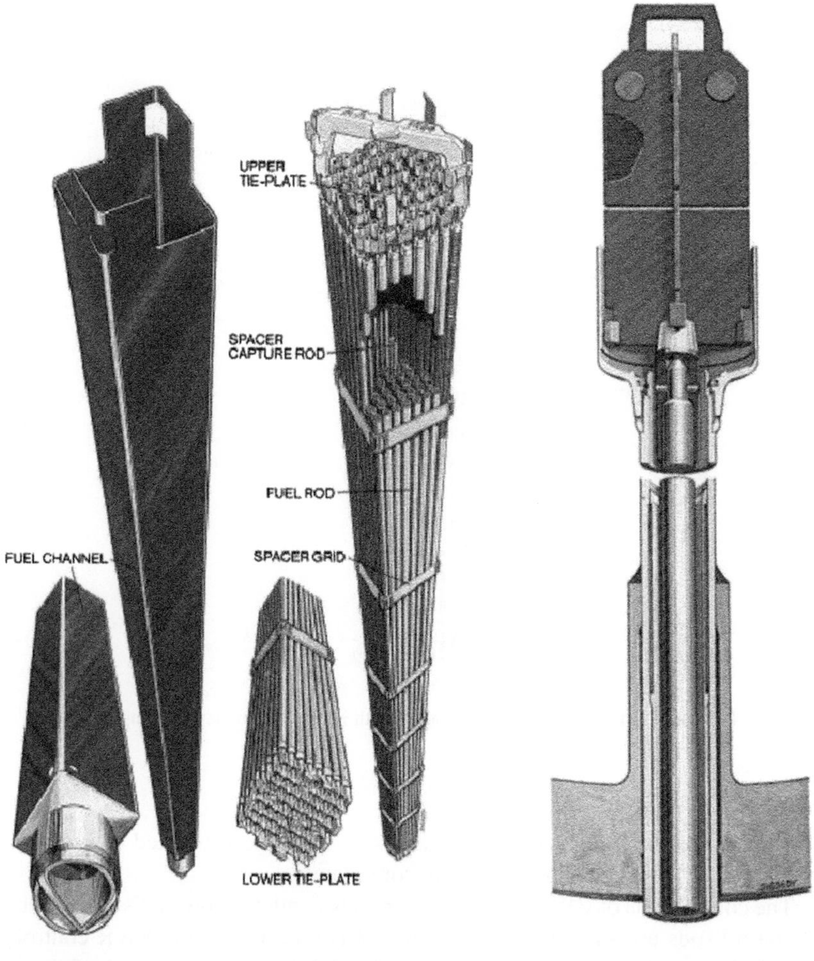

FIGURE 2.32 Left—Fuel assembly and channel of a BWR. Right—Control rod guide tube, housing, and stub tube on a BWR lower head.

TABLE 2.7 Main Design Features That Affect the BWR versus PWR Differences in In-vessel Accident Progression

Items	PWR	BWR
Core	No canister for fuel assembly, radial mixing of coolant possible	A square canister for fuel assembly serves as a coolant channel
	Control rod materials: Ag-In-Cd	Control rod material: B_4C
	Guide tubes in each assembly	Four fuel assemblies around a control rod forms a fuel module
	Core plate provides core support	Core plate provides lateral support. CRGTs provide vertical support
Lower plenum	Core support structure, Instrumentation penetrations	CRGTs and IGTs, Drainage plug, Internal pumps, CRGT cooling
Upper plenum	Control rod housings	Steam separators, steam dryers

- Boron carbide (B_4C) is employed as the absorber of BWR control rods. Thus, during a severe accident, the reaction of boron carbide occurs with steam or other materials. The melting temperature of the B_4C is at 1200°C to 1400°C, which is significantly higher than that of Ag-In-Cd used in control rods of PWRs.
- The cruciform control rods of a BWR are situated in the gaps between the fuel assemblies, while the control rods of a PWR are inserted into the fuel assemblies.
- The control rods of BWR have an independent cooling system.
- The BWR lower plenum features a lower head with the numerous penetrations that hold CRGTs and instrument guide tubes (IGTs). The CRGTs tubes enable the insertion of the control rods from beneath the core (see Figure 2.24). Much of the volume immediately beneath the core plate is occupied by the control rod guide tubes and instrument guide tubes. This is very distinct from PWR lower head where only a few penetrations for instrumentation are present.
- There is a forest of drive mechanisms of the control rods situated under the lower head of a BWR.
- The upper plenum of BWR is much bigger than that of PWR. It takes up almost half of the total vessel volume, and is mainly occupied by the steam separators and driers.

- Most BWRs are equipped with automatic depressurization system (ADS), which diminishes the likelihood of high-pressure scenarios of severe accident.

2.6.2. Accident Progression

The physical picture of in-vessel accident progression in a BWR is generally similar to that of the PWR. The main phenomena that govern core degradation, melt relocation, and vessel failures have already been discussed in the preceding sections and therefore are not repeated here. The following remarks highlight aspects where the BWR-specific features exert the most influence on the in-vessel accident progression and consequences.

2.6.3. Core Degradation and Melt Relocation

The core degradation is affected by the specific configuration of a BWR core due to the following aspects:

- The two-phase flow (instead of single-phase flow as in PWRs) in the core region prior to accident is important to thermal-hydraulics and core degradation (e.g., the swollen and collapsed levels of water in the core are important for timing of the dry-out and core uncovery).
- There is no radial mixing between the core channels of a BWR, while the mixing is possible in a PWR core since there are no such core channels there.
- The heat transfer of the core also depends on its channel and gap configuration. Special consideration is needed for radiation and conduction heat transfer of the canisters.
- The melt (debris) relocation path strongly depends on the channel and gap configuration. Before the canister is melted through by ablation, the debris in the channel is separated from the debris in the gap.
- After the canister partially fails, the debris in the channel are possible to relocate to two destinations: channel and gap. If the canister is entirely melted, the channel and gap will merge together finally. After the failure of the canister wall, the flow conditions change and the situation is similar to the PWR configuration.
- The presence of B_4C in the control rods affects the corium physicochemistry, formation of eutectics, and their relocation.
- The core plate failure does not lead to the collapse of fuel modules (and the core) in a BWR, while the failure of the supporting plate of a PWR core results in the collapse of the fuel and debris above it. The core collapse in a BWR depends on the integrity of the CRGTs in the lower plenum.

- The designs of fuel channels (with canisters), CRGTs and large core dimension have a significant influence on the dynamics of core degradation and melt relocation.
- Consequently, special care should be taken for BWR when coming to models of severe accident phenomena.

2.6.4. Melt Progression in the Lower Head

The BWR lower head is geometrically complex, having a large contact surface area with CRGTs and IGTs. The CRGTs are also functioning as obstacles to melt progression. The relocated debris is therefore more easily quenched for BWR. After the remaining water boils off, the CRGTs are heated up and melted. Thus, the great number of CRGTs not only affects the progress of melt relocation, but also contributes to the metal mass in corium. This is why the corium of BWRs has a higher percentage of metal mass (resulting from the larger inventory of both stainless steel and zircaloy).

2.6.5. Melt-coolant Interactions

The constrained geometry of coolant in the lower plenum (due to the forest of vessel penetrations, i.e. the CRGTs and IGTs) limits the fuel-coolant premixing and consequently diminishes the potential for energetic interactions upon melt relocation to the lower head. Thus, in-vessel steam explosion is not expected to be of risk importance for BWR plants [92].

2.6.6. In-vessel Melt/Debris Coolability

For the same power rating, the diameter of a BWR lower head (6.4 m in inner diameter) is larger than that (4 m) of a PWR. This means that the BWR lower head has a larger external surface area. Thus, the BWR has a better chance for in-vessel coolability and retention, should the external vessel cooling be available. On the other hand, since the lower drywell (the cavity below the lower head) of a BWR is much larger in depth and diameter than the reactor pit of a PWR, it will take a much longer time for the water to reach the lower head by flooding the cavity. So, the answer to in-vessel coolability of a BWR via external vessel cooling as a SAM measure is not straightforward [100]. More research work is needed in this respect.

The CRGT cooling system of a BWR provides a new SAM avenue for in-vessel coolability and retention. Since the cooling flow rate (~60 g/s in each CRGT and ~10 kg/s in total) is so small, the coolant flow can be made possible by a battery-driven pump. Powered by the independent source, coolant flow through CRGTs can be used to remove decay heat during the accident, thus serving as a mitigative measure (reducing thermal load on the lower head structures; delaying the time of vessel failure; reducing the amount of melt

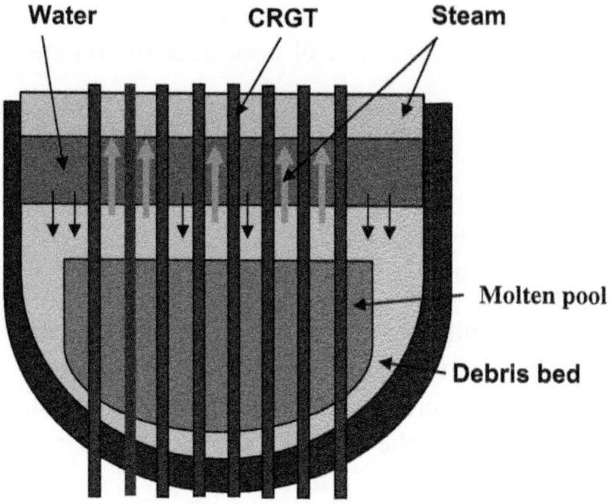

FIGURE 2.33 Configuration of corium pool in a BWR lower head with CRGT cooling.

available for discharge upon the RPV failure; and potentially providing coolability and retention).

An extensive study on the efficiency of CRGT cooling as a SAM measure has been conducted at the Royal Institute of Technology [75] [76] [77] [78]. Figure 2.33 shows a conceptual configuration of corium pool in a BWR lower head with CRGT cooling and top flooding. A portion of the debris bed is heated up and remelted, leading to formation of a melt pool in the middle of the lower plenum. Two cooling mechanisms are considered here for heat removal from the corium pool: the first is the cooling by water flows in CRGTs (CRGT cooling), and the second is the cooling by water on top of the melt pool. External vessel cooling is not considered here since as mentioned above it is unavailable at the early stage (<8 hrs) of a severe accident.

To be able to simulate melt pool heat transfer in a complex geometry such as in the BWR lower head with a forest of penetrations (CRGTs and IGTs), a suitable simulation tool with an effective convectivity model (ECM) and phase-change ECM (PECM) was developed and validated. The ECM and PECM were then applied to assess the efficiency of CRGT cooling in a BWR lower head. The simulation results show that for a debris bed (melt pool) less than 0.7 m thick formed in the lower plenum, the CRGT cooling at the nominal water flow rate is sufficient for removing the decay heat and thus is likely adequate to ensure the integrity of the reactor pressure vessel. With a thicker debris bed, the vessel wall is predicted to fail in the section close to the uppermost region of debris bed (melt pool) [76]. The simulation with a metal layer shows that in a stratified melt pool (0.2 m metal layer in 1.0 m total thickness) the focusing effect is ameliorated, thanks to the CRGT cooling. However, upon formation of the stratified melt pool, the transient heat flux to CRGTs submerged in the metal layer may be high enough to cause dry-out inside the CRGTs, leading to potential failure due to its high temperature.

In this case, increase of flow rate in the CRGT cooling is necessary to ensure the CRGT integrity and debris coolability. More results and insights of the study can be found in [79].

2.6.7. Vessel Lower Head Failure

The BWR lower head wall (150 mm) is thinner than the PWR's (200 mm). However, the pressure in a BWR is lower than that in a PWR, given the actuation of the ADS. There should not be a big difference in creep failure between the two vessels, but the ablation of a BWR vessel wall may be faster.

It is believed that the most probable failure mode for a BWR lower head is the penetration failure (melt-through or drop-away of the guide tubes). IGTs are considered as the more vulnerable to earlier failure, due to the IGT's relatively small size (and hence thermal capacity) and lack of external support. Melt attack on a penetration may disable the weld between the penetration housing and the stub tube on the vessel lower head. Molten corium can enter the penetration's interior space and cause the penetration to fail.

2.6.8. Hydrogen Production and Combustion

Production of hydrogen as a noncondensable gas in a large quantity (due to BWR's large inventory of materials) can influence the system pressure. The majority of operating BWRs employ inerted containment, making the issue of hydrogen combustion irrelevant, particularly in consideration of in-vessel accident progression.

In case of containment isolation failure or hydrogen leakage to the reactor building (not inerted) sourrounding the BWR containment, hydrogen combustion is possible and can cause damage. But this is out of the scope of this chapter (in-vessel phenomena).

2.7. VVER (Eastern PWR)

(Z. Hozer)

2.7.1. Design

The VVER is a pressurized light water reactor of Soviet design. It operates on the same principles as a Western PWR reactor and uses similar technological systems. The first VVER reactors were constructed in the 1960s and later more than 40 units were built in the Soviet Union and in several other countries.

The core of VVER reactors is characterized by hexagonal geometry: the fuel assemblies have hexagonal form, and the assemblies are arranged on the basis of hexagonal symmetry. The fuel rods are arrayed on a triangular lattice. The VVER-440 type assemblies are covered by zirconium shroud, and they are

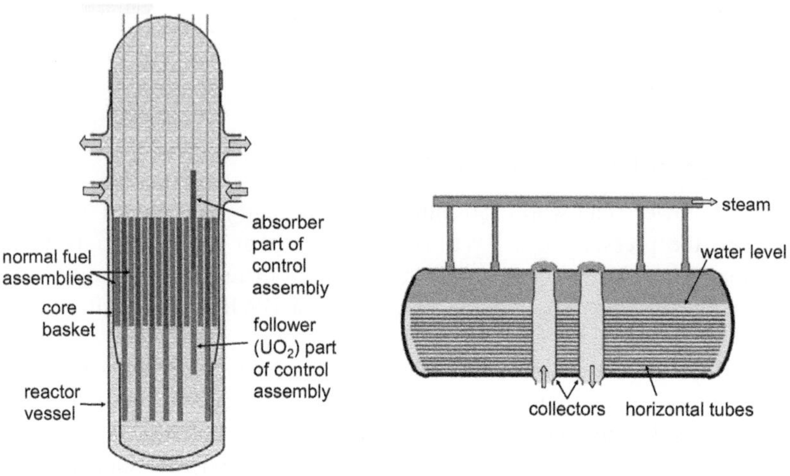

FIGURE 2.34 VVER-440 core design (left) and steam generator (right).

more BWR-like than PWR-like assemblies. The VVER reactor cores are constructed of slightly different materials from those of PWRs. Probably the most significant difference is the cladding material: E110 (Zr1%Nb) alloy in VVERs and Zircaloy (ZrSn alloy) in PWRs.

The VVER-440 reactors are equipped with control assemblies that are twice as long as the standard assemblies (Figure 2.34). Its upper part is a hexagonal boron steel (2% B, 20% Cr, 16% Ni) absorber. The lower part of the control assembly is the *follower*, which consists of fuel rods and which is similar to the normal fuel assembly. In normal operation most of the control rods are in their upper position: the followers are in the core, and the absorbers are above the core. During a reactor scram each control assembly moves down: the follower goes below the core, and the absorber part enters the core from the top. Special steel guide tubes are located in the lower part of the reactor vessel to house the control assembly followers.

The VVER-1000 control rods are similar to PWR clusters. In the original VVER-1000 design, B_4C control rods were used, but today some units use Ag-In-Cd rod segments, too.

The VVERs have horizontal steam generators that contain large water volumes on the secondary side (Figure 2.34). The heat transfer surface is provided by thousands of horizontal tubes that are connecting the hot and cold collectors and are located below the water level of the secondary side coolant. The VVER-440 reactors have valves in the hot and cold legs to isolate the loops in case of leakage from steam generator tubes.

The most important design features that might affect the in-vessel severe accident progression in VVERs compared to PWRs are summarized in Table 2.8 [5].

TABLE 2.8 Comparison of Some Characteristics of VVER and PWR Reactors

	VVER-440	Westinghouse 2 loop	VVER-1000	Westinghouse 4 loop
Thermal power (MW)	1375	1192	3000	3400
Number of loops	6	2	4	4
Steam-generator	horizontal	vertical	horizontal	vertical
Primary pressure (MPa)	12.5	15.5	15.7	15.5
Fuel mass (t)	42.	31.7	91.8	101.
Fuel lattice	triangular	square	triangular	square
Fuel assembly type	hexagonal with shroud	square	hexagonal	square
Cladding material	E110	Zircaloy-4	E110	Zircaloy-4
Control rod type	control assembly	cluster	cluster	cluster
Absorber material	boron steel	AIC	B_4C/AIC	AIC
Bottom head form	elliptical	hemispherical	elliptical	hemispherical

2.7.2. Accident Progression

The initial events leading to core melt are practically the same in VVERs and in PWRs. Due to the special arrangement of collectors in the steam generators (Figure 2.34), the leakage or rupture of SG collector cover can result in a primary-to-secondary break, which generally has a larger break size than the SG tube rupture.

Accident progression in VVERs in general is similar to that in PWRs. However, the differences due to geometry and materials can result in differences in the accident sequence and in the timing of events. The large amount of water on the secondary side of steam generators and in the lower plenum of VVER-440 reactors might provide a long period before core damage in some accident scenarios.

2.7.3. Core Degradation

There are slight differences between the high-temperature oxidation and hydrogen uptake of the zirconium alloys used in PWRs and VVERs. The alpha-to-beta phase transition takes place at about a 100 K lower temperature in E110 than in Zircaloy, and this affects the thermo-mechanical phenomena (such as plastic deformation and burst of fuel rods) taking place between 1100 and 1300 K (Figure 2.35) [80]. The large mass of Zr shroud of VVER-440 provides a significant additional hydrogen production source.

The complex and heterogeneous geometry of VVER-440 core influences the steps of core degradation and melt progression. The power generated in the lower part of control assemblies, which are located below the core, can produce

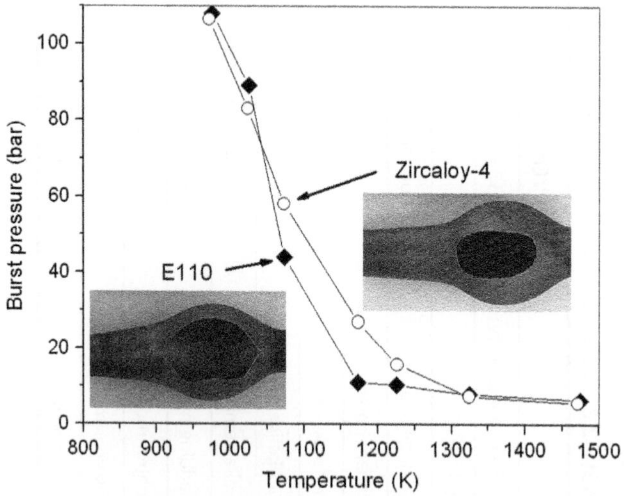

FIGURE 2.35 Burst pressure of VVER (E110) and PWR (Zircaloy-4) cladding tubes.

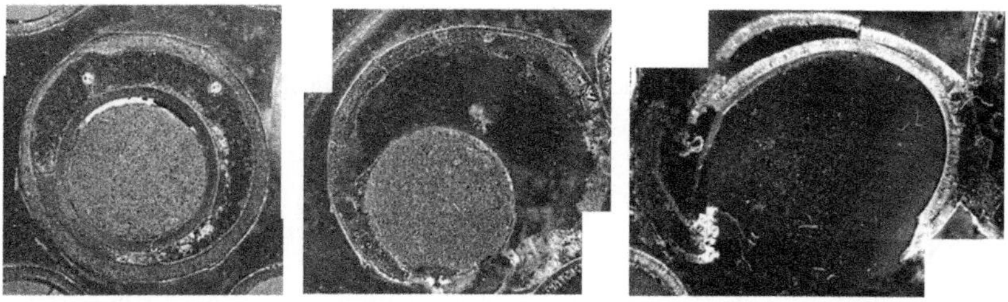

FIGURE 2.36 Steps of VVER-1000 type B_4C control rod degradation: melting of SS cladding (left), oxidation of B_4C (middle), oxidation of Zr guide tube (right).

steam even in dry core conditions and result in some acceleration of the failure of the lower tie plate. The shroud of fuel assembly may hamper the radial spreading of molten materials. The melting of the absorber part of control assembly can create large channels for a downward flow of melts, while the followers and the structures below the core facilitate blockage formation. If the frozen material in the lower part of the assemblies creates complete blockage, then the most likely relocation mechanism will be the side penetration through core barrel.

The loss of rod-like geometry in VVER-1000 starts with the melting of the stainless steel cladding of control rods and the oxidation of boron-carbide (Figure 2.36) [81]. The high-temperature oxidation of B_4C in steam is an exothermic reaction that produces high amounts of hydrogen, gaseous carbon, and boron-containing species [82]. Similar phenomena take place in BWRs having the same kind of control rods.

2.7.4. In-vessel Melt Retention

Both VVER-440 and VVER-1000 type reactor vessels have an elliptical (or more precisely torospherical) bottom head, while the PWR reactors have a hemispherical bottom head. In the elliptical bottom head, the molten pool can extend up to the cylindrical part of the vessel, and the corium creates heat flux not only on the bottom of reactor vessel, but on the sidewall as well. The structure and the convective behavior of the molten pool in the VVER bottom head and in the hemispherical PWR bottom head seem to be similar: only slight differences are expected in the heat transfer profile.

The only operating reactor in the world that has the in-vessel retention of corium as an approved severe accident management measure is the VVER-440 Finnish Loviisa plant. The approach selected takes advantage of the unique features of the plant, such as low power density, a reactor vessel without penetrations and ice-condenser, which ensures a flooded cavity [83].

The latest design of VVER-1000 reactors includes a unique core catcher system under the reactor vessel for the reception, localization, and cool-down of the molten corium in the course of severe accidents [84].

ACKNOWLEDGMENTS

The authors of Section 2.2 wish to acknowledge B. Adroguer (IRSN, retired) for his support to the writing of this Section.

REFERENCES

[1] S. Kinnersly, et al., In-Vessel Core Degradation in LWR Severe Accidents: A State of the Art Report. NEA/CSNI/R (1991) 12.

[2] T. Haste, B. Adroguer, U. Brockmeier, P. Hofmann, K. Müller, M. Pezzilli, In-Vessel Core Degradation in LWR Severe Accidents, State of the Art Report. Update January 1991–June 1995, EUR 16695 EN (January 1996).

[3] T. Haste, B. Adroguer, R.O. Gauntt, J.A. Martinez, L.J. Ott, J. Sugimoto, K. Trambauer, 1996, In-vessel core degradation validation matrix. NEA/CSNI/R(95)21, OCDE/GD(96)14 (January 1996).

[4] K. Trambauer, T. Haste, B. Adroguer, Z. Hózer, D. Magallon, A. Zurita, 2000, In-vessel core degradation code validation matrix, update 1996–1999. OECD/CSNI report NEA/CSNI/R (2000). 21, February 2001.

[5] Z. Hózer, K. Trambauer, J. Duspiva, Status Report on VVER-Specific Features Regarding Core Degradation. NEA/CSNI/R(98)20 (September 1998).

[6] G. Bandini, Status of Degraded Core Issues—A Synthesis Paper. NEA/CSNI/R (2001). 5, OECD 2001.

[7] T. Haste, B. Adroguer, N. Aksan, C. Allison, S. Hagen, P. Hofmann, V. Noack, Degraded Core Quench: A Status Report. November 1995; NEA/CSNI/R(96)14, OCDE/GD(97)5 (August 1996).

[8] T. Haste, K. Trambauer, Degraded Core Quench: Summary of Progress 1996–1999. NEA/CSNI/R(99)23 (February 2000).

[9] NEA/CSNI, Technical Opinion Paper on Fuel-Coolant Interaction. NEA/CSNI/R(99)24 (November 1999).

[10] D. Magallon, et al., MFCI Project Final Report. European Commission report INV-MFCI(99)-P007 (1999).

[11] M. Steinbrück, M. Große, L. Sepold, J. Stuckert, Synopsis and outcome of the QUENCH experimental program, Nuclear Engineering and Design 240 (2010) 1714–1727.

[12] H. Tuomisto, V. Strizhov, B.R. Sehgal, A. Behbahani, R. Gonzalez, B. Sanderson, K. Trambauer, Application of the OECD Rasplav Project Results to Evaluations at Prototopic Accident Conditions, OECD RASPLAV Seminar, Munich, Germany, November 2000.

[13] H. Helf, et al., Catalogue of Generic Plant States leading to Core Melt in PWRs, includes Appendix 1: Detailed Description of Sequences Leading to Core melt. NEA/CSNI/R (1996). 18, November 1996.

[14] O. Nusbaumer, Decay heat in nuclear reactors.http://www.ontronic-efi.com/decay_heat.htm.

[15] B. Adroguer, et al., Core loss during a severe accident (COLOSS Project), Nuclear Engineering and Design 235 (2005) 173–198.

[16] B.C. Oberlander, M. Epeland, M.O. Solum, PIE results from the high burn-up (92 MWd/kg) PWR segment after LOCA testing in IFA 260-4, Proc. of enlarged Halden Programme group meeting, F2-7, Loen (Norway), 2008.

[17] Z. Hózer, 2002, Summary of the Core Degradation Experiments CODEX, Eurosafe Fo-rum 2002, November 4–5, http://www.eurosafe-forum.org/grs/de/ed 2002.html (2002).

[18] Y.G. Dragunov, V.V. Shchekoldin, I.I. Fedik, N.Y. Parshin, Computational Analysis of PARAMETER Facility Experiments, NURETH-11, International Topical Meeting on the Nuclear Reactor Thermal-Hydraulics, October 2–6, 2005, Avignon, France.
[19] M.S. Veshchunov, K. Mueller, A.V. Berdyshev, Molten corium oxidation model, Nuclear Engineering and Design 235 (Issue 22) (November 2005) 2431–2450.
[20] I. Shepherd et al., Investigation of Core Degradation, COBE project, FISA-99 Symposium, Luxembourg, November 1999, EUR 19532 EN.
[21] E.W. Coryell, 1994, Summary of Important Results and SCDAP/RELAP5 Analyses for OECD LOFT. Experiment LP-FP2, NUREG/CR-6160, NEA-CSNI-R(94)3 (1994).
[22] I. Shepherd, et al., Oxidation Phenomena in Severe Accidents (OPSA) Final Report. EUR 19528 EN (2000).
[23] B. Adroguer et al., Corium Interactions and Thermochemistry, CIT project, FISA-99 Symposium, Luxembourg, Novermber 1999, EUR 19532 EN.
[24] R. Silverii, D. Magallon, 1998, FARO-LWR programme. Test L-26S Data Report, Technical Note I.98.229 JRC Ispra.
[25] B. Cheynet, P.Y. Chevalier, E. Fischer, Thermodynamic modelling and data banking: NUCLEA, MASCA Seminar, Aix-en-Provence, France, June 10–11, 2004.
[26] J.M. Seiler, K. Froment, Material effects on multiphase phenomena in late phase of severe accidents of nuclear reactor, Multiphase Science and Technology 12 (2000).
[27] V. Dauvois, S. Goldstein, C. Gueneau, K. Froment, J.M. Seiler, Boundary conditions for liquid corium in thermalhydraulic steady state and experimental validation, Proceedings of ICONE 8, April 2–6, 2000, Baltimore, MD USA.
[28] S.S. Abalin, I.P. Gnidoi, A.I. Surenkov, V.F. Strizhov, Data base for 3rd and 4th series of RASPLAV salt tests. OECD RASPLAV Report (1998).
[29] X. Gaus-Liu, A. Miassoedov, T. Cron, T. Wenz, In-vessel Melt Pool Coolibility Test-Description and Results of LIVE Experiments, Nuclear Engineering and Design 240 (2010) 3898–3903.
[30] M. Barrachin, F. Defoort, Thermophysical properties of In-Vessel Corium, MASCA Seminar Aix-en-Provence, France, June 10–11, 2004.
[31] J.M. Seiler, A. Fouquet, K. Froment, F. Defoort, Theoretical analysis for corium pool with miscibility gap, Nuclear Technology 141 (3) (2003) 233–243.
[32] T.G. Theofanous, C. Liu, S. Angelini, O. Kymaäläinen, H. Tuomisto, S.L. Additon, Experience from the first two integrated approaches to In-Vessel retention through external cooling, OECD/CSNI/NEA Workshop on large molten pool heat transfer CEA Grenoble, March 9–11, 1994.
[33] T.G. Theofanous, C. Liu, S.L. Additon, S. Angelini, O. Kymäläinen, T. Salmassi, In vessel coolability and retention of a core melt. DOE/ID-10460 (July 1995).
[34] F.A. Kulacki, A.A. Emara, Steady and transient convection in a fluid layer with uniform volumetric energy sources, J. Fluid Mech. 83 (part 2) (1977) 375–395.
[35] U. Steinberner, H. Reineke, Turbulent buoyancy convection heat transfer with internal heat sources, Proc. of 6th Int. Journal of Heat and Mass Transfer Conf., Toronto, Canada, August 7–11, 1978.
[36] T.G. Theofanous, C. Liu, Natural convection experiments in a hemisphere with Rayleigh numbers up to 10^{15}, ANS National Conf. on Heat Transfer. Portland, 1995 USA, 349–365.
[37] T.G. Theofanous, M. Maguire, S. Angelini, T. Salmassi, 1996, The first results from the ACOPO experiments, Nuclear Engineering and Design 169 (1997) 49–57.
[38] J.M. Bonnet, J.M. Seiler, OECD RASPLAV In-Vessel Corium Pool Thermalhydraulics for the Bounding Cases. RASPLAV seminar, Münich (November 13, 2000).

[39] M. Jahn, H.H. Reinecke, Free convection heat transfer with internal heat sources. Calculations and measurements, Proc. of the 5th Int. Conf. on Heat Transfer, vol. III, Tokyo, September 3–7, 1974.

[40] S.J. Asfia, V.K. Dhir, An experimental study of natural convection in a volumetrically heated spherical pool bounded on top with a rigid wall, Nuclear Engineering and Design 163 (1996) 333–348.

[41] S.W. Churchill, H.H.S. Chu, Correlating Equations for Laminar and Turbulent Free Convection from a Vertical Plate, International Journal of Heat and Mass Transfer 18 (1975) 1323–1329.

[42] T.C. Chawla, S.H. Chan, Heat Transfer from Vertical/Inclined Boundaries of Heat Generating Boiling Pools, In. J. Heat Mass Trans. 20 (5) (1977) 499–506.

[43] Kim, et al., Effect of nanoparticles on CHF enhancement in pool boiling of nano-fluids, Int. J. of Heat Mass Transfer 49 (2006) 5070–5074.

[44] J.H. Jeong, R.J. Park, K.H. Kang, S.B. Kim, H.D. Kim, Experimental study on CHF in a Hemispherical Narrow Gap, OECD CSNI Workshop on In-Vessel Core Debris Retention and Coolability Garching, Germany, March 3-6, 1998.

[45] V. Asmolov, L. Kobzar, V. Nickulshin, V. Strizhov, Experimental study of heat transfer in the slotted channel at CTF facility, OECD/CSNI Workshop on In-Vessel Core Debris Retention and Coolability, Garching, Germany, March 3-6, 1998.

[46] O. Herbst, H. Schmidt, W. Köhler, W. Kastner, Experimental contributions on the Gap cooling process crucial for RPV integrity during the TMI-2 accident, 7th Int. Conference on Nuclear Engineering Tokyo, Japan, April 19–23 (ICONE 7) (1999).

[47] M. Murase, T. Kohriyama, K. Kawabe, Y. Yoshida, Y. Okano, Heat transfer in narrow gap, Proceeding of 9th International Conference on Nuclear Engineering (ICONE 9), Nice, France, April 8–12, 2001.

[48] J.M. Seiler, Analytical model for CHF in narrow gaps in vertical and hemispherical geometries, Nucl. Eng. & Des. 236 (2006) 2211–2219.

[49] T.G. Theofanous, S. Syri, The coolability limits of a lower reactor pressure vessel head, NURETH 7, Nuclear Engineering and Design 169 (1997) 59–76.

[50] T.N. Dinh, T. Salmassi, T.G. Theofanous, Limits of coolability of the AP 1000 ULPU 2400 configuration V facility. CRSS-03/06 (June 30, 2003).

[51] F.B. Cheung, Limiting factors for external vessel cooling, 10th Int, Topical Meeting on Nuclear Reactor Thermalhydraulics (NURETH 10) (2003).

[52] T.G. Theofanous, S. Syri, T. Salmassi, Critical heat flux through curved, downward facing, thick walls, OECD/CSNI/NEA Workshop on large molten pool heat transfer—Grenoble, France, March 9–11, 1994.

[53] S. Rougé, SULTAN test facility; Large scale vessel coolability in natural convection at low pressure, NURETH 7, Saratoga Springs, September 10–15, 1995.

[54] K.H. Kang, R.J. Park, S.D. Kim, H.D. Kim, Simulant pool experiments on the coolability through external vessel cooling strategy, ICAPP 05 conference Seoul, Korea, May 15–19, 2005.

[55] J.M. Seiler, B. Tourniaire, F. Defoort, K. Froment, Consequences of physico-chemistry effects on in-vessel retention issue, 11th Topical Meeting on Nuclear Reactor Thermal-hydraulics (NURETH 11) (2005).

[56] H.G. Willschuetz, E. Altstadt, B.R. Sehgal, F.P. Weiss, Coupled thermal structural analysis of LWR vessel creep failure experiments, Nuclear Engineering and Design 208 (2001) 265–282.

[57] H.G. Willschuetz, E. Altstadt, B.R. Sehgal, F.P. Weiss, Simulation of creep tests with French or German RPV-steel and investigation of a RPV-support against failure, Annals of Nuclear Energy 30 (2003) 1033–1063.

[58] C. Sainte Catherine, Tensile and creep tests material characterization of pressure vessel steel (16MND5) at high temperatures (20 up to 1300°C), Report SEMT/LISN/RT/98-009/A, CEA, France, (experimental data files), 1998.

[59] Ph. Mongabure (2000), Creep tests on LHF and OLHF SA533B1 steel, Report SEMT/LISN/RT/00-060/A, CEA Saclay – DMS2, December 2000.

[60] B.R. Sehgal, A. Theerthan, A. Giri, A. Karbojian, H.G. Willschütz, O. Kymäläinen, S. Vandroux, J.M. Bonnet, J.M. Seiler, K. Ikkonen, R. Sairanen, S. Bhandari, M. Bürger, M. Buck, W. Widmann, J. Dienstbier, Z. Techy, R. Taubner, T.G. Theofanous, T.N. Dinh, Assessment of reactor vessel integrity (ARVI), Nuclear Engineering and Design 221 (2003) 23–53.

[61] E. Altstadt, H.G. Willschuetz, B.R. Sehgal, F.P. Weiss, Modelling of in-vessel retention after relocation of corium into the lower plenum. Scientific Reports of Forschungszentrum Dresden-Rossendorf, N°FZR-437, ISSN 1437-322X (2005), http://www.fzd.de/publications/007712/7712.pdf.

[62] H.G. Willschuetz, E. Altstadt, B.R. Sehgal, F.P. Weiss, Recursively coupled thermal and mechanical FEM-analysis of lower plenum creep failure experiments, Annals of Nuclear Energy 33. 2 (2006) 126–148.

[63] T.Y. Chu, M.M. Pilch, J.H. Bentz, J.S. Ludwigsen, W.Y. Lu, L.L. Humphries, Lower Head Failure Experiments and Analyses, Report NUREG/CR-5582, SAND98-2047, Sandia National Laboratories, Albuquerque, NM, USA (1999).

[64] L.L. Humphries, T.Y. Chu, J. Bentz, R. Simpson, C. Hanks, W. Lu, B. Antoun, C. Robino, J. Puskar, Ph. Mongabure, OECD Lower Head Failure Project Final Report, Sandia National Laboratories, Albuquerque, NM, (2002) 87185-1139.

[65] L. Nicolas, M. Durin, V. Koundy, E. Mathet, A. Bucalossi, P. Eisert, J. Sievers, L. Humphries, J. Smith, V. Pistora, K. Ikonen, Results of benchmark calculations based on OLHF-1 test, Nuclear Engineering and Design 223 (2003) 263–277.

[66] S. Bechta, et al., Experimental Study of Interactions Between Suboxidized Corium and Reactor Vessel Steel, Proceedings of ICAPP '06, Reno, NV USA, June 4–8, 2006, Paper 6054.

[67] V.E. Denny, B.R. Sehgal, Analytical Predictions of Core Heat-Up, Liquefaction, Slumping Proc. International Mtg., LWR Severe Accident Evaluation, vol. 1, p. 5.4, Cambridge, MA, USA (August 28 to September 1, 1983).

[68] V.E. Denny, The CORMLT Code for the Analysis of Degraded Core Accidents, EPRI NP-3767 CCM EPRI, Palo Alto, CA, USA (1984).

[69] B.R. Sehgal, V.E. Denny, W.A. Stewart, B.-J. Chen, Effects of Natural Convection Flows on PWR System Temperatures during Severe Accidents, Proceedings of National Heat Transfer Conference, Denver, CO, USA (August 1985).

[70] W.T. Sha, et al., COMMIX-1A: A Three Dimensional Transient Single Phase Component Computer Program for Thermal-Hydraulic Analysis, NUREG/CR-0785, ANL-77-96, Argonne National Laboratories (1978).

[71] W.A. Stewart, A.T. Pieczynski, V. Srinivas, Natural Convection Experiments for PWR High-Pressure Accidents, EPRI-TR102815; EPRI Licensed Report (August 1993).

[72] B. Pershagen, Light water reactor safety, Pergamon Press, 1989.

[73] J.R. Lamarsh, A.J. Baratta, Introduction to nuclear engineering, Prentice Hall, 2001.

[74] A. Carlen, N. Garis, et al., Störnings-handboken – BWR, SKI Report 03 (2003) 02.

[75] C.T. Tran, T.N. Dinh, The effective convectivity model for simulations of melt pool heat transfer in a light water reactor pressure vessel lower head. Part I: physical processes, modeling and model implementation, Progress in Nuclear Energy 51 (2009) 849–859.
[76] C.T. Tran, T.N. Dinh, The effective convectivity model for simulations of melt pool heat transfer in a light water reactor pressure vessel lower head. Part II: Model assessment and application, Progress in Nuclear Energy 51 (2009) 860–871.
[77] C.T. Tran, T.N. Dinh, Simulation of core melt pool formation in a reactor pressure vessel lower head using an effective convectivity model, Nuclear Engineering and Technology 41 (7) (2009) 929–944.
[78] C.T. Tran, P. Kudinov, T.N. Dinh, An approach to numerical simulation and analysis of molten corium coolability in a BWR lower head, Nuclear Engineering and Design 240 (9) (September 2010) 2148–2159.
[79] C.T. Tran, P. Kudinov, The Effective Convectivity Model for simulation of molten metal layer heat transfer in a Boiling Water Reactor lower head, Proceedings of ICAPP'09, Shinjuku Tokyo, Japan, May 10–14, 2009.
[80] Z. Hózer, C. Győri, M. Horváth, I. Nagy, L. Maróti, L. Matus, P. Windberg, J. Frecska, Ballooning experiments with VVER cladding, Nuclear Technology 152 (2005) 273–285.
[81] Z. Hózer, L. Maróti, P. Windberg, L. Matus, I. Nagy, Gy . Gyenes, M. Horváth, A. Pintér, M. Balaskó, A. Czitrovszky, P. Jani, A. Nagy, O. Prokopiev, B. Tóth, Behaviour of VVER fuel rods tested under severe accident conditions in the CODEX facility, Nuclear Technology 154 (2006) 203–317.
[82] M. Steinbrück, M.S. Veshchunov, A.V. Boldyrev, V.E. Shestak, Oxidation of B_4C by steam at high temperatures: New experiments and modelling, Nucl. Eng. Des. 237 (2007) 161–181.
[83] O. Kymäläinen, H. Tuomisto, T.G. Theofanous, In-vessel retention of corium at the Loviisa plant, Nucl. Eng. Des. 169 (1997) 109–130.
[84] S.V. Svetlov, V.V. Bezlepkin, I.V. Kukhtevich, S.V. Bechta, V.S. Granovsky, V.B. Khabensky, V.G. Asmolov, V.B. Proklov, A.S. Sidorov, A.B. Nedozerov, V.F. Strizhov, V.V. Gusarov, Yu.P. Udalov, Core Catcher for Tianwan NPP with VVER-1000 reactor. Concept, Design and Justification, 2003, ICONE11-36102.
[85] L.A. Stickler, J.L. Rempe, S.A. Chavez, G.L. Thinnes, S.D. Snow, R.J. Witt and M.L. Corradini, Vessel Investigation Project - Calculations to Estimate the Margin-to-Failure in the TMI-2 Vessel, 1993, OCDE-NEA TMI-2, TMI V(93)EG01 report.
[86] Nuclear Technology, Special edition TMI-2, 87 (1989).
[87] Three Mile Island reactor pressure vessel investigation project, Proc. of an Open Forum Sponsored by OECD NEA and USNRC (1993).
[88] K. Atkhen, G. Berthoud, Experimental and numerical investigations based on debris bed coolability in a multidimensional and homogeneous configuration with volumetric heat source, Nuclear Technology 42 (June 2003).
[89] P. Hofmann, Reactions- und Schmelzverhalten des LWR-Corekomponenten UO_2, Zircaloy and Stahl wärhend der Abschmelzperiode, KfK 2220 (July 1976).
[90] J.M. Seiler, K. Ikkonen, R. Sairanen, S. Bhandari, M. Bürger, M. Buck, W. Widmann, J. Dienstbier, Z. Techy, P. Kostka, R. Taubner, T. Theofanous, T.N. Dinh, Assesment of reactor vessel integrity, Nuclear Engineering and Design 221 (2003) 21–53.
[91] J.M. Bonnet, J.M. Seiler, In-vessel corium pool thermalhydraulics for the bounding case, Salt Expert Group Meeting RASPLAV Seminar (2000).
[92] V.A. Bui, Phenomenological and Mechanistic Modelling of Melt-Structure-Water Interactions in a Light Water Reactor Severe Accident, Doctoral Thesis at Royal Institute of Technology, Stockholm (1998).

[93] C. Müller, Review of debris bed cooling in the TMI-2 accident, Nuclear Engineering and Design 236 (2006) 1965–1975.

[94] V. Koundy, F. Fichot, H.G. Willschuetz, E. Altstadt, L. Nicolas, J.S. Lamy, L. Flandi, Progress on PWR lower head failure predictive models, Nuclear Engineering and Design 238 (2008) 2420–2429.

[95] F. Fichot, F. Duval, N. Trégourès, C. Béchaud, M. Quintard, The impact of thermal non-equilibrium and large-scale 2D/3D effects on debris bed reflooding and coolability, Nuclear Engineering and Design 236 (2006) 2144–2163.

[96] W. Hering, Ch. Homann, Degraded core reflood: Present understanding and impact on LWRs, Nuclear Engineering and Design 237–24 (December 2007) 2315–2321.

[97] R.B. Pond, J.E. Matos, Nuclear Mass Inventory, Photon Dose Rate and Thermal Decay Heat of Spent Research Reactor Fuel Assemblies. http://www.rertr.anl.gov/FRRSNF/TM26REV1.html.

[98] D.A. Powers, L.N. Kmetyk, R.C. Schmidt, A Review of the Technical Issues of Air Ingression during Severe Reactor Accidents, NUREG/CR-6218 (September 1994).

[99] G. Bandini, R.O. Gauntt, T. Okkonen, K.Y. Suh, I. Shepherd, T. Linnemann, Molten Material Relocation to the Lower Plenum: A Status Report, September 1998, NEA/CSNI/R(97)34 (September 1998).

[100] S.A. Hodge, J.C. Clevelend, T.S. Kress, M. Petek, Identification and Assessment of BWR In-Vessel Severe Accident Mitigation Strategies, NUREG/CR-5869 (October 1992).

[101] P. Roux, F. Fichot, S. De Pierrepont, D. Gobin, D. Goyeau, M. Quintard, Modelling of binary mixture phase change: assessment on RASPLAV Salt Experiments, NURETH 11, Avignon, October 2–6, 2005.

[102] H. Combeau, Appolaire, J.M. Seiler, Interface temperature between solid and liquid corium in severe accident situations: a comprehensive study of characteristic time delay needed for reaching liquidus temperature, OECD/NEA/CSNI MASCA2 seminar, Cadarache, France, October 11–12, 2007.

[103] B. Tourniaire, J.M. Seiler, K. Froment, F. Defoort, F. Fichot, M. Barrachin, C. Journeau, The French interpretation of the results of MASCA programme and reactor applications, Transactions of the American Nuclear Society, Anaheim (USA), 2008.

Chapter 3

Early Containment Failure

Chapter Outline

Nuclear Safety in Light Water Reactors. DOI: 10.1016/B978-0-12-388446-6.00003-4

3.1. HYDROGEN BEHAVIOR AND CONTROL IN SEVERE ACCIDENTS

(I. Kljenak, A. Bentaib, and T. Jordan)

It was March 28, 1979, at the Three Mile Island 2 (TMI-2) nuclear power plant near Harrisburg, Pennsylvania (USA), the second day of the accident that would become one of the (unfortunate) milestones in the history of civilian use of nuclear power. On this day an intense hydrogen burn occurred in the containment [4]. Hydrogen had been generated during the degradation of the reactor core by zirconium–water interaction and had been accumulating in the containment. Hydrogen was released to the containment from the reactor cooling system through the pressurizer and reactor coolant drain tank at a vent. The hydrogen burn occurred 9 h and 50 min after the initiation of the accident. Analyses performed after the accident asserted that, by the time the burn occurred, the generated hydrogen had become well mixed throughout most of the containment. What triggered hydrogen combustion might never be known with certainty: a plausible hypothesis is that it was initiated by an electric spark. Be that as it may, the hydrogen burn did not cause any severe damage: the pressure spike, which lasted less than a few seconds, reached an approximate maximum value of 2.8 bar (Figure 3.1). The entire burn lasted approximately 12 s. In fact, this combustion of approximately 350 kg of hydrogen caused the only serious pressure load during the whole accident. Since the TMI-2 containment withstood this load, there were no consequences for the

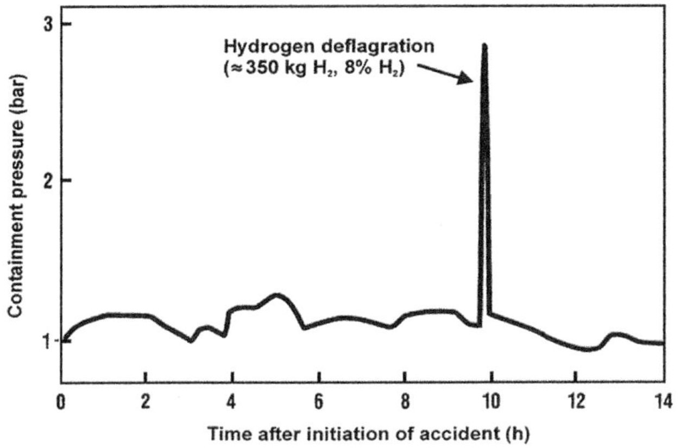

FIGURE 3.1 Containment pressure during TMI-2 accident [4].

environment. But the hydrogen burn at the TMI-2 plant revealed a serious gap in knowledge of nuclear safety: although it was not unknown, the risk of hydrogen combustion had not been given sufficient attention before. The TMI-2 hydrogen combustion was thus a key event that, apart from other research in nuclear safety, also initiated intensive research on hydrogen combustion, that is, its prevention and mitigation during hypothetical severe accidents in nuclear power plants. Both of the following severe accidents in nuclear reactors that had drastic consequences for the environment—Chernobyl in 1986 and Fukushima in 2011— confirmed the importance of proper hydrogen risk management.

3.1.1. The Hydrogen Combustion Threat in Nuclear Power Plant Containments

Hydrogen is neither toxic for humans (however, like other gases, it might displace oxygen to a level where asphyxiation could happen) nor acts as a corrosive medium. The main threat caused by the presence of a large amount of hydrogen in a nuclear power plant is its possible combustion. As the TMI-2 and Fukushima accidents revealed, hydrogen combustion could cause high pressure spikes (mechanical loads) and high temperatures (thermal loads). These explosive loads can cause immediate failure of the containment or can damage the equipment - some of it perhaps being used to mitigate the consequences of the accident – and/or can, as demonstrated in Fukushima, cause the failure of the nuclear power plant (NPP) containment, thus breaking the last safety barrier and allowing the release of fission products that have accumulated in the containment to the environment.

The present chapter provides an overview of the issue of hydrogen combustion risk in an NPP containment during a severe accident.

Specifically, the chapter is meant to provide answers to the following questions:

- How is hydrogen generated in a LWR NPP during a severe accident?
- How can the generated hydrogen form flammable mixtures in the reactor's subcompartments or in the containment?
- At what conditions is hydrogen combustion expected to occur during a severe accident in an NPP?
- What are the different modes of hydrogen combustion and the respective threats that they create?
- What are the currently employed safety systems, which are expected to mitigate the hydrogen combustion risk in NPPs during a hypothetical severe accident?
- What experimental investigations have been performed to date on distribution and combustion of hydrogen?
- Which theoretical methods are being used to predict hydrogen distribution and combustion during a hypothetical severe accident?

The topic of hydrogen combustion risk in severe accidents is a wide one. First, it covers many different fields, starting from hydrogen generation, then continuing through hydrogen distribution and combustion, to conclude with hydrogen mitigation measures. Specifically, hydrogen behavior in core meltdown accidents is influenced by many physical processes that occur on different time and length scales. The most important phenomena are the following:

- Release of hydrogen and steam at the leakage location of the primary circuit,
- Hydrogen distribution and mixing with the air–steam atmosphere in the containment,
- Hydrogen ignition when a burnable mixture is formed,
- Slow quasi-laminar deflagration,
- Fast turbulent deflagration,
- Detonation,
- Pressure and temperature loads on the reactor building.

Second, apart from the most elementary facts, most physical aspects of hydrogen behavior in nuclear power plants are still being investigated (especially experimentally), which shows that knowledge of hydrogen behavior is far from complete. The present chapter will mostly be limited to stating the basic acknowledged facts about the hydrogen combustion risk in an NPP containment during a severe accident. It will also acquaint the reader with the most important ongoing investigations and provide references where more information about specific issues may be found.

The treatment of the present chapter is limited to those aspects of basic phenomena that are relevant for the understanding of hydrogen behavior during severe accidents in NPPs. Also, this chapter does not provide sufficient material that would enable the readers to apply theoretical methods in their own work on hydrogen behavior or to perform safety analyses of NPPs, since only an overview of the main existing theoretical approaches is presented. To obtain additional information necessary for the actual application of the theoretical methods (that is, to determine the expected hydrogen combustion mode, or to model and simulate hydrogen mixing and combustion), the reader will have to consult additional references, provided at the end of the chapter. In other words, the present chapter provides starting points from which readers may continue their studies of specific topics.

3.1.2. Basic Physics: Hydrogen Properties, Mixing, Combustion, Flammability, and Flame Propagation

This section presents some general basic physics about hydrogen, its mixing and combustion. Specific descriptions pertaining to phenomena in a nuclear power plant are presented in later sections.

Hydrogen Properties

The hydrogen atom consists of a nucleus, formed of a single proton, and of a single electron. The hydrogen molecule consists of two hydrogen atoms bounded together. The molar mass of hydrogen is 2.016 g/mole. Hydrogen is thus much lighter than other gases likely to be found in the atmosphere of a nuclear power plant containment during a severe accident (steam, air consisting mostly of oxygen and nitrogen, as well as carbon monoxide and carbon dioxide generated by molten core—concrete interaction). Thus, when hydrogen is part of a heterogeneous atmosphere in a closed compartment, it is likely to accumulate in the upper parts due to its buoyancy.

The dynamic viscosity of hydrogen at 0 °C is $8.4 \cdot 10^{-6}$ Pa·s, which is somewhat lower than the dynamic viscosity of saturated steam at 1 bar ($12 \cdot 10^{-6}$ Pa·s), and about one-half of the dynamic viscosity of air at 0 °C ($17.4 \cdot 10^{-6}$ Pa·s). For the specific situation of a severe accident, this means that when hydrogen is flowing upward, air entrainment would be less intensive than in a comparable situation with steam, but not fundamentally different.

The diffusivity of hydrogen in air at pressure of 1 atmosphere and temperature of 25 °C is $0.41 \cdot 10^{-4}$ m^2/s, whereas the diffusivity of steam in air at the same conditions is $0.26 \cdot 10^{-4}$ m^2/s. Thus, during a severe accident, greater diffusion of hydrogen than of steam may be expected. This is relevant for predicting the formation of regions with high hydrogen concentration, which makes hydrogen ignition and subsequent combustion more likely.

Hydrogen Mixing

When considering the phenomenon of mixing, one is usually implying the mixing of liquids instead of the mixing of gases. If two miscible liquids are mixed together (say, water and ethanol), they will form a homogeneous mixture and not separate if left to themselves in a container. In fact, miscible liquids dissolve into each other. However, if two immiscible liquids are mixed (say, water and cooking oil) and left to themselves, they will not form a homogeneous mixture but will eventually separate into two horizontally stratified layers after a certain time. These different behaviors are related to the nature of intermolecular forces between molecules of different liquids.

In principle and in the sense that is usually applied to liquids, all gases are miscible (the expression "immiscible gases," which appears in the literature, refers to a different phenomenon). A particular situation occurs if two gases are mixed together, but one has an initial momentum and/or higher initial temperature. Although transfer of momentum and/or energy between the gases will occur, this initial difference may eventually promote a separation of the components. The expected behavior would depend on the exact conditions and circumstances, but one should bear in mind this aspect when considering a specific situation.

When hydrogen flows and diffuses through an air–steam atmosphere in a large enclosure (such as an NPP containment), it may be considered that, for some time, an air–steam–hydrogen mixture cloud exists in some parts of the enclosure. However, on a longer time scale, in the absence of other driving forces, hydrogen will gradually separate, flow upward, and accumulate in the upper parts of the enclosure. This process is called stratification. If hydrogen has an initial momentum or a higher temperature than the surrounding atmosphere, some of the momentum or heat will of course be lost to the surrounding gas, but the separation of hydrogen will be faster.

Hydrogen Flammability and Combustion

Hydrogen is a burnable gas, which means that it reacts chemically with oxygen to form water:

$$2H_2 + O_2 \rightarrow 2H_2O$$

This chemical reaction releases energy in the form of heat. The lower heat of combustion amounts to 120 kJ per g (gram) of hydrogen.

In an accident situation in a nuclear power plant, combustion will usually occur in a premixed "cloud" consisting of hydrogen, air, usually steam, and eventually other gases. Although hydrogen is a burnable gas, it does not ensure that it will burn immediately when mixed with oxygen. One of the issues that have to be understood is the following: what are the necessary conditions for sustainable hydrogen combustion to occur?

First, combustion is a physical phenomenon that has to be started by some initiating event, that is, ignition. Theoretically, it would be possible to have a concentration of a burnable gas mixed in air or oxygen much above the threshold necessary for combustion to be sustainable, without combustion occurring. The reaction rate at ambient temperatures is negligible. However, it can be accelerated by a catalyst or by raising the temperature.

Second, even if hydrogen combustion is initiated, it needs favorable conditions to be sustained. If conditions are not favorable, combustion will die out. The main physical condition that defines the conditions in a (gas) cloud, necessary for the sustainability of hydrogen combustion, is its composition. The pressure and the temperature also influence flammability, although their influence is secondary. The ranges of species concentrations within which the cloud is burnable are called "flammability limits". These limits are usually determined by standard combustion experiments. There is a "lower flammability limit" (that is, a necessary minimum concentration of burnable gas), as well as a "higher flammability limit" (that is, a maximum concentration of burnable gas, as the mixture should also contain a sufficient amount of oxidant) [43].

Hydrogen Flame Propagation—Standing, Deflagration, and Detonation

A flame is caused by a self-propagating (with respect to the unburned gas) exothermic reaction that usually has a luminous reaction zone associated with it. When combustion of hydrogen is initiated, the flame may stay at a fixed location or may propagate. If the flame stays at a fixed location, the phenomenon is called "standing diffusion flame combustion". This may be observed for non-premixed hydrogen inventories released into air. Otherwise, and in particular for premixed situations, the flame can propagate in basically two different modes through the flammable parts of the cloud:

- If the flame propagates at subsonic velocity, the phenomenon is called "deflagration"; deflagration may be "slow" (laminar) or "fast" (mostly turbulent);
- If the flame propagates at supersonic velocity, the phenomenon is called detonation.

In the detailed structure of the flame, the temperature must increase smoothly from the initial to the final state. The product concentration will increase similarly, whereas the concentration of fuel and oxidant must show a corresponding decrease (Figure 3.2) [47].

In the topic of flame propagation, the flame speed is defined as the velocity of the flame relative to a stationary observer. The burning velocity is the velocity of the flame front with respect to the unburned (moving) gas immediately ahead of the flame. Figure 3.3 shows photographs of a laminar deflagration, a turbulent deflagration, and a quasi-detonation.

The deflagration mode of flame propagation is the most common. The velocity of the unburned gas ahead of the flame is produced by the expansion of the combustion products. The flame speed and, as a consequence, the

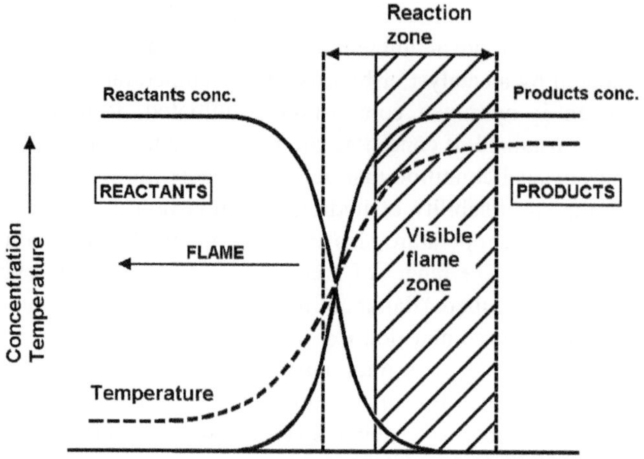

FIGURE 3.2 Concentration and temperature profiles associated with a one-dimensional, premixed, adiabatic flame.

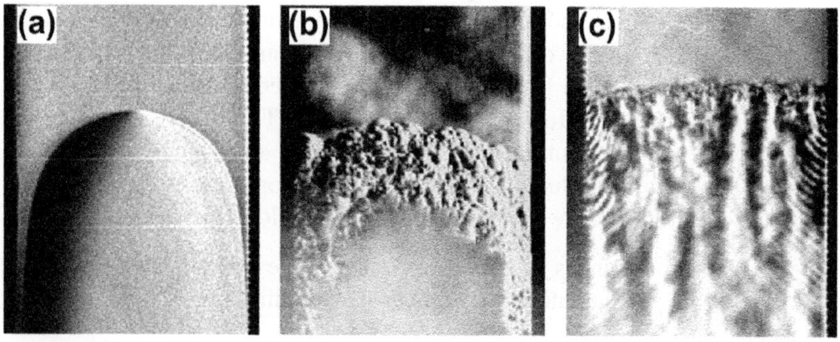

FIGURE 3.3 Combustion regimes - (a) laminar deflagration; (b) turbulent deflagration; (c) detonation.

combustion pressure will strongly depend on the mixture composition, the cloud size, and the geometrical conditions within the cloud (i.e., equipment, piping, etc.) and geometries confining the cloud (i.e., buildings, containments, etc.). If the hydrogen concentration is greater, the released combustion energy will be higher, causing a faster expansion of combustion products and thus a higher flame speed. Typical flame speeds in the deflagration mode are from the order of 1 to 1000 m/s. The pressure may reach values of several bars, depending on the flame speed.

When the cloud is ignited by a weak ignition source (such as a spark or a hot surface), the flame starts as a laminar flame, for which the basic mechanism of propagation is molecular diffusion of heat and mass. This diffusion process into the unburned gas is relatively slow and the laminar flame will propagate with a velocity of the order of 1–10 m/s. In most accidental combustions, the laminar flame will accelerate and transit into a turbulent deflagration (i.e., turbulent flame). The acceleration is usually caused by instabilities and by the flow field ahead of the flame front, which becomes more and more turbulent by the interaction with equipment, piping, structures, and so on. In this way, the flame velocity may attain values of up to 1000 m/s. For strong deflagrations, shock waves may propagate ahead of the deflagration (i.e., the flame). One of the mechanisms causing the increased combustion rate in turbulent deflagrations is the wrinkling of the flame front by large turbulent eddies. When the turbulent integral length scale is of the order of the thickness of the flame front or lower, the flame becomes a thick turbulent flame brush. In this regime, turbulence increases diffusion of heat and mass and thereby causes a high combustion rate, which in turn increases further the flow velocity and the turbulence ahead of the flame. This strong positive feedback mechanism causes flame acceleration and high explosion pressures, and, in some cases, transition to detonation.

However, turbulence induced by obstacles in the displacement flow does not necessarily enhance the combustion rate. High-intensity turbulence may also lower the overall combustion rate by excessive flame stretching and by rapid mixing of the burned products and the cold unburned mixture. If the temperature of the reaction zone is lowered to a level that can no longer sustain continuous propagation of the flame, the flame can be extinguished locally. The quenching by turbulence becomes more significant as the velocity of the unburned gas increases. This can set a limit to the positive feedback mechanism and, in some cases, lead to the total extinction of the flame. Hence both the rate of flame acceleration and the eventual outcome depend on the competing effects of turbulence on combustion.

The behavior of the turbulent deflagration also depends on the comparative magnitudes of the chemical time scales, which determine the reaction rate, and the time scales of turbulent flow. Chemical time scales may be shorter, comparable, or longer than the characteristic time scales of turbulence. However, chemical reactions introduce additional time scales but no viscous effects [52].

Formally, a detonation is defined as a combustion wave propagating at supersonic velocity relative to the unburned gas immediately ahead of the flame; that is, the velocity is greater than the speed of sound in the unburned gas. Detonation is the most devastating form of gas combustion: in a hydrogen–air cloud, a detonation wave would propagate at a velocity of 1500–2000 m/s, and the peak pressure would be typically 15 to 20 bar (but could also reach values up to 40 bar or more due to reflections and superpositioning of shock waves). An actual detonation is a three-dimensional shock wave immediately followed by a reaction zone (flame). The shock wave and the combustion wave are in this case coupled: the shock compression heats the gas and triggers the combustion. The leading shock consists of curved shock segments. At the

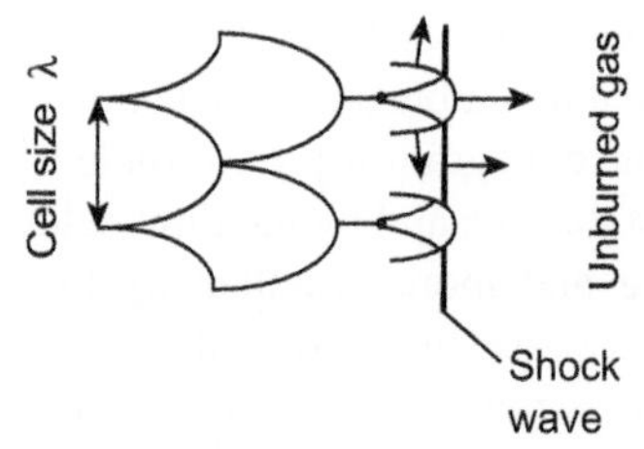

FIGURE 3.4 Pattern of shock wave structure during flame acceleration [40].

detachment lines between these shock segments, the shock wave interacts in a Mach stem configuration (Figure 3.4).

The size of the "fish shell" pattern generated by the triple point of the shock wave is a measure of the reactivity of the mixture and represents a length scale that characterises the overall chemical reaction in the wave. This length scale, λ, is often called the "cell size" or the "cell width". The more reactive the mixture is, the smaller the cell size is. The cell size is measured experimentally and there are some variations in the reported results (variations of a factor of two are not uncommon). Figure 3.5 shows an example of the dependence of the detonation cell size as a function of hydrogen concentration, obtained from experiments.

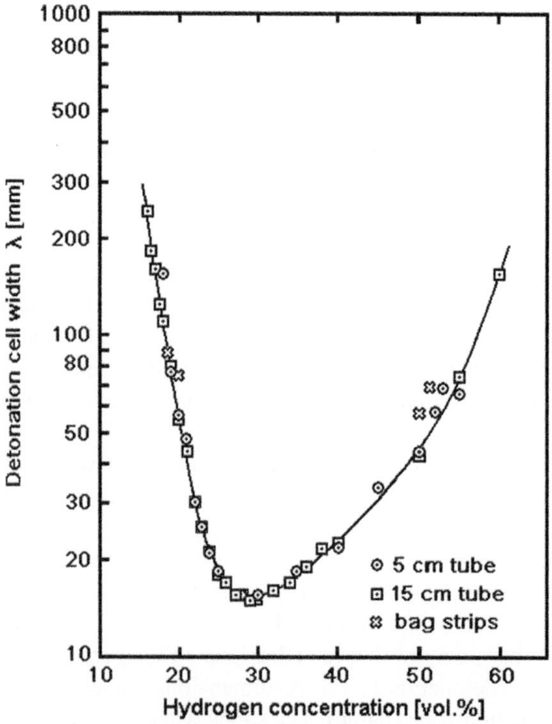

FIGURE 3.5 Experimental measurement of detonation cell size for hydrogen–air mixtures at atmospheric pressure [56].

The mechanism of transition to detonation is still not fully understood. Presently, apart from some criteria, there is no theory (in the strict sense of the term) that can reliably predict conditions for deflagration to detonation transition. To some extent, the transition from deflagration to detonation can be evaluated based on knowledge of the cell size of the combustion mixture and the characteristic size of the premixed cloud. From a practical point of view, it is important to recognize that transition to detonation will cause extremely high pressures in the area where the transition takes place.

3.1.3. Hydrogen Generation During a Severe Accident in a Light Water Reactor (LWR)

Before burning, hydrogen must be somehow produced. In this section, we will see how hydrogen is generated during a severe accident, at what time after the start of the accident different modes of hydrogen generation occur, as well as what quantities of generation may be expected [6] [7].

In principle, there are several possible sources of hydrogen in the course of a hypothetical severe accident:

- Zirconium—steam reaction (zircaloy oxidation),
- Boron carbide—steam reaction,
- Uranium—steam reaction,
- Steel—steam reaction,
- Molten corium–concrete interaction (MCCI),
- Water radiolysis,
- Corrosion of metallic compounds such as zinc, aluminium, or iron,
- Interaction between corium debris and containment atmosphere.

In severe accidents, hydrogen production is a fast process due to zirconium oxidation by steam with a magnitude from 0.1 to 5.0 kg/s (degradation and reflood of the overheated core). Because this kinetics is at least ten times higher than the other sources, the hydrogen production induced by the other sources may be neglected in a first approach. In general, the determination of the time-dependent hydrogen source for a given accident is still subject to considerable uncertainties.

In the literature, hydrogen generation is sometimes divided into in-vessel phenomena (that is, phenomena occurring in the reactor pressure vessel), and ex-vessel phenomena. The following sources will be considered here in more detail: zirconium oxidation (see also the chapter 2 of the present book) and molten core–concrete interaction.

Zirconium Oxidation

Zirconium (Zr) is the main constituent of the fuel cladding. When a reactor core is uncovered during a severe accident, the hot zircaloy (zirconium alloy) comes into contact with steam and oxidises according to the reaction:

$$Zr + 2H_2O \rightarrow ZrO_2 + 2H_2$$

Thus, hydrogen generation by Zr oxidation starts right after the core is uncovered. If the core starts degrading by melting, Zr oxidation naturally continues. The relocation of molten material following clad failure exposes new surfaces to the reaction. In the case of a typical PWR reactor core, some 150–200 kg of hydrogen may be generated in the first few hours of an accident. For large BWRs the generated hydrogen mass may be even a factor 2 to 5 higher.

If the reactor core is reflooded, the oxidation rate of zircaloy (and, consequently, the production rate of hydrogen) is increased due to the large quantities of steam that are generated. As a result, local hydrogen production rates may increase by an order of magnitude relative to the rates during the initial heating and melting of the core. Thus, a mitigation measure that is meant to prevent severe long-term consequences contributes to the increase of the hydrogen threat in the short term.

In principle, hydrogen generation during the early in-vessel phase of the accident is the phenomenon that is known best. Nevertheless, predictions of hydrogen generation for the same accident scenario can vary significantly, particularly during fuel rod destruction or displacement. The uncertainties concerning the rate of hydrogen generation generally increase with the progression of the accident, since the initial and boundary conditions for the metal–water reaction (such as temperature and surface of the metal exposed to steam, or the steam supply) as well as the physical phenomena occurring, become increasingly uncertain.

The TMI-2 accident has been analyzed in detail regarding the hydrogen that was generated during core degradation. The analysis of available data indicated that the hydrogen generation rate peaked at more than 20 kg/min shortly after 6:54 (the accident started at 4:00), when the hot reactor core was quenched. Just prior to that period, the zirconium-water reaction may have been steam-limited. Approximately 400 kg of hydrogen were generated between 6:12 and 7:00, and approximately 460 kg of hydrogen had been generated by 7:48. Similar inventories should have been produced in the Fukushima accidents.

If the core degradation continues further without successful cooling in the reactor pressure vessel, the entire reactor core will presumably melt, and the molten core will eventually accumulate in the vessel lower plenum. The oxidation of metals present in the corium pool also generates hydrogen.

Molten Corium–Concrete Interaction (MCCI)

The accumulation of the core melt in the reactor pressure vessel lower plenum may eventually lead to the rupture of the vessel and spilling of the melt in the reactor cavity. Cavities are made of concrete. Whether it is better that the cavity be flooded with water beforehand, so that the melt starts cooling faster, but at the risk of triggering a steam explosion, is still being debated and will not be considered in this section. Be that as it may, the core melt will come into contact with the concrete in the cavity, which will lead to concrete ablation. The phenomenon, which is known as MCCI, is complex, all the more so as the melt may stratify into a metallic and an oxide layer. The phenomenon of MCCI is considered in detail in chapter 4 of the present book. Basically, hydrogen production is due to the oxidation of metals by the steam released during the interaction. A significant part of hydrogen is produced during the early phase of corium–concrete interaction, while Zr is being oxidized. After depletion of Zr and its follow-on products, long-term hydrogen release during MCCI is governed by Fe oxidation with typical release rates of 2 mol/s (4 g/s). However, this release is also accompanied by a large rate of steam flow, which reduces the flammability of the atmosphere.

The following facts about MCCI are important in any consideration of the hydrogen combustion risk:

- The expected rate of hydrogen generation during early phases of MCCI is of the order of 1–2 kg/s. The amount of generated hydrogen is mostly limited by the available Zr that remains after much of it was presumably oxidized during the early in-vessel phase of the accident.
- Apart from hydrogen, carbon monoxide (CO) and carbon dioxide (CO_2) are also expected to be generated, depending on the type of concrete.
- The threat posed by the generation of gases is due not only to flammability, but also to possible pressurization of the containment, as H_2, CO, and CO_2 are noncondensable.

During a hypothetical severe accident, failure of the reactor pressure vessel, spilling of the core melt in the reactor cavity, and start of MCCI are expected to occur a few hours after the start of the accident. MCCI could theoretically go on (if measures taken are unsuccessful in cooling the melt) for a few days, when basement melt-through could occur.

3.1.4. Hydrogen Distribution in an NPP Containment

The TMI-2, Chernobyl, and Fukushima accidents demonstrated that hydrogen combustion can occur during a severe accident in a nuclear power plant. Some risk studies have shown that severe accidents can involve the release of up to 2 kg/s of hydrogen during core degradation, yielding even more than 1000 kg of hydrogen in the containment in the first 7 hours of the accident. Due to various physical mechanisms in the atmosphere (that we shall consider later), hydrogen is usually not uniformly distributed, so that high local concentrations may occur, creating favorable conditions for combustion.

The issues that are related to the occurrence of high local hydrogen concentrations, where hydrogen combustion is likely, are:

- Why is such a situation likely to occur?
- Which passive measures, that is, features that are built in the containment, may be used to prevent it?
- Which active measures, that is, systems that are either automatically or manually actuated, may be used to prevent it?

This section addresses the first issue, while the second and third issues are covered in Section 3.1.7 on hydrogen mitigation. The first issue is related to the phenomenology of multicomponent atmosphere mixing and, as a consequence, hydrogen distribution in the atmosphere. Experimental and theoretical investigations that have been and are still being carried will also be briefly described.

Mechanisms of Hydrogen Distribution

If hydrogen is distributed uniformly in the containment of a LWR, the concentration is in principle insufficient to pose a threat to the containment integrity. However, if the concentration is non-uniform, the threat of detonation or transition to detonation can exist. The important matter is whether a high local concentration of hydrogen may occur, that is, how hydrogen is distributed in the containment. This is why hydrogen distribution is important and why it is being intensely investigated, both experimentally and theoretically.

How can local conditions prone to hydrogen combustion occur? These conditions may occur due to the combination of various physical mechanisms that influence the mixing of the containment atmosphere and thus determine the spatial distribution of different species (including hydrogen) in the containment. These basic mechanisms are:

- gas flow (that may occur for various reasons),
- molecular diffusion,
- heat transfer between various containment structures (walls, pipes, platforms, stairs, tubes, other equipment, etc… mostly made of steel and concrete) and the containment atmosphere,
- mass transfer (that is, in the case of a severe accident, steam condensation).

The flow of gas in the containment atmosphere may be caused by the following driving forces: inertial forces and buoyant forces.

Gas flow caused by inertial forces may occur due to the injection of steam from the reactor cooling system, following a pipe break. The inertial flow may be the flow of the steam itself, or the flow of other gas components in the containment atmosphere, entrained by the flow of steam. At the stage of the accident when hydrogen enters the containment, the convective motion in the atmosphere should in principle already have been established.

Gas flow caused by buoyant forces occurs due to the differences in density of different components: gases with lower density flow upward. These differences in density may be due either to the intrinsic gas property (molecular mass) or to the difference in gas temperature (in which case, the density differences are expected to attenuate in the course of the gas flow). As hydrogen is a light gas, the upper containment zone, after a certain time, is expected to contain a mixture relatively rich in hydrogen. In an actual containment, the expected time scale of hydrogen accumulation in a (more or less) horizontal layer is of the order of hours.

Heat transfer occurs between the structures of the containment (walls, platforms, pipes, equipment, etc.) and the atmosphere. The temperature of the containment structures may be either higher or lower than the temperature of the atmosphere, so that structures could cause either heating or cooling of the nearby gas. The heating or cooling directly influences the density of the atmosphere, first of the adjacent gas and then of the gas further away, thus creating a contributing force for upward or downward buoyant flow.

Mass transfer in the containment during an accident in a nuclear power plant consists essentially of steam condensation. Most of the steam condensation is supposed to occur on the surface of structures. However, some experimental evidence suggests that steam condensation in the bulk of the atmosphere is also likely to occur. Be that as it may, steam condensation affects the gas transport within the containment atmosphere in the following ways:

- Steam condensation affects the local gas density, thus promoting upward or downward flow due to buoyancy (for instance, local steam condensation within a steam-air mixture causes the mixture to become heavier in comparison to the neighboring gas, thus inducing a secondary downward flow along the condensing structures that brings hydrogen from higher to lower locations);
- Steam condensation lowers the local steam concentration, thus promoting steam diffusion and Stefan flow toward regions of low steam concentration. The Stefan flow may also entrain hydrogen.

The presence of hydrogen (as well as other noncondensable gases, such as CO and CO_2) also influences steam condensation on the structures, as it increases resistance to steam diffusion.

The relative importance of the enumerated physical phenomena may be different in different accident circumstances, at different stages of the accident, as well as in different parts of the containment. For instance, in the case of a pipe break, inertial effects near the break definitely dominate the buoyant effects (not to mention molecular diffusion), at least during the very early stages of the accident. However, later on, or in some parts of the containment removed from the location of the break, buoyant effects may be dominant.

We have seen why hydrogen distribution is important and which basic phenomena influence it. As to what will be the result of the interaction of these phenomena, a general answer cannot be provided. The outcome of the interaction depends on the specific case, that is:

- Location of the hydrogen source and mass generation rate;
- Geometry of the containment, including the volume and disposition of different compartments, and the openings between them;
- Spatial arrangement of structures, as it influences the heat transfer and steam condensation;
- Influence of the hydrogen mitigation systems;

- Initial conditions (pressure, atmosphere temperature, atmosphere composition) at the start of hydrogen release into the containment;
- Boundary conditions, namely: direction and mass flow rate of the steam injection from the reactor coolant system (in the case of a pipe break), and temperature of the structures in the containment.

To gain physical insights and develop methods, which are able to answer the question of whether high local hydrogen concentration may be expected in specific situations, both experimental and theoretical investigations are being carried out.

Experimental Investigations of Hydrogen Distribution

The specific purposes of experimental investigations of hydrogen-related phenomena (both hydrogen distribution and hydrogen combustion) are:

- to develop the understanding of the investigated phenomena and their interaction,
- to develop and validate theoretical models, which could then be used for assessment of the hydrogen combustion threat in actual plants.

These assessments are usually performed within the so-called safety analyses of nuclear power plants.

So far, the major unresolved issue of experimental investigations of the hydrogen distribution and combustion is the scaling-up of experimental findings from experimental facilities to actual plants. Namely, even volumes of so-called large-scale containment experimental facilities are close to one-hundredth to one-thousandth of the volume of actual plant containments. Thus, although qualitatively similar behavior of hydrogen mixing and combustion may be expected in actual plant containments as observed in experiments, much uncertainty still remains.

Be that as it may, experiments still provide valuable insight. Experimental investigations of hydrogen distribution are in fact investigations of atmosphere mixing in a large enclosure. Experiments are performed in containment experimental facilities. Usually, initial conditions of some kind are established. These may consist either of homogeneous or non-homogeneous (usually stratified) atmosphere, consisting either of air and steam, air and helium (which is usually used instead of hydrogen for safety reasons, as it is expected to behave similarly, but is not burnable), or air, steam, and helium. Then, a gas is injected (usually steam, to simulate the flow from the primary system through a pipe break), either in a horizontal or vertical direction, either with low momentum (plume) or high momentum (jet), at specific locations (upper or lower part of the facility, centrally or not), and the ensuing evolution of the atmosphere is observed, typically over a period of several hours. The observations include measurements of pressure, atmosphere temperature (at different locations, to observe the thermal stratification), wall temperatures, atmosphere composition (also at different locations, to observe the non-homogeneity of the atmosphere), and flow velocities (to observe the gas circulation pattern). The purpose of these investigations is to find out how a "cloud" of hydrogen that would be released in a containment during an accident would behave under various circumstances and influences.

The description of methods used in experiments on atmosphere mixing is beyond the scope of this section and will not be presented here. Readers may find information on that topic in the References listed at the end of the chapter.

The most important experimental containment facilities that have been employed to date may be roughly divided into three groups:

- Recently employed single-compartment facilities: TOSQAN (IRSN) and MISTRA (CEA) before the modification into a simple multicompartment facility,
- Simple multicompartment facilities: MISTRA (after the modification), PANDA (PSI), THAI (Becker Technologies GmbH, Germany),
- Complex multicompartment facilities: Heissdampf Reaktor or HDR (Kahlstein, Germany), Battelle Model Containment or BMC (Battelle Ingenieurtechnik, Germany), NUPEC (Japan).

Although one would expect the opposite, first experiments on hydrogen distribution were performed in complex multicompartment facilities. The state of development of experimental methods at the time made it difficult to observe experimentally a non-homogeneous atmosphere in a single compartment. That is, these older facilities were not equipped with experimental devices that would allow detailed measurements on the local scale (for example, temperature or species concentration measurements at different locations over a large volume). However, these experiments allowed the observation of the evolution of the atmosphere composition in a network of compartments, which was the primary purpose.

With the development of measurement techniques, the attention focused later on the observation of non-homogeneous atmosphere in single-compartment facilities. These experiments allowed elimination of the influence of flow obstruction on mixing, thus making it possible to dissociate convective and diffusive mixing from other influences.

The next step in complexity concerns experiments in simple multicompartment facilities. Actually, these facilities have little similarity to actual containments, but are designed in such a way that phenomena expected in an actual plant will occur during experiments, including the interaction (friction, heat transfer) between the atmosphere and the structures. The benefits of the experiments in these facilities are the following:

- The containment is subdivided into compartments, so that the conditions are more realistic than in a single-compartment facility,

- The relatively simple arrangements make it easier to dissociate physical causes and effects, so that the basic mechanisms of atmosphere mixing and stratification are easier to understand.

The main characteristic of newer experimental programs (both single and multicompartment) is that they are being carried out in smaller facilities than the above-mentioned older facilities, but with more sophisticated measurement methods, which allow the observation of variables at a local scale in relatively large compartments—that is, the characterization of a non-homogeneous atmosphere. Experimental observations include measurements of local temperature, local species concentrations, and sometimes local flow velocities. Of course, intrusive methods that are sometimes employed disturb the flow in the atmosphere, causing perhaps some minor influence on the atmosphere mixing. Nevertheless, even the smaller facilities are large enough for the influence of measurement devices on the observed phenomena to be considered as negligible.

The remaining part of this section briefly describes the above-mentioned experimental facilities. However, a general overview of all experimental results is beyond the scope of this section. The interested reader may find further information in the References at the end of the chapter.

The TOSQAN Experimental Facility

The TOSQAN facility (Figure 3.6) consists of a cylindrical steel vessel with an internal volume of 7 m^3 (not including the sump) [10] [21] [28]. The diameter of the main cylindrical part is 1.50 m, whereas the overall height is 4.80 m. The wall temperature may be controlled separately in different parts, so that the temperature may be set to different values to either promote or prevent steam condensation on different parts of the wall. The TOSQAN vessel was first used to study the mixing and stratification of air–steam and air–steam–helium atmosphere following the injection of steam, and later the influence of containment sprays on atmosphere depressurization and mixing.

As an illustration of experiments, tests TOSQAN 101 and TOSQAN 113 were performed to investigate the effects of containment sprays on atmosphere depressurization and mixing. In test 101, the vessel was filled with superheated steam and the vessel walls were kept at a sufficiently high temperature to prevent steam condensation. Sprays were then actuated and the atmosphere depressurization was observed. In test 113, helium was injected just below the top of the vessel to create a horizontal layer. Sprays were then actuated and the gradual atmosphere homogenization was observed.

The MISTRA Experimental Facility

The MISTRA facility (Figure 3.7) is a large cylindrical vessel, with a volume of 98 m^3. The diameter of the main cylindrical part is 4.25 m, whereas the overall

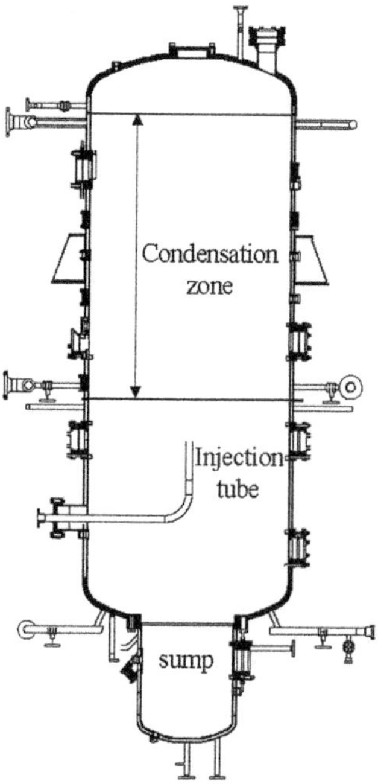

FIGURE 3.6 Vertical cross section of the TOSQAN experimental facility (courtesy of IRSN).

height is 7.0 m [10] [31]. The containment is thermally isolated and equipped with three internal condensers which are thermoregulated. The condensers have the shape of cylindrical walls (without the base plane faces) and are located one above the other, concentric to the containment walls, with a narrow gap between the internal surface of the containment wall and the external surface of the condensers. The approximate height of the condensers is 1.5 m, so that the three condensers together cover roughly the upper 5 m of the internal cylindrical wall. The separate regulation of the condensers' temperature enables the setting of values that will either promote or prevent steam condensation, or cool or heat the atmosphere and thus promote natural circulation. The gap between the containment wall and the condensers is sufficiently wide so that steam condensation also occurs there. This gap is a somewhat awkward feature of the facility, as it does not correspond to the arrangement in an actual containment.

The MISTRA facility was originally a single-compartment vessel. It was later transformed into a multicompartment vessel by installing a vertical internal cylinder, closed at the bottom, and a dividing horizontal annular plate at the level between the lowest and the middle condenser, in the outer space between the newly installed inner cylinder and the vessel wall. At the inner edge, the plate is welded to the cylinder, but the outer edge does not extend right to the wall, so that the atmosphere may circulate through a narrow annular gap.

FIGURE 3.7 The MISTRA experimental facility (courtesy of CEA).

While the vessel had a single compartment, the MISTRA facility was used to investigate the non-homogeneous air–steam and air–steam–helium atmosphere. Experiments on containment sprays were also performed. After the vessel was compartmentalized, experiments on atmosphere mixing, influence of steam plume and jet, natural convection, and containment sprays were performed again. As an illustration, in some tests, a helium layer in the upper part of the vessel was established first. Then, the influence of steam plumes and jets with different momentum on the helium layer was observed to assess the possibilities of breaking up eventual atmosphere stratification during an accident in a real nuclear power plant and thus reduce the local hydrogen concentration.

The PANDA Experimental Facility

PANDA is a large-scale thermal-hydraulics test facility designed and used for investigating containment system behavior and phenomena for different, mainly BWR, designs and for large-scale separate-effects tests [12] [25] [26] [27] [34]. The containment compartments and the reactor pressure vessel are simulated by six cylindrical pressure vessels. In addition, the facility includes four water pools with four condensers. The temperature of the walls is not controlled. The height of the facility is 25 m, and the total volume of the vessels is 460 m^3. As the facility is flexible, tests on hydrogen (represented by helium) distribution were performed employing only the two upper cylinders (with a total volume of 183 m^3), which represent the containment compartments. These cylinders are interconnected by a pipe of about 1 m diameter. Similar as in the MISTRA facility, in some tests, a helium layer in the upper part of one of the cylinders was established first. Then, the influence of various effects, such as vertical or horizontal steam jets or buoyant flows created by heating or cooling devices, on the helium layer was observed.

The THAI Experimental Facility

The THAI experimental facility (Figure 3.8) is located at Becker Technologies GmbH in Eschborn, Germany [10]. The main component of the facility is

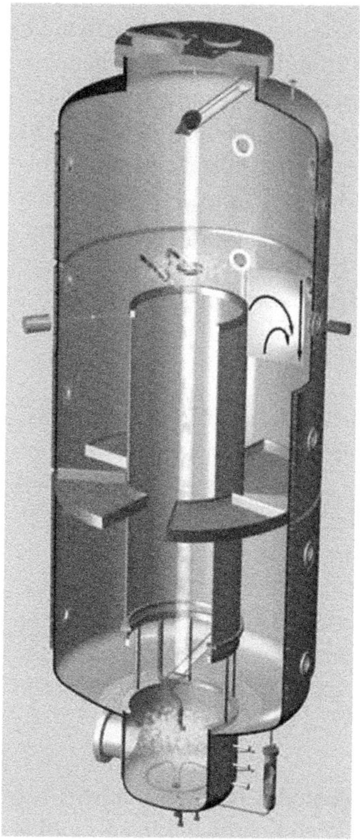

FIGURE 3.8 View of the THAI experimental facility (courtesy of Becker Technologies).

a cylindrical steel vessel with a total volume of 60 m^3, of 9.2 m total height (including the sump compartment at the lower end), and of 3.2 m diameter. The outside walls of the containment are thermally insulated. The vessel space is subdivided by an open inner cylinder and a horizontal separation plane in the annular region between the wall and the inner cylinder, with vent openings. The separation plane is located a little below the middle level and consists of four equally spaced condensate-collecting trays that span from the inner cylinder wall to the vessel wall. Each tray covers one-sixth of the annular cross section. The outer cylindrical wall has oil cooling / heating jackets. As an illustration of experiments, in some tests, a helium layer in the upper part of the vessel was established first, and the influence of steam jets at different locations on the helium layer was observed. Experiments on hydrogen combustion were also performed (see the next section).

The Heissdampf Reaktor

The containment of the Heissdampf Reaktor (HDR) integral experimental facility had a circular cross section with a hemispherical dome [32]. The containment was 60 m high with an approximate free volume of 11,300 m^3. The containment's lower and middle parts were divided into 70 compartments, whereas the large upper dome (approximately 4800 m^3) was not subdivided. Two diametrically opposite staircases (spiral and straight) represented the main vertical gas flow paths in the containment. External sprays were located in the dome upper part, between the outer concrete wall and the inner containment steel shell. The sprays were designed to wet the external surface of the hemispherical part of the steel shell. The facility was used to perform experiments on hydrogen (represented by helium) distribution, as well as hydrogen combustion. The most widely known test is the experiment E11.2 on helium distribution, which was used later for the OECD International Standard Problem No. 29.

The NUPEC Experimental Facility

The NUPEC experimental facility was designed to perform hydrogen distribution tests (using helium) [8] [29] [30]. It represents a linearly 1:4 scaled-down Westinghouse standard containment, with a free volume of about 1300 m^3 (height: 17.4 m, inner diameter: 10.8 m). The vessel is partitioned into 25 compartments, connected between them with 66 flow paths. The facility is divided into three main floors. The gap between the first floor and the containment wall is sealed, whereas there is a narrow cylindrical gap between the outer edges of the other floors and the containment wall. Apart from the usual features, the NUPEC facility also has a containment spray supply system.

The Battelle Model Containment

The Battelle Model Containment (BMC) represented a linear 1:4 downsized model of a German PWR containment [33]. The compartmentalized facility, which was built of unlined concrete, had a variable geometry (which means that some openings between compartments could be closed for particular experiments) and a total volume of about 640 m^3 (diameter: 12 m, height: 10 m). It consisted of nine rooms, but some of them were closed for some experiments. The main compartments were the inner cylindrical room and the outer ring, both open at the top to the containment dome area. Among other experiments, the BMC facility was used to perform tests on hydrogen distribution, deflagration, and mitigation.

Theoretical Investigations of Hydrogen Distribution

As already mentioned, the distribution of hydrogen is also being investigated theoretically. Basically, the purposes of the theoretical investigations are:

- to understand the mechanisms of hydrogen distribution from basic physical principles, that is, from the laws governing mass, momentum, and energy transport,
- to develop theoretical models, which will be able to predict the hydrogen distribution in the containment (that is, the time and space dependent composition of the atmosphere) of an actual plant during a severe accident.

In the field of nuclear engineering research, these investigations consist mostly in simulating experiments that were performed in containment facilities (some of them are described just above) with computer codes. The differences between experimental and theoretical results are then analyzed (one should note that experimental results are almost always considered as being correct, allowing only for the uncertainties of experimental methods), in order to:

- detect deficiencies of the models in the code that are the main cause of the discrepancies, and thus the incorrect physical assumptions on which the models may be based,
- correct the models.

Sometimes, simulations are performed for accident scenarios in actual plants, for which there are no corresponding experimental results. They are applications of developed methodologies. These applications are not only valuable, but in fact they are the essential purpose of the methodologies (codes) developed.

The models used in computer codes may use different levels of description; that is, they may solve the transport equations of fluid mechanics at different spatial scales. In the field of containment atmosphere modeling, two types of codes are basically used: Lumped-Parameter (LP) codes and Computational Fluid Dynamics (CFD) codes.

The present section not only compares different LP and CFD codes, but also presents their basic principles and acquaints the reader with different methodologies that are available. Although the most widely used codes are mentioned, the reader should consult the relevant literature (i.e., mostly code user manuals) for further details. In particular, differences between results obtained with LP and CFD codes are well illustrated in OECD International Standard Problem N°47 (2002–2005) [10].

Modeling with Lumped-Parameter Codes

Lumped-Parameter (LP) codes describe an NPP containment as a network of control volumes (sometimes called "cells" or "zones"), connected with junctions (sometimes called "vents" or "flow paths"). The number of control volumes in LP containment simulations typically ranges from 10 to 100. Control volumes contain so-called heat structures: these include all the structures such as walls, pipes, platforms, and equipment. These structures are referred to as heat structures because in the modeling of heat and mass transfer in LP codes, they act basically as heating or cooling reservoirs and provide surfaces for steam condensation. In each control volume, conditions are modeled as homogeneous, so that they are described by single values of pressure, temperature, species concentrations, and other variables. Since only scalar equations are solved in LP models, control volumes are repositories of mass and energy, but do not contain information of momentum direction.

Junctions between control volumes connect fictive "control points" within each volume. Junctions are not repositories of mass or energy, and do not contain heat structures. The mass flow rate in junctions is determined using various forms of one-dimensional momentum equations. Generally in LP codes, connections between control points are only modeled via a single flow path, so that no simultaneous counterflow is possible. Turbulence effects are not modeled at all, but its influence might be taken into account via prescribed flow-loss coefficients in the momentum equation.

LP codes were devised basically to enable predictions of so-called thermal-hydraulic conditions in the containment (that is, pressure, atmosphere temperature, and temperatures of structures) at accident conditions. In principle, each control volume should correspond to an actual containment compartment. Thus, the volumes of these control volumes could be from a few cubic meters up to a few hundred or even thousand cubic meters.

LP codes are also sometimes used to model the non-homogeneous atmosphere in large compartments (e.g., in the dome region of a large containment). This is achieved by subdividing large volumes into smaller control volumes and connecting them with "fictive" junctions (fictive, because they do not exist as such in the containment). The cross-sectional area of these junctions is equal to the area of the dividing surface between the connected volumes. The direction of the flow is implicitly defined by the spatial arrangement of the zero-dimensional control points. That is, the convection paths are predetermined, or at least restricted, by the user due to the chosen subdivision of compartments into control volumes (this subdivision is usually referred to as "nodalization").

The following LP codes are employed primarily in the modeling of hydrogen distribution (see the chapter 8 for more details on ASTEC, MAAP and MELCOR codes):

- ASTEC (i.e., the CPA module) (IRSN-GRS),
- COCOSYS (GRS) [80],
- CONTAIN (SNL, USA) [81] [82], with the last version CONTAIN 2.0 released in 1997;
- MAAP (i.e., the module that deals with the containment) (Fauske and Associates, USA);
- MELCOR (i.e., the module that deals with the containment) (SNL, USA),
- TONUS-LP (IRSN-CEA).

The reader is referred to the References section at the end of the chapter for more details on the results obtained so far with LP codes.

Modeling with Computational Fluid Dynamics codes

Computational Fluid Dynamics (CFD) codes is a term that has become commonly used for computer codes that numerically solve the transport equations of fluid mechanics (continuity, momentum, and energy), using a local instantaneous description. In current CFD codes, the influence of turbulence is usually modeled with the k-ε model. With the current computers, 100,000 to several million cells with a similar number of discrete points can be used nowadays in a CFD calculation to obtain the actual solution.

The main advantages of CFD codes are as follows:

- They are applicable to a wider range of conditions than LP codes, that is, also to conditions where inertial effects play an important role.
- Providing that an adequate numerical grid and other numerical parameters are used, the solution should not (in principle) depend on the user space subdivision (that is, on the numerical grid) and other settings strongly relying on the user experience.

The main drawbacks of CFD codes are:

- They require long computation times, as sufficiently fine numerical grids (hundreds of thousands of cells) may be required for simulations in containments.
- The development and testing of complex input models, with complicated numerical grids, may require long preparation times.

Besides, due to the actual state of containments, numerical grids, even very complex ones, still represent idealized models. Thus, approximations are involved.

CFD codes that are being used for simulating containment phenomena may be divided into two groups: multipurpose codes and codes specifically developed for simulations of phenomena in NPP containments.

The most well-known multipurpose CFD codes that have been used so far to simulate atmosphere mixing and stratification in a containment ([11] [13] [14] [16] [17] [18] [19] [22] [33]) are CFX (ANSYS Inc. in the USA), STAR-CD (CD-ADAPCO in UK), and FLUENT (ANSYS Inc. since 2004).

Codes that are being specifically developed for applications to atmosphere behavior in a containment usually include models of phenomena that are expected to occur during accidents and models for efficiently simulating the special equipment of a NPP containment like sprays, recombiners, and rupture discs. In addition, these codes have usually been validated using relevant experiments, so one may use them with more confidence for simulating containment phenomena. Besides, as they are meant to be used for simulations in large volumes, they may be devised for coarse numerical grids, with subgrid phenomena taken into account with adequate constitutive relations.

The most well-known containment-specific CFD codes are:

- GASFLOW, which was first developed at Los Alamos National Laboratory (USA) and is now being developed at KIT (Germany) [83],
- GOTHIC, developed by Numerical Applications Inc. (USA), now owned by EPRI (USA) [84] [85] [86],
- HYKA3D (SNU, Korea) [87],
- NEPTUNE-CFD, developed by EDF (France) [23],
- TONUS-3D (IRSN-CEA) [88].

It should be pointed out that GOTHIC is not strictly a CFD code, as it uses a relatively coarse grid. Thus it allows the simulation of atmosphere mixing in large volumes, but still solves the equations of fluid mechanics on a smaller length scale than LP codes.

The main drawback of containment-specific CFD codes is that they have not attained the level of development (in terms of user-friendliness) where they could be used confidently without support from the code developers. As they are not commercially viable, their development depends mostly on the resources of the developing research organizations and funding of research agencies. Due to limited funds, the development of these codes usually focuses on the adequate simulation of commonly expected physical phenomena, and they are not being tested on a wide range of possible scenarios.

The reader is referred to the References section at the end of the chapter for more details about the results obtained so far with CFD codes.

To conclude this section on theoretical modeling of hydrogen distribution, it should be emphasized that code validation is a continuous task. The ability of codes to predict complex containment phenomena that govern hydrogen distribution is assessed time and again with data obtained from new experiments. Internal and international benchmark exercises are often organized with this purpose in mind.

3.1.5. Hydrogen Combustion in an NPP Containment

In Section 3.1.2, the fundamentals of hydrogen combustion were described. In this section, we discuss the specifics of hydrogen combustion in an NPP containment.

Many general facts concerning experimental and theoretical investigations, different experimental facilities, and different computer codes are stated in the previous section on hydrogen mixing, so that they are not repeated in this section.

Specific Considerations for Hydrogen Combustion in NPPs

Flammability

Hydrogen combustion will presumably occur in the containment atmosphere during a severe accident if the following conditions are satisfied:

- The cloud, that is, the gaseous mixture containing hydrogen, must be flammable, with adequate physical conditions (hydrogen concentration in the mixture, pressure, temperature) for sustainable hydrogen combustion.
- The gaseous mixture must be ignited.

Once hydrogen is ignited, the combustion process is basically uncontrollable until all the flammable hydrogen mixtures are consummed.

In the TMI-2 accident (which is mentioned at the beginning of this chapter), the hydrogen burn occurred throughout essentially all the containment during a period of approximately 12 seconds. According to analyses performed after the accident, less than 5% of the combustion took place in the first 6 seconds, less than 40% during the next 3 seconds, and more than half of the combustion occurred during the last 3 seconds. There was no detonation. Even though the hot burned gas was losing heat to the unburned gas and surrounding surfaces after it burned, compressive heating was dominant and significantly increased the gas temperature until the pressure peaked.

The sustainability of hydrogen combustion is in general inferred from the well-known Shapiro diagram (Figure 3.9) [53]. In this triangular diagram, the three sides represent the volumetric concentrations of air, steam, and hydrogen, which are likely to be the dominant components of the atmosphere of an NPP containment during a severe accident (at least before the occurrence of molten core–concrete interaction, which is expected to generate large quantities of CO and CO_2 as well). The Shapiro diagram was determined empirically. The diagram indicates the region in which hydrogen combustion is sustainable, as well as the limits between different modes of hydrogen combustion. The limits in the diagram are not intrinsic characteristics of the gas mixture; in principle, they are only valid for the geometry in which they were obtained. However, the diagram is being widely used for various geometries, as it provides basic information about the likelihood of hydrogen combustion. The curve "BURN LIMIT" indicates the limits of the region in which sustainable hydrogen combustion would occur. A smaller region within this region is indicated by the

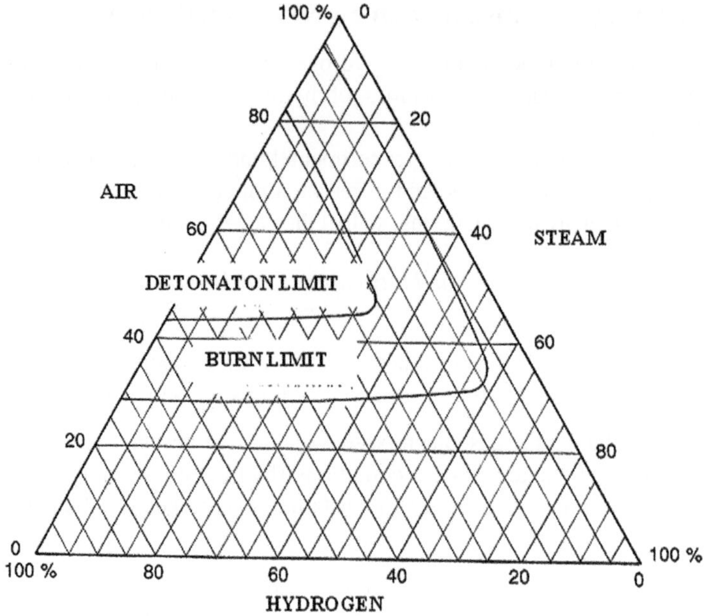

FIGURE 3.9 Shapiro diagram: detonation and burn limit for hydrogen–air–steam mixtures [53].

curve "DETONATION LIMIT." Within this region, hydrogen detonation could occur. The limits are also dependent, though weakly, on temperature and pressure. As might have been expected, there are minimum threshold concentrations of hydrogen and air necessary for combustion to occur. Steam acts as an inerting gas.

As an interesting fact, the hydrogen concentration in the TMI-2 containment before the occurrence of the combustion probably averaged just below 8 vol.% (according to analyses performed after the accident).

As already mentioned, the Shapiro diagram is not strictly valid because of many remaining uncertainties: apart from the uncertainties of the concentrations, there are also uncertainties pertaining to the geometrical characteristics—dimensions, shape—of the vessel or enclosure in which combustion is supposed to occur. Nevertheless, the diagram provides useful information about the likelihood of hydrogen combustion and the prevailing combustion regimes.

Ignition

Ignition sources, which may initiate hydrogen combustion, can be divided into random and intentional ignition events. Random ignition sources are sparks, flames, and hot surfaces. Intentional ignition sources are hydrogen igniters (see discussion about hydrogen mitigation). Although accidental ignition is a random event (therefore not certain), past experience with industrial accidents has shown that, conservatively, the presence of an ignition source should be assumed when performing risk analyses and safety assessments.

The differentiation of weak and strong iginition refers to the ignition energy and timing. Strong ignition, like a primary explosion, usually not present in a containment, may immediately initiate detonation, whereas weak ignition will kick off a deflagration (see below the subsection "Combustion modes").

A large number of potential ignition sources can be identified during a severe accident in a nuclear power plant, as, for example, electrical systems, bursting pipes, or hot core melt particles. The controlling factor for occurrence of hydrogen combustion is then the flammability of the gaseous mixture.

Combustion Modes

As mentioned earlier, hydrogen combustion may occur in different modes, be it as a standing flame, or a moving flame with different modes (deflagration or detonation). If oxygen and ignition sources are present in the vicinity of the hydrogen release (e.g., near the pipe break where hydrogen generated in the reactor core flows out), then hydrogen will ignite and could burn as a standing flame at the release location. Standing diffusion flames are capable of exerting thermal loads on the containment via three different heat transfer mechanisms:

- Direct contact with the structures,
- Heat transfer via radiation (mainly infrared radiation on steam),
- Convection of hot combustion products followed by gas/structure heat transfer.

After a weak ignition, which is the most likely ignition in an accidental scenario, hydrogen combustion is expected to start as a slow, quasi-laminar, premixed hydrogen–air–steam deflagration. Slow combustion regimes do not endanger the containment in terms of pressure loads, but rather by temperature loads. Combustion will propagate preferably in the direction with the highest flame speed, which means generally toward richer and/or dryer mixtures and into regions with high turbulence.

In a flammable mixture, in particular when the mixture is close to stoichiometric conditions, combustion may be initiated by an energy source of a few mJ. Consequently, in the presence of electrical power sources, it appears probable that ignition would occur rapidly once the flammable cloud reaches the point of ignition. If hydrogen concentration is below 10 vol.%, a much more powerful energy source (at least 100 kJ) is required to trigger a stable detonation immediately. This explains why direct detonation in an NPP containment can be ruled out for practical purposes. Thus, the only mechanism considered likely to provoke detonation is flame acceleration and deflagration-to-detonation transition.

Geometrical Considerations

If the gas mixture cloud is truly unconfined and unobstructed (i.e., no equipment or other structures are engulfed by the cloud), the flame is not likely to accelerate to velocities of more than 20–25 m/s, and the overpressure will be negligible. However, this is usually not the case of combustion in an NPP containment. First, although these are large enclosures and much different from vessels or tubes, they still act as confinements (just notice that a homogeneous cloud of 1 kg hydrogen mixed with air to give a 10% volumetric concentration will fill a volume of more than 100 m^3 at atmospheric pressure and 20 °C). Second, the flame is likely to be obstructed by some equipment (such as tubes or platforms).

In fact, the geometry of the combustion volume is the most important and the most complex parameter for flame acceleration in an NPP containment. The geometry of a single combustion compartment and the arrangement of a multicompartment containment will influence the flame acceleration. The three main parameters may be summarized as size of obstacles, distance between two successive obstacles, and degree of confinement (that is, geometrical discontinuities on the combustion path). In an actual NPP geometry, data such as blockage ratio (size of obstacles) or distance between successive obstacles cannot always be defined because of their complexity. The main characteristics of such geometry are the very irregular arrangements compared to well-defined experimental conditions. Thus, in real geometries, contrary to many experiments, combustion processes are two or three dimensional.

Influence of Containment Sprays

Containment sprays are engineered safety devices that spray liquid water in the containment atmosphere in the form of droplets. The main purpose of containment sprays is to decrease the containment pressure by causing steam condensation on spray droplets.

Liquid water is sprayed through nozzles, located in the upper region of the containment dome. The characteristics of nozzles determine the size of the droplets. If the liquid is distributed in smaller droplets, then the available interface area between the steam in the atmosphere and the water is larger, which increases the condensation rate. The size of droplets also influences their velocity. Slower droplets will promote condensation, as they will be longer in contact with the containment atmosphere. Faster droplets will promote atmosphere mixing through more intensive momentum transfer (i.e., entrainment) from the droplets to the surroundings. Usually, the nozzle vendors provide the expected size distribution of droplets.

The mechanisms of condensation and entrainment have the opposite consequences on hydrogen combustion. On one hand, the atmosphere mixing due to the entrainment by falling droplets (that is, momentum transfer from the droplets to the atmosphere) causes a more uniform hydrogen concentration in the containment. On the other hand, steam condensation on droplets causes a decrease of steam concentration in the atmosphere, and a corresponding increase of hydrogen concentration.

Hydrogen Concentration Gradient

Apart from obstacles, another physical condition, which apparently influences flame acceleration and is likely to occur in an NPP containment, is the hydrogen concentration gradient. Namely, due to the mode of release of hydrogen in the containment and the interaction of different phenomena in the containment atmosphere (inertial or buoyant flow within the containment, heat transfer between atmosphere and structures that influences buoyancy effects, steam condensation that acts as an atmosphere sink), hydrogen concentration within flammable mixture clouds is likely to be non-homogeneous. Due to various circumstances, the hydrogen concentration gradient along the flame propagation direction may be positive as well as negative. Recently, experimental investigations were performed on flame propagation in atmospheres with hydrogen concentration gradients. However, no definitive conclusions may be stated yet. Besides, some of these experiments involved flame propagation in tubes, so that the observed phenomena are mostly one-dimensional, and the application of experimental findings to actual NPP containments remains to be established. Of course, the application of these findings to actual NPP containments would also require a reliable prediction of the non-homogeneous containment atmosphere, which is another open issue (see Section 3.1.4 on hydrogen distribution).

Flame Acceleration

Under the effect of hydrodynamic instabilities, acceleration and turbulence (caused primarily by obstacles in the flame's path), as well as due to a hydrogen concentration gradient, an initially laminar deflagration may accelerate. This feedback mechanism can cause a transition from a slow laminar to a fast turbulent deflagration.

If the flame accelerates during hydrogen deflagration and sufficient space and time for acceleration are available, then combustion may very well evolve into detonation. From the practical point of view, the main quantitative difference between deflagration and detonation is the order of magnitude of the resulting pressure increase, which is much higher in the latter case and could very well threaten the integrity of the containment. Thus, in nuclear safety, hydrogen detonation is a phenomenon that should definitely be prevented.

Two transition criteria are used to determine the expected combustion mode:

- criterion for flame acceleration (slow to fast deflagration);
- criterion for transition from deflagration to detonation.

These transition criteria, which are described next, are not based on elementary theoretical treatments of the combustion process, which are described later (see page 222). They are merely "ad-hoc" or empirical criteria to determine the expected mode of hydrogen combustion. These criteria are commonly applied nowadays, but could very well be replaced by more elaborate criteria in the future.

Criterion for Flame Acceleration

For a conservative estimation of the transition from slow to fast deflagration, the so-called σ criterion is commonly used. This criterion was derived from experiments with different hydrogen-oxygen-diluents mixtures in obstructed tubes of different scales. The expansion ratio σ of the mixture is defined as the ratio of the density of unburned mixture to the density of burned mixture at constant pressure. It is an intrinsic continuous property of the mixture.

A flame is considered unstable if thermal diffusion may overcome hydrogen diffusion and lead to quenching. First of all, the combination $\beta(Le - 1)$ is a defining parameter for thermal-diffusion flame instability, where:

- $\beta = E_a(T_b - T_u)/(RT_b^2)$: Zeldovich number (non-dimensional measure of the temperature sensitivity of the reaction rate),
- Le: Lewis number (ratio of thermal to molecular diffusivity),
  where E_a is effective activation energy, T_u initial flame temperature, and T_b maximum flame temperature.

The stability boundary corresponds to $\beta(Le - 1) = -2$. Flames are stable with $\beta(Le - 1) > -2$ and unstable with $\beta(Le - 1) < -2$. The necessary (but not sufficient) conditions for development of a fast combustion regime are then:

- $\sigma > (3.5–4)$, for mixtures with $\beta(Le - 1) > -2$
- $\sigma > \sigma^*(\beta)$, for mixtures with $\beta(Le - 1) < -2$

The function $\sigma^*(\beta)$ is represented as a limiting boundary between experimental results corresponding to different combustion regimes (Figure 3.10).

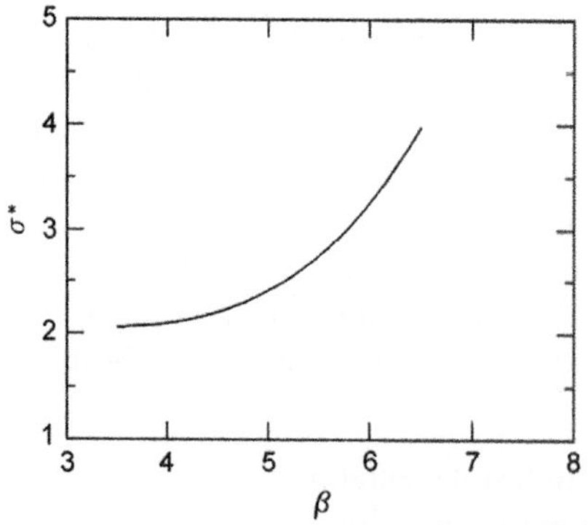

FIGURE 3.10 Function $\sigma^*(b)$ determined from experiments, valid for $\beta(Le - 1) < -2$ – a fast combustion regime may develop if $\sigma > \sigma^*(b)$ [36].

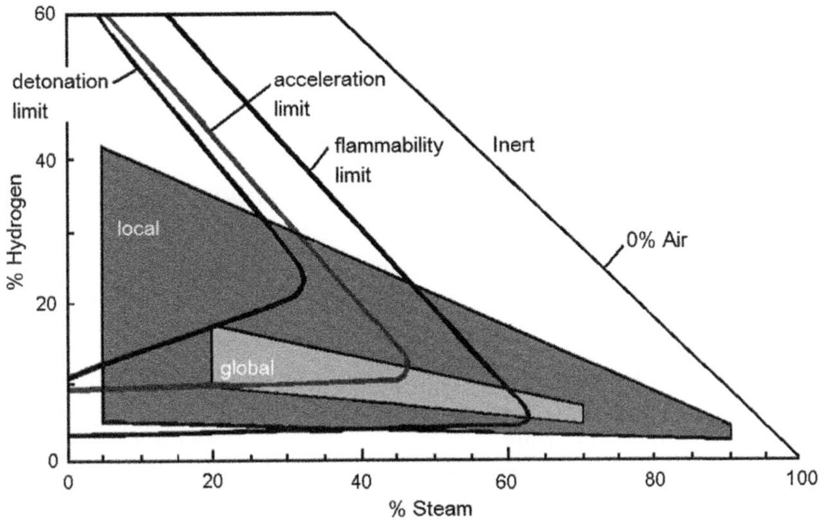

FIGURE 3.11 Example of combustion regime limits (determined from experiments in HDR facility) [36].

These conditions are based on the results of numerous experiments at various scales and in various geometries. However, other requirements should be met as well so that flame propagation can result in formation of fast combustion regimes. The most important are the requirements of a large-enough scale (flame propagation distance) and favorable geometry (obstructions) for effective flame acceleration.

To calculate β, the values of effective activation energy E_a and T_b in addition to T_u are required. Thermodynamic calculations provide data on T_b for each particular mixture (equilibrium temperature of combustion products at constant pressure). The effective activation energy E_a can be estimated from dependence of laminar flame speed on T_b. For hydrogen-lean mixtures, such estimates give an average value of $E_a/R = 9800$ K, and for hydrogen-rich mixtures, $E_a/R = 17700$ K.

Limits between different combustion regimes may also be determined from experiments, as shown on Figure 3.11.

Criterion for Deflagration to Detonation Transition

For predicting the deflagration to detonation transition (DDT), the so-called λ-criterion is commonly used: it may be expressed as $D/\lambda > 7$, where D is the characteristic geometrical size of the reacting mixture and λ is the detonation cell size of the average mixture composition (see Section 3.1.2 and Figure 3.5). The overall analysis of experimental data shows that the detonation onset requires a certain minimum scale of the reacting cloud.

These criteria for flame acceleration and deflagration-to-detonation transition, together with the analysis of hydrogen distribution in the containment building and the facility's geometry, can be used to identify potentially dangerous situations during an accident in an actual NPP.

Experimental Investigations of Hydrogen Combustion in Containment

Similarly, as stated before for experimental investigations of hydrogen distribution, the specific purposes of experimental investigations of hydrogen combustion are:

- to develop a phenomenological understanding of the hydrogen combustion processes, which may be used for qualitative prediction of hydrogen combustion in actual plants,
- to develop and validate theoretical models, which may then be used for quantitative assessment of the consequences of hydrogen combustion in actual plants (that is, calculate the expected pressure and temperature increase).

Unlike the atmosphere behavior in large enclosures (which includes hydrogen distribution in the containment), which is not a widely researched topic, the investigations of gas combustion form a wide research field of their own. However, in the present section, the description of investigations, both experimental and theoretical, will be limited to those studies pertaining to hydrogen combustion in NPP containments.

When performing an experiment on hydrogen combustion, a mixed atmosphere (homogeneous or not, usually composed of air and hydrogen, or air, steam and hydrogen) is usually established in a closed vessel. Then, the mixture is ignited, and the resulting hydrogen combustion is observed. The measured physical variables, as well as the number of measuring locations, depend on the experimental facility. Usually, pressure and temperature are measured. The flame propagation may be inferred from temperature measurements or visual observations through windows. A specific feature of these experiments is that they typically last at most a few seconds.

Due to the complexity and to the safety requirements of hydrogen combustion experiments, many experiments are being carried out in facilities that have little resemblance to actual containments (such as, for example, wide tubes). Nevertheless, such experimental results are still valuable for understanding the phenomenology of hydrogen combustion. On the other hand, some of the experiments are indeed being performed in vessels of containment-like shape. These experiments are likely to produce more realistic results (that is, results that would match more closely what could be expected in an actual plant).

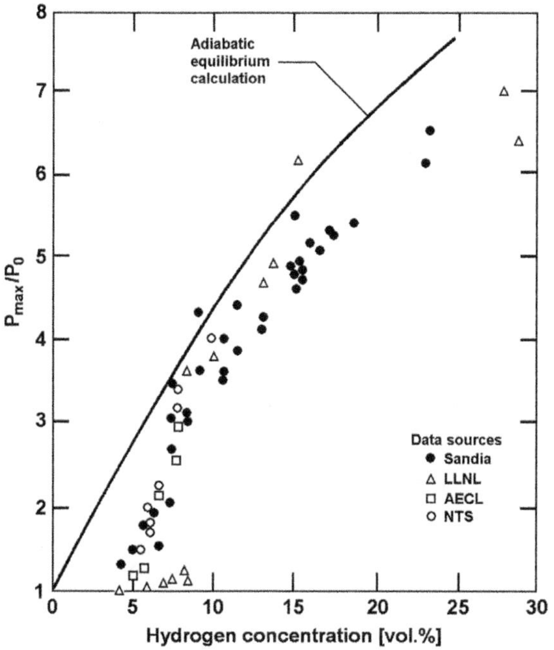

FIGURE 3.12 Peak combustion pressures, measured during large-/small-scale experiments [56].

Hydrogen combustion experiments were started in the United States, with single and multicompartment facilities [5]. The main findings of these experiments were that burning of lean mixtures was found to occur somewhat more readily in large scale than small scale, and multiple pressure spikes have not been observed under any conditions involving continuous hydrogen injection with preactivated igniters. Figure 3.12, which shows the measured change in pressure ratio (peak pressure vs. initial pressure) with hydrogen concentration obtained from these experiments, illustrates this aspect of the physics of hydrogen combustion.

The most well-known recently used facilities used for experiments in hydrogen combustion are briefly described below (further information is available in the References at the end of the chapter).

- HYKA experimental platform (KIT) that consists of several facilities. The following are the most relevant for the topic of the present chapter:
  - A1 horizontal cylindrical vessel (volume 98 m^3, length 12 m, inner diameter 3.3 m), designed for static pressure up to 11 MPa, equipped with several windows for visual observation,

FIGURE 3.13 HYKA A2 experimental facility (courtesy of KIT).

- A2 vertical cylindrical vessel (volume 220 m^3, height 8 m, inner diameter 6 m), designed for static pressure up to 2 MPa, also equipped with windows for optical access (Figure 3.13),
- A3 vertical cylinder (volume 33 m^3, height 8 m, inner diameter 2.5 m), designed for static pressure up to 6 MPa, which can be connected to the A1 vessel,
- A8 horizontal cylindrical vessel (volume 8.8 m^3, length 3.7 m, inner diameter 1.8 m), designed for static pressure up to 12 MPa, also equipped with glass windows for visual observation,
- Test Cell TC rectangular chamber (volume approximately 160 m^3, dimensions 8.53 × 5.5 × 3.3 m).

- ENACCEF experimental facility (CNRS/Orléans in France) (Figure 3.14) that consists of a vertical acceleration tube (3.2 m long and 154 mm i.d., total volume 0.062 m^3) and a cylindrical dome (1.7 m long, 738 mm inner diameter, total volume 0.658 m^3), mounted centrically above the tube [39]. In the tube, obstacles of various geometries can be inserted. The facility is equipped with devices to measure the flame propagation velocity (using photomultiplier tubes) as well as the pressure build-up inside the vessel during the combustion. In some tests, a hydrogen concentration gradient (positive or negative) along the tube was established first. The air–hydrogen mixture was then ignited near the tube bottom and the flame propagation was observed.

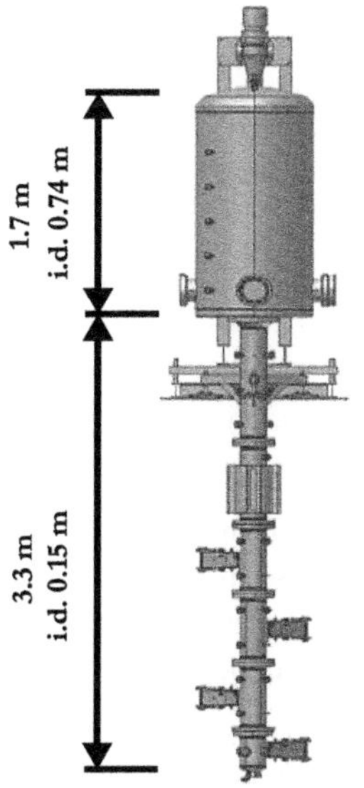

FIGURE 3.14 ENACCEF experimental facility (courtesy of IRSN).

- The RUT facility (Kurchatov Institute in Moscow, Russia) is basically a large duct with variable cross section, which can be subdivided into a number of compartments [42]. A channel (35 m long, with a volume of about 180 m^3) with obstacles is connected to a block of three compartments (of approximate volume 60 m^3 each, divided by a wall with a blockage ratio) and then to another channel (also of approximate volume 60 m^3). The measurement system includes photodiodes to observe the flame propagation.
- THAI experimental facility that was already described in the Section 3.1.4 on experiments on hydrogen distribution [37]. To perform experiments on hydrogen combustion, the internal cylinder and condensate trays are removed, so that the vessel is effectively empty. The pressure and the temperature are measured with a sufficient number of devices, so that not only the vertical propagation of the flame, but also its propagation in the radial (i.e., horizontal) direction is observed. In some tests, a homogeneous air–steam–hydrogen mixture was established first. The mixture was then ignited near the bottom of the vessel, and the flame propagation was observed.
- The combustion test facility or CTF (AECL Whiteshell Laboratories in Canada) is a 10.7 m^3 vertical cylindrical 10 MPa pressure vessel, with a diameter of 1.5 m and a height of 5.7 m [57]. The flame position may be detected using an array of thermocouples and pressure transducers mounted near the vessel's vertical axis.

Theoretical Investigations of Hydrogen Combustion in Containment

Similar to what was stated in Section 3.1.4 for hydrogen distribution, the purposes of theoretical investigations of hydrogen combustion in nuclear power plants are:

- to understand the mechanisms of hydrogen combustion from basic physical principles,
- to develop theoretical models, which will be able to predict the hydrogen combustion in the containment of an actual plant during a severe accident.

As with hydrogen distribution, the investigations of hydrogen combustion consist mostly in simulating experiments that were performed in containment facilities (some of them were described in the previous section) with computer codes. The differences between experimental and theoretical results are then analyzed in order to:

- detect deficiencies of the models in the code that are the main cause of the discrepancies,
- correct the models.

As for hydrogen distribution, simulations of accident scenarios in actual plants are sometimes performed, for which there are no corresponding experimental results. They may be considered as applications of theoretical investigations, which are of course not only valuable, but they are in fact the essential purpose of the theoretical investigations.

The models used in computer codes may use different levels of description; that is, they may solve the transport equations of fluid mechanics on different spatial scales. As in the field of containment atmosphere modeling, two types of codes are used as well in the field of hydrogen combustion: Lumped-Parameter (LP) codes and Computational Fluid Dynamics (CFD) codes.

The essential characteristics of both types of codes have already been described in Section 3.1.4 on hydrogen distribution. As before, the purpose of the present section is not to compare different LP and CFD codes, but merely to present their basic principles and acquaint the reader with different possibilities that are available. Although the most widely used codes are mentioned, the reader should consult the relevant literature (i.e., mostly code user manuals) for further details. The differences between results obtained with LP and CFD codes are well illustrated in the recent OECD International Standard Problem No. 49 (2009-2010).

Modeling with Lumped-Parameter Codes

Models used in Lumped-Parameter codes to simulate hydrogen combustion in a containment have two specific features:

- Although actual flame propagation is fast, it is still not instantaneous. Thus, although conditions in control volumes that make up the containment in a model for an LP code are homogeneous, the codes still assume that the propagation of the flame over cells takes some time. Flame propagation velocities are usually calculated with physical models implemented in the code or may sometimes be prescribed by the user.
- Flame propagation from a control volume to the next control volume in the direction of the propagation is not automatic, but must fulfill certain criteria, again determined by the model used in the code. Thus, the subdivision of the containment into control volumes may significantly influence the results.

The following LP codes are mostly used in the modeling hydrogen combustion:

- ASTEC (i.e. CPA module) (IRSN-GRS),
- COCOSYS (GRS) [80],
- CONTAIN (SNL in USA), with the last version 2.0 released in 1997 [81] [82],
- MELCOR (i.e., the module that deals with the containment) (SNL in USA),
- TONUS-LP (IRSN-CEA).

The reader is referred to the References for more details on results obtained so far with LP codes.

Modeling with Computational Fluid Dynamics Codes

CFD codes that have been mentioned in the previous section on hydrogen distribution also have built-in models that allow them to simulate hydrogen combustion. However, flame propagation is not obtained implicitly from the solution of basic equations of fluid mechanics, but additional models have to be included in the system (similar to the k-ε model, which is used to take into account the influence of turbulence). For instance, the CFX4 code has two models for deflagration: the eddy-break-up model and the mixed-is-burnt model. The TONUS code uses the eddy-break-up model as well, whereas the REACFLOW code uses the eddy-dissipation model. New combustion models suitable for modeling large-scale combustion events are under development. The COM3D code offers the aforementioned models and additionally some new developments such as the CREBCOM and KYLCOM models. It should also be pointed out that some CFD codes have only models for deflagration, whereas others have only models for detonation.

Multipurpose CFD codes that have been used so far to simulate hydrogen combustion in the containment [38] [51] are:

- CFX (ANSYS Inc. in the United States),
- FLUENT, which has also been developed by ANSYS Inc. since 2004,
- STAR-CD (CD-ADAPCO in UK).

The most well-known containment-specific CFD codes are:

- COM3D (KIT) for the modeling of turbulent deflagration,
- DET3D (KIT) for the modeling of detonation, for lean premixed systems only [55],
- GOTHIC (Numerical Applications Inc. in the United States), now owned by EPRI (USA) [50] [84] [85] [86].
- REACFLOW, which was developed at the EU Joint Research Center and allows the simulation of hydrogen deflagration [38],
- TONUS-3D (IRSN-CEA) [88].

The reader is referred to the References at the end of the chapter for more details about the results that have been obtained so far with CFD codes.

To conclude this section on theoretical modeling of hydrogen combustion, it should be pointed out that code validation is a continuous task. The ability of codes to predict hydrogen combustion, including flame propagation, is assessed time and again with the performing of new experiments and availability of new experimental results. Internal and international benchmark exercises are often organized for this purpose.

3.1.6. Mitigation of Hydrogen Combustion Risk

Mitigation of hydrogen combustion risk [58 to 80] designates the prevention of severe consequences (such as damage to the safety equipment, or failure of the containment)—that is, the prevention of severe thermal and pressure loads, and not the prevention of hydrogen combustion. Thus, in principle, hydrogen combustion may be accepted. It may even be started deliberately early in the accident, if it is believed that this might be the best option to prevent more severe consequences if combustion would start later, when the hydrogen concentration would be much higher.

At present, the following options are considered to mitigate the hydrogen risk in a nuclear power plant containment during a severe accident:

- inert the containment atmosphere, that is, remove or dilute oxygen,
- mix the containment atmosphere to prevent high local concentrations of hydrogen,
- consume hydrogen by recombining or deliberate ignition (the last option being adequate when hydrogen concentration is still low enough).

Concerning the adequacy of different options, there are differences of opinion; there are also diverse strategies available, all of which may be equally adequate.

This section presents the different hydrogen mitigation measures.

Inerting of the Containment Atmosphere

Inerting of the containment atmosphere by injecting an inert gas (such as nitrogen or carbon dioxide) is an option that is employed for all existing BWR containments. Complete inerting (i.e., combustion suppression at all hydrogen concentrations) is possible only when the carbon dioxide or steam concentration exceeds approximately 60 vol.% in the atmosphere; inerting with nitrogen requires in excess of 75 vol.% [58]. Inerting by dilution assumes that the diluent is thoroughly mixed with the atmosphere in the containment by an appropriate diluent distribution system.

Mixing of the Containment Atmosphere

As discussed earlier, the necessary condition for hydrogen combustion to occur is that an adequate composition of a gaseous mixture is established. Although the thresholds of different component concentrations cannot be defined strictly, it is still reasonable to expect hydrogen to be flammable if its concentration is much higher than a threshold value, with a steam concentration much lower than another (relevant) threshold value. Given the large volume of a PWR containment, conditions for hydrogen combustion quite possibly would not be met if all the generated hydrogen would be distributed uniformly. Thus, it makes sense to prevent eventual high local hydrogen concentrations and distribute the hydrogen more uniformly throughout the containment by mixing the atmosphere.

For a mixing of the containment atmosphere to be achieved, the arrangement of compartments and the disposition of various equipments might prove important. Containment compartments should have sufficiently large openings to allow venting. Walls and equipment should not be disposed in such a way as to disturb the expected flow paths that will promote mixing. Mixing may be achieved in two ways: by promoting passive mixing through the design of the containment, and by actively mixing the atmosphere through use of safety systems.

Passive mixing may occur due to the gas flow inside the containment. The gas flow may be induced by:

- the break flow from the reactor cooling system,
- the buoyancy effects, due to density differences caused by different temperatures (also induced by the break flow from the reactor cooling system),
- the buoyant flow, caused by the recombination of hydrogen in passive autocatalytic recombiners.

We do not consider the spontaneous buoyant flow of hydrogen as passive mixing of the atmosphere, as the lightness of hydrogen effectively promotes its accumulation in the upper parts of the containment, achieving a result contrary to what is being sought.

In current existing or designed nuclear power plants, cooling fans may be used for atmosphere mixing. Basically, fans are meant to be used for design-basis accidents. Also, their basic purpose is the cooling of the atmosphere, not its mixing. Nevertheless, fans promote the mixing of the atmosphere, as they induce the flow of gas.

Preventing High Local Hydrogen Concentration by Hydrogen Consumption

Passive Autocatalytic Recombiners

Passive autocatalytic recombiners (PARs) are devices that usually consist of a casing and catalyst-coated substrates (metallic plates coated with platinum or ceramic pellets compound coated with palladium). The casings have a lower and an upper opening. The hydrogen-rich gas mixture normally enters the device at the bottom and flows upward. Hydrogen molecules coming into contact with the catalyst surface react with oxygen in the air. Namely, the chemical reaction between hydrogen and oxygen to form water (steam) is an exothermic reaction:

$$2H_2 + O_2 \rightarrow 2H_2O + 120\ \text{kJ/g (of hydrogen)}$$

The reaction starts only after overcoming the required activation energy. This energy can be significantly reduced by the use of catalysts, so that the reaction can start at low temperatures without propagation to the surrounding atmosphere.

Catalytic recombiners are described as passive because such devices are self-starting and self-feeding, require no external energy, and need no action by operators. A catalytic recombiner comes into action spontaneously as soon as the hydrogen concentration begins to increase at the recombiner's inlet. In some experimental tests, catalytic recombiners start up with hydrogen concentration equal to 1–2 vol.%, subject to design-dependent minimum conditions on temperature and humidity.

The heat produced due to the exothermic reaction would enhance buoyancy and thus increase transport of the gas mixtures that leave the recombiner at the top. At low hydrogen concentration, energy from the recombination of hydrogen with oxygen is released at a relatively slow (but continuous) rate into the containment. The heat produced creates strong buoyancy effects that increase the influx of the surrounding gases to the recombiner. These natural convective flow currents promote the mixing of the surrounding containment atmosphere.

Consumption of hydrogen by recombiners is limited due to the speed of natural circulation flows and diffusion, and the availability of reacting species. Thus, PARs are possibly not sufficient for massive hydrogen releases. Besides, hydrogen transport to the recombiners cannot be assured in a sufficient way.

A PAR-system consists of catalytic recombiners—from 30 to 60 for a typical PWR— distributed in the containment to accommodate a wide range of hydrogen release scenarios. The arrangement of the individual catalytic recombiners inside the reactor building is determined by the projected hydrogen release rate, location and distribution, the geometry and operational constraints on maintenance areas, and the accessibility. The disposition of PARs is very important: first, they will cause hydrogen recombination in their vicinity, and, second, they will influence the flow pattern in the containment. Thus, the efficiency of the PARs could be very much dependent on their locations in the containment. For this, simulations of gas flow in the containment atmosphere may be performed to determine the optimal disposition of PARs.

It should also be noted that, in principle, the exothermal reaction in a PAR could lead to an overheating of the catalyst elements and consequently cause an unintended ignition of the hydrogen/air mixture. Investigations of this possibility are also being carried out.

Hydrogen Igniters

Igniters, as their name suggests, are devices used to purposely ignite a flammable mixture. The basic logic in the use of hydrogen igniters is to deliberately cause hydrogen combustion while the hydrogen concentration is still relatively low. In this way, hydrogen will be consumed, while the resulting temperature and pressure increase will be relatively low (in any case much lower than they would be at a later stage, when the hydrogen concentration would be higher). With an optimized arrangement of igniters (in the sense that combustion starts as soon as possible), the combustion rate should be low (slow deflagration mode) and the excess pressure loads should be moderate (lower than 0.3 bar). In principle, deliberate ignition does not increase the severity of an accident because a flammable mixture would deflagrate sooner or later in any case, as there are certainly uncontrolled ignition sources in the containment.

To prevent high local concentrations of hydrogen, igniters should be installed at locations where significant hydrogen release or flow and steam condensation (which also causes the increase of hydrogen concentration) could be expected. Some studies have shown that about 150 igniters are needed in a typical German PWR to ensure reliable hydrogen reduction at low concentration levels.

Two types of igniters are mostly used: spark and catalytic igniters. The spark igniter initiates hydrogen combustion by means of a high-energy spark. Components involved in producing the spark are enclosed in a metal housing. When a severe accident starts, pressure and/or temperature or hydrogen concentrations actuate the electronics of the battery-fed igniter which will produce high-voltage sparks.

The catalytic igniter initiates combustion with the aid of a special catalyst made of a precious metal, also enclosed in the housing. In the event of a severe accident, the gas mixtures flow into the housing, and the catalyst recombines the hydrogen. The gas mixture is heated by the energy release in the exothermic reaction and ignited when it has reached the self-ignition temperature.

There are aso other types of igniters, such as the diesel engine glow plugs used in U.S. PWRs, which do not cause a spark but use a high surface temperature.

3.2. DIRECT CONTAINMENT HEATING (DCH)

(L. Meyer and C. Caroli)

3.2.1. Introduction and Background

In case of core meltdown accident in a PWR, a liquid corium containing metals and oxides may relocate in the lower head of the reactor vessel. If the lower head fails, the molten corium is released into the reactor vessel cavity along with steam. The location, time, shape, and size of the breach influence the melt release. However, the most influential parameter determining the mode and consequences of the melt release is the in-vessel pressure when the vessel fails. If failure occurs with an in-vessel pressure of the same level or only slightly above the pressure in the containment, the molten part of the core inventory will flow into the reactor vessel cavity primarily by gravity. This may lead to MCCI (see Chapter 4) but does not threaten the integrity of the containment in the short term. If the vessel failure occurs with an in-vessel pressure much higher than the containment pressure, high-pressure melt ejections (HPMEs) occur. In this case the melt will be forcefully ejected into the reactor vessel cavity, finely fragmented, and transported out of the reactor vessel cavity. During this phase the metallic part of the melt may be oxidized by steam producing additional hydrogen and heat. An efficient heat exchange between the finely fragmented melt particles and the containment atmosphere will heat up and pressurize the containment since it is a constant volume system. Combustion of hydrogen produced during the melt dispersion or previously released into the reactor building may also contribute to pressurize the containment. These processes, referred to as direct containment heating (DCH), may then endanger the integrity of the containment and are likely to induce an early containment failure.

If the occurrence of DCH is strictly related to the primary circuit pressure at vessel failure, its consequences depend also on the breach characteristics, on the amount and characteristics (degree of oxidation, temperature) of the molten mass released, and on the layout of the reactor vessel cavity and reactor building. The evaluation of the consequences of DCH is therefore plant dependent.

The TMI-2 accident could have led to a DCH event if the vessel had failed.

3.2.2. Importance from a Risk Perspective

DCH is particularly feared because it is liable to induce an early containment failure and thus to produce an early and not filtered radioactive release from the reactor containment to the environment. Therefore in case of DCH there may be very little time available to take actions to protect or evacuate populations living near the plant.

DCH may occur only in the event of high-pressure vessel failure. The LWR design base loss of coolant accident (LOCA) sequences, except small LOCA sequences, are characterized by a rapid primary circuit depressurization and are normally not liable to induce DCH unless late core reflooding occurs (a mitigating action undertaken to enable in-vessel retention) when water vaporization may repressurize the primary circuit. Core meltdown accidents at high primary circuit pressure are sequences such as complete loss of feedwater to steam generators, complete loss of cooling source, or station blackout. These may lead to induce DCH. However, in order to reduce the probability of occurrence of DCH, mitigating actions are normally undertaken in these conditions by the emergency response team. In particular, intentional depressurization of the primary cooling circuit is foreseen by the emergency operating procedure using the primary circuit safety valves. In some case, as for example in EPR, a special safety valve is installed on the primary cooling circuit in order to secure the primary circuit depressurization in these situations and lower the risk to have a high-pressure failure of the vessel. The possibility of DCH is thus limited to those situations where, in spite of the mitigating actions undertaken by the operating personnel, core meltdown occurs under high-pressure conditions. However, in such cases the whole primary circuit is normally pressurized and overheated, and its failure may occur in different locations. If the first failure occurs in the surge line or in the legs between reactor pressure vessel and steam generators, the following primary circuit depressurization avoids the occurrence of DCH and the consequences of the accident are expected not to be severe for the containment. If the first failure occurs in the steam generator tubes, the following primary circuit depressurization should also avoid the occurrence of DCH. However, steam generator tube failures themselves constitute a bypass of the containment and may be responsible for large radioactive release into the environment (refer to Chapter 2 for a description of these accidents). Finally, if the first failure takes place in the vessel lower head, DCH occurs.

In the unlikely case that DCH occurs, the consequences for the reactor containment integrity, and thus the potential radioactive release in the environment, strongly depend on the design and layout of the reactor vessel cavity and reactor building. Indeed, as it has been shown by the R&D work carried out up to now on this issue (see the following Sections), the DCH-induced containment pressurization depends mainly on the efficient heat exchanges between dispersed corium and containment atmosphere, by exothermic reactions between corium and steam (metallic corium oxidation) and by the concurrent hydrogen combustion during the melt dispersion (other phenomena may contribute to the containment pressurization, but they are not believed to be as important). The experimental work carried out up to now has shown that the fraction of corium dispersed outside the reactor vessel cavity strongly depends on the cavity layout and more generally on the containment design. On the other side experimental and theoretical R&D work has also shown that heat transfer between dispersed corium and containment gas increases when the dispersed corium reaches large containment subcompartments. DCH-induced pressurization of the containment is thus strongly dependent on the reactor layout. It should be mentioned that, based on this observation and on some experimental results on corium dispersion, the initial layout of the EPR reactor vessel cavity was slightly modified in order to reduce potential consequences of DCH in this reactor. Finally, it should be noted that, based on the analyses of experimental results, it is now believed that DCH may occur only in case of vessel failure with a primary pressure above a threshold value. Evidence exists, however, that the cutoff pressure varies from design to design, and it does not seem realistic to consider the same cutoff pressure for all the existing plants.

3.2.3. DCH Phenomenology

High-pressure ejection of corium and steam out of the vessel is characterized by three stages: single-phase liquid corium jet, two-phase corium and steam jet, and gaseous steam jet. The duration of these stages depends on the mass of corium relocated to the lower head, the sectional area and location of the break, and the pressure in the reactor coolant system. Pressurized ejection initially causes fragmentation of corium in liquid droplets. A flow of steam and corium through the cavity is then established. This very complex flow is strongly influenced by the cavity geometry. It is also subject to various phenomena (Figure 3.15): projection of corium onto the cavity walls and formation of

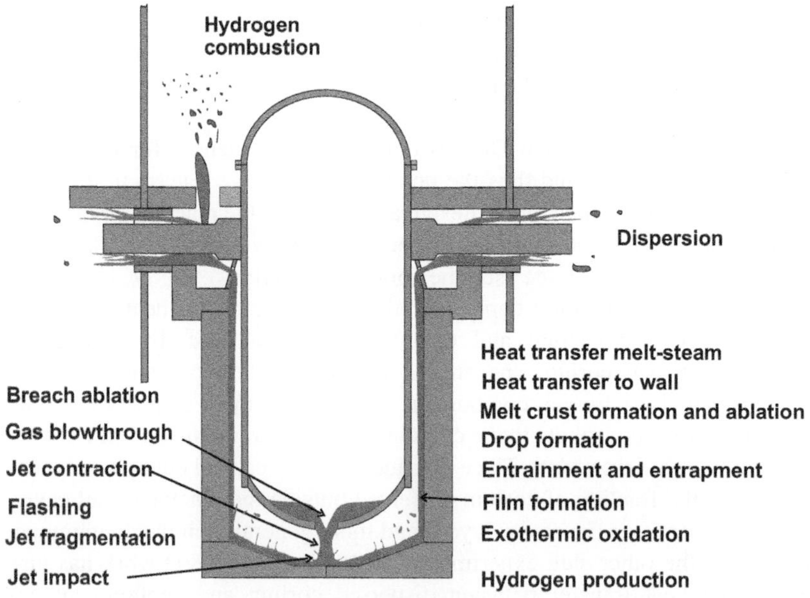

FIGURE 3.15 Phenomena occurring during a forceful melt ejection.

a liquid film along these walls, entrainment and fragmentation of the film by the flow of steam and formation of corium droplets, coalescence and/or fragmentation of these droplets, and so on. As a result, part of the corium entrained by the steam is transported to the areas adjoining the cavity, whereas a portion remains trapped in it. During this entrainment phase, the steam and the droplets interact thermally and chemically. Consequently, steam temperature and cavity pressure increase considerably.

Hydrogen is produced by the exothermic reaction between the corium metals and steam. However, hydrogen combustion is not possible in the vessel cavity because its atmosphere contains little oxygen, having been swept out by the steam flow. When hot gases and corium particles enter the containment, they contribute to the overheating and rapid pressurization of its atmosphere. The larger the mass of corium dispersed and the finer its fragmentation, the larger the increase in heat transfer and containment pressure. The distribution of corium in the various containment volumes and the duration of discharge also play an important role in the degree of pressurization. Moreover, when the extremely hot gases and corium particles enter the containment, they will provoke more or less rapid hydrogen combustion. This combustion is very complex because it combines the characteristics of a diffusion flame (combustion of a hydrogen jet entering the containment) with the characteristics of a premixed flame (combustion of the initially present hydrogen already mixed with air). It will increase the pressure spike in the containment if the characteristic combustion times are close to those of corium discharge and dispersion. However, determining characteristic combustion times is complicated by the complex nature of combustion during DCH.

The pressure rise is governed by the competing processes, of addition of heat energy to the containment atmosphere and heat losses to structures and the containment wall. Since all transfer processes are rate-limited, the peak pressure rise will be high if the heat transfer processes from the different sources to the containment atmosphere are fast and coherent.

3.2.4. Vessel Failure Modes

The location, time, shape, and size of the breach greatly influence the melt release. They depend on the temperature distribution in the vessel wall (which in turn depends on the condition of the molten core, i.e., temperature, composition, distribution, and mass of the melt), the design and the location of penetrations in the lower head, the pressure in the vessel, and the vessel steel properties at high temperature. Experimental information about vessel failure modes comes mainly from the LHF (lower head failure) and FOREVER (Failure Of REactor VEssel Retention) experiments [91] [92] [93] (see also the Chapter 2). Two basically different failure modes can occur, depending on the core design. If the in-core instrumentation is through the bottom head, these penetrations can be weakened or enlarged by contact with melt and act as a location for failure initiation, or the whole instrument guide tube can be ejected. The second mode is initiated by a weakening of the vessel wall caused by an internal melt jet during melt relocation or a hot spot occurring within the area of the highest heat fluxes. Both events can cause a failure in the vessel bottom. If a corium pool forms in the lower head, the pool may be stratified with a metal layer on top of an oxide pool and the highest heat flux from the pool to the vessel wall in the locations of contact of the vessel with the metal layer (focusing effect). This edge-peaked heat flux can lead to a side failure. The rip can span only a small part of the vessel perimeter or can progress to a fast unzipping of the lower head. In the LHF experimental program (and its continuation OECD Lower Head Failure (OLHF)) an unzipping of the vessel was observed in two out of the eight tests carried out. If in one case this result seems to be related to a sudden pressure increase in the experimental procedure, however, in the second case it was not expected. Further R&D work carried out on this issue (for example, by IRSN and CEA) indicated that this behavior could be related to the metallurgical characteristics of the steel and the influence on its high-temperature behavior.

Larger breach sizes generally give rise to larger melt dispersion. Melt fragmentation may, however, be bound because not all the ejected melt mass can be fragmented within the short duration of the blowdown in case of vessel unzipping.

3.2.5. Discharge Phenomena

Hole Ablation

The molten corium is significantly hotter than the melting point of the vessel steel wall, and therefore the initial hole size in the lower head will enlarge due to heat transfer from the flowing melt to the wall [94] [95] [96]. Ablation of the hole will occur when the interface temperature between steel wall and flowing melt reaches the melting point of the steel. A simplified model for the ablation rate may be derived balancing the convective heat transfer to the wall with the energy required to heat and melt the wall:

$$\dot{D}_h = \frac{dD_h}{dt} = \frac{2h_{d,w}(T_c - T_{mp,w})}{\rho_w[c_{p,w}(T_{mp,w} - T_w) + h_{fw}]}$$

where

D_h = *diameter of breach*	T_w = *temperature of steel wall*
$h_{d,w}$ = *heat transfer coefficient*	ρ_w = *density of steel wall*
T_c = *temperature of corium*	$c_{p,w}$ = *specific heat of steel*
$T_{mp,w}$ = *melting point of steel wall*	$h_{f,w}$ = *heat of fusion of wall*

The key parameter is the heat transfer coefficient, which depends above all on the question of whether a crust is formed at the interface, and whether that crust is stable or floating on the ablating steel and is stripped from the interface by the flowing corium. The final size of the hole is governed by the duration of melt discharge. The *time to eject all the melt* from the reactor pressure vessel (RPV) with a constant hole size is:

$$\tau_M = \frac{M^o}{C_d A v_M \rho_M}$$

where

M^o = *initial melt mass* v_M = *velocity of melt jet*
C_d = *discharge coefficient* ρ_M = *density of melt*
A = *area of breach*

However, due to the increasing hole size, the actual discharge time will be shorter. This time can be determined taking into account the ablation rate. Generally, for small initial diameters the time for the ablation process is longer and the hole growth will be larger than for initially larger holes, since the total melt discharge time is larger for smaller initial hole sizes. This is generically shown in Figure 3.16.

This means that different small initial breach sizes will be growing to a more or less identical final size, depending mainly on the available melt mass, while large initial holes will not grow much more. A code HAMISA [96] was written to provide modeling of the hole ablation and the discharge behavior from a vessel breach.

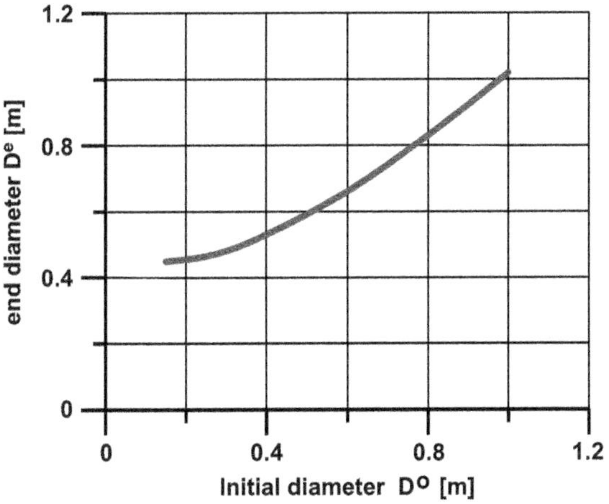

FIGURE 3.16 Generic behavior of breach diameter with wall ablation.

Discharge Stages

Vessel Bottom Failure

Three distinct stages of the outflow can be distinguished in case of bottom failure of lower head of the RPV (Figure 3.17):

1. the single-phase liquid flow,
2. the blowthrough with two-phase flow,
3. single-phase gas flow.

The single-phase gas flow can be again subdivided in choked flow (for high ratios between RPV pressure and cavity pressure) and subsonic single-phase gas flow. The duration of the individual stages depends on the hole size, the height of the liquid in the lower head, the pressure, and the liquid density.

The single-phase liquid discharge progresses until, at some critical level, a funnel-like depression forms over the outlet hole. This is termed gas blowthrough. Pilch and Griffith [97] have done a detailed literature survey with regard to the application to the DCH-problem. For the critical liquid level at the onset of blowthrough, they recommend using the Gluck correlation [98] [99]:

$$\frac{h_b}{d} = 0.43\frac{D}{d}\tan\mathrm{h}\left(Fr^{1/2}\frac{d}{D}\right)$$

with h_b, the height of liquid at start of blowthrough, D, the inner diameter of the lower head, d, the diameter of the breach, and the Froude number $Fr = v_L/(gd)^{1/2}$, where v_L is the liquid velocity at the breach and g the gravity acceleration.

For large D/d ratios it limits to $\frac{h_b}{d} = 0.43 \cdot Fr^{0.5}$

The time of blowthrough is then a function of the initial liquid height, the hole diameter, respectively, the diameter as function of time, the pressure in the RPV, and the properties of the steam and the melt.

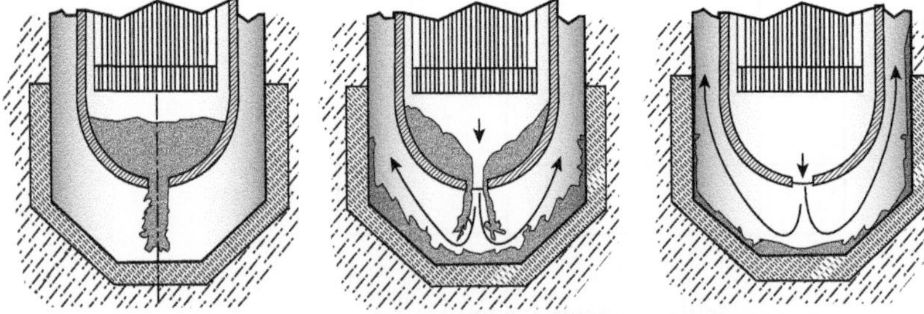

FIGURE 3.17 Stages of melt discharge and blowdown.

Most important for the DCH processes is the interaction of corium and steam during the two-phase stage of the blowdown but also thereafter. The velocity of the jet increases with decreasing liquid fraction up to the beginning of single-phase gas flow when the gas velocity at the breach will be velocity of sound (choked flow), as long as the vessel pressure is higher than the critical pressure (that is roughly equal to twice the cavity pressure). The turbulent blowdown process causes intensive mixing of melt and steam, especially in cavity geometries with a large distance between lower head and cavity floor. Here a strong recirculation of the flow may occur.

Vessel Side Failure

A side failure is possible with edge-peaked heat flux. The level of the melt pool will be near or above the failure site. At the start of the blowdown melt is forcefully ejected out of the breach, hitting the wall of the cavity. Blowthrough occurs when the liquid level at the breach sinks below the upper edge of the failure. A stratified flow at the breach forms, where gas entrains melt out of the hole until the liquid level drops below a critical distance below the lower edge of the breach [100] [101]. The gas velocity, the density ratio of gas and liquid, the surface area of the liquid pool, the duration of the blowdown, and the height-to-width ratio of the breach govern this entrainment process and thereby this critical distance. Therefore, the entrained liquid fraction can be higher with a small breach than with a large one because the blowdown time is longer while the maximum velocity may be the same. Also, the fraction of melt dispersed out of the reactor cavity may be larger with smaller lateral breaches than with very large ones. Lateral breaches generally lead to a smaller dispersion of melt than central bottom failures, even if the dispersed fraction is related to the liquid mass that has been ejected out of the RPV. The main effect is probably the circumferential component of the velocity in the cavity [104]. The melt mass remaining in the lower head does not contribute significantly to DCH, except for some oxidation at the pool surface of the remaining melt pool in the vessel.

Jet Types and Fragmentation

A forcefully ejected liquid jet is generally fragmented. Only in special conditions can a compact jet be obtained over a short distance. These conditions do not exist at vessel failure. If gas is dissolved in the melt, the gas will expand when the pressure drops and break up the jet. Also, without this effect the jet boundary will not be smooth, and the velocity difference between jet and surrounding gas will lead to the fragmentation of the jet. The governing parameter is the Weber number:

$$We = \frac{d\rho_G v_{rel}^2}{\sigma} \geq 12$$

FIGURE 3.18 (left) Single-phase jet, (right) two-phase jet in DISCO-C experiment with water.

where

d = *droplet diameter*
ρ_G = *density of gas in cavity*
v_{rel} = *relative velocity between gas and jet*
σ = *surface tension*

If the Weber number is higher than a critical value (approximately 12), the liquid is fragmented into smaller droplets. This effect is more important for large distances between vessel and cavity floor. For short distances the jet impact at the cavity bottom may be the predominant cause of fragmentation.

After gas blowthrough, the fragmentation of the jet by two-phase flow is very intense (Figure 3.18). The Weber number will be much larger due to the high gas velocity, and the drop size consequently will be smaller. The enlarged contact area between steam and melt promotes chemical reactions (oxidation and hydrogen generation), heat transfer, and release of radionuclides. The fragmented droplets or particles may be easily transported out of the cavity.

3.2.6. Cavity Phenomena

Discussion of phenomena in the cavity should always be accompanied by the details of the cavity geometry. The important characteristics for melt dispersal and DCH are the flow paths out of the cavity and their destinations, the existence of melt retention features, and the height below the pressure vessel. Table 3.1 and Figure 3.19 show those characteristics for different reactor plants.

Film Formation, Entrainment, and Debris Transport

During single-phase liquid jet discharge, the melt hits the cavity bottom and can be further fragmented. The larger part moves as a film along the floor and is

TABLE 3.1 Characteristics of Different Reactor Cavities

	Zion	Calvert Cliffs	EPR	P'4	Konvoi	VVER-1000
Direct flow path into containment dome	no	yes	no	yes	no	no
Flow path from lower cavity into compartments	yes	no	negligible	yes	no	yes
Flow path along main cooling lines into pump and steam generator rooms	small	yes	yes	yes	yes	negligible
Melt retention features	no	no	no	no	yes	no
Cavity height below pressure vessel	high	low	low	high	low	low

redirected into vertical direction at the wall. It moves up the wall by inertia as a closed film or breaks up after a certain distance. If the initial velocity was high enough, melt can leave the cavity during this stage, usually as coarse fragments and large drops. During the two-phase blowdown, the film is accelerated and torn apart, and droplets are entrained by the much faster gas (steam) flow. A criterion as to whether entrainment takes place is given by the Kutateladze number:

$$Ku = \frac{\rho_G\, u_G^2}{(\rho_L\, g\, \sigma)^{1/2}}$$

Different values of the threshold for debris dispersal are given for different cavities. For the Zion cavity Ku of 6 – 13 was stated. The results of the DISCO experiments in EPR geometry could be correlated by

$$f_d = 0.4 \log_{10}(Ku)$$

where f_d gives the mass fraction of melt dispersed out of the cavity, with the limit value of 0.76 ([102] [103] [104] [105]). Many such dispersal correlations were devised for different cavity designs [89], with the objective of defining a *low-pressure cutoff,* such that the RCS pressure is insufficient to cause significant dispersal from the cavity. Different ways to tackle the problem have been taken. On the one side there are the pure correlations with dimensionless groups fitted to measured results, and on the other side there are phenomenological models with parameters adapted to the experimental results. All are valid only for the investigated cavity geometries, and some cannot be adapted

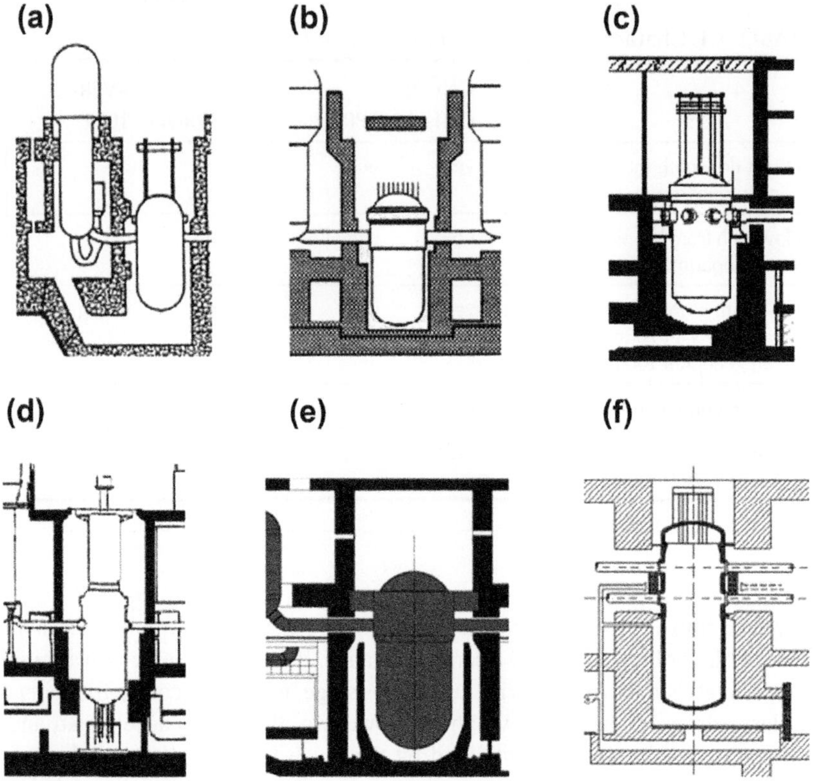

FIGURE 3.19 Different cavity designs: (a) Zion, (b) Calvert Cliffs, (c) EPR, (d) P'4, (e) Konvoi, (f) VVER-1000/320Kozloduy.

to prototypic conditions. Great care has to be taken when using these correlations for reactor application.

Considering the differences in the cavity designs listed in Table 3.1 and shown in Figure 3.19, we can understand that it is not only important to know how much melt is carried out of the cavity, but also where its final location is and what the size of the debris is. Figure 3.20 shows an overview of typical results from scaled experiments with steam and iron–alumina melt for five different reactor plants. These results may vary with RCS pressure and breach size, but the relations between the different cavity designs will persist. In all reactor cavitys some melt remains just because of crust formation, so even higher pressures would not blow out more than approximately 80%. In the Zion plant all dispersed melt is transported via the large chute into a relatively small compartment, which limits the consequences for DCH at a low level, in spite of the high dispersal rate. The Calvert Cliffs cavity has a direct flow path into the containment dome: here the potential for DCH is high. The EPR cavity is

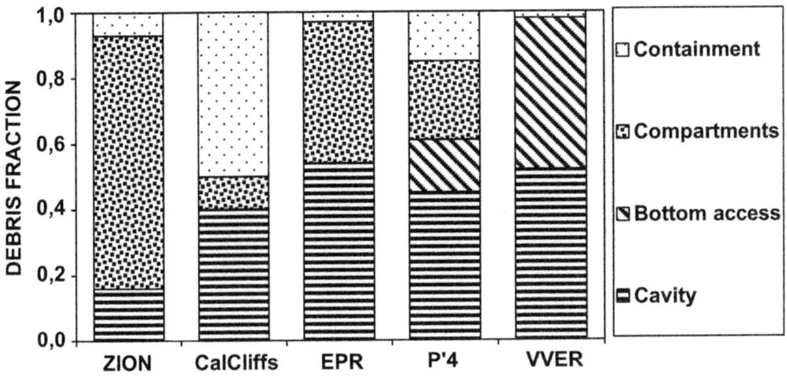

FIGURE 3.20 Typical dispersal fractions in different reactor designs.

similar to the Calvert Cliffs design, with the important difference that no direct path leads to the containment dome. Therefore, more melt is retained in the cavity, and little is dispersed into the containment dome. Again, DCH is limited to a low level.

The Konvoi cavity is similar to the EPR cavity with two exceptions. The flow cross section from the cavity into a reactor room above the pressure vessel is closed by sealing plates during normal operation, which may break down when the overpressure in the cavity is above 2 bars. Further, there is a biological shield, and there are eight pressure-venting flaps behind it, which may be open during an accident, in which case water would enter up to a level of the center of the flaps. If both flow paths remain closed, the melt dispersal fractions are similar as in the EPR geometry. If any one of the flow paths is open, the DCH effect is mitigated, either because melt is trapped in a small compartment (first case) or some melt is quenched in water behind the biological shield (second case). The French 1300 MW reactor plant P′4 has a very high cavity, where mixing of steam and melt is very strong, leading to efficient heat transfer from melt to blowdown steam and a high rate of oxidation and hydrogen production. This, together with the fraction transported directly to the containment dome, can lead to relatively high DCH. However, this is mitigated by the high melt retention capability in the cavity and the fact that the melt fraction transported to the bottom access chamber adds little to DCH. The cavity of the VVER-1000 plant is almost completely closed, connected to the outside only by venting lines and a door that opens at 0.4 MPa overpressure. In the event of vessel failure, the pressure in the cavity will quickly reach 0.4 MPa, and the door will yield. So half of the melt will be trapped in the bottom access chamber, and the other half will remain in the cavity. Although the bottom access chamber is connected to the containment dome, the melt is not finely fragmented and does not reach much of the containment. Therefore little DCH occurs.

The debris found in the cavity is mostly in the form of crusts, the remains of melt films. In the bottom access chambers and in the compartment of the Zion plant, the melt enters as film and droplets, which hit a wall after a relatively short flight path. Melts ending up in the pump and steam generator rooms have to change direction several times during their way from the vessel to their final destination. A part gets trapped in the bends along their way and will freeze on the wall; the rest is fragmented and may hit walls in the rooms. The degree of fragmentation varies during the blowdown; it is highest during the single-phase gas blowdown. Also the breach size governs the degree of fragmentation by determining the mass flow rate of the steam and thereby the maximum velocity in the cavity and in the flow cross section along the main cooling lines. Generally, the mean size of the debris found in the subcompartments is larger (1–3 mm) than that in the containment dome (0.2–1 mm).

Debris–Gas Interaction

The most intense interaction between fragmented melt and steam occurs in the cavity. Both heat transfer and chemical reactions heat up the steam and increase the pressure locally. Depending on the cavity geometry, pressure peaks may reach high values (over 1 MPa). These peaks can even stop the melt ejection out of the reactor cavity temporarily.

The heat transfer from the debris to the gas phase is by convection and radiation. The heat transfer rate is given by

$$\dot{Q} = h A_d (T_d - T_g)$$

where h is the effective heat transfer coefficient for both convective and radiative heat transfer, and A_d is the debris–gas interface area. The interface area of a given mass varies with the particle diameter as d^{-1}. Generally, gas phase convective heat transfer dominates the radiative contributions in the cavity and can be expressed by

$$Nu = a + b \ \mathrm{Re}^{0.5} \ \mathrm{Pr}^{0.33}$$

for flow over a sphere, with $Nu = h\,d/k$, where k is the gas thermal conductivity. These correlations imply that the heat transfer rate varies with drop diameter approximately as $d^{-1.5}$, if radiation and heat transport inside the melt drop are neglected. The latter is of course true only for small drops. Another restriction to this conclusion is the effect of the velocity difference that plays a role in the Nusselt number correlation, via the Reynolds number. Smaller particles follow the gas velocity closer than large particles, and consequently the Reynolds number is smaller.

Analogue considerations with respect to the diffusion coefficient lead to similar conclusions for the chemical reaction rates. In terms of the characteristic time scales of these processes, they are roughly between 0.003 and 3 seconds for drop diameter between 0.1 and 10 mm. Since the particle size depends mainly on the velocity (by the Weber number), and the velocity is not much affected by facility scale, the particles have similar sizes in small-scale experiments and at reactor scale (provided they have the same physical properties). However, the particle residence times vary linearly with scale. Therefore, the reliability of experimental results has to be ensured taking these time scales into account.

Since the most intense interaction occurs during the period of simultaneous steam blowdown with melt fragmentation, that is, during the two-phase flow and single-phase steam flow with melt entrainment, the ratio of this characteristic dispersal time to the total blowdown time is used in some analysis methods [89] and is termed the coherence ratio $R\tau = \tau_{disp}/\tau_{blowdown}$, where τ_{disp} is the time for melt dispersal and $\tau_{blowdown}$ is the time during which blowdown occurs. However, these analysis methods were developed on the basis of results from experiments in Westinghouse reactors (Zion and Surry) and hence are applicable to these reactor geometries only.

The metal part of the corium can be oxidized by the oxygen in the cavity atmosphere and by the blowdown steam. A significant amount of oxygen will only be present during the period of single-phase melt ejection; later the air atmosphere will be blown out of the cavity by the steam. If un-oxidized metal droplets or particles are still present in the subcompartment or the containment dome, the reaction with air–oxygen can continue there. During the metal-steam reaction, hydrogen is produced. The most important metals in the corium that can be oxidized are zirconium, chromium, and iron. The main oxidation processes are as follows.

with oxygen:	Zirconium	$Zr + O_2$	$\rightarrow ZrO_2$	−1082 kJ/mole
	Chromium	$2Cr + 3/2O_2$	$\rightarrow Cr_2O_3$	−1138 kJ/mole
	Iron	$Fe + ½\ O_2$	$\rightarrow FeO$	−244 kJ/mole
with steam:	Zirconium	$Zr + 2H_2O$	$\rightarrow ZrO_2 + 2H_2$	−598 kJ/mole
	Chromium	$2Cr + 3H_2O$	$\rightarrow Cr_2O_3 + 3H_2$	−414 kJ/mole
	Iron	$Fe + H_2O$	$\rightarrow FeO + H_2$	−2 kJ/mole
and the hydrogen reaction is		$H_2 + ½O_2$	$\rightarrow H_2O$	−242 kJ/mole

The reaction enthalpy is temperature dependent: the data given are for reaction at 2500 K. The reactions with zirconium and chromium are highly exothermic and can go to near completion. For the iron–steam reaction, chemical equilibrium between H_2 and H_2O is reached at a H_2/H_2O-ratio of approximately 2. The reactions with steam can be limited by the amount of available steam or limited by time. The limitation by time would rather arise in small-scale experiments than in reactor scale. If the reaction is not completed in the cavity, it can continue in the compartments and the containment dome, with oxygen from either air or steam. Considering the heat release by the two ways—(1) the reaction with oxygen from the containment atmosphere and (2) the reaction with steam, producing hydrogen and then the hydrogen combustion with the air-oxygen in the containment—the total amount of heat release is the same, as can be seen from the numbers given above. However, it may have a different impact on the temperature and pressure increase in the containment, where and when the heat is released. This will be discussed in Section 3.2.7.

It should be noted that the blowdown steam can already contain hydrogen from the metal–steam reaction inside the RPV.

Debris–Water Interaction

The effect of water in the cavity or co-ejected from the vessel with the melt can be quite diverse, depending on the amount of water and the cavity geometry. If the cavity is filled with water, steam explosions are possible. Moderate amounts of water can either mitigate or augment DCH loads. Mitigation effects arise by the quenching of debris, suppression of complete chemical reaction, and, in the containment dome, suppression of hydrogen combustion by steam inerting and cooling by aerosol water. Augmentation effects include increasing the supply of steam for thermal and chemical reactions (in case of blowdown steam limited conditions), and increasing the velocity for dispersal processes. Experiments generally did not show a significant effect of water on the DCH load. Co-ejected saturated water, undergoing flash vaporization as it left the vessel, had no particular impact. For the non-prototypic case with water at 300 K, which does not vaporize following depressurization, a considerable reduction (30%) in the pressure loads was observed. Tests with a small quantity of water in the cavity did not produce a significant impact.

Debris–Insulation Interaction

Only small experimental evidence is available about the effect of the interaction between RPV insulation and debris [122]. It is seen that insulation material is ablated by the hot debris and entrained by the steam. No major plugging of flow paths can be expected. The sheet metal will be oxidized and is an additional source of hydrogen, though quite small.

3.2.7. Phenomena in the Containment Dome

Heat Transfer

Debris particles can enter the containment dome either on the direct path through the annular space between RPV and cavity wall and openings at the top of the reactor cavity, or via subcompartments and connections between them and the dome. In the second case, larger particles cannot follow the change of direction and will hit walls. The same considerations to the interaction between melt droplets and atmosphere apply as described in the previous Section. Since the particle sizes reaching the containment dome are small and the flight paths long, it is more likely that interactions with the containment atmosphere can reach equilibrium.

Heat transfer to structures limits and reduces the peak pressure. When the melt hits the structures, it will quickly freeze and transfer its energy to the structures. Radiation *heat transfer to structures* may contribute to the heat loss and mitigate DCH loads. The mean free path of thermal radiation in a steam atmosphere is similar to characteristic containment dimensions. However, with aerosols present in the containment atmosphere the free path is much shorter, and part or all of the radiation energy can be deposited in the atmosphere.

The main heat sink and limiting effect on the peak pressure is the *gas-to-structure heat transfer*. The heated blowdown steam and atmosphere lose energy to structures in the subcompartments and the containment dome by convection. The height of the pressure peak is determined by the time scales of the heat transfer to gas and the heat transfer from gas to structures. High containment pressures are reached if the processes of energy transfer to the atmosphere are faster than the relatively slow processes of heat losses to the structure.

Hydrogen Combustion

The blowdown steam together with produced hydrogen enters the upper dome of the containment through several flow paths from the cavity or subcompartments. The hydrogen comes from the metal–steam reaction in the RPV and in the cavity (150 – 800 kg). The jet entrains the atmosphere, which contains oxygen and preexisting hydrogen. The hydrogen concentration in the containment atmosphere at vessel failure depends on the accident scenario and on the presence/absence of recombiners. However, hydrogen concentrations well above the flammability limit (4% molar) may not be excluded. In these conditions, hot debris particles act as igniters. If the temperature of the jet is high enough (~800°C), autoignition may occur. The hydrogen burns as a diffusion flame, and the rising hot combustion products increase the mixing of the atmosphere entraining fresh oxygen and hydrogen rich gas into the flame. Depending on the hydrogen and steam content and the temperature, volumetric combustion can also take place. If these processes happen on the same time scale as the debris dispersal process, they contribute to peak pressure. The pressure increase due to hydrogen combustion can be higher than that by other processes. Because of the flammability limits for hydrogen–air–steam mixtures, not all available hydrogen will burn. Between 30 and 90% of the available hydrogen has been observed to burn in experiments. The highest fractions burned were found on both ends of the spectrum regarding the preexisting hydrogen before the blowdown: if no or very little hydrogen existed in the containment atmosphere, up to 90% of the produced hydrogen burned; on the other hand, if the concentration was 4% or higher, the burned fraction was around 80% of total available hydrogen. For similar reactor geometries the peak pressure rise correlates quite well with the total hydrogen burned per volume unit ($\Delta p = 23 \pm 4$ kPa/[g/m^3]).

No significant scaling effect on the peak pressure was found for geometries similar to the Zion reactor plant. Also, recent separate effects tests on hydrogen combustion under DCH conditions in a simplified geometry (no compartments) in two different scales (1:18 and 1:7) showed no effect of scale [123].

Pressure Increase

The issue of the maximum containment pressure due to DCH is predominant besides the issue of final corium location that concerns long-term cooling considerations. Concerning the load on the containment building, the DCH peak pressure is quasistatic, lasting several seconds, contrary to the short pressure spikes caused by steam explosions. Containment buildings are generally designed for pressure loads between 0.5 and 1.0 MPa. Figure 3.21 shows some typical experimental results for a cavity design, similar to the Calvert Cliffs (open) and the EPR (closed) design but otherwise identical geometries and initial pressures. The initial hydrogen concentration in case A and C was 2.6%. However, the produced amount of hydrogen in case C was only 60% of that in case A. Both time and linear dimension scale in this case by a factor of 18.

Here the different paths of heat release due to metal oxidation play a role. If oxidized by air, heat is mainly deposited in the corium droplets, which may not have the chance to transfer all heat to the atmosphere because of hitting structures, trapping, and agglomeration. On the other hand, hydrogen may or may not burn at the time scale of heat transfer from the corium to the atmosphere, and the fraction of burned hydrogen can vary depending, among other parameters, on the amount of preexisting hydrogen and the compartmentalization of the containment building. Key data for the examples are as follows:

	A	B	C
Melt fraction dispersed into containment dome	0.50	0.65	0.02
Hydrogen fraction burned	0.82	-	0.43

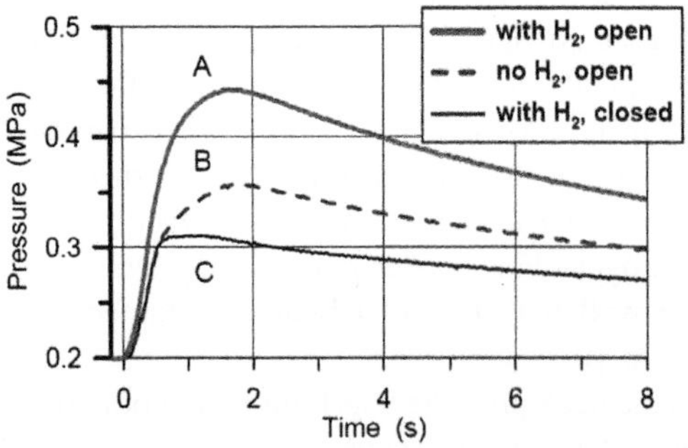

Pressure increase in containment for three cases:

A. steam driven with hydrogen production and combustion (open)
B. nitrogen driven, no hydrogen production, direct path (intense heat transfer)
C. steam driven, but no direct flow path to containment (closed, little concurrent combustion)

FIGURE 3.21 Pressure increase in DISCO tests.

3.2.8. Experimental Database

Experiments to investigate DCH phenomena are performed in mock-ups, which provide small-scale reproduction of the main geometrical characteristics of the reactor containments at different levels of detail. Based on the results of these tests, models and correlations are developed for simulation codes, which are then used to study reactor-scale consequences of a DCH event. The fact that all experiments are conducted in scaled-down test facilities with simulant fluids calls for a careful examination of questions related to scaling, in order to ensure that the important processes of interest are well scaled and to assess the effects of possible distortions in relation to a nuclear power plant [116] [117]. The experiments fall in two main categories: tests using simulant liquids at low temperature and tests employing material with high melting temperature.

The experiments of the first category focus on the fluid dynamic aspects of DCH. They are separate-effect tests that aim at characterizing corium dispersion and establishing correlations for entrainment of corium droplets out of the cavity toward the containment and the adjacent compartments. These correlations are a function of the experimental parameters, which generally include the size of the lower head break, vessel pressure, and physical properties of simulant materials and the carrier gas exiting the vessel. Several cavity geometries have been studied, and various simulant materials have been used for this type of test: water, oils, Wood's metal, and gallium. The advantage of these last two metallic alloys is that they possess properties (density, viscosity, surface tension) similar to those of corium. The key results of these tests are characteristic blowdown times and the distribution of the dispersed material, as well as the pressure changes over time in the vessel and the cavity.

The second group of experiments provides insight into thermal and chemical aspects in addition to the dynamic aspects. The simulant material used for this type of test generally is a mixture of iron and alumina (Al_2O_3), the product of a thermitic reaction, at ejection temperatures between 2000 and 2300°C. In some tests, chromium was included to better model the chemical reactivity of prototypic molten corium that may contain highly reactive zirconium metal. A small number of tests employed a mixture of uranium dioxide, zirconia, and steel, materials representing the molten fuel closer to the reactor conditions. If steam is used as driving gas, hydrogen production and combustion occur. Otherwise, if a chemically inert gas (nitrogen or argon) is used, only the thermal effects can be studied.

Historically, DCH investigations can be divided into two phases.

HPME Experimental Program for U.S. Plants

A large database was established in the years 1985 to 1997 by a concerted action in the United States with partners in the United Kingdom, Spain, and Korea. It was initialized by the Zion Probabilistic Study in 1981, where it was argued that a high-pressure melt ejection (HPME) would sweep out the melt into the containment, where the melt would settle into a coolable geometry on the containment floor. Thus HPME was regarded as a positive process to terminate the accident. However, simple experiments and calculations gave rise to the fear that the accompanying pressure increase could endanger the integrity of the containment.

Because of the large number of experiments performed, only an overview shall be given here. A more detailed description of all experiments of the first stage can be found in the book article on DCH [89], the topical issue on DCH of Nuclear Engineering and Design [90] and the State-of-the-Art report [102], and of course in the respective research reports listed in these references.

Experiments with cold simulant melts to investigate fluid dynamic phenomena were conducted by Brookhaven National Laboratory (BNL), Purdue University, UKEA Winfrith, Korea Advanced Institute of Science and Technology (KAIST), and the Korea Atomic Energy Research Institute (KAERI). Water and bismuth–lead alloy were used as melt simulants in most cases. These tests were done in scales varying from 1:40 to 1:25. Many tests were done in the whole pressure range, from low (RPV pressure lower than 0.3 MPa) to high pressure, in order to find the 'low-pressure cutoff' for different breach sizes and simulant liquids. Another objective was to determine the droplet size distribution in order to understand the entrainment mechanisms and obtain data for the melt–gas interaction correlations [121]. A comparative experimental study has been performed by KAERI between the Zion cavity without a direct pathway to the dome and the French 900 MWe PWR, which has this path along the annular gap around the RPV and has the large instrument channel similar to Zion [118]. These tests showed that, for the case of water as simulant material, when vessel pressure is above 1.5 MPa, 80% of the water is entrained through the annulus and then ejected into the containment. On the other hand, at pressures below 0.2 MPa, more than 90% of the simulant material remains in the cavity.

Experiments with high-temperature chemically reactive melts were conducted by Sandia National Laboratory (SNL), Argonne National Laboratory (ANL) and Fauske and Associates Inc. (FAI). Generally, iron–alumina melts were used; only ANL performed a few experiments with prototypic melt, containing UO_2, ZrO_2, Zr and steel. In most cases steam was used as driving gas, and in some experiments nitrogen was applied. Most of the tests were performed at high RPV pressure in the range of 3 to 8 MPa and relatively small breach sizes at the center of the lower head.

Separate-effect tests and integral effects tests have been conducted. The investigations focused on Westinghouse plants (geometries similar to Zion and Surry) and related ones in United Kingdom, Spain, and Korea. These reactors have an in-core instrumentation through the lower vessel head and a large instrumentation chute connecting the lower cavity to a subcompartment. The

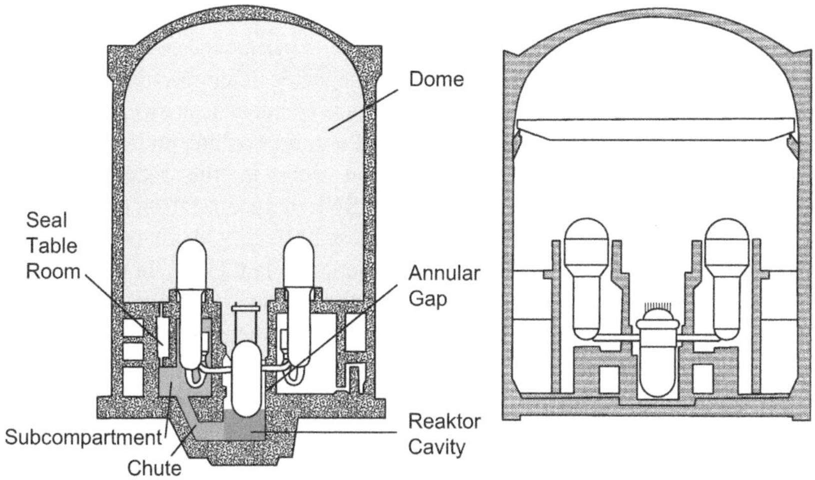

FIGURE 3.22 Configuration of Westinghouse plant (Zion: left) and combustion engineering plant (Calvert Cliffs: right) (drawings from [120] and [122]).

Zion geometry is referred to as "closed", with no direct pathway toward the dome but an indirect pathway between the cavity and the dome via the instrument tunnel and the intermediate compartments. The Surry geometry is referred to as "open": there is a direct pathway toward the dome, via the annular section around the vessel (Figure 3.22). However, the ratio of the flow cross sections, cavity exit to annular gap, is in the order of 12:1 to 5:1, depending on the choice of relevant parameters. Because of this large ratio, both geometries fall into the same category, where most of the melt is dispersed into the intermediate compartments, and very little into the containment dome, in contrast to the Combustion Engineering (CE) plant geometries (Calvert Cliffs) (Figure 3.22). Thus we have to distinguish geometries with large instrumentation chutes and relatively small paths directly to the containment dome and those without a chute. In both cases there are "open" and "closed" cavities.

The modeling of reactor specific details was stepped up during the campaign, noting their significance for the DCH processes. The problem of scaling was met by doing counterpart experiments, at different scales. The largest facility (SNL CTTF) represented the Surry plant in a scale 1:5.75. The SNL Surtsey facility had a scale of 1:10, the FAI facility was 1:20, and the ANL facility scaled the Zion plant 1:30 and 1:40. Data from 58 DCH relevant tests with hot melt are available relating to the Westinghouse geometry. Counterpart tests in different scales did not reveal any strong effect of physical scale. Tests performed with iron alumina on the one hand and with prototypic corium compositions on the other hand did not show significant differences either. Tests performed with uranium-based melt showed somewhat lower pressurization than the equivalent iron–alumina melt tests.

The data obtained served as the basis for developing analytical models (see Section 3.2.9) that were used to extrapolate the experimental results to prototypic conditions. The models predicted manageable loads for the Zion and Surry containments for the high-pressure severe accident scenario.

A small series of tests was performed for the geometry of a CE plant, which does not have an instrumentation duct in the reactor cavity (Figure 3.22). These tests were performed at SNL in a scale 1:10 with hot melt (CE and CES tests [120]) and at Purdue in a scale 1:20 with air or helium as driving gas, and water or Wood's metal as melt simulant [121]. In these types of plants, debris can be transported directly to the dome (average of 58% vs. 10% in Surry), and trapping of debris in subcompartments is absent. Refined analyses considering containment failure probability curves together with high-pressure melt ejection probabilities resulted in a conditional containment failure probability near or below the success criteria for DCH (<0.01), also for CE plants.

The main conclusion of this research program was as follows: if high-pressure melt ejection by depressurization of the RCS is prevented, then the pressure load due to DCH will probably stay below the design pressure of the containment.

Low-pressure Melt Ejection Experiments for European Plants

In 1998 two experiments were performed at SNL in cooperation with FZK, IRSN, and US NRC, with a cavity geometry similar to a CE plant but adapted to characteristics of the EPR cavity and with a direct path into the containment dome [103]. These tests were performed with iron–alumina melt, steam, and a prototypic containment atmosphere in a scale 1:10. The blowdown pressures were lower, 1.1 and 1.5 MPa versus approximately 8.0 MPa in the CE tests, and the breach diameter was 0.1 and 0.04 m versus 0.05 and 0.04 m in CE tests. The melt mass was larger with 62 kg versus 33 kg in CE tests. The melt dispersal into the containment dome was large, and the maximum pressure increase was 0.4 MPa, the same as the maximum pressure found in the CE experiments with 8 MPa RPV pressure. The conclusion was that, with a hole size equal to or larger than 0.5 m diameter (in reactor scale) in the lower head, even at low blowdown pressures the containment load can be high.

In 1998 an experimental and analytical program started at FZK to investigate melt ejection scenarios for typical European reactor designs. Experiments have been performed in two facilities, with a 1:18 scaled EPR geometry, characterized by a narrow cavity without exit other than through the annular space between pressure vessel and cavity wall, leading either directly to the upper containment or into the pump and steam generator rooms along the flow path around the main cooling lines. Since the system pressure in core melt accidents will be low due to compulsory system depressurization, the vessel failure pressures were kept between 0.8 and 2.2 MPa in these experiments.

Over 40 experiments were performed with cold simulant fluids in a test facility DISCO-C to study the fluid-dynamic processes with different failure modes of the bottom head under low-pressure conditions. The fluids employed were water or Woods metal instead of corium, and nitrogen or helium instead of steam. The main results from the cold experiments were as follows: large holes ($\varnothing \geq 0.5$ m, scaled) at the bottom of the bottom head led to high dispersed melt fractions (>50%). The maximum dispersed fraction for such breaches was reached already at pressures below 2.0 MPa. A certain amount was trapped in the reactor cavity depending on its geometry. With breaches at the side of the lower head, the dispersed melt fraction was lower and even with unzipping of the bottom head the dispersed fraction was smaller than with central holes [104].

In a second facility (DISCO-H), six experiments in the same scale were performed with an iron–alumina melt, steam, and a prototypic atmosphere in the containment. To determine the contribution of hydrogen combustion, one test was performed with nitrogen and air only. All breaches were holes at the center of the bottom head, which led to higher melt dispersal, but might be less likely than breaches at the side of the lower head [105] [108].

Also, experiments have been carried out in the DISCO facilities with the geometry of the French P'4 plant at linear scale 1:16. Fourteen tests with cold simulant melts (water and gallium alloy) and five experiments with iron–alumina melt were performed with similar initial conditions as in the preceding program [106]. The cavity of the 1300 MWe plant P'4 is characterized by a large distance between RPV lower head and cavity floor, a niche in the cavity, a relatively large flow path out of the lower cavity into a cavity bottom access room, and the upward directed flow paths through the space between RPV and cavity wall, leading either directly into the upper containment building or along the main cooling lines into the pump and steam generator rooms (see Table 3.1 and Figure 3.19).

One experiment was conducted with a geometry of the cavity of the Bulgarian VVER-1000 plant (Figure 3.19) [107] [108].

The phenomenon of hydrogen combustion during a DCH event has grown in importance since it was found that, for high preexisting hydrogen contents in the containment atmosphere, more than half the pressure increase can be due to combustion [108]. The available models are not adequate to extrapolate from small scale to reactor scale. Therefore, separate-effect tests were performed in the DISCO facility (scale 1:18) and in a larger test facility (scale 1:7), where hydrogen-steam jets from the reactor cavity are ejected into simplified models of EPR or P'4 reactor containments that have an air–steam–hydrogen atmosphere [123]. Dedicated CFD combustion codes (e.g., COM3D) can be validated on the results of these experiments and can also be used for extrapolation [119].

3.2.9. Modeling Tools

The ultimate goal of all modeling is the application to reactor scale and conditions, since all experiments have been performed in small scale and most of them with simulant fluids, and their results cannot simply be applied to the real plant. To reach this goal, modeling is essential to analyze experimental results, achieve insights in processes, identify sensitive parameters, and assess scaling characteristics.

There are three classes of modeling: simple analytical models, system codes with increasing level of detail in mechanistic models, and multidimensional CFD codes.

Analytical Models

Simple physics models attempt to represent or bound only the dominant processes contributing to DCH loads. They can be applied economically many times with different initial conditions and variation of uncertain parameters to create a probability distribution of containment loads.

The *Single Cell Equilibrium (SCE)* model [89] [112] is a useful tool to calculate the upper bound of pressure increase in the containment. It takes the entire containment volume as a single control volume, all thermal and chemical interactions between debris and containment atmosphere are at equilibrium, and heat loss to structures is excluded.

The containment pressurization ΔP is given by:

$$\frac{\Delta P}{P^0} = \frac{\Delta U}{U^0} = \frac{1}{U^0}\frac{\sum \Delta E_i}{1+\Psi} \quad \text{or}$$

$$\Delta P = \frac{\kappa - 1}{V}\frac{(\Delta E_b + \Delta E_t + \Delta E_r + \Delta E_{H2} + \Delta E_w)}{1+\Psi}$$

where

P^0 = *Initial containment pressure*
U = *Internal energy*
κ = *Ratio of gas specific heats*
V = *Containment volume*
Ψ = *Ratio of heat capacity between dispersed debris and containment atmosphere*

ΔE_b = *Energy of RCS steam and water*
ΔE_t = *Latent and sensible heat of debris*
ΔE_r = *Metal oxidation*
ΔE_{H2} = *Combustion of produced and preexisting hydrogen*
ΔE_w = *Water vaporization (energy sink)*

Experiments have shown that the SCE model yields a pressure rise too high by a factor 2 to 5.

The mitigating processes that should be considered are:

- Hydrodynamic processes limiting the amount of corium that mixes efficiently with the entire containment atmosphere (trapping in compartments, melt film, etc.)

- Kinetic processes limiting the thermal energy transfer on the DCH time scale (droplet size, surface effects, freezing, etc.)
- Processes limiting the amount of hydrogen that can be produced (the above plus steam limitation, equilibrium states, etc.)
- Processes limiting the combustion rate of produced or preexisting hydrogen (stratification, threshold temperature, oxygen starvation, etc.)
- Heat transfer to structures and vaporization of water acting as heat sinks

The Two Cell Equilibrium Model (TCE) [112] takes most of these limiting processes into account. Specifically, it treats the thermal and chemical interactions separately in two different locations, cavity plus compartments and the dome, and limits the interactions by applying interaction times through flow processes. All DCH processes are determined in the two volumes, with $\Delta U = \Delta U_1 + \Delta U_2$. The individual contributions are expressed as the product of an efficiency and the maximum internal energy change based on the single cell model:

$$\Delta U = (\eta_1 + \eta_2)\,\Delta U_{\text{SingleCell}} \quad \text{and the pressurization is then}$$

$$\frac{\Delta P}{P^o} = (\eta_1 + \eta_2)\left[\frac{\Delta P}{P^o}\right]_{\text{SingleCell}}$$

The efficiencies account for the dominant mitigating effects of compartmentalization and non-coherence of dispersal and blowdown. The amount of blowdown steam participating in DCH is limited by the coherence factor (see Section 3.2.6). The debris dispersal is partitioned according to the ratio of the flow cross sections to these volumes. The energy deposition in small volumes is limited by reaching temperature equilibrium between debris and atmosphere faster than in larger volumes. The model needs as input correlations for the fractions of melt ejected and dispersed from the reactor cavity and the coherence time. The TCE model was validated against experiments and was applied to plant analysis. It worked best for "closed" cavities of the Westinghouse plants, but showed discrepancies for tests that had reactive atmospheres or water in the cavity. Care must be taken if it is applied to other geometries.

The *Convection-Limited Containment Heating Model (CLCH)* [113] treats the hydrodynamic, thermal, and chemical processes sequentially in the RPV, cavity, and containment, and limits the interactions by applying interactions times through flow processes, similar to the TCE model. The model consists of nine simple equations, which can be put into dimensionless form. The solution, that is, the pressure increase, can be expressed as a function of eight independent dimensionless groups:

$$\frac{\Delta P}{P^o} = f\left[\frac{P_{0,v}}{P^o}, \frac{V_m}{V}, \frac{T_{0,v}}{T}, \frac{M_{Zr}}{M_{total}}, \frac{M_{ss}}{M_{total}}, \frac{m_{H2,v}}{m_{total,v}}, \frac{m_{H2,c}}{m_{total,c}}, \frac{\tau_m}{\tau_s}\right]$$

with ratios of: (1) initial steam pressure in primary system to containment, (2) volume of melt to volume of containment, (3) initial steam temperature to containment atmosphere temperature, (4) zirconium mass fraction, (5) steel mass fraction, (6) molar hydrogen fraction in steam, (7) molar hydrogen fraction in containment, and (8) melt dispersal/steam blowdown time. The unknown parameter No. 8 must be derived from experiments. The model is a good tool for analyzing the effect of different parameters. All results attain an asymptotic behavior at $\tau_m/\tau_s \sim 1$, with $\Delta P/P_o \sim 3$; implying thereby that for the Zion-like geometry the containment DCH pressure will be limited to 3 times the pressure prevailing before the HPME event. The CLCH model is restricted to Zion-like geometry.

Models for Use in Risk Codes

The second class of models belongs to control-volume codes (or Lumped-Parameter) where DCH processes are modeled in different levels of detail (see Chapter 8 for all references on MAAP, MELCOR, and ASTEC). There are the U.S. code systems MAAP, MELCOR, and CONTAIN [114]. A more specialized lumped parameter code is COCOSYS (Containment-Code-System) [115], which is a GRS code system based on mechanistic models for the comprehensive simulation of all relevant phenomena, processes, and plant states during severe accidents in the containment of LWR; its DCH models are similar as in CONTAIN. ASTEC, developed by IRSN and GRS, is a modular code system with dedicated DCH modules: RUPUICUV for the processes in the cavity and CPA for the processes occurring in the containment. These codes are fast-running codes but rely on validated correlations and/or analytical models for each reactor design. They are not aimed at studying and understanding phenomenology, but they reflect the level of knowledge for a given process.

The DCH models in the three main integral codes are briefly described as follows:

- MELCOR: The dispersed debris are distributed among compartments under user control. Heat transfer to the atmosphere and to surfaces, chemical reactions, and hydrogen combustion are included, and the user must select the appropriate model parameters. The various parameters and time constants are difficult to select for geometries where experimental data are not available.
- MAAP: The DCH module has many versions that use different correlations for the evaluation of the melt dispersal from the reactor cavity as a function of initial state (vessel pressure and break diameter). The droplets are assumed to be in dynamic equilibrium with the gases. The distribution of droplets between the various cavity outlets thus depends on gas flow rates. Once the debris is dispersed from the cavity, it attains thermal equilibrium in the first containment compartment. Only steam is consumed by oxidation of the dispersed corium droplets (Zr, Cr, Fe, and Ni successively). Neither

the oxygen in the air nor the oxidation of spread corium is taken into account. The code yields reasonable representation of containment pressurization found in early SNL-Surtsey tests.

- ASTEC: The pressure loads during DCH are assessed using the RUPUICUV and CPA modules (plus the CORIUM simple module that calculates the heat exchanges between corium droplets and containment atmosphere). The cavity phenomena are addressed by the RUPUICUV module. The total dispersed fraction is given by a correlation that is based on tests representative of a specific geometry. The user defines the partition of ejected fuel into various compartments. The droplets are assumed to be in thermal equilibrium with the gas in the cavity. Containment heat exchanges are assessed using simple convection models in the CPA module. Beyond a threshold temperature, steam oxidation of corium metals becomes possible, producing hydrogen and heat. Pressure and temperature loads as well as hydrogen combustion are also modeled.
- CONTAIN: it is possible to describe corium debris transport, flows between containment building compartments, debris trapping by various structures, chemical reactions (corium oxidation), heat transfers between debris and the atmosphere (through convection or radiation), and combustion of hydrogen produced by DCH as well as preexisting hydrogen. The CONTAIN 0-D code is the most advanced of its kind, offering an impressive collection of computational options and backed by a solid qualification database. However, validation was limited to U.S. reactors with Zion and Surry geometries, and recent use of the code by the GRS for DISCO tests on EPR and P′4 geometries did not produce satisfactory results. This may also be due to the code's complexity: due to the many options, it requires a high level of expertise from the user.

A general comment on code systems is that, if the use of simplified modeling allows these codes to perform fast-running computations, the complexity of phenomena involved in DCH is hardly well represented in these codes, and they often need tuning of their parameters to adapt them to the different conditions. For instance, the user must define some critical parameters such as the heat transfer coefficient between droplets and containment atmosphere, the droplet flight duration, the repartition of the dispersed mass.... Moreover, the modeling was based on experiments with specific geometries and their full applicability to other reactors is not demonstrated.

The validation and development is still ongoing for the ASTEC and COCOSYS codes (the latter in close analogy with the main CONTAIN modeling), which should improve some of the current weaknesses.

CFD Codes

The third class of modeling is done with two- or three-dimensional CFD codes, with at least three phases and three velocity fields for the different components, liquid melt and solid debris appearing as droplets or film, water, and the gas (steam, air, etc.). Mechanistic models treat each phenomenon locally in each computational cell. One of the most advanced and validated codes is AFDM (Advanced Fluid Dynamics Modeling) which belongs to the internationally developed SIMMER-code family [109] [110]. It is a multiphase, multicomponent 2-D code with three velocity fields, and special DCH models have been incorporated and validated against SNL and FZK experiments. The other code is MC3D, a 3-D multiphase IRSN code, originally developed for fuel-coolant interaction. This code is still in its development stage for the application to DCH [111]. The analysis with CFD codes gives insight into details that cannot be measured and helps to devise appropriate experiments. Once validated against different experiments, these codes should give the highest confidence of all three modeling tools, when applied to reactor scale. These codes, however, have long computing times, and their application in Probabilistic Safety Assessment (PSA) studies is therefore limited.

A recent trend is the application of commercial codes (e.g., CFX or STAR-CD) where new models are implemented or existing ones are qualified to calculate separate processes, such as the melt flow in the reactor cavity or the hydrogen combustion in the containment.

3.2.10. Validation of the Models

For all three classes of modeling the challenges are, first, the extrapolation from experimental scale to reactor scale, with the problem of the different effects of length, area, and volume scales on processes such as droplet fragmentation, flight times, mixing rates, combustion rates, and other; second, the extrapolation from simulant melts to corium, with different temperatures (absolute, freezing point, etc.), different oxidation potential, different specific energy content, and different density. Appropriate experiments with real corium for validation purposes are not available.

For future safety assessments fast-running codes are necessary, and here the biggest challenge is the transfer of the detailed models used in CFD codes or their approximation to fast-running system codes.

3.2.11. Conclusion

The experimental database shows that the consequences of DCH are essentially related to cavity geometry, and to the pathways between it and other containment areas. In particular, it is now widely accepted that the consequences of DCH are limited in reactors with no direct pathway between the cavity and the containment dome. In contrast, the situation is less clear for reactors that do have a direct pathway between the cavity and the containment; tests show that substantial fractions of corium may be dispersed into the containment in such cases if pressure in the reactor coolant system is relatively high at the time of DCH.

Combustion of the hydrogen produced by oxidation as well as the hydrogen initially present, if its concentration is high (above ~4%), appears to be the dominant phenomenon for containment pressurization. It should be noted that DCH oxidation is difficult to extrapolate to reactor scenarios based on tests with simulant materials, which are much less reactive than corium components, especially zirconium.

As to DCH modeling, it has proven complicated; the complexity and diversity of the phenomena involved do not lend themselves to simplified modeling. Multiphase simulation software appears to be the most promising option, but existing solutions are not yet able to satisfactorily predict outcomes on a reactor scale, due to problems in modeling oxidation and combustion in particular.

The impact of co-ejected water or water in the cavity on DCH has not really been characterized either; a better understanding is needed. For this point as well, simulation software may provide new insight.

For a more detailed overview on the DCH issue, references [89] and [102] are recommended, which reflect the state of the art as of 1996.

3.3. STEAM EXPLOSION IN LIGHT WATER REACTORS

(D. Magallon)

3.3.1. Introduction

During a severe accident, large amounts of molten core materials (corium) containing a large fraction of molten fuel may enter into contact with the cooling water generating a so-called fuel-coolant interaction (FCI). If a number of conditions are realized, transfer to the water of a significant part of the energy stored in the corium melt may take place so rapidly that explosive vaporization of the water occurs. This process, called steam explosion, can result in mechanical damage of the reactor structures. Evaluation of steam explosion risk requires being able to quantify the level of the generated loads to verify whether reactor structures are capable of withstanding them, and to design appropriate countermeasures. This is particularly important for the containment, which in LWRs is the last barrier to the release of radioactive fission products to the environment.

Figure 3.23 shows the FCI situations that are considered susceptible to resulting in a steam explosion with potential damage to the structures in LWR severe accidents.

After melting of a certain amount of core materials, a breach eventually forms in the crust or structure that holds the melt within the core, and the corium melt starts flowing down toward the lower head. The first FCI situation that occurs is when the melt interacts with the water in the RPV lower head. A steam explosion occurring during this FCI may induce failure of the lower head and/or a liquid slug playing against the uppermost structures of the RPV,

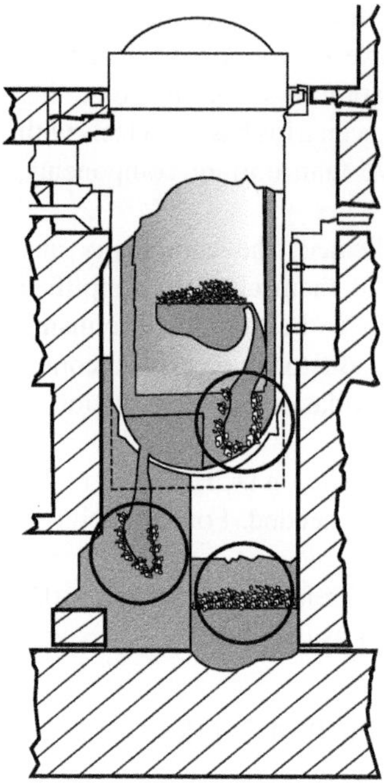

FIGURE 3.23 Different possible FCI scenarios in an LWR accident.

accelerating them up to hitting the upper head with a significant amount of kinetic energy. This energy is maximised if the lower head does not fail and may induce rupture of the bolts that hold the upper head into place. The upper head is in turn accelerated upward up to the containment roof like a missile and may cause failure of the containment wall. This mode of failure of the containment, known as "alpha-mode failure," received the most attention in the past because it was believed to be the highest probability cause of containment failure due to a steam explosion and because of its early occurrence in the accident sequence. A second international steam explosion review performed in 1996 [124], complemented by the experimental program BERDA performed at FZK [125], allowed eliminating this issue as a risk for containment failure.

Studies on in-vessel steam explosion have focused on the risk of lower head failure. In existing reactors, lower head failure is expected due to thermal loading of the corium debris, and accident management procedures are designed accordingly.[1] But a steam explosion inducing an early vessel failure during core melt slump into water would challenge those procedures. For the

1. A particular case is the TMI-2 accident where no vessel failure occurred. The reasons for that are addressed in Chapter 2.

new generation of reactors based on in-vessel retention of the corium debris, it is obvious that the risk of lower head failure due to a steam explosion has to be evaluated. Theofanous and colleagues performed such an analysis for AP-600-like reactor design and concluded that the lower head was strong enough to withstand the loads generated by a steam explosion with a very low ($<10^{-3}$) conditional probability of failure [126]. The recent international OECD SERENA program is moving toward the conclusion that in-vessel steam explosion would not induce lower head failure in LWRs in general (see Section 3.3.3).

After vessel melt-through, the molten corium slumps into the cavity. Either because it is part of the accident management strategy to flood the cavity before vessel failure for quenching the melt, or because it is very likely that water will collect on the cavity bottom during the earlier phase of the accident, ex-vessel steam explosion has to be considered as a potential risk. There are no phenomenological differences between in- and ex-vessel steam explosion, but the different initial and boundary conditions can significantly impact on the energy yield of the explosion. These differences are essentially the pressure of the system (a few bar ex-vessel versus up to 20 bar in-vessel at the time of vessel melt-through), the temperature of the water (subcooled ex-vessel vs. saturated in-vessel), the melt composition (e.g., higher steel content ex-vessel), the melt delivery (e.g., single large pour ex-vessel vs. multi-jet pour in-vessel), the depth of the water (7–9 m ex-vessel in Swedish BWR vs. 2 m in-vessel in PWR). In addition, the structural capacity of the reactor cavity is much lower than the capacity of the lower head. Large uncertainties exist on the level of loads generated on the reactor cavity wall by ex-vessel steam explosion, and most of the current efforts aim at resolving this issue.

Another FCI situation that can lead to a steam explosion in LWRs is flooding of debris after melt collection in RPV lower head or cavity, leading to what is in general referred to as “stratified steam explosions”. These situations lead to explosions that are less energetic than those associated with melt pour because of the limited quantity of melt that can mix with water before the explosion (mixing around the interface melt–water) and because the system is less constrained, thus allowing early venting of the explosion. They received far less attention than melt pour situations and will not be considered here. The interested reader can refer to [127] for experimental studies and [128] for theoretical considerations and modeling of stratified steam explosion.

It has to be noted that a power excursion accident may induce steam explosion in some types of reactors. A fast power increase induces melting of the fuel and its high-pressure ejection from the fuel element claddings into the coolant, resulting in fast vapor production. Steam explosions of this type occurred in the Chernobyl accident. This type of accident is not considered an issue for LWRs, since the western LWRs in general reduce their reactivity as the coolant boils.

The remainder of this section is devoted to describing steam explosion resulting from melt pouring into water, as being the most challenging generic

FCI situation for LWRs. Except when otherwise mentioned, the "system" will always mean the components in interaction plus the surrounding structures.

3.3.2. Some Definitions

In this section, fuel–coolant interaction or corium–coolant interaction, designated under the term of FCI, will refer to the generic process in which molten core material (corium) is put in close contact with water coolant. As an example, an FCI occurs as soon as corium melt is poured in water (or vice versa) during a severe accident, independently of the evolution of the system. A steam explosion is a particular type of FCI: "steam explosion" will be used each time we speak of an FCI with energy so rapidly transferred from the melt to the water that pressure peaks of large amplitude and short duration (dynamic pressure) are produced. What is behind "large" and "short" is a question of scale. An event resulting from fast fragmentation of a single melt drop interacting with water into very fine fragments ("explosion of the drop") and producing peaks of order 1.0 MPa and duration 10^{-6} s is a steam explosion. It is of little concern for reactor safety, but holds mechanisms of fundamental fragmentation and heat and mass transfer that govern larger scale explosions. In addition, such a local explosion occurring during FCI involving reactor-scale amounts of corium melt and water can be the initiator of an explosion involving tons of melt and having milliseconds duration.

It can be shown, and it has been experimentally verified, that large quantities of melt penetrating at once into water cannot produce a one-step explosion like a single drop does.[2] Pre-fragmentation of the melt occurs, and the steam explosion is the result of a process that, starting from the explosion of a single fragment or small group of fragments, will amplify and propagate to all or part of the pre-fragmented melt. This sort of propagating event, which can generate pressure peaks of the order of 100 MPa and milliseconds duration, is challenging for the reactor structures and is therefore an issue for reactor safety. The self-sustained character of such explosions makes them comparable to chemical detonations: they are called thermal detonations.

The "conversion ratio" of the explosion is usually the ratio of the mechanical energy output to the total thermal energy content of the corium mixed with water at the time the explosion occurs. "Efficiency" is the conversion ratio expressed as a percentage of the melt thermal energy. It is generally used as an indicator of the violence, intensity, and strength of the explosion for a given melt–water system. Care must be taken with using and extrapolating this notion from small (experimental) to large corium masses ("melt mixed with water" difficult to evaluate for large masses) and to other

2. Actually, a single-drop explosion is not a true one-step process, as the drop is completely fragmented after two to three cycles of growth and collapse of the steam bubble surrounding the drop.

geometries or situations than those used to establish it. Damage that may be induced by an explosion depends, among other things, on the quantity of participating melt (see the footnote associated with the paragraph "Conclusions on Triggering" in Section 3.3.4). Nevertheless, the notion is largely used as a measure of the explosivity of a melt–coolant system.

3.3.3. Conceptual Description of the Steam Explosion Process

On the basis of experimental observations and analyses, a large-scale steam explosion associated with melt pour can be represented conceptually as in Figure 3.24. First, the corium melt penetrates and breaks up into water in a film boiling regime: it creates a melt–water–steam mixture (premixing). Typically, the melt fragments that form during this phase are of cm size (coarse mixing). Due to the presence of a vapor film between melt and water, heat transfer from corium to water is relatively low and the corium remains essentially in a liquid state. After a certain amount of melt has penetrated into the water, an event (the trigger) may initiate the steam explosion process itself somewhere in the premixture, generally when the melt front hits the bottom of the housing structure (here vessel or cavity). The trigger induces vapor film collapse locally, and liquid (melt)–liquid (water) contact occurs, giving rise to fine fragmentation of the melt (typically fragments <100 μm) allowing rapid heat transfer to the water and subsequent high pressurization. While expanding, this zone produces further fragmentation around it, and this process escalates and propagates to all the premixture. Propagation is a self-sustained process that can reach supersonic velocities depending on the premixture characteristics. In this case, the

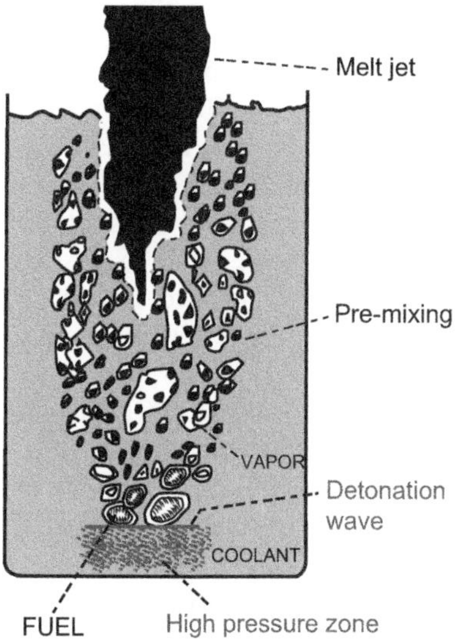

FIGURE 3.24 Conceptual picture of a steam explosion associated with melt pour (partly from [137]).

propagation front is associated with a shock that reaches the water surface in a few milliseconds and leaves behind it a high-pressure region inducing dynamic loading of the surrounding structure (pressure of 120 MPa and impulse loading of 75 kPa.s have been measured in 1-D experiments with 1.5 kg of alumina melt [129]). It can be considered similar to a chemical detonation, where fragmentation plays the role of chemical reaction. Analogy with a chemical detonation together with a model for 1-D geometry was first proposed by Board et al. in a famous letter to *Nature* in 1975 [130]. They called it a "thermal detonation". Pressure relief of the exploded zone can induce the generation of missiles leading, for example, to alpha-mode failure and/or to high static pressurization of the housing structure if the volume is closed.

The four steps—premixing, triggering, propagation, and expansion—and the reasons why an explosion may or may not occur will now be described in more detail.

3.3.4. Description of the Various Steps of a Steam Explosion

Premixing

A significant amount of melt must be involved in a steam explosion in order for it to have a "chance" to damage the reactor structures. If a steam explosion occurs as soon as melt contacts the water in a surface explosion, the melt mass involved is very small and so is the explosion strength. Actually, a significant amount of melt, if not all, can penetrate into the water before or without any steam explosion. As the melt descends through the water, it fragments due to hydrodynamic forces and releases energy to the water. Steam is produced and a melt–water–steam mixture forms. Given the fragmentation process of the melt and the boiling regime of water, the rate of heat transfer is governed by the mass pouring rate of the melt. The partly quenched debris eventually settle on the bottom where they form a debris bed and/or re-melt due to decay heat forming a molten pool. This is what occurred in the TMI-2 accident, where about 20 tons of molten corium relocated in the lower head in approximately 2 min while the vessel was full of water. It resulted in a vessel pressure increase of 2 MPa due to steam production, but no steam explosion occurred. In steam explosion terminology, this mixing is called premixing because it precedes a possible steam explosion, which is another type of mixing process. An example of two premixing configurations, as obtained with two different melts in the KROTOS facility (illustrated in Figure 3.25), is shown in Figure 3.26.

Figure 3.27 shows typical sizes of particles collected after FCI without steam explosion in KROTOS (5-kg-scale corium melt [131] and FARO (100-kg-scale corium melt [132]) experiments. Premixing with alumina melts in KROTOS produced fragments of mean size of the order of 10 mm as compared to 1 mm with corium.

Another condition to be verified in order for a steam explosion to occur is that during the mixing phase most of the corium remains in molten state. This is

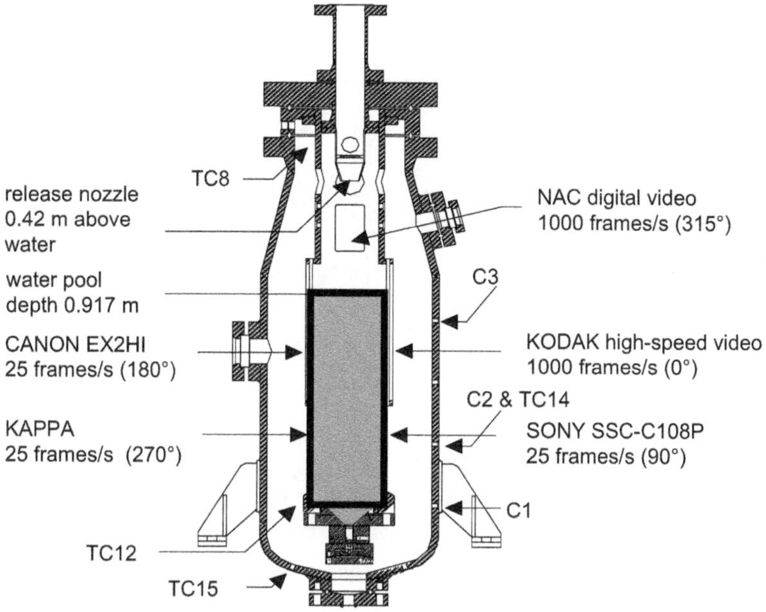

FIGURE 3.25 KROTOS test vessel and visualization devices. Melt poured in water via the release nozzle. Water pool diameter 0.2 m, depth ~1 m.

possible when melt and water mix together in a film boiling regime of water, because heat transfer to water through the film is low. Film boiling is always the boiling regime of water in LWR core meltdown accidents due to the very large difference existing between the temperature of the melt (above 2000°C) and the boiling temperature of the water (even at elevated pressure). Therefore the necessary condition for a steam explosion occurrence is always verified in LWR severe accident conditions[3]. But this condition is not sufficient: a trigger is required.

Triggering

Film boiling is an unstable situation that can be disrupted by events that occur during melt descent through water. Collapse of the vapor film somewhere in the premixture may initiate a fine fragmentation of the melt and a rapid heat transfer to water, which, extending to all the premixture in a few milliseconds, may give rise to high pressurization of the system. The initiating event is called the trigger.

3. Provided, of course, that liquid water remains present in the premixture. If all liquid water is swept out due to, for example, intense vapor counterflow, no steam explosion can occur.

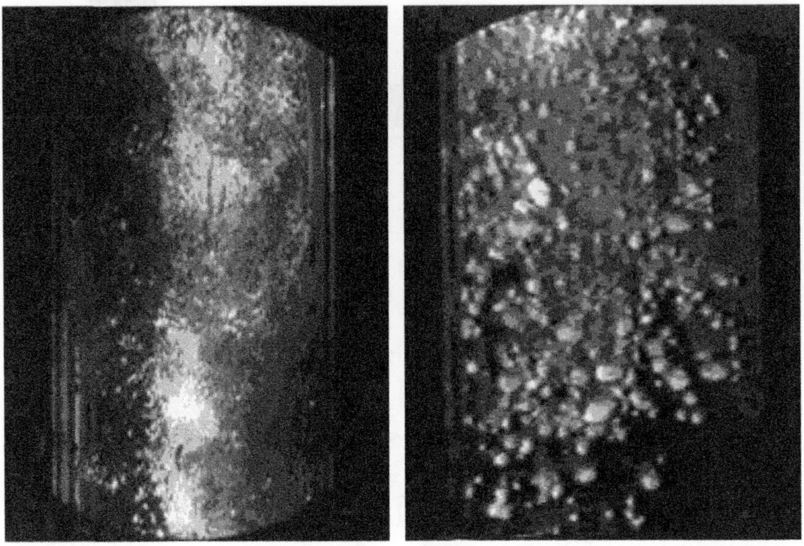

FIGURE 3.26 Premixing configurations in KROTOS tests. Left: Corium melt. Right: Alumina melt. Viewing area 100 × 200 mm, CANON video camera (from [131]).

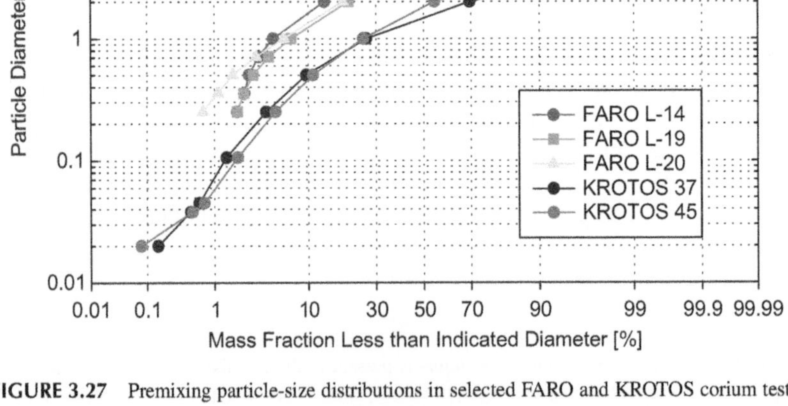

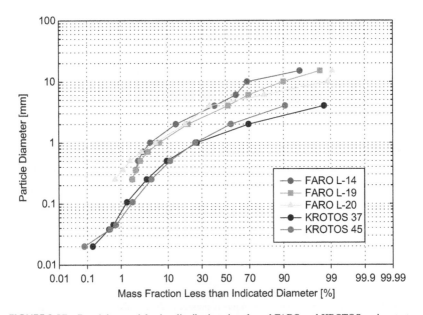

FIGURE 3.27 Premixing particle-size distributions in selected FARO and KROTOS corium tests.

"Spontaneous" Trigger

The experimental database allows establishing the conditions that are more or less favorable to "spontaneous steam explosions". A spontaneous explosion does not require any external event: the trigger is generated by the system itself. It has been observed that, when a steam explosion occurs spontaneously, it often triggers when the melt contacts the structure containing water, which generally occurs when the melt front reaches the structure bottom. The reasons for that are not clear, but the change of boiling regime due to melt impact on the structure (transition from film boiling to transition or nucleate boiling) may generate localized pressure pulses that are capable of triggering the explosion. Another possibility is water entrapment by melt, followed by rapid vaporization of the entrapped water. Installing a thermoplastic liner on steel structures has resulted in suppressing spontaneous steam explosion in some experiments, which indicates that the change of boiling regime was the most likely cause of the explosion in those cases. Unfortunately, this behavior cannot be generalized.

Steam explosions have been found more difficult to trigger spontaneously when the system pressure is high. The reason is that stability of the vapor film increases with the pressure, and thus more energy is required to collapse it. Large void (steam) fraction in the premixture can also prevent spontaneous occurrence of steam explosion. Here, it is the large steam fraction (large film thickness) that makes liquid–liquid contact difficult. For instance, it is more difficult to trigger explosions in saturated water than in subcooled water. The presence of noncondensable gases in the mixture has a cushioning effect that hinders film collapse. Noncondensable gases can result from entrainment of cover gas by the melt when penetrating into the water and/or from chemical reaction occurring during the mixing process itself—for example, hydrogen production due to corium oxidation by steam. In the latter, the exothermic oxidation reaction in turn enhances steam production, which combines with hydrogen generation to further increase void fraction and thus resistance to film collapse.

More recently, on the basis of experimental results obtained in the TROI facility (South Korea [133]), it has been suggested that corium of non-eutectic compositions may be more resistant to steam explosion than eutectic corium melts. However, no clear confirmation has been obtained so far, and investigations are still ongoing.

No steam explosion was observed in the TMI-2 accident, which had most of the conditions that are unfavorable to a spontaneous steam explosion: high pressure (above 10 MPa), saturated water (large void fraction), hydrogen production (Zr oxidation during quenching), and non-eutectic composition of corium, including fission products (which are believed to substantially increase the gap between liquidus and solidus temperatures). Similarly, no spontaneous steam explosion has been observed in any of the FARO and KROTOS tests performed at JRC-Ispra with up to 175 kg of 80 w% UO_2-20 w% ZrO_2 melts (~20–50 K interval between liquidus and solidus temperatures) at pressures between at 0.1 and 5.0 MPa in either saturated or subcooled water. But spontaneous steam explosions are easily obtained in a TROI facility with 70 w% UO_2-30 w% ZrO_2 melts (0 K between liquidus and solidus).

"External" Trigger

An event external to the system may trigger an explosion. Experimentally, artificial triggering of steam explosions has been obtained by applying an external shock on the steel container housing the FCI (e.g., hammer blow), a shock wave generated inside the system by the rupture of a high-pressure gas capsule or by a chemical explosion, a rapid water motion generated by a piston. Figure 3.28 shows the pressure signature in pure water of two different triggers used in the KROTOS facility, where the trigger is located at the bottom end of the test tube. Pressure signals correspond to six elevations evenly distributed over the water column on the test tube wall. In both cases, the energy of the trigger is similar and around a few hundred Joules, but one has small amplitude and "long" duration while the other has large amplitude and short duration.

The experimental database shows that a steam explosion may be obtained when an external trigger is applied to those situations that exhibit a certain resistance to spontaneous steam explosion (e.g., high system pressure, thermoplastic liner on structures), and rather easily. Both triggers of Figure 3.28 have been able to set off steam explosions in KROTOS corium/water experiments in conditions where no spontaneous explosions were observed [131]. The explosive trigger of Figure 3.28 was able to cause an explosion in the only attempt performed in the FARO facility (FARO L-33 performed with 25 kg of corium melt at pressure 0.4 MPa and water at ambient temperature [134]).

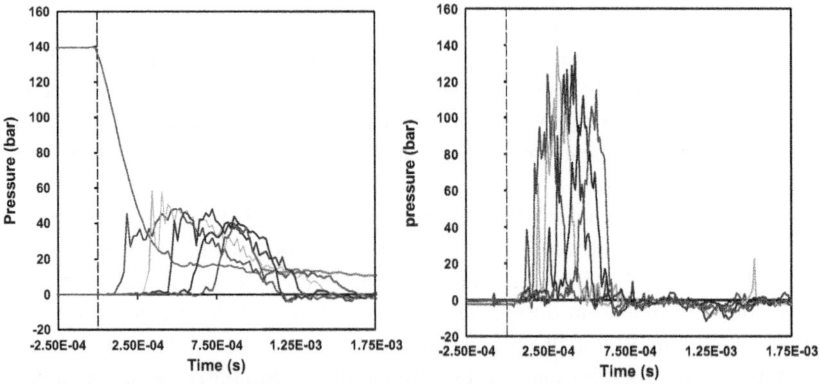

FIGURE 3.28 Pressure wave generated in pure water by two different types of trigger used in the KROTOS facility. Left: Rupture of a 14 MPa gas capsule at bottom of test section. Right: Explosion of 0.17 g of PETN at bottom of test section (pressure transducers located on the wall of the test tube at different elevations).

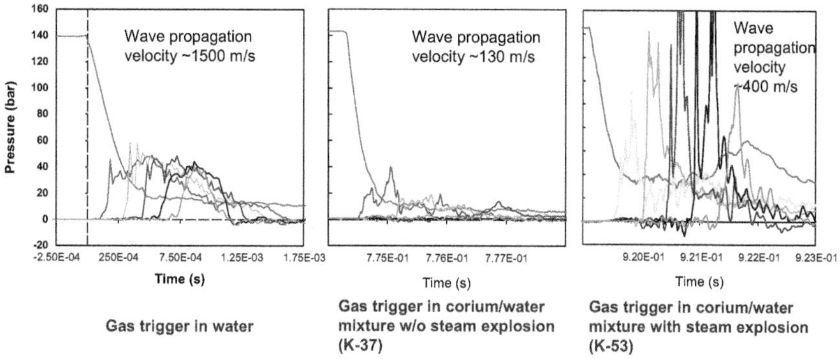

FIGURE 3.29 Possible evolution of premixtures exposed to the same trigger.

Figure 3.29 shows the pressure response to the gas trigger of Figure 3.28 for two corium–water systems in the KROTOS facility. In the first case (K-37), the trigger was not sufficient to induce an explosion. The pressure wave was damped with respect to the pure water case and propagated at a velocity typical of the premixture characteristics (130 m/s). In the second case (K-53), the same trigger was sufficient to induce an explosion, and the propagation front moved at a velocity of 400 m/s into the premixture. The main difference between the tests was the system pressure (0.10 MPa in K-37 vs. 0.36 MPa in K-53). In K-37 large void (steam fraction) in the premixture due to the low system pressure could inhibit the explosion even with an external trigger. The slightly higher pressure in K-57 was enough to reduce the void sufficiently for a steam explosion to occur with the same trigger.

Conclusion on Triggering

From the preceding discussion, it can be concluded that it is practically impossible to predict whether or not a steam explosion will trigger in a real situation. This makes the definition of countermeasures to prevent steam explosions very difficult. Actually, any melt/coolant mixture involving high-temperature melt such that film boiling is the dominant boiling regime of the coolant may generate a steam explosion if a sufficiently energetic trigger is provided[4]. The problem is establishing how energetic is energetic enough for

4. Note that there is no experimental evidence that it would be possible to trigger an explosion when pouring molten corium into water for very high system pressures such as in the TMI-2 accident. The maximum system pressure used to trigger an explosion with UO_2-based melt was 1. 0 MPa (SUW series performed at Winfrith with a UO_2-Mo mixture at 3600K [135]). Triggering the explosion was not easy, but the results showed an enhanced melt quantity participating in the explosion with respect to tests performed at lower pressures in the same facility. Thus, while the efficiency based on participating melt was lower than in a similar test performed at lower pressure, it induced more damage to the facility due to the higher melt mass participating.

a given system and/or situation, and whether energies of external triggers necessary to generate an explosion in experimental systems can be found in internal events occuring during core meltdown in reactor accident sequences. Past studies have been inconclusive in those respects, and, according to the present situation of the FCI research, little progress is expected in this area in the near future. It is the reason why present analyses of the steam explosion risk postulate that an explosion always occurs during an FCI (triggering probability equal to 1) and assess the consequences it may have for the surrounding structures. This helps to design structures that can withstand a steam explosion and define severe accident management strategies accordingly.

Propagation

The trigger is a local event initiator of a process that will propagate to all the premixture and lead to rapid transfer to water of a significant part of the thermal energy stored in the melt. The characteristic time of propagation is the order of a few milliseconds at reactor scale. One may say that propagation is the steam explosion phase where thermal energy of the melt is converted into thermal energy of the coolant.

It is generally admitted that the trigger (e.g., melt–bottom impact) causes the vapor film around the neighbor particles to collapse, resulting in a "liquid–liquid" contact (liquid water–liquid melt), inducing rapid fine fragmentation of the melt. This process is called thermal fragmentation. One possible way this process may occur is illustrated in Figure 3.30. Film collapse induces liquid water jet impingement on the melt drop surface (top picture) and water penetration inside the drop (middle picture). Pressure generated inside the drop by heat transfer and vaporization of the entrapped water causes the drop to fragment and the fragments to be expelled out of the drop (bottom picture), as well as the vapor film to grow and collapse again, inducing further fragmentation of the drop. This phenomenon, occurring simultaneously at several locations at the surface of the drop, induces its rapid destruction (often called "explosion of the drop") and generation of high-pressure locally by the thermal energy transfer. It has been experimentally observed that a molten tin drop in water is totally fragmented (exploded) after a number of film growth/collapse cycles lasting a few ms, while the very nature of the way the drop fragments has not been observed so far (see [136] as the most recent experimental contribution on this subject).

The explosion of a drop (or a few drops) may in turn induce neighboring drops to explode in the same way, and this phenomenon may propagate step by step to all the premixture. Propagation velocity depends on the conditions in the premixing region, that is, for a given system, essentially the component distribution (melt–water–steam). It may correspond to the propagation velocity of disturbances in the premixing region, resulting in sequential ignition of the mixture. In this case there is no steam explosion, and the pressurization of the system is relatively limited, slow, and uniform, without generation of shock

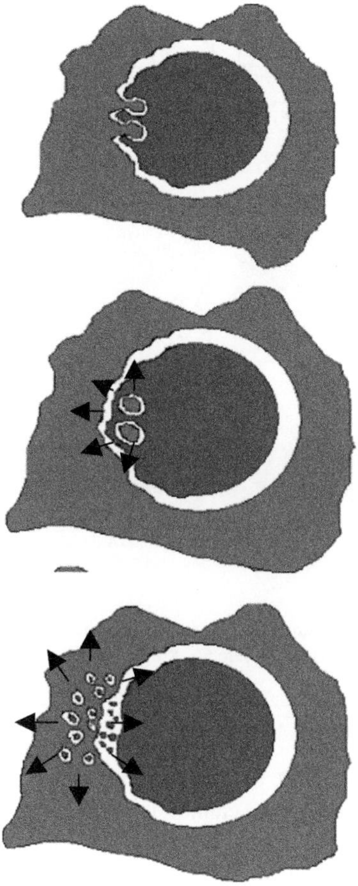

FIGURE 3.30 Conceptual illustration of thermal fragmentation induced by film collapse (from [137]).

waves. Typical velocities in this case are of the order of some tens of m/s. An example of sequential ignition of explosions of drops not having the characters of a thermal detonation is given in Figure 3.31 [138]. In this experiment, a first drop was triggered on the left by a spark discharge, and the explosion propagated from left to right at a velocity around 8 m/s, not sufficient to induce generation of shock waves and a global pressurization of the system. A series of pressure picks of amplitude less than 2.0 MPa and duration 10^{-5} s corresponding to the explosion of each drop was recorded at the base of the test rig. This did not represent any danger for the thermoplastic test container.

If the mixture conditions are favorable (e.g., large concentration of melt drops and sufficient liquid water), the propagation velocity may rapidly escalate up to reaching a quasi-steady state presenting all the characters of a detonation: a shock wave sustained by energy released in the zone immediately following

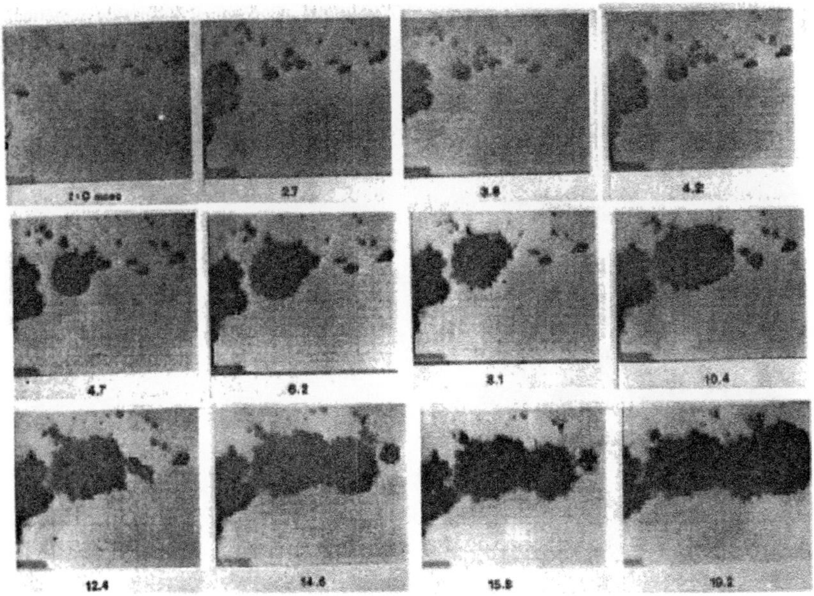

FIGURE 3.31 Propagation of vapor explosion in a multiple drop system (from [138]). Tin drops 1073 K; Water 327 K; Drop spacing ~2 cm; Measured propagation velocity 8 m/s.

the shock front propagating at supersonic velocity in the premixture. It is generally believed that thermal fragmentation alone cannot produce such a detonation wave, but is progressively substituted or complemented by hydrodynamic fragmentation induced by differential velocity between melt and coolant created by passage of the shock front. An illustration of this process is given in Figure 3.32. As a comparison with the sequential ignition of drops of Figure 3.31, a thermal detonation of a tin/water mixture issued from a tin jet penetrating in water propagates at a velocity of order 100 m/s and produces a single peak of order 10 MPa pressure at the wall and 10^{-3} s duration. Tin/water explosions are generally mild events.

Depending on the conditions, the premixture can "burn" more or less completely before any pressure relief can take place, creating very high pressures behind the front. Supercritical explosions with dynamic pressures of order 100 MPa and impulses in excess of 100 kPa.s have been obtained in Al_2O_3/water systems in quasi 1-D geometry (Figure 3.33). In one of these tests, the 40-mm-thick steel bottom of the test tube was severely damaged, and the tube itself was permanently deformed. Note that in multidimensional systems like in reactor accident, "venting" may reduce the pressurization at the walls and mitigate the explosion effects with respect to one-dimensional situations.

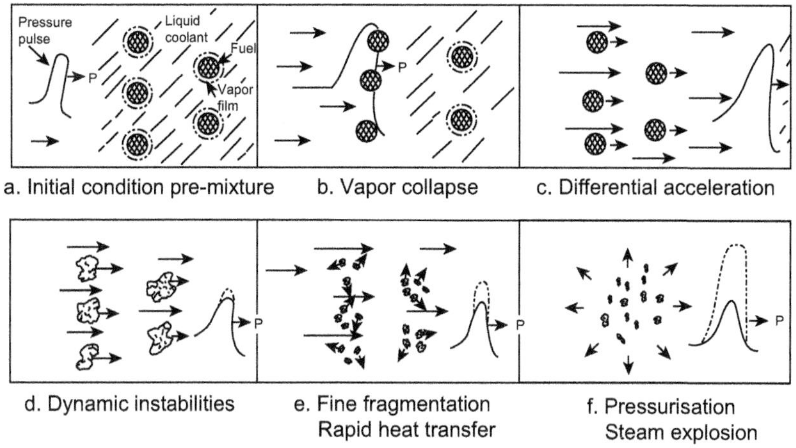

FIGURE 3.32 Illustration of a thermal detonation (from [139]).

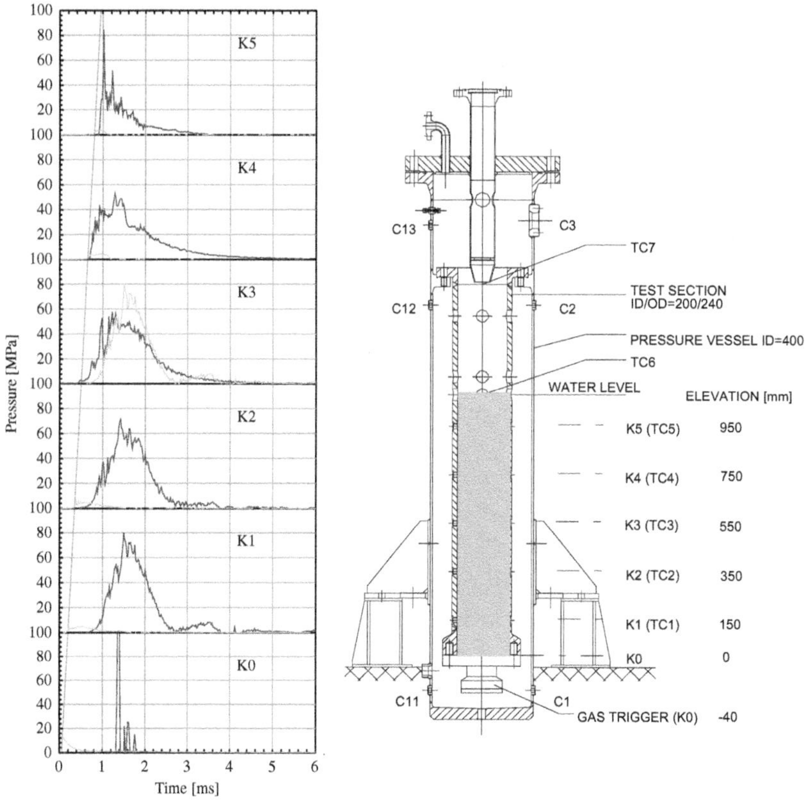

FIGURE 3.33 Explosion pressures measured at test section wall at different elevations in KROTOS K-49: 1.47 kg of alumina melt at 2700 K in 1-m-depth water at 294 K.

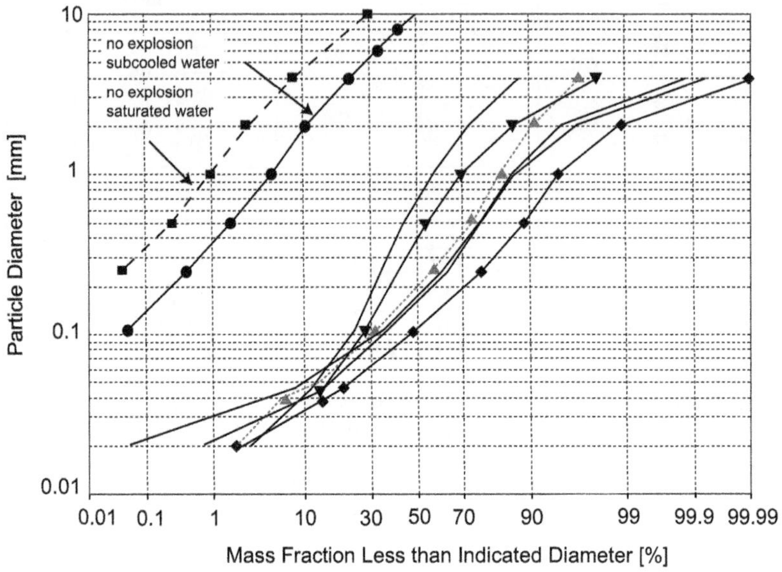

FIGURE 3.34 Debris size in selected KROTOS alumina tests with steam explosion compared to debris size without steam explosion (top two curves).

Fragments of size < 100 μm are produced during this process. Figure 3.34 shows particle-size distributions of debris from steam explosion with alumina melt as compared to debris from experiments without steam explosion.

Expansion

Expansion is the phase in which thermal energy of the coolant is converted into mechanical energy, resulting in possible damage to the surrounding structures. Keeping in mind the analogy with a chemical detonation and considering a (stationary) detonation wave progressing in the premixture from the closed end to the open end of a very long tube (one-dimensional situation with the explosion triggered at the closed end), one can see that:

- The region behind the detonation front (the "burn gases" region) is an expansion zone where elevated pressures are reached depending on the premixture characteristics (see, e.g., cases of tin and alumina above),
- The premixing zone ahead of the detonation wave is not "seeing" anything before being reached by the wave. The same holds for the space downstream to the premixing region.

If the tube is sufficiently strong to withstand the impulse resulting from high pressure behind the detonation front, all the space surrounding the tube is not affected by what is occurring in the tube itself until all the premixture has entirely "burned" (i.e., the detonation wave has reached the downstream boundary of the premixture). Then pressure relief against the surroundings will occur. Suppose the tube is vertical and the explosion occurs under water. Tube pressure relief will induce ejection of materials from the tube and upward acceleration of the water slug and possible other overlying materials (possibly leading to alpha-mode failure as described earlier). In a closed system, pressurization of the surroundings occurs. This situation can be considered to occur in the quasi one-dimensional KROTOS configuration using a 20-mm-thick steel test tube (Figure 3.33).

If the tube is not strong enough to withstand dynamic loading, part of the mechanical energy will be consumed for deforming and rupturing the tube, inducing a venting of the explosion, and a reduction of the energy transferred to the upward motion of the slug is expected.

Transposition of this ideal picture to a steam explosion occurring in the vessel lower head in case of core meltdown accident could lead to alpha-mode failure and/or lower head failure, which are the two major issues for in-vessel FCI. In reality, FCI in reactor conditions is likely to be a 3-D situation and the separation between propagation on one hand and expansion on the other hand is not so straightforward. The development of multidimensional FCI codes allows a representation of this interdependency.

Summary of the steps of steam explosion

Corium melt poured into water experiences coarse fragmentation (fragments of order some mm) in a film boiling regime due to hydrodynamic effects. This phase is called premixing and is unstable (meta-stable). Destabilization of the system locally induces drop-size explosion with production of 100 μm-size fragments, which starts propagating through the premixture (triggering). If conditions are favorable (unfavorable from the safety point of view), escalation to supersonic propagation velocity may occur, inducing a self-sustaining process similar to a chemical detonation extending to all or part of the premixture. This process occurring in the RPV lower head induces dynamic loading of the lower head walls with possible failure of the lower head. Recent studies (in particular, the international OECD SERENA program) have reached consensus that failure of the lower head exposed to this type of loading is near to be ruled out from the risk perspective (see Section 3.3.7). A slug (missile) made of water and/or overlaying materials may be accelerated by a steam explosion occurring in the lower head and act against the upper head and induce failure of the upper head, becoming in turn a missile that may fail the the containment roof (alpha-mode failure). This mode of failure of the containment has been definitely ruled out as an issue from the risk perspective. A steam explosion occurring in the cavity beneath the RPV in case of rupture of the lower head by either mechanical or thermal loading may damage the cavity wall, challenging containment integrity and/or may induce displacement of the vessel with possible further damage to the primary circuit and bypass sequences. At present, this issue remains unresolved (see Section 3.3.7).

3.3.5. Global Estimates of Steam Explosion Energetics

Independently of the amount of melt that can be (pre-)mixed with water and of the geometry, it is possible to calculate, by thermodynamic methods, the work done during the expansion phase as a function of the ratio of the initial volumes fractions of melt and water. In 1965 Hicks and Menzies proposed a thermodynamic estimate giving an effective upper limit of the energy conversion for fast reactors [140]. This was done by considering instantaneous and constant-volume mixing of melt and coolant to thermodynamic equilibrium, followed by isentropic expansion of the system to its end state (atmospheric pressure). The maximum conversion ratio (mechanical energy output divided by the thermal energy of the melt) is found for almost equal volumes of melt and coolant. This maximum is in the range 50–60% for corium–water and alumina–water systems, and around 27% for tin–water systems. The supercritical explosions observed with alumina in the quasi 1-D geometry of KROTOS (Figure 3.33), which produced one of most violent explosions ever observed[5], correspond to a conversion ratio of less than 3% (considering all the melt present in water at the time of the explosion) inducing severe damage to the test tube, and those performed with tin and corium less than 0.5%. This is also true for all experiments performed since Hicks-Menzies' estimates.

It should be noted that the original Board and Hall thermal detonation model, which, contrary to the Hicks-Menzies model, gives a description of the dynamic processes involved in a steam explosion, results in similar efficiencies as shown in Figure 3.35.

Clearly, the above methods are not applicable for reactor safety assessment. In the real world, premixtures are not so ideal, and the number of dissipation processes limits energy conversion. Models have been subsequently developed to take into account these effects parametrically, such as limiting the amount of participating melt or considering the transient nature of the fragmentation. They were useful to help identifying some key processes that limit explosion efficiency, but they are not sufficient to address the complexity of the reactor situations. To address these situations, mechanistic codes have been developed, which are able to calculate complete FCI sequences and structure loading.

3.3.6. Insight into Modeling of Steam Explosion in CFD Codes

As can be understood from the above sections, steam explosion is a multi-component, multiphase, multi-time-scale phenomenon with heat and mass transfer. There are basically two components: corium and water. In some cases, inert gases are also present—for example, when oxidation occurs during

5. Values around 10% were reported for the FITSB series performed at Sandia National Laboratories (USA), but the method the experimental team employed to obtain these values has been criticized in particular by Farawila et al. [143], who obtained values around 2%.

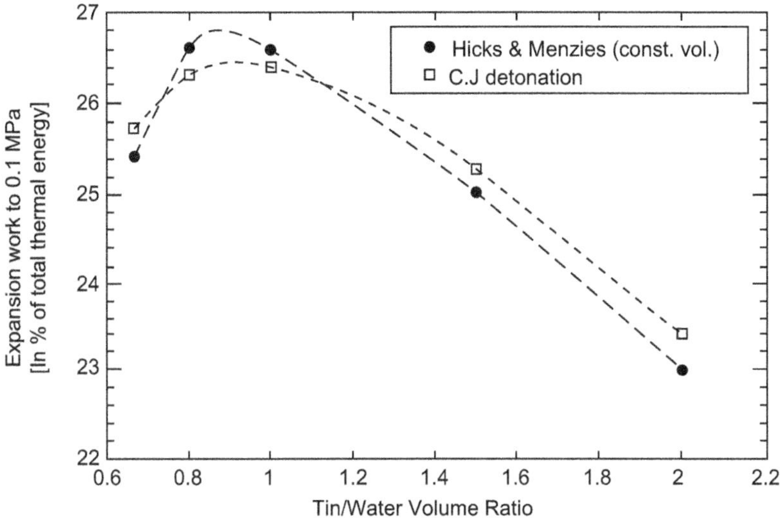

FIGURE 3.35 Comparison of expansion work following a constant-volume process (Hicks-Menzies process) and a Chapman-Jouguet detonation (Board-Hall process) for tin at 1773 K and 50% initial void fraction (from [141]).

melt–water interaction, producing hydrogen. There are two phases for the water (liquid and vapor) and two phases for the melt (liquid and solid), with phase change for each component. It takes seconds or minutes to (pre-) mix melt and water together depending on the accident scenario and instant of triggering, while propagation occurs at a millisecond time scale.

Describing all these processes in a unified structure is a tremendous task. It has been achieved by developing FCI deterministic codes, which describe all these processes with different degrees of approximation and several uncertainties that will now be discussed. The codes have to resolve a huge system of equations of mass, momentum, and energy for each component and/or phase according to their degree of details. The closure of the system is obtained through constitutive laws for the various phenomena, in particular melt fragmentation, heat and mass transfer, and friction laws for both the premixing and the explosion phase. Most of the codes are 2-D axisymmetric, and the real 3-D situations are approximated by various artifacts.

Modeling of the Premixing Phase

Since this gives the initial state of the medium through which the explosion will propagate, it is of fundamental importance to calculate this phase properly. Initial state means essentially geometry and distribution of water, gas(es), and corium in the premixture. For the corium, it is further important to know which proportion freezes during melt penetration, as essentially liquid corium can be finely fragmented by the explosion wave and deliver its energy to the water during the time scale of the propagation/expansion phase.

If we consider a corium melt jet penetrating into water, we may distinguish between jet break-up into particles (sometimes called primary break-up) and, possibly, fragmentation of those particles into smaller particles (secondary break-up). But in a reactor accident, it is very unlikely that a nice corium jet forms, so that usefulness of addressing jet break-up for reactor applications is still a matter of debate. In some codes, the melt is considered entering into water as a collection of drops of given sizes. The initial given size(s) of the particles is (are) deduced from experimental data or other considerations that are more closely related to the code structure. Note that in case jet break-up is calculated, the dimension (diameter) of the particles generated from the break-up process is also an input datum, in general. But unlike the former case, they are formed as the jet descends through the water according to jet break-up laws, which may be different from one code to another. The lack of detailed experimental data in representative conditions has made it difficult to converge toward a common understanding on this issue so far.

The particles generated from the jet or given *a priori*, experience, in principle, further fragmentation as they progress down into water until cohesive forces (surface tension) balance the disruptive forces (inertia). Again here different models are used, none of them having been fully validated so far.

The downward progression and spread of jet and drops are governed by the coolant two-phase flow generated by heat transfer from the corium to the water. Heat exchange and steam generation are established by using film-boiling correlations resulting in different flow regimes as a function of the local gas fraction, from bubbly flow (water is the continuous phase) to droplet flow (gas is the continuous phase) or churn turbulent flow.

As melt pour into water may last minutes when there is a reactor accident, part of the corium may solidify before an explosion occurs. This limits, for a given scenario, the quantity of melt that may be involved in the explosion. For instance, if an explosion had occurred in TMI-2 at the end of melt relocation in the lower head (which lasted ~2 minutes), part of the relocated 20 tons would certainly have been solidified, and thus would not have been available for the explosion. Another process that may limit the amount of premixture susceptible to sustain propagation is water depletion due to entrainment of water out of the zone by escaping steam (no liquid water means no steam explosion). At the very beginning of FCI research, attempts were made to quantify this "limit to premixing"—that is, to find out the maximum amount of melt that could be mixed with water before fluidization of liquid water by steam. Depending on the hypotheses and parameter choices, the quantity varied by several orders of magnitude, which makes the method inapplicable for safety assessment.

In summary, deterministic FCI codes are presently used for calculating the premixing phase. They allow establishing the time-dependent flow maps associated with melt penetration into water, and thus the time-dependent properties of the mixture a steam explosion may propagate through. They resolve systems of mass, momentum, and energy equations for coolant and melt, depending on their degree of approximation. They vary essentially by the description of the melt jet (continuous jet or collection of drops), drop break-up processes, film boiling, coolant flow, partitioning of energy release between steam and water. Their differences highlight a lack of understanding of these phenomena and of their importance for calculating real situations. This is due essentially to the difficulty in getting the required data in the extreme conditions testing must be performed. This issue is addressed further in Section 3.3.7.

Modeling of the Trigger

The apparently stochastic character of the triggering of a steam explosion does not allow predicting where and when it will occur. So in the codes, the trigger time and position are always imposed by the user. It has been observed in experiments that spontaneous explosions often trigger when the melt contacts the system bottom, and it is generally assumed in numerical simulations of reactor accidents. However, this might not give the strongest explosion. So, for risk assessment, the effect of changing the time and position of the trigger on the explosion strength may be necessary, which requires a large amount of calculations.

Several ways are used to simulate a trigger, all resulting in a local over-pressurization in a elementary cell of the computational mesh. For instance, applying 15 MPa in the centerline-lowermost cell of the lower head or cavity (similar to what is done in KROTOS tests) is generally sufficient to start the explosion process.

Modeling of the Explosion Phase

With respect to the premixing phase, the differences in the treatment of the explosion phase lie in the description of the generation of fine fragments and rapid heat release to the coolant.

Today two models are used in the codes to describe the explosion phase from a given or calculated premixture: the micro-interaction model and the nonequilibrium model.

In the micro-interaction model used in ESPROSE and IDEMO codes, a certain amount of water, steam, and melt fragments are mixed together and in thermal equilibrium, forming a *m-fluid* whose expansion is the cause of the pressure build-up. Figure 3.36 illustrates the concept according to Yuen and Theofanous as compared to the ideal Board and Hall type of propagation (referred to as "classical concept" in the Figure).

In the non-equilibrium concept used in MC3D and TEXAS codes, part of the heat from the fragments is used to produce vapor, which is the cause of the pressure build-up.

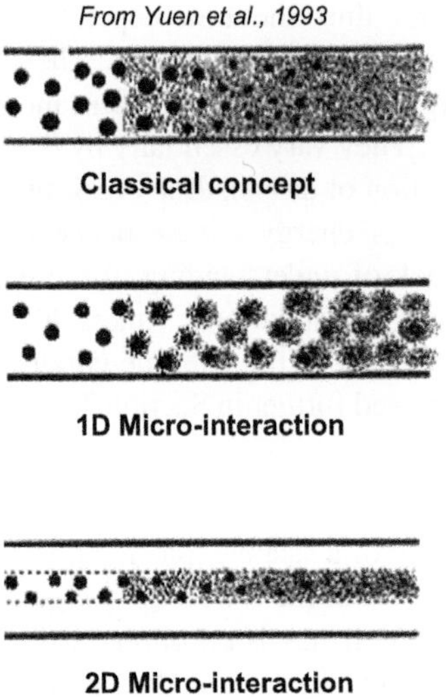

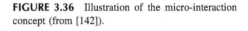

FIGURE 3.36 Illustration of the micro-interaction concept (from [142]).

Again, the lack of direct observation due to technical difficulties makes it difficult to reach convergence on the fundamental processes that actually govern this phase.

Status of Understanding

The diversity of the approaches reveals a lack of understanding of some fundamental aspects of FCI and steam explosion phenomena. This is due not only to the complexity of the phenomenon itself as already mentioned (1-D two-phase flow is already difficult to model!), but also to the difficulties of direct observation in turn due to the extreme conditions in which testing must be performed. Low-temperature experiments or/and experiments performed at high temperature with "simulant" melts, for example, alumina, have been used to develop and validate the models. But when applied to experiments performed with prototypical corium melts in realistic conditions, they have difficulties reproducing the results. In particular, they cannot reproduce the low explosivity of corium melts as found in recent tests (KROTOS and TROI already mentioned) in comparison to alumina melt: in KROTOS explosion efficiency is divided by 10 when passing from alumina to 80 w% UO_2 –20 w% ZrO_2 melts. It is not clear yet whether the differences in the premixing configurations or in the material properties are the main causes of the differences in the explosion energetics.

If the differences are due to the premixing characteristics (essentially component fraction and drop size), it may be an artifact of the experimental geometry and conditions. The main question is whether or not those premixing configurations that produced strong explosions with alumina are possible for other situations, and in particular prototypic reactor situations.

If the differences are due to the melt properties, then the question is whether those properties are a common characteristic of all possible corium compositions. This would allow to generalise the relatively mild character of explosions with corium melts. The main differences in melt properties/behavior that can affect premixing and propagation are:

- Density, specific energy,
- Freezing during premixing, which both reduces the amount of melt available for fine fragmentation and makes it more difficult for a partially frozen droplet,
- Viscosity, especially for materials with a large solidus-liquidus intervals, that is, non-eutectic mixtures,
- Physico-chemical behavior of the materials,
- Hydrogen production.

The FCI codes are not presently modeling these "material effects", except for hydrogen production in some of them.

3.3.7. OECD SERENA Program

So, what to do? Do we need to resolve all these uncertainties for reactor safety assessment purposes? In order to evaluate it, an international program performed under the auspices of OECD/NEA has been achieved (SERENA Phase1). The scope was to make a status of the FCI codes' capabilities to calculate steam explosion in reactor situations and to find out what would be strictly needed to bring understanding and predictability of FCI energetics to sufficient levels for risk management.

The codes presently used by different organizations throughout the world for safety assessment were involved in the program (see Table 3.2). The predictive capabilities of the codes were assessed on two typical configurations, found to be of common major concern for in- and ex-vessel situations, respectively. Figure 3.37 shows the two selected configurations and the initial conditions. A trigger was applied when the melt front contacted the bottom.

Despite the differences in modeling and approaches, and the large scatter of hypothesis and results for both premixing and explosion, the reactor predictions indicate clear tendencies as shown in Figure 3.38 (presented here for illustrative purpose only):

- For the in-vessel case, the calculated loads are below the capacity of the defined-model intact vessel (estimated to be above 300 kPa.s),
- For the ex-vessel case, the calculated loads, even though low, are above the capacity of the defined-model cavity walls (estimated to be a few tens of kPa.s). The scatter of the results raises the question of quantifying the containment safety margin for ex-vessel steam explosion.

TABLE 3.2 Partnership in SERENA and Associated Codes

Partners	Codes
CEA/IRSN	MC3D
GRS (IKE)	IKEMIX, MC3D/IKEJET, IDEMO
FZK	MATTINA
KAERI/KMU	TEXAS V, TRACER II
NRC (UCSB, UW)	PM-ALPHA, ESPROSE, TEXAS
KINS	IFCI 6.2
JAERI	JASMINE
NUPEC	VESUVIUS
EREC	VAPEX

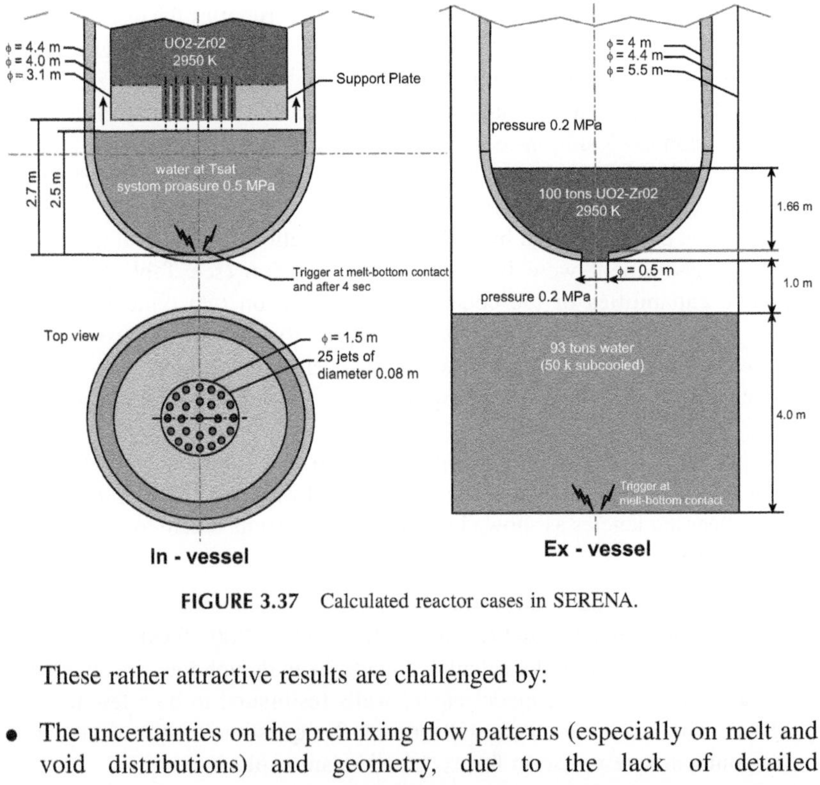

FIGURE 3.37 Calculated reactor cases in SERENA.

These rather attractive results are challenged by:

- The uncertainties on the premixing flow patterns (especially on melt and void distributions) and geometry, due to the lack of detailed

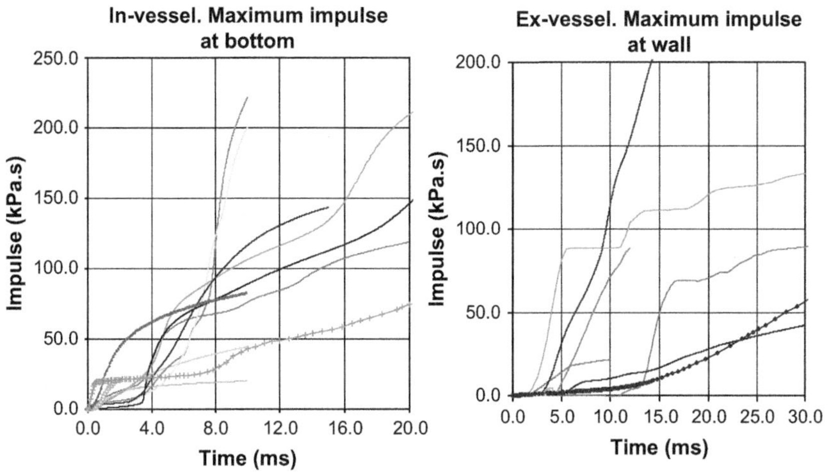

FIGURE 3.38 In- and ex-vessel calculated loads by the various codes used in the SERENA exercise.

experimental data. It is believed in particular that the codes overestimate void fraction to such high levels that it acts as a mitigator of steam explosion.

- The uncertainties on material influence on the energetics. Some data are available for two close corium oxidic compositions. In some cases, "reduced" parameters have been used to model the explosion without firm physical reasoning.
- The scatter of the results for ex-vessel situation does not allow quantifying the containment safety margins.

In other words, if it could be confirmed by providing sufficient data on premixing zone internal structure that void is properly calculated, in-vessel steam explosion issue could be considered definitively resolved from the perspective of risk of rupture by dynamic loading. In the same way, if the reduced energetics could be confirmed for a large spectrum of corium compositions (possibly finding bounding compositions), the problem of quantification of containment safety margins for ex-vessel steam explosion would reduce to narrowing the scatter between the codes predictions.

3.3.8. Next Steps

We arrive at the important conclusion that even though large uncertainties still exist as to fundamental aspects of FCI and steam explosion processes, resolving the issue for reactor applications depends on resolving a very limited number of uncertainties, namely, level of void in premixture and material effect on steam

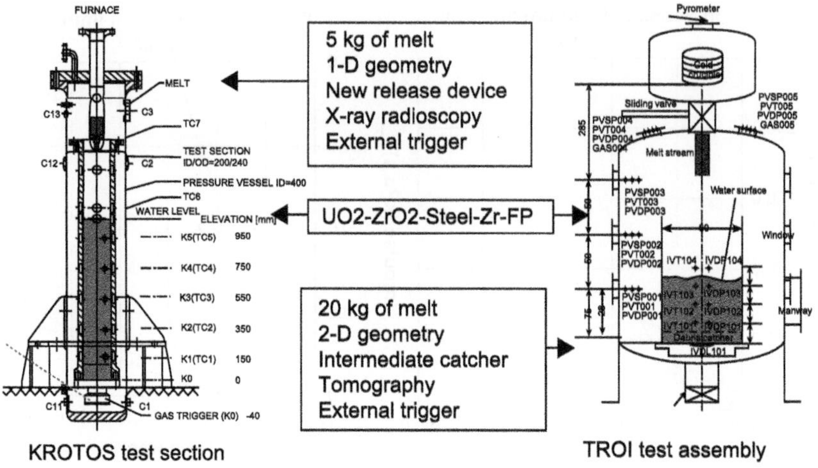

FIGURE 3.39 KROTOS and TROI facilities to be used in SERENA-2.

explosion energetics. This led to the proposal for a Phase 2 of SERENA program aiming at obtaining the complementary data on these issues and reducing the scatter of the predictions.

The proposal is based on use of two complementary corium facilities to obtain the required data (Figure 3.39):

- KROTOS (CEA, France), being more suited for investigating intrinsic FCI characteristics of the melts and equipped with a high-energy X-ray radioscopy device to "visualize" premixing,
- TROI (KAERI, Korea), being more suited for testing more reactor-like conditions by having more mass and 3-D capabilities, and equipped for tomography.

It is proposed that five complementary tests be performed in each facility, varying the geometry and corium compositions.

The tests are accompanied by analytical activities aiming at:

- Determining, through pre- and posttest calculations, what is needed to be improved in the models to adequately calculate the premixing flow patterns and corium explosion yields,
- Identifying the reasons for the significant spread of the results in the calculations of Phase 1,
- Integrating the findings in the codes and verifying the progress made in reducing the scattering of the predictions of the reactor cases of Phase 1 of the project.

The four-year program started in October 2007.

3.3.9. Summary and Conclusions on Steam Explosion

Steam explosion occurring during corium melt pouring into water has received the most attention so far as it is considered the most challenging for LWR structures. The accident conditions in a LWR are such that when molten corium penetrates into water it first fragments into centimeter-size particles, while remaining essentially molten. A "coarse mixture" or premixture of corium drops, steam, and liquid water forms. This is an unstable situation, and in some cases very fine fragmentation of the corium drops occurs, inducing a very rapid release of the remaining internal energy of the corium to the water, with shock wave formation and subsequent expansion of the system giving rise to dynamic loading of the surrounding structures. This process is called a steam explosion.

Conceptually, a steam explosion may be considered a propagating event similar to a chemical detonation, as originally suggested by Board et al. (thermal detonation concept): a coarse mixture of melt, steam, and water exposed to a triggering event may detonate, that is, be the place where a shock wave develops and propagates at supersonic speed in the mixture sustained by energy release in the zone immediately behind the shock front. As far as reactor safety is concerned, the main challenge is to establish the characteristics of the mixture that can form, for a given accident scenario, in terms of melt–water–steam fractions and melt quantity, and to evaluate the level of dynamic loading to the structure in case a steam explosion takes place in this mixture.

Experiments performed under a large variety of conditions have shown that maximum energy conversion ratios (mechanical energy release divided by thermal energy of the melt in the premixture) were one order of magnitude lower than those predicted by the thermodynamic model of Hicks-Menzies. The reasons for that were found in the heterogeneity of actual premixtures and energy dissipation processes in the propagation wave, which are not taken into account in either the Hicks-Menzies thermodynamic or Board et al. thermal detonation models. Deterministic codes have been developed that describe these processes in a detailed manner, thus obtaining more realistic estimates of mechanical energy release and corresponding dynamic loads.

Although these codes still have a number of parameters that are not totally validated because of the difficulty in obtaining sufficiently detailed data due to the extreme conditions in testing, they have reached a certain level of maturity and are able to calculate reactor situations as shown in Phase 1 of the OECD SERENA program. The large scattering of the predictions reflects the lack of understanding of a number of FCI processes, but, as far as reactor application is concerned, the main uncertainties lie on the void level in premixture and generic explosion behavior of corium melts.

Reducing these uncertainties to a sufficient level for reactor safety assessment is within the scope of Phase 2 of SERENA, whose achievement could be a definite step toward resolving the steam explosion issue for LWRs. If it is confirmed that void was properly calculated in Phase 1, we could reasonably consider closed the issue of bottom vessel failure by steam explosion, which is especially important for the new generation of reactors based on in-vessel retention (IVR) concepts (e.g., AP600, AP1000, etc.). For the ex-vessel case, reducing the scatter of the predictions would probably not eliminate the risk of cavity damage, but will allow specifying the level of the loads. A problem is then to analyze it in terms of impact on the containment and on severe accident management strategies, which is to a large extent design dependent.

Another problem is that we are not presently able to propose solutions to prevent steam explosion. Adding substances in the water to change its properties (e.g., increase surface tension and/or viscosity) has been studied in the past but did not provide reliable results. Putting igniters (controlled artificial triggers) as for hydrogen is an appealing possibility, but hardly feasible and applicable in practice.

Lastly, besides the issue of dynamic loading, dispersion and production of very fine debris by the explosion(s) may impact the debris cooling and retention strategies, which directly affect the issue of stabilization and termination of severe accidents.

3.4. INTEGRITY OF CONTAINMENT STRUCTURES

(P. Eisert and J. Sievers)

3.4.1. Introduction

The main role of the containment of a nuclear power plant (NPP) is to protect the public and the environment from a radioactive release, especially in case of a severe accident. For this reason, the integrity of the containment structure and the leak tightness must be ensured in every event that could possibly occur. The events that can threaten the integrity of the containment can be distinguished in external events, such as earthquake, airplane crash, or gas cloud explosion, and internal events, such as station blackout or loss of coolant accidents (LOCAs) caused by a leak or a break of a coolant line. A containment affected by external events is loaded mainly by induced vibrations, impact loads, or external pressure loads. In contrast, internal events mainly cause increasing pressure and temperature inside the containment. Depending on national regulations, the safety design basis for containments of NPPs are internal loads and external loads.

In addition, the regulations state that the effects of severe accidents (e.g., caused by a core melt scenario) have to be taken into account in the framework of level 2 probabilistic safety assessment (PSA). This kind of accident generates internal pressures and temperatures that can exceed the design values for containments of NPPs built in the past, and these may cause catastrophic failure of the containment. Further dynamic loads due to rapid hydrogen combustion may possibly occur. So knowledge of the limit on the containment's load-carrying capacity, which may be much greater than the design values, is an essential concern for accident management. Taking the limit load-carrying capacity of the containment into account may allow an intentional release of part of the containment atmosphere to prevent catastrophic failure of the containment. Alternatively, such a release can be delayed as far as possible. Delaying the release of part of the containment atmosphere leads to a decrease of the release of fission products into the environment since the aerosol deposition on containment surfaces with additional time reduces the fission products in the containment atmosphere. It has to be noted that these loads generally cause development of considerable plastic strain in the containment materials. The design of third-generation NPPs such as the European Pressurized Reactor (EPR) takes into consideration the effect of such kinds of accidents.

Beginning with a representative overview of different containment types, the following Sections focus on the behavior of NPP containments loaded beyond design loads caused by severe accidents like core melt. Experiments concerning this question are represented, as well as analysis methods to determine the limit load-carrying capacity of NPP containments. In connection with this issue, an insight on failure criteria is given, which can be employed for assessing the integrity of the different types of containment structures.

3.4.2. Overview of Different Types of Containment Structures

Single- and double wall-containment designs provide differently for the requirements that have to be fulfilled by the containment. Figure 3.40 shows a single-wall reinforced concrete containment of an American PWR, and Figure 3.41 shows a Russian prestressed concrete PWR containment. In countries where an airplane crash has to be considered as an external event, its load is normally carried by the outer wall of a double-wall containment like that in the French PWR type "N4" and the German PWR type "KONVOI" (see Figures 3.42 and 3.43). Both outer containment walls consist of a reinforced concrete shell. The different wall thickness (N4: 0.55 m, KONVOI: 1.8 m) is due to the different load assumptions that were taken into account in France and Germany for an airplane crash.

The role of a single-wall containment or the inner wall of a double-wall containment is to carry the pressure load caused by a design accident—that is, a double-ended break of a coolant line under normal operation conditions. In this case, the pressure within the inner containment wall should rise only up to the design pressure of the inner containment wall. In addition, sufficient leak tightness against the release of fission products is required from the inner containment wall. Leakage of the internal atmosphere is collected in the so-called annular space of double-wall containments between the inner and outer containment wall that is operated under slight vacuum conditions.

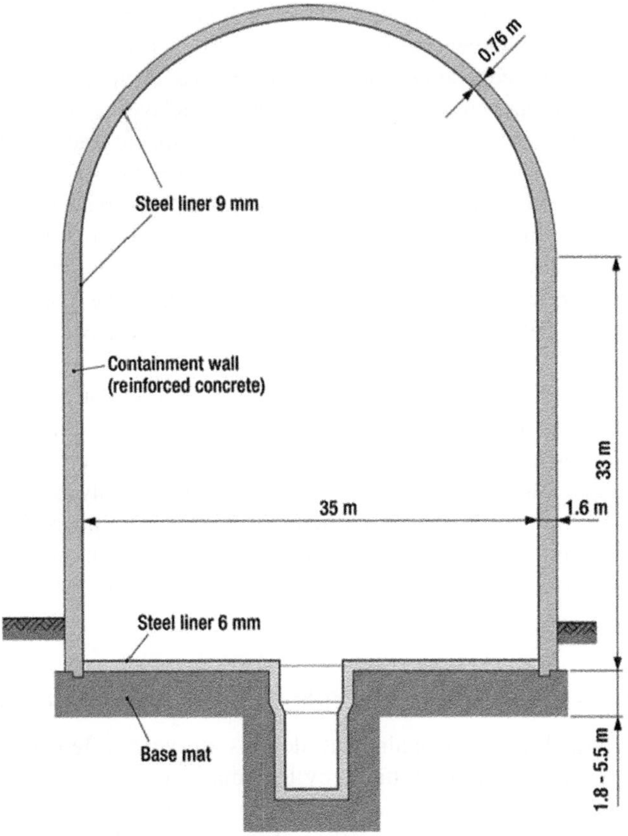

FIGURE 3.40 Single-wall reinforced PWR-Containment of an American NPP.

The inner containment wall of the French PWR (N4) consists of a prestressed concrete shell of about 1.3 m thickness, while the German PWR (Konvoi) consists of a spherical steel shell of about 38 mm thickness. The leak tightness of the French inner containment is solely ensured by the compressive stress in the concrete. No liner is implemented. Other inner walls of similar type of containments are equipped with a steel liner.

An example of a single-wall containment of a PWR is the prestressed concrete containment of the Russian PWR (VVER 1000). To ensure sufficient leak tightness, the containment wall is covered by a steel liner implemented at the inner surface of the containment (Figure 3.41). The load caused by internal pressure in case of an incident is carried mainly by tendons as in the inner containment wall of the French double-wall containment. The tendons are conducted in sheaths that are implemented in the wall space before concreting

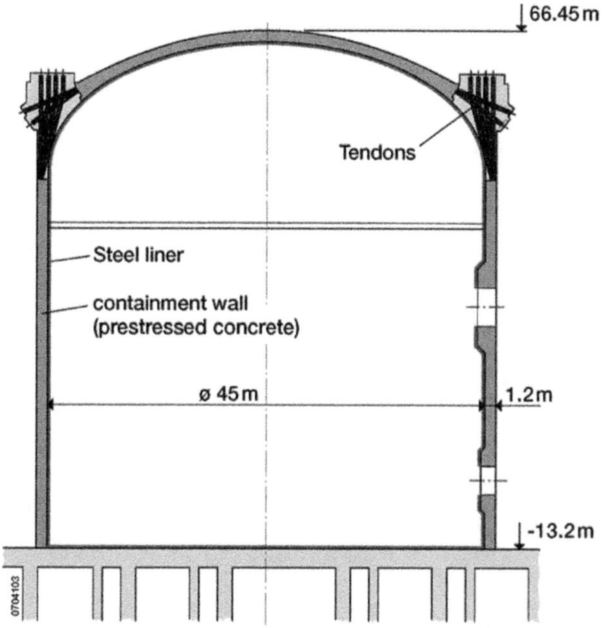

FIGURE 3.41 Single-wall PWR-Containment of a Russian NPP, type VVER 1000.

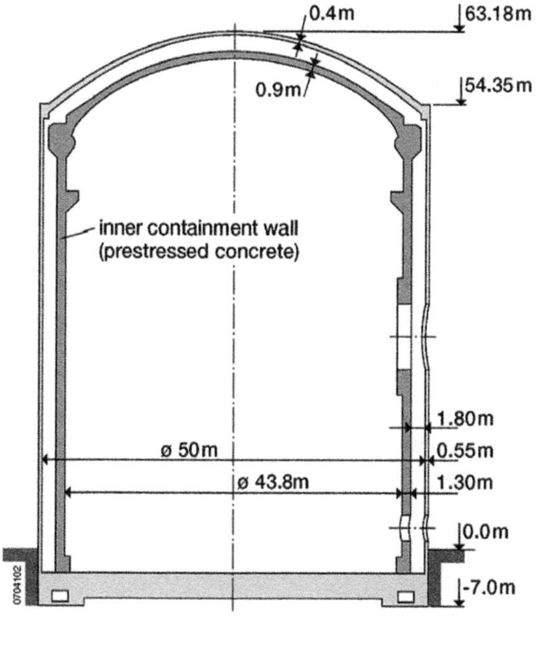

FIGURE 3.42 Double-wall PWR-containment of a French NPP, type N4.

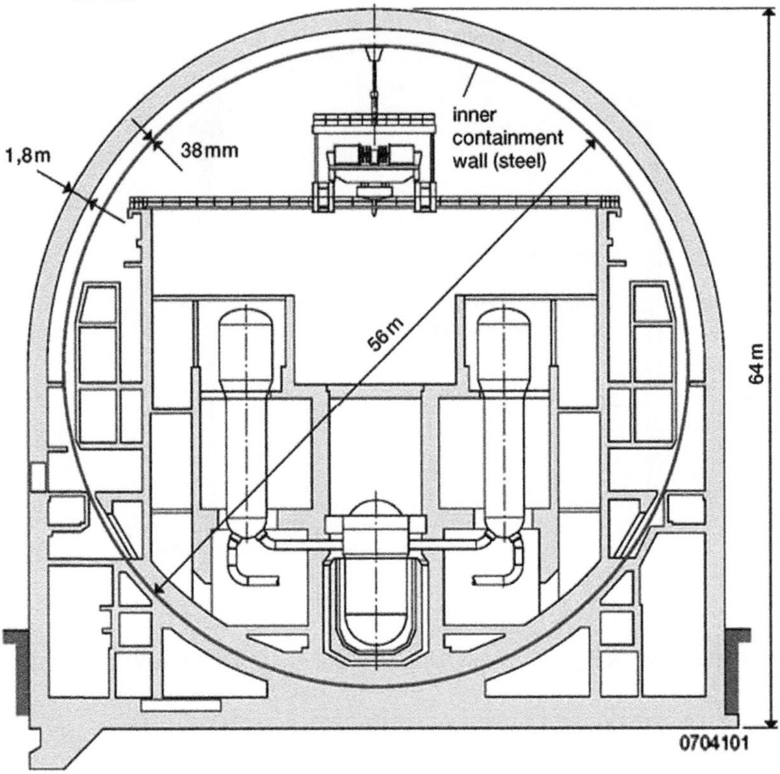

FIGURE 3.43 Double-wall PWR-containment of a German NPP, type KONVOI.

the wall. While the tendons in French prestressed concrete containment walls are grouted (the sheaths with tendons inside are filled with a special mortar after prestressing), the tendons in Russian containments are ungrouted. To prevent the tendons from corrosion under these conditions, they are protected by grease. Only ungrouted tendons can be replaced in case of failure or re-tensioned in case of loss of the prestressing force during the operation time (or aging).

While PWR containments, known as dry containments, are designed with the capacity to contain the release of the energy of the whole volume of the primary coolant in case of a LOCA, BWRs are designed with a pressure suppression system. In BWRs the containment is therefore divided into two main compartments, drywell and wetwell. Consequently, PWR containments are large in size, and because of the pressure suppression system the containments of BWRs are significantly smaller. The inner containment of a BWR type "SWR 72" consists of different parts. The cylindrical part (rising wall) consists of prestressed concrete. The bottom and the upper ceiling are both manufactured of reinforced concrete, and the upper closure of the inner containment consists of a steel shell located directly above the reactor pressure vessel. The inner surface of the inner containment is covered by a steel liner.

3.4.3. Experiments on the Integrity of NPP Containments

In the past, several experiments concerning the integrity of containments of NPPs due to internal pressure beyond design have been performed [144]. The range of the tests envelopes steel-, reinforced- and prestressed-concrete containment models in different scales. For example, the latest tests concerning a 1:4 scaled prestressed concrete containment vessel (PCCV) with liner are described in the following. The objective was to extend the understanding of capacities of actual containment structures, taking in account results from this test and other previous research [144].

The prototype for the scaled test model is the containment building of Unit 3 of the Ohi NPP in Japan, which is shown in Figure 3.44. This figure also shows locations of interest where failure is expected in case of overpressurization.

Failure is expected in areas where larger changes of the stiffness of the structure occur. This meets in particular areas of larger penetrations, for example, main steam-line penetration and equipment hatch as well as changes in the wall thickness of the containment vessel (all locations with notation "e" in Figure 3.44) or the intersection between basemat and containment wall (location b). Also, the buttress where the tendons are anchored is an area of interest (location d). Areas of changes of the stiffness in a structure cause strain concentrations in the material, which leads to earlier failure than in an undisturbed area (location a).

The dimension of the 1:4 scaled model of the PCCV is shown in Figure 3.45. To ensure a comparable behavior of the prototype and the model of the containment vessel also, the arrangement of the reinforcement of the concrete wall as well as the design of the liner has been adjusted. Figure 3.46 shows the arrangements of the prototype and of the model of the PCCV for comparison. The liner at the inner surface of the containment wall is adumbrated on the left outer sides of the two shown cross sections of the reinforced concrete wall. It is fixed to the concrete of the PCCV prototype as well as of the model by studs, respectively T-bars.

For observation of the deformation behavior of the containment model due to increasing internal pressure, the tendons, reinforcement, and liner were equipped by adequate sensors (e.g., strain gauges), which for the most part were concentrated in areas where failure is expected (Figure 3.46).

In addition, acoustic sensors were used to detect rupture of single wires of the tendons or loss of leak tightness by flow off noise at generated cracks in the liner (for further information see [145]). The increase of pressure in the

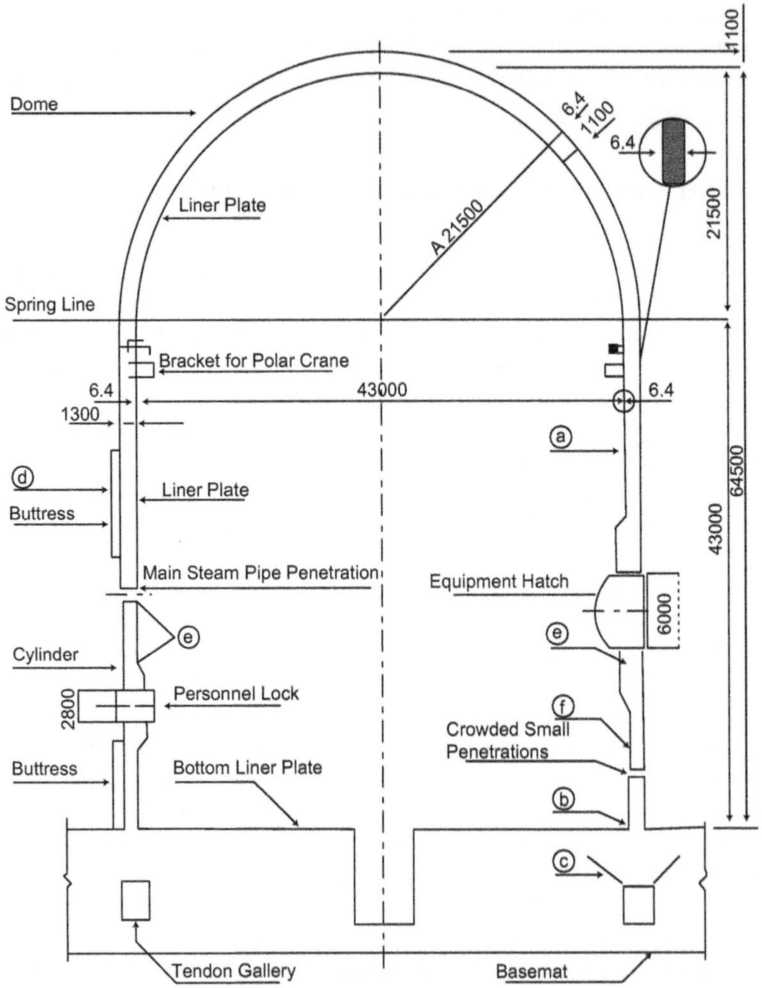

FIGURE 3.44 Actual Prestressed Concrete Containment Vessel (PCCV) - Dimensions (measurements in mm) and indication of potential failure locations a until f (from [145]).

model containment vessel was generated by the input of nitrogen gas. Figure 3.47 shows the completed model of the PCCV.

Selected Test Results of the 1:4 Scaled PCCV-model

After several preparatory tests concerning mainly the structural integrity and the leak tightness of the PCCV, the containment was loaded by slowly increasing internal pressure to perform the limit state test (LST). The load was increased in intervals alternating with dwell times in order to observe the respective time-dependent state of the internal pressure. The goal of this test

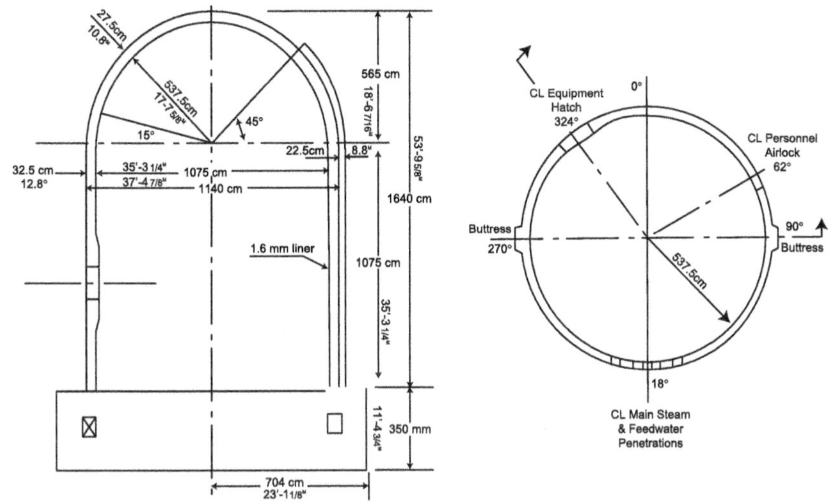

FIGURE 3.45 1:4 scaled design of the Prestressed Concrete Containment Vessel model - vertical and horizontal section (from [145]).

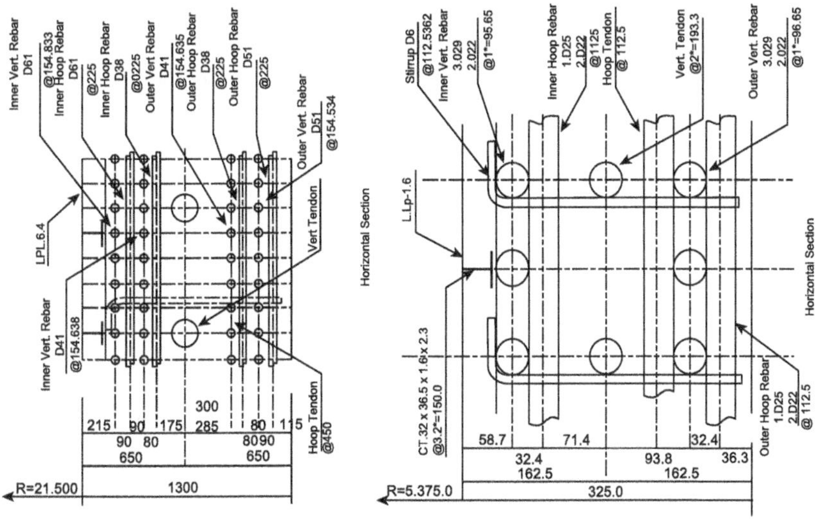

FIGURE 3.46 Comparison of reinforcement-, tendon-, and liner arrangement of the real PCCV (left side) and the 1:4 PCCV model (right side) (from [145]).

was to determine at which pressure the limit of the structural integrity of the PCCV is reached. This limit is expressed in the multiple of the design pressure P_d, which is 0.39 MPa.

At a pressure level of about 2.4 P_d (~0.94 MPa), the first decrease in pressure was observed, which exceeds the specification of the leak tightness

FIGURE 3.47 Completed 1:4 scaled model of the PCCV (from [145]).

considerably (1.86% mass/day). In this case the first crack in the liner was located in the vicinity of the equipment hatch by the acoustic sensors (see Figure 3.48). This location is consistent with the predicted failure location "e" in Figure 3.44. Strain gauges in the vicinity of the crack indicate a strain of about 1% in the liner, which confirms plastic deformation.

At a pressure level of about 3.1 P_d (~1.21 MPa), the leak rate is about 100% mass/day caused by an increasing amount of cracks in the liner. At a pressure of

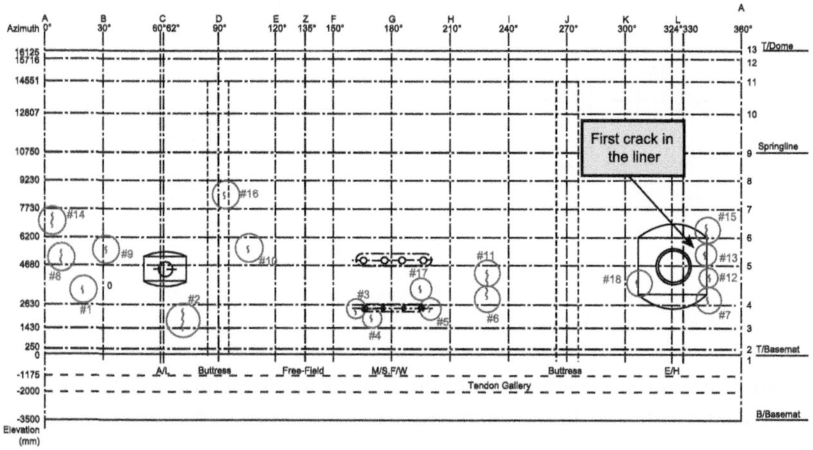

FIGURE 3.48 Crack locations in the liner after termination of the limit state test (LST) indicated by cycles #X (from [145]).

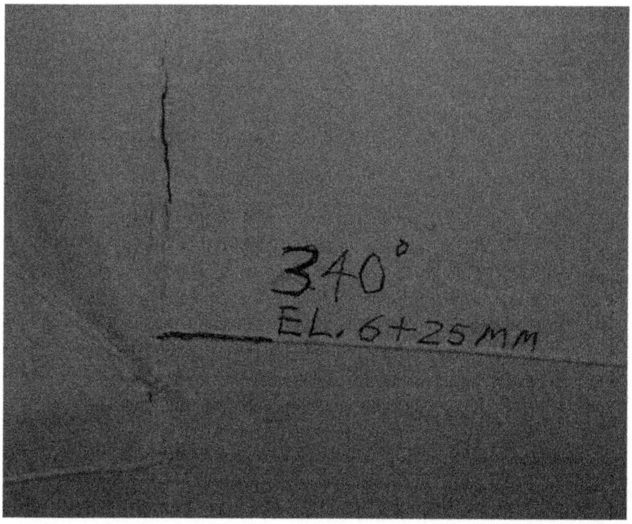

FIGURE 3.49 Crack #15 in the liner localized at azimuth ~340° (see Figure 3.48) after the limit state test (LST) (from [145]).

about 3.3 P_d (~1.3 MPa), the leak rate exceeds the capacity of the pressurization system, and so the test was terminated. For this pressure the leak rate was estimated to be of the order of about 900%/day.

Figure 3.48 shows the crack pattern in the liner that was formed during the LST. Eighteen cracks were accumulated during the test. As can be seen, the cracks preferably occur in the environment of disturbed areas of the structure. Figure 3.49 shows a photograph of the crack at position 15.

In a further experimental phase, the structural failure mode test (SFMT), it was shown that the model containment failed by a long axial crack at a pressure of 3.63 P_d (~1.42 MPa). The failure process was initiated by the failure of circumferential tendons. Details are given in [145].

The crack pattern in the outer concrete surface generated during the LST is shown in Figure 3.50.

Figure 3.51 shows a photograph of a cut-out of the crack pattern on the outer surface of the concrete wall (see highlighted area at the left side in Figure 3.50).

3.4.4. Structural Mechanics Analyses

In order to be able to give a realistic description of the behavior of complex structures under the influence of extreme loading, the nonlinear load response requires that a sophisticated analysis technique be employed. In this connection the finite element method (FEM) is an adequate analysis method to simulate the structural response of reactor containments.

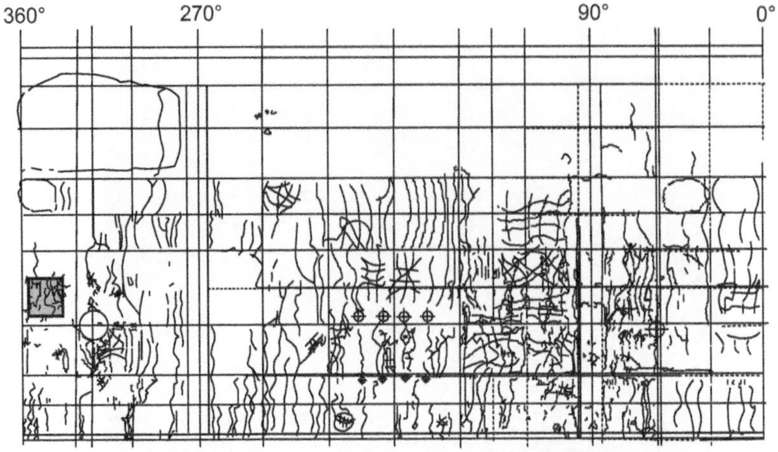

FIGURE 3.50 Crack pattern in the concrete at the outer surface of the PCCV model wall generated up to termination of the limit state test (LST) (from [145]).

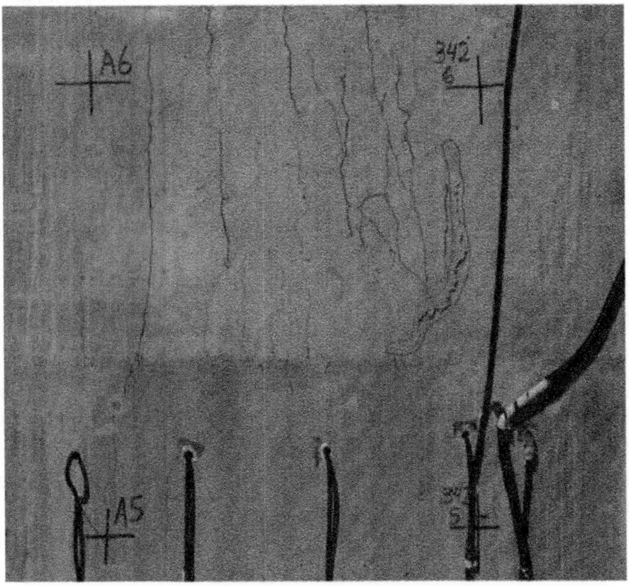

FIGURE 3.51 Crack pattern in the outer surface of the concrete wall in the vicinity of cable feed-through at azimuth 350° after the limit state test (LST) (from [145]; see also Figure 3.50).

Benchmark Activities in the Framework of the International Standard Problem ISP-48

In parallel to the experimental efforts concerning the behavior of the 1:4 scaled PCCV under increasing internal pressure beyond design, among other things

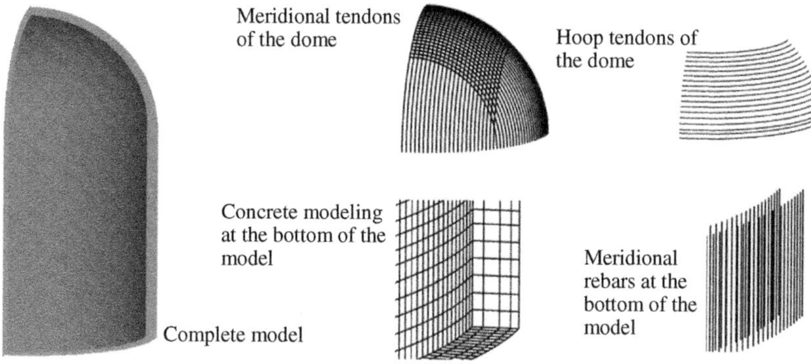

FIGURE 3.52 3D-finite element model of the prestressed reinforced concrete wall of the 1:4 model containment (PCCV) 90° section of undisturbed area (from [146]).

benchmark activities have been created in the framework of the OECD/CSNI International Standard Problem ISP-48 [147] to compare the observed behavior of the PCCV, with results generated by analyses with FEM codes.

The prestressed reinforced concrete wall with liner requires complex finite element models (see Figure 3.52). Due to stiffness deficit in the wall in areas of penetrations this deficit will be compensated by local thickening of the wall and increase of reinforcing steel. Also changes in the layout of the tendons are necessary. All these aspects have to be considered.

Based on such kinds of analysis techniques, the structural mechanics behavior of the 1:4 PCCV model was calculated in the framework of the ISP-48 activities by eight organizations. At selected positions of the model containment, the global deformation and the strains in the liner, rebars, tendons, and concrete were compared between the calculations and the experimental measures.

Structural Mechanics Phenomena in Severe Accident Loading

During the progression of severe accident scenarios, the atmosphere in the containment, which is in contact with the inner surface of the containment wall, could be heated up to several hundred degrees Celsius. As an example, Figure 3.53 shows the development of pressure and temperature in the containment as it is assumed for a severe accident scenario caused by a station blackout in the framework of ISP-48. In consequence of a zirconium–water reaction caused by core melting, hydrogen combustion is assumed at about 4.5 hours after initiation of the accident. In the considered case, a pressure peak of about 0.8 MPa and a temperature peak of about 620°C occur due to deflagration. The pressure as well as the temperature rises abruptly within about 500 ms. After 15 minutes, the pressure decreases to about 50% of the peak pressure, and the temperature decreases to about

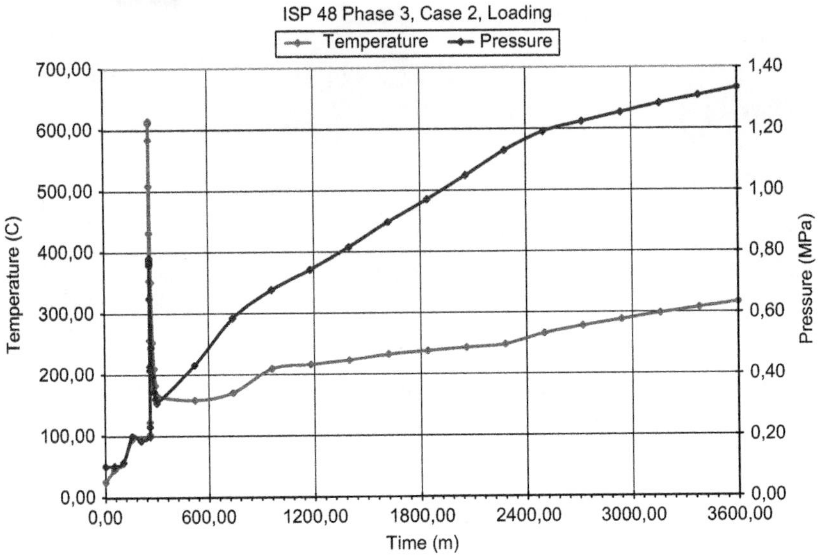

FIGURE 3.53 Pressure and temperature time history assumed in case of a severe accident [147].

one-third of the peak temperature. After occurance of these pressure and temperature transients, pressure and temperature in the containment increase continuously due to further energy input generated by the core melt.

Depending on the ratio of the gas mixture, hydrogen combustion can lead to detonation with duration of a few milliseconds or to deflagration with duration up to about some 100 milliseconds, with a corresponding rise of pressure and temperature. Particularly for deflagrations, the effects on the loading conditions for the containment are severe for several minutes. Investigations concerning the behavior of the steel containment of a PWR in case of severe accidents have shown [148] that the limit load-carrying capacity in terms of internal pressure can be considerably larger for loading due to hydrogen detonation compared with hydrogen deflagration. This is also confirmed by [149].

In connection with this issue, the eigenfrequencies and the eigenmodes of the containment that essentially depend on the mass and stiffness characteristics of the containment play a decisive role. In particular, low eigenfrequencies that spend a large contribution to the dynamic response of the containment can be excited by the dynamic load of a hydrogen deflagration. For thin-walled steel containments, special attention has to be taken if the duration of the decrease of temperature and pressure after hydrogen deflagration needs several minutes as described above. Because steel has good heat conduction, the temperature in the wall needs only a short time for heating the shell. Due to strength reduction of the steel caused by increasing temperature, this could be a problem concerning the containment integrity.

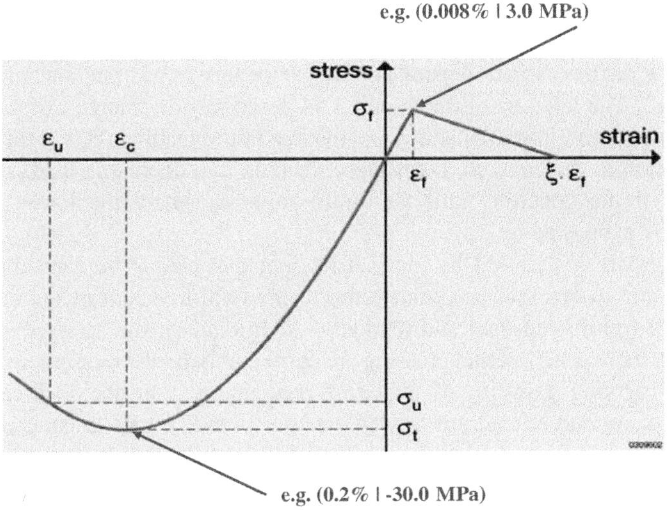

FIGURE 3.54 Stress-strain characteristic of concrete.

Therefore special attention has to be taken with regard to the accuracy of the loading conditions of both pressure and temperature in case of severe accidents.

For thick-walled prestressed concrete containments, this question is not of essential importance because of the low heat conduction of the concrete and the relatively large distance of the load-carrying tendons from the inner surface of the containment wall. In this case, the tendons are not affected significantly by increasing temperature inside the containment. Due to the temperature difference between the inner and outer surface of the containment, a corresponding temperature gradient arises within the containment wall. The thermal expansion of the inner surface region is constrained, which leads to compressive stresses. Consequently, the outer surface region is loaded by tensile stresses, which lead to concrete cracking because concrete is not able to carry more than about 5 to 10% of the compressive strength in tension (see Figure 3.54). Therefore cracks in the concrete are formed at a very low stress level in the area loaded by tensile stresses. In prestressed concrete containments, this phenomenon can only appear when the compressive stresses induced in the concrete by the tendons are completely removed.

Figure 3.54 shows some fundamental concrete material characteristics in terms of a uniaxial stress strain relation that can be used in finite element calculations. The triangular stress-strain progress describes the behavior of concrete under tensile conditions. At the peak, position crack formation occurs according to the smeared crack approach. The occurrence of cracks in concrete does not affect sudden total loss of tensile stresses. The decrease of the tensile capacity caused by increasing strain is approximated by linear decline of the curve by a factor ξ, which depends on the characteristics of the concrete to be considered. The left side of the Figure 3.54 describes the behavior of concrete under compressive conditions. Beyond the maximum loading (σ_t/ε_c), increased deterioration of the concrete takes place, leading to a decreased load-carrying capacity of the concrete until the strain value ε_u where the load-carrying capacity is exhausted (σ_u).

Tensile stresses caused by thermal loading that cannot be carried by the concrete due to cracking are transferred to the reinforcement of the containment wall (reinforced bars and tendons). Additional tensile stresses that are caused by increasing internal pressure are carried only by the reinforcement. So the load-carrying capacity of the reinforcement governs the load-carrying capacity of reinforced and prestressed concrete containments in general.

Assessment of the Load-carrying Capacity of Containments Loaded by Severe Accidents

The load-carrying capacity of containments is determined in general by finite element calculations because this analysis method allows a detailed description of the real structure concerning shape and material characteristics as well consideration of complex loading conditions. This is of crucial importance because disturbed areas of the containment wall like pipe ducts (main steam pipe or feedwater line) or hatches (equipment hatch, personal airlock) cause changes in the local stiffness of the containment shell that lead, for example, to strain or stress concentrations, which generally results in a significant reduction of the load-carrying capacity of the containment. With regard to prestressed concrete containments with liner, the experiments show cracks in the liner, particularly in the vicinity of disturbed areas (see Figure 3.48).

Calculations based on simple models (e.g., analytical approaches like boiler formula or finite element models describing the undisturbed containment shell only) can give only rough information about the global behavior of the structure, which generally differs significantly from the results at disturbed areas generated by detailed finite element models.

The assessment of the results of finite element calculations concerning the load-carrying capacity of a structure is based on failure criteria. Depending on the kind of failure, the following different failure criteria can be used for assessing the integrity of steel containments:

Kind of Failure	Criteria
Ductile fracture	Equivalent stress $\leq$ engineering flow stress
	Equivalent strain $\leq$ allowable strain
Brittle fracture	Maximum principle stress $\leq$ reduced ultimate strength

For the assessment of steel containment structures like thin shells or thick-walled components like nozzles, the allowable strain may be determined based on the maximum uniform elongation of stress-strain curves generated by a uniaxial loaded test specimen. In the uniaxial test specimen, necking occurs after exceeding the strain at maximum uniform elongation, which cannot be transmitted to multiaxial loaded stress states. For consideration of the influence of multiaxial stress conditions, the allowable local strain limits are dependent on local stress conditions. This can be considered by reducing the maximum uniform elongation by a so-called triaxiality factor.

For the integrity assessment of reinforced or prestressed concrete containment structures, the strains in the reinforcement and the tendons (if present) can be limited by the maximum uniform elongation because these components show a one-dimensional character as the uniaxial test specimen from which the stress-strain curves are generated. Strains beyond the maximum uniform elongation should not be used in the sense of safety reasons.

Also, the effects of manufacturing defects during construction such as weld defects have to be taken into consideration. These effects, combined with application of the failure criteria mentioned above, lead to additional reduction of the allowable stress or, strain values of the failure criteria.

In conclusion, for the assessment of the limit load-carrying capacity of containment structures loaded by severe accidents, multiple influencing variables have to be noticed:

- Strains in steel components like rebars and tendons are limited by the maximum uniform elongation.
- The allowable strain of structures like plates, shells, and nozzles is reduced by multiaxial stress conditions.
- Failure criteria refer to ductile or brittle failure modes of structures.
- Manufacturing defects generated during construction have to be considered.

ACKNOWLEDGMENTS

The authors of Section 3.1 wish to acknowledge Dr. Breitung (KIT, retired) for his support to the writing of this Section.

REFERENCES

[1] International Atomic Energy Agency (IAEA), Approaches and Tools for Severe Accident Analysis for Nuclear Power Plants, Safety Reports Series No.56, Vienna, Austria (2008).

[2] U. Bielert, W. Breitung, A. Kotchourko, P. Royl, W. Scholtyssek, A. Veser, A. Beccantini, F. Dabbene, H. Paillère, E. Studer, T. Huld, H. Wilkening, B. Edlinger, C. Poruba, M. Movahed, Multi-dimensional simulation of hydrogen distribution and turbulent combustion in severe accidents, Nuclear Engineering and Design 209 (2001) 165–172.

[3] W. Breitung, P. Royl, Procedure and tools for deterministic analysis and control of hydrogen behavior in severe accidents, Nuclear Engineering and Design 202 (2000) 249–268.

[4] J.O. Henrie, A.K. Postma, Lessons Learned from Hydrogen Generation and Burning during the TMI-2 Event, U.S. Department of Energy, 1987. Report GEND-061.

[5] F. Rahn, Technical Foundation of Reactor Safety—Knowledge Base for Resolving Severe Accident Issues, Rev.1, Final Report No.1022186 (2010).

[6] OECD Nuclear Energy Agency, In-Vessel and Ex-Vessel Hydrogen Sources, Report by NEA Groups of Experts, Report NEA/CSNI/R(2001)15(2001).

[7] G. Schanz, B. Adroguer, A. Volchek, Advanced treatment of zircaloy cladding high-temperature oxidation in severe accident code calculations, Part I. Experimental database and basic modeling, Nuclear Engineering and Design 232 (2004) 75–84.

[8] OECD Nuclear Energy Agency, Final Comparison Report on ISP-35: NUPEC Hydrogen Mixing and Distribution Test (Test M-7-1), OECD/NEA/CSNI, Report OECD/GD(95) 29(1994).

[9] OECD Nuclear Energy Agency, SOAR (State-Of-The-Art) on Containment Thermal-hydraulics and Hydrogen Distribution, Prepared by an OECD/NEA Group of Experts, OECD/NEA/CSNI, Report CSNI-R(99)16 (1999).

[10] H.-J. Allelein, K. Fischer, J. Vendel, M. Malet, E. Studer, S. Schwarz, M. Houkema, H. Paillère, A. Bentaib, International Standard Problem ISP-47 on Containment Thermal Hydraulics, OECD/NEA/CSNI, Report NEA/CSNI/R(2007)10(2007).

[11] M. Andreani, K. Haller, M. Heitsch, B. Hemström, I. Karppinen, J. Macek, J. Schmid, H. Paillère, I. Toth, A benchmark exercise on the use of CFD codes for containment issues using best practice guidelines: A computational challenge, Nuclear Engineering and Design 238 (2008) 502–513.

[12] O. Auban, R. Zboray, D. Paladino, Investigation of large-scale gas mixing and stratification phenomena related to LWR containment studies in the PANDA facility, Nuclear Engineering and Design 237 (2007) 409–419.

[13] M. Babić, I. Kljenak, B. Mavko, Prediction of light gas distribution in experimental containment facilities using the CFX4 code, Nuclear Engineering and Design 238 (2008) 538–550.

[14] M. Babić, I. Kljenak, B. Mavko, Simulations of TOSQAN containment spray tests with combined Eulerian CFD and droplet-tracking modelling, Nuclear Engineering and Design 239 (2009) 708–721.

[15] Y.S. Choi, U.J. Lee, G.C. Park, Study on local hydrogen behaviors in a subcompartment of the NPP containment, Nuclear Engineering and Design 208 (2001) 99–116.

[16] M. Heitsch, R. Huhtanen, Z. Techy, C. Fry, P. Kostka, J. Niemi, B. Schramm, CFD evaluation of hydrogen risk mitigation measures in a VVER-440/213 containment, Nuclear Engineering and Design 240 (2010) 385–396.

[17] M. Heitsch, D. Baraldi, H. Wilkening, Simulation of containment jet flows including condensation, Nuclear Engineering and Design 240 (2010) 2176–2184.

[18] M. Houkema, N.B. Siccama, J.A. Lycklama à Nijeholt, E.M.J. Komen, Validation of the CFX4 CFD code for containment thermal-hydraulics, Nuclear Engineering and Design 238 (2008) 590–599.

[19] I. Kljenak, M. Babić, B. Mavko, I. Bajsić, Modeling of containment atmosphere mixing and stratification experiment using a CFD approach, Nuclear Engineering and Design 236 (2006) 1682–1692.

[20] J. Lee, G.C. Park, Experimental study on hydrogen behavior at a subcompartment in the containment building, Nuclear Engineering and Design 217 (2002) 41–47.

[21] J. Malet, E. Porcheron, J. Vendel, OECD International Standard Problem ISP-47 on containment thermal-hydraulics—Conclusions of the TOSQAN part, Nuclear Engineering and Design 240 (2010) 3209–3220.

[22] J.M. Martin-Valdepeñas, M.A. Jiménez, F. Martín-Fuertes, J.A. Fernández, Improvements in a CFD code for analysis of hydrogen behaviour within containments, Nuclear Engineering and Design 237 (2007) 627–647.

[23] S. Mimouni, J.-S. Lamy, J. Lavieville, S. Guieu, M. Martin, Modelling of sprays in containment applications with a CMFD code, Nuclear Engineering and Design 240 (2010) 2260–2270.

[24] F. Niu, H. Zhao, P.F. Peterson, J. Woodcock, R.E. Henry, Investigation of mixed convection in a large rectangular enclosure, Nuclear Engineering and Design 237 (2007) 1025–1032.

[25] D. Paladino, R. Zboray, P. Benz, M. Andreani, Three-gas mixture plume inducing mixing and stratification in a multi-compartment containment, Nuclear Engineering and Design 240 (2010) 210–220.

[26] D. Paladino, R. Zboray, M. Andreani, J. Dreier, Flow transport and mixing induced by horizontal jets impinging on a vertical wall of the multi-compartment PANDA facility, Nuclear Engineering and Design 240 (2010) 2054–2065.

[27] D. Paladino, R. Zboray, O. Auban, The PANDA tests 9 and 9bis investigating gas mixing and stratification triggered by low-momentum plumes, Nuclear Engineering and Design 240 (2010) 1262–1270.

[28] E. Porcheron, P. Lemaitre, A. Nuboer, V. Rochas, J. Vendel, Experimental investigation in the TOSQAN facility of heat and mass transfers in a spray for containment application, Nuclear Engineering and Design 237 (2007) 1862–1871.

[29] D.W. Stamps, CONTAIN Assessment of the NUPEC Mixing Experiment. Report SAND94-2880, Sandia National Laboratories, Albuquerque, NM, USA, 1995.

[30] D.W. Stamps, Analyses of the thermal-hydraulics in the NUPEC ¼-scale model containment experiments, Nuclear Science and Engineering 128 (1998) 243–269.

[31] E. Studer, J.-P. Magnaud, F. Dabbene, I. Tkatschenko, International standard problem on containment thermal-hydraulics ISP47, Step 1—Results from the MISTRA exercise, Nuclear Engineering and Design 237 (2007) 536–551.

[32] L.A. Valencia, Hydrogen distribution tests under severe accident conditions at the large-scale HDR facility, Nuclear Engineering and Design 140 (1993) 51–60.

[33] H. Wilkening, D. Baraldi, M. Heitsch, CFD simulations of light gas release and mixing in the Battelle Model-Containment with CFX, Nuclear Engineering and Design 238 (2008) 618–626.

[34] R. Zboray, D. Paladino, Experiments on basic thermalhydraulic phenomena relevant for LWR containments: Gas mixing and transport induced by buoyant jets in a multi-compartment geometry, Nuclear Engineering and Design 240 (2010) 3158–3169.

[35] OECD Nuclear Energy Agency, Flame Acceleration and Transition to Detonation in Hydrogen/Air/Diluent Mixtures, State-of-the-Art Report by an NEA Group of Experts, Report NEA/CSNI/R(92)3 (1992).

[36] OECD Nuclear Energy Agency, Flame Acceleration and Deflagration-to-Detonation Transition in Nuclear Safety, State-of-the-Art Report by a Group of Experts, Report NEA/CSNI/R(2000) 7.

[37] OECD Nuclear Energy Agency, OECD/NEA THAI Project, Hydrogen and Fission Product Issues Relevant for Containment Safety Assessment under Severe Accident Conditions, Report NEA/CSNI/R(2010) 3.

[38] D. Baraldi, M. Heitsch, H. Wilkening, CFD simulations of hydrogen combustion in a simplified EPR containment with CFX and REACFLOW, Nuclear Engineering and Design 237 (2007) 1668–1678.

[39] A. Bentaib, Explosion d'hydrogène—Synthèse des connaissances et méthode d'évaluation du risque. Report IRSN/DSR/SAGR/2006-128, Fontenay-aux-Roses, France, 2006.

[40] D. Bjerketvedt, J.R. Bakke, K. van Wingerden, Gas Explosion Handbook, Journal of Hazardous Materials 52 (1997) 1–150.
[41] W. Breitung, R. Redlinger, Containment pressure loads from hydrogen combustion in unmitigated severe accidents, Nuclear Technology 111 (1995) 395–419.
[42] W. Breitung, S. Dorofeev, A. Kotchourko, R. Redlinger, W. Scholtyssek, A. Bentaib, J.-P. L'Heriteau, P. Pailhories, J. Eyink, M. Movahed, K.G. Petzold, M. Heitsch, V. Alekseev, A. Denkevits, M. Kuznetsov, A. Efimenko, M.V. Okun, T. Huld, D. Baraldi, Integral large scale experiments on hydrogen combustion for severe accident code validation—HYCOM, Nuclear Engineering and Design 235 (2005) 253–270.
[43] H.F. Coward, G.W. Jones, Limits of Flammability of Gases and Vapors, Bulletin 503, U.S. Bureau of Mines, 1952.
[44] S.B. Dorofeev, M.S. Kuznetsov, V.I. Alekseev, A.A. Efimenko, A.V. Bezmelnitsyn, Yu.G. Yankin, W. Breitung, Effect of Scale and Mixture Properties on Behavior of Turbulent Flames in Obstructed Areas. Preprint IAE-6127/3, RRC Kurchatov Institute, Report FZKA 6268, Forschungszentrum Karlsruhe, Germany, 1999.
[45] S.B. Dorofeev, V.P. Sidorov, M.S. Kuznetsov, I.D. Matsukov, V.I. Alekseev, Effect of scale on the onset of detonations, Shock Waves 10 (2000) 137–149.
[46] S.B. Dorofeev, M.S. Kuznetsov, V.I. Alekseev, A.A. Efimenko, W. Breitung, Evaluation of limits for effective flame acceleration in hydrogen mixtures, Journal of Loss Prevention in the Processes Industries 14 (2001) 583–589.
[47] J.F. Griffiths, J.A. Barnard, Flame and Combustion, Blackie Academic & Professionals, 1995.
[48] M. Kuznetsov, V. Alekseev, A. Bezmelnitsyn, W. Breitung, S.B. Dorofeev, I. Matsukov, A. Veser, Yu. Yankin, Effect of Obstacle Geometry on Behavior of Turbulent Flames. Preprint IAE-6137/3, RRC Kurchatov Institute, Report FZKA 6328, Forschungszentrum Karlsruhe, 1999.
[49] M. Kuznetsov, V. Alekseev, S.B. Dorofeev, Comparison of critical conditions for DDT in regular and irregular cellular detonation systems, Shock Waves 10 (2000) 217–224.
[50] J.L. Lee, J.J. Lee, G.C. Park, Assessment of the GOTHIC code for prediction of hydrogen flame propagation in small scale experiments, Nuclear Engineering and Design 236 (2006) 63–76.
[51] M. Manninen, A. Silde, I. Lindholm, R. Huhtanen, H. Sjövall, Simulation of hydrogen deflagration and detonation in a BWR reactor building, Nuclear Engineering and Design 211 (2002) 27–50.
[52] N. Peters, Turbulent Combustion, Cambridge University Press, 2000.
[53] Z.M. Shapiro, T.R. Moffette, Hydrogen Flammability Data and Application to PWR Loss-of-Coolant Accident. Report WAPD-SC-545, Bettis Plant, Pittsburgh, Pennsylvania, USA, 1957.
[54] V.P. Sidorov, S.B. Dorofeev, Influence of initial temperature, dilution, and scale on DDT conditions in hydrogen-air mixtures, Archivum Combustionis 18 (1998) 87–103.
[55] R. Redlinger, DET3D—A CFD tool for simulationg hydrogen combustion in nuclear reactor safety, Nuclear Engineering and Design 238 (2008) 610–617.
[56] L.B. Thompson, J.J. Haugh, B.R. Sehgal, Large scale hydrogen combustion experiments, Proc. Int. Conf. on Containment Design (1984). Toronto, Ontario, Canada.
[57] D.R. Whitehouse, D.R. Greig, G.W. Koroll, Combustion of stratified hydrogen-air mixtures in the 10.7 m3 Combustion Test Facility cylinder, Nuclear Engineering and Design 166 (1996) 453–462.
[58] International Atomic Energy Agency (IAEA), Mitigation of Hydrogen Hazards in Water Cooled Power Reactors, IAEA-TECDOC-1196, Vienna, Austria (2001).

[59] International Atomic Energy Agency (IAEA), Implementation of Hydrogen Mitigation Techniques and Filtered Containment Venting, Working Material IAEA-J4-TC-1181, Vienna, Austria (2001).

[60] OECD Nuclear Energy Agency, Hydrogen Management Techniques in Containment, Report by a Group of Experts, Report NEA/CSNI/R(93)2 (1993).

[61] OECD Nuclear Energy Agency, Proceedings of the OECD/NEA/CSNI Workshop on the Implementation of Hydrogen Mitigation Techniques, Winnipeg, Manitoba, Canada, May 1996, Report AECL-11762, NEA/CSNI/R(96)8, Whiteshell Laboratories, Pinawa, Manitoba, Canada (1997).

[62] OECD Nuclear Energy Agency, The Implementation of Hydrogen Mitigation Techniques, Summary and Conclusions, OECD Workshop, Winnipeg, Manitoba, Canada, May 13–15, 1996, NEA/CSNI/R(96)9(1996).

[63] OECD Nuclear Energy Agency, Implementation of Hydrogen Mitigation Techniques During Severe Accidents in Nuclear Power Plants, Report OECD/GD(96)195 (1996).

[64] E. Bachellerie, F. Arnould, M. Auglaire, B. de Boeck, O. Braillard, B. Eckardt, F. Ferroni, R. Moffett, Generic approach for designing and implementing a passive autocatalytic recombiner PAR-system in nuclear power plant containments, Nuclear Engineering and Design 221 (2003) 151–165.

[65] T.K. Blanchat, D.W. Stamps, Deliberate Ignition of Hydrogen-Air-Steam Mixtures in Condensing Steam Environments. Report NUREG/CR-6530, SAND94-1676, Sandia National Laboratories, Albuquerque, NM, and University of Evansville, Evansville, IN, USA, 1997.

[66] T.K. Blanchat, A. Malliakos, Analysis of hydrogen depletion using a scaled passive autocatalytic recombiner, Nuclear Engineering and Design 187 (1999) 229–239.

[67] P. Bröckerhoff, W. von Lensa, E.-A. Reinecke, Innovative devices for hydrogen removal, Nuclear Engineering and Design 196 (2000) 307–314.

[68] J. Deng, X.W. Cao, A study on evaluating a passive autocatalytic recombiner PAR-system in the PWR large-dry containment, Nuclear Engineering and Design 238 (2008) 2554–2560.

[69] F. Fineschi, M. Bazzichi, M. Carcassi, A study on the hydrogen recombination rates of catalytic recombiners and deliberate ignition, Nuclear Engineering and Design 166 (1996) 481–494.

[70] K. Fischer, Qualification of a passive catalytic module for hydrogen mitigation, Nuclear Technology 112 (1995) 58–62.

[71] K. Fischer, P. Broeckerhoff, G. Ahlers, V. Gustavsson, L. Herranz, J. Polo, T. Dominguez, P. Royl, Hydrogen removal from LWR containments by catalytic-coated thermal insulation elements (THYNCAT), Nuclear Engineering and Design 221 (2003) 137–149.

[72] R. Heck, G. Kelber, K. Schmidt, H.J. Zimmer, Hydrogen reduction following severe accidents using the dual recombiner-igniter concept, Nuclear Engineering and Design 157 (1995) 311–319.

[73] M. Heitsch, Fluid dynamic analysis of a catalytic recombiner to remove hydrogen, Nuclear Engineering and Design 201 (2000) 1–10.

[74] H.J. Kim, C.H. Sohn, S.H. Chung, H.D. Kim, Laser Rayleigh measurement of mixing processes and control of hydrogen combustion using quenching meshes, Nuclear Engineering and Design 187 (1999) 291–302.

[75] W.E. Lowry, B.R. Bowman, B.W. Davis, Final Results of the Hydrogen Igniter Experimental Program. Report NUREG/CR-2486, Lawrence Livermore National Laboratory, Lawrence, CA, USA, 1982.

[76] E.-A. Reinecke, I.M. Tragsdorf, K. Gierling, Studies on inovative hydrogen recombiners as safety devices in the containments of light water reactors, Nuclear Engineering and Design 230 (2004) 49–59.

[77] E.-A. Reinecke, A. Bentaib, S. Kelm, W. Jahn, N. Meynet, C. Caroli, Open issues in the applicability of recombiner experiments and modelling to reactor simulations, Progress in Nuclear Energy 52 (2010) 136–147.
[78] P. Royl, H. Rochholz, W. Breitung, J.R. Travis, G. Necker, Analysis of steam and hydrogen distributions with PAR mitigation in NPP containments, Nuclear Engineering and Design 202 (2000) 231–248.
[79] S.Y. Yang, S.H. Chung, H.J. Kim, Effect of pressure on effectiveness of quenching meshes in transmitting hydrogen combustion, Nuclear Engineering and Design 224 (2003) 199–206.
[80] H.-J. Allelein, S. Arndt, W. Klein-Hessling, S. Schwarz, C. Spengler, G. Weber, COCOSYS: Status of development and validation of the German containment code system, Nuclear Engineering and Design 238 (2008) 872–889.
[81] K.K. Murata, D.C. Williams, J. Tills, R.O. Griffith, R.G. Gido, E.L. Tadios, F.J. Davis, G.M. Martinez, K.E. Washington, Code Manual for CONTAIN 2.0: A Computer Code for Nuclear Reactor Containment Analysis. NUREG/CR-65, SAND97-1735, Sandia National Laboratories, Albuquerque, New Mexico, USA, 1997.
[82] K.K. Murata, D.W. Stamps, Development and Assessment of the CONTAIN Hybrid Flow Solver. Report SAND96-2792, Sandia National Laboratories, Albuquerque, New Mexico, USA, 1996.
[83] J.R. Travis, P. Royl, R. Redlinger, G. Necker, J.W. Spore, K.L. Lam, T.L. Wilson, B.D. Nichols, C. Müller, GASFLOW-II: A Three-Dimensional-Finite-Volume Fluid-Dynamics Code for Calculating the Transport, Mixing, and Combustion of Flammable Gases and Aerosols in Geometrically Complex Domains, Theory and Computational Model vol. 1, (1998) Reports FZKA-5994, LA-13357-MS.
[84] H. Holzbauer, L. Wolf, GOTHIC verification on behalf of the Heiss Dampf Reaktor hydrogen-mixing experiments, Nuclear Technology 125 (1999) 166–181.
[85] M. Andreani, D. Paladino, Simulation of gas mixing and transport in a multi-compartment geometry using the GOTHIC containment code and relatively coarse meshes, Nuclear Engineering and Design 240 (2010) 1506–1527.
[86] M. Andreani, D. Paladino, T. George, Simulation of basic gas mixing tests with condensation in the PANDA facility using the GOTHIC code, Nuclear Engineering and Design 240 (2010) 1528–1547.
[87] Y.S. Choi, U.J. Lee, J.J. Lee, G.C. Park, Improvement of HYCA3D code and experimental verification in rectangular geometry, Nuclear Engineering and Design 226 (2003) 337–349.
[88] S. Kudriakov, F. Dabbene, E. Studer, A. Beccantini, J.-P. Magnaud, H. Paillère, A. Bentaib, A. Bleyer, J. Malet, E. Porcheron, C. Caroli, The TONUS CFD code for hydrogen risk analysis: Physical models, numerical schemes and validation matrix, Nuclear Engineering and Design 238 (2008) 551–565.
[89] M.M. Pilch, M.D. Allen, D.C. Williams, Heat Transfer During Direct Containment Heating, in: G.A. Greene (Ed.), Adv. in Heat Transfer, vol. 29, Academic Press, 1997.
[90] Nucl. Eng. Des., 164, 1996 (Topical Issue on DCH).
[91] T.Y. Chu, M.M. Pilch, J.H. Bentz, J.S. Ludwigsen, W.Y. Lu, L.L. Humphries, Lower Head Failure Experiments and Analyses. Report, NUREG/CR-5582, SAND98-2047, Sandia National Laboratories, Albuquerque, NM, USA, 1999.
[92] B.R. Sehgal, R.R. Nourgaliev, T.N. Dinh, A. Karbojian, FOREVER experimental program on reactor pressure vessel creep behavior and core debris retention, In: Proceedings of the 15th International Conference on Structural Mechanics in Reactor Technology (SMiRT-15), August 15–20, 1999, Seoul, Korea.

[93] OECD/CSNI Workshop on In-Vessel Core Debris Retention and Coolability, Summary and Conclusions, NEA/CSNI/R(98)21. http://www.nea.fr/html/nsd/docs/1998/csni-r98–21.pdf, 1999.

[94] J.J. Sienicki, B.W. Spencer, Superheat effects on localized vessel breach enlargement during corium ejection, Trans. ANS 52 (1986) 522–524. Abstract-INSPEC.

[95] M.M. Pilch, Continued enlargement of the initial failure site in the reactor pressure vessel, in: Appendix J, NUREG/CR-6075, SAND93-1535, Sandia National Laboratories, Albuquerque, NM, 1993.

[96] T.N. Dinh, V.A. Bui, R.R. Nourgaliev, T. Okkonen, B.R. Sehgal, Modelling of heat and mass transfer processes during core melt discharge from a reactor pressure vessel, Nuclear Engineering and Design 163 (Issues 1-2) (June 1996) 191–206.

[97] M.M. Pilch, R.O. Griffith, Gas blowthrough and flow quality correlations for use in the analysis of high pressure melt ejection (HPME) events. SANDIA Report, SAND91–2322, UC-523, Sandia National Laboratories, 1992.

[98] D.F. Gluck, J.P. Gillie, D.J. Simkin, E.E. Zukowski, Distortion of the liquid surface under low g conditions, in: D.J. Simkin (Ed.), Aerospace Chemical Engineering, 62, No. 1, AIChE, New York, 1966, p. 150.

[99] D.F. Gluck, J.P. Gillie, E.E. Zukowski, D.J. Simkin, Distortion of a free surface during tank discharge, J. Spacecraft 3 (1966) 1691.

[100] C. Smoglie, J. Reimann, Two-phase flow through small branches in a horizontal pipe with stratified flow, Int. J. Multiphase Flow 12 4 (1986) 609–625.

[101] C. Smoglie, J. Reimann, U. Müller, Two-phase flow through small breaks in a horizontal pipe with stratified flow, Nuclear Engineering and Design 99 (1987) 117–130.

[102] High-pressure melt ejection (HPME) and direct containment heat (DCH), State-of-the-Art Report, OCDE/GE(96)194, NEA/CSNI/R(96)25, CSNI OECD Nuclear Energy Agency (1996).

[103] T.K. Blanchat, M.M. Pilch, R.Y. Lee, L. Meyer, M. Petit, Direct Containment Heating Experiments at Low Reactor Coolant System Pressure in the Surtsey Test facility, Sandia National Laboratory report, NUREG/CR-5746, SAND99-1634, 1999.

[104] L. Meyer, M. Gargallo, Low Pressure Corium Dispersion Experiments with Simulant Fluids in a Scaled Annular Cavity, Nuclear Technology 141 (2003) 257–274.

[105] L. Meyer, G. Albrecht, D. Wilhelm, Direct Containment Heating Investigations for European Pressurized Water Reactors, NUTHOS-6, Nara, Japan, October 4–8, 2004

[106] R. Meignen, D. Plet, C. Caroli, L. Meyer, D. Wilhelm, Direct containment heating at low primary pressure: Experimental investigation and multi-dimensional modelling, NURETH-11, Paper: 164, Avignon, France, 2005.

[107] S. Kisyoski, I. Ivanov, D. Popov, et al., DISCO-L2 Test Report, EC-SAM-LACOMERA-D12, January 2006.

[108] L. Meyer, G. Albrecht, C. Caroli, I. Ivanov, Direct containment heating integral effects tests in geometries of European nuclear power plants, Nucl. Eng. Des. 239 (2009) 2070–2084.

[109] W.R. Bohl, D. Wilhelm, The Advanced Fluid Dynamics Model Program: Scope and Accomplishment, Nuclear Technology 99 (1992) 309–317.

[110] D. Wilhelm, Analysis of a Thermite Experiment to Study Low Pressure Corium Dispersion, Forschungszentrum Karlsruhe, Report FZKA 6602 (2001).

[111] G. Berthoud, M. Valette, Development of a multidimensional model for the premixing phase of a fuel–coolant interaction, Nucl. Eng. Des. 149 (1994) 409–418.

[112] M.M. Pilch, A two-cell equilibrium model for predicting direct containment heating, Nucl. Eng. Des. 164 (1996) 61–94.

[113] H. Yan, T.G. Theofanous, The prediction of direct containment heating, Nucl. Eng. Des. 164 (1996) 95–116.

[114] K.E. Washington, D.S. Stuart, Comparison of CONTAIN and TCE calculations for direct containment heating of Surry, Nucl. Eng. Des. 164 (1996) 201–210.

[115] H.-J. Allelein, W. Klein-Hessling, N. Reinke, K. Trambauer, ATHLET-CD und COCOSYS: die mechanistischen Rechenprogramme der GRS zur Simulation schwerer Störfälle (ATHLET-CD and COCOSYS: the mechanistic computer codes of GRS for simulating severe accidents), Atw. Atomwirtschaft-Atomtechnik - Internationale Zeitschrift für Kernenergie, 49, no. 5 (2004) 301, 337–341.

[116] H.-J. Allelein, K. Neu, J.-P. Van Dorsselaere, K. Müller, P. Kostka, M. Barnak, P. Matejovic, A. Bujan, J. Slaby, European validation of the integral code ASTEC (EVITA), Nucl. Eng. Des. 221 (2003) 1–3, 95–118.

[117] N. Zuber, et al., An integrated structure and scaling methodology for severe accident technical issue resolution: Development of methodology, Nuclear Eng. Design 186 (1998) 1–21.

[118] S.B. Kim, R.J. Park, H.D. Kim, C. Chevall, M. Petit, Reactor Cavity Debris Dispersal Experiment with Simulant at Intermediate System Pressure, SMIRT15 Post Conference Seminar on Containment of Nuclear Reactors, Seoul, Korea, August 23–24 (1999).

[119] L. Meyer, A. Kotchourko, Separate Effects Tests on Hydrogen Combustion during Direct Containment Heating Events in European Reactors, SMIRT-19, August 12–17, 2007, Toronto, Canada.

[120] T.K. Blanchat, M. Pilch, M.D. Allen, Experiments to Investigate Direct Containment Heating Phenomena with Scaled Models of the Calvert Cliffs Nuclear Power Plant. NUREG/CR-6469, SAND96-2289, Sandia National Laboratories, Albuquerque, NM, February 1997.

[121] M.L. De Bertodano, A. Becker, A. Sharon, R. Schnider, DCH Dispersal and Entrainment Experiment in a Scaled Annular Cavity, Nucl. Eng. Des. 164 (1996) 271–285.

[122] T.K. Blanchat, M.D. Allen, M.M. Pilch, R.T. Nichols, Experiments to Investigate Direct Containment Heating Phenomena with Scaled Models of the Surry Nuclear Power Plant. NUREG/CR-6152, SAND93-2519, Sandia National Laboratories, Albuquerque, NM, 1994.

[123] L. Meyer, G. Albrecht, Experimental study of hydrogen combustion during DCH events in two different scales, NURETH-14, 2011, Toronto, Canada.

[124] S. Basu, T. Ginsberg, (Eds.), Proceedings of the Second Steam Explosion Review Group (SERG-2), NUREG-1524 (1996).

[125] R. Krieg, et al., Load carrying capacity of a reactor vessel head under a corium slug impact from a postulated in-vessel steam explosion, Nucl. Eng. Des. 202 (2000) 179–196.

[126] W.W. Yuen, T.G. Theofanous, S. Angelini, Lower Heat Integrity Under Steam Explosion, Nuclear Engineering & Design 189 (1999) 7–57.

[127] D.L. Frost, B. Bruckert, G. Ciccarelli, Effect of boundary conditions on the propagation of a vapour explosion in stratified molten tin /water systems, Nucl. Eng. Des. 155 (1995) 311–322.

[128] S. Picchi, G. Berthoud, MC3D modelling of stratified explosions, Proc. Workshop on Severe Accident Research in Japan, Tokyo, Japan, 4-6 November 1998 (JAERI-Conf 99-005), pp. 370–375.

[129] I. Huhtiniemi, D. Magallon, H. Hohmann, Results of recent KROTOS FCI tests: alumina versus corium melts, Nucl. Eng. Des. 189 (1999) 379–389.

[130] S.J. Board, R.W. Hall, R.S. Hall, Detonation of fuel coolant explosions, Nature 254 (1975) 319–321.

[131] I. Huhtiniemi, D. Magallon, Insight into steam explosions with corium melts in KROTOS, Nucl. Eng. Des. 204 (2001) 391–400.

[132] D. Magallon, I. Huhtiniemi, Corium melt quenching tests at low pressure and subcooled water in FARO, Nucl. Eng. Des. 204 (2001) 369–376.

[133] J.H. Song, et al., Fuel-Coolant Intraction in TROI Using a UO_2-ZrO_2 Mixture, Nucl. Eng. Des. 222 (2003) 1–15.

[134] D. Magallon, I. Huhtiniemi, Energetic event in fuel-coolant interaction test FARO L-33, ICONE-9, Nice, France, April 8–12, 2001.

[135] M.J. Bird, An experimental study of scaling in core melt/water interaction, 22nd National Heat Transfer Conf, Niagara Falls (1984).

[136] R.C. Hanson, et al., Dynamics and preconditioning in single-droplet vapour explosion, Nucl. Tech. 167 (2009) 223–234.

[137] M.L. Corradini, B.J. Kim, M.D. Oh, Vapour Explosion in Light Water Reactors: A Review of Theory and Modelling, Prog. Nucl. Energy 22 (1988) 1–117.

[138] D.L. Frost, G. Ciccarelli, Propagation of Explosive Boiling in Molten Tin-Water Mixtures, Proc. of 1988 ASME/AIChE National Heat Transfer Conference, Houston, HTD-96, vol. 2 (1988) 539–548.

[139] A.W. Cronenberg, R. Benz, Vapour Explosion Phenomena with respect to Nuclear Reactor Safety Assessment, Advances in Nuclear Science and Technology 12 (1980) 247–335.

[140] E.P. Hicks, D.C. Menzies, Theoretical Studies on the Fast Reactor Maximum Accident, ANL-7120 (1965) 654–670.

[141] D.L. Frost, J.H.S. Lee, G. Ciccarelli, The use of HugoniotAnalysis for the Propagation of Vapor Explosion Waves, ShockWaves 1 (1991) 99–110.

[142] W.W. Yuen, T.G. Theofanous, The prediction of 2D thermal detonations and resulting damage potential, Nuclear Engineering and Design 155 (1995) 289–309.

[143] Y.M. Farawila, S.I. Abdel-Khalik, On the calculation of steam explosion conversion ratios from experimental data, Nuclear Science and engineering 104 (1990) 288–295.

[144] M.F. Hessheimer, R.A. Dameron, Containment Integrity Research at Sandia National Laboratories An Overview. NUREG/CR-6906, U.S. Nuclear Regulatory Commission, 2006.

[145] M.F. Hessheimer, E.W. Klamerus, L.D. Lambert, G.S. Rightley, R.A. Dameron, Overpressurization Test of a 1:4-Scale Prestressed Concrete Containment Vessel Model, NUREG/CR-6810, U.S. Nuclear Regulatory Commission, 2003.

[146] H. Grebner, J. Sievers, Structure Simulation of Prestressed Concrete Containment Structure, 18th International Conference on Structural Mechanics in Reactor Technology, Beijing, 2005.

[147] OECD Nuclear Energy Agency, International Standard Problem No. 48— Containment Capacity Synthesis Report, NEA/CSNI/R 5 (2005) 2005.

[148] Gesellschaft für Reaktorsicherheit, Deutsche Risikostudie Kernkraftwerke Phase B, Verlag TÜV Rheinland, 1990.

[149] R. Krieg, B. Dolensky, B. Goller, W. Breitung, R. Redlinger, P. Royl, Assessment of the load–carrying capacities of a spherical pressurized water reactor steel containment under a postulated hydrogen detonation, Nucl. Technol. 141 (2003) 109–121.

BIBLIOGRAPHY ON STEAM EXPLOSION

M. Berman, D.F. Beck, Steam Explosion Triggering and Propagation: Hypotheses and Evidence, Proc. 3rd Intern. Seminar on Containment of Nuclear Reactors Peaceful, UCLA, Los Angeles, August 10-11, 1989 (SNL report SAND89-1878C).

G. Berthoud, Vapor Explosions, Annu. Rev. Fluid Mech., 32 (2000) 573–611.

A.W. Cronenberg, R. Benz, Vapour Explosion Phenomena with respect to Nuclear Reactor Safety Assessment, Advances in Nuclear Science and Technology, 12 (1980) 247–335, Pergamon Press.

D.F. Fletcher, R.P. Anderson, A Review of Pressure-Induced Propagation Models of the Vapour Explosion Process, Prog. Nucl. Energy, 23 (1990) 137–179.

D. Magallon, I. Huhtiniemi, P. Dietrich, G. Berthoud, M. Valette, W. Schütz, H. Jacobs, N. Kolev, G. Graziosi, B.R. Sehgal, M. Bürger, M. Buck, E. von Berg, G. Colombo, B.D. Turland, G.P. Dobson, D. Monhardt, MFCI Project—Final Report, INV-MFCI(99)-P007, EUR 19567 EN, 2000 (Work performed in the frame of the 4th Framework Programme of the European Commission under contract FI3S-CT92-0004).

B.D. Turland, G.P. Dobson, Molten Fuel Coolant Interactions: A state-of-the-art report, EUR 16874 EN, 1996 (Work performed in the frame of the 4th Framework Programme of the European Commission under contract FI3S-CT92-0004).

Chapter 4

Late Containment Failure

Chapter Outline

4.1. DEBRIS FORMATION AND COOLABILITY

(M. Bürger and B.R. Sehgal)

4.1.1. Introduction

The previous chapter described the dynamic loads that could be generated during the progression of a severe accident and that could lead to an early

Nuclear Safety in Light Water Reactors. DOI: 10.1016/B978-0-12-388446-6.00004-6

failure of the containment. In this chapter we will describe the loads, which could fail the containment later in time, that is, considerably after the failure of the vessel. The loads we consider are due to the thermal attack of the core melt discharged from the vessel on the containment structures. In particular, we consider the attack of corium melt on the concrete basemat in the containment, which leads to (a) the basemat concrete erosion and (b) the generation of hydrogen, CO, steam, CO_2 from concrete erosion, which pressurize the containment and could, in time, fail it. These dual consequences (containment failure and basemat melt-through) can be altogether avoided, if the corium melt can be cooled to a temperature below that for concrete ablation.

It should be emphasized here, again, that the late failure of containment allows a large fraction of the fission product aerosols to agglomerate and deposit on the containment surfaces, where they can dissolve in water, which may be there due to the containment sprays or steam condensation. They are then carried to containment sumps, as they were found in the TMI-2 containment after the accident. Clearly, the late failure of containment is of much smaller consequence for the public than an early failure of the containment, except for the consequence of ground and water contamination that could possibly result from the basemat melt-through by the core melt.

Returning to the phenomenology of the corium melt attack on the containment concrete basemat (described later in this chapter) and its mitigation by cooling the melt, there are two preferred methods employed for cooling the melt: (1) the melt discharged from the failed vessel is released onto the basemat in the normally dry vessel cavity, which is then flooded with water as soon as possible after the arrival of the melt, and (2) the vessel cavity is filled with water, before the arrival of the melt, and the melt discharged from the vessel falls into the pool of water maintained in the vessel cavity. The first accident management measure is employed in most PWRs and BWRs, except that the Nordic BWRs employ the second measure. The Westinghouse Severe Accident Management Guidelines (SAMGs) advise the utility company to add water in the vessel cavity at least to the level of the equator of the lower head of the vessel. The depth of water pool in the Westinghouse PWR vessel cavity could be of the order of 3–4 meters and of 7–11 meters in the vessel cavities of the Nordic BWRs. The French and German PWRs do not fill their cavities with water and actually design the containment to avoid entry of even the condensed water in the containment. Thus there are three different vessel cavities to consider: initially dry, initially with water pools of relatively small depth, and those with a deep pool of water. Water for each of these cases would be substantially subcooled.

In this section we will consider the cases of dry and wet cavities and the potential for cooling the melt either with flooding of water from the top; or with melt dropped into shallow and deep pools of water. Our aim is to assess whether the melt can be cooled to temperatures lower than the concrete ablation temperature, in order to avoid (1) further generation of fission products, (2) erosion of concrete leading to basemat melt-through and (3) containment failure caused by generation of gases from concrete ablation and from any other processes that accompany the interaction of core melt and water, as for example a possible steam explosion, when melt is dropped into a water pool.

It should be recognized that, in order to stabilize and terminate a severe accident, cooling of the fuel, whether it is still intact as rods, or in the form of particulate debris, or in the form of melt, has to be performed and the heat transported to a heat sink. This has been most evident in the Fukushima accident, but it was also necessary in the TMI-2 accident, where the presence of water in the vessel and the natural circulation flow to the steam generators, which acted as the heat sinks, stabilized the accident and cooled the melt and the particulate debris present in the vessel to sufficiently low temperatures not to threaten vessel failure and to reach the termination of the accident later on. Similar efforts were undertaken at the Fukushima plants, except that due to lack of electric power and the damage to pumps the heat sinks could not be established early enough to prevent damage to the core. The venting of the vessel and the containment leads to the release of radioactivity to the environment. Thus, for stabilization and termination of a severe accident in which core melt reaches the containment, the quenching of melt and/or particulate debris is needed quite readily. Time is of essence, since the gas generation and the containment pressurization are a function of time, and so are any accident mitigation measures that may be instituted by the operators. In this context the containment volume and the heat removal rate from the containment are important parameters.

The two methods of achieving melt coolability in the containment, that is, by pouring water on top of a melt pool in a dry cavity; or pouring melt into a water pool in a "wet cavity," have different phenomenology for melt cooling. The melt–water interaction in the former case is relatively benign after the initial contact as an insulating crust is formed all around the melt pool. In the latter case the oxidic-ceramic melt breaks into melt particles of varying sizes, resulting in orders of magnitude increase in the surface area of contact between the original melt and the water. This can lead to a steam explosion or to generation of a large volume of steam. The particles formed can accumulate as a heaped debris bed on the basemat surface, which again decreases the surface area of contact. However, if the porosity of the debris bed is not very low, water entry in the bed becomes possible and the bed could be coolable.

The operative difference between the two methods of cooling is the melt fragmentation process, which increases the surface area for heat removal from the melt. Thus, the phenomenology for the two methods of cooling is different but has similarities. In this very context there is a combination developed at FZK (now KIT, Karlsruhe), called the COMET concept in which water is added from the bottom of a melt layer. The rapid and high-volume steam generation at the bottom of the melt pool creates interconnected porosity in the melt pool through which water ingression takes place and cools the melt. This process of inducing fragmentation in a melt layer is a relatively controlled process, since the amount of water added into the melt layer is controlled. It has been found that this method of inducing fragmentation for melt cooling is more efficient than that of adding water from top on a melt layer.

In the following subsections we will describe the experiments performed, the understanding achieved, and the methodologies developed to date for the various melt coolability processes.

4.1.2. Coolability of ex-Vessel Melt Pools

Water Flooding of a Melt Pool in the Initially Dry Vessel Cavity

This is the situation in many of the Generation II PWRs and BWRs. The accident management scheme in these plants is that after the melt discharged from the vessel is deposited in the initially dry vessel cavity, the cavity is flooded with water to cool the melt. It is the natural process of accident management that is inherent in all Generation II LWRs, except the Nordic BWRs.

Experimental Results

Recognizing that ex-vessel coolability of a core melt resulting from a severe accident in which a vessel failure occurs is needed to protect the integrity of the containment and to prevent a large release of fission products to the environment, a research program named Melt Attack and Coolability Experiments (MACE) was initiated [1] at Argonne Natural Laboratory (ANL) in the late 1980s under the sponsorship of an international consortium led by the Electric Power Research Institute (EPRI). The aim of the MACE Program was to demonstrate the coolability of a core melt pool under prototypic conditions. It was believed that pouring water over a corium melt pool would cool it and that type of accident management which can be achieved in most PWRs and BWRs would be sufficient to stabilize and terminate a severe accident.

The MACE Program constructed a large-scale test facility shown in Figure 4.1 in which a melt pool was created from powders of natural UO_2, ZrO_2, and some concrete products through resistance heating. The size of the first experiment [2] was 30 cm × 30 cm × 15 cm depth, which was interacting with a limestone common sand concrete basemat. The corium melt was prepared by resistance heating and was heated at the decay power level with the same electrodes. The corium powder, is nonconductive, but becomes conductive after melting temperature is reached. The first experiment (30 cm × 30 cm × 15 cm melt pool), after water flooding, was found to develop a crust on its cooled surface which attached itself to the cavity walls and prevented the contact of water with the corium melt. The latter kept attacking the concrete basemat, and as it followed the surface of the concrete, it separated itself from the crust that remained attached to the side walls, preventing the contact of the water overlayer with the melt. The crust was strong enough to support the load of water pool on top of the melt in the vessel

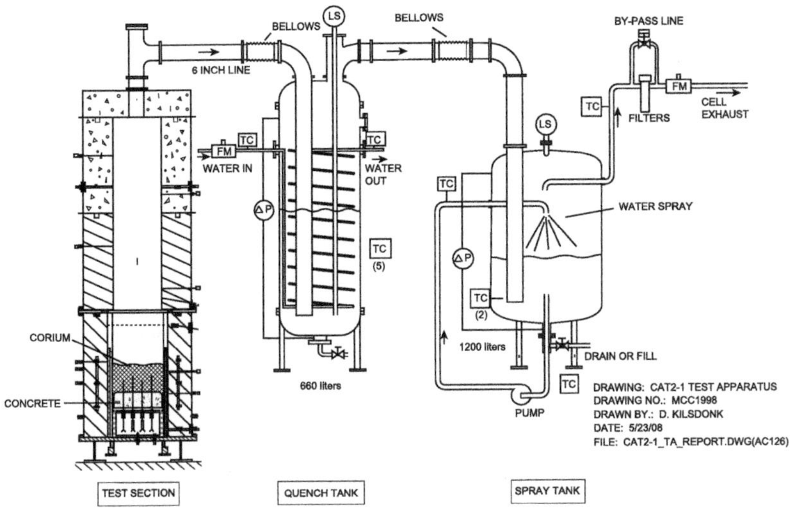

FIGURE 4.1 Schematic of the MACE large melt coolability facility.

of the experiment. The experiment was shut down when the concrete erosion reached a specified value.

The argument was voiced that the span of the melt pool in the vessel cavity of the containment would be about 6 meters and that such a wide crust would not be structurally stable in a PWR containment. The size of the MACE experiment was increased to 50 × 50 × 25 cm depth and later to 120 × 120 × 20 cm size employing approximately 2 tons of melt in order to experience a crust failure and the resulting enhancement in heat removal from melt zone. These experiments employed a different technique for generating the corium melt; that is, a corium thermite consisting of UO_2 + Zr + chrorium + a few percent concretes, was ignited to provide sufficient chemical energy from the oxidation of zirconium to generate a melt pool within tons of seconds. The melt generated was heated with resistance heating with electrodes placed on two sides of the melt pool.

Unfortunately, the large experiments, of 50 × 50 × 25 cm melt pool and the 120 × 120 × 20 cm melt pool, flooded with water on top, displayed the same behavior; that is, a crust was formed, which allowed the passage of concrete ablation gases through it, but not the passage of water through it to allow intimate contact of water with the melt. The melt, though reduced in temperature, kept attacking the concrete base mat, albeit with a very low erosion rate. The heat removal rate at the top surface of the melt pool was measured in all three experiments. It was formed to be $\approx 2\ MW/m^2$, but later as the crust thickness increased, it decreased to $\approx 0.1\ MW/m^2$, which was less than the decay heat input to the melt pool. In the 120 × 120 × 20 cm experiment, considerable water ingression was noted, since the post-test examination showed that the crust thickness of ≈ 10 cm; that is, approximately 50% of the melt pool volume had been quenched. The 10 cm thickness of melt, however, kept eroding the concrete and separating from the crust, which was attached to the side walls.

A 50 cm × 50 cm × 25 cm depth integral melt coolability test with siliceous concrete was also performed, whose results were approximately the same as for the tests employing limestone common sand concrete. Actually the siliceous concrete has much less gas content than the LCS concrete, and its volume reduction is also less than that of LCS concrete. These characteristics are important for one of the four heat-removal modes identified and described in the next paragraph.

Four modes of heat removal from the melt pool were identified: (1) the bulk boiling during the initial melt–water contact; (2) water ingression into the crust and conduction; (3) melt eruptions into water, when the heat generated in the melt is greater than that removed by conduction or water ingression through the crust, and (4) local crust breakup, or fractures, leading to renewal of melt–water contact, which may form another crust underneath.

The integral test program was modified to investigate the modes of heat transfer through separate-effect tests. The intent was to develop validated models that could be employed for the evaluation of prototypic coolability configurations. A test simulating the melt eruption was performed in the MACE Program, in which the gas injection rate at the bottom of the melt pool was varied and melt eruptions into overlaying water were generated. Data on the entrainment coefficient were obtained.

The MACE Project was followed by the MCCI-1 and MCCI-2 Projects [3], also conducted at the Argonne National Laboratory from 2002 to 2010, under the auspices of the Nuclear Energy Agency (NEA) of the Organization for Economic Cooperation and Development (OECD). The objective of the MCCI-1 Project was to perform experiments at reasonable scale, employing prototypic melt material, to understand the physics of the different coolability mechanisms and to provide data for validation of the models developed for these mechanisms. In addition to the separate effect tests, some integral tests were also performed to observe if a melt pool could be cooled by flooding water from the top.

The MCCI-1 Project constructed the experimental facility called Small Scale Water Ingression and Crust Strength (SSWICSS) [4] to study the water ingression phenomenon, which should be a dominant mechanism for cooling the melt pool as also observed in the cooling of molten lava masses. The SSWICS facility (see Figure 4.2) employed a corium melt mass of approximately 100 kg, generated by a thermite reaction, with water added immediately on the top surface of the generated melt pool. The SSWICS tests employed melts with different fractions of concrete to model the various states of corium melt that occur in a prototypic coolability scenario. The MCCI-1 Project also measured the strength of the crusts that were formed during the coolability process. The corium employed in the SSWICS tests was not heated with electrodes; the heat removed in the SSWICS tests was the sensible heat of the melt generated by the thermitic process. The integral tests in the MCCI-1

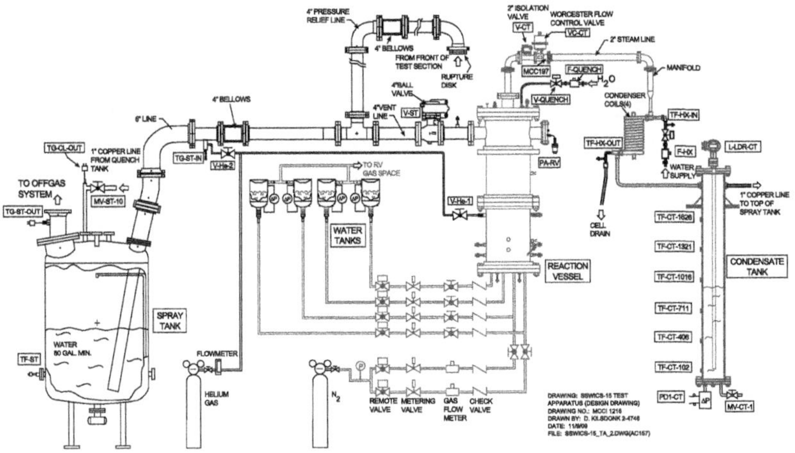

FIGURE 4.2 The SSWICS melt quench facility.

Project were of much larger size (50 × 50; 70 × 70) and the corium in these tests was heated electrically to simulate the decay heat generation. They were performed in the large melt coolability facility (Figure 4.1), as were the MACE tests performed earlier.

The integral tests performed in the MCCI-1 program were basically a continuation of the 2-D corium concrete interaction (CCI) experiments that were flooded with water after the specified concrete ablation was obtained. These experiments thus represented integral tests with late addition of water. The CCI experiments are described in the chapter on corium–concrete interaction (Section 4.3).

The data obtained in the MCCI-1 Project [3] showed that the addition of concrete to the melt reduced the rate of water ingression. The rate reached a plateau; perhaps a certain rate of water ingression will always occur in the melt pool crust, as long as the dry-out condition is not reached. It was clear that deterioration of water ingression will occur as more concrete is added, since its addition to the melt tends to produce a mixture material with greater fracture toughness.

The measured crust strength was low, and that would indicate that a 6–8 meter diameter crust would break under the loading of the water pool. However, any break in crust and the cooling of the melt underneath would lead to formation of a new crust. A cascading failure of crusts formed would be required to enhance the rate of melt coolability significantly. Another observation made during the tests was that the attempts to break the crusts with a loaded sharp-pointed tool were not successful.

The cooling of integral corium concrete interaction (CCI) tests by water flooding was not successful, except for one case with the LCS concrete [3]. The concrete ablation continued, though at a somewhat lower rate, but the melt pool temperature under the top crust did not decrease below the concrete ablation temperature.

The MCCI-2 Project [3], also performed under the auspices of the OECD, was a continuation of the MCCI-1 Project. Its objectives with respect to melt coolability were to (1) to obtain more data on the water ingression mechanism, in particular with gas injection from the bottom to represent the gas generation that would occur with the ablation of concrete; (2) to perform tests incorporating methods to enhance coolability of the melt, for example, the COMET concept (described later in this section) and the cooling plate at the bottom face of the melt pool, as employed in the Gen. III plant: EPR and (3) to perform integral tests for validation of melt coolability models. Another major objective was to further develop the CORQUENCH code; incorporating the models describing the various coolability mechanisms developed with the understanding achieved in the MCCI and MACE projects.

The data obtained on the water ingression cooling mechanism showed that with gas injection its rate improved considerably [3]. These data were used to modify the previous model of water ingression in the CORQUENCH code (described in the next subsection, *The CORQUENCH Code and Its Predictions*). The integral tests were performed in the MACE large-melt coolability facility (see Figure 4.1). The test performed with a cooling plate at the bottom of the melt pool did not indicate substantial heat transferred to the cooling water flowing in the cooling plate. Perhaps an inclined cooling plate would be more effective since the bubble departure velocities at the downward-facing heat transfer surface would increase substantially and the heat transferred would increase.

A test was performed in the SSWICS facility to test the COMET concept. The approximately 100 kg of corium generated through the thermitic reaction was flooded with water, but water was injected from the bottom of the melt pool with nozzles, with and without concurrent gas injection. It was found that the corium pool cooled down at a much faster rate. It was also found that the gas addition along with water did not affect the cooling rate. It should be noted that the corium in the SSWICS tests is not heated electrically.

The integral test with COMET-type water addition into the bottom surface of the melt pool was not performed due to an operational decision. Thus, there is no data from the MCCI Project on the enhancement of melt coolability of a heated corium melt pool with the COMET concept. Incidentally, the integral test employed 700 kg of corium melt with the melt pool dimension of 70 cm × 70 cm × 20 cm. This test was started by employing a UO_2-ZrO_2 (corium) thermite mixture without any concrete addition. It was found that this melt material would fracture sufficiently to provide a greater water ingression rate, which was able to cool the melt in a few hours. This was an unusual event, since in all of the previous tests that employed a small fraction of concrete in the corium melt mixture (representing the small concrete ablation that would take place before the operator in a typical plant would flood the melt pool with water), the melt could not be cooled completely and the concrete ablation did not terminate. Concrete addition to the melt will occur in the melt pool, since the melt jet descending from the vessel onto the concrete basemat would ablate some concrete in the stagnation zone and incorporate it in the melt pool. Thus, the successful coolability occurrence in the last test conducted in the MCCI-2 Project may not be prototypic.

In summary, the data obtained from the MACE and the MCCI-1 AND MCCI-2 integral test program did not show a clear tendency that a corium melt pool interacting with the concrete basemat in a severe accident will be coolable with water flooding on top. These experimental programs employed prototypic corium melt mixture material, concrete-basemat compositions of LCS and siliceous mixtures, prototypic decay power input, and melt temperatures.

There was partial success in the test employing the large (120 × 120 cm) melt pool, in which about 50% of the melt pool was cooled, through the water ingression mode of heat removal. In the test employing the LCS concrete some melt eruptions occurred, which cooled a considerable fraction of the melt due to the fragmentation of the erupting melt on contact with the overlying water. The last integral test in the MCCI-2 program cooled down since the starting melt composition did not include any concrete constituents and water ingression was at its maximum.

The MACE and the MCCI program also performed separate effect experiments to obtain understanding of the various modes of heat transfer that occur during the coolability process when a core melt pool is flooded with water from above. Those separate effect tests provided the data for validation of the models developed for the different modes of heat removal. The models developed are described in the next subsection.

Modeling of the Coolability of Ex-vessel Melt Pools by a Water Overlayer

The MACE and the MCCI Projects identified four different cooling mechanisms for the corium melt layer flooded with water from the top. There were (1) bulk cooling (2) water ingression (3) melt eruptions, and (4) the crust breach. Data were obtained from the tests performed in the large melt coolability facility for the MACE and the MCCI integral and separate effect tests and in the SSWICS facility for the water ingression mechanism. These experiments were helpful in reaching an understanding of these heat removal mechanisms and the data obtained in the experiments was employed to validate the models developed for each of these heat removal processes. The models developed for each of these processes are described below.

Bulk Cooling

The bulk cooling process is perhaps the best understood. This process occurs before a stable crust is formed on the upper surface of the melt layer. The high gas generation in the attack of the concrete by the melt pool is responsible for the bulk cooling phase through the enhancement of the interfacial surface area of melt in contact with water. The large bubble generation and the steaming rate also contribute to this enhancement. The effective heat transfer coefficient is a function of the gas sparging rate, bubble diameter, melt/crust thermophysical properties, coolant properties, containment pressure, and the crust fracture strength. The bulk cooling heat transfer coefficient is directly proportional to the gas sparging rate, which in turn is inversely proportional to the system pressure.

The bulk cooling phase continues until conditions are reached for the formation of a stable crust. These are: (1) reduction in the temperature of the melt surface to the melt solidus temperature so that a crust can be formed and (2) a reduction in the gas sparging rate so that the crust formed is stable. Models have been developed for the critical superficial gas velocity [5], at which a stable crust can be formed and for the melt–water heat transfer rate during the bulk cooling phase [6]. They have been incorporated in the code CORQUENCH [7]. The predicted heat fluxes have been compared quite favorably to the MACE data.

Water Ingression

The formation of the stable crust isolates the liquid melt from intimate contact with water, and the only mode of heat transfer will be the heat conduction from the melt to the water through the crust, which unfortunately is a very low conductivity material. The crust, however, has porosity or cracks, since the gases produced by the melt attack on concrete are vented through the crust to the water layer above. These cracks / porosity also provide pathways for water to ingress into the crust and move the heat removal surface closer to the liquid melt. The process of water ingression is also responsible for quenching the crust into its interior, which leads to development of additional cracks that in turn provide a further passageway for water to ingress further into the crust layer. Thus, the water ingression through the porosity and cracks inside the crust can be considered as thinning the thermal boundary layer within the crust at the crust melt interface and augmenting the otherwise conduction-limited heat transfer. This mechanism is a strong function of the crack propagation process within the crust or in effect of the generated crust permeability.

The water ingression process has been observed in the cooling of lava and hot rocks, flooded with water. Lister [8] developed a theory of a cracking front that separates a convective region in cracked porous rock from the conductive boundary layer. The cracks within the material propagate when the tensile stress exceeds the overburden pressure. Further cooling and shrinkage widen the cracks to allow water percolation with the bulk permeability becoming a strong function of crack spacing and temperature. Lister developed equations for the bulk permeability, which allowed the evaluation of the dry out heat flux. He applied his theory to the cooling of the lava fields and rocks.

Recently, Epstein [9] developed a purely thermal steady-state model of the water ingression process. He later combined the thermal model with Lister's cracking front model to obtain an approximate solution for the dry-out heat flux and for the rate of downward cooling and solidification of the crust. With his combined model, Epstein could predict the field data on the water flooding of the lava flow. Lomperski and Farmer [4] [10] [11] further extended the Epstein model to apply to the case of corium and derived a correlation for the corium dry-out heat flux with one empirical constant to be determined. They determined this empirical constant by employing the data on water ingression obtained in the SSWICS tests. The resulting model for the water ingression into corium melt is included in the CORQUENCH code.

Melt Eruptions

The melt eruptions were observed in all of the MACE tests performed with limestone common sand (LCS) concrete. The phenomenon is that of an eruption like that of a volcano of the liquid melt enclosed by the crust at the top. The gas generated by the interaction of liquid melt with concrete, in spite of its flow through the cracks in the crust, could create sufficient pressure under the crust to erupt through some holes in the crust. The gas flow would entrain liquid melt, which first increases the size of hole in the crust and then comes in contact with water region above the crust. There, it fragments into a particle bed, which is cooled in the water overlayer. Thus, a melt eruption event can transfer much heat from the liquid melt pool under the crust to the water pool above the crust. The melt eruption events generally occur when the heat removal rate from the liquid melt falls below the decay heat addition rate to the melt. This leads to increase of the liquid melt temperature, more rapid erosion of the concrete below the liquid melt, and a faster gas generation rate, which cannot be accommodated by the flow through the porosity formed in the crust.

The melt entrainment coefficient is the parameter of prime importance for evaluating the process of heat transfer through melt eruption. This was measured in the experiments performed in the PERCOLA [12] program conducted by CEA in Grenoble, France. The tests employed simulant materials of different viscosities to determine the dependence of the entrainment coefficient on viscosity. This was needed since the viscosity of the liquid melt keeps changing as concrete products are incorporated into the melt with the ablation of concrete. A single-phase flow model for liquid ejection through a hole in the crust and a two-phase flow jetting eruption models were developed by Tourniaire and Seiler [13]. The friction for the liquid flow through a small hole in the crust and the pressure drop through the crust layer influence the flow rate of the liquid melt. This model did not represent the solidification of corium melt, as could occur during its flow through the colder crust, and did not provide a hole density in the crust. These deficiencies were corrected in a model extension by Farmer [14], which considered the thermal hydraulics, pressure drop, and the freezing process of liquid melt flow through a permeable crust. The model developed by Farmer is being extended for the two-phase flow effects in a PhD thesis completed [15] very recently at the University of Wisconsin. It is clear that even for entrainment coefficients as low as 0.1 to 0.01% considerable liquid melt could be transformed

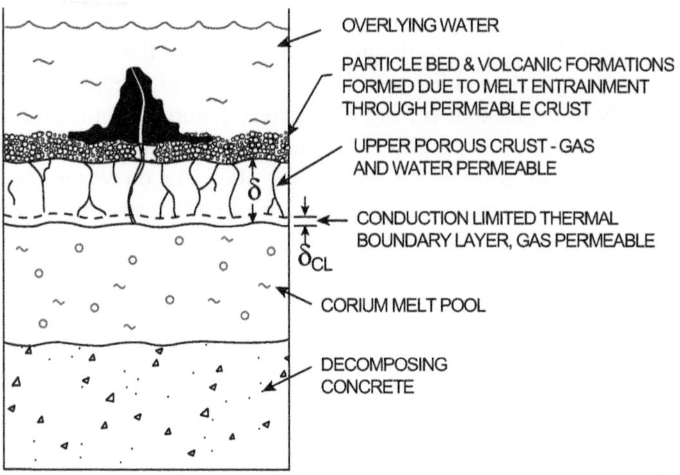

FIGURE 4.3 Schematic of CCI with water ingression and melt eruption cooling mechanisms.

into a quenched debris bed on top of the crust barrier. The water ingression and the melt eruption process are shown in Figure 4.3.

Crust Breach

Thus far a crust breach has not been observed in any of the melt coolability experiments conducted. However, if we consider a large-span-bridged crust, which could be formed in the vessel cavity of the containment of a prototypic PWR, crust breach is almost a certainty as the water pool increases in height above the bridged crust. The crust breach will renew the contact of water with liquid melt, which will then again go through the processes of bulk cooling, formation of another stable crust, water ingression, and melt eruption. This repetition of the cycle implies extraction of much heat content of the liquid melt, and the process could repeat itself further until all the liquid melt is cooled and solidified. The final debris configuration may be a set of layers of particles/porous corium with layers of permeable crust in between.

The above scenario applies to suspended crust. If, however, a crust were floating on top of a liquid melt pool, there would not be any issue of crust breach. The more likely configuration may be that crust has attached itself to the sidewalls of the cavity; however, in the center region, the crust is floating on the liquid melt pool. This could lead to crust breach in the middle region of the melt pool, as the liquid melt moves down as it ablates concrete.

The suspended crust can be modeled as a flat plate loaded with a uniform weight of water plus the weight of the crust itself. There are correlations available for stress in a flat plate as a function of plate area, thickness, and loading. The fracture strength of the crusts formed in the MACE and MCCI tests has been measured, and that would determine whether or not a particular crust configuration would fail. Analyses have been performed to determine the thickness of a stable crust in a 6 meter diameter cavity. The indicated thickness is 20 to 30 cm, which implies that about one-third to one half of the melt from a bounding accident could be in the crust formed from the crust survival considerations. A thick bridge crust most probably will not form, since it would take considerable time to develop one. A floating crust would be the most likely situation, with some fractured and cooled crust attached to the walls. In a floating crust configuration, the operative mechanisms for achieving coolability of the melt pool would be bulk cooling, water ingression, and melt eruptions.

The CORQUENCH Code and Its Predictions

The modeling developed for the heat transfer processes of bulk cooling, water ingression, melt eruption, and the crust breach was incorporated in the CORQUENCH code [7], under development for several years by Farmer at Argonne Natural Laboratory. The CORQUENCH code can perform the analysis for coolability of a corium melt pool interacting with concrete, be it siliceous (SIL) or limestone common sand (LCS).

The CORQUENCH code was applied to scope an approximate melt coolability envelope for both the SIL and LCS concretes. Vessel cavities in the containments of prototypical reactor plants were modeled, and the time of water flooding after the discharge of the melt on the cavity floors was varied. The selected parameter was the content of concrete in the corium melt at the time of cavity flooding, since it was known that the water ingression mechanism is a function of this parameter. This parameter also affected the fracture strength of concrete, which is a parameter in determining the cooling achieved by the melt eruption mechanism. The accident was considered as stabilized when the melt temperature dropped below the temperature for concrete ablation; that is, the melt pool was sufficiently cooled that no more ablation occurred or fission products and concrete ablation gases were released. The accident scenario assumed that the collapsed depth of the core melt in the vessel cavity would be 40 cm, which for a 6 meter diameter cavity would be $\approx$90 tons of melt. This may be less than that for a bounding accident in which a whole core melts and melts all the steel structure in the lower head. A floating crust configuration was assumed for this analysis. Figures 4.4 and 4.5, taken from the publication [3] of Farmer and colleagues, show the differences in the coolability behavior of the LCS and the SIL concretes. The LCS concrete with its greater gas generation contributes much more coolability potential from the melt eruption mechanism than the SIL concrete, which has quite a low concentration of limestone, the CO_2 generating material. The figures also show that indeed a considerable depth of basemat could be ablated before accident (melt) stabilization occurs; in particular if water is not added at the very instant when the corium–concrete interaction starts with the arrival of melt onto the basemat of the vessel cavity. A 40 cm deep melt pool attacking the LCS concrete could stabilize with basemat erosion to 1 meter depth in almost 10 hours, if the initial concrete content in the corium melt is less than 15%. Greater melt depth

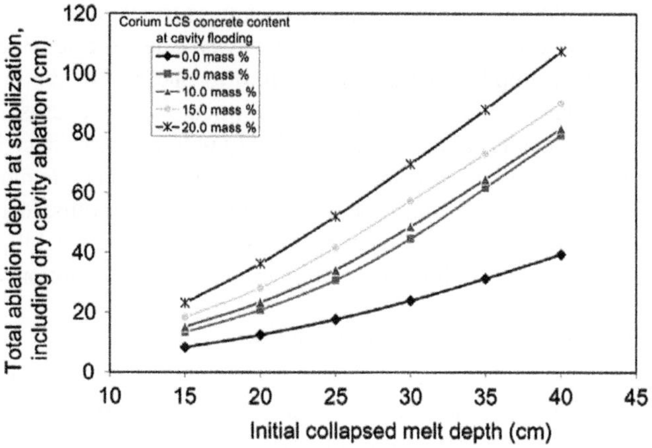

FIGURE 4.4 Prediction of basemat ablation depth at melt stabilization for a LCS concrete.

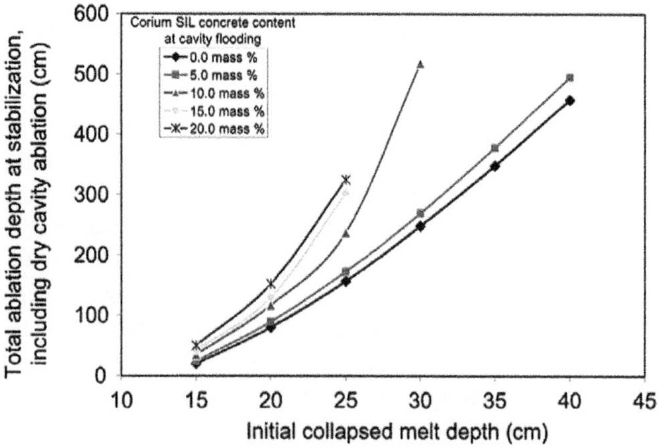

FIGURE 4.5 Prediction of basemat ablation depth at melt stabilization for a siliceous concrete.

than 40 cm and/or greater initial concrete content in the melt would make it very difficult to claim accident (melt) stabilization, with water flooding.

The SIL concrete calculations with CORQUENCH code assume that there will be melt eruptions, but with the lower gas flow rates resulting from the lower gas generation rates from the SIL concrete. Figure 4.5 shows that even with 5% concrete content in the melt, the basemat may be eroded to the depth of 5 meters, in about one week, if the initial melt depth is 40 cm. Any further delay in flooding of water or melt depths greater than 40 cm would not lead to accident stabilization with water flooding from the top. Melt depth of about 20 cm may be feasible since the stabilization erosion depth would be ≈ 1.5 meters with $<5\%$ concrete content in the corium.

Conclusions on the Coolability of Ex-vessel Melt Pools

Since the containment is a constant volume system and the concrete ablation generates large quantities of noncondensable gases, the containment pressure will increase unless it is vented. Venting without filtering is not a good idea, as we know from the recent experience with the Fukushina accident, since the fission products resident in the containment (the source term) will also be released to the environment. Many PWR and BWR plants in Europe have filtering capability, but in general it is a good idea to stabilize the melt/accident without any venting, filtered or otherwise.

Melt coolability with water flooding on top has been found to be not as effective an accident management measure as originally believed. This assessment does not even take into account the many uncertainties that are inherent in the experiments and in the models that have been developed from the knowledge gained from the experiments. It appears that additional measures would be needed to achieve melt stabilization and quenching readily, in order to preclude the failure of the containment or to resort to venting of the containment to preserve its integrity. One such measure is the COMET concept for melt/debris coolability, which will be presented in the next subsection.

4.1.3. Cooling of a Melt Pool by Water Injection from the Bottom

The COMET Concept and Experimental Results

Since melt coolability may be hard to achieve with water flooding from the top, alternative and innovative means have been explored to cool and quench the melt. Experiments have been performed [16] [17] at the COMET facility in Forschungszentrum Karlsruhe (FZK) in which water is injected from the bottom of a melt pool at a slight overpressure. It has been found that melt cooling and quenching is quite readily (in about one hour) achieved. No steam explosion occurred even with the Al_2O_3 melt, which has been found to explode spontaneously when dropped into a pool of water. It appears that the addition of sacrificial concrete to the Al_2O_3 melt reduces its explosivity. Another reason may be that water injection into the melt is at a very slow rate and its rapid evaporation does not provide the conditions for forming a premixture, leading to a steam explosion.

Three variants of the COMET design exist for the basic processes of passive flooding and cooling, which have been evaluated by experiments:

- The first variant uses an array of plastic tubes, embedded in a horizontal concrete layer and connected to a water reservoir pressurized by a static overhead (Figure 4.6). Water is fed into the melt through the plastic tubes after erosion of the sacrificial concrete layer.
- The second variant replaces the array of tubes by a layer of porous, water-filled concrete (CometPC = COMET Porous Concrete). A uniform

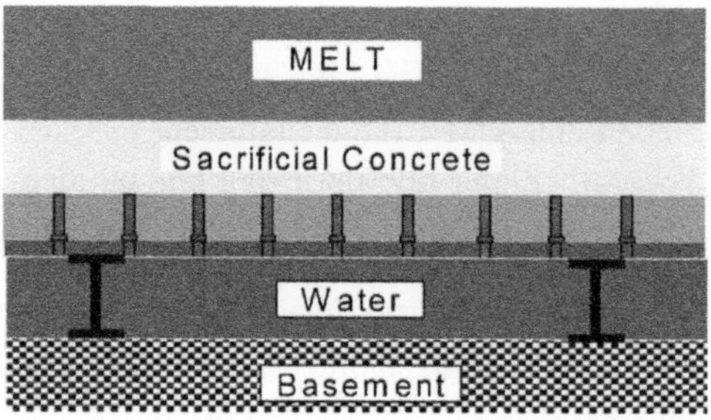

FIGURE 4.6 Sketch of COMET bottom plate for water injection through an array of plastic tubes.

horizontal distribution of the water may be achieved by a second, high porosity concrete layer underneath.

- In the CometPCA cooling concept (Figure 4.7), water channels with a regular pattern have been inserted in the porous and water-filled concrete layer, so that the melt is penetrated more homogeneously by the water. This modified concept combines the advantages of the original COMET concept with water nozzles and those of the CometPC concept.

Different solutions can be envisaged for the location of the COMET coolant device in the containment. In the spreading area of the EPR, it can be used as an alternative retention concept. It had originally been designed for this. It can also be applied in the vessel cavity, which may be enlarged laterally to reduce the height of the melt. Optionally, the coolant water level that starts to build up after onset of cooling may be designed to rise to a sufficiently high level, in order to cool residual core debris directly in the RPV, thus reducing and possibly eliminating further melt releases. Especially the simple CometPCA concept could be implemented in existing GEN II plant containments and thus provide the means to achieve melt coolability in those plants.

Several experimental series have been performed by FZK to test and optimize the functioning of the different variants of the COMET concept. Major series were COMET-T, COMET-U, and COMET-H for the first concept with melting plugs and CometPC-H and CometPCA-H for the concepts with porous concrete. Besides COMET-U, in which cooling of UO_2-rich oxide melts

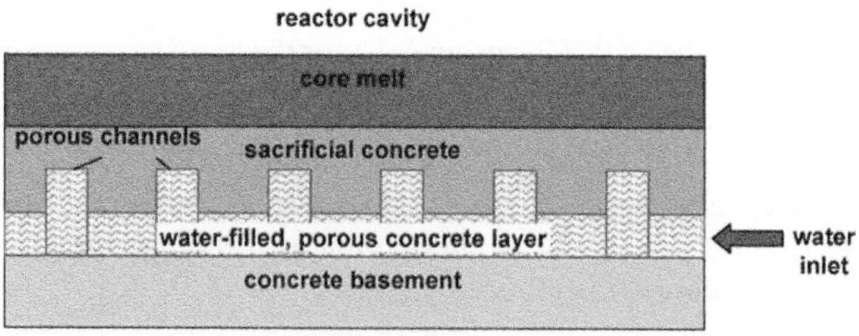

FIGURE 4.7 The CometPCA concept as a combination of melt plugs and porous concrete layer.

was successfully demonstrated, thermite generated, high-temperature melts of iron and aluminium oxide were used, with the addition of 35 wt.% CaO to the oxide. More details are given in [17].

Simulant melt experiments were performed at KTH, in a series of Debris Coolability with Bottom Injection at High Temperature (DECOBI-HT) experiments [18] with the binary oxide mixtures $CaO–WO_3$ and $MnO_2–TiO_2$. In contrast to the previous experiments with $CaO–B_2O_3$ melt, which is a glass-type material with high viscosity, the experiments with the ceramic type melts showed substantial mixing of the melt with the coolant leading to rapid quenching of the complete melt. More detailed information is given by Sehgal et al. [19].

The investigations, performed at FZK, were used to optimize the COMET concepts and to define the range of applicability under reactor conditions. The COMET concept with the injection tubes is considered to be mature for reactor application. Also for the CometPCA concept, recent investigations have demonstrated its technical feasibility for corium layers up to 0.5 m high. Thus, the COMET bottom injection concept is able to guarantee safe arrest and cooling of the melt under ex-vessel conditions in the existing GEN II plants for melt layers up to at least 50-cm depths. Further experimentation would be needed to certify the COMET concept for melt depths greater than 50 cm. Alternatively, melt spreading may be employed to reduce the depth of the melt pool.

Modeling of Porosity Formation and of Cooling by Bottom Injection of Water

The large number of successful cooling experiments for the bottom flooding concept shows the high potential to achieve rapid quenching and safe long-term cooling of the melt in various scenarios. The crucial process for successful cooling is sufficient breakup of the compact corium layer and the formation of a porous structure. It is therefore especially important to understand the processes that determine the porosity formation.

A general pattern of processes is assumed, as sketched in Figure 4.8, describing breakup of melt and porosity formation from the beginning of water injection. This picture corresponds to the observations in the experiments that initially all water evaporates (possible only by strong lateral spreading of water, which also produces the lateral distribution of porosities obtained at the end) and that the strong steam outflow at the top of the melt occurs from the beginning.

In the lower region, the injected water evaporates, thereby forming porosity and quenching the melt. Lateral distribution of porosity within this region occurs due to the strength of fuel-coolant interaction (FCI), which provides the resulting pressure buildup driving lateral expansion and spreading of water. In the case of too little pressure buildup from evaporation (or gas injection), the gas would mainly flow upward, thus forming a limited region of porosity as an upward channel around the water or gas injection location. From a laterally extended steam production region, the steam produced flows upward through many channels and by this method generates interconnected porosity in the

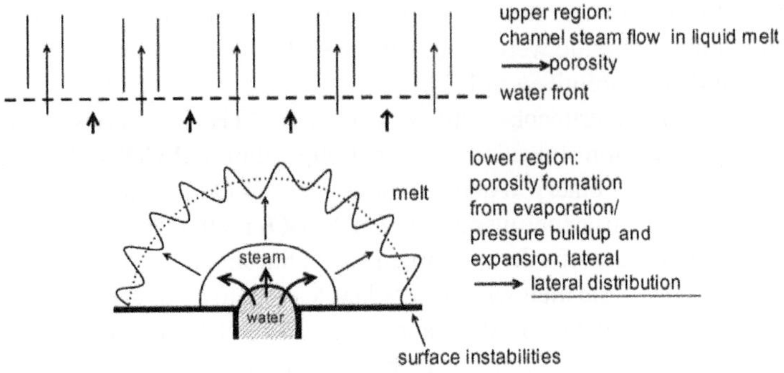

FIGURE 4.8 Pattern of processes involved in the COMET concept.

upper region. In later stages, the water progresses further upward and may there also come into contact with still liquid melt, producing additional lateral porosity. But the major effect would be quenching; and thus freezing, the porosities (channels) formed in the liquid state. Generally, porosity formation by evaporation with resulting local pressure buildup should become weaker when the steam can better escape through porosities. On the other hand, pressure buildup is favored by high friction hindering the steam flow upward.

In particular, the rapid lateral expansion of the interaction between melt and injected water must be explained, in contrast to a formation of thick channels supporting axial upflow of steam and water. As the only mechanism yielding the observed phenomena, a strong local pressure buildup with subsequent strong expansion is envisaged here. This produces lateral motions as compared to only axial ones from buoyancy. Furthermore, these lateral expansions produce oscillations (expansion/collapse effects), giving rise to enhanced fragmentation. Thus, local pressure buildup is considered a key feature of the process of porosity formation with lateral extension.

To obtain an overall picture of the processes in the melt layer, including hindrances to porosity formation, or building only a few paths of steam escape, or too rapid solidification in the entrance region or around such paths, the WABE code has been adapted [17] for local porosity formation, depending on local pressurization that exceeds the hydrostatic head of melt. The inverse process of porosity disappearance is also included. Freezing stops both processes. This and temperature-dependent viscosity of the melt are included as effects in the correlation of local porosity formation rates. For starting the analysis, very small initial porosity and hydraulic diameters are assumed.

Within the validation process, the COMET-T experiments of FZK were modeled. The pattern of processes sketched above resulted naturally, simply based on the local pressurization leading to formation of porosity and its growth. Also, the quantitative features, for example, of rapid increase of the evaporation rate to a maximum, were well met, and approximate agreement was obtained with the experimental results not only about this and maximum evaporation rates depending on water inflow pressure, but also with respect to porosities formed, first appearance of (liquid) water at top, overall quenching times, and so on [17].

Furthermore, related to the DECOBI experiments of KTH and the COMET experiments of FZK, influences of melt viscosity and water inflow rate have been checked. Understanding could be provided by the calculations about an increase of lateral extension of porosity formation with lower viscosity, in contrast to a tendency of single-channel formation (dominance of buoyancy) with higher viscosity. Increased lateral extension of porosity formation in COMET experiments by a higher water injection rate, rather than increased porosities themselves, was also obtained, as well as, in contrast, no such effect with a DECOBI experiment with a lower melt height and lower temperature (thus promoting the single-channel effect by favoring buoyancy against pressurization effects).

Thus, the results obtained from the WABE-COMET Code support not only the modeling, but also the effectiveness of the COMET concept. However, the validation process is evolving, including model improvements, to assure the applicability of the model over a wide range of conditions. Applications to reactor-related cases have been performed to support the applicability of the COMET concept, including optimization, for example, to reduce steaming rates without reducing sufficient porosity formation and coolability. The CometPC(A) concept with porous concrete, which may be implemented more easily in the reactor cavities of existing plants, should be further investigated.

Finally, the results of recent calculations for the VULCANO-COMET experiment [20], performed at CEA with corium melt, are described here. These experiments represented the CometPCA concept (Figure 4.9). The results of precalculations with WABE-COMET in Figure 4.10 show the porosities and temperatures at 200 s, as well as the water and steam-flow patterns. A comparison of the measured and calculated steam mass flow rate and total mass is given in Figure 4.11. The short quenching times from the experiment as well as the calculation support the present understanding of mechanisms. In view of uncertainties of initial conditions for the calculation (e.g., initial melt temperature at the time of water entry after concrete ablation) and the complexity of processes, the comparison is quite promising. An explanation of the stronger steam production from the model may also be that in the experiment concurrent opening of the two inlets did not occur, which was assumed in the calculation.

4.1.4. Coolability by Dropping Melt into a Water Pool

This is the second preferred accident management measure to achieve coolability of the melt pool that would be discharged into the containment. The action here is that, instead of maintaining a dry cavity under the vessel, a water

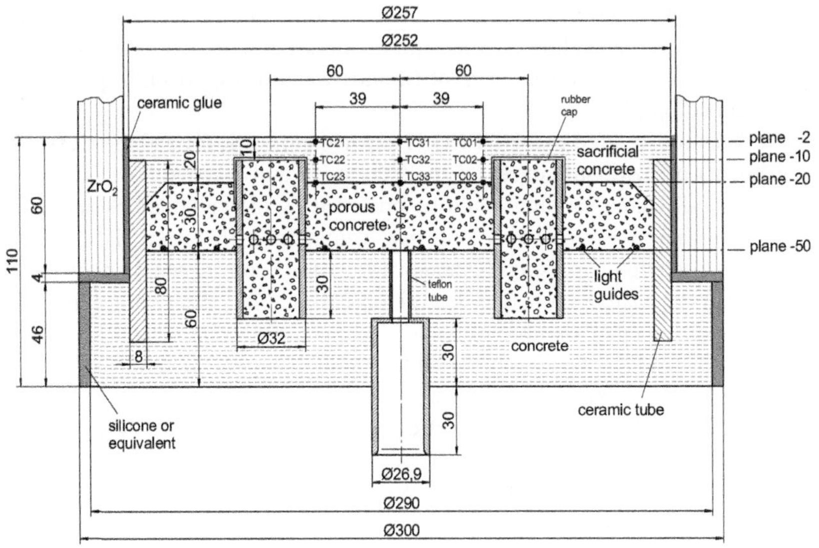

FIGURE 4.9 Sketch of the core catcher element tested in VULCANO.

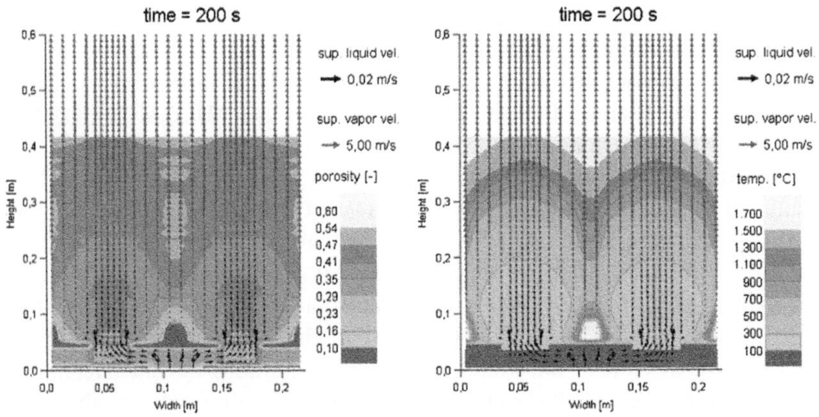

FIGURE 4.10 (left) Porosity after 200s; (right) temperature (1800 K: solidus temperature) after 200s.

pool should be established in that cavity before the failure of the vessel. This results in the discharge of the melt jet (from the vessel on its failure) into that water pool. As mentioned in the introduction, the phenomenology of such a fuel-coolant interaction would result in the processes of melt jet break-up, further fragmentation, and formation of a particulate porous debris bed, generating decay heat. This bed generates decay heat and has to be cooled and

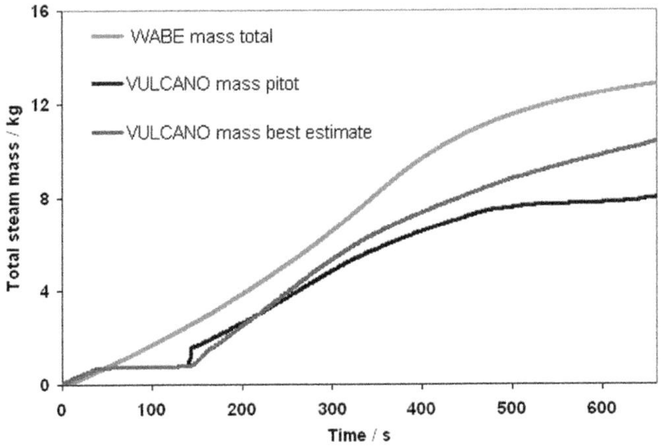

FIGURE 4.11 Comparison of measured and calculated steam mass flow rate and total mass.

kept cool in order to avoid its remelting and interaction with the concrete basemat of the vessel cavity. The whole process of the fuel-coolant interaction (FCI), the jet break-up and fragmentation, the formation of the debris bed, possible remelting, and the achievement of the coolability of the melt/particulate debris bed is the subject of this section. It is perhaps evident that these processes depend on the initial conditions of the melt coming from the vessel and are therefore dependent on both the in-vessel and the ex-vessel scenarios that may occur in the course of the severe accident before its termination by achieving the coolability of the melt discharged into the vessel cavity.

Scenarios of Melt Release into the Cavity

RPV failure initiates scenarios of melt release into the cavity. With respect to the influences on melt behavior and especially coolability, such scenarios differ in:

- the kind of release: massive break and melt gush, pressurized outflow under dispersion (DCH), or smoother, continuous outflow from limited holes and in limited rate,
- the duration of release: continuous melt outflow over a limited time or outflow in batches with significant time periods in between,
- the water level in the wet cavity,
- the geometrical conditions (reactor cavity design), especially the depth of cavity and lateral space for spreading of melt,
- the existence of dedicated measures, for example, sacrificial material to be melted and mixed with the corium, spreading and cooling in core catchers.

Focusing on generic aspects, at first variants between the extremes of a completely dry cavity and of a deep water pool in the cavity are to be considered here. In the accident management (AM) concepts for BWRs, water pools of 7–11 m depth are foreseen, while in existing PWRs only about 1–2 m (or even less) cavity depth are available, and mostly the cavities are considered to be dry at the onset of melt release from the RPV. This depends on AM actions followed by the individual plant owners. The German and the French reactor plants deliberately do not flood their PWR vessel cavities or BWR drywells, while the Swedish BWRs and the Westinghouse PWRs fill the cavities with water. Westinghouse has adopted the guideline of adding water into the vessel cavity to cool the lower head from outside in order to prevent vessel failure and, if this fails, to form a particle bed in the cavity that is easier to cool.

A deeper water pool favors the formation of coolable particle debris. Furthermore, the kind of melt release is important. A large melt pour can generally be considered as less favorable for breakup than a limited diameter of outflow from the RPV. Further information on favorable melt release conditions can be expected from a better understanding of breakup processes. The question that arises then is which release modes are to be considered as realistic.

Assuming an already developed large melt pool in the lower head of the RPV from a bounding accident, it is expected from natural convection of the melt with internal heat generation that vessel melt-through occurs at the lateral top edge, the hottest region (highest temperature of vessel wall). Then, a hole in this range should form, through which melt flows out. When this occurs at one location, the outflowing melt immediately reduces the pool level, thus tending to exclude formation of other openings with outflowing melt. Only by a practically simultaneous failure at different locations, multiple streams or jets of melt would be released or even a circumferential break yielding a large pour could occur. But this appears to be possible only by forcing such a simultaneous failure with a high pressure in the vessel. Excluding this by performing depressurization measures should also exclude such events or at least give them only a low probability. Uncertainties in the size and size development of outflow holes remain, but in general the size will also be limited due to the inherent restriction of thermal load (pool level) by outflow. Some fish-mouth-like break may occur under the influence of moderate pressures and some hole increase due to ablation of vessel wall [21] during outflow, for example, as in the FOREVER tests [22]. Size ranges of up to ~30 cm hole diameter may be concluded for reactor conditions. The hole progression will follow the pool-level decrease. Due to outflow at the top of the pool, the outflow velocity will remain small, and thus also the mass flow.

In a BWR, the penetrations of guide tubes for the control rods through the lower head are considered to be weak locations that may give rise to downwards leakage. Then, a melt pool may fail downward through several holes. This should not occur if a stable crust could develop during formation of a large pool. Then, again, a failure may be expected at the top boundary. If premature failure of guide tubes were to occur, only limited melt would be in the lower head, thus again yielding only small pouring rates and limited pour diameters. The same is valid for designs including a dewatering nozzle as the weakest location.

Premature failure possibilities during the phases of formation of a melt pool are also to be considered for a PWR. Lower regions of the RPV may be weakened already due to earlier melt deposition and during buildup of the melt pool. Downward failure may not be completely excluded. But again the trend would be that either failure occurs with still relatively low amounts of melt or that outflow at top of a developed pool is favored. Further, crust formation at the boundaries acts to stabilize a pool during the buildup phase. The latter is even more valid for cases with water in the cavity. On the other hand, uncertainties come into play by possible material layering in a melt pool. During pool formation, a premature failure would mean limited melt mass and failure at the top due to a metal layer with a low-outflow rate. Other types of layering may be possible and need further research and consideration.

From consideration of these possible scenarios, it may be concluded that rather limited diameter melt streams (smaller than 40 cm) would flow into the cavity for most PWR and BWR scenarios. With (partly) water-filled cavities, inflow of such melt jets into water pools would then have to be considered as the next event in the scenario. A similar conclusion has been drawn in GAREC analyses [23]: it was based on arguments that at vessel failure only part of the core may be in a liquid state and that after depressurization the vessel will not fail with a large opening in a very short time.

Water Pools in the Cavity

The following subsection considers the scenario sketched in Figure 4.12, with the major processes of jet breakup indicated in Figure 4.13. A possible adverse effect in this AM concept is the possibility of the occurrence of steam explosions. Strong steam explosions could fail or damage vessel cavity and the containment. Strong explosions with corium fuel interacting with water have been discounted (see Chapter 3) due to the physio-chemical properties of corium. However, the core melt discharging from a BWR vessel would contain a substantial amount of steel, whose FCI behavior is not similar to that of corium. Production of rather fine particles even by mild steam explosions has to be considered as challenging the coolability of particle beds.

Experiments on Formation of Particulate Debris in Water Pool

The breakup of melt jets penetrating into water pools has been investigated experimentally with corium melt in the FARO experiments [24], using jets of 5–10 cm diameter with about 5 m/s initial penetration velocity into 1–2 m deep water. Up to 177 kg melt mass was released, mostly into saturated water under different system pressures from 5 – 0.2 MPa. In cases with low system pressure (0.2 – 0.4 MPa), subcoolings of water of up to 100 K were prevalent. A major outcome of these experiments was that significant breakup occurs with corium melt jets flowing under realistic conditions into water. Further, similar particle-size distributions were obtained for the fragmented parts, in spite of the

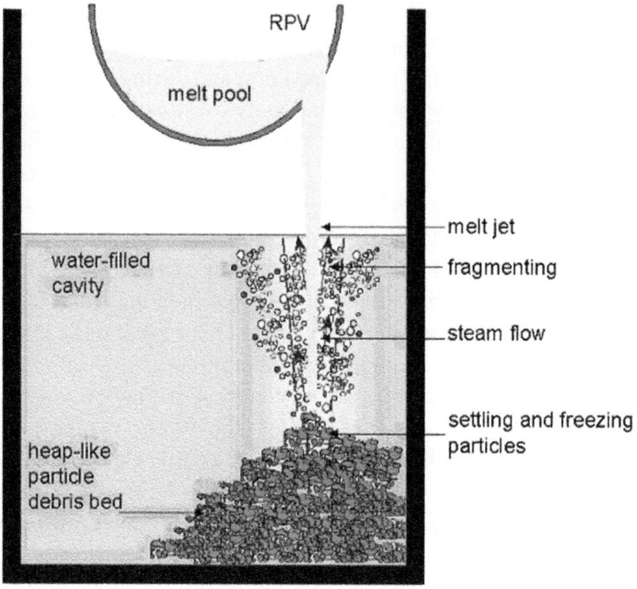

FIGURE 4.12 Sketch of scenario of melt outflow from RPV and formation of particulate debris in water-filled cavity.

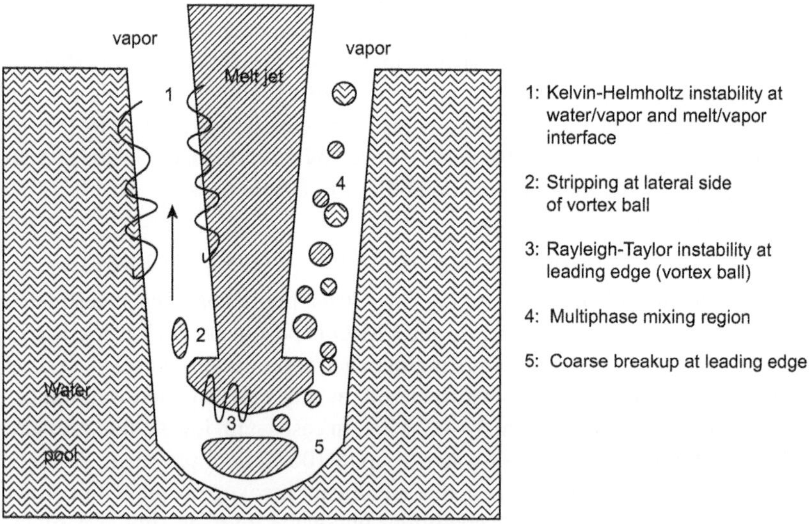

FIGURE 4.13 Processes of breakup of melt jet in water pool.

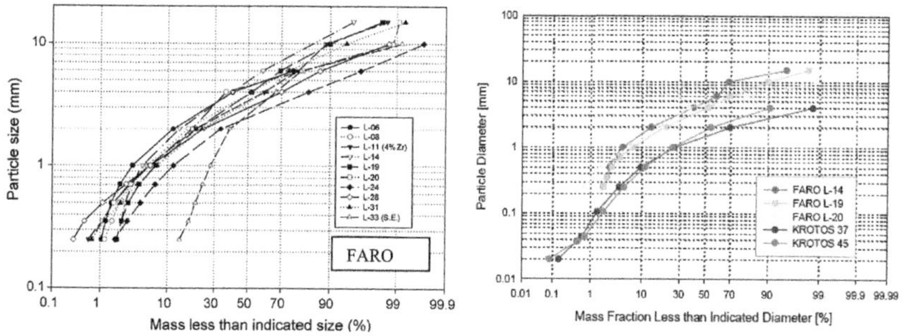

FIGURE 4.14 Particle-size distributions of FARO and nonexplosive KROTOS tests with corium.

significant variation of conditions (see Figure 4.14). On the other hand, there were differences in the part of loose debris (taken for size distribution) and hard debris (cake), for example, FARO L-14 (10 cm melt release diameter, 5 MPa system pressure, saturated water, 2.05 m water depth) yielded 20 kg cake and 105 kg loose debris; FARO L-28 (5 cm diameter, 0.5 MPa system pressure, saturated water, 1.44 m water depth) yielded 77.5 kg cake and 84.5 kg loose debris; and FARO L-31 (5 cm diameter, 0.2 MPa, subcooling of 104 K, 1.45 m water depth) yielded no cake and 83 kg loose debris.

The remarkably similar size distribution of the fragmented parts in L-28 and L-31 with the same absolute mass, for the same conditions besides the system pressure and especially subcooling, indicates similar breakup processes. Measured penetration data (gained from thermocouples) show a significant decrease of penetration velocity in the range of 0.7–0.8 m below water level for both experiments, which may be taken as an indication of the coherent jet length during the first penetration. Then, complete breakup may be concluded in both cases, and the cake part resulting in L-28 would be interpreted as a result of re-agglomeration or even coalescence of still liquid, not sufficiently quenched drops separated from the jet, as coherent jet arrives at bottom. Complete breakup in L-31 could then be understood by more effective quenching due to the higher subcooling. It is of interest here that a smaller cake part (16%) resulted for L-14 than for L-28 (48%) in spite of the thicker jet (10 cm diameter), but with deeper water of 2.05 m. Nearly the same result was obtained with L-24 at 0.5 MPa system pressure instead of 5 MPa with L-14 (saturated water in both cases). This supports the interpretation of the difference between complete jet breakup and cake mainly resulting from the degree of quenching of resulting drops of melt formed during settling. Indeed, closer inspection of the debris in the FARO experiments shows that the cakes are constituted from agglomerated particles, which supports the above interpretation.

Other important properties of particle beds concern the spatial distribution of particle sizes in the bed, regions of cake parts, layering of small particles above larger ones, degree of local mixing of particles of different sizes, as well as the shape of the whole bed, laterally uniform or heap-like. The porosities must also be considered. These depend on the local mixing of particles of different size and of heterogeneities in the bed structure. Unfortunately, bed porosities were not measured in the FARO experiments.

The FARO experiments are the most relevant experiments, involving prototypic corium material, thick jets, large masses, and resulting long pours. Other experiments on breakup of corium jets have been performed at ANL in the CCM series with jets of 2–5 cm diameter, but much less mass: maximum 12 kg [25]. Uncertainties about interpretation of results of the experiments exist, due to mass jet lengths being in the range of the breakup lengths. Nevertheless, the correlation of jet breakup length L per diameter D given as (L/D ~10–19) from earlier studies is similar to that obtained from FARO experiments. The same holds for the mass mean particle sizes of ~1–5 mm, which were also obtained in the FARO experiments. Debris sintered together was obtained in an experiment with lower water depth and higher water temperature, in contrast to other experiments, again indicating that the quenching conditions are decisive for particle bed formation with only small cake parts.

In the KROTOS experiments with corium [24], even less mass of ~3 kg was released to water, with jet diameters of ~3 cm. Somewhat smaller mass mean particle sizes than with FARO resulted throughout (~1 mm instead of 3–4 mm; see also the KROTOS particle-size distribution in Figure 4.14). This indicates an additional, different breakup mechanism under these conditions. With alumina, much larger particles resulted in KROTOS experiments (around 15 mm), which is discussed in [26] to be probably caused by breakup already at onset of melt penetration into water, due to large-impact velocities at the intermediate catcher, small mass of 1.5 kg, and low melt density. But even in the PREMIX experiments [27], large particles of around 9 mm resulted with alumina, although the released mass was up to 60 kg and experimental conditions were considered as similar between FARO-L28 and PREMIX-16.

Modeling of Formation of Particulate Debris in Water Pool and Validation

While in the initial phases of melt jet penetration breakup processes are expected at the leading edge of the jet, due to Rayleigh-Taylor (RT) instabilities as a result of strong deceleration or due to deformation processes yielding vortex formation and stripping of material, these processes should become less effective with penetration becoming smoother and with an increasing jet part not involved. Processes of stripping of melt along the sides of the jet should then become dominant (Figure 4.13). Such stripping processes have already been attributed to Kelvin-Helmholtz (KH) instabilities in the relative parallel flow of ambient medium within the THIRMAL model [28].

In a resulting surrounding mixture of melt drops with water, the ambient medium flow is essentially a multiphase flow with steam driving upward along with the surrounding water (buoyancy) and melt and water droplets falling down or being entrained in the upward flow. In initial phases with still relatively small steam content, the relative water flow, that is, melt jet velocity and water density, is decisive for KH wave formation on the lateral melt jet surface ("thin-film" behavior according to Epstein and Fauske [29]). However, a thick multiphase mixture flow with large parts of steam flow around the melt jet would subsequently determine the melt breakup by wave production and stripping along the jet sides. Then, the steam velocity and density become decisive. Due to the effect of water density in "thin-film" wave stripping, smaller fragments may be obtained in earlier phases of jet penetration. With smaller mass of melt released, this part may then have a significant influence and thus explain the average of smaller particles obtained in KROTOS experiments.

Doubts exist concerning the KH model of wave growth as the basis for the stripping processes. Especially, it is known that too small wavelengths result, indicating too small drops being stripped, as well as too rapid wave growth, that is, too rapid erosion of the jet. This behavior can be understood from the underlying approach of a velocity jump profile in the ambient flow. In spite of attempts (see, e.g., [30]) to develop a more detailed model based on velocity profiles and also nonlinear instability theory, the present approaches in codes are still using linear KH instability modeling, partly by adapting model parameters to experimental results. Such an adaptation is, for example, done in the code IKEJET [31] by assuming a reduction factor on the relative ambient velocity in order to account for velocity profiles. Based on this, the wave growth and the wave stripping (from an additional stripping criterion) is calculated, that is, the jet erosion and the fragment sizes. For more details, see, e.g., [31].

The formation of particulate debris then concerns especially the settling and cooling processes of the drops of melt resulting from jet breakup. This includes further breakup processes for melt drops falling in water. Models and correlations exist for such processes [32]. It appears, however, in case of FARO, that the jet breakup by stripping yields essentially the final particle sizes, as, for example, obtained from analyses with IKEJET/IKEMIX (combined jet breakup and settling code [31]). Key parameters for melt breakup, and also for settling as particle bed, are steam production and steam release from the mixture. The resulting void in mixture should strongly influence the steam and mixture flows along the jet via the pressure distribution. These flows along the jet influence the breakup. Furthermore, cooling and crust formation during the settling of drops will be strongly influenced by the void in mixture. By reducing heat transfer, a higher void will hinder crust formation and thus settling as stable particles. Instead, agglomeration and coalescence may become important.

In the OECD-SERENA project on steam explosion [33], a conclusion about the present modeling of corium/water mixing processes was that significantly too high a void was calculated in the mixture. This was especially the experience of the posttest calculation of FARO L-28, which encountered the problem that with rapid high-void buildup the heat transfer is reduced, then insufficient steam can be produced further to maintain the experimentally obtained long and practically constant pressure increase over the melt pouring time of 5–6 s. As a solution, it was considered [31] that more rapid steam removal should occur in the models. This refers to the friction laws, especially water-steam interfacial friction, and respective flow patterns. As an example for attempted improvements, the approach with IKEJET/IKEMIX is mentioned here: it introduces different steam velocities in transition flow patterns between water and steam continuous phases, considered to be composed by limiting configurations of both types. A significantly improved agreement with the experimental pressure development in FARO L-28 can be seen in the Figure 4.15. A transition from an initial "thin-film" to a final "thick-film" behavior is not yet included here. Also the particle sizes are in the experimental range, which is the most important result for the particle bed formation. However, due to the feedback of mixture properties, especially void, with melt breakup and quenching, the other features of the model also become important.

For clarification about the adequacy of the modeling, the breakup into much larger particles with alumina must also be explained. Using the experimental jet penetration velocities from PREMIX experiments, calculations with IKEJET/IKEMIX also yielded (with the same model parameters as applied for FARO) large particle sizes in the range of the experimental ones. This can be understood partly from material effects directly affecting KH instabilities (especially the small alumina density), but a major effect comes from the relatively low jet

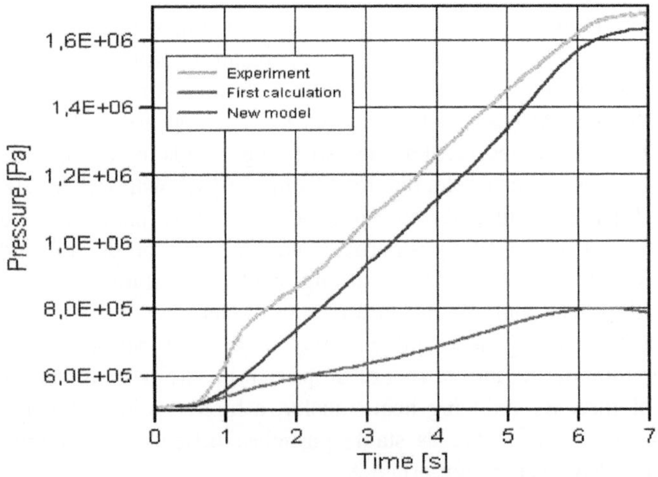

FIGURE 4.15 Pressure in FARO L-28 experiment (green line) versus calculations performed with IKEJET/IKEMIX under variation of interfacial friction modeling (first calculation: unreduced friction model; new model: two-velocity model in transition region).

penetration velocities that tend to yield shorter jets, thus smaller relative steam velocities (increasing with jet length) and by the larger drops. The reduced penetration velocity detected in the experiment could not, however, be reproduced by the model.

A further key phenomenon to be understood is the striking similarity of particle-size distributions in FARO experiments over a large spectrum of variation of conditions (Figure 4.14). The jet diameter and the water depth may indeed not basically change the breakup mechanism if dominance of the "thick-film" KH mechanism is established and the jet is thick enough (Weber and Reynolds numbers sufficiently high) to exclude coarse breakup modes as varicose or sinuous type breakup [26]. Then the atomization mechanism (Figure 4.13) prevails over most of the breakup history, and the jet diameter and water depth may have greater influence on the fractions of the fragmented and nonfragmented (or re-agglomerated) parts than the fragment distribution, at least if the jet length does not vary too strongly between breakup length and water depth as limits. However, that the results do not depend on pressure and subcooling appears puzzling, since the mixture properties and especially the void should be strongly affected. Independence may only be understood if an external water head essentially determines the steam upflow along the jet, yielding roughly a kinetic energy of steam proportional to the water head. In the KH approach, the kinetic energy term of ambient flow is also decisive for wave growth. This would mean that the high void buildup in the mixtures, for example, in FARO L-28, does not significantly influence the steam flow along the melt jet. For the cases with high subcooling, such as FARO L-31 or -33, a sufficiently thick region with steam should exist around the jet supporting the "thick-film" KH mechanism. Presently, the conclusions cannot be clearly based on calculations, and the modeling features need further checking.

Particle beds can only be formed if the drops of melt from jet breakup can sufficiently cool down and develop stable crusts. Otherwise re-agglomeration, the sticking together of particles or the coalescence of liquid parts, occurs. Even with complete jet breakup, the final debris would contain significant cake parts, if cooling is not sufficient. Modeling of crust formation at the drops is complicated due to a large number of different single drops to be followed passing through various conditions, deformation and eventually breakup of falling drops, complicated processes of crust formation with noneutectic corium, which may progress via mushy states. Thus, in the frame of the overall modeling, sufficient simplifications are required as, for example, introduced in IKEMIX [31]. This requires validation and adaptation. The goal must be to cover a large range of conditions by approximations employed.

It remains to determine the bed structure and especially the bed porosity. While an approximate determination of characteristic particle sizes appears to be feasible according to the above outline and with less surety also an evaluation of sufficiently crusted parts, this appears presently to be impossible concerning the bed structure and porosities. Experimental clarification about

FIGURE 4.16 Debris from DEFOR-07.

major trends and influences is required. For this purpose, the DEFOR experiments at KTH are being performed [34]. Six experiments have been performed with eutectic CaO-B_2O_3 melt, one experiment with WO_3-CaO melt, released as 2 cm diameter jets into water (falling height in air of 60 cm, water depth varied between 40 and 65 cm). Initial melt temperatures were between 1200°C and 1350°C, and initial water temperatures varied from 7°C to 88°C (0.1 MPa system pressure). In more recent experiments, the melt material with higher density WO_3-CaO (6500 kg/m^3) has been employed instead of CaO-B_2O_3 with 2500 kg/m^3. The major results from DEFOR show very high measured porosities (up to 65%) with the higher density WO_3-CaO melt.

Figure 4.16 shows the debris from one experiment, with the heap-like shape obtained. Larger particles resulted with the low-density material CaO-B_2O_3 than with the higher-density one (WO_3-CaO), as shown in Figure 4.17.

Coolability of Porous and Particulate Debris

Experimental Results and Overall Modeling Requirements

Heap-like bed configurations are to be expected as a result of melt breakup in a deep water pool in the cavity, with a nonhomogeneous internal structure of low-porosity regions, re-agglomerated drops of melt or unfragmented melt forming so-called cake parts, and also of regions with higher porosity. The latter

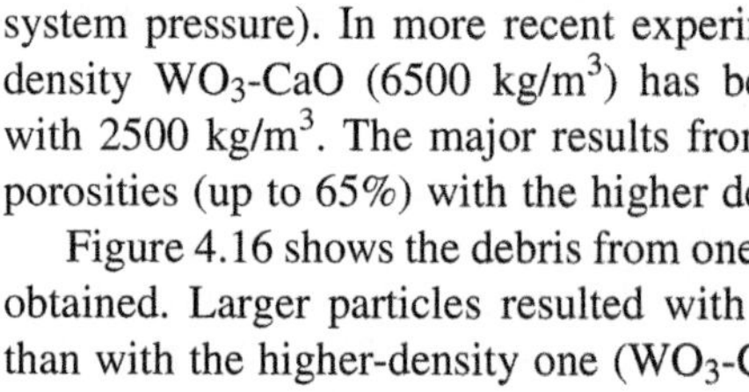

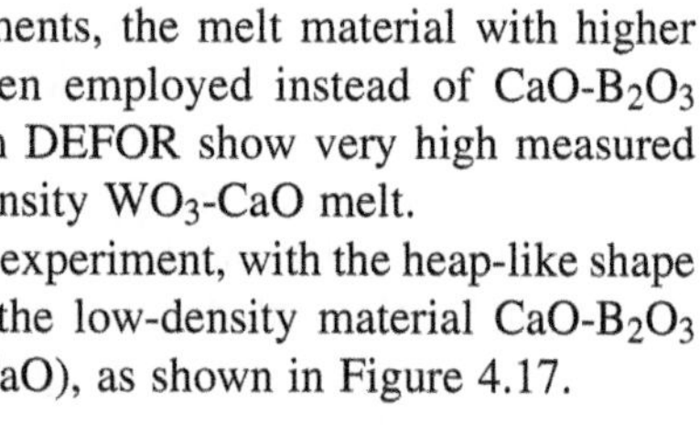

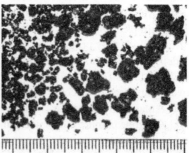

FIGURE 4.17 Comparison of debris from DEFOR: low-density (left) and higher density melt.

may in general be favored by the settling process, which does not support local mixing of particles of different size. Further, nonspherical particles tend to increase the local porosity due to their hindering the embedding of smaller particles. On the other hand, very small particles may occur, for example, from weak steam explosions. They could better be mixed into the coarser particle bed. However, smaller particles can also be expected to settle more slowly, that is, at top of the bed. Cake parts are correspondingly expected at bottom. But liquid melt may also enter the bed only in later phases, forming crusted areas in upper regions when the bed has already grown in height (thus the falling length in water is reduced) and the void in the mixture has increased.

Classical analyses mainly considered the coolability of 1D homogeneous beds under top flooding—that is, a water pool above a particle bed. For dry-out analyses, the bed is assumed to be initially filled with water and in a quenched state (saturation temperature). Then, the question is whether the decay heat can be removed and a steady state of water loss by evaporation and water supply can be established, instead of the development of a dry zone and continued heat-up and extension of such a zone. It is obvious that water supply from top against upstreaming steam (countercurrent) and via a region with accumulated steam can become difficult with increasing power and steam production. Better coolability should be reached by water injection from below where the water flows into the bed via water-rich regions and co-current with upflowing steam. Indeed, experiments [35] showed much better coolability with bottom than with top flooding. More than twice the dry-out heat flux (DHF), that is, the heat removed at top per unit cross section of bed, was obtained with bottom injection from a lateral water column of the same height as the bed.

With heap-like, heterogeneous beds inflow of water from top will be most favorable for the smaller heights (smaller steam accumulation and flow) at the edges of the beds. Further, the lateral pressure differences of water–steam mixture columns will produce a natural convection flow of water, driven into the bed from the sides, especially at bottom, then upflowing in the bed with vapor generation. Thus, through this mechanism, cooling, via (partly) bottom inflow of water, is realized.

In order to realistically evaluate the coolability for various reactor scenarios, it is thus necessary not just to use DHF values, but also to analyze the real configurations. This can only be done by validated 2D/3D computer codes which are capable of treating lateral water inflow. In addition to general requirements for the 2D/3D codes, as the separate description of water and steam flows in a multiphase approach, the constitutive laws for friction and heat transfer need special attention. Approximations and adaptations to experiments are indispensable. However, the level of detail (the physical mechanisms taken into account) is important for enabling adaptation to a large bandwidth of conditions, that is, applicability to various reactor scenarios.

The classical approach of Ergun for friction laws of single-phase flows in porous media has been adapted to multiphase flows by introducing relative

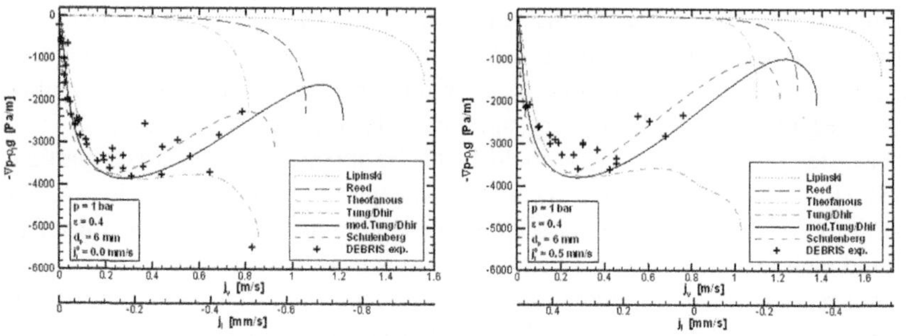

FIGURE 4.18 Pressure gradients, depending on j_g, j_l combinations (j_g increasing with height z), for case with top flooding (left = $j_{l,0}$ = 0 mm/s, right = $j_{l,0}$ = 0.5 mm/s).

permeabilities and passabilities of the phases [36]. Interfacial friction between steam and water is thus included via respective friction of the fluids against the solid, depending on the void fraction. This is not sufficient in general, if interfacial friction between steam and water plays a role. A unified description for co- and countercurrent flows requires a separate interfacial friction term. In fact, calculations on the experiments of Hofmann [35] failed to reproduce DHF values of the bottom-fed beds with the description adapted to top-flooding (see [36]). The DHF values obtained for bottom flooding were significantly too low.

The need of an interfacial friction term has also been shown by the analysis of the DEBRIS experiments [37], which especially address the constitutive laws. Section-wise axial pressure differences were measured in a debris bed of about 60 cm height, composed of uniform spherical particles (3 mm and 6 mm), heated inductively, allowing derivation of quasi-local pressure gradients. Figure 4.18 shows examples that mainly demonstrate that the adaptation of friction laws employing only the relative permeabilities and passabilities, in order to fit the measured DHF, yields no agreement with the measured pressure drops. With the classical friction models, without explicit interfacial friction, a significant deviation from static pressure drop occurs for larger (superficial) steam velocities j_g, close to DHF (curves at the top in Figure 4.18). This is due to the small water velocities j_l that nevertheless yield high friction with low water content. But the experiments show significant effects for even small j_g. These can only be obtained by explicit inclusion of interfacial friction, as in the approaches of Tung and Dhir and Schulenberg [36] [37]. Further clarification is still required [36].

So far, realistic debris and consequences for modeling have been considered with respect to the overall configuration of particulate debris beds, especially water inflow conditions. This requires 2D/3D models as well as specific modeling of friction laws (interfacial friction) to account for various conditions. Boil-off under decay heat with initially saturated conditions as well as quenching of hot debris are to be analyzed. While the DEBRIS experiments

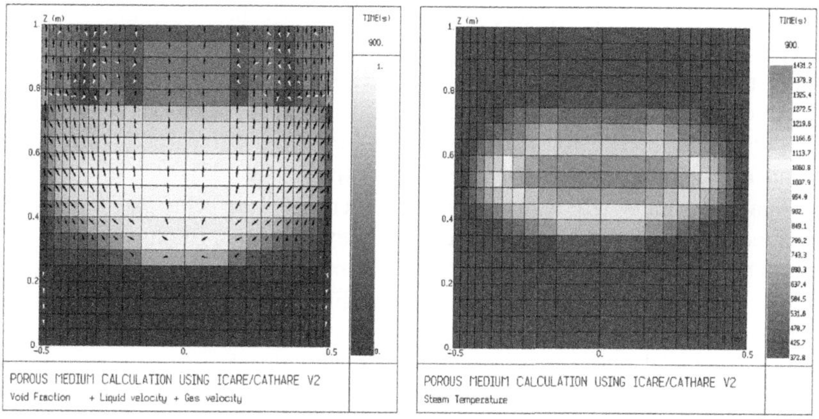

FIGURE 4.19 ICARE/CATHARE simulation of a planned experimental configuration with heated debris bed.

were oriented at local constitutive laws, experiments as POMECO at KTH addressed the effect of conditions of water inflow with different kinds of downcomers [38]. Enhancement of cooling was shown and could be reproduced by calculations (see also below in the next subsection, Experiments with Particle Mixtures and Validation).

Comparisons with earlier quenching experiments and similar experiments in DEBRIS have been performed [39], employing the WABE module of the code ATHLET-CD and ICARE/CATHARE codes. Simplifications of heat transfer models may be possible since small quenching zones can be expected due to slow water ingression in the debris beds. Higher water inflow rates found in the experiments of Tutu et al. [36] [39] [40] yield more complicated patterns with extended quenching zones, but may be unrealistic. New quenching experiments addressing 2D/3D effects are planned at IRSN. A pre-calculation with ICARE/CATHARE is shown in Figure 4.19 [39]. Water inflow from bottom via lateral downcomers into a nonhomogeneous dry bed at about 1300 K initial temperature can be seen. The steam flow (black arrows) produced through the bed provides cooling of the debris bed heated by the decay heat (right curve).

Experiments with Particle Mixtures and Validation

Another important aspect of prototypic debris concerns the local distribution of particle sizes of nonspherical shape. This has been addressed by POMECO [38] and STYX [41] experiments, both using sands with mixed particle sizes. In the STYX experiments, the particle-size distribution obtained in FARO experiments is simulated. Rather small DHF values resulted in these top flooding experiments, to be expressed in significantly smaller effective particle diameters than obtained from surface or even mass averaging (0.8 mm resulted, compared to mass average diameters of 3–4 mm from FARO). The question remains whether the well-mixed particles in the experiment correspond to the local bed structure obtained in reality under settling conditions. This issue is being investigated by the DEFOR experiments (see the previous subsection, Modeling of Formation of Particulate Debris in Water Pool and Validation). In the present DEBRIS program, mixtures of nonspherical particles are also included, partly with debris from PREMIX experiments [27].

Analyses with WABE on these boil-off experiments are being performed for continued validation and also for checking of experimental results (e.g., of the effects of nonhomogeneous heating in STYX by heating wires). As examples, at first, with assumed homogeneous heating, Figure 4.20 shows results of the DHF values for the STYX 2 series, and Figure 4.21 shows a corresponding development of the "saturation nose" (saturation = local water part in pores). The latter demonstrates that, with increasing bed power, conditions are reached, here with 618 kW/m^3 (371 kW/m^2 as total power per unit cross section at top), where a steady state is no longer possible and a slow dry-out development starts, even though at the last power level of 590 kW/m^3 a steady state was still established with >30% water phase everywhere in the bed. This indicates a drastic change of conditions, in spite of the slow dry-out development. The underlying instability is due to the countercurrent flow limitation under top flooding. In the friction laws of Reed applied here, no explicit interfacial friction is included, but a drastic increase of friction with decreasing water phase [36]. Figure 4.21 shows the developing "nose" of saturation (due to loss of sufficient water supply from top) moving downward until reaching zero saturation, that is, complete dry-out, only at bottom. Water, still in the bed, moves downward and covers deficits of supply from the top in lower regions, thus maintaining the profile there up to the time when the nose reaches the

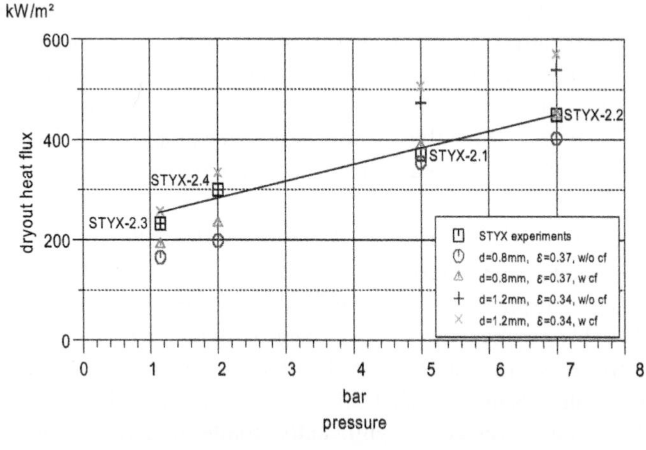

FIGURE 4.20 WABE-2D calculations of STYX test series 2.1–2.4 with different assumptions on particle diameter d, porosity ε and with (w) or without (w/o) including capillary forces (cf).

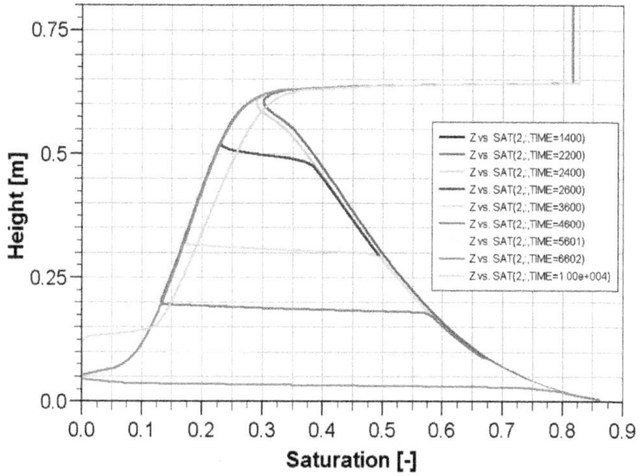

FIGURE 4.21 WABE calculation of STYX-2.1. Axial saturation profiles at different times. Last power level with steady saturation profile is 590 kW/m^3 (1400 s–2200 s: red curve covers black). With power level 618 kW/m^3, the saturation profile becomes transient (2400 s). After 10,000 s all water up to 12 cm from bottom is evaporated.

bottom. The nose development results from this loss of supply from the top and loss to the bottom regions. If the limiting power level is just surpassed, the loss of supply from the top must result in a first dry zone at bottom, and higher powers will yield the first dry spots in higher regions.

Figure 4.22 shows a result from a calculation modeling nonhomogeneous heating in STYX. Water downflow occurs here between the heating wires, which are heated with an increased power to yield the same total heating power as with assumed uniform power. Here, the first dry zone appears at the top of the bed, which is typical for bottom inflow of water (dry zone first at location with highest steam accumulation, in contrast to breakdown of water

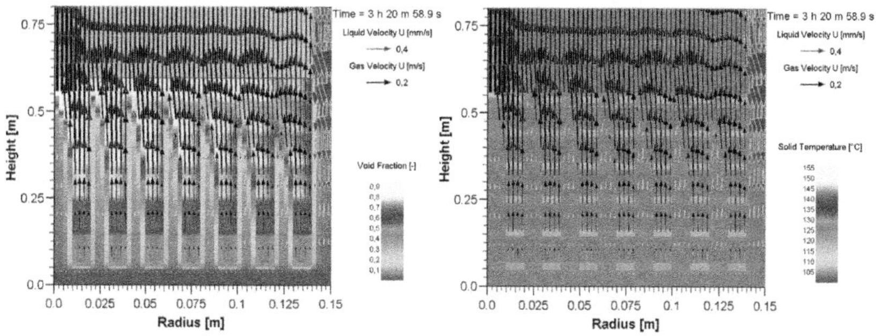

FIGURE 4.22 WABE calculation for STYX-2.3 with nonhomogeneous heating.

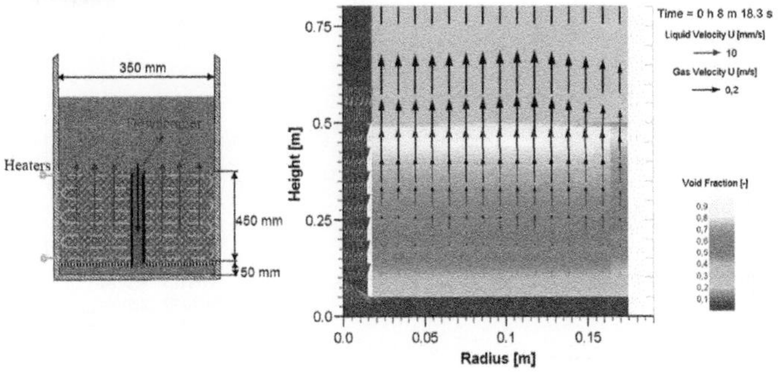

FIGURE 4.23 POMECO experiment (left) WABE calculation for POMECO (right).

supply from top with lower void under top flooding). Figure 4.22 also shows a temperature increase in the dry zones. Nevertheless, a steady state is established due to cooling in the steam flow. However, in the STYX experiments, the first dry zone was obtained at the bottom, which may be taken as indication of a uniform behavior, in spite of nonhomogeneous heating. Perhaps lateral spreading of steam is underestimated in the WABE code.

A similar behavior is obtained for a case of a POMECO experiment with downcomer, calculated by KTH with WABE [38]. Figure 4.23 shows the experimental setup and the calculation results (half-vessel) demonstrating the natural-circulation-driven cooling, due to the lateral pressure differences from pure water and water/steam mixture hydrostatic heads. Here, the calculation agrees with the experimental result very well with a DHF of 346 kW/m^2 from the calculation compared to the measured value of 331 kW/m^2 measured in the experiment.

Analysis Results for Reactor Scenarios

Analyses for reactor scenarios have been performed in order to check the applicability of the model and especially the influences of multidimensional flows and bed inhomogeneities on coolability. A significantly better coolability is expected, for example, from heap-like debris configurations, as outlined above. In the frame of SARNET, especially calculations on in-vessel debris cooling have been performed for the core region as well as the lower plenum with ICARE-CATHARE (IRSN) and ATHLET-CD or separately with the module WABE (IKE) in quenching and boil-off modes. A significant improvement of coolability by lateral water inflow has been shown [36] [39] [40]. Including cooling of dry, hot regions by steam, coolable states of a particle bed in the lower head (about 1.3 m deep bed in center, 2 mm diameter particles, porosity 0.4, system pressure 0.3 MPa) were obtained in WABE calculations until reaching a power of about 350 W/kg, although a first dry zone appears already with 200 W/kg. Under 1D top flooding, a dry zone would already

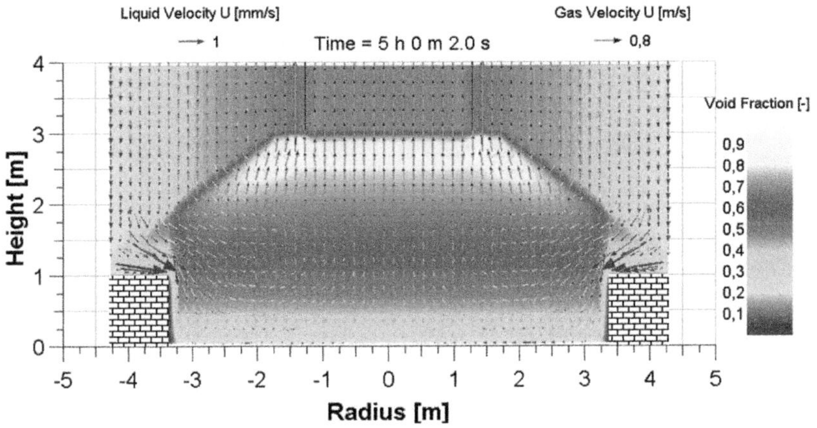

FIGURE 4.24 WABE calculation on quenching of a debris bed in a BWR cavity (initial temperature: 1300 K, total mass: 516 tons, mass of UO_2: 157 tons, Volume 103 m^3, uniform particle diameter of 2 mm, power 505 kW/m^3, porosity 0.4, system pressure: 0.3 MPa).

develop with about 140 W/kg. Thus, DHF values from 1D top flooding configurations strongly underestimate the coolability of real configurations. Calculations with cake parts underlined the cooling potential. Similar results were obtained by ICARE/CATHARE calculations.

Finally, with respect to the present subject of debris coolability in the cavity, a calculation with WABE is shown in Figure 4.24 for a heap-like debris assumed initially dry and at a temperature of 1300 K. The figure shows the lateral inflow of water and mostly upflow of water streams inside the bed. This pattern yields coolability in spite of the bed of 3 meters depth.

4.1.5. Conclusions on Debris Bed Formation and Coolability

The major conclusions from the above model development and analyses on debris formation and coolability for melt dropped into a water pool are as follows:

- Melt release from a failing RPV can be expected to mostly occur with relatively small (<30 cm) breaches and moderate flow rates, but with extended duration.
- Such release modes favor breakup of melt and formation of particulate debris formation in deep water pools, established in BWRs as an accident management measure. Models of these processes exist (IKEJET/IKEMIX, MC3D) and have been shown to describe the essential effects consistently under various conditions. Further confirmation by continued validation is required. Proposals for simplified modeling in ASTEC can then be derived. Features of the debris bed formation not described in

the models are the debris shape and porosities, depending on details of the settling process. Experimental support of such data is expected especially from the DEFOR experiments. First results indicate significantly higher porosities than those assumed previously and, as expected, settling as heap-like configurations. Further model development is especially necessary on solidification of the settling melt drops in order to distinguish between parts of loose particles and compact "cakes" of frozen or crusted melt.

- Coolability of particulate debris is strongly supported by inflow of water from the sides, especially over bed regions of lower height. Heap structures are favorable for this, yielding inherently such flows of natural convection type due to the lateral differences in water/steam composition and resulting pressure differences. Classical approaches only considering top flooding in 1D configurations (uniform height) strongly underestimate coolability. In contrast to such configurations, water inflow into heaps occurs via water-rich regions and mostly in co-current flow with steam.
- Models to describe this extended coolability must calculate the multidimensional fluid flows in complex, heterogeneous structure. The available models, such as WABE-2D (implemented in ATHLET-CD) and ICARE/CATHARE, have been shown to have this capability. Continued validation and especially application to reactor cases are required to confirm the capabilities. Proposals for modeling in ASTEC are to be derived.
- With only shallow water pools in PWRs, breakup would be reduced greatly and be incomplete, thus the production of coolable debris. Nevertheless, as compared to top flooding of melt in an initially dry cavity, there should still be advantages in retardation of basement penetration. More efficient cooling can in principle be reached by bottom injection of water into melt, for example, via porous concrete layers.
- Bottom injection of water into melt has been elaborated as an AM measure in the COMET concept of FZK, primarily for alternative use in the EPR core catcher concept. Experiments showed the effective porosity formation and cooling, thus the applicability of the concept. This finding is further supported by the DECOBI small-scale experiments at KTH and modeling at IKE (WABE extension to describe porosity formation). Validation of WABE-COMET has been performed successfully by FZK experiments, DECOBI experiments, and recently a CometPCA experiment in the VULCANO facility of CEA applying corium melt. Further checking and elaboration of the model is required, as well as application to reactor conditions, especially concerning the CometPC(A) concept (easily implemented) to enhance the coolability of melt pools / particulate debris beds that may be formed in the vessel cavities of existing PWRs and BWRs. Alternatively, downcomers of ≈ 1 meter height may be installed, which are covered by

water. These downcomers would be effective in breaching the melt jet, into particulate melt. The main role of downcomers is to bring water from the top of the bed to the bottom of debris for injection from below. This downcomers scheme has been developed at KTH [19]. The downcomers established a natural circulation flow, between the water pool above the downcomers and the reservoir of water at the bottom of downcomers to cool the particulate debris bed. The dry-out heat flux with the downcomers was found in POMECO experiments to be two to three times the DHF with water flooding only at the top of the bed.

4.2. CORIUM SPREADING

(C. Journeau, C. Spengler, J.J. Foit, J.C. Latché, B. Spindler, and J.M. Veteau)

Nomenclature for Section 4.2

D	width
F	force
h	thickness
H	enthalpy
L	spreading length
T	temperature
U	characteristic velocity
V	volume
x_f	position of the spreading front
h(x,t)	shape of a gravity current
g	acceleration due to gravity
R	space of real numbers
(x,t)	space-time coordinates

Greek letters

α	power exponent describing the flow rate law.
μ	dynamic viscosity
ν	kinematic viscosity
ρ	density
σ	surface tension
τ	characteristic time
ϕ	solid fraction
φ	heat flux
δ_T	thermal boundary layer thickness
δ_W	kinematic boundary layer thickness
$\delta(x)$	Dirac distribution
$\Gamma(x)$	Euler Gamma function

4.2.1. Introduction

Spreading under gravity has received much attention in the field of nuclear safety research. It is the basis of the EPR core catcher concept [42] [43]. In the case of a postulated meltdown accident with a failure of the EPR pressure vessel, the ex-vessel melt will first be collected in the reactor cavity up to the

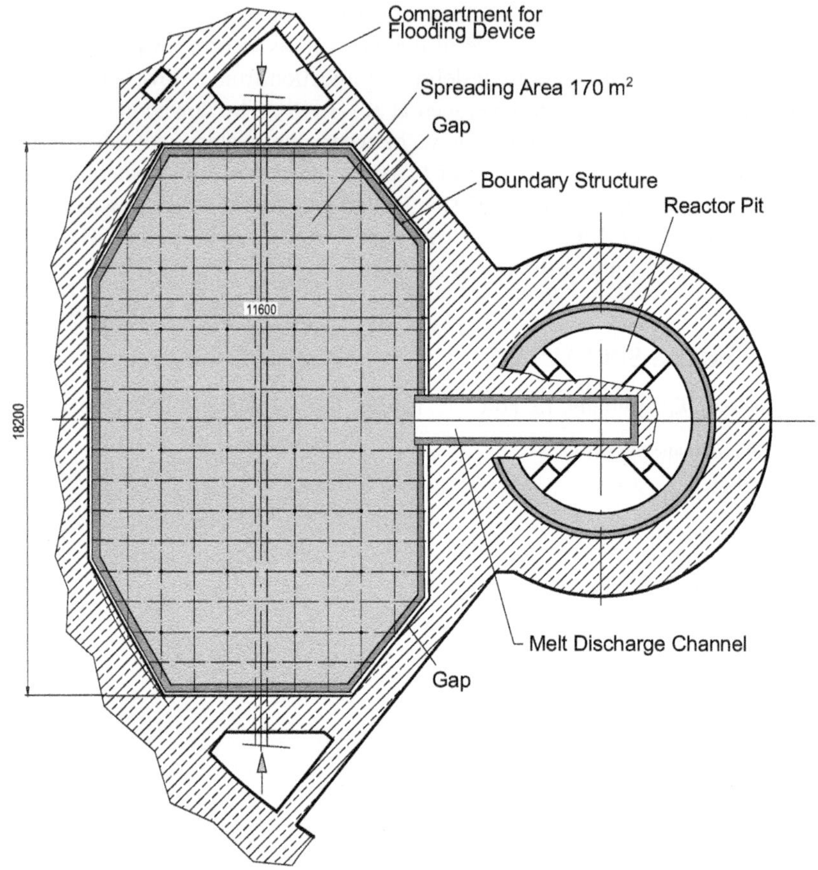

FIGURE 4.25 Top view of the EPR spreading section [42].

melt-through of a sacrificial concrete-covered gate. Afterward, the gate opens, and the corium melt is released into a spreading compartment. The EPR core catcher (Figure 4.25) concept relies on a sufficiently homogeneous spreading of the corium–concrete mixture in order to ensure its coolability after passive initiation of flooding by water.

Spreading is also important for the discharge of corium in large reactor pits and in core-catchers such as that of the Tian Wan VVER-1000 [44].

In the first section, the physics of spreading will be described, followed by the major spreading experiments. The key objective with respect to reactor safety is to verify the coolability of the finally immobilized melt layer that requires the prediction of its thickness h given, as an average, by the ratio of the volume V divided by the area A on which the melt is spread. The objective of detailed numerical models, which will be described in a third section, is to solve the appropriate force balance equations describing the movement of the melt volume with time, while considering mass conservation and heat transfer from the melt to the environment. Finally, spreading codes and their applications to reactor case will be discussed.

4.2.2. Spreading Phenomenology

When an isothermal fluid is spread over an inclined plane (Figure 4.26 left), or on a horizontal plane (Figure 4.26 right), the flow is generally governed by a competition between inertia, viscous, and capillary forces (depending on initial conditions and extent of flow) in balance with gravity. These flows have been thoroughly studied in the field of volcanological lava flows. Griffiths and Fink [49] have studied the effect of cooling for various flow regimes, which will be detailed in Section 4.2.4.

The characteristic of lava and corium spreading flows is that the flowing material cools while it flows. Cooling and spreading are coupled by the rheological properties. The apparent viscosity increases as its temperature decreases, especially in the solidification range where solids are formed in the melt. One of the specificity of corium–concrete mixtures is the very large (of the order of 1000°C) interval between liquidus (the appearance of the first solid) and solidus (the disappearance of the last liquid) temperatures.

Except for the trivial case in which the whole spreading area is filled, the spreading flow stops either when the spread thickness reaches the capillary limit (surface tension overcomes the driving forces) or when the melt cools so that the melt apparent viscosity is so large that the flow is practically stopped.

These phenomena will be described in more detail in Section 4.2.4. First, the experimental results will be described and their main lessons will be summarized.

4.2.3. Major Experimental Investigations

A variety of spreading experiments were performed within various projects, such as CORINE (CEA, France), see Figure 4.27 [65], KATS (FZK, Germany) [70], ECOKATS (FZK) [45] [71], VULCANO (CEA, France) [72], FARO (JRC, Italy) [73], and COMAS (Siempelkamp, Germany) [74] and S3E (KTH, Sweden). The experiments were performed on ceramic or concrete surfaces

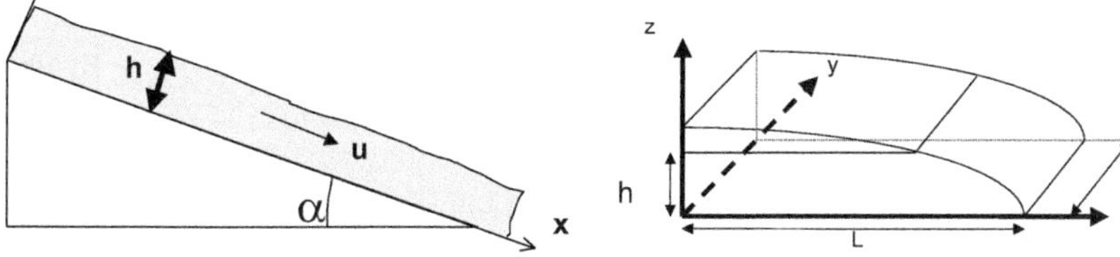

FIGURE 4.26 Spreading over an inclined plane (left) and over a horizontal plane (right).

FIGURE 4.27 Spread of a low-temperature metallic alloy in the CORINE facility (note the spread height sensors, above the spreading plane) ©CEA.

under various conditions, for example, melt overheat and melt release rates with either a thermite oxide melt, based on iron-alumina (KATS, ECOKATS) or prototypic corium melts (COMAS, VULCANO, FARO).

Table 4.1 gives an overview of the major spreading experimental programmes using simulant materials. In Table 4.2 the experiments with realistic corium melts are summarized.

All these experiments had two objectives: to provide phenomenological observations on the spreading processes and to contribute to an experimental database against which the spreading models and codes have been validated.

CORINE

In the CORINE experiments, performed at CEA Grenoble [65], low-temperature melt volumes of about 40 l were spread into an angular sector. The CORINE program featured three groups of experiments on the spreading of liquids due to gravity: a first group of experiments focused on the hydrodynamics of spreading using isothermal water-glycerol or water–hydroxyethyl cellulose mixtures so as to reach fluid viscosities as high as 2 Pa.s. Inertia-gravity controlled flows as well as gravity viscous flows were realized. These experiments served as the validation of the hydrodynamic modeling of spreading over a horizontal surface.

In a second group of experiments, eutectic and noneutetctic Bi-Sn or Bi-Pb alloys were used as melt simulants with a solidification range between 138°C and 192°C. These tests showed that mixtures presenting noncongruent[1] melting could spread better than those with a congruent melting (pure substances or

1. Melting over a range of temperature

TABLE 4.1 Experimental Programs with Simulant Materials

Program	Lab.	Material	Scale (spread volume)	Geometry	Initial and boundary conditions
CORINE [65]	CEA (France)	Low-temperature simulants (water, glycerol, low fusion point metallic alloys)	~50 liters	19° sector	Flow rate Material (viscosity, pure compound-noneutectic alloys) Top/bottom cooling Effect of sparging gas through the substrate
Greene [67]	BNL (USA)	Lead	~1 liter	Square section	Mass Overheat Water height
S3E [68]	KTH (Sweden)	Low and medium (1200°C) temperature simulants	5–20 liters	1-D channel Rectangular section	Flow rate Overheat Material Substrate (effect of concrete) Dry/wet spreading
SPREAD [69]	Hitachi Energy Res. Lab. (Japan)	Steel	1–15 liters	Rectangular channel Half-disk	Spread mass Overheat Flow rate Inlet geometry

(*Continued*)

TABLE 4.1 Experimental Programs with Simulant Materials—cont'd

Program	Lab.	Material	Scale (spread volume)	Geometry	Initial and boundary conditions
					Substrate Water level
KATS [70]	KIT (Germany)	Aluminium Thermite (Al_2O_3 + Fe) around 2000°C	~60 liters	Rectangular channel	Spread mass Flow rate Substrate Addition of concrete decomposition products Oxide and/or metal phase Reflooding
ECOKATS [45] [71]	KIT (Germany)	41 wt.-% Al_2O_3, 24 wt.-% FeO, 19 wt.-% CaO and 16 wt.-% SiO_2 at 1600 °C	Up to 850 liters	Channel leading into a rectangular 3 m × 4 m section	Viscous/inertial regime Ceramics and concrete substrate

TABLE 4.2 Prototypic Corium Experimental Programs

Program	Lab.	Material	Scale (spread volume)	Geometry	Parameters
COMAS [74]	Siempelkamp (Germany)	Corium–concrete-Iron mixtures Liquidus around 1900°C	20–300 liters	Rectangular channels 45° sector	High flow rates (>150 kg/s) Effect of silica fraction Substrate (ceramic/metal/concrete)
FARO [73]	JRC Ispra (European Commission)	UO_2_ZrO_2 Liquidus around 2700°C	~20 liters	19° sector	Presence/absence of a shallow water layer Metallic substrate
VULCANO [72]	CEA (France)	UO_2_ZrO_2 + concrete products Liquidus from 1900 to 2700°C	3–10 liters	19° sector	Flow rate Composition Nature of substrate (ceramic, concrete, steel)

eutectics). This was attributed to the fact that efficient spreading continues below the liquidus for noncongruent melting compositions.

In a third group of experiments, other effects (spreading of a melt onto an already encrusted melt, spreading of a melt in the presence of sparging gas) were studied. To investigate the effects of gas release from the substrate caused by thermal erosion of the underlying concrete by a corium melt on spreading, a series of experiments using water and a hydroxyethyl cellulose (HEC)-water mixture were carried out also in a 19° angular geometry [62]. A strong increase in viscosity of water in the presence of a gas flow through the bottom plate during spreading was observed. The spreading experiments performed with HEC-water mixtures show a complex influence of bubbles on rheology. The HEC-water mixture is a non-Newtonian, shear-thinning fluid. In the presence of bubbles during spreading, shear stress is applied to a small thickness of liquid between neighbouring bubbles. This leads to an increase in viscosity. On the other hand, the increase in the shear rate in the interfacial layers will result in a viscosity decrease due to the shear-thinning behaviour of the HEC mixture. If the latter effect is dominant, the effective viscosity will decrease [52].

S3E

Several spreading experiments with binary melts were performed at KTH Stockholm (Sweden) [68]. The binary mixtures used were an oxide mixture ($CaO - B_2O_3$) and a salt mixture ($NaNO_3 - KNO_3$) with volumes up to 20 l and initial temperatures at about 1100°C (oxides) and 300°C (salts). The melts were spread into 1D and 2D test sections made of concrete and steel, respectively. Although different phenomena were observed for the cases of spreading on steel and concrete (smooth spreading in case of steel, violent interactions and gas bubbling in the melt in case of concrete), similar final spreading lengths were obtained. (See Figure 4.28.)

KATS

At KIT a range of experiments were performed with thermite melt on the basis of alumina and iron [70]. This facility provided a large experimental database concerning large masses of melt. The oxidic and metallic parts of the melt were spread into separate sections, for which 1D channels and 2D rectangular sections were used. For a widening of the solidification temperature range of the oxidic melt, approximately 10 wt.-% of silica was mixed into the alumina. The matrix covered experiments on concrete as well as on ceramics. The final spreading length in the KATS 13 and VULCANO VE-U7 experiments was approximately 30% shorter than that observed in the analogous experiments on ceramic substrates. Spreading on concrete was also studied with some water present on the surface of the substratum and with an epoxy coating of the concrete. The presence of a shallow water layer had no effect on the spreading length.

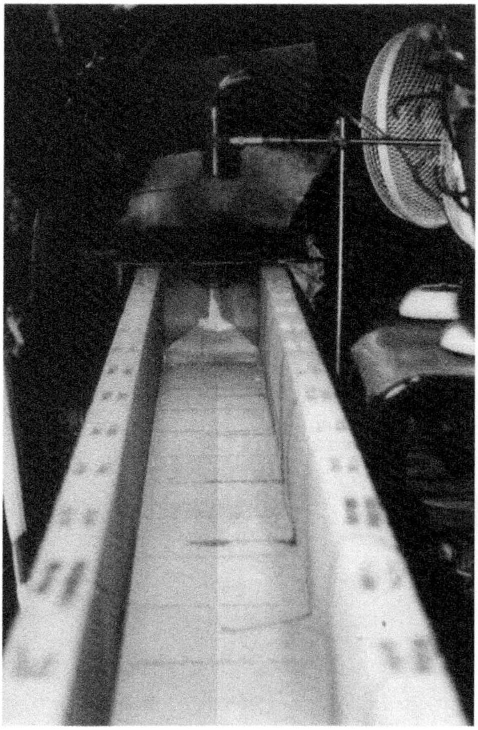

FIGURE 4.28 Oxidic (30% BaO 70% B_2O_3) 1D spread in S3E facility ©KTH.

ECOKATS

The ECOKATS-V1 and ECOKATS-1 large-scale experiments [45] were performed to investigate spreading under conditions relevant to an ex-vessel melt release during a core melt accident, using a highly suitable thermite-generated multicomponent oxide melt with a sufficiently wide solidification range (Figure 4.29). The initial temperature of the melt was chosen to be approximately 70°C above the liquidus temperature to allow solidification to start during the melt pouring process. After the onset of solidification, the front advanced by deforming the crust, pushing it aside and generating breakouts. In the ECOKATS-1 experiment, 550 kg of an alumina-concrete mixture were spread over a concrete test section.

Successful spreading is important to minimize the melt height and therefore to enhance the cooling process when water is added to the melt. These are the conditions that were realized in the ECOKATS-2 experiment (see Figure 4.30) [71]. Shortly after completion of the spreading process, the melt was flooded from the top, a process that could be the result of an accident management procedure in case of an accident. But efficient fragmentation and cooling was restricted to the upper few centimeters of the oxide melt, and this is not adequate to extract sufficient decay power from the bulk of the melt. Some

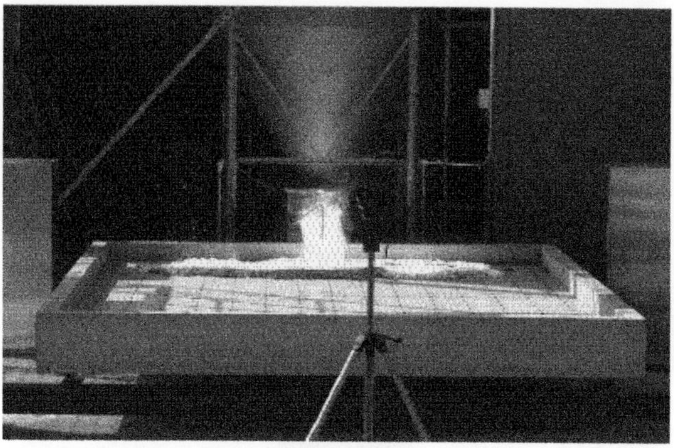

FIGURE 4.29 Onset of the thermite spread over ECOKATS-1 test section [45].

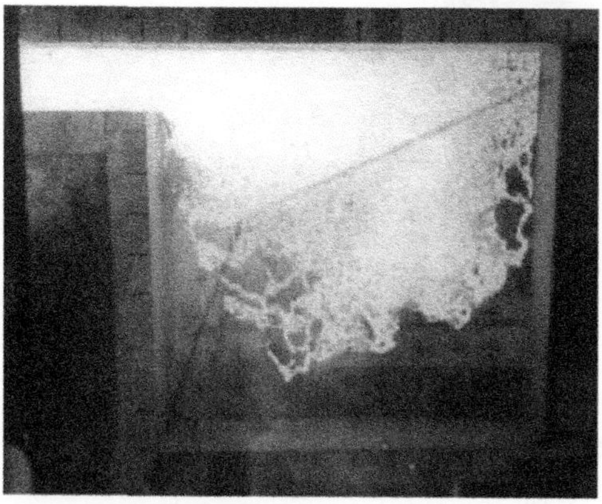

FIGURE 4.30 Test ECOKATS-2, 7.4 s after spreading start [71].

uncertainties, however, persist from the present experiment as no decay heat could be simulated in the melt. Also, the properties of the melt were different, especially with respect to the oxide density and the level of the freezing temperature, and the spreading dimension in a reactor, which is typically 6 m instead of 2 m. It is unlikely, however, that this changes the sequence and importance of the observed coolant phenomena.

COMAS

In the COMAS project conducted at Siempelkamp (Krefeld, Germany), representative corium melts for the EPR were spread at temperatures around 2000°C [74]. The experimental matrix included spreading of melt into 1D-channels (40 cm wide) with different substratum materials (iron, concrete, ceramics) and spreading into a 2D section made of iron. The maximum mass of melt that was spread in a single experiment was nearly 2 tons. Maximum volume fluxes

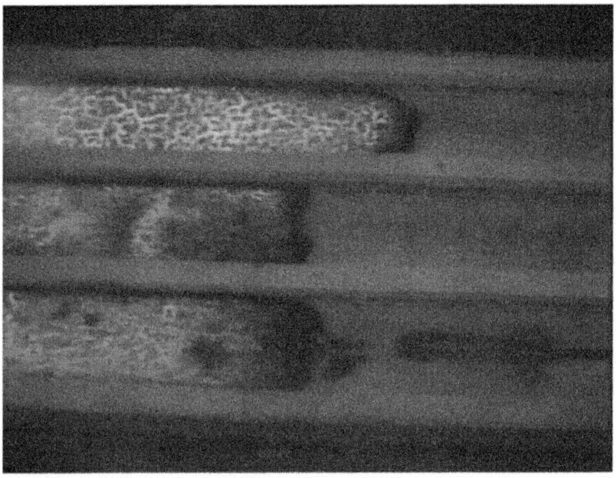

FIGURE 4.31 COMAS test EU-2B; spreading over three different substrates.

achieved in the experiments were approximately 60 l/s, leading to front velocities of the melt of about 2 m/s. The average thickness of the spread melt was between 3 and 5 cm (including pores). Temperature differences between the interior and the crust surface of about 400°C were observed. For the finally obtained length of the spread melt, no influence of the substratum material was observed because the spreading was dominated by gravity-inertial forces; therefore, the viscous forces were of minor importance. The experiments (see Figure 4.31) showed an efficient spreading of the melt at the initial and boundary conditions selected.

FARO

Binary mixtures of UO_2 and ZrO_2 were studied in dedicated spreading experiments in the FARO facility at JRC ISPRA (Italy) [73], see Figure 4.32. For such melts that are representative of spreading without prior mixing with sacrificial concrete, the temperature region for solidification is only ~50°C wide.

Melt masses of approximately 100 kg were spread into a sector-like 2D geometry with volume fluxes at about 2 l/s. Initial temperatures were reported to be close to liquidus of the melt (2640°C), but the uncertainty of the

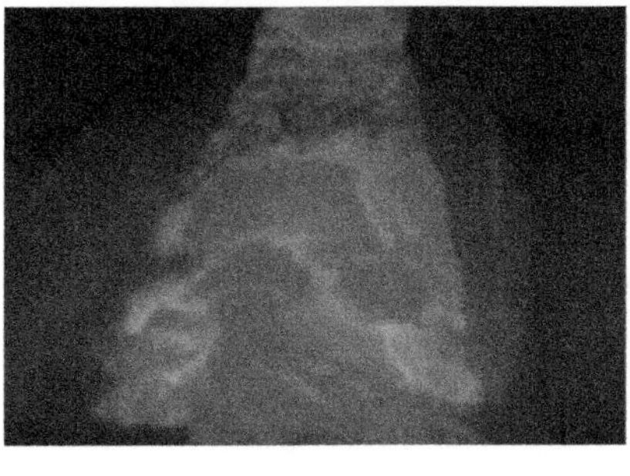

FIGURE 4.32 Infrared thermography of the FARO L32S spread (crust cracks appear brighter).

temperature measurement (±50°C) compared to the temperature range of solidification (50°C) is not negligible. In these experiments a rather inhomogeneous spreading was observed. The spreading was characterized by irregular local stops of the melt, which can be attributed to the impact of crusts at the surface, followed by some continuation of melt progression due to local failure of crusts/melt overcoming the crusts. Spreading on a wet substratum surface was also studied but showed no effect.

VULCANO

At CEA Cadarache, spreading experiments were conducted with simulant melts and with representative corium melts for the EPR in a 2D sector-like geometry [72]. The experiments were performed with small-volume fluxes (<1 l/s) and with small integral melt masses (<100 kg). For the case of the simulant melt, urania was replaced by hafnia. The experiments with corium melts showed that for the selected initial and boundary conditions the spreading was sufficiently efficient, due to the large temperature range for solidification of the melts. A substantial contribution of crusts to the retardation of the flow was not observed. The stopping and immobilization of the melt was, rather, attributed to the increasing viscosity of the melt with proceeding solidification. For corium–concrete mixtures, there is rather a smooth viscous skin on the corium surface, which is generally folded (see, e.g., Figure 4.33). It has been shown [76] that these folds are due to instabilities caused by the rapid variations of viscosity with depth at the upper boundary layer.

The major effect of concrete decomposition products is that it enlarges substantially the solidus-liquidus interval from 50 to 100°C for (U, Zr)O_2 mixtures to more than 1000°C. The amount of heat to extract before the crust is fully solid is more than twice for the corium–concrete mixtures than for in-vessel corium including only urania and zirconia. Thus, during spreading of corium–concrete mixtures, the surface remains viscoplastic while it may be completely solidified for (U, Zr)O_2 mixtures.

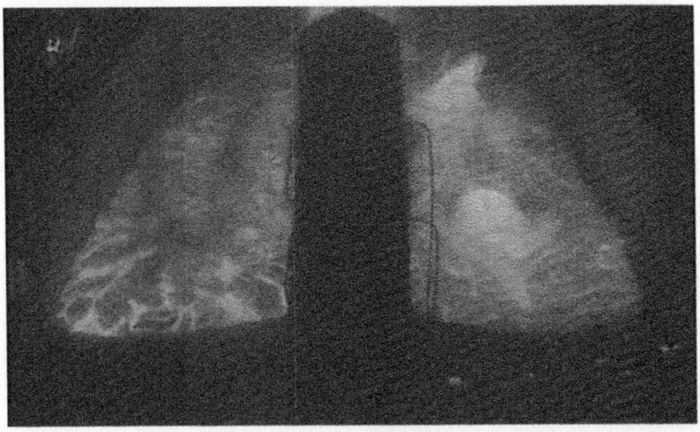

FIGURE 4.33 VULCANO VE-U7 test over a ceramic (left) and a concrete (right) spreading section.

4.2.4. Analysis and Model Development

In a first subsection, useful self-similar solutions will be presented for the gravity-viscous spreading of a constant volume of liquid or a volume of liquid that increases as a power law of time. In the second subsection, the effects of the cooling and solidification process on fluid flow will be discussed. Thereafter the available computer codes and their validation status will be presented.

Basic Analysis of Isothermal Spreading Flows

A melt dome spreading in a horizontal channel of width D is considered as shown by Figure 4.26 right. The average melt height (thickness) is given by $h(t)$, the extent of the spread melt is represented by $L(t)$ and the velocity of the front in the x-direction is denoted by $U(t)$, which is of the order of L/t. If the force of gravity were not counterbalanced by other forces acting on the mass of the melt, the front of the melt dome would be accelerated (increase of U, decrease of h) due to gravity. In this case, the spreading would be limited at last by the surface tension of the melt. This is the case, when the hydrostatic pressure at the final average melt thickness h_{cap} of the shallow melt layer equals the corresponding capillary pressure, which is of the order of magnitude of:

$$h_{cap} \sim \sqrt{\frac{\sigma}{\rho g}}$$

The final extent L_{max} of a melt spread under isothermal conditions could then be easily evaluated from this relation using the expression $L_{max} = V/(h_{cap}\ D)$. In contrast to the capillary thickness (order of some millimeters), the results of important spreading experiments with prototypic corium melts show larger melt thicknesses after immobilization of the melt (e.g., in ECOKATS-1 [45]: ~3.5 cm). This indicates that there is a strong influence of cooling on real spreading processes, which has to be considered adequately by analytical models.

Considering an isothermal flow in a channel, the different forces acting on the melt are given with time-dependent reference quantities as reported in Table 4.3.

Previous research work on the spreading of gravity currents under isothermal conditions [46] [47] has shown that the influence of retarding forces

TABLE 4.3 Forces and Reference Quantities of an Isothermal Flow in a Channel

Quantity	Volume V	Front velocity U	Gravity F_g	Inertial forces F_i	Viscous forces F_v
Scaling	$\sim hLD$	$\sim L/t$	$\sim \rho g h^2 D$	$\sim \rho U^2 h D$	$\sim \eta ULD/h$

(e.g., viscous forces F_v) increases together with the melt-covered area ($=L\cdot D$ in Figure 4.26 right; see also Table 4.3) and that there is a sharp transition from an initial spreading regime controlled by a balance of inertial forces F_i and gravity F_g toward a spreading regime controlled by a balance of viscous forces F_v and gravity F_g. Considering isothermal flows, these two regimes can be modelled by the simplified force balance equations

$$\begin{aligned} F_i &= F_g \qquad \text{for } F_i \gg F_v\,, \\ 0 &= F_g + F_v \quad \text{for } F_i \ll F_v \end{aligned}$$

leading to power law relations for the time-dependent position of the melt front $L(t)$ [46] [47] [48].

Spreading of a Constant Volume

When the lubrication theory for low Reynolds number flows is used, the following evolution equation for the fluid shape h(x, t) holds [46] [47]:

$$\frac{\partial h}{\partial t} - \frac{1}{3}\frac{g}{\nu}\frac{\partial}{\partial x}\left(h^3\frac{\partial h}{\partial x}\right) = 0, \tag{1}$$

where ν is the kinematic viscosity and g the gravity constant.

The spreading of a fluid droplet of the shape $h_0(x)$ and with the volume, $\int_{\mathbf{R}} h_0(x)dx$ is described by Eq. (1) and the initial data,

$$h(x,0) = h_0(x). \tag{2}$$

There is a well-established mathematical theory for solution of equations (1) and (2).

One of the most important differences between the solutions of Eq. (1) and the solutions of the linear diffusion equation is the existence of solutions for which the front propagates at finite speed except possibly at $t = 0$.

The instantaneous release of the volume (per channel width), v, of the fluid is described by the initial data given by

$$h(x,0) = \mathrm{v}\delta(x) \tag{3}$$

where $\delta(x)$ denotes the Dirac delta distribution. A self-similar solution of the above problem was obtained by Barenblatt [47] and later by Huppert [46].

The solution of Eq. (1) with the initial data given by Eq. (3) can be written as

$$h_\delta(x,t) = \begin{cases} \left(\frac{3}{10}\right)^{\frac{1}{3}} \left(\frac{3\mathrm{v}^2\nu}{g}\right)^{\frac{1}{5}} \xi_f^{2/3}\, t^{\frac{-1}{5}} \left[1 - \left(\frac{3\nu}{g\mathrm{v}^3}\right)^{\frac{2}{5}} \xi_f^{-2} x^2 t^{-\frac{2}{5}}\right]^{1/3}, & |x| \le x_f \\ 0 & |x| > x_f \end{cases} \tag{4}$$

This empirical correlation has been validated [58] [60] against experimental data on semisolid metallic alloys, corium mixtures, molten salts, and solidifying lavas and basalts. However, for solidification during spreading, a thermodynamical equilibrium generally cannot be assumed.

As observed in the KATS, ECOKATS, and VULCANO experiments, spreading that is governed by gravity-viscous forces is influenced by the gas release from the substrate caused by melt-induced thermal erosion of the underlying concrete, which changes viscosity and mixes the surface boundary layers with the interior of the spreading oxide melt. The latter process cools the bulk of the melt. The final spreading length in the two-dimensional experiments KATS 13 [61] and VULCANO VE-U7 was approximately 30% shorter than that observed in the analogous experiments on ceramic substrates. Analysis of the CORINE experiments [62] with water and shear-thinning hydroxyethyl cellulose (HEC)-water mixtures shows that the front motion can be described in terms of approximate self-similar gravity-viscous solutions with a modified viscosity [52]. A different approach using a concept of friction two-phase multiplier was proposed in [62]. The rheology of bubbly suspensions depends on the volume fraction of bubbles and the amount of deformation ([63] [64]) and is still not well understood.

Recognizing the phenomenological complexity of physico-chemical and multi-phase transport processes present in high-temperature corium spreading, and the practical irreducibility of large uncertainties in measuring spreading melt's time- and spatially-resolved characteristics and properties, Dinh et al (KTH) [68] pursued an integral approach for the assessment of efficacy of corium spreading as a strategy to manage ex-vessel corium risk. Within the KTH risk-oriented framework, the safety's figure of merit is the spread corium layer thickness (and decay heating in that layer), which determines the corium long-term coolability. Along this line, experimentation, modelling and analysis all shifted focus from measuring, describing and computing details of corium spreading dynamics to quantification of the final spread melt thickness.

To characterize melt spreading efficacy, the KTH group uses a dimensionless length scale L of spread melt thickness scaled to isothermal capillary thickness. Using mass and energy conversation equations and channel flow analytical solutions in limiting inviscid and viscous regimes, the KTH scaling equation is obtained, relating L to a dimensionless time T in a square-root relation

$$L = CT^{1/2}.$$

Here T is ratio of time scales of physics that govern corium melt spreading, namely hydrodynamics τ_{hydro} and solidification $\tau_{solidif}$. The solidification time scale is evaluated using a melt immobilization criterion, by which the bulk corium enthalpy reaches a certain "mushy" degree β. The faster the solidification (the smaller $\tau_{solidif}$), the thicker the immobilized melt layer (the larger L).

By benchmarking the scaling equation against measured data from a large number of simulant-material and prototypic-corium one-dimensional melt-spreading experiments, C and β are found to be unity and half, respectively.

Notably, KTH team applied the scaling relation to make successful pre-test predictions for high-temperature binary-oxidic and corium experiments [68]. The method was extended to two-dimensional and open spreading, although the method qualification in these geometries remains limited due to lack of relevant data. Readers are referred to [68] for details of derivation and validation.

More broadly, designer's and regulator's confidence in methods for predicting corium spreading efficacy in reactor scenarios depends largely upon quality of experimental data used for the predictive model calibration and code validation. The question about relevance and scalability of certain data sets looms particularly large when the validation experiments are performed with relatively small corium quantities, or using simulant materials under conditions far from prototypic reactor conditions. KTH method is most conducive for evaluating all experiments, from real-corium to simulant-materials, low to high temperature, on a unified scaling basis relative to the reactor scales, allowing one to appropriately weight the information value of various data sets. Thus, instead of relying on hard-to-measure parameters (often "unknowable unknowns") for calibration of high-fidelity, computationally-expensive, multi-dimensional dynamic models, KTH analysis method achieves its predictive capability through a balance between modelling, experimentation, and validation.

4.2.5. Spreading Codes

Code Development for Nonisothermal Corium Spreading

Because of the enormous computer resource requirements, a range of approaches evolved using various degrees of simplification of the transport equations or of the numerical discretization scheme or of averaging material properties for finitely-small mesh sizes. Those approaches, which have received sufficient discussion and validation, are presented with their specific features in Table 4.4.

Validation of Spreading Codes

Each code has been validated against its own validation matrix, from analytic cases and isothermal spreading to prototypic corium experiments. The final validation of the spreading codes was performed in the frame of two European benchmarks:

- A benchmark on the simulant material spreading experiment ECOKATS-1 [77] (Figure 4.35).
- A benchmark on the VULCANO VE-U7 spreading benchmark with prototypic composition close to that which is expected at the EPR gate melt through [78].

These exercises show a good estimation of the spreading surface (or thickness) with uncertainties in the order of 20%.

TABLE 4.4 Comparison of Characteristics of the Spreading Codes

Feature	MELTSPREAD	CORFLOW	THEMA	LAVA	CROCO
Reference to basic literature	Farmer, 1993 [80]	Wittmaack 1997 [81]	Spindler, Veteau 2004 [82]	Allelein 2000 [83] Spengler 2002 [84]	Piar 1999 [85], Michel 1999 [86], Vola 2004 [87]
Forces considered in transport equation	F_g, F_i, F_v	F_g, F_i, F_v	F_g, F_i, F_v	F_g, F_v	F_g, F_i, F_v
Material law (stress-strain-relation)	Newtonian	Newtonian, Bingham	Newtonian, (Bingham)	Newtonian, Bingham	Newtonian, non-Newtonian
Spatial discretization	x	x, y, z	x, y	x, y	x, z or r, z
Rheological properties dependent on temperature	yes	yes	yes	yes	yes
Vertical temperature profiles $T(z)$ at (x, y)	no	yes	no	no	yes
Vertical velocity profiles $w(z)$ at (x, y)	only average value solved, based on vertical integration of transport equation	yes	only average value solved, based on vertical integration of transport equation	only average value solved, based on analytical approximation of velocity profiles	yes
Bottom crust top crust	yes yes	yes yes	yes yes	no yes	simulated by viscosity/ yield stress increase

(*Continued*)

TABLE 4.4 Comparison of Characteristics of the Spreading Codes—cont'd

Feature	MELTSPREAD	CORFLOW	THEMA	LAVA	CROCO
Downward heat flux from melt to substratum surface	forced convection and/or effect of bubble agitation	$\lambda\, dT(z)/dz$ (based on temperature profile $T(z)$), incl. crust formation	Nusselt correlation for horizontal flow along a plate	empirical heat transfer coefficient or Nusselt correlations to be provided in input file	$\lambda\, dT(z)/dz$ (based on temperature profile $T(z)$), incl. crust formation
Heat conduction in the concrete substratum	yes, incl. phase change	yes, no phase change	yes, incl. phase change	yes, no phase change	yes, incl. phase change
Upward heat transfer from the melt to its free surface	Nusselt correlation	$\lambda\, dT(z)/dz$ (based on temperature profile $T(z)$), incl. crust formation	Nusselt correlation	Nusselt correlation in combination with top crust formation	$\lambda\, dT(z)/dz$ (based on temperature profile $T(z)$), incl. crust formation
Radiation from the free surface	gray body radiation law at T_{int}	gray body radiation law at T_{int}	gray body radiation law at T_{int}	gray body radiation law at T_{int}	gray body radiation law at T_{int}
effect of gas from concrete melting	on heat transfer only	increase of initial viscosity in input	on friction factor only	increase of initial viscosity in input	bubbles-induced modification of transport properties

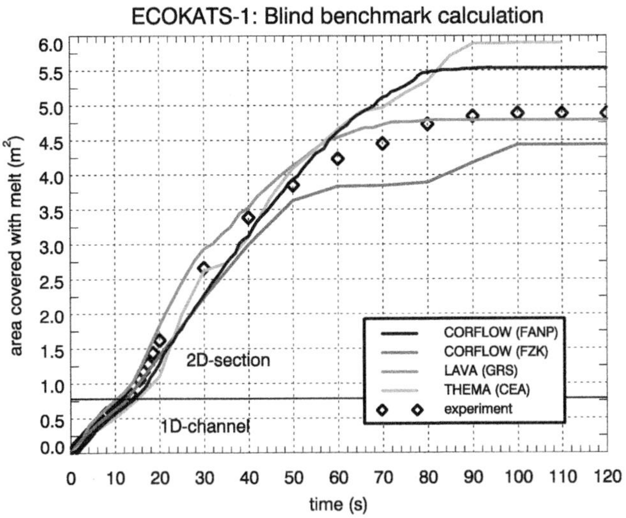

FIGURE 4.35 Code results for the spreading area covered with melt versus time in the blind benchmark for ECOKATS-1.

As an illustration, the results obtained in the framework of the European ECOSTAR Project are presented here. Nonisothermal spreading of an oxide melt on ceramic and concrete 2D surfaces (fed by the way of a 1D channel) is studied in the large-scale ECOKATS-1 experiment, which was performed at FZK [45] [77]. Blind calculations performed in the frame of a dedicated benchmark action with some of the spreading codes (see Table 4.4) served to validate the codes' capabilities to predict the spreading of a melt with well-known material properties. Concerning coolability of the core melt, it is sufficient to assess the capabilities of the code with respect to

- predicting the surface covered by the spreading melt;
- the similarity of the distribution of the melt thickness compared to the experimental findings;
- the calculated temperatures of the melt.

Therefore, for the assessment of the codes' results, the major quantity, which is compared here, is the extent (spreading area) of the melt spread as a function of time. The major result of the benchmark was that the uncertainty for the codes having participated in the benchmark was shown to be less than approximately ±20% – provided there is sufficient knowledge of the material properties of the melt.

With view to assessing, with an integral code like ASTEC, the proper operation of the ex-vessel melt retention and coolability concept of the EPR for a wide parameter range of initial and boundary conditions, a much faster approach in terms of computational costs compared to the detailed codes is required in the first instance of detecting problematic spreading scenarios. With regard to this objective, Dinh et al. [68] used the idea of a criterion to estimate the end of spreading due to cooling.

For implementation into ASTEC, Spengler proposed using an alternative criterion [79] which follows more closely the time evolution of spreading according to the isothermal approximate self-similar solutions for oxidic melts [52] in combination with an empirical stopping criterion to estimate the final extent of spreading. In certain boundary cases and due to the shortcomings of such parametric approaches, the application of the more detailed code of Table 4.4 might nevertheless be necessary to assess further the consequences for the specific accident sequence under investigation.

4.2.6. Reactor Applications

With these validated codes and models, it is possible to extrapolate to reactor case. Steinwarz et al. [88] indicate that, even for very conservative situations, spreading is not a problem.

For instance, Wittmaack [81] simulated corium spreading in the EPR core-catcher with the CORFLOW code. He considers that 370 t (59 m^3) of corium (250 t of a corium-sacrificial concrete mixture and 120 t of metal spread on a 178 m^2 area (10 m long) is modeled by 27 × 13 × 14 meshes. In this reference case, corium is released in 8 s through a 2.4 m^2 gate, that is, a 7.4 m^3/s average flow rate. Analysis of the corium–concrete interaction in the reactor pit [89] indicates that the melt initial temperature is around the liquidus temperature, implying a small solid fraction and thus a low viscosity, which is favorable to spreading. However, for the prediction of melt temperature during corium–concrete interaction by analytical tools, an uncertainty range of some hundred °C must be considered.

The oxidic melt liquidus temperature is estimated around 2300°C, and CORFLOW calculations show that for an initial temperature above 1830°C, spreading flows rapidly, reaches the spreading section wall in less than 7 s, and covers the wall surface in less than 12 s. After a few minutes, the free surface is almost flat. Radial velocities reaching 4.7 m/s have been computed.

Similar results have been obtained from THEMA calculations [90]. Actually, the main uncertainty is related to the size of the breach in the gate, which controls the corium flow rate. If this rate were extremely small, spreading would be less efficient [81]. CORFLOW calculations performed by Foit [91] show that the gate opening must be larger than 0,17 m^2 to ensure a homogeneous spreading. Azarian et al. [92] consider as lower bound of the EPR design a flow rate of 1 t/s (150 l/s). Therefore, as all calculations predict a large margin on spreading, it can be considered that a satisfactory spread will be achieved in all the considered configurations.

4.3. CORIUM CONCRETE INTERACTION AND BASEMAT FAILURE

(H. Alsmeyer, M. Cranga, J.J. Foit, C. Journeau, J.M. Seiler, and B. Tourniaire)

4.3.1. Introduction

During a severe accident with loss of heat removal and subsequent failure of the reactor pressure vessel, the molten corium would discharge into the reactor cavity and interact with the concrete of the basemat. Controlled by the decay power of the fission products in the corium, the long-lasting process of MCCI, namely, the gradual ablation of the basemat and the walls of the reactor pit, may jeopardize the integrity of the containment.

These processes and their consequences are of essential importance for the safety analyses of existing Gen II reactors. Also, MCCI defines the initial and boundary conditions to control the corium melt in different core retention concepts, which are under consideration and development for Gen III reactors. In the EPR, for example, the core melt would interact with concrete during its temporary retention in the reactor pit [89].

In a typical commercial PWR of 1300 MWe, about 120 t of oxides (UO_2, ZrO_2 ...) and 80 t of metal (Fe, Cr, Ni, Zr) could relocate to the reactor cavity. For a 6 m diameter cavity, this leads to a melt height of the order of 1 meter. This height would be reduced in the case of spreading of the melt on larger surfaces. The corium is heated by the internal decay power (typically, 30 MW initially, decreasing to 10 MW after 10 days), which is predominantly generated in the oxide fraction of the corium. Without additional measures, the long-lasting decay power may lead to melt-through of the basemat over a period of several days, depending on its thickness. Further on, the release of steam, hydrogen and other noncondensible gases from concrete decomposition increases the containment pressure and may also result in the accumulation of flammable hydrogen concentrations, with the possible consequence of containment overpressurization. It should also be noted that the production of aerosols during MCCI causes changes in the behavior of the already existing aerosols in the containment and hence in the source term.

The consequences of MCCI can be catastrophic with (a) the large fission product release and (b) the land contamination. The first would be due to the containment rupture by the pressurization caused by the large gases generated in the MCCI, and the second would be due to the basemat melt-through. However, cooling of the melt should stop the MCCI and prevent the two consequences. The topic of melt-coolability has been treated in Section 4.1.2.

4.3.2. Phenomenology of MCCI

MCCI is characterized by an imposed power, and not by an imposed temperature. If heat removal is not sufficient, the temperature of the melt will increase until melting or dissolution of adjacent structures does occur. This precludes the "simple" use of refractory materials, for example, as a liner without sufficient capacity of heat extraction.

More precisely, corium heats up until reaching at least partial melting (or even total melting) of oxide materials (UO_2, ZrO_2) and of metals released from the reactor vessel (temperature around 2200°C) causing the buildup of a corium pool. The corium pool reaches a temperature higher than the melting onset temperature of concrete surrounding it. Heat released by fission products is convected to corium pool boundaries and triggers the partial melting of reactor pit concrete walls and their loss of integrity by concrete decomposition, the so-called ablation process. Concrete ablation occurs between around 1200°C and 1450°C for most concretes. So, after an initial heating phase MCCI generates a continuous ablation of concrete walls. During this later phase, the decay power generated within the pool is transmitted mostly to corium/concrete interfaces and for a minor part by radiation between the corium pool surface and surrounding walls in particular in dry conditions.

The corium pool behavior during MCCI depends on:

- Transport and thermochemical properties, which depend on corium composition and thus on concrete fraction;
- Heat flux at pool interfaces depending on decay power dissipation, corium pool volume, and also on ablated concrete brought into the pool;
- Superficial gas velocity (volumetric gas flowrate per unit surface) along the corium/concrete interface, which is proportional to heat flux and concrete gas content (H_2O, CO_2), but inversely proportional to the reactor containment pressure above the pool.

It is then convenient to define several phases of MCCI characterized by the concrete oxide mass fraction in the corium pool and the gas superficial velocity value:

- The short-term phase corresponding to a concrete oxide mass fraction lower than around 25%, that is, the first hours of MCCI in the reactor case, with physical and thermochemical properties closer to those of initial corium inventory with a high heat flux toward pool interfaces and an intense gas bubbling corresponding to an elevated gas superficial velocity with an order of magnitude of a few tens of cm/s;
- The medium-term phase corresponding to a higher concrete oxide mass fraction up to around 50%, concerning in the reactor case the period between around 5 and 15 hours after MCCI onset, with physical and thermochemical properties altered, compared to those of the initial corium inventory with a still high heat flux toward pool interfaces leading to a gas superficial velocity of a few cm/s. It is mainly during this phase that the remaining metallic zirconium and steel oxidizes, generating extra heat; it is likely that several successive corium pours out of the vessel

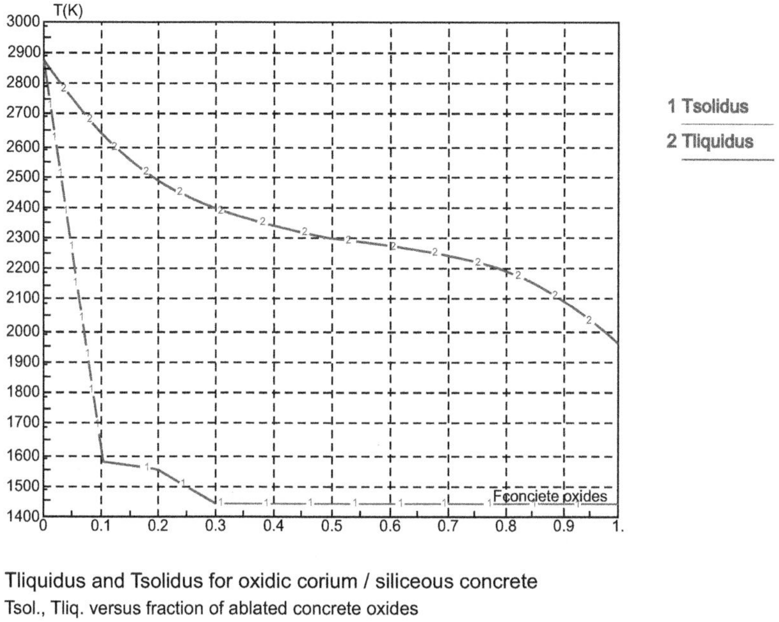

FIGURE 4.36 Evolution of solidus and liquidus temperatures of a mixture of oxidic corium and siliceous concrete versus ablated siliceous concrete fraction.

will take place during this phase, but considering that the total inventory is released instantaneously it can be considered as a bounding case (for the evaluation of the basemat ablation delay).

- Finally, the long-term phase corresponding to a prevailing oxidic concrete mass fraction, concerning in the reactor case the period beyond roughly 15 hours after MCCI onset, with a reduced heat flux toward pool interfaces, an increased corium viscosity because of the silica influx due to concrete ablation and a low gas superficial gas velocity around, or below, 1 cm/s.

As an example, the evolution of liquidus and solidus temperatures for the oxidic corium/concrete mixture versus the mass fraction of concrete included into the pool is plotted in case of one particular siliceous concrete in Figure 4.36, showing the large evolution of thermochemical corium properties during concrete ablation.

The research carried out in relation to MCCI aims to provide the necessary understanding and modeling to make an acceptable prediction of the kinetics of the axial and radial ablation of the basemat and walls of the reactor pit, together with gas and aerosol release and composition during an MCCI. Predictions of present codes for plant behavior are presented at the end of this chapter, and the uncertainties and remaining questions are discussed.

4.3.3. R&D Approach on MCCI

MCCI is characterized by the intense coupling of many complicated phenomena, such as:

- High-temperature concrete behavior and its decomposition;
- Thermal hydraulics and heat transfer from the corium pool agitated by gas bubbles;
- Physico-chemistry of the multicomponent melt as influenced by changes in material composition through admixture of molten concrete, oxidation of metals, or onset of solidification;
- Partial solidification of the melt and the behavior of slurries or interfacial crusts;
- Exothermal oxidation of metallic species.

Consequently, identification, understanding, and quantification of these phenomena with respect to their importance for accident analysis requires an iterative process of integral experiments, separate effect tests (see Section 4.3.4), and development of models and their validation (see Section 4.3.5). Finally, their integration into sophisticated computer codes should allow predictions for the course of the accident (see Section 4.3.7).

4.3.4. Experimental Results

This section presents some of the most signficant experimental data obtained on MCCI in the last 20 years.

Behavior of Concrete under Extreme Temperatures

Concrete is a complex mixture of cement, water, and aggregates that, for most plants, consists of variable proportions of silica (SiO_2) and limestone ($CaCO_3$). Under extreme temperatures of the corium melt, which are between about 2500 and 1500 K, it will decompose into an oxidic, lava-like melt and gases, namely, steam and carbon dioxide [93]. The gases bubbling through the corium pool partly oxidize the metals in the pool and produce hydrogen and carbon monoxide, accumulating in the containment. The liquid decomposition products of concrete dissolve in the oxidic phase of corium and increase its volume.

As described by several authors ([94], [95], [96]), decomposition of concrete during heat-up starts with evaporation of physically bound water around 100°C. Dehydration of chemically bound water occurs up to 550°C. Decarbonation of $CaCO_3$ ($CaCO_3 \Rightarrow CaO + CO_2$) from the cement and carbonate aggregates occurs approximately between 700°C and 900°C. Liquid phases start to form between 1100°C and 1450°C. Figure 4.37 illustrates the gas release from siliceous concrete during heat-up as measured by the weight loss of a specimen.

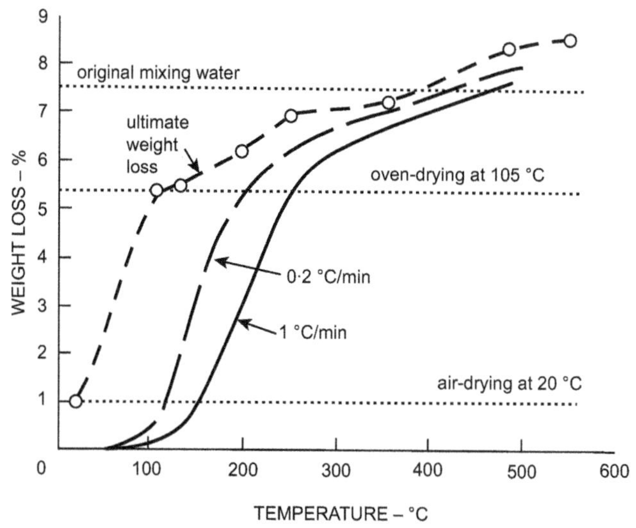

FIGURE 4.37 Decomposition of Siliceous Concrete during Heat-up.

The amount of water that is present in concrete as chemically bound, physically bound, and free water is typically around 7 weight-% or 217 m^3 STP m^3 per m^3 of concrete. The corresponding steam volume that will percolate through the corium melt at some 2000 K is quite high, namely, about 1000 m^3 per 1 m^3 of decomposed concrete. This highlights the strong effect of gases on agitation and heat transfer processes in the melt.

The decomposition enthalpy to bring concrete from room temperature to total melting is generally between 1.8 and 2.5 MJ/kg and includes the different chemical reaction enthalpies. The range of solidus and liquidus temperatures for several types of concretes deduced from available data (in particular Roche et al [97]) is indicated on Table 4.5.

It is essential to note that decomposition of concrete by the thermal attack of the high-temperature corium melt is reasonably described as a melting process

TABLE 4.5 Solidus and Liquidus Temperatures for Different Types of Concrete (Note that a wide range of concrete composition exists, which may affect their properties.)

	Siliceous	Lime-siliceous	Limestone
Solidus Temperature	~1100°C	~1100°C	~1200°C
Liquidus Temperature	1250/1450°C	1250/1450°C	~2300°C

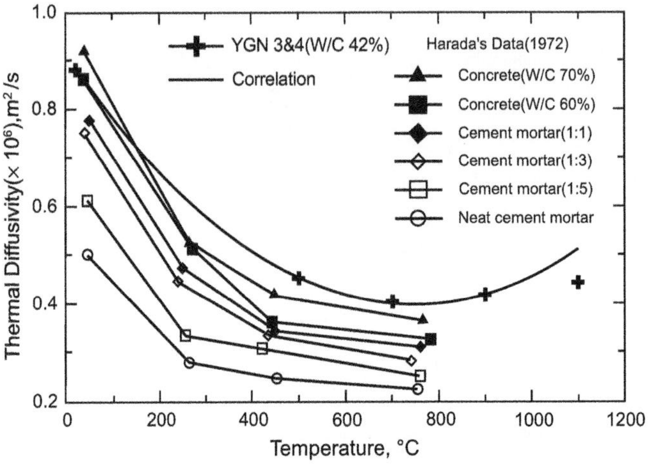

FIGURE 4.38 Typical concrete thermal diffusivity [98] data from YoungGwang Nuclear plant concrete, compared to literature data. Correlation: $\alpha = 9.1639\ 10^{-7}\ T(°C)^2 - 0.00136982\ T + .909062$ (in $10^{-6}\ m^2/s$).

with simultaneous gas release. Spalling of concrete plays a minor role. However, transfer of unmelted aggregates (silica rich) into the melt was observed as a result of the deterioration of the cement concrete matrix in some VULCANO real material MCCI experiments [118].

During heat-up of the concrete, its thermal properties change, mainly due to the loss of water [93] [98]. Figure 4.38 shows the thermal diffusivity versus temperature for a typical basaltic (amorphous silica aggregates) concrete. Water content mainly affects the thermal conductivity. Its reduction at higher temperatures reduces the ability of long-term heat conduction through the structural concrete. Concrete heat conduction is low enough to have a negligible impact on MCCI in case of thick concrete walls (thickness around 1 m or more), which is true in the reactor case for the major part of the MCCI phase. Furthermore, above ~400°C, the concrete becomes brittle and may form particles, which leads to a further decrease of the thermal conductivity. However, concrete heat conduction plays a significant role either at the very beginning of MCCI in the reactor case due to transient effects, thus delaying the concrete ablation onset, or in experiments of smaller scale with a much thinner wall, inducing significant heat losses across the concrete walls.

Concrete density ranges between 2 and 2.5 g/cm^3 and it decreases with temperature. The maximum thermal elongation can reach 1.5%. Concrete-specific heat ranges between 900 and 1400 J.kg^{-1}.K^{-1}. The Eurocode related to the effects of fire on concrete structures [99] recommends values of concrete physical and mechanical properties against temperature.

Early Large-scale Experiments

Table 4.6 summarizes the major experiments that were conducted in the early days of MCCI research, mostly using simulant oxides instead of UO_2. Many experiments were conducted in the years 1970–1990 at Sandia Nat. Lab.—SS, LS, BURN, COIL, TURC, SWISS, SURC, FRAG, HOT SOLID, WITCH, GHOST— and will not be detailed here ([170], [171], [172], [173], [174]). These tests mainly analyzed the behavior of concrete during the ablation process, the ablation kinetics, and also the release of fission products.

Most of the experiments use an electrical heating technique to simulate the internal decay heat and to allow studies of long-term behavior—either induction heating when a metal melt is present, or direct Joule heating in case of pure oxidic melts (see the sketch of ACE facility in Figure 4.39). Except for the BETA experiments, all of these were devoted to 1D geometries. An overview of these tests and their use in code development and application is given in [100]. The 1D tests showed what was expected: the ablation rate is, in the first order, the ratio of the heat flux to the enthalpy needed to heat and melt a unit volume of concrete. All experiments show the strong release of gases bubbling through the melt and generating a well-stirred melt pool. After a first intense period of interaction upon melt release, controlled by the initial melt superheat and the chemical oxidation processes, the typical long time erosion rates are several centimeters per hour. The 2D BETA experiments showed pronounced axial erosion by the metal phase, depending on the power generated in the melt. Figure 4.40 shows the section of the siliceous BETA V2.1 crucible and that of the BETA V3.2 crucible, which was made of pure limestone according to American specifications. The large amount of CO_2 released during the lime-burning process dominated the gas release. The CaO has been found in a zone some millimeters ahead of the melting concrete interface. The crucible showed clear erosion by concrete spalling. Lateral erosion in the central part of the crucible is much more pronounced than in comparable tests with siliceous concrete.

This result must not be directly applied to the oxidic melt behavior but requires careful consideration of the fact that decay heat was provided only in the metal (at the opposite of the reactor case). Moreover the adequate transport properties of the oxide must be evaluated using adequate models for fluids of high Prandtl number [108] and taking into account further experiments with oxides, which are available nowadays.

The liquid decomposition products accumulate in the oxide phase of the corium melt and add mainly SiO_2 and CaO. This extends the solidification temperature interval since the solidus temperature of the oxide mixture drops down to about 1100°C.

The gas release is higher for limestone concrete—because of the decomposition of $CaCO_3$—than for siliceous concrete and is linked to the ablation rate. A substantial fraction of carbon dioxide and steam percolating through the corium melt oxidizes the metal fraction in the melt basically in the

to present at least after some transient phase an anisotropic ablation pattern (more efficient ablation of the sidewalls compared to downwards ablation). It is the case both in the CCI3 experiment with a symmetrical square cavity geometry with two lateral ablatable walls and the CCI5 experiment (except in the first part of the test) with a similar geometry but a larger transversal dimension and a single lateral ablatable wall. At the opposite, the tests with limestone-rich concrete (e.g., CCI2) show a more isotropic ablation (see Figure 4.44), except perhaps in the first part of the test. The same type of influence of concrete nature on the ablation pattern was observed in VBU4, VBU5, and VBU6 VULCANO experiments. Besides recent VULCANO tests, VBES-U2 [118] with a "cement clinker" concrete similar to a limestone concrete but with a low CO_2 content and VB-U7 with iron oxide EPR concrete) also leads to a prevailing lateral ablation; these tests permitted to get a more precise view of the impact of concrete nature. Indeed, according to these recent results, the larger lateral ablation would not be caused by the low-gas content of siliceous concrete, facilitating the buildup of a resistive accumulation or crust at the cavity bottom, but rather by the existence of stable aggregates still solid at the ablation temperature, leading to a preferential digging or slumping and subsequent ablation of lateral concrete walls. This picture remains to be confirmed by additional experiments and further analysis.

In case of water flooding, COTELS tests (with siliceous concrete) indicated that water penetration in the degraded concrete walls limited sideward ablation and, thus, favored downward ablation. A satisfactory understanding of the role of crusts, their stability, removal, and potential regeneration is, however, still lacking. Gas release from the decomposing concrete as well as its mechanical stability—both are quite different for siliceous and limestone type of concrete—has been considered to influence crust formation, with different behavior for horizontal or inclined concrete walls. The concrete melting temperature (Table 4.5) may also influence crust stabilities and the observed anisotropy.

In the long-term accident situation, the corium melt is thought to form a layered configuration in which the denser steel layer is located beneath the oxide melt. This situation is investigated in the COMET-L experiments (Figure 4.47) [114] [115] [116]—however, with simulated high-temperature corium melts and an imperfect simulation of the distribution of the decay power, since power is injected only into the metal layer instead of both oxidic and metallic layers in the reactor case. It is observed that the lower steel melt forms a crust on the concrete bottom. The crust is partially removed from time to time with periods of higher and lower gas release and varying melt agitation. Late flooding of the upper (oxidic) melt surface is also investigated to quantify the role of flooding in concrete erosion.

In the VULCANO facility, a series of experiments have been started to study MCCI with prototypic oxidic and metallic corium in stratified configurations, in which sustained heating is prototypically applied for more than 90%

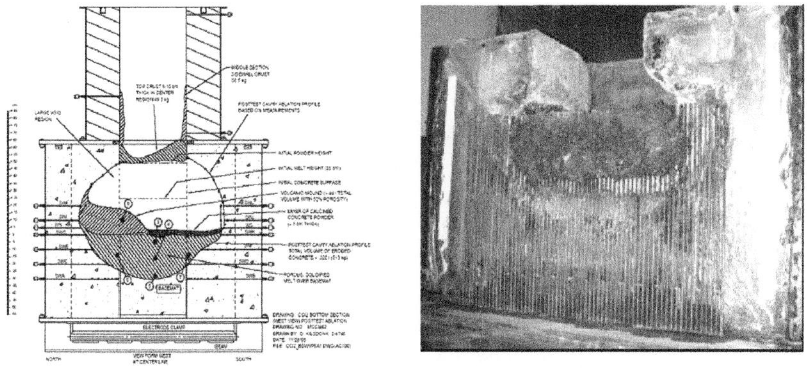

FIGURE 4.44 CCI experiments—left: sketch of CCI2 test with limestone-rich concrete, isotropic ablation—right: posttest photograph of CCI3 cavity in siliceous concrete, mostly sideward erosion.

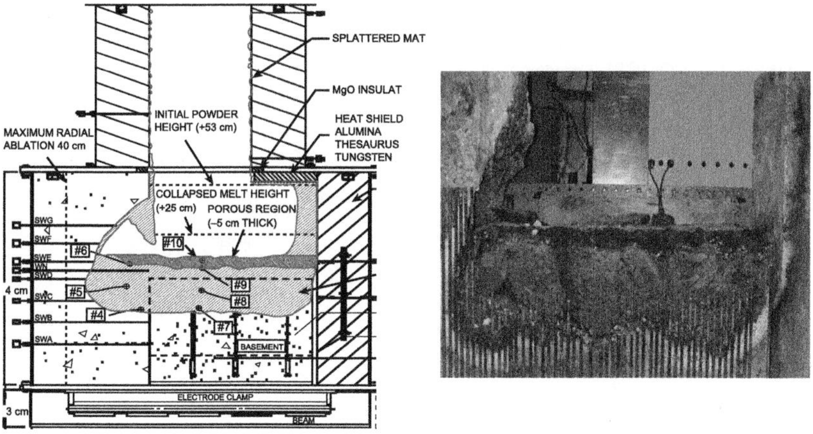

FIGURE 4.45 CCI experiments—left: sketch of larger scale CCI5 test siliceous concrete, with only one ablatable lateral wall—right: posttest photograph of the CCI5 cavity, mostly sideward erosion.

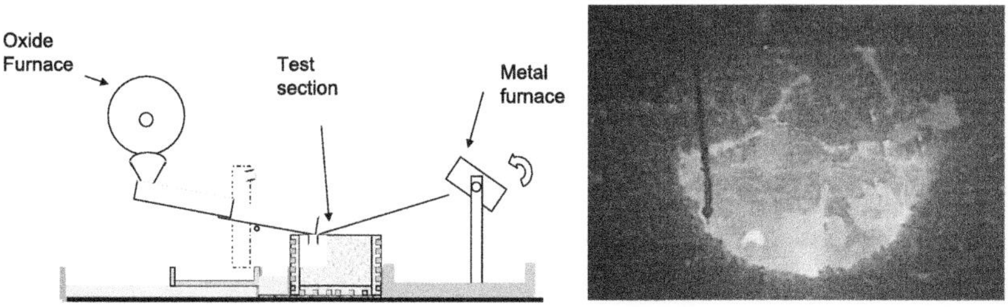

FIGURE 4.46 VULCANO experiments—left: sketch of the facility with furnace, pouring device and test section—right: top view of bubbling corium pool during VB-U4 test.

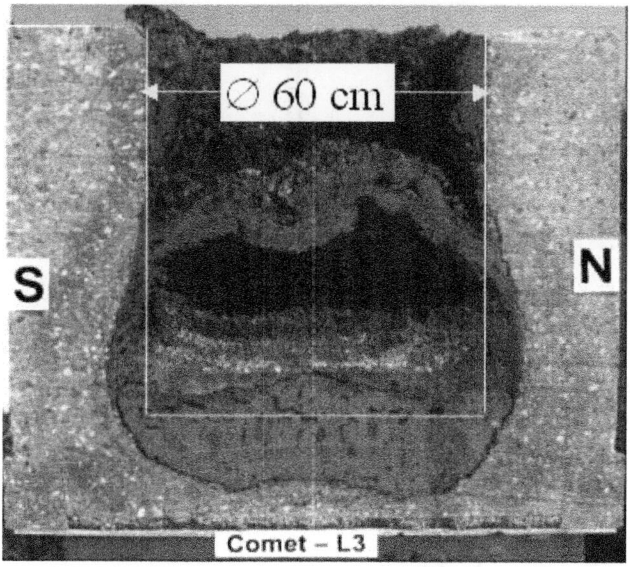

FIGURE 4.47 COMET-L3 eroded cavity—white lines show the initial cavity: voiding of cavity occurred by melt eruption during the quenching phase.

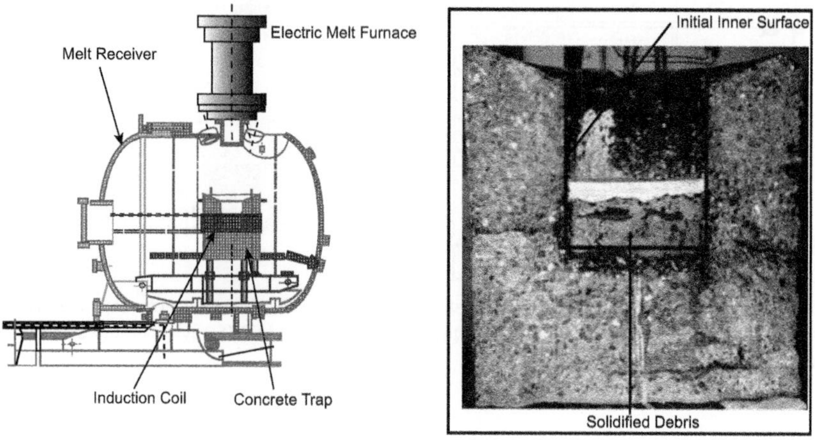

FIGURE 4.48 COTELS—left: scheme of the facility—right: cut of the corium and concrete after test.

to the lighter oxides [109] [110]. First, oxide-metal VULCANO experiments did not demonstrate the pool stratification into two separate oxide and metal layers except at the end of the VBSU1 test with limestone-sand concrete where most of the metal was oxidized (see Figure 4.49). However a segregation (and not a real stratification) between oxidic and metallic phases was observed at some parts of the pool in the VBS-U2 and VBS-U3 tests with siliceous

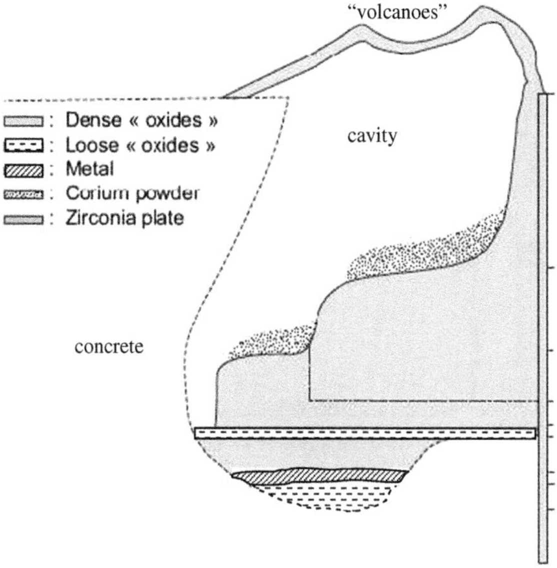

FIGURE 4.49 Posttest view of the VBS-U1 test section: the large void at the top is evidence of the large corium swelling during the experiment.

concrete, where the oxidation of metals remains limited. Further work is underway to understand the oxide-metal phenomenology.

Separate Effects Experiments

Many of the correlations and models used in the computer codes were deduced from results of separate effects experiments that study the specific phenomena in analytical tests. Based on physical arguments, such as similarity analysis or detailed modeling, and mostly checked by large-scale experiments described above, these results are transferred to the situation and materials, which exist during the anticipated reactor accident. This section summarizes experiments performed to study heat transfer between volumetric heated pool and porous wall, heat transfer between immiscible liquids, mixing and stratification of immiscible liquids, and pool/concrete interface temperature.

Heat Transfer between a Heated Pool and a Porous Wall with Gas Injection

Numerous separate effects tests [122] [147] [150] seeking to determine the heat exchange coefficient between a liquid pool and a porous wall with gas injection have been carried out during the past 30 years. An analysis of the results of these tests shows that the physical properties of the liquids used are often close to those of water, and that the available data mainly relate to horizontal walls. In the case of water, the measurements carried out as part of the various experimental

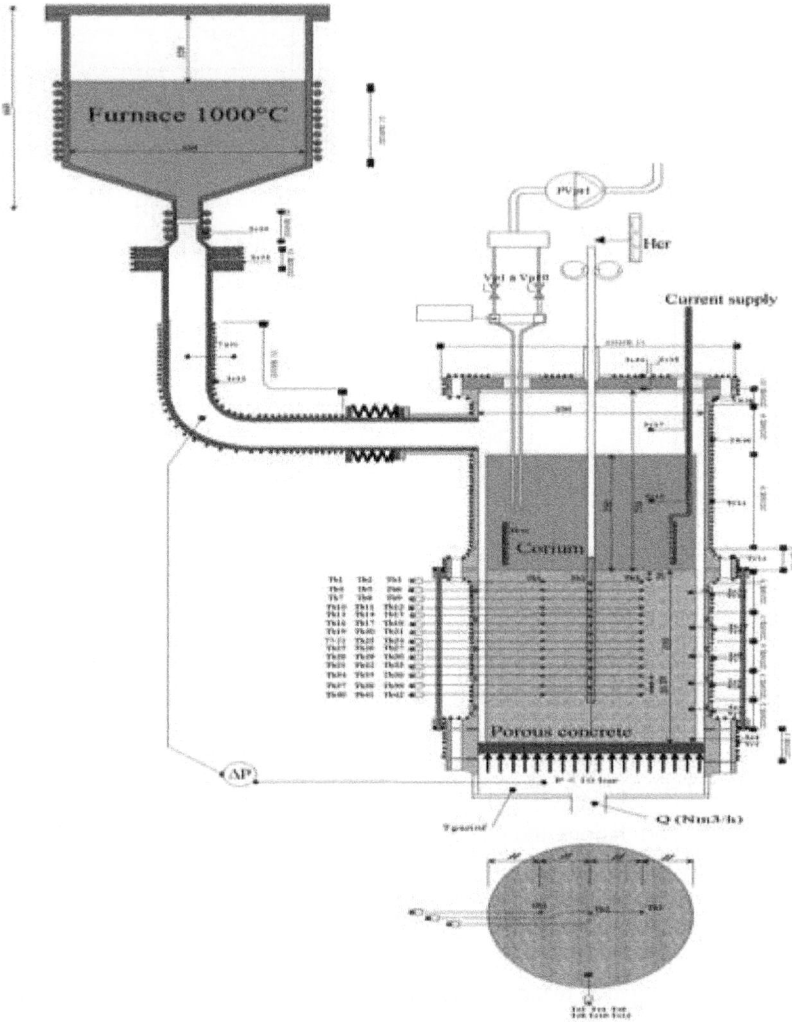

FIGURE 4.50 Scheme of the 1D ARTEMIS facility: because of the zero gas content in the stimulant concrete, gas is injected through the porous concrete.

programmes all give very similar results for a given superficial gas velocity. Data relating to viscous liquids (comparable with concrete-enriched corium) and vertical walls are relatively rare [123]. It would, however, appear that the heat exchange coefficients for vertical and horizontal walls are broadly similar when water is used [123] [124]. These experiments led to the development of correlations reported in Table 4.8. They are expressed in the form Nu = f (Re, Pr) or equivalently St = f (Re, Pr, Fr), or Nu = f (Ra, Pr), according to whether the heat transfer is assumed to be controlled by forced convection due to the gases from the

TABLE 4.8

Name [Reference]	Correlation	
BALI	$Nu = 19.67\left(\frac{\rho_{melt} j_G^3}{\mu_{melt} \cdot g}\right)^{0.136} \times Pr_{melt}^{-0.22}$ downwards	$Nu = 24.75\left(\frac{\rho_{melt} j_G^3}{\mu_{melt} \cdot g}\right)^{0.073} \times Pr_{melt}^{-0.29}$ upwards
Blottner	$Nu = \left[\frac{gL^3}{\nu_{melt}\alpha_{D,melt}}(0.00274\beta\Delta T + 0.4\alpha_{void})\right]^{1/3}$	
Deckwer	$St = 0.1(ReFrPr^2)^{-0.25}$	
Kutateladze-Malenkov	$\frac{3.10^{-5}}{\sqrt{\frac{j_G \cdot \mu_{melt}}{\sigma}}}\left(\frac{c_{p,melt}P.j_G}{\lambda_{melt} \cdot g}\right)^{2/3}$ for $4.10^{-4} < j_G \cdot \mu_{melt}/\sigma_{melt} < 10^{-2}$	$3.10^{-4}\left(\frac{c_{p,melt}Pj_G}{\lambda_{melt} \cdot g}\right)^{2/3}$ for $j_G \cdot \mu_{melt}/\sigma_{melt} < 10^{-2}$

Where:
$c_{p,melt}$ is the melt heat capacity,
j_G is the superficial gas velocity,
g is the gravity,
L is the pool size,
Nu is the Nusselt number,
P is the pressure,
Pr is the Prandtl number,
St = Nu/Re.Pr is the Stanton number,
$\alpha_{D,melt}$ is the heat diffusivity in the melt,
α_{void} is the void fraction in the melt,
β is the expansion coefficient of the melt,
λ_{melt} is the expansion coefficient of the melt,
ρ_{melt} is the mass density of the melt,
ρ_{melt} is the mass density of the melt,
μ_{melt} is the dynamic viscosity of the melt,
ν_{melt} is the cinematic viscosity of the melt,
ΔT is the temperature difference across the corium pool

decomposition of the concrete, or by free convection. The limited range of physical properties of the fluids used in the tests employing simulants makes it difficult to evaluate these models in spite of the fact that they produce results that are in relatively good quantitative agreement in these tests. Although these models give identical results on the basis of available experimental data, there is a wide variation in the results of the correlations when they are used with parameters representative of the reactor case [125]. This scattering probably indicates that some of the models do not fully reflect all of the physical phenomena and the relative weightings of the various parameters. For this reason, a model [149] based on a more phenomenological approach was recently suggested. This model gives satisfactory results for the available experimental data (horizontal and vertical walls) but requires additional validation for viscous liquids [124], especially for vertical gas injection.

Heat Transfer between Two Immiscible Liquids with Gas Bubbling

Results related to the heat exchange coefficients between two immiscible liquids through which a gas is passing (the same configuration as in a stratified corium melt) are even rarer (Greene [127], Werle [145]). Analysis of the results of these tests reveals a wide scattering in the measurements (by a factor of around 5), a lack of relevance of the systems being studied (absence of solidification at the interface layer), a wide gap between simulant fluid physical properties and those of the reactor case, in particular for viscosity and surface tension values, and required improvements of experimental procedures.

Following the experimental work, Greene and Irvine [144] proposed a correlation to calculate the heat transfer coefficient between two immiscible liquids with gas bubbling. The main feature of this correlation is that it is independent of the viscosity of both liquids. When used in reactor calculations, it might lead to a prevailing axial ablation in a stratified configuration with metal at the bottom. Recent ABI simulant experiments performed at CEA Grenoble confirm the high value of the oxide/metal convective heat transfer coefficient and agree within a factor of 3 to 5 with Greene's data. Conclusions to be drawn from this short review of experimental data and models are the following: the oxide/metal convective heat transfer coefficient is surely much larger than that at the oxidic corium/concrete interface enhancing possibly the axial ablation kinetics if a stratified configuration with metal at the bottom is maintained [156]. Therefore, even if no fully validated correlation is available for describing the oxide/metal convective heat transfer in the reactor case, sufficiently conservative applications to the reactor case can be performed using a formulation giving an upper bound of this convective heat transfer.

Interface Temperature

The ARTEMIS program [121] is the only program so far to have determined the temperature at the boundaries of a corium pool during MCCI. The program made use of simulants, including salts (LiCl and $BaCl_2$) with phase diagrams similar to those of reactor constituents, to investigate the relationships between physico-chemistry and thermohydraulics when solidification starts to occur. Tests carried out using one-dimensional configurations (horizontal corium–concrete interface) have confirmed that, under test conditions believed to be representative of a reactor, the temperature at the free interface of the crust was close to the liquidus temperature of the pool but showed some possible deviation from the liquidus, and that the pool temperature decreased in accordance with the decrease in liquidus temperature as the concrete content of the pool increased. The results of these experiments are only in partial agreement with a model assuming a pool/crust thermodynamic equilibrium and show a deviation from the equilibrium due to the buildup of a possible thick solid particle accumulation [157] or layer [158] at the corium/concrete interface in case of a high gas superficial velocity or fast ablation. The second phase of the program (ARTEMIS 2D) expands the studies to include multidimensional configurations.

Mixing and Stratification of Immiscible Liquids with Gas Bubbling

Several studies of the mixing and stratification of immiscible liquids with gas bubbling have been carried out with the aim of predicting the configurations (mixed or stratified) of a corium pool during MCCI. Most of the experimental work on this topic has been carried out using simulants [126] [129] [130], and was restricted to hydrodynamic considerations only (no effects of crusting at the interface). This work aimed to determine the mixing and separation thresholds in terms of the superficial gas velocity (or void ratio) as a function of the difference in densities between the two liquids. A review of the data reported in [129] indicates a degree of variance, which can be marked between the results of the various tests. This is partly due to the different physical properties of the liquids.

The results of tests using simulants have been used as a basis for the development of experimental correlations [129] in which the variation in the results is similar to that in the measurements. Greene [127] has developed a more mechanistic approach based on the modeling of the entrainment of a heavy liquid into a lighter one in the wake of bubbles but not leading to a usable stratification criterion. These models and correlations have not been validated against tests with prototypic materials; they are very conservative because the stratification criterion is chosen equal to the threshold for reaching the metal/oxide complete mixing and ignore a possible influence of the ratio of metal to oxide layer volumes. Improvement of these models and of their validation would be useful because their application for reactor conditions determines the evolution of pool configuration and can impact strongly on the axial ablation kinetics during MCCI [156].

4.3.5. Overview of Existing Models and Codes

General Code Aspects

The early MCCI experiments clearly demonstrated that the heat transfer between melt and concrete controls its erosion [128]. The heat transfer mechanisms are nevertheless strongly dependent on the (varying) composition of the melt and concrete.

The heat flux φ_i through a given interface i is given by:

$$\varphi_i = h_i(T_{pool} - T_i) \quad (20)$$

where h_i is the convective heat transfer coefficient (mixed convection due to the sparging gases) and T_i is the interface temperature for the considered boundary.

The heat balance on the ablating concrete boundary gives:

$$\rho.\mathrm{v}_{\mathrm{abl}}.\Delta \mathrm{H}_{\mathrm{concrete}} = \varphi_{\mathrm{i}} \tag{21}$$

where $\Delta H_{concrete}$ is the decomposition enthalpy necessary to heat and melt a unit mass of concrete, ρ the mass density of concrete, and v_{abl} the local ablation velocity on the interface *i*. It must be noted that this mass must be divided into a volatile part (especially important in presence of limestone, $CaCO_3$) and a part that will form the corium–concrete mixture.

The closure of this system of equations is given by the global conservation of energy:

$$\sum \varphi_i \cdot S_i = \dot{Q} \tag{22}$$

Basically in all models, metallic and ceramic phases are to be distinguished, with a large variety of composition. In the early phases of MCCI, mixtures of uranium and zirconium dioxides are slightly heavier than the metallic phases (mainly molten steel and zircaloy). As concrete decomposition products will mix with the oxide phase of corium, it will become lighter during the course of the accident, than the metallic phase. Thus, it may be expected to have an emulsion of oxides and metals in the early time and, later, a transition to a layered configuration with metal at the bottom. The threshold between these two configurations may be determined from the BALISE correlation [129]: the pool is an emulsion if the gas superficial velocity J_g (in cm/s) satisfies

$$\mathrm{J_g} \geq \mathrm{bHS}\,\frac{\rho_{\mathrm{H}} - \rho_{\mathrm{L}}}{\rho_{\mathrm{L}}} \tag{23}$$

where $\mathrm{bHS} = 0.054\,\mathrm{m/s}$ and ρ_H and are, respectively, the densities of the heavy and light phase of the melt-concrete mixture or if, as proposed by Epstein et al. [130], the density of the mixture of the two liquid phases and bubbles is lower than the density of the light phase:

$$\frac{\rho_H(1-\alpha)}{1+V_l/V_H}\left[1+\frac{\rho_l V_l}{\rho_H V_H}\right] \leq \rho_l. \tag{24}$$

The first computer codes to describe the corium-concrete interaction were CORCON (developed by Sandia National Labs [131] from 1981 to 1993) and WECHSL, the development of which started at FZK in 1981 [134]. WECHSL-Mod 3 was released in 1995 [135]. Other codes, TOLBIAC-ICB [140], COSACO [89] [141], and ASTEC/MEDICIS [143], were developed more recently.

These codes aim at modeling the physical and chemical phenomena governing the molten core-concrete interaction in a severe reactor accident, when the molten core has penetrated the pressure vessel. Figure 4.51 shows the main

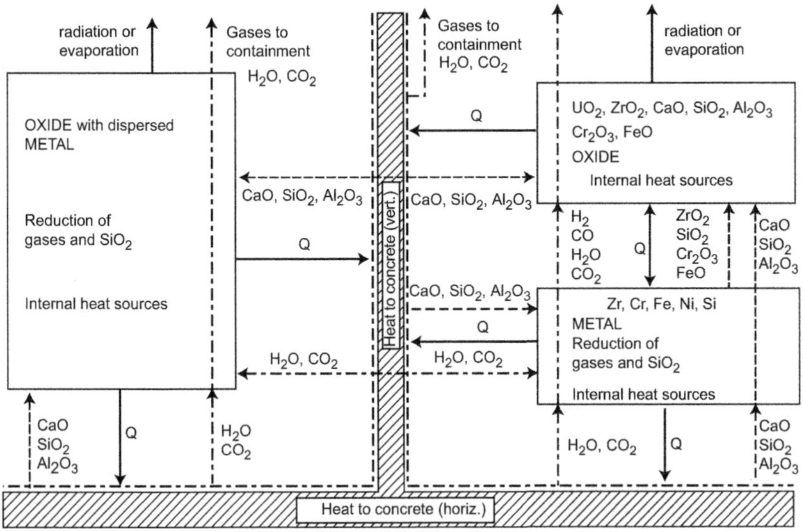

FIGURE 4.51 Sketch of the main mass and heat transfers described by a MCCI code.

mass and heat transfers that are modeled. The left side shows models for a homogeneous oxide melt into which the metal phase is homogeneously dispersed in the form of droplets, whereas the right side shows the layered melt configuration with the metal layer at the bottom, overlaid by the oxidic corium layer in the initially cylindrical concrete cavity.

Figure 4.52 gives a general picture of the corium pool description performed by MCCI codes. The metallic melt may contain Zr, Cr, Fe, Ni, and Si. The oxide layer is composed of UO_2, ZrO_2, CaO, SiO_2, Al_2O_3, Cr_2O_3, and FeO. Internal energy can be produced by decay heat or by exothermic chemical reactions. The mass and composition of the melt change as liquid concrete is

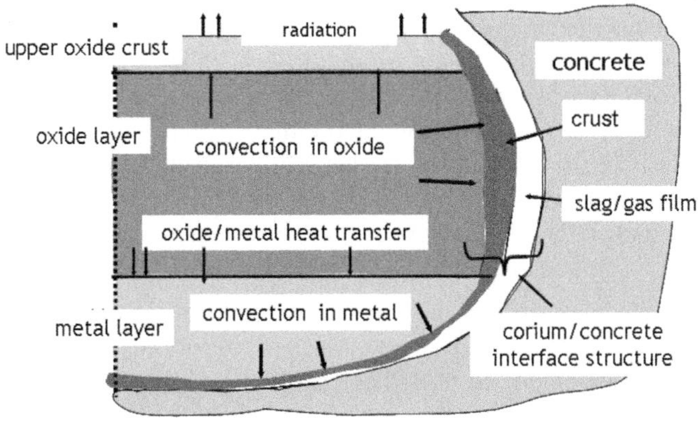

FIGURE 4.52 Sketch of the corium pool described by a MCCI code.

mixed into the oxidic corium melt and as the chemical oxidation of some of the metals takes place. Energy is lost to the melting concrete and to the upper containment by thermal radiation or by evaporation of sump water, possibly flooding the surface of the melt.

During cool-down of the melt, crust formation at the pool/concrete interfaces and at the upper pool interface is modeled. Crusts are assumed to be permeable to gases. The codes perform calculations from the time of initial contact of a hot molten pool until long-term basemat erosion over several days with the possibility of basemat penetration.

Differences between MCCI codes concern used convective heat transfer correlations taken from literature (in particular for the convective oxide/metal heat transfer), thermochemistry data for the corium, the treatment of chemical reactions, the pool stratification models, the cavity ablation model, and the pool/concrete interface model. These differences can be summed up as follows:

- Former codes (WECHSL and CORCON) use explicit descriptions of the thermochemical properties of corium/concrete mixtures, while more recent codes (TOLBIAC, COSACO, MEDICIS) compute the thermodynamic equilibrium to obtain these data;
- Oxidation reactions between metals and concrete gases are described more precisely by CORCON, COSACO, and TOLBIAC codes using thermodynamic equilibrium between condensed and gasous phases, while in WECHSL and MEDICIS they are described in a simplified way using an oxidation priority order for the successive oxidation of different metals;
- Pool stratification is described in most codes using the criterion similar to that derived from BALISE data and mentioned above; an exception is CORCON, which can describe the progressive mixing of metal and oxide layers using more mechanistic models for settling and entrainment phenomena, but without sufficient validation up to now;
- The local cavity ablation velocity is determined by the heat flux continuity (above equation (21)) in all codes, but the local ablation direction is determined differently by codes assuming a direction more or less normal to the ablation front, thus possibly leading to different cavity shapes;
- A last but not the least difference between available codes concerns the treatment of heat transfer from the bulk pool toward the pool/concrete interface and especially the detailed structure of this interface at the vicinity of the concrete ablation interface (see Figure 4.52).

Different models have been proposed to characterize the interface condition between melt and the adjacent concrete, which is possibly covered by a layer of (partly) solidifying melt. In the classical model, the pool/crust interface temperature is assumed to be at the melt solidus temperature (see Appendix 1 for more details on chemical thermodynamics and phase diagrams) or at the so-called softening temperature [135], or at the concrete melting[2] temperature [131]. At the concrete wall a "mushy zone" exists in which the concentration of solid melt particles increases as the temperature profile comes down from the bulk temperature in the melt to the interface temperature at the wall.

In another modeling proposed by Seiler and Froment [132] [133], the boundary temperature that is taken into account is the liquidus temperature of the melt pool. This second model gives very different results since the difference between the liquidus and solidus temperature in corium–concrete mixtures can be as much as 1000 K as seen in Figure 4.36). The underlying assumption is the existence of a thermodynamic quasi-steady state during solidification of a multicomponent melt: the more refractory species of the oxidic melt solidify forming a solid solution of mainly UO_2 and ZrO_2, and are assumed to deposit as crust at the concrete walls. This changes the composition of the remaining liquid corium melt as it is depleted of the refractory components. This process is called fractional crystallization and is related to phase segregation. During the ablation, the melt is enriched in species from concrete, which leads to a decrease of the liquidus temperature, and thus, to the decrease of the melt temperature.

The content of the different interface models and their implementation into the available codes are detailed hereafter in next 2 sub-chapters. Main code features are displayed on Table 4.9.

Classical Pool/Concrete Interface Models

In WECHSL code [135], the modeling of the heat transfer from the melt to the concrete is based on the following ideas. The heat transfer from the melt bulk to the concrete is characterized by processes forming boundary layers at the melt pool surface facing concrete. The most important process that governs the heat transfer phenomena is the release of large-volume fluxes of gases from the decomposing concrete. If the superficial velocity of the gases being released from the concrete is sufficiently high, a stable gas film is formed between the melt and concrete. If the superficial gas velocity drops below a limiting value, the heat transfer will be governed by a nucleate boiling type due to the discrete bubble gas release. As the melt is intensively stirred by the gases released during concrete ablation, the bulk of each layer of the melt is assumed to be isothermal. Because of the high thermal conductivity and the low viscosity of the metallic phase, the temperature drop across the boundary layer of the metal layer is very small. For an oxide melt with a high Prandtl number, $Pr > 1$, the temperature drop across the boundary layer is quite significant. Due to

---

2. Strictly speaking, the concrete, being a multicomponent mixture, does not have a defined melting temperature but a melting range (see Table 4.1). The concrete melting temperature generally corresponds to the temperature at which concrete loses its geometry, that is, to a liquid fraction of the order of 30 vol.%.

TABLE 4.9 Summary of Major Characteristics for Most Available MCCI Codes

Feature	ASTEC	TOLBIAC	COSACO	CORCON	WECHSL
Thermochemistry database	interface with thermochemistry module	coupling with thermochemistry module	coupling with thermochemistry module	correlations	correlations, input table
Pool/crust temperature	user-parameter (linear interpolation or threshold molten fraction)between Tsol and Tliq	$T_{liquidus}$	specific model between T_{sol} and T_{liq}	$T_{solidus}$	$T_{sol} \leq T_{freez} \leq T_{immob}$
h_{conv} at pool interfaces	main available correlations	main available correlations	Bali/Kutateladze	Kutateladze	
h_{conv} at oxide/ metal interface	Greene x user's factor	Bali	Bali (oxide) + Kutateladze (metal)	Greene	Werle
Configuration evolution	yes (simple criteria)	yes (simple criteria)	yes (simple criteria)	yes (detailed models)	no
Corium quenching	yes	yes detailed	no	no	no
Flexibility of interface model	large	medium	medium	low	medium

cool-down of the melt, the temperature in the melt boundary layer facing the concrete may drop far below the liquidus temperature, which characterizes the onset of solidification, because of the wide solidus-liquidus interval. The initially thin viscous layer will grow, and the solid volume fraction will increase with a further temperature decrease. The strong variation of the viscosity with temperature in the temperature boundary layers at the interfaces of the melt considerably reduces the heat transfer from the pool. A correlation proposed in [135] [136] is used for correcting the appropriate nondimensional heat transfer parameter, that is,

$$Nu = Nu_0 \left[0.645 \left(\frac{\mu_i}{\mu_b} \right)^n + 0{,}356 \right], \quad n = 2$$

Nu is the corrected Nusselt number and Nu_0 is the constant property solution. The viscosity μ_i is the viscosity at the interface temperature, while μ_b is evaluated at the bulk temperature [137] [138]. The most pronounced rheological changes with temperature occur at temperatures for which the solid volume fraction, f_{immob}, passes through the critical range of 50–80 vol.%. This temperature is defined to be the freezing temperature, T_{freez}, $T_{sol} \leq T_{freez} \leq T_{immob}$, of the oxide melt at which a stable crust starts to form. Both the crusts and the liquid melt have the same composition. The viscosities of the oxide melt below the liquidus temperature are assumed to follow the modified Pinkerton-Stevenson correlation. The viscosity measurements performed for the melt used in the KATS tests [139] are in good agreement with the modified correlation. For the oxide phase and the dispersed melt, the solidus and liquidus temperatures are determined either from a quasi-binary phase diagram or from a user input table. The solidus and liquidus temperatures of the metallic phase in WECHSL are calculated from the chromium-nickel-iron, zirconium-iron, and silicon-iron phase diagrams. Between the crust and the concrete a gas film exists. The heat transfer between the melt bulk and the inside of the crust is determined by a discrete bubble type heat transfer mechanism, with the driving temperature difference determined by the bulk temperature and the freezing temperature, T_{freez}, of the melt layer at the crust interface. A steady-state heat conduction is assumed for thin crusts, whereas in the case of thick crusts the one-dimensional transient heat conduction is modeled.

Pool/Concrete Interface Model with Thermal-Hydraulic-Physico-Chemistry Coupling

In the more recent TOLBIAC-ICB [140] and COSACO [89] [141] codes, there is an explicit coupling between thermal-hydraulic and physico-chemistry. In TOLBIAC-ICB, a crust is assumed to exist at the concrete wall; its interface temperature to the residual melt is the liquidus temperature of the melt (as derived from the quasi-steady-state hypothesis [142]). Under these circumstances, the heat transfer to the concrete is determined by processes in the pool,

For the heat transfer between two stratified layers, two competing heat transfer models have been proposed. According to Greene [144], the heat transfer at the interface between oxide and metal layer can be written as:

$$Nu = 1.95\,\mathrm{Pe}^{0.72} \text{ or } Nu = 1.95 \cdot (\rho_{melt} \cdot j_G \cdot c_{p,melt})^{0.72}$$

Where:

$c_{p,melt}$ is the melt heat capacity,
j_G is the superficial gas velocity,
Nu is the Nusselt number,
Pr is the Prandtl number,
$Pe = Re.Pr$ is the Peclet number,
λ_{melt} is the expansion coefficient of the melt,
ρ_{melt} is the mass density of the melt,

As mentioned in previous sections, this correlation provides quite different results compared to the BALI correlation [146] or Werle's correlation [145] or other correlations applicable to the pool/concrete interface and listed in Table 4.8.

Coolability Models During MCCI

Models have been built to describe the main cooling mechanisms observed in real material MCCI experiments in the case of top quenching such as mentioned above, mainly water ingression and melt eruption, the second being surely the more efficient mechanism. As far as water ingression is concerned, the extracted heat flux due to water penetration through the upper crust cracks is determined by the crust permeability itself resulting from crust cracking; a detailed model was proposed some years ago for describing this crust cracking, depending on crust mechanical properties and indirectly on the concrete fraction included within the corium and validated against SSWICS experiments [179]. The efficiency of water ingression decreases with increasing ablated concrete fraction and can be high only in the early MCCI phase. As far as melt eruption is concerned, several models were successively proposed. A first very simple model assumes that the volumetrical flow rate of ejected corium is proportional to the gas flow rate [176]; the proportionality factor (entrainment factor) can be deduced from an entrainment factor correlation deduced from simulant experiments on liquid ejection [180]. A second model (so-called PERCOLA model) describes in greater detail the hydrodynamics of the corium ejection through holes in the upper crust [181]. This model was validated against PERCOLA experiments and should permit evaluation in a more realistic way of the entrainement factor. However this supposes that features of these holes (diameter and area density), which are input data of the model, are better known. A model developed recently allows evaluation of the size and density of ejection holes in the upper crust [182]

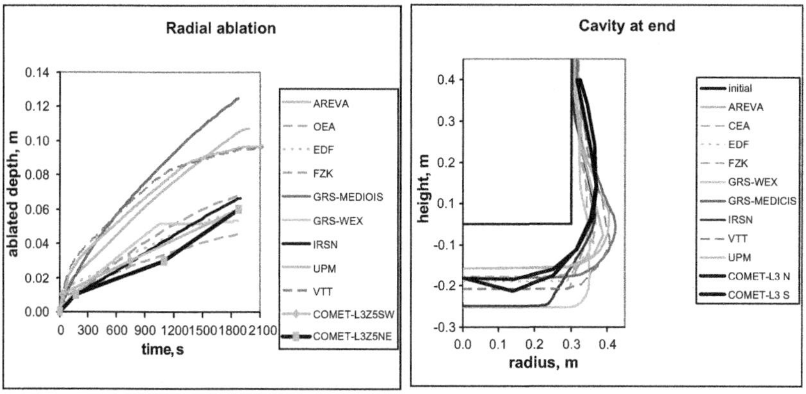

FIGURE 4.59 COMET-L3: (left) calculated and measured lateral ablation versus time—(right) calculated and measured final cavity boundaries.

however, is essential for the long- time cavity formation in the region of the metal melt.

The problem that exists is probably related to the formation of metal crusts at the concrete interface and their stability. Accumulation of molten concrete under the steel crust leads to eruptions that were observed in the experiment but are not modeled in any code. For the horizontal crust, the eruptions are probably driven by the weight of the melt. For the lateral crusts, the gas generated by the ablating concrete along the lateral interface may widen the gap between crust and concrete, thus increasing the resistance to heat transfer.

The results of this benchmark exercice performed on the most recent oxide/metal experiments available point out the discrepancies between codes and then the uncertainties in the prediction of the axial ablation in case of a stratified pool. However, no extrapolation of these results to the reactor case is possible even as a trend: indeed, the nonprototypical power injection only in the metal layer might simulate the case where the decay power injected into the oxide layer is focused on the metal layer because of the likely high oxide/metal convective heat transfer. However, the much lower ratio of metal thickness compared to axial ablation depth in the reactor case and compared to that of the experiment should lead to a lower ratio of radial ablation to an axial one because of geometrical effects. This aspect concerning the impact of stratification on 2D ablation in the reactor case will be illustrated in the next section (see Figure 4.63).

4.3.7. Code Application to Plant Analysis

WECHSL Application to Plant Analysis

The current level of agreement with experimental data would appear to be good enough to justify use of the MCCI codes for risk assessment studies, provided the uncertainties in the predicted results are taken into account. Nevertheless, plant calculations still require extrapolation beyond the existing experimental database. In particular, the treatment of long-term radial ablation by the oxidic corium fraction layered over the metallic melt fraction is not yet well established, and is expected to affect the prediction for axial ablation by the metal melt.

The various accident sequences, to be considered in safety analysis, result in a broad variation of the initial conditions at the time of RPV failure. Additionally, there are uncertainties in the knowledge of the properties of corium melts, including the viscosity and the solidus liquidus temperatures. Consequently, the initial conditions for the MCCI analysis are not well known. The uncertainties are related to:

- Time of RPV failure;
- Initial melt temperature;
- If melt stratification takes place;
- Initial melt composition;
- Melt configuration (layered/mixed);
- Slumping rates of the melt from the RPV.

The concrete decomposition products from the basemat erosion would be incorporated quickly into the heavy oxide layer, leading to a rapid reduction in the density of that layer. When the densities of metal and oxide melt fraction are nearly equal, gas flow from the decomposing concrete would lead to mixing and entrainment of the metal into the continuous oxide phase. If thereafter melt stratification takes place the oxide density continues to decrease due to the influx of light concrete oxides, the oxide layer will relocate above the metal layer.

This is the situation calculated (Figure 4.60) with the WECHSL code using its layered melt model employing Werle's correlation [145] for convective heat transfer, causing a lower heat transfer than that proposed by Greene [144]. The calculated erosion profile in a 6 m thick basemat is given for a 1200 MWe PWR that was investigated in the German Risk Study B [153]. In the early interaction phase, basemat erosion is dominated by the metal phase driven mainly by the initial melt overheat and the exothermal oxidation in the metal phase, especially that of Zr and Cr. When the internal heat source is restricted to the long-term decay power, the metal layer starts to form a crust at the concrete interface, and erosion by the metal layer slows down significantly. Subsequently, the erosion occurs mainly through the oxidic melt that releases the majority of the decay power. Propagation of the oxide is sidewards, and downward oxide erosion develops around the partly solidified metal layer, forming an annulus that propagates faster than the metal layer in the central part. The step in the axial ablation profile characterizes the transition from metal to oxide erosion. Eventually, penetration of the basemat occurs by the oxide melt and is calculated to occur after 150 hours (6.3 days).

Substantial uncertainties exist as to the timing and rate of the separation of metal and oxide melts into the layered configuration. Therefore, as a limiting

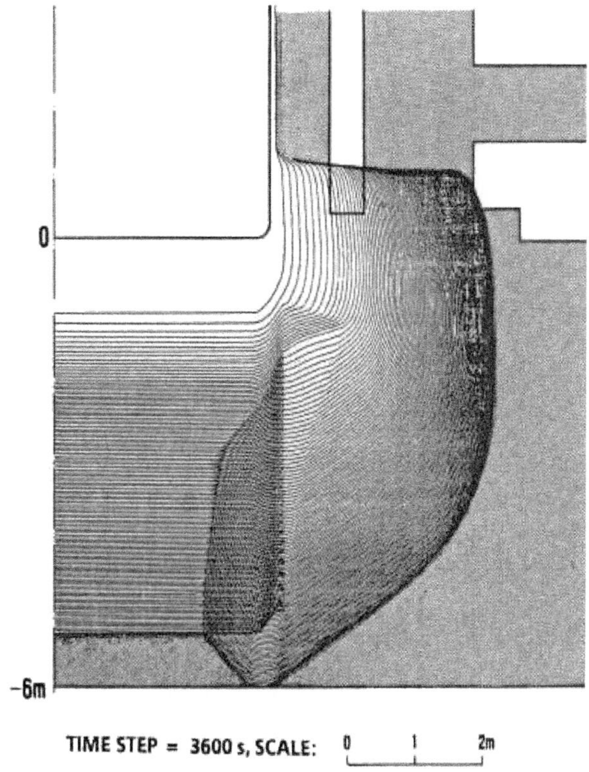

FIGURE 4.60 German Risk Study, WECHSL calculation predicts basemat penetration after 150 h.

case, calculations were performed with WECHSL and CORCON, assuming the completely mixed melt configuration, with the metal droplets homogeneously dispersed in the oxide melt. This calculation was performed within the MCCI Program of the Commission of European Communities under the Reinforced Concerted Action on Nuclear Fission Safety Research [154] and is given in Figure 4.61. In comparison with the layered melt situation (Figure 4.60), the different partition of the radial and axial heat fluxes to the concrete leads to somewhat different WECHSL predictions for the mixed configuration. Slower penetration of the 6 m thick basemat (within 188 h), higher eroded mass of concrete, and earlier sump water ingression are the main consequences of a more pronounced lateral erosion and slower radial erosion rates calculated for the mixed melt. For the same mixed melt situation, CORCON predicts more radial and less axial ablation, and therefore, a longer time to basemat penetration than WECHSL by more than a factor of 2. Moreover the CORCON version under consideration predicts a greater ratio of upward to downward heat than WECHSL, which enhances the differences between both codes. The predicted cavity shape (Figure 4.61) has disclosed a weakness in the CORCON

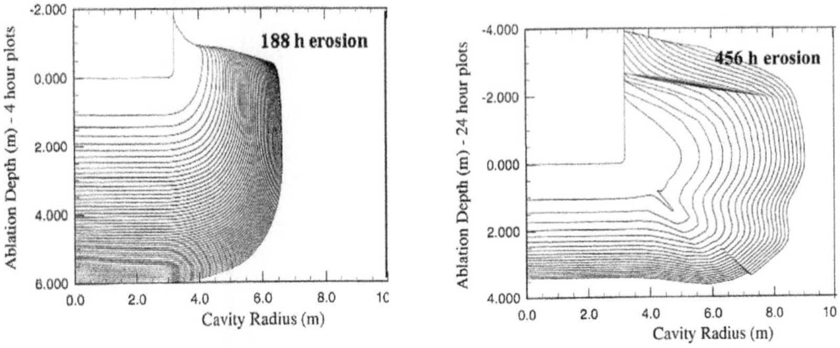

FIGURE 4.61 Cavity-calculated profiles for mixed melt with the codes WECHSL (left) and CORCON (right).

model for cavity growth. Water ingression to the melt surface that would occur after 3 to 5 hours has a minimal impact on the erosion. Both codes model a stable surface crust, and no enduring penetration of water into the corium melt is taken into account.

ASTEC Application to Plant Analysis

This section presents an application of the ASTEC code (MEDICIS module) for a typical 900 MWe PWR [143] [151] [156]. The main data used for the reactor geometry and material compositions are given in the Table 4.10.

Main assumptions, values of key physical parameters, and choice of boundary conditions are listed in the Table 4.11. The heat transfer models are the same as those in the ACE and CCI2 calculations. The heat transfer coefficient at the pool outer interface is not dependent on the interface orientation. In agreement with the interpretation work performed with MEDICIS of past MCCI experiments (see Section 4.3.6), there is a solidification temperature slightly below $T_{liquidus}$ equal to $0.8T_{liquidus} + 0.2T_{solidus}$.

The temperature of the reactor pit wall above the corium pool is set to a constant value near the melting point of steel, which is higher than the concrete

TABLE 4.10 Initial Conditions and Reactor Pit Geometry

Initial oxidic corium inventory (t)	UO_2 mass: 82, ZrO_2 mass: 19.5
Initial metallic corium inventory (t)	Zr: 4.8, Fe: 35, Ni: 4, Cr: 6
Reactor pit radius	3 m
Basemat thickness	3 m to 4 m
Concrete characteristics	siliceous concrete with 6% Fe

TABLE 4.11 Choices of Assumptions and Values of Key Parameters for Reactor Calculations

Pool configuration	homogeneous, stratified, with configuration evolution
Heat convective coefficients	BALI's correlation at the outer layer
	$h_{slag} = 1000$ W/m^2/K
	Greene's correlation [144] along the oxide/metal layer interface
Solidification temperature	from $T_{solidus}$ to $T_{liquidus}$ (γ between 1 to 0)
Concrete type	reinforced siliceous concrete with 6% Fe 'standard concrete'
Time after scram, decay power (W/kg U)	1 h: 283, 3 h: 227, 7 h: 190, 15 h: 157, 20 h: 145, 50 h: 108, 9 d: 63
Initial corium temperature	2673 K (the oxide phase is solid)
Pit wall temperature	1700 K

ablation temperature. As the temperature of concrete surface should not exceed its melting temperature, this assumption is conservative because it will underestimate the power radiated from the corium pool to the reactor pit walls. No corium quenching by flooding is taken into account in the present calculations.

Large uncertainties still exist on the pool configuration and its possible evolution. Therefore three very different scenarios concerning the pool configuration are considered.

Scenario 1: The assumption of a homogeneous oxide-dominated melt configuration maintained during the whole MCCI phase (see Figure 4.62) leads to a rather slow concrete erosion with melt-through times around 9 days for a basemat thickness of 4 m). The reason for the slow erosion in this case is the uniform heat transfer coefficient distribution along all pool interfaces, leading to spatially uniform ablation depths. This slows down the axial erosion because of the large ablated concrete volume, as it appears in Figure 4.62.

Scenario 2: The assumption of a fixed stratified configuration used as a very conservative boundary case gives faster concrete erosion, with melt-through times ranging between 14 hours only and 3.4 days, depending on the choice of solidification temperature and on the basemat thickness. The reason for this configuration's faster erosion is the high heat transfer coefficient at the oxide/metal layer interface compared to that along the oxide layer/concrete interface (see Table 4.10). This leads to focusing the heat toward the metal layer; as the

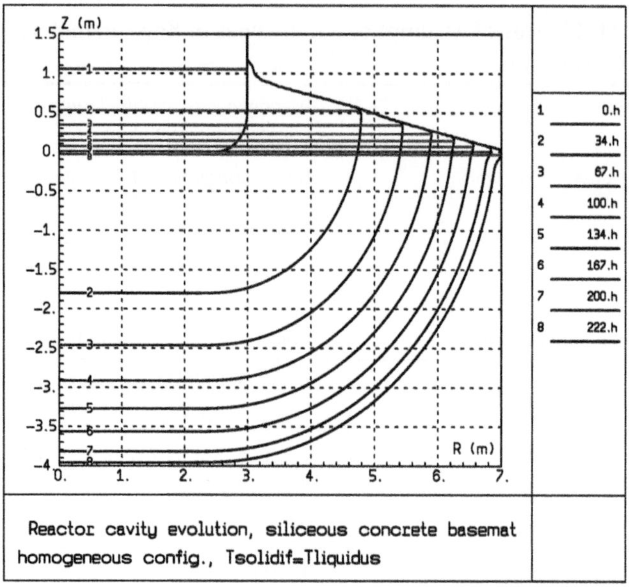

FIGURE 4.62 Cavity erosion with homogeneous configuration; melt-through at 222h (9.2 days).

thickness of the metal layer is rather thin compared to the axial ablation depth, the effective lateral heat flux from the metal layer to the lateral concrete wall is low because of the short contact between this metal layer and the lateral concrete wall, and consequently most of the decay heat is focused on the bottom pool interface, thus leading to an axial erosion much faster than the lateral (see Figure 4.63).

Scenario 3: Finally, a more realistic scenario is assumed with four successive phases (see Figure 4.64). In this scenario, the pool configuration is initially stratified with the oxide layer below the metal layer because of the higher initial oxide density, followed by a homogeneous pool. Then, the pool is again stratified but with the metal layer below, and finally again homogeneous after disappearance of the metal layer due to oxidation. The subsequent evolution of the pool configuration is evaluated using pool configuration switch criteria consistent with BALISE experiments [129]. The consequence of modeling this sequence of pool configurations is the large delay of the predicted melt-through time by more than 24 hours, compared to that obtained in the case of a steady stratified metal/oxide configuration (compare Figure 4.64 and Figure 4.63).

The results obtained with the more realistic pool configuration scenario are still very conservative because pessimistic boundary conditions are chosen (high wall temperature around the steel-melting temperature, total core inventory in reactor pit at initial time). Moreover, the BALISE stratification criterion is too conservative because it corresponds to the complete oxide/metal mixing occurrence and also ignores the decrease of the superficial gas velocity

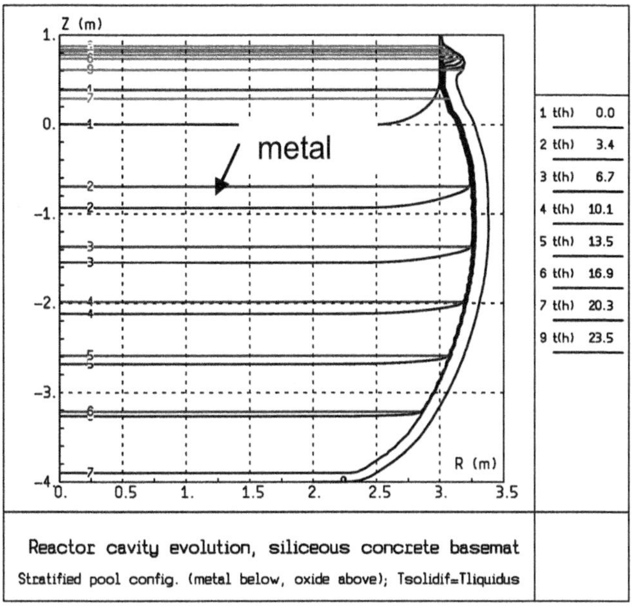

FIGURE 4.63 Reactor cavity erosion with stratified configuration (melt-through: 23.5h).

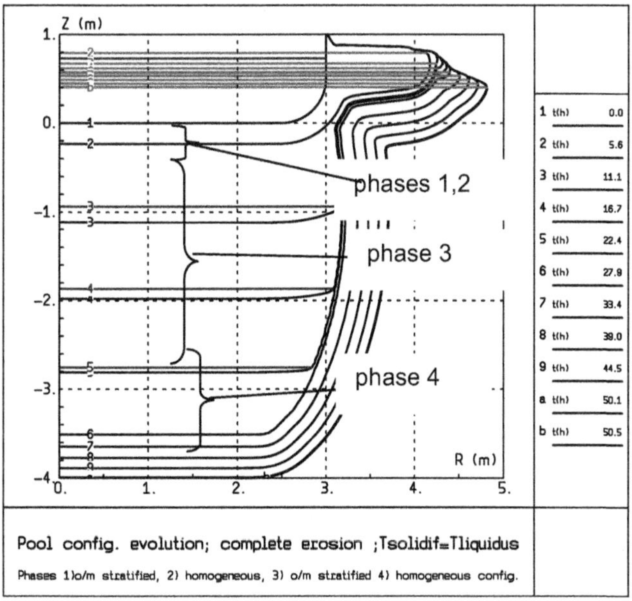

FIGURE 4.64 Case with pool configuration evolution in four phases until basemat melt-through at 50.5h.

threshold with decreasing ratio of metal to oxide volume and corium quenching is ignored. Furthermore, the heat transfer at the pool/concrete interface is assumed to be independent of the interface orientation for oxide melts. However, in the case of a siliceous concrete and for a homogeneous oxide pool, the heat transfer might be smaller at the bottom interface than at the lateral one according to some recent CCI-OECD results (CCI3 test [112]); a model taking into account such results would lead to further delay the axial ablation compared to the radial one. However the observed behaviour in CCI experiments is not yet well understood.

Less pessimistic results (i.e., with a slower axial ablation) are obtained in the case of limestone-common sand concrete because of the higher superficial gas velocity due to the higher concrete gas content reducing the time duration of the stable stratified configuration phase or even suppressing it, and of the higher ablation enthalpy causing a slower ablation velocity.

The large axial ablation results partially from the use of Greene's correlation for the heat transfer over the horizontal metal-oxide interface. Some authors criticize the use of this correlation. For example, in the TOLBIAC-ICB code, it is currently recommended that the same correlation be used for the heat transfer at each interface, including the metal-oxide interface, because of uncertainties in the extrapolation of the different existing correlations to the reactor case [155]. However, there is no commonly agreed reason for using the same heat transfer correlations for these two different interfaces. Indeed, gas bubbling through the interface probably induces a local mixing of metal droplets with the overlaid oxide pool and hence different types of efficient heat transfer. Moreover recent analytical experiments on the oxide/metal heat transfer confirmed the high level of oxide/metal heat transfer as mentioned above [156].

Recent studies [156] also showed that the axial ablation kinetics remains fast even if the oxide/metal heat transfer is reduced by a factor of 5. Consequently, the impact of the uncertainty on this heat transfer is rather moderate, provided that the oxide/metal heat transfer remains much larger than the convective heat transfer from the bulk oxide layer to the lateral concrete interface, which is likely. The same studies highlighted the influence of the stratification criterion, which appears to be the major uncertainty for MCCI in a stratified pool configuration, as shown in Figure 4.65.

A more realistic stratification criterion might be, for example, to use a twice lower threshold superficial gas velocity (corresponding to the entrainment threshold from the bottom layer in BALISE experiments) combined with a minimal metal thickness of 5 cm instead of 1 cm permitting the stratification (in order to maintain a separate metal layer only if it is thicker than a few gas bubbles). This stratification criterion would lead to a melt-through of five days (see Figure 4.65) instead of around two days in the reference case mentioned above (see Figure 4.64).

The reactor benchmark exercise performed in the frame of the European SARNET network [161] permitted an evaluation of the discrepancies between

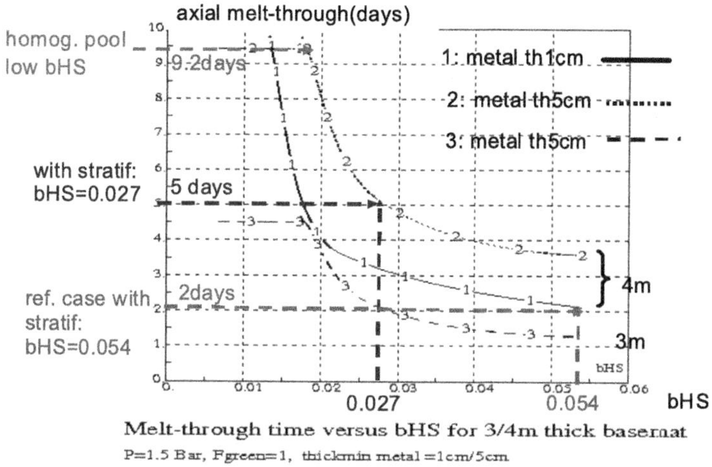

FIGURE 4.65 Melt-through for a 3 or 4 m thick basemat versus stratification criterion (proportionality factor bHS between Jg threshold and relative layer density difference) and, minimal metal thickness allowed for stratification in case of siliceous concrete.

codes in reactor applications. Codes used are: TOLBIAC [140] by CEA, WECHSL [135] by KIT, ASTEC/MEDICIS [143] by GRS and by IRSN, and MELCOR/CORCON [131] by UPM and CORQUENCH [7] (containing quenching models but very close to CORCON code in dry conditions and assuming a fixed homogeneous configuration) by VTT. The agreement between most codes for the axial ablation of the first 3 meters (see Figure 4.66a) or the first four days of MCCI (see Figure 4.66b) is rather good if assuming the pool configuration remains homogeneous during the whole MCCI phase.

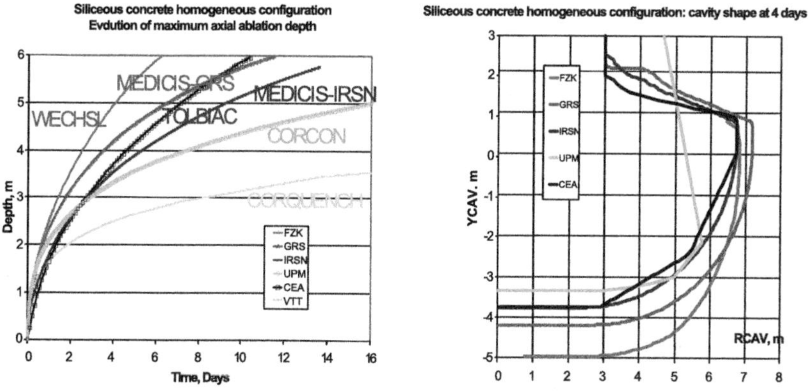

a/ axial ablation depth versus time (days) b/ cavity shape at 4 days interaction

FIGURE 4.66 SARNET benchmark reactor applications with homogeneous pool configuration in case of siliceous concrete.

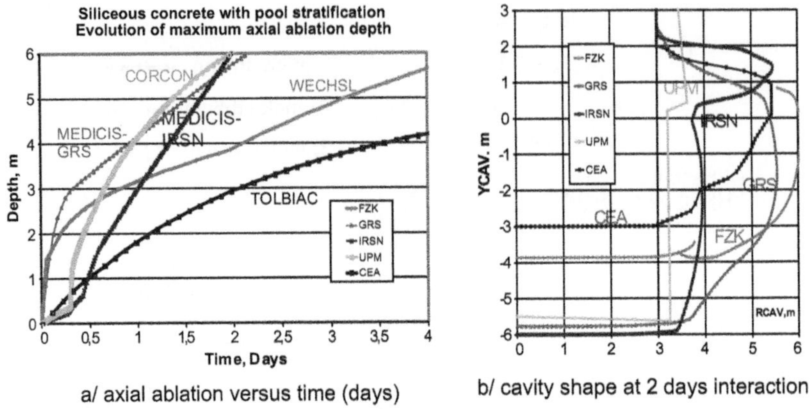

FIGURE 4.67 SARNET benchmark reactor applications with pool configuration evolution or fixed stratification (MEDICIS-GRS) in case of siliceous concrete (left) axial ablation versus time (days)—(right) cavity shape at 2 days interaction.

The reason of this broad agreement is that most of the codes use assumptions for the convective heat transfer within a mixed homogeneous pool and the pool/concrete interface structure independent of the interface orientation, leading to a more or less isotropic 2D ablation. Deviations increase progressively versus time due to the impact of different pool/concrete interface models leading to different pool temperatures. The deviations would change drastically if codes take into account the possible prevailing lateral ablation observed in experiments with siliceous concrete: the trend should be to get a significant further delay of the melt-through time but strongly dependent on the detailed modeling of the anisotropic 2D heat flux distribution in each code.

Large discrepancies between codes even for the axial ablation of the first meters (see Figure 4.67a) or the first days of MCCI (see Figure 4.67b) appear, assuming that the pool configuration evolves during MCCI and a stratified pool phase is possible.

The reasons for the code discrepancies in case of pool stratification are mainly the differences for the used oxide/metal convective heat transfer, the chosen stratification criteria, and, the assumption of a fixed stratified pool configuration kept during the whole MCCI phase in case of MEDICIS-GRS and WECHSL calculations thus leading to a very fast axial ablation kinetics during the first 12 hours. The two completely different codes, CORCON and MEDICIS, using similar assumptions for the oxide/metal convective heat transfer and for the stratification criterion, give rather close predictions. The TOLBIAC code predicts the slowest axial ablation kinetics: the main reason is the use of the BALI correlation for the oxide/metal convective heat transfer leading to a very low heat transfer value, whereas the TOLBIAC calculation uses the same stratification criterion as the MEDICIS-IRSN calculation leading to a much faster axial ablation kinetics because of the high oxide/metal convective heat transfer.

4.3.8. Remaining Uncertainties

Comparison of code results shows remarkable differences in the prediction of cavity formation, especially in the ratio of radial to axial erosion as a long-term process: predictions for the time of basemat melt-through in the containment may differ by more than a factor of 3 for a four meter thick basemat. The reason for these discrepancies is that models are different between codes and the impact of the model differences is enhanced in the long term. Parametric reactor calculations using the advanced MCCI codes point out the most important items that have a strong impact on the time for axial basemat penetration: the long-term distribution of the heat convection to the concrete surface and the structure of pool/concrete interfaces, the pool stratification and mixing criteria, and the convective heat transfer between oxide and metal layers with gas bubbling.

From a general point of view, the reliability and validation of the MCCI models is insufficient. These lacks and uncertainties result from the difficulties encountered in solving the following issues.

Determination of the 2D Ablation in the Case of a Homogeneous Pool

This requires modeling both of boundary layers of a viscous melt stirred by gas bubbling and of formation of interfacial crusts that may disappear intermittently in particular due to gravity along the lateral interface or due to dissolution at the bottom interface. Recent 2D MCCI experiments also pointed to the strong influence of the concrete type and in particular the aggregate behavior (molten or still intact) at the concrete ablation threshold on the pool/concrete interface structure. This leads to a complex structure of pool/concrete interface depending on the interface orientation and impacting on the heat flux distribution at the pool boundaries and consequently on the 2D ablation anisotropy. The interpretation of 2D MCCI experiments permitted deriving some model trends for the pool/concrete interface structure, but additional work is required to confirm the modeling and to make it sufficiently reliable for extrapolation at the reactor scale.

Existence of Pool Stratification in a Realistic Situation

The stratified situation with metal at the bottom was not up to now demonstrated, although it appears likely in the long-term MCCI phase of the reactor case, at least in case of a low or moderate concrete gas content (case of siliceous concrete). As decay heat is mainly released in the oxide, the interfacial heat transfer from the oxide to the steel layer, which is probably high, will focus the decay power downward and speed up the axial ablation as long as enough metal is available and stratification remains stable. Moreover, it must be kept in mind that a pure oxidic melt may not be the most likely situation in the accident, even in the very long-term MCCI phase, as melting of rebars is a steady source of steel that will accumulate at the bottom as long as oxidation of metals does not counterbalance the iron in-flux. First, representative MCCI experiments with oxide and metal have started only recently [110]. Further experiments on this situation are required but are extremely difficult. Nevertheless, performing an experiment with real material and aiming at demonstrating whether the pool stratification exists during MCCI would be valuable because it would help settle the issue of oxide/metal stratification during MCCI.

Stratification of Metal and Oxide Melts under Gas Bubbling

This complex phenomenon remains a real challenge: it is influenced by the density differences, by the melt viscosities, by the size distribution of the metal droplets, and, of course, by the convection processes in the melt induced by the gas release from concrete. However, additional analytical experiments on pool stratification and mixing thresholds with enough prototypical material properties should permit more realistic and reliable models for predicting pool stratification in the reactor case.

In spite of persisting uncertainties, some rather positive trends can be clearly outlined for the reactor case predictions *in dry conditions*:

- The use of a more realistic and validated stratification criterion will permit excluding any early melt-through (before a few days) for an at least 3 m thick basemat;
- If the influence of concrete type is confirmed at a longer time scale (beyond around 10 hours), for some concrete types (in particular the siliceous concrete) it will lead to a prevailing lateral ablation, which will strongly delay the axial melt-through time. The consequence might be to counterbalance the impact of stratification, which is promoted by the low gas content in case of siliceous concrete, and to delay the melt-through largely beyond 10 days for a thick enough basemat (at least around 4 m);
- In case of concrete with higher gas content such as limestone-sand concrete, stratification becomes unlikely or even excluded, and the melt-through time for a 4 m thick basemat will also reach around 10 days.

However, MCCI will not definitely stop in dry conditions even using more realistic but still conservative models: it points out the importance of mitigating MCCI using existing or new devices involving top and/or bottom water injection. (This topic is discussed in Section 4.1, Debris Formation and Coolability.)

ACKNOWLEDGMENTS

This chapter used the bibliographical synthesis of Matthieu Guillaumé (CEA Grenoble) [159] as a useful source of information.

REFERENCES

[1] B.R. Sehgal, B.W. Spencer, D.J. Kilsdonk, D.R. Armstrong, R.W. Aeschlimann, ACE Program Phase D: Melt Attack and Coolability Experiment (MACE) Program, OECD CSNI Specialist Meeting on Core Debris Concrete Interactions, Karlsruhe, Germany, April 1–3, 1992.

[2] B.W. Spencer, M. Fischer, M.T. Farmer, D.R. Armstrong, MACE Scoping Test Data Report, EPRI/MACE-TR-D03 (1991).

[3] M.T. Farmer, D.J. Kilsdonk, R.W. Aeschlimann, Corium Coolability under Ex-Vessel Accident Conditions for LWRs, Nucl. Eng. and Tech. (June 2009) 41–45.

[4] S. Lomperski, M.T. Farmer, Experimental Evaluation of the Water Ingression Mechanism for Corium Cooling, Nuclear Eng. Design 237 (2006) 905.

[5] M.T. Farmer et al., Modeling and Database for Melt-Water Interfacial Heat Transfer, 2nd CSNI Specialist Meeting on Core Debris-Concrete Interactions, Karlsruhe, Germany, April 1–3, 1992.

[6] M.T. Farmer, J.J. Sienicki, B.W. Spencer, CORQUENCH: A Model for Gas Sparging-Enhanced, Melt-Water, Film Boiling Heat Transfer, ANS Winter Meeting on the Thermal Hydraulics of Severe Accidents, Washington, D.C. USA, November 11–15, 1990.

[7] M.T. Farmer, Modeling of Ex-Vessel Corium Coolability with the CORQUENCH Code, Proceedings 9th Int. Conf. on Nucl. Eng. (April 8–12, 2001). Nice, France.

[8] C.R.B. Lister, Qualitative Theory on the Deep End of Geothermal Systems, Proceedings 2nd UN Symposium on Development and Use of Geothermal Resources, San Francisco, CA USA, May 20–29, 1975.

[9] M. Epstein, Dryout Heat Flux During Penetration of Water into Solidifying Rock, J. Heat Transfer 128 (2006) 847.

[10] S. Lomperski, M.T. Farmer, S. Basu, Experimental Investigation of Corium Quenching at Elevated Pressure, Nuclear Eng. Design 236 (2006) 2271.

[11] S. Lomperski, M.T. Farmer, Corium Crust Strength Measurements, Nuclear Eng. Des. 239 (11) (November 2009) 2551–2561.

[12] B. Tourniaire, J.M. Seiler, J.M. Bonnet, M. Amblard, Liquid Ejection through Orifices by Sparging Gas—The PERCOLA Program, Proceedings 10th Int. Conf. on Nucl. Eng. (April 2–6, 2000). Arlington, VA USA.

[13] B. Tourniaire, J.M. Seiler, Modeling of Viscous and Inviscid Fluid Ejection through Orifices by Sparging Gas, Proceedings ICAPP '04 (June 13–17, 2004). Pittsburgh, PA USA.

[14] M.T. Farmer, Phenomenological Modeling of the Melt Eruption Cooling Mechanism During Molten Corium Concrete Interaction (MCCI). Paper 6165, Proceedings ICAPP '06 (June 6–8, 2006). Reno, Nevada USA.

[15] K.R. Robb, M.L. Corradini, Experimental and Theoretical Investigation of Melt Eruptions during MCCI, OECD/NEA/IRSN Sponsored MCCI Seminar 2010, Cadarache, France, November 15–17, 2010.

[16] H. Alsmeyer, C. Spencer, W. Tromm, The COMET concept for cooling of ex-vessel corium melts, ICONE-6 Conf., San Diego, USA, 1988.

[17] W. Widmann, M. Bürger, G. Lohnert, H. Alsmeyer, W. Tromm, Experimental and theoretical investigations on the COMET concept for ex-vessel core melt retention, Nucl. Eng. Des. 236 (2006) 2304–2327.

[18] D. Paladino, B.R. Sehgal, DECOBI: investigation of melt coolability with bottom coolant injection, Prog. Nucl. Energy 40 (2) (2002) 161–206.

[19] B.R. Sehgal, et al., Phenomenological studies on melt coolability by bottom injection during severe accidents, Royal Institute of Technology, Stockholm, 2001. EU-Report. Report SAM-ECOSTAR-P11.

[20] C. Journea, H. Alsmeyer, Validation of the COMET Bottom-Flooding Core-Catcher with Prototypic Corium, ICAPP'06, Reno, NV, USA, June 4–8, 2006.
[21] B.R. Sehgal, T.N. Dinh, J.A. Green, D. Paladino, Experimental Investigation on Vessel-Hole Ablation During Severe Accidents, SKI Research Report 97 (December 1997) 44. 91 p.
[22] B.R. Sehgal, Stabilisation and termination of severe accidents in LWRs, Nucl. Eng. Des. 236 (2006) 1941–1952.
[23] GAREC working Group, J.M. Seiler, GAREC analyses in support of ex-vessel retention concept, OECD workshop on Ex-vessel Debris Coolability, Karlsruhe, Germany, November 15–18, 1999.
[24] D. Magallon, Characteristics of corium debris bed generated in large-scale fuel-coolant interaction experiments, Nucl. Eng. Des. 236 (2006) 1998–2009.
[25] S.K. Wang, C.A. Blomquist, B.W. Spencer, L.M. Mc Umber, J.P. Schneider, Experimental Study of the Fragmentation and Quench Behaviour of Corium Melts in Water, 25th Nat. Heat Transfer Conf. (July 24–27, 1988). Houston, Texas.
[26] M. Leskovar, R. Meignen, C. Brayer, M. Bürger, M. Buck, Material Influence on Steam Explosion Efficiency: State of Understanding and Modelling Capabilities, ERMSAR 2007 Conf. (June 12–14, 2007). Karlsruhe, Germany.
[27] A. Kaiser, W. Schütz, H. Will, PREMIX Experiments PM12-PM18 to Investigate the Mixing of a Hot Melt with Water, Report FZKA 6380, FZ Karlsruhe GmbH, 2001.
[28] C. Chu, et al., Ex-vessel melt-coolant interactions in deep water pool: studies and accident management for Swedish BWRs, Nucl. Eng. Des. 155 (1995) 1–2. 159–214.
[29] M. Epstein, H.K. Fauske, Steam Film Instability and the Mixing of Core-Melt Jets and Water, Nat. Heat Transfer Conf. (August 1985). Denver, Colorado.
[30] M. Bürger, Particulate debris formation by breakup of melt jets: Main objectives and solution perspectives, Nucl. Eng. Des. 236 (2006) 1991–1997.
[31] G. Pohlner, Z. Vujic, M. Bürger, G. Lohnert, Simulation of melt jet breakup and debris bed formation in water pools with IKEJET/IKEMIX, Nucl. Eng. Des. 236 (2006) 2026–2048.
[32] M. Bürger, E.V. Berg, M. Buck, U. Fichter, A. Schatz, Fragmentation and Film Boiling as Fundamentals in Premixing, Multidisciplinary Internat. Seminar on Intense Multiphase Interactions, Santa Barbara, California, June 9–13, 1995.
[33] D. Magallon, Results of Phase 1 of OECD Programme SERENA on Fuel - Coolant Interaction, ERMSAR-2005, Aix-en-Provence, France, November 14–16., 2005.
[34] W. Ma, Status and Progress of DEFOR Program at KTH, 3rd Annual Review Meeting of SARNET, Garching, Germany, January 29–February 2, 2007.
[35] G. Hofmann, On the location and mechanisms of dryout in top-fed and bottom-fed particulate beds, Nucl. Technol. 65 (1984) 36–45.
[36] M. Bürger, M. Buck, W. Schmidt, W. Widmann, Validation and application of the WABE code: Investigations of constitutive laws and 2D effects on debris coolability, Nucl. Eng. Des. 236 (2006) 2164–2188.
[37] P. Schäfer, M. Groll, R. Kulenovic, Basic investigations on debris cooling, Nucl. Eng. Des. 236 (2006) 2104–2116.
[38] W. Ma, T.N. Dinh, M. Buck, M. Bürger, Analysis of the effect of bed inhomogeneity on debris coolability, 15th Int. Conf. on Nuclear Engineering (April 22–26, 2007). Nagoya, Japan.
[39] M. Bürger, Core and Debris Coolability during Reflooding, 3rd Annual Review Meeting of SARNET, Garching, Germany, January 29–February 2, 2007.
[40] F. Fichot, F. Duval, N. Trégourès, C. Béchaud, M. Quintard, The impact of thermal non-equilibrium and large-scale 2D/3D effects on debris bed reflooding and coolability, Nucl. Eng. Des. 236 (2006) 2144–2163.

[41] I. Lindholm, et al., Dryout heat flux experiments with deep heterogeneous particle bed, Nucl. Eng. Des. 236 (2006) 2060–2074.

[42] M. Fischer, The severe accident mitigation concept and the design measures for core melt retention of the European Pressurized Reactor (EPR), Nucl. Eng. Des. 230 (2004) 169–180.

[43] F. Bouteille, G. Azarian, D. Bittermann, J. Brauns, J. Eyink, The EPR overall approach for severe accident mitigation, Proc. 13th Int. Conf. Nucl. Eng. (2005). Beijing, China, Paper ICONE13-50018.

[44] S.V. Svetlov, V.V. Bezlepkin, I.V. Kukhtevich, S.V. Bechta, V.S. Granovsky, V.B. Khabensky, V.G. Asmolov, V.B. Proklov, A.S. Sidorov, A.B. Nedoresov, V.F. Strizhov, V.V. Gusarov, Yu.P. Udalov, Core Catcher for TianWan NPP with VVER-100 reactor. Concept, Design and Justification, Proc. 11th Int. Conf. Nucl. Eng. (2003). Tokyo, Japan, Paper ICONE11-36102.

[45] H. Alsmeyer, T. Cron, J.J. Foit, G. Messemer, S. Schmidt-Stiefel, W. Häfner, H. Kriscio, Test Report of the Melt Spreading Tests ECOKATS-V1 and ECOKATS-1, SAM-ECOSTAR-D15/FZKA 7064, Karlsruhe 2004.

[46] H.E. Huppert, The propagation of 2 dimensional and axisymmetric viscousgravity currents over a rigid horizontal surface, J. Fluid Mech. 121 (1982) 43–58.

[47] G. Barenblatt, On some unsteady motions of a liquid and gas in a porous medium (in Russian), Akad. Nauk SSSR Prikl. Mat. Mech. 16 (1952) 67–78 and 679–698.

[48] N. Didden, T. Maxworthy, The Viscous Spreading of Plane and Axisymmetric Gravity Currents, J. Fluid Mech. 121 (1982) 27–42.

[49] R.W. Griffiths, J.H. Fink, Effects of Surface Cooling in the Spreading of lava flows and domes, J. Fluid Mech. 252 (1993) 667–702.

[50] D. Bercovici, A theoretical model of cooling viscous gravity currents wits temperature-dependent viscosity, Geophys. Res. Lett. 21 (1994) 1177–1180.

[51] A. Friedman, S. Kamin, The asymptotic behaviour of gas in an n-dimensional porous medium, Trans. Amer. Math. Soc. 262 (No. 2) (1980) 551–563.

[52] J.J. Foit, Spreading Under Variable Viscosity and Time-Dependent Boundary Conditions: Estimate of Viscosity from Spreading Experiments, Nucl. Eng. Des. 227 (2004) 239–253.

[53] J.J. Foit, Large-scale ECOKATS experiments: Spreading of oxide melt on ceramic and concrete surfaces, Nucl. Eng. Des. 236 (2006) 2567–2573.

[54] H. Pinkerton, L. Wilson, Factors controlling the lengths of channel fed lava flows, Bull. Volcanol 56 (1994) 108–120.

[55] M.C. Flemmings, Behavior of Metal Alloys in the Semisolid State, Mettallurgical Trans. 22A (1991) 957–981.

[56] F.J. Ryerson, H.C. Webb, A.J. Pijinski, Rheology of subliquidus Magmas 1. Picritic compositions, J. Geophys. Res. 93 (1988) 3421–3436.

[57] J.-M. Seiler, J. Ganzhorn, Viscosities of corium–concrete mixtures, Nucl. Eng. Des. 178 (1988). 259–268.

[58] M. Ramacciotti, F. Sudreau, C. Journeau, G. Cognet, Viscosity models for Corium Melts, Nucl. Eng. Design 204 (2001) 377–389.

[59] S. Arrhenius, The viscosity of solutions, Biochem J. 11 (1917) 112–133.

[60] C. Journeau, G. Jeulain, L. Benyahia, J.-F. Tassin, P. Abélard, Rheology of mixtures in the solidification range, Rhéologie 9 (2006) 28–39.

[61] J.J. Foit, Spreading on ceramic and concrete substrates in KATS experiments, Proceedings of the Annual Meeting on Nuclear Technology, Stuttgart (2002) 211–215.

[62] J.-M. Veteau, B. Spindler, G. Daum, Modelling of two-phase friction from isothermal spreading experiments with gas fed from the bottom and application to spreading

accompanied by solidification, Proc. 10th Int. Top. Mtg. Nuclear Reactor Thermal Hydraulics (2003). NURETH-10, Seoul, Korea.

[63] M. Manga, J. Castro, K.V. Cashman, M. Loewenberg, Rheology of bubble-bearing magmas, J. Volcanol. Geotherm. Res. 87 (1998) 15–28.

[64] E.W. Llewwellin, H.M. Mader, S.D. Wilson, The rheology of bubbly liquid, Proc. R. Soc. Lond A 458 (2002) 987.

[65] J.-M. Veteau, R. Wittmaack, CORINE experiments and theoretical modelling, in: G. Van Goetem, W. Balz, E. Della Loggia (Eds.), FISA 95 EU Research on severe accidents, Office Official Publ. Europ. Communities, Luxembourg, 1996, pp. 271–285.

[66] B. Spindler, J.-M. Veteau, Simulation of spreading with solidification: assessment synthesis of THEMA code, CEA public report CEA-R6053 (2004).

[67] G.A. Greene, C. Finrock, J. Klages, C.E. Schwarz, S.B. Burton, Experimental Studies on Melt Spreading, Bubbling Heat Transfer and Coolant Layer Boiling, Proc. 16th Water Reactor Safety Meeting (1988) 341–358. NUREG/CP-0097.

[68] T.N. Dinh, M.J. Konovalikhin, B.R. Sehgal, Core Melt Spreading on a reactor Containment Floor, Progr. Nucl. Energy 36 (4) (2000) 405–468.

[69] H. Suzuki, T. Matsumoto, I. Sakaki, T. Mitadera, M. Matsumoto, T. Zama, Fundamental experiment and analysis for melt spreading on concrete floor, Proc. 2nd ASME/JSME Nucl. Eng. Conf. 1 (1993) 403–407.

[70] G. Engel, G. Fieg, H. Massier, U. Stiegmaier, W. Schütz, KATS experiments to simulate corium spreading in the EPR code catcher concept, OECD Wkshp Ex-Vessel Debris Coolability, Karlsruhe, Germany (1998).

[71] H. Alsmeyer, T. Cron, G. Messemer, W. Häfner, ECOKATS-2: A Large Scale Experiment on Melt Spreading and Subsequent Cooling by Top Flooding, Proc. ICAPP'04 (Int. Conf. Advances in nuclear Power Plants), Pittsburg, PA, Communication no. 4134 (2004).

[72] C. Journeau, E. Boccaccio, C. Brayer, G. Cognet, J.-F. Haquet, C. Jégou, P. Piluso, J. Monerris, Ex-vessel corium spreading: results from the VULCANO spreading tests, Nucl. Eng. Des. 223 (2003) 75–102.

[73] W. Tromm, J.J. Foit, D. Magallon, Dry and wet spreading experiments with prototypic materials at the FARO facility and theoretical analysis, Wiss. Ber. FZKA 6475 (2000) 178–188.

[74] W. Steinwarz, A. Alemberti, W. Häfner, Z. Alkan, M. Fischer, Investigations on the phenomenology of ex-vessel core melt behaviour, Nucl. Eng. Des. 209 (2001) 139–146.

[75] B. Eppinger et al., Simulationsexperimente zum Ausbreitungsverhalten von Kernschmelzen: KATS-8 bis KATS-17, Wissenschaftliche Berichte FZKA 6589 (2001).

[76] C. Journeau, F. Sudreau, J.-M. Gatt, G. Cognet, Thermal, physico-chemical and rheological boundary layers in multi-component oxidic melt spreads, Int. J. Therm. Sci. 38 (1999) 879–891.

[77] C. Spengler, H.-J. Allelein, J.J. Foit, H. Alsmeyer, B. Spindler, J.-M. Veteau, J. Artnik, J., M. Fischer, Blind benchmark calculations for melt spreading in the ECOSTAR project, Proc. ICAPP '04 (Int. Conf. Advances in nuclear Power Plants), Pittsburg, PA, Communication no. 4105 (2004).

[78] C. Journeau, J.-F. Haquet, B. Spindler, C. Spengler, J.J. Foit, The VULCANO VE-U7 Corium spreading benchmark, Progr. Nucl. Energ., 48 (2006) 215–234.

[79] C. Spengler, A Fast Running Method for Predicting the Efficiency of Core Melt Spreading for Application in ASTEC, Jahrestagung Kerntechnik, Berlin, 2010.

[80] M.T. Farmer, J.J. Sienicki, C.C. Chu, B.W. Spencer, The MELTSPREAD-1 computer code for the analysis of transient spreading and cooling of high temperature melts, Report EPRI TR-103413 (1993).

[81] R. Wittmaack, CORFLOW: A code for the numerical simulation of free-surface flow, Nucl. Technol. 116 (1997) 158–180.
[82] B. Spindler, J.-M. Veteau, The simulation of melt spreading with THEMA code, Nucl. Eng. Design 236 (2006) 415–441.
[83] H.-J. Allelein, A. Breest, C. Spengler, Simulation of core melt spreading with LAVA: Theoretical background and Status of Validation, Wiss. Ber. FZKA 6475 (2000) 189–200.
[84] C. Spengler, 2002, Simulation der Ausbreitung von Kernschmelzen mit einem Binghamschen Fließmodell unter Berücksichtigung von Phasenübergängen, PhD thesis, Ruhr-Universität Bochum, Germany.
[85] B. Piar, B. Michel, F. Babik, J.-C. Latché, G. Guillard, J.-M. Ruggieri, CROCO: A Computer Code for Corium Spreading, Proc. 9th International Topical Meeting on Nuclear Thermal Hydraulics (NURETH-9), San Francisco, CA (1999).
[86] B. Michel, B. Piar, F. Babik, J.-C. Latché, G. Guillard, C. De Pascale, Synthesis of the validation of the CROCO-V1 spreading code, Proc. OECD Workshop on Ex-Vessel Debris Coolability, Karlsruhe (1999).
[87] D. Vola, F. Babik, J.-C. Latché, On a numerical strategy to compute gravity currents of non-Newtonian fluids, Journal of Computational Physics 201 (2004) 397–420.
[88] W. Steinwarz, W. Koller, W. Häffner, C. Journeau, J.M. Seiler, K. Froment, G. Cognet, S. Goldstein, M. Fischer, S. Hellmann, M. Eddi, H. Alsmeyer, H.J. Allelein, C. Spengler, M. Bürger, B.R. Sehgal, M.K. Koch, Z. Alkan, J.B. Petrov, M. Gaune-Escart, F.P. Weiss, G. Bandini, Ex-vessel core melt stabilization Research (ECOSTAR), in: G. Van Goethem, A. Zurita, J. Martin Bermejo, P. Manolatos, H. Bischoff (Eds.), FISA 2001: EU research in reactor safety, Office Official Publ. Eur. Communities, Luxembourg, 2002, pp. 274–285.
[89] M. Nie, Temporary melt retention in the Reactor Pit of the European Pressurized water Reactor (EPR), Doctoral Thesis, University of Stuttgart, 2004.
[90] B. Spindler, C. Brayer, M. Cranga, L. de Cecco, P. Montanelli, D. Pineau, J.M. Veteau, Assessment of the THEMA code against spreading experiments, in: H. Alsmeyer (Ed.), OECD Wkshp on ex-vessel coolability, Wiss. Ber. FZKA 6475 (2000) 221–234.
[91] J.J. Foit, C. Spengler, N. Dyllong, Ausbreitung von Schmelzen, Corium Workshop, Kompetenzverbund Kerntechnik, GRS Cologne, February 2006.
[92] G. Azarian, H.M. Kursawe, M. Nie, M. Fischer, J. EyinK, R.H. Stoudt, EPR Accident threats and Mitigation, Proc. Int. Congress Advances nucl. Power Plants (ICAPP'04) (2004). Pittsburgh, PA.
[93] S. Malaval, C. Journeau, A. Smith, J.M. Bonnet, Thermal Study of a typical concrete from a nuclear power plant, 13th Int Conf Nucl. Eng. (2005). Beijing, China.
[94] L. Alarcon-Ruiz, et al., The Use of Thermal Analysis in assessing the Effect of Temperature on a Cement Paste, Cement and Concrete Research 35 (2005) 609–613.
[95] G.A. Khoury, Effect of Fire on Concrete and Concrete Structures, Prog. Struct. Engng Mater. 2 (2000) 429–447.
[96] A. Noumowe, Effet de Hautes Températures (20–600°C) sur le béton: cas particulier du béton à hautes performances, Ph.D in Institut National des Sciences Appliquées, Lyon, France (1995).
[97] M.F. Roche, L. Leibowitz, J.K. Fink, L. Baker Jr., Solidus and Liquidus Temperatures of Core-Concrete Mixtures, U.S. Nuclear Regulatory Commission Report NUREG/CR-6032, Argonne National Laboratory Report ANL-93/9 (1993).
[98] K.Y. Shin, S.B. Kim, J.H. Kim, M. Chung, P.S. Jung, Thermophysical properties and transient heat transfer of concrete at elevated temperatures, Nucl. Eng. Des. 212 (2002) 233–241.

[99] AFNOR, Eurocode 2: Design of concrete structures, Part 1–2 general rules—Structural fire design, European Standard, NF EN 1992-1-2 (2005).

[100] H. Alsmeyer, et al., Molten corium/concrete interaction and corium coolability—a state of the art report, Report EUR 16649, European Commission (1995).

[101] D.H. Thompson, M.T. Farmer, J.K. Fink, D.R. Armstrong, B.W. Spencer, Compilation, analysis and interaction of ACE Phase C and MACE experimental data, Report ACEX TR-C-14, Argonne National Laboratory, Chicago, IL, USA, 1997.

[102] E.R. Copus, Sustained Uranium dioxide Concrete interaction tests: The SURC test series, 2nd OECD (NEA) Spec. Mtg. on Molten Core Debris—Concrete Interactions, Karlsruhe, Germany (1992).

[103] E.R. Copus, R.E. Blose, J.E. Brockmann, R.B. Simpson, D.A. Lucero, Core-Concrete Interactions Using Molten Urania With Zirconium on a Limestone Concrete Basemat, The SURC-1 Experiment, Report NUREG/CR-5443, Sandia National Lab. (1992).

[104] E.R. Copus, R.E. Blose, J.E. Brockmann, R.B. Simpson, D.A. Lucero, Core-Concrete Interactions using Molten UO2 with Zirconium on a Basaltic Basemat: The SURC-2 Experiment, (1990) Sandia Nat. Lab. Report NUREG/CR-5564, SAND90–1022 (1990).

[105] E.R. Copus, R.E. Blose, J.E. Brockmann, R.D. Gomez, D.A. Lucero, Core-Concrete Interactions using Molten Steel with Zirconium on a Basaltic Basemat: The SURC-4 Experiment, In: NUREG/CR-4994, SAND87–2008, Sandia National Laboratories (1989).

[106] R.E. Blose, D.A. Powers, E.R. Copus, J.E. Brockmann, R.B. Simpson, D.A. Lucero, Core-Concrete Interactions with Overlying Water Pools - The WETCOR-1 Test, Sandia Nat. Lab. Report NUREG/CR-5907, SAND92-1563 (1993).

[107] M.T. Farmer, B.W. Spencer, D.J. Kilsdonk, W. Aeschlimann, Status of large scale MACE Core Coolability experiments, OECD Workshop on Ex-Vessel Debris Coolability, Karlsruhe, Germany (1999).

[108] J.J. Foit, Modelling oxidic molten core-concrete interaction in WECHSL, Nucl. Eng. Des. 170 (1997) 73–79.

[109] C. Journeau, P. Piluso, J.F. Haquet, S. Saretta, E. Boccaccio, J.M. Bonnet, Oxide-metal corium–concrete interaction test in the VULCANO facility, ICAPP'07, Int Conf Advances Nucl. Power Plants (May 13–18, 2007). Nice.

[110] C. Journeau, J.M. Bonnet, L. Ferry, J.F. Haquet, P. Piluso, Interaction of Concretes with Oxide + Metal Corium: The VULCANO VBS Series, ICAPP'09, Int Conf Advances Nucl. Power Plants (May 10–14, 2009). Tokyo.

[111] M.T. Farmer, S.W. Lomperski, S. Basu, The results of the CCI-2 reactor material experiment investigating 2-D core-concrete interaction and debris coolability, Proc. 11th International topical meeting on nuclear reactor thermal hydraulics (Nureth 11) (2005). Avignon, France.

[112] M.T. Farmer, S.W. Lomperski, S. Basu, The Results of the CCI-3 Reactor Material Experiment Investigating 2-D Core-Concrete Interaction and Debris Coolability with a Siliceous Concrete Crucible, Proc. Int. Cong. Advanced Nuclear Power Plants (ICAPP'06) (2006). Reno, NV.

[113] C. Journeau, P. Piluso, J.F. Haquet, Behaviour of nuclear reactor pit concretes under severe accident conditions, Proc. CONSEC '07, Concrete under Severe Conditions (2007). Tours, France.

[114] H. Alsmeyer, A. Miassoedov, M. Cranga, R. Fabianelli, I. Ivanow, G. Doubleva, The COMET-L1 experiment on long-term concrete erosion and surface flooding, Proc. 11th International topical meeting on nuclear reactor thermal hydraulics (NURETH1-1) Avignon, France (2005) and FZKA 7213, SAM-LACOMERA D14 (2006).

[115] G. Sdouz, R. Mayrhofer, H. Alsmeyer, T. Cron, B. Fluhrer, J.J. Foit, G. Messemer, A. Miassoedov, S. Schmidt-Stiefel, T. Wenz, The COMET-L2 experiment on long-term MCCI with steel melt, FZKA 7214, SAM-LACOMERA D15 (2006).

[116] H. Alsmeyer, T. Cron, B. Fluhrer, G. Messemer, A. Miassoedov, S. Schmidt-Stiefel, T. Wenz, The COMET-L3 Experiment on long-term melt-concrete interaction and cooling by surface flooding, FZKA 7244 (2006).

[117] Y. Maruyama, Y. Kojima, M. Tahara, H. Nagasaka, M. Kato, A.A. Kolodeshnikov, V.S. Zhdanov, Yu.S. Vassiliev, A study on concrete degradation during molten core/concrete interactions, Nucl. Eng. Des. 236 (2006) 2237–2244.

[118] C. Journeau, L. Ferry, P. Piluso, J. Monerris, M. Breton, G. Fritz, T. Sevon, Two EU-funded tests in VULCANO to assess the effect of concrete nature on its ablation by molten corium, 4th European Review Meeting on Severe Accident Research (ERMSAR-2010), Bologna (Italy), May 11–12, 2010.

[119] T. Sevon, T. Kinnunen, J. Virta, S. Holmström, T. Kekki, I. Lindholm, HECLA experiments on interaction between metallic melt and hematite—containing concrete, Nucl. Eng. Des. 240 (2010) 3586–3593.

[120] M. Cranga, L. Ferry, J.F. Haquet, C. Journeau, B. Michel, P. Piluso, G. Ratel, K. Atkhen, MCCI in an oxide/metal pool: lessons learnt from VULCANO, Greene, ABI and BALISE experiments and remaining uncertainties, 4th European Review Meeting on Severe Accident Research (ERMSAR-2010), Bologna (Italy), May 11–12, 2010.

[121] J.M. Veteau, Experimental investigation of interface conditions between oxidic melt and ablating concrete during MCCI by means of simulating material experiments: the Artemis program, Proc. 11th International topical meeting on nuclear reactor thermal hydraulics (NURETH-11) (2005). Avignon, France.

[122] M.R. Duignan, G.A. Greene, T.F. Irvine Jr., Heat transfer from a horizontal bubbling surface to an overlying water pool, Chem. Eng. Comm. 87 (1990).

[123] D.K. Felde, H.S. Kim, S.I. Abdel-Khalik, Convective heat transfer correlations for molten core debris pools growing in concrete, Nucl. Eng. Des. 58 (1980).

[124] B. Tourniaire, O. Varo, Assessment of two-phase flow heat transfer correlation for molten core-concrete interaction study, Proc. Of ICAPP'06, Reno, 2006.

[125] D.R. Bradley, Modelling of heat transfer between core debris and concrete, ANS Proceedings of the 1988 Nat. Heat Tansfer Conf (1988).

[126] J.L. Casas, M.L. Corradini, "Study of void fractions and mixing of immiscible liquids in a pool configuration by an upward gas flow", Nucl. Technology 99 (1992).

[127] G.A. Greene, Heat, mass and momentum transfer in a multifluid bubbling pool, Advances in Heat Transfer 21 (1991).

[128] H.J. Allelein, M. Bürger, Considerations on ex-vessel corium behavior: Scenarios, MCCI and coolability, Nucl. Eng. Des. 236 (2006) 2220–2236.

[129] B. Tourniaire, J.M. Bonnet, Study of the mixing of immiscible liquids by sparging gas: results of the BALISE experiments, Proc. 10th Int. Topical Mtg on Nuc. Reactor Thermal Hydraulics (NURETH 10) (2003). Seoul, Korea.

[130] M. Epstein, D.H. Petrie, J.H. Linehan, G.A. Lambert, D.H. Cho, Incipient stratification and mixing in aerated liquid-liquid or liquid-solid mixtures, Chem. Eng. Sci. 36 (1981) 84–86.

[131] D.R. Bradley, D.R. Gardner, J.E. Brockmann, R.O. Griffiths, CORCON-Mod3: An integrated computer model for analysis of molten core-concrete interactions, USNRC Report NUREG/CR-5843 (1993).

[132] J.M. Seiler, Phase segregation model and pool thermal hydraulics during molten core concrete interaction, Nucl Eng. Des. 166 (1996) 259–267.

[133] J.M. Seiler, K. Froment, Material effects on multiphase phenomena in late phases of severe accidents of nuclear reactors, Multiphase Sci. Technol. 12 (2000) 117–257.

[134] M. Reimann, W.B. Murfin, The WECHSL-code: A computer program for the interaction of a core melt with concrete, Kernforschungszentrum Karlsruhe, KfK 2890 (1981).

[135] J.J. Foit, M. Reimann, B. Adroguer, G. Cenerino, S. Stiefel, The WECHSL-Mod3 Code: a computer program for the interaction of a core melt with concrete including the long term behavior; Model descriptions and User's manual, FZKA 5416 (1995).

[136] J.J. Foit, A. Miassoedov, Modelling of Viscosity and Heat Transfer of Complex Oxidic Melts in WECHSL, Wiss. Ber. FZKA 5507 (1995).

[137] A. Hackel, W. Gröll, Zum Wärmeübergangsverhalten zähflüssiger Öle, Verfahrenstechnik 3 (1969) 141–145.

[138] H. Hausen, Bemerkung zur Veröffentlichung von A. Hackel, W. Gröll, Zum Wärmeübergangsverhalten zähflüssiger Öle, Verfahrenstechnik 3 (1969) 355.

[139] B. Eppinger et al., Simulationsexperimente zum Ausbreitungsverhalten von Kernschmelzen: KATS-8 bis KATS-17, Wissenschaftliche Berichte FZKA 6589 (2001).

[140] B. Spindler, B. Tourniaire, J.M. Seiler, Simulation of MCCI with the TOLBIAC-ICB code based on the phase segregation model, Nucl. Eng. Des. 236 (2006) 2264–2270.

[141] M. Nie, M. Fischer, G. Lohnert, Advanced MCCI modelling based on stringent coupling of thermalhydraulics and real solution of thermochemistry in COSACO, Proc. ICONE10, Arlington, VA (2002).

[142] J.M. Seiler, K. Froment, Material Effects on Multiphase Phenomena in Late Phases of Severe Accidnets of Nuclear Reactors, Multiphase Sci. Technol., 12 (2000) 117–257.

[143] M. Cranga, R. Fabianelli, F. Jacq, M. Barrachin, F. Duval, The MEDICIS code, a versatile tool for MCCI modelling, Proceedings of ICAPP '05, Seoul, KOREA (2005).

[144] G.A. Greene, T.F. Irvine, Heat transfer between stratified immiscible liquid layers driven by gas bubbling across the interface, ANS Proc. Nat. Heat Transfer Conf (1988).

[145] H. Werle, Enhancement of heat transfer between two horizontal liquid layers by gas injection at the bottom, Nucl. Technol. 59 (1982) 160–164.

[146] J.M. Bonnet, Thermalhydraulic phenomena in corium pools for ex-vessel situations: The BALI experiment, Wiss. Ber. FZKA 6475 (2000).

[147] F.G. Blottner, Hydrodynamics and Heat Transfer Characteristics of Liquid Pools with Bubbles Agitation, Sandia Laboratories Report NUREG CR-0944 (1979).

[148] W.D. Deckwer, On the mechanism of heat transfer in bubble column reactors, Chem. Eng. Sc. 35 (1980) 1341–1346.

[149] B. Tourniaire, A heat transfer correlation based on a surface renewal model for molten core concrete interaction study, Nucl. Eng. Des. 236 (2006) 10–18.

[150] S.S. Kutateladze, I.G. Malenkov, Boiling and Bubbling Heat transfer under the conditions of free and forced convection, 6th Int. heat Transfer Conf. (1978). Toronto, Canada.

[151] C. Spengler, H.J. Allelein, M. Cranga, F. Duval, J.P. Van Dorsselaere, Assessment and development of Molten Corium Concrete Interaction Models for the Integral ASTEC code, EUROSAFE Forum, Brussels, November 7–8, 2005.

[152] B. Spindler, K. Atkhen, M. Cranga, J.J. Foit, M. Garcia, W. Schmidt, T. Sevon, C. Spengler, Simulation of Molten Corium Concrete Interaction in a Stratified configuration: the COMET-L2–L3 Benchmark, ERMSAR-07, June 2007.

[153] Deutsche Risikostudie Kernkraftwerke, Phase B, Verl. TÜV Rheinland (1990).

[154] J.J. Foit, L.D. Howe, Plant Application of CORCON and WECHSL, in: H. Alsmeyer et al., Molten Corium/Concrete Interaction and Corium Coolability—A State of the Art Report, EUR 16649 EN (1995), 130–179.

[155] B. Spindler, Thermo-physical Properties Needs for Severe Accident Mechanistic Codes: MCCI with TOLBIAC-ICB, International Workshop on Thermo Physical properties of Materials, Paris, October 2006.

[156] M. Cranga, L. Ferry, J.F. Haquet, C. Journeau, B. Michel, C. Mun, P. Piluso, G. Ratel, K. Atkhen, MCCI in an oxide/metal pool: lessons learnt from VULCANO, Greene, ABI and BALISE experiments and remaining uncertainties, ERMSAR-10, Bologna (Italy), May 11–12, 2010.

[157] B. Michel, M. Cranga, Interpretation and calculations for the first series of tests for the ARTEMIS program (corium–concrete interaction with simulating materials), Nucl. Eng. Des. 239 (2009) 600–610.

[158] M. Guillaumé, H. Combeau, J.M. Seiler, An improved interface model for molten corium–concrete interaction, Nucl. Eng. Des. 239 (2009) 1084–1094.

[159] M. Guillaumé, Modélisation de l'interaction entre le coeur fondu d'un réacteur à eau pressurisée et le radier en béton du bâtiment réacteur, Ph.D. thesis, INP-Lorraine, Nancy, France (2008).

[160] V. Strizhov, V. Kanukova, T. Vinogradova, E. Askenov, An assessment of the CORCON-Mod3 Code, Part I: thermohydraulic calculations, NUREG/IA-0129, published by NRC, 1996.

[161] M. Cranga, B. Spindler, E. Dufour, D. Dimov, K. Atkhen, J. Foit, M. Garcia-Martin, T. Sevon, W. Schmidt, C. Spengler, Simulation of Corium Concrete Interaction in 2D geometry, Progr. Nuclear Energy Journal 52 (2010) 76–83.

[162] J.J. Foit, L.D. Howe, Thermal Hydraulic codes, in: H. Alsmeyer, et al., Molten Corium/Concrete Interaction and Corium Coolability—A State of the Art Report, EUR 16649 EN (1995) 130–179.

[163] H. Alsmeyer, BETA Experiments in Verification of the WECHSL Code: Experimental results on the melt-concreteinteraction, Nucl. Eng. Des. 103 (1987) 115–125.

[164] C. Renault, J.J. Foit, J. Poubllan, WECHSL-Mod3 Assessment Report, Note Technique Semar 92/93 KfK 5164, 1992.

[165] C. Renault, J.J. Foit, Assessment Status of the WECHSL-Mod3 Code, 2nd OECD (NEA) CSNI Specialist Meeting on Molten Core Debris-Concrete Interactions, KfK 5108, Karlsruhe, Germany, 1992.

[166] M. Firnhaber, H. Alsmeyer, International Standard Problem No.30: BETA V5.1 Experiment on Melt-Concrete Interaction, OECD Nuclear Energy Agency NEA/CSNI/R(92), 9 (1992).

[167] J.J. Foit, Improved WECHSL Models Including Zirconium Oxidation and its Verification by New BETA Experiments, 2nd OECD (NEA) CSNI Specialist Meeting on Molten Core Debris-Concrete Interactions, KfK 5108, Karlsruhe, Germany, 1992.

[168] C.Y. Paik, R.W. Reeves, W. Luangdilok, R.E. Henry, Q. Zhou, Current status of molten corium concrete interaction modellling in MAAP, MCCI-OECD seminar, Cadarache, France, 2010.

[169] G. Cenerino, V. Crutel, ACEX: amendment n°1 to Contract W03425-02: condensed Phase chemistry - Task3: Plant Application of GEMINI/WECHSL, Report DPEA/SEAC/96 (1996).

[170] D.A. Powers, D.A. Dahlgren, J.F. Muir, W.D. Murfin, Exploratory study of molten core material/concrete interactions, SAND (1977) 77–2042.

[171] D.A. Powers, F. Arellano, Large-Scale, Transient Tests of the Interaction of Molten Steel with Concrete, SAND (1981). 81–1753 NUREG/CR/2282.

[172] J.A. Cronager, Suo-Antilla, D.R. Bradley, J.E. Brockman, TURCl: Large Scale Metallic Melt-Concrete Interaction Experiments and Analysis, NUREG/CR/4420, printed in January 1986 SAND 85–0707, 1986.

[173] E.R. Copus, Sustained URanium-Concrete interaction: the SURC experiments, OCDE/SCIN Specialist MCCI meeting Palo-Alto, CA, USA, September 3–5, 1986.
[174] E.R. Copus, D.R. Bradley, Interaction of Hot Solid Core Debris with Concrete, NUREG/CR/4558 SAND 85–1739, printed in June 1986.
[175] S. Levy, Summary of coolability studies undertaken by ACE / MACE / ACEX, MACE project document (2002).
[176] J.M. Bonnet, J.M. Seiler, Coolability of Corium Spread Onto Concrete Under Water, the PERCOLA Model, 2nd CSNI Specialists Meeting on Core Debris-Concrete Interactions, OECD/NEA/CSNI R(92)10 (1992).
[177] K.R. Robb, M.L. Corradini, Experimental and Theoretical Investigation of Melt Eruptions during MCCI, MCCI Seminar, November 15–17, 2010, Cadarache, France.
[178] M.T. Farmer, A Summary of Modeling Activities Related to Debris Coolability and Core-Concrete Interaction, MCCI Seminar, November 15–17, 2010, Cadarache, France.
[179] S. Lomperski, M.T. Farmer, Experimental evaluation of the water ingression mechanism for corium cooling, Nuclear Engineering and Design 237 (December 2006) 905–917.
[180] F.B. Ricou, D.B. Spalding, Measurements of entrainment of axisymmetrical turbulent jets, J. Fluid Mechanics 11 (1961) 21–32.
[181] B. Tourniaire, J.M. Seiler, J.M. Bonnet, M. Amblard, Experimental study and modelling of liquid ejection through orifices by sparging gas, Nucl. Eng. Des. 236 (2006) 2281–2295.
[182] M.T. Farmer, Phenomenological Modeling of the Melt Eruption Cooling Mechanism during Molten Corium Concrete Interaction (ICAPP06), Reno, NV, USA, June 2006.
[183] K.R. Robb, M.L. Corradini, Towards understanding melt eruption phenomena during molten corium concrete interactions, Proc. of ICONE 18 Conference, Xi'an, China, May 2010.

Chapter 5

Fission Product Release and Transport

Chapter Outline

Nuclear Safety in Light Water Reactors. DOI: 10.1016/B978-0-12-388446-6.00005-8

5.1. INTRODUCTION

(C. Housiadas, M. Kissane, and R. Sehgal)

The term *source term* is widely used in industrial safety studies connected with risk and environmental impact assessment analyses. Generally, the term refers to the amount of hazardous material released to the environment from a facility following an accident. The evaluation of this quantity is of great importance because it is the first essential step in the whole risk assessment chain. Clearly, the source term is the necessary input for assessing the distribution of hazardous material in the various environmental media (air, water, etc.), and by that, assessing the exposure, doses and, ultimately, possible health effects to workers and public.

The specificity of a severe nuclear reactor accident in relation to an accident in a nonnuclear industrial facility lies in the potential release of radioactive products into the environment and the resulting radiological consequences. These radioactive products include fission products (FPs), activation products (APs) from core structural materials, and heavy nuclei (HN) such as uranium and transuranium elements. The evaluation of the source term is a long-standing issue in severe accident research. Still, the term *source term* is not included in the glossary of terms in nuclear science and technology of the American Nuclear Society. In many cases the term is used to describe the release in the external environment. However, it is often also used to refer to the release in the containment, considering that the total inventory in the containment may potentially be released to the environment. In this book the following definition of source term will be used: the quantity, time, history, and chemical and physical form of radionuclides released to the environment, or present in the containment atmosphere, during the course of a severe accident.

As is evident, this definition does not mention the term *fission products*, but instead replaces it by the more general phrase *chemical and physical form of radionuclides*. This is because the FPs (and structural materials) released from a degraded core are drastically transformed, both physically and chemically, prior to becoming the "ingredients" of the source term. This is because of the vigorous physical and chemical processes that take place as the released materials transcend the primary system and the containment.

The release phases of an accident are conveniently summarized in [1].[1] In the case of a postulated pipe rupture, activity can first be released to the containment atmosphere associated with the very small amount dissolved in the coolant water itself; for a large- break LOCA, this has been evaluated [1] to take place over 25 seconds for Westinghouse PWRs and over about 13 seconds for Babcock and Wilcox BWRs. Consideration of this release is outside the scope of this chapter. The second step, the gap release phase, starts when the fuel cladding ruptures, allowing release of the more volatile radionuclides such as noble gases (xenon and krypton), iodine, and caesium, which have been released from the fuel during normal operation, and accumulated in the gap between the fuel and the cladding. Typically, this is a small percentage of the fuel inventory of these elements, and the duration of the release is typically about half an hour. In this phase, most of the FPs are still retained in the fuel itself. This phase ends when the temperature reaches such a level that significant amounts of FPs can no longer be retained within the fuel.

The main releases occur during the early in-vessel, ex-vessel, and late in-vessel phases. In the in-vessel phase, the temperatures of fuel and structural materials reach such a level that the initial reactor core geometry is no longer maintained and core materials melt and relocate to the bottom of the reactor vessel. Typically, this phase lasts some tens of minutes depending on the type of reactor and accident sequence. This phase, in which several tens of percent of volatile FPs and smaller quantities of less volatile elements can be released, ends when the bottom head of the pressure vessel fails, allowing core melt to enter the cavity. Reactions between the molten corium and concrete allow further FP release in this ex-vessel phase, which can last a few hours, ending when the debris has cooled to a low enough temperature. The late in-vessel stage starts at vessel failure, and may involve further release from the fuel and revolatilization of material already deposited in the reactor coolant system (RCS). This phase can last up to about 10 hours.

Figure 5.1 shows schematically the source term pathway from release from the fuel through release to the environment. Accordingly, the section is organized by considering the distinct, separate steps in the shown pathway.

The mode of production and the main characteristics of FPs are described in Section 5.2. During the core degradation phase in the vessel, these products are released from the fuel, mainly in the form of vapor or gas (Section 5.3). The vapors first cool down in the upper part of the vessel and then in the primary cooling system, which leads to significant retention according to the different mechanisms described in Section 5.4. Having arrived in the containment in the form of aerosols or gas via the break, new mechanisms of retention or

1. Note: At the time of writing, this reference was under review and was likely to be revised; such a revision might result in changes in the timings mentioned below.

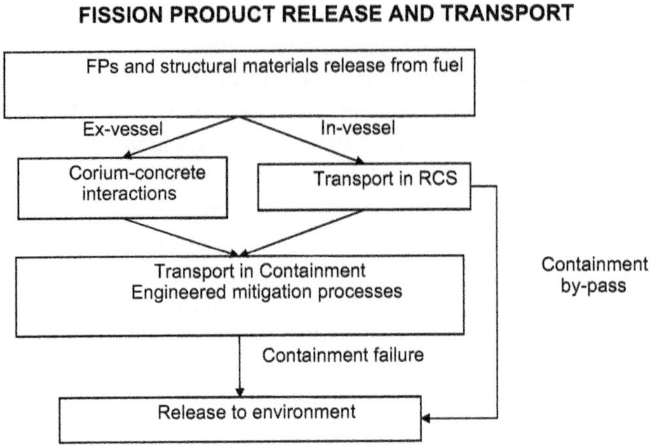

FIGURE 5.1 Schematic representation of the source term pathway.

reemission come into play, particularly through exchanges between the containment atmosphere, its walls and the sump (Section 5.7). In the case of vessel failure, there may be an additional source of release following interaction between the corium and the reactor pit concrete (Section 5.6). Finally, direct releases into the environment can occur in the case of containment bypass (Section 5.5). The release and transport of these radioactive elements into the environment are described in other parts of the book (Chapter 7).

5.2. FISSION PRODUCT INVENTORY AND VARIATIONS

(G. Ducros)

5.2.1. How Fission Products Are Produced

The production of radionuclides and the variation in their inventory during and after reactor irradiation are governed by the Bateman equations [2]. For each FP,[2] the concentration variations of N_i in the fuel can be described by three generic expressions:

- Generation from heavy nuclei by fission: $\phi \cdot \sum_{hn} \sigma_f^n \cdot y_n^i \cdot N_{hn}$;
- Generation from one (or several) parent by radioactive decay or by neutronic capture from one (or several) parent: $\sum_{j,k} (\alpha_i^j \cdot \lambda_{Pj} \cdot N_{Pj} + \beta_i^k \cdot \sigma_c^{Pk} \cdot \phi \cdot N_{Pk})$;
- Disappearance by radioactive decay and/or by neutronic capture: $(\sigma_a^i \cdot \phi + \lambda_i) \cdot N_i$

2. Equations are similar for variations in HN and AP inventories without an expression for production by fission.

The general equation used to calculate variations is expressed as follows:

$$\frac{dN_i}{dt} = \phi \cdot \sum_{hn} \sigma_f^n \cdot y_n^i \cdot N_{hn} + \sum_{j,\,k} (\alpha_i^j \cdot \lambda_{Pj} \cdot N_{Pj} + \beta_i^k \cdot \sigma_c^{Pk} \cdot \phi \cdot N_{Pk}) - (\sigma_a^i \cdot \phi + \lambda_i) \cdot N_i \quad (1)$$

where:

ϕ is the neutron flux ($n \cdot cm^{-2} \cdot s^{-1}$)
σ_f^n is the microscopic fission cross section of each heavy nucleus (hn)
y_n^i is the fission yield of each heavy nucleus (hn) for the FP_i
N_{hn} is the concentration of each heavy nucleus (hn) in the fuel
α_i^j and β_i^k are the branching ratios (generally equal to 1) of the parents P_j and P_k to the FP_i
N_{pj} and N_{pk} are the concentrations of the parents P_j and P_k in the fuel
λ_{pj} is the radioactive decay constant of the parent P_j
σ_c^{Pk} is the microscopic capture cross section of the parent P_k
σ_a^i is the microscopic capture cross section of the FP_i
λ_i is the radioactive decay constant of the FP_i

This system of equations can be taken into account numerically by using specific computer codes, such as ORIGEN2 [3] developed in the United States by ORNL and widely used in the international scientific community, or DARWIN-PEPIN [4] developed in France by CEA. These computer codes require coupling with neutronic codes to calculate variations in the different cross sections over time.

As an example, Table 5.1 summarizes the inventory of each FP and each heavy nucleus, as well as the variations in the total core activity, from reactor shutdown up to one month afterward. This calculation was obtained using the DARWIN-PEPIN code for a 900 MWe PWR at the end-of-life with quarter core loading, that is, initial load of 72.5 tons of uranium having reached a respective burn-up of 10.5, 21, 31.5 and 42 GWd/t for the four quarters.

The analysis of Table 5.1 makes it possible to underline various characteristics of several elements, such as:

- The mass inventory is 10 times more abundant for caesium than for iodine, the two main volatile FPs from the short-term radiological standpoint. This means that though caesium iodide (CsI) is one of the most probable chemical forms of iodine in PWR fuel from a thermodynamic viewpoint, it hardly represents 10% of the caesium inventory, leaving this element with the possibility of chemically binding with other FPs;
- In a similar manner, the mass inventory of the fission gases is 10 to 15 times higher for the noble gas xenon than for the noble gas krypton.

TABLE 5.1 Inventory and Variation in the Activity of Fission Products in a 900 MWe PWR after Shutdown

		Activity (% of total activity)			
Fission products	Total mass (kg at shutdown)	Shutdown	1 Hour	1 Day	1 Month
As	**7.39E-03**	0.20%	0.01%	0.00%	0.00%
Se	**3.14E+00**	0.58%	0.02%	0.00%	0.00%
Br	**1.16E+00**	1.17%	0.20%	0.00%	0.00%
Kr	**2.21E+01**	2.32%	1.46%	0.03%	0.06%
Rb	**2.03E+01**	3.22%	0.84%	0.01%	0.00%
Sr	**5.51E+01**	4.50%	3.85%	2.57%	6.10%
Y	**2.89E+01**	5.84%	5.11%	3.40%	8.16%
Zr	**2.10E+02**	4.73%	3.83%	4.63%	10.30%
Nb	**3.24E+00**	7.09%	5.68%	5.93%	13.18%
Mo	**1.84E+02**	4.28%	2.28%	2.90%	0.01%
Tc	**4.52E+01**	4.82%	2.50%	2.77%	0.01%
Ru	**1.37E+02**	1.85%	3.11%	3.67%	10.27%
Rh	**2.36E+01**	2.30%	3.42%	4.96%	10.26%
Pd	**5.93E+01**	0.19%	0.33%	0.18%	0.00%
Ag	**3.97E+00**	0.14%	0.11%	0.12%	0.05%
Cd	**4.00E+00**	0.03%	0.02%	0.01%	0.00%
In	**8.20E-02**	0.13%	0.03%	0.01%	0.00%
Sn	**2.65E+00**	0.66%	0.15%	0.02%	0.01%
Sb	**8.98E-01**	1.76%	0.68%	0.17%	0.06%
Te	**2.62E+01**	3.85%	4.16%	2.88%	0.69%
I	**1.27E+01**	5.70%	8.94%	6.39%	0.65%
Xe	**3.07E+02**	4.33%	3.60%	5.12%	0.41%
Cs	**1.61E+02**	3.82%	1.27%	0.46%	1.61%
Ba	**8.21E+01**	4.67%	3.75%	3.46%	3.45%
La	**6.99E+01**	4.71%	5.22%	3.57%	3.25%

TABLE 5.1 Inventory and Variation in the Activity of Fission Products in a 900 MWe PWR after Shutdown—cont'd

		Activity (% of total activity)			
Fission products	Total mass (kg at shutdown)	Shutdown	1 Hour	1 Day	1 Month
Ce	**1.63E+02**	3.61%	5.04%	7.41%	16.01%
Pr	**6.21E+01**	3.10%	4.63%	5.49%	11.76%
Nd	**2.07E+02**	0.68%	1.07%	1.25%	0.82%
Pm	**1.24E+01**	0.65%	1.22%	1.65%	1.48%
Sm	**3.57E+01**	0.21%	0.46%	0.54%	0.00%
Eu	**8.90E+00**	0.08%	0.19%	0.29%	0.36%
Actinides					
U	**6.99E+04**	9.37%	3.91%	0.00%	0.00%
Np	**3.15E+01**	9.37%	22.76%	29.86%	0.02%
Pu	**5.89E+02**	0.05%	0.11%	0.19%	0.80%
Am	**6.18E+00**	0.00%	0.00%	0.00%	0.00%
Cm	**2.09E+00**	0.01%	0.03%	0.06%	0.21%
Total activity (Bq)		**5.91E+20**	**2.42E+20**	**1.39E+20**	**3.30E+19**

For most FPs, the main term expressing their production comes from the fission of heavy nuclei via the associated fission yield (the first term of the general equation (1)), whereas the main term for the disappearance of radioactive FPs comes from their radioactive decay. Equation (1) can therefore be simplified by the following equation:

$$\frac{dN_i}{dt} = y^i_{eq} \cdot F - \lambda_i \cdot N_i \tag{1a}$$

where:

y^i_{eq} is the equivalent fission yield for all of the heavy nuclei

$F = \phi \cdot \sum_{hn} \sigma_f^n \cdot N_{hn}$ represents the number of fissions/second and can be simplified by the ratio P/E_f of the irradiation or thermal power (P) over the mean fission energy (E_f) equivalent to about 200 MeV.

This equation can be easily integrated, which makes it possible to perform design calculations manually.

$$N_i = \frac{y_{eq}^i \cdot F}{\lambda_i} \cdot (1 - e^{-\lambda_i \cdot t}) \tag{2}$$

Two boundary conditions can also be deduced from this simplified equation:

- For stable FPs or radioactive FPs with long half-lives in terms of the irradiation time ($\lambda_i \rightarrow 0$), the inventory increases linearly as a function of time:

$$N_i = y_{eq}^i \cdot F \cdot t \tag{3}$$

- For radioactive FPs with short half-lives in terms of the irradiation time ($\lambda_i \rightarrow \infty$), the inventory is limited by the saturation value and therefore stabilizes at this value when the irradiation time exceeds three times that of the FP half-life:

$$N_i = \frac{y_{eq}^i \cdot F}{\lambda_i} \tag{4}$$

These simplified expressions clearly show the preponderant influence of the fission yield. The fission yield is distributed according to a two-peak curve, which depends only on the second order of the fission type (thermal for ^{235}U or ^{239}Pu and fast for ^{238}U, etc.). The highest yields concern the mass numbers ranging from 85 to 105 and from 130 to 150 (see Figure 5.2). There is nevertheless a slight shift in this curve toward higher masses for plutonium fissions, which results in a higher production of the elements Ru, Rh, and Pd. In particular, the fission yields of ^{103}Ru and ^{106}Ru are, respectively, two times and ten times higher when fission occurs with ^{239}Pu in comparison with ^{235}U.

5.2.2. Specificity of Stable Fission Products

The total mass inventory of the FPs therefore mainly concerns stable FPs and several long-lived radioactive FPs. This inventory increases almost linearly as a function of the burn-up at about 75 kg/(GWd/t), which is equivalent to about 2 tons of FPs produced in a 900 MWe PWR core in equilibrium.

The stable FPs control the physico-chemical variations in the fuel under accident conditions. For this reason, they have a negative effect on core degradation for two reasons:

- The potential formation of UO_2-FP eutectics[3] tends to lower the melting temperature [5], with the FP concentration in the fuel being significant at high burn-ups (higher than 10 atomic % above 50 GWd/t);

3. ZrO_2 from cladding oxidation can also play a role in reducing the melting temperature of the corium mixture, as also discussed in [5].

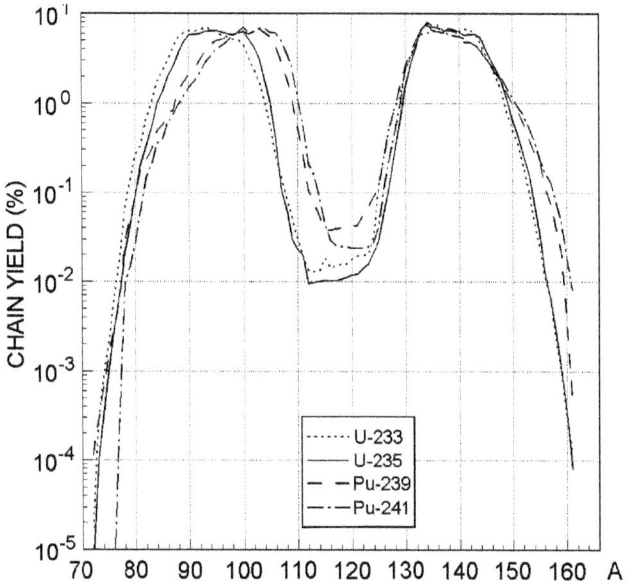

FIGURE 5.2 Yield of fission products according to the mass number. *(from [153] with permission from Elsevier.)*

- The action of pressure from the fission gases can cause the grain boundaries to fracture and result in the formation of fuel debris.

5.2.3. Specificity of Radioactive Fission Products

Though negligible in terms of mass, radioactive FPs are responsible both for most of the core decay heat and the radioactivity that could be released into the environment.

The activity of each isotope decreases exponentially after reactor shutdown[4] according to the following equation:

$$A_i(t) = \lambda_i \cdot N_i(t) = A_{i,t0} \cdot e^{-\lambda_i \cdot \Delta t} \tag{5}$$

By intercomparison, the impact of these radioactive isotopes reaches a maximum when the decay time $\Delta t = t - t_0$ is equal to the order of magnitude of the half-life of each FP.

4. Except for daughter FPs, that is, products with a parent radionuclide that has a higher or equivalent half-life such as $^{140}Ba/^{140}La$, $^{132}Te/^{132}I$, and $^{95}Zr/^{95}Nb$ for which the activity variation of the daughter product is also related to that of the parent.

For example, the radiological impact of iodine, which only has short-lived isotopes[5] (the longest being 8 days for ^{131}I), covers a period of about one month following an accident. Conversely, the radiological impact of caesium (two long-lived isotopes: 2 years for ^{134}Cs and 30 years for ^{137}Cs) is minor in the days following an accident and only becomes preponderant after one year.

In schematic terms, the impact of radioactive FPs under accident conditions is dependent on their degree of volatility:

The most volatile FPs are released from the vessel, transported and partially deposited in the primary cooling system and then in the containment where they can thus contaminate the environment via leakage paths. In this case, the most radiotoxic isotopes are ^{133}Xe, ^{132}Te, ^{132}I, and ^{131}I in the short term, and ^{134}Cs and ^{137}Cs in the long term. Ruthenium (^{103}Ru in the medium term and ^{106}Ru in the long term) may also be a hazard in specific accident scenarios with air ingress in the core.

Nonvolatile FPs remain associated with the fuel (^{239}Np and ^{140}La in the short term, and $^{95}Zr/^{25}Nb$ and ^{144}Ce in the medium and long term). Owing to the decay heat that continues to be released, they can heat and melt the core if the latter is not cooled sufficiently.

Table 5.2 lists the main radioactive FPs and specifies their scope of action in the short, medium, and long term, as well as their degree of volatility.

Figure 5.3 shows the variations in the decay heat of a 900 MWe PWR after its shutdown by indicating the contribution of the different classes of radionuclides [6]. This variation decreases very quickly in time: it represents about 5% of the total reactor power at the time of shutdown, and then only 1% after one hour and 0.1% after one month. The significant contribution of volatile FPs in the short term comes from the iodine isotopes and lanthanide isotopes (in green), which cover a longer half-life owing to ^{140}La. Actinides (in red) also greatly contribute in the short term due to the isotope ^{239}Np with a half-life of 2 days.

5.2.4. Physico-chemical State of Fission Products in the Fuel

Under nominal irradiation conditions in a PWR, the FPs will accumulate in the fuel matrix according to different chemical states [7]. Based on a degree of decreasing volatility, the following categories can be distinguished:

- Dissolved oxides, which represent almost half of all FPs, particularly Sr, Y, Zr, La, Ce, and Nd;
- When the limit of solubility is reached, some of these FPs will precipitate into oxides. This mainly concerns Ba and Nb;
- As the oxygen potential is imposed by the oxide fuel, other FPs form metallic precipitates: Tc, Ru, Rh, Pd, and Mo partially;

5. Other than ^{129}I with a half-life of $1.6.10^7$ years, which can be considered as a practically stable element and whose significance is more so related to "waste disposal" rather than its potential impact under accident conditions.

TABLE 5.2 Main Radioactive Fission Products

Active in the short term		Active in the medium term		Active in the long term	
Fission product	Half-life	Fission product	Half-life	Fission product	Half-life
Kr88	2.8 h	Zr95/Nb95	64 d/35 d	***Kr85***	10.7 y
Sr91/Y91m	9.5 h/0.8 h	Mo99	2.8 d	Sr90	28.6 y
Sr92/Y92	2.7 h/3.7 h	Ru103	39 d	Ru106	1.0 y
Y93	10.5 h	Sb127	3.8 d	Ag110m	0.7 y
Zr97/Nb97	17 h/1.2 h	**I131**	8.0 d	Sb125	2.8 y
Ru105/Rh105	4.4 h/35.5 h	**Te131m**	1.2 d	**Cs134**	2.1 y
I133	20.8 h	**Te132/I132**	3.2 d/2.3 h	**Cs137**	30.1 y
I134	0.9 h	***Xe133***	5.2 d	Ce144	284 d
I135	6.6 h	***Xe133m***	2.2 d	Eu144	8.6 y
Xe135	9.1 h	Ba140/ La140	12.8 d/1.7 d		
Ce143	1.4 d	Ce141	32 d		
		Ce143	1.4 d		
		Nd147	11.1 d		
		<u>Np239</u>	2.43 d		
Key	***XXX***	Fission gas	XXX	Non-volatile FP	
	XXX	Volatile FP	<u>XXX</u>	Actinide	

- To date, the chemical state of <u>volatile FPs</u> (Br, Rb, Te, I, Cs) has not been fully explained. Most are probably in the form of dissolved atoms. Above a certain temperature, they are found in a gaseous form (especially in the center of the pellet) and can migrate radially to condense in cooler areas, such as upon contact with the cladding. It has often been hypothesized that compounds such as CsI, Cs_2MoO_4, and Cs_2Te or caesium uranates are formed, but this hypothesis remains to be confirmed experimentally;
- Lastly, <u>fission gases</u> (xenon and krypton) exist as atoms dissolved in the UO_2 matrix, or in the form of gas bubbles located in inter- and intragranular

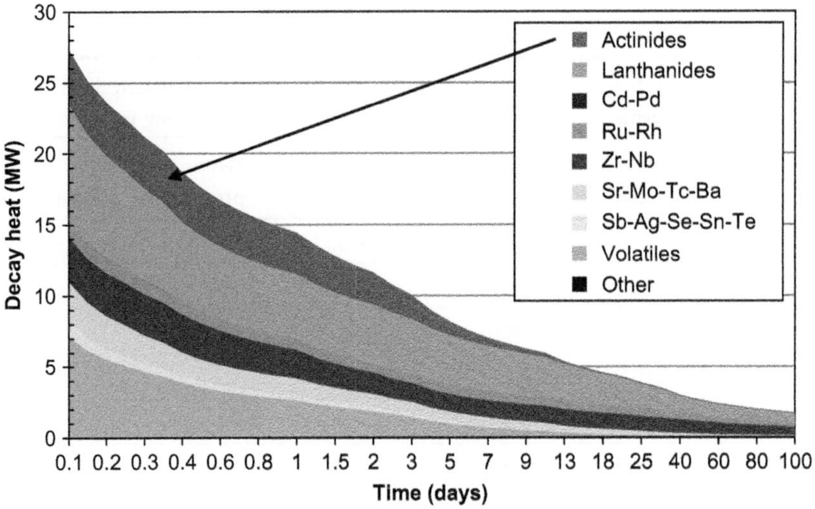

FIGURE 5.3 Variations of decay heat in a 900 MWe PWR between 0.1 and 100 days.

spaces, with the gas solubility limit being low. The fraction of gas accumulated in the grain boundaries is most likely to be released under accident conditions, in particular under LOCA conditions.

The chemical state of the FPs from the first three categories is not stable: some can move from one category to another depending on the operating temperature, the oxygen potential in the fuel (which increases with the burn-up since fissions have an oxidizing nature), and the burn-up (increase in the concentration of FPs in the matrix). This is particularly true for molybdenum, which mainly precipitates in a metallic form but which can also be found in an oxidized state (particularly on the periphery of the pellets or in MOX fuel), or for niobium and strontium whose oxides can be partially dissolved and partially precipitated.

Figure 5.4 was taken from reference [7] and summarizes these different chemical states.

5.3. IN-VESSEL FISSION PRODUCT RELEASE

(M. Kissane, G. Ducros, and M. Koch)

5.3.1. Phenomenology of Fission Gas Release

Fission Gases

During normal irradiation in a PWR, fission gases form in the grains of UO_2. These atoms of gas either diffuse toward the grain boundaries or precipitate into intergranular bubbles, slowing down their rate of migration

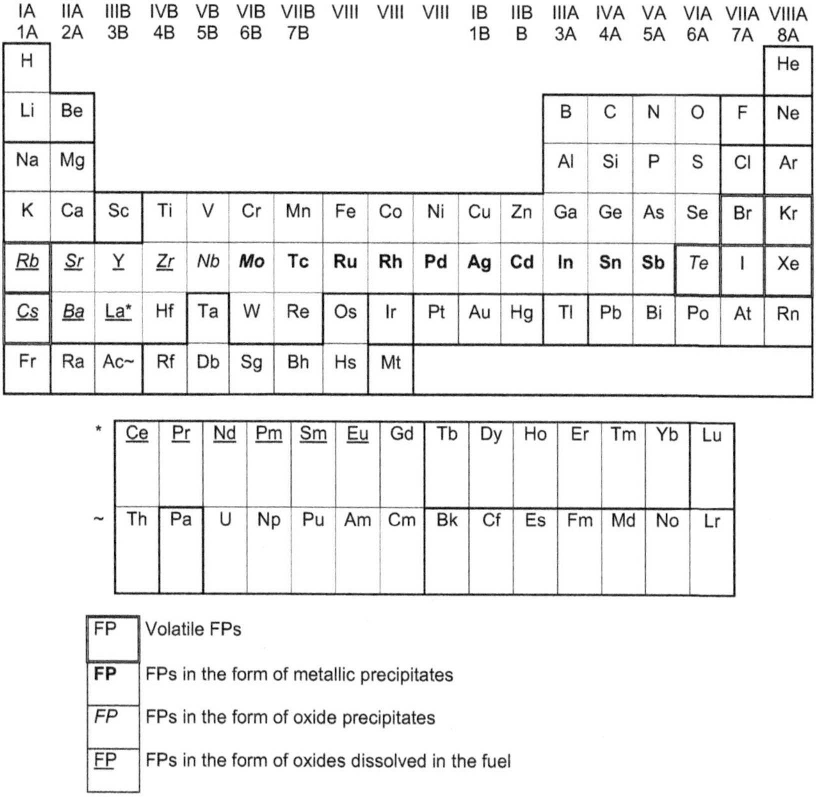

Chapter | 5 Fission Product Release and Transport

IA 1A	IIA 2A	IIIB 3B	IVB 4B	VB 5B	VIB 6B	VIIB 7B	VIII	VIII	VIII	IB 1B	IIB B	IIIA 3A	IVA 4A	VA 5A	VIA 6A	VIIA 7A	VIIIA 8A
H																	He
Li	Be											B	C	N	O	F	Ne
Na	Mg											Al	Si	P	S	Cl	Ar
K	Ca	Sc	Ti	V	Cr	Mn	Fe	Co	Ni	Cu	Zn	Ga	Ge	As	Se	Br	Kr
<u>Rb</u>	<u>Sr</u>	Y	Zr	Nb	Mo	Tc	Ru	Rh	Pd	Ag	Cd	In	Sn	Sb	Te	I	Xe
<u>Cs</u>	<u>Ba</u>	La*	Hf	Ta	W	Re	Os	Ir	Pt	Au	Hg	Tl	Pb	Bi	Po	At	Rn
Fr	Ra	Ac~	Rf	Db	Sg	Bh	Hs	Mt									

*	Ce	Pr	Nd	Pm	Sm	Eu	Gd	Tb	Dy	Ho	Er	Tm	Yb	Lu
~	Th	Pa	U	Np	Pu	Am	Cm	Bk	Cf	Es	Fm	Md	No	Lr

FP	Volatile FPs
FP	FPs in the form of metallic precipitates
FP	FPs in the form of oxide precipitates
<u>FP</u>	FPs in the form of oxides dissolved in the fuel

FIGURE 5.4 Chemical state of fission products in the fuel *(from [7])*.

toward the grain boundaries. The bubbles may then re-dissolve under the influence of fission spikes, phenomena that speed up the rate at which gas is supplied to the grain boundaries. Once on the grain surface (mainly by atomic diffusion, but also by bubble migration), the fission gases accumulate to a point where they coalesce to form larger bubbles and fill the boundaries. These bubbles are then capable of moving into the free volume of the rod.

During the accident, the gas populations can be classified into four categories:

- Gas atoms dissolved in the matrix;
- Intragranular gas bubbles with little mobility;
- Gas accumulated in the grain boundaries;
- Gas released in the gap and in the upper plenum of the fuel rod.

Their release is governed by a number of mechanisms. The first release phase (often referred to as the "burst release") corresponds to the release of gases accumulated in the intergranular spaces. The fraction already released in the rod plenum during normal irradiation must also be taken into account, ranging from a few percent to 10%, depending on the burn-up, the irradiation power, and the fuel type. Such releases occur at the beginning of the temperature rise at around 1000°C, though this is sometimes lower for high burn-up fuels. The second phase involves the release of intragranular gases via a thermally activated diffusion process that begins with dissolved atoms. The gases trapped in the intragranular bubbles are the last to be released, which generally occurs when the fuel melts.

It is therefore important to correctly quantify the respective fractions of these four gas populations when modeling gas releases, which depend on their radial position in the pellet (different temperatures, HBS[6] microstructure on the periphery, etc.) and the fuel type (heterogeneous MOX fuels have a higher intergranular fraction). It is also important to point out that these fractions depend on the radioactive half-life of the gases, which has a positive impact on the source term in as much as short-lived gases have had the time to decay after having migrated toward the grain boundaries. This effect is mainly beneficial for LOCA- or RIA-type accidents as the total gas inventory is generally released during a severe accident.

Other Fission Products

It is generally accepted that the release of FPs follows a two-phase process: (1) the FPs in solution in the matrix (or the precipitates when the solubility limit has been reached) diffuse as far as the grain boundaries, and then (2) a mechanism of mass vaporization transfers the FPs from the grain surface outside of the fuel matrix. This mechanism also involves a number of physico-chemical aspects: potential formation of defined compounds (CsI, molybdates, zirconates, and uranates of caesium, barium, strontium, etc.), or the oxidation or reduction of precipitates by steam and/or hydrogen. These chemical reactions have a significant impact on the volatility of some elements. The basic high-temperature thermodynamic parameters governing the formation and destruction of these species are currently poorly understood, which is problematic when it comes to the mechanistic modeling of these processes.

Other than releases outside the fuel matrix, potential chemical interactions with the cladding and/or the core structural elements can also reduce the volatility of some elements through the formation of more refractory species. Last of all, once released from the core, a significant fraction of the FPs condenses in the colder areas of the upper core structures, before even reaching the primary cooling system or the containment. This is especially true for low-volatile FPs.

6. High Burn-up Structure, or rim structure.

Qualitatively speaking, the main physical parameters influencing the release of FPs are as follows:

- The temperature is the main parameter, at least until loss of the core geometry;
- The oxidizing-reducing conditions have a significant impact on the fuel. The release kinetics of volatile FPs are particularly accelerated under oxidizing conditions. Furthermore, the overall release of certain FPs is very sensitive to the oxygen potential. For example, the release of Mo increases in steam, whereas that of Ru can be very high in air. Conversely, the release of Ba (as for Sr, Rh, La, Ce, Eu, and Np) increases under reducing conditions;
- The interactions with the cladding and/or the structural elements can play a major role. For example, the presence of tin in the cladding delays the emission of the volatile elements tellurium and antimony. Barium significantly contributes to the decay heat (via its daughter product ^{140}La) and is also partially trapped in both the cladding (probably due to the formation of zirconates) and the structural steels;
- The burn-up accentuates releases, in terms of both the kinetics of volatile FPs and the release amplitude of low-volatile species such as Nb, Ru, La, Ce, and Np;
- The fuel type also seems to have a significant impact: MOX releases tend to be higher than those of UO_2. This phenomenon is probably related to its heterogeneous microstructure, with the presence of plutonium-rich agglomerates where the local burn-up can be very high;
- Last of all, the state of the fuel during its in-vessel degradation has a significant influence: the transition from a "degraded rod" geometry to a "debris bed" geometry also involves an increase in releases via the increase in the surface-to-volume ratio. Conversely, the transition from a debris bed to a molten pool slows down the release of FPs as a solid crust forms on the surface of the molten pool.

Degree of Volatility

We owe our current state of knowledge to numerous analytical or separate-effects programs performed in various countries, not to mention several integral-type programs such as the international program Phébus FP.[7] These different programs (summarized in the following section) made it possible to

7. Here we use the definitions "integral" and "separate effects" as used in the CSNI Code Validation Matrix for core degradation experiments [8]. "Integral" tests are performed to simulate simultaneously different phenomena and processes and to study the interactions amongst them. Facility and boundary conditions are oriented on real reactors and possible conditions during postulated transients. "Separate-effects" tests, on the other hand, are performed to study single phenomena, to support model development and to derive quantitative correlations for the simulation of those processes.

schematically classify the FPs and fission gases into four categories of decreasing volatility, with the following characteristics:

- *Fission gases* (xenon and krypton) and *volatile FPs*: iodine, caesium, bromine, rubidium, tellurium, antimony, and probably silver. Almost all of these products are released as of 2300°C, even before reaching molten pool conditions. The release kinetics of these elements are accelerated under oxidizing conditions and are slightly delayed for Te and Sb due to interactions with tin in the cladding,[8] which has little impact on the source term in the end;
- *Semivolatile FPs*: molybdenum, rhodium, barium, palladium, and technetium. They are characterized by high levels of release, which are sometimes equivalent to those of volatile FPs, that is, almost the totality, while being highly sensitive to the oxidizing-reducing conditions and resulting in significant retention in the upper core structures;
- *Low-volatile FPs*: strontium, yttrium, niobium, ruthenium, lanthanum, cerium, and europium. They are characterized by low but significant levels of release, ranging from a few percent to 10% of the initial inventory during the fuel rod degradation phase (prior to loss of core geometry). Nevertheless, these releases can reach much higher levels for fuels with very high burn-ups (15 to 30% releases were measured for Nb, Ru, and Ce in UO_2 fuel at 70 GWd/t) or under specific conditions; for example, the release of Ru can be almost total in air, whereas reducing conditions will favor the release of Sr, La, Ce, and Eu. Significant retention of these elements is nevertheless expected in the upper core structures;
- *Nonvolatile FPs*: zirconium, neodymium, and praseodymium. To date, no significant release of these three elements has been demonstrated experimentally. These are the most refractory FPs.

Lastly, actinides can be subdivided into two categories. The first category includes uranium and neptunium whose releases can reach 10% before the fuel melts. This behavior is similar to that of low-volatile FPs. Uranium releases are higher in oxidizing conditions, whereas neptunium releases are favored by reducing conditions. Plutonium releases are always very low—typically below 1%—which means they tend to behave more like nonvolatile FPs. Plutonium releases can be higher under reducing conditions.

These volatility classes have been clearly demonstrated through online gamma spectrometry measurements performed on fuels irradiated in analytical experiments. Figure 5.5 (drawn from the VERCORS 5 test [9], [10]) illustrates these four categories:

- volatile FPs (represented by ^{137}Cs);
- semivolatile FPs (represented by ^{140}Ba);

8. For Zircaloy cladding containing 1 to 2% tin.

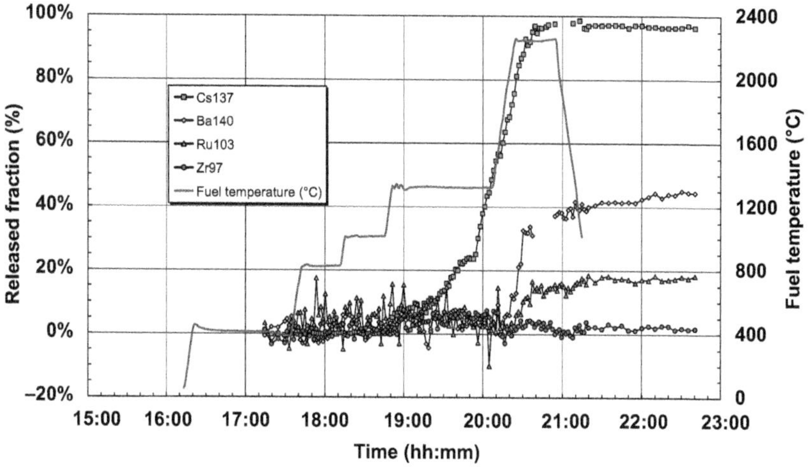

FIGURE 5.5 The four classes of volatility for fission products according to the VERCORS 5 test *(from [10])*.

- low-volatile FPs (represented by ^{103}Ru);
- nonvolatile FPs (represented by ^{97}Zr).

5.3.2. Experimental Programs Devoted to Fission Product Release

The experimental programs on in-vessel FP releases have mainly relied on out-of-pile analytical experiments performed on sections of irradiated fuels. They were eventually supplemented by in-pile integral tests, particularly in terms of coupling core degradation with FP releases. A detailed up-to-date bibliography of such programs is available in reference [11].

Analytical Experiments

Six major analytical programs have been performed since the 1970s: SASCHA in Germany [12]; HI/VI in the United States [13]; CRL in Canada [14]; HEVA/VERCORS in France [10], [15], [16], [17]; VEGA in Japan [18]; and QUENCH-VVER [19] in Russia.

The **SASCHA** program was the first analytical experiment of its kind. A total of about 50 tests were performed, including two tests on molten corium–concrete interaction. It was performed on unirradiated UO_2 fuel with Zircaloy cladding mixed with steel (plus Ag-In-Cd control rod material and Inconel). The UO_2 pellets were manufactured with simulant FP tracers (simfuel) at a concentration equivalent to a burn-up of 44 GWd/t. This combination of cladding fuel and steel weighed about 150 g and was heated at temperatures between 1600°C and 2300°C under various atmospheres, including argon, air, and steam. Eutectics formed, and the mixture melted completely at 2300°C because of the presence of steel. Furthermore, the method of implanting additives during pellet sintering to simulate the FPs in the grain boundaries was not really representative of the real location of the FPs during irradiation. Nevertheless, this program did provide a series of preliminary estimates for the main FPs, especially iodine and caesium up to 2000°C, as well as establishing the coefficients of the CORSOR laws (see Section 5.3.3). Figure 5.6 illustrates the SASCHA test facility.

ORNL conducted the **HI/VI program** between 1981 and 1993 on behalf of the NRC. This groundbreaking program was considered a reference for a long time. The fuel samples were 15 to 20 cm long sections of UO_2 fuel (100 to 200 g) sealed at the ends and irradiated in a power reactor. A hole was drilled in the cladding at midplane height. It was heated in an induction furnace. A horizontal configuration was implemented in six tests from the HI series (burn-up 10–40 GWd/t), while a vertical configuration was used in the seven tests from the VI series (burn-up 40–47 GWd/t). Figure 5.7 illustrates the VI and HI test loops.

The FPs were collected downstream of the furnace using three devices that operated in sequence. Each device was equipped with a thermal gradient tube (TGT) to collect condensable gases, a series of filters designed to trap aerosols, and active carbon cartridges to trap the volatile forms of iodine and fission gases. NaI detectors were set up for the online measurement of rare gases condensed in the cold traps (^{85}Kr), and the ^{134}Cs and ^{137}Cs were deposited in the TGTs and filters. From the VI-4 series onward, an additional online gamma

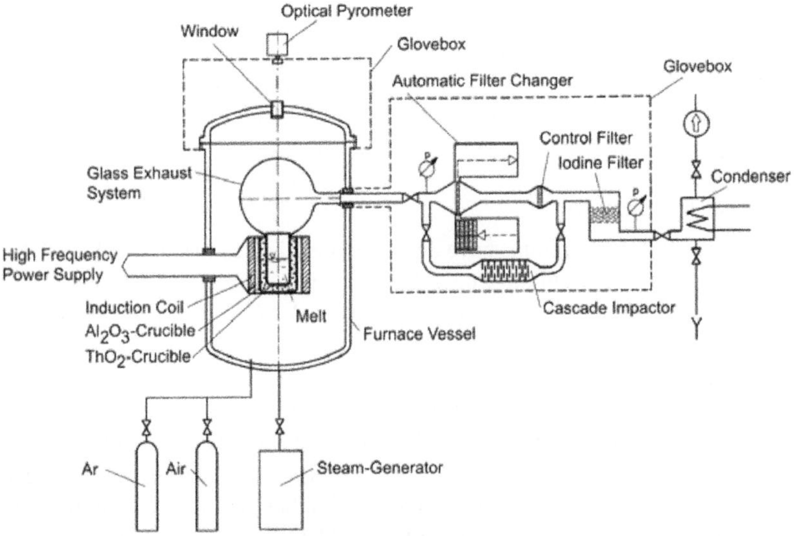

FIGURE 5.6 SASCHA test facility for core-melt release studies *(from [154])*.

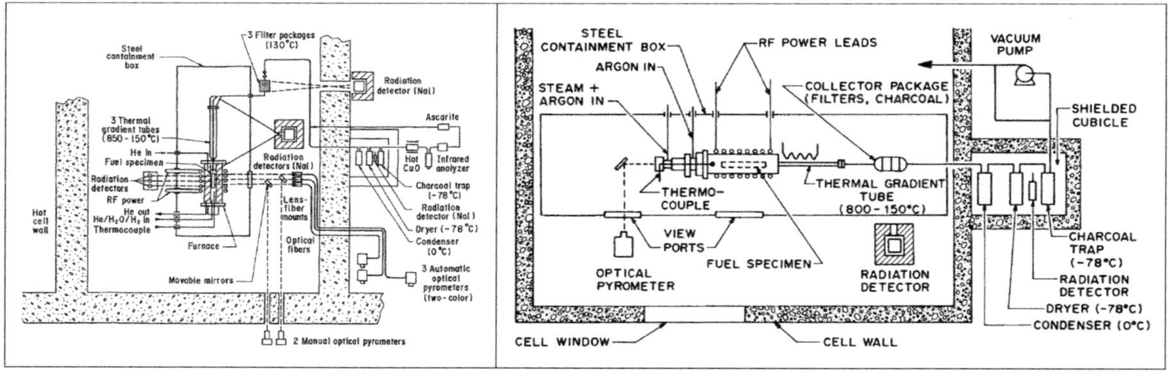

FIGURE 5.7 Experimental loops VI (left) and HI (right) *(from [155], [156])*.

spectrometry device was installed at the bottom and top of the fuel in order to monitor fuel rod collapse. This program provided highly representative results on the release of FPs, but only in relation to long-lived FPs (mainly ^{85}Kr, ^{106}Ru, ^{125}Sb, ^{134}Cs, ^{137}Cs, ^{144}Ce, and ^{154}Eu) as the samples were not re-irradiated prior to the tests.[9]

The **CRL program** is a highly analytical program and is still in progress. It consists of many tests (more than 300) on fragments of irradiated UO_2 fuel (100 mg to 1 g) and on short cladded fuel sections. The burn-up is rather low—characteristic of the CANDU reactor technology—and ranges between 5 and 20 GWd/t.[10] Some samples are re-irradiated before the tests to measure the short-lived FPs. Tests are performed under a mixture of Ar/H_2, steam or air at temperatures limited to 2100°C due to the use of a resistance furnace. These tests are greatly contributing to the understanding of the effect of the oxygen potential on the release of low-volatile FPs. The first program in particular revealed the very high release of ruthenium in air, including at low temperatures of about 1000–1500°C. Certain tests on cladding-free fragments have also revealed the very high volatilization of the fuel at high temperature and under extremely oxidizing conditions, associated with a high release of very low-volatile or nonvolatile FPs, via a "matrix stripping" process: 50% for ^{95}Nb and ^{140}La, and 30% for ^{95}Zr.

The **HEVA/VERCORS program** was conducted by the CEA for IRSN and EDF, and aimed at quantifying the release of FPs and heavy nuclei (kinetics and total release) from irradiated nuclear fuel under conditions representative of a severe accident. These tests were performed in a high-activity cell on different types of fuel samples irradiated in a PWR (around 20 g of fuel) under a range of experimental conditions. Most of the samples were re-irradiated for a few days at low power in an experimental reactor in order to build up an inventory of the short-lived FPs responsible for the worst radiological effects. These samples were then heated in an induction furnace under a variable atmosphere of steam and hydrogen. Fission product releases were measured online by gamma spectrometry during the accident sequence, with a direct view of the fuel. Other complementary online gamma spectrometry equipments were installed in view of specific parts of the loop (thermal gradient tube, filters, gas capacity). Quantitative gamma spectrometry measurements of all test loop components were then taken after dismantling. Twenty-five tests were carried out between 1983 and 2002 in three phases: (1) eight HEVA tests (release of volatile and some semivolatile FPs up to 2100°C), (2) six VERCORS tests (volatile, semivolatile, and some low-volatile FPs up to 2300°C, the limit for fuel collapse),

9. Additional measurements were performed by chemical analysis after rinsing the test loop components. This was needed to quantify the release of other elements (Sr, Mo, Te, I, Ba, U, and Pu), but the results were very imprecise in comparison with the spectrometry measurements of the radioactive fission products.

10. Except for two tests performed with PWR fuel at 54 GWd/t.

and (3) eleven HT/RT tests involving all types of FPs until the melting point was reached. The HT test loop (see Figure 5.8) was the most complex and the most instrumented loop. It was also designed with the goal of studying the transport of FPs in a PWR primary cooling system and their potential interactions with the neutron absorbing components of PWRs (Ag, In, Cd, and B).

One advantage of this program was the extensive scope of the experimental database compiled, particularly thanks to the optimized fuel re-irradiation conditions that made it possible to quantify FP releases over a wide range of half-lives, from around 10 hours up to 30 years. The parameters investigated during these tests included the temperatures reached (above or below the melting point of the fuel), the oxidizing-reducing conditions of the atmosphere, the burn-up (up to 72 GWd/t), the fuel type (usually UO_2 though two tests used MOX) and the initial fuel geometry (intact or fuel debris to simulate the final phase of a severe accident).

The **VEGA program** (Verification Experiment of Gas/Aerosol Release) was very similar to the VERCORS program, especially the VERCORS HT series. This program was performed by JAEA between 1999 and 2004, including a total of 10 tests: eight on UO_2 fuel and two on MOX fuel. The samples—heated to high temperatures in helium or steam—were irradiated pellets with or without their original cladding. The experimental device was very similar to the one used in the VI program, particularly with TGTs and filters downstream of the furnace operating in a sequential mode (Figure 5.9).

Some samples were re-irradiated prior to the tests but under less than optimal conditions (irradiation time too short and decay time too long) when measuring very short-lived FPs. A unique feature of this program was the addition of tests at 1 MPa, which demonstrated the reduction in caesium releases.

The **QUENCH-VVER program** is the only analytical program studying FP releases under reflood conditions. This program was financed by the International Science and Technology Centre (ISTC) and was performed by the NIIAR in Dimitrovgrad. The fuel samples (15 cm long sections of VVER UO_2 fuel with original cladding at burn-ups of 50–60 GWd/t) were pre-oxidized at 1300°C in steam prior to testing and then heated to reflood temperature. Reflood was simulated by quickly flooding the sample in a tank of water at ambient temperature in the furnace (Figure 5.10).

The fission gases released were measured online by gamma spectrometry (^{85}Kr) and mass spectrometry (stable xenon isotopes). Volatile FPs (mainly ^{137}Cs because there was no re-irradiation) were measured after testing in the reflood water and in the fuel sample. About 20 tests were carried out, with variable cladding pre-oxidation rates and reflood temperatures (maximum of 1700°C).

Integral Tests

When combined with analytical tests, integral tests are very useful in compiling a comprehensive database. They mainly help representing the

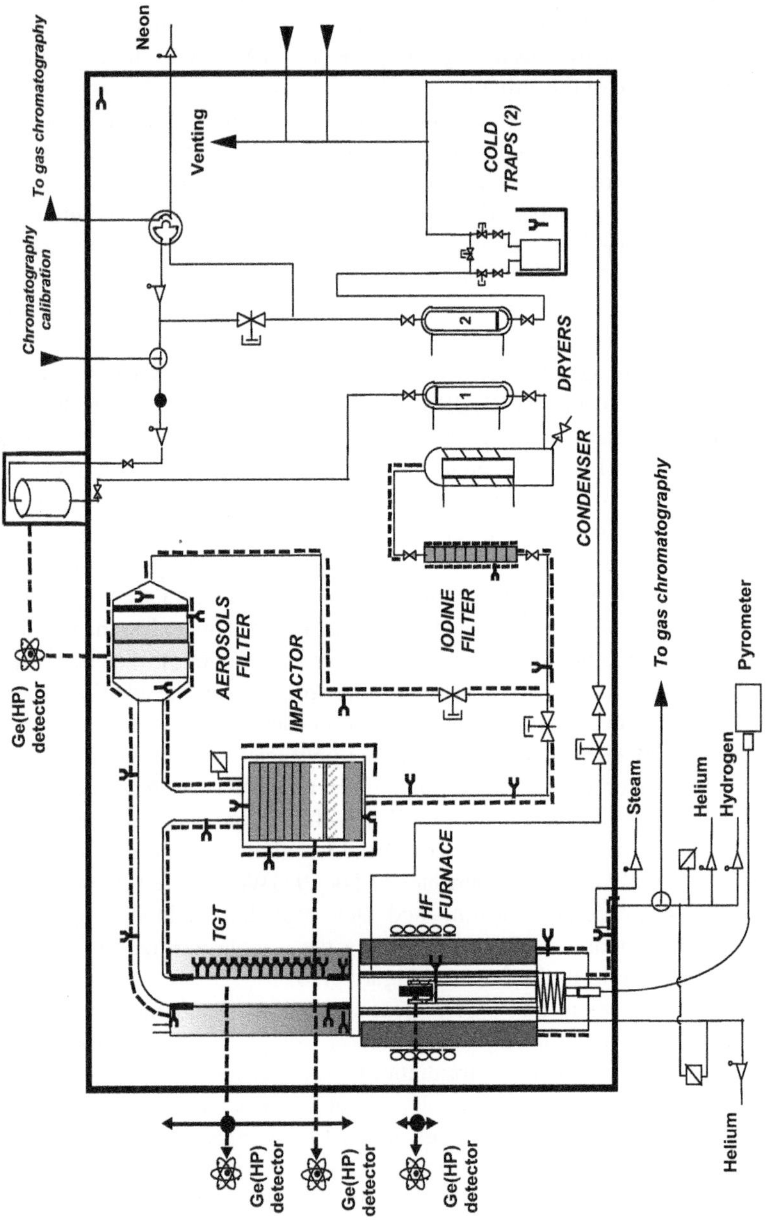

FIGURE 5.8 VERCORS HT loop.

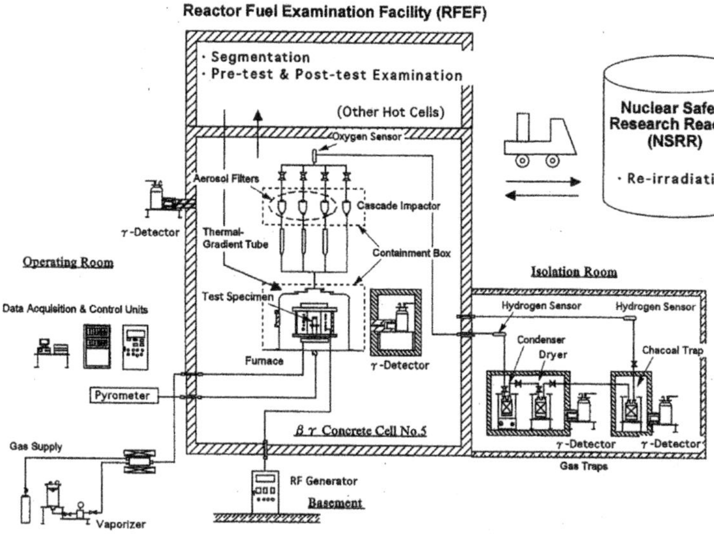

FIGURE 5.9 VEGA loop *(from [157])*.

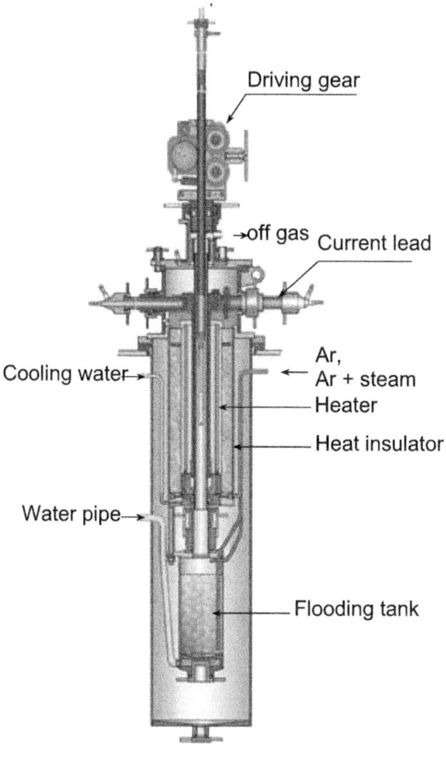

FIGURE 5.10 QUENCH-VVER device installed in a hot cell *(from [19])*.

physical phenomena coming into play during a severe accident. Other than studying in-vessel FP releases, they are also used to study FP transport in the primary cooling system and their behavior in the reactor containment in a more representative way. Performed with larger quantities of fuel, integral tests make it possible to lift potential bias from very small-scale analytical tests.

Generally performed in an experimental reactor, these tests nevertheless have trouble precisely quantifying FP releases, especially since it is impossible to set up online measurement systems very close to the fuel. Integral tests do, however, provide complementary information such as on the coupling of core degradation and FP releases. They are also used to study the effect of material interactions upon the drop in the release of certain FPs.

Numerous programs were launched after the TMI-2[11] accident, which were continued for over two decades between 1980 and 2000.

The following examples are worth mentioning:

- The ACCR-ST program [21] (1985–1989) involved two tests performed on a mini-bundle of four rods irradiated down 15 cm. These tests were used to compare releases obtained by reactor heating with those recorded during the VI analytical tests using furnace heating;
- The PBF-SFD program [22] (1982–1985) involved four tests performed on a bundle of 32 rods 1 meter in length;
- The FLHT program [23] (1985–1987) involved four tests on a bundle of 12 real-length rods (3.7 meters);
- The LOFT-LP program [24] (1984–1985) was performed in a test reactor with a fuel assembly of 100 rods 1.7 meters in length. It included a full PWR loop (steam generator, pressurizer, and primary pumps). The second test (FP2) included a reflood phase and was one of the most representative tests in terms of a PWR severe accident.

More recently, the Phébus FP program [25], [26] has probably been the most instructive thanks to its highly developed FP measuring instrumentation. The tests have covered all accident phases, from the start of core degradation to the formation of a molten pool in the reactor vessel. The in-pile test section was composed of a bundle of 20 irradiated fuel rods 1 meter in length, except for the FPT0 test (20 fresh fuel rods) and the FPT4 test (debris bed configuration). A central fuel rod was used to simulate a control rod (SIC or B_4C). The FPs released were transported through temperature-regulated lines in order to represent the transport conditions in the primary cooling system. The system was connected to a 10 m^3 tank simulating a PWR reactor containment

11. TMI-2 provided a lot of useful information on fission product releases in reactor vessels [20] .

and equipped with a sump in the lower part, which contained pH-regulated water (Figure 5.11).

Five tests were performed between 1993 and 2004:

- The FPT0 test [27] (1993) was performed with fresh fuel and a silver/indium/cadmium (SIC) control rod under steam-rich conditions with an acid sump;
- The FPT1 test [28] (1996) was identical to FPT0, but the fuel was irradiated to 24 GWd/t;
- The FPT2 test [29] (2000) used the same fuel and control rod as in test FPT1, but in a less oxidizing environment with an alkaline sump that was evaporating at the end of the test;
- The FPT3 test [30] (2004) was performed under the same conditions as FPT2, but with a boron carbide (B_4C) control rod and an evaporating acid sump;
- The FPT4 test [31] (1999) used an initial debris bed configuration to simulate the final phase of a severe accident.

In terms of in-vessel FP releases, the Phébus FP program made a huge contribution by identifying:

(a) The coupling between degradation and barium releases, with these releases being much less than those recorded in the analytical tests. The difference in behavior was explained by Ba interactions with the cladding, and possibly iron, which reduced its volatility;
(b) The low releases from the molten pool.

Otherwise, the results of FP releases were consistent with those from the analytical tests.

More than FP release results, this program made its highest contribution on FP transport in the RCS and behavior in the containment, particularly for the understanding of iodine chemistry, such as its physical and chemical forms in the circuit (e.g., partition between aerosol and gaseous/vapor forms), and retention in the sump water and reaction with paints in the containment. The latter points are particularly relevant concerning the amount of volatile iodine that remains in the containment atmosphere, which could potentially be released to the environment in the case of containment failure.

5.3.3. Calculation Models and Codes

Two approaches are traditionally used to model FP releases and to integrate these models into the computer codes: (1) an empirical approach uses simplified models that can be easily integrated into the integral codes, and (2) a mechanistic approach describes models in a very detailed way with all the physical phenomena coming into play.

Among the simplified models, the first to be implemented were the so-called CORSOR models, which were followed by slightly more elaborate models, based on the "limiting principle" phenomenon: the ELSA module of

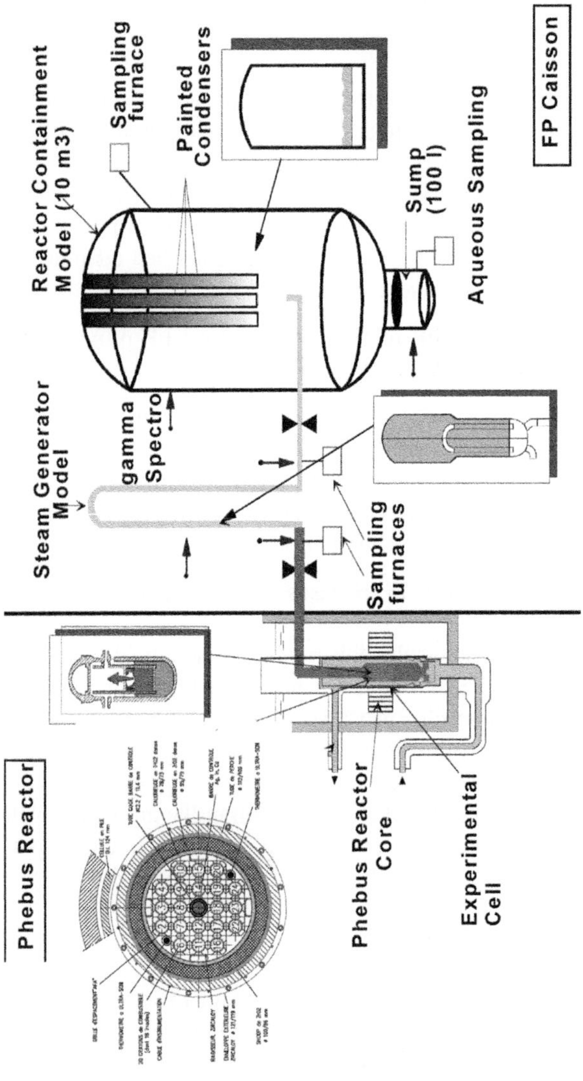

FIGURE 5.11 Phébus FP facility.

the ASTEC code is a good illustration of this. Lastly, today the MFPR code can be considered today as a reference in mechanistic codes.

CORSOR Models

These models are integrated into most international integral codes, such as MELCOR or MAAP, or in the ATHLET-CD code. The FPs are grouped together in around 10 volatility categories. The variation of the N_i concentration of each group is given by a simple analytical expression, associated with a release rate K, which is related to the radioactive decay [32]:

$$\frac{dN_i}{dt} = -K \cdot N_i \tag{6}$$

where:

$$N_i = N_{i,0} \cdot e^{-K \cdot t} \tag{7}$$

The release (R) at each instant (t) is thus given by the expression:

$$R = 1 - \frac{N_i}{N_{i,0}} = 1 - e^{-K \cdot t} \tag{8}$$

The release rate K depends on the fuel temperature. Two mathematical formulations of K have been established, respectively, within the CORSOR and CORSOR-M models.

The original CORSOR model calculates K according to the following expression:

$$K = A_i \cdot e^{B_i \cdot T} \tag{9}$$

where A_i and B_i are the constants determined experimentally for each FP category and adjusted in different temperature ranges, and T is the temperature.

The CORSOR-M model calculates K according to an Arrhenius-type expression:

$$K = K_0 \cdot e^{-\frac{Q}{R \cdot T}} \tag{10}$$

where K_0 and Q are constants resulting from adjustments to experimental results.

The A_i, B_i, K_0, and Q constant coefficients were mainly established on the basis of SASCHA and HI/VI analytical experiments for both types of model.

A third CORSOR-BOOTH model was implemented with the aim of improvement and concern for a better physical representation of the phenomena [33]. It is based on the diffusion of atoms in solids according the classical Fick's law [34]:

$$\frac{\partial C}{\partial t} = D \cdot \nabla^2 C \tag{11}$$

where C represents the FP concentration at a given point in the fuel and at a given instant, and D is the diffusion coefficient of the species considered in the fuel.

To simplify this expression, the Booth hypothesis consists of assuming that the fuel is made up of spherical grains with radius "a" and considering that the concentration is nil on the surface of the grain, that is, that release is total as soon as the FPs have reached the grain boundaries [35]. Variations in the released fraction *Fr(t)* as a function of time can thus be described by:

$$Fr(t) = 6\cdot\left[\frac{\mathrm{D}\cdot\mathrm{t}}{\pi\cdot\mathrm{a}^2}\right]^{1/2} - 3\cdot\frac{D\cdot t}{a^2} \text{ for } \frac{D\cdot t}{a^2} \leq \frac{1}{\pi^2} \tag{12}$$

$$Fr(t) = 1 - \frac{6}{\pi^2}\exp\left[-\frac{\pi^2 D\cdot t}{a^2}\right] \text{ for } \frac{D\cdot t}{a^2} > \frac{1}{\pi^2} \tag{13}$$

with:

$$D = D_0 \exp\left(-\frac{Q}{R\cdot T}\right) \tag{14}$$

D_0: pre-exponential term of the Arrhenius law
Q: activation energy
R: perfect gas constant

The coefficients D_0 and Q were established experimentally on the basis of the HI/VI and HEVA/VERCORS experiments [36].

Semimechanistic Models of the ASTEC Computer Code

In the ASTEC integral code, the ELSA module calculates the release of FPs from the fuel. It models FP releases according to three categories, based on the simple "limiting phenomenon" principle [37].

- So-called volatile FP releases (Xe, Kr, I, Br, Cs, Rb, Sb, and Te) are governed by the diffusion mechanism in the fuel grain, according to an improved Booth model in which the diffusion coefficient is a function of the fuel stoichiometry (as well as the temperature and grain size). This coefficient is the same for all the FPs, excepting Sb and Te for which a release delay is applied to take into account their retention in the cladding, as long as the latter has not oxidized completely;
- So-called semivolatile FP releases are governed by the mass transfer induced by their vaporization at the grain boundaries. The steam pressures are tabulated by thermodynamic correlations carried out with GEMINI2 (Sr, Ru, Ba, and La) or FACT (Mo, Ce, Eu) solvers. It is the same mass-transfer mechanism that is then applied to all the FPs released from the molten pool;

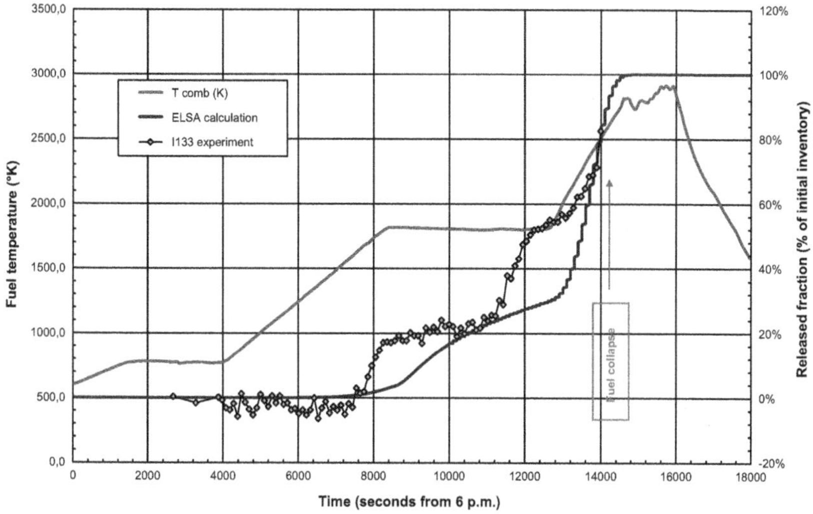

FIGURE 5.12 Iodine release kinetics during the VERCORS HT1 test *(from [38])*.

- Nonvolatile FP releases are governed by the vaporization of UO_2 when it becomes overstoichiometric and oxidizes until UO_3 forms. This category also covers U, Np, Pu, Am and Cm actinides.

As an example, Figure 5.12 compares volatile FP releases (shown here by ^{133}I) as calculated by ASTEC/ELSA and measured online during the VERCORS HT1 test (T_{comb} represents the fuel temperature). The final releases are well reproduced by the calculation. However, there is a slight underestimation of the releases at intermediate temperature (between 1000 and 1500°C) because the intergranular inventory was not taken into account in this model. More accurate modeling therefore requires a mechanistic model, similar to that implemented in MFPR.

Mechanistic Approach of the MFPR Computer Code

MFPR is a 0-D mechanistic code designed to model releases from solid UO_2 fuel [39], [40], [41]. The FPs—considered stable—are incorporated into the fuel matrix in the form of atoms or oxides. Two types of modeling are performed: one involving fission gases and the other dedicated to FPs.

Modeling of the fission gases includes all the previously described physical phenomena: intragranular diffusion of atoms and bubbles to the grain boundaries, with the bubble formation (nucleation, growth) and destruction mechanisms (return to solution form) being taken into account. Releases occur from the grain boundaries after coalescence and interconnections of the gas bubbles.

At present, FP modeling involves 13 elements: Cs, I, Te, Mo, Ru, Sb, Ba, Sr, Zr, La, Ce, Nd, and Eu. They are assumed to diffuse to the surface of the grain boundaries, with some oxidation occurring along the way. They then form three distinct phases: a metallic phase, a ternary phase (known as the gray phase and composed of FP oxides), and a specific phase for CsI. The releases are then governed by the thermodynamic equilibrium of these three phases, with the gas contained in the grain boundaries.

Figure 5.13 illustrates all of the mechanisms modeled in the MFPR computer code.

5.3.4. Conclusion on Release of Fission Products and Future Requirements

The experimental database of analytical tests performed on irradiated fuel sections is relatively broad in the case of UO_2 fuels with average burn-ups. It is supplemented by integral Phébus FP-type tests. These experiments have helped improve our understanding of the different parameters influencing releases, such as the temperature, oxidizing-reducing conditions, interactions with structural materials (especially the cladding), the burn-up, the nature of the fuel (UO_2 or MOX) and its state (solid or liquid fuel).

These results have made it possible to elaborate and validate two types of models. Some mechanistic models are used to describe most of the interactions in the fuel and above all to interpret the tests. Simplified models can be derived from these mechanistic models, which describe the main phenomena and can be used in the release modules of integrated codes.

The hypotheses formulated to interpret the tests make it possible to reproduce correctly the influence of the different parameters, temperatures, burn-ups and atmospheres on releases. They are based mainly on the physico-chemical transformations inside the fuel. Apart from predicting the behavior of fission gases, the MFPR code is used to describe variations in the composition of the different phases containing FPs inside the fuel, as well as the chemical species of the elements involved. However, these hypotheses still need a certain amount of validation, which requires a program focusing on the microanalysis of fuel subjected to release tests.

The experimental database currently available needs to be extended to cover high burn-up UO_2 fuels, MOX fuel, and future advanced fuels. It also needs to cover specific accident scenarios such as the accidents involving air ingress, highly reducing conditions, and the effect of reflooding on FP releases.

5.4. FISSION PRODUCT TRANSPORT IN THE REACTOR COOLANT SYSTEM

(M. Kissane and C. Housiadas)

5.4.1. Physico-chemical Effects

During a severe accident in a nuclear power plant, fission products, actinides, and structural materials are released as gases or vapors from the degrading core

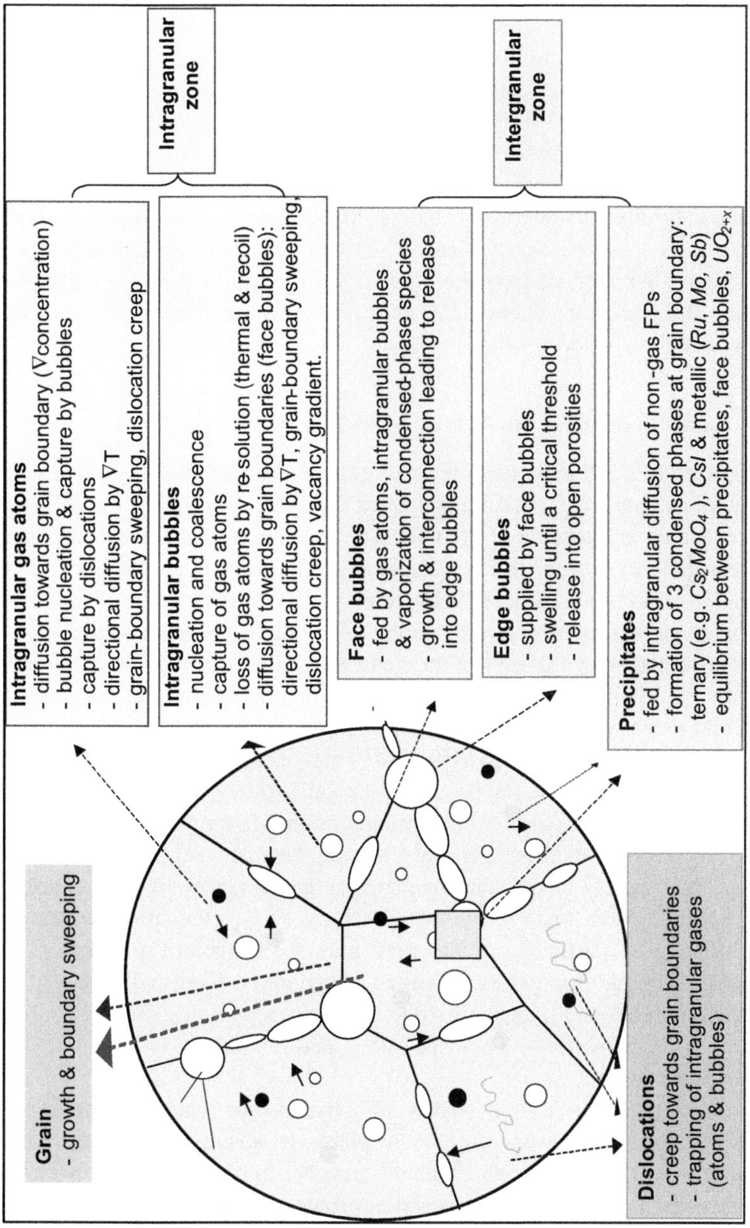

FIGURE 5.13 Schematic representation of the models of the MFPR code *(from [41])*.

into the reactor coolant system. These are then swept, in general, by a steam-hydrogen gas mixture toward the breach/leak path in the RCS. A number of important physico-chemical processes occur between the point of release from the core and release via the breach of still-suspended materials into the containment (or into the auxiliary building in the case of a containment-bypass sequence). The physical effects taking place involve primarily aerosol physics and dynamics. We present the theory on aerosols, with emphasis on describing in more detail the most relevant processes in severe accident phenomenology, by also providing the necessary theory to describe them mathematically. We also review briefly the chemistry in the RCS, mainly conditioned at high temperature by gas phase processes, which can be calculated using thermodynamic laws. We then address the phenomenology in the primary circuit by presenting the important phenomena and trends, with reference to the basic processes previously addressed.

5.4.2. Basic Processes in Aerosol Physics and Dynamics

The vaporized FPs and structural materials experience a significant reduction in temperature as they leave the core and enter the RCS. As a result, the released species condense and take the form of aerosol particles suspended and entrained by the gas (steam) flow, with the notable exception of iodine and ruthenium, which can in some circumstances largely remain in gaseous form for ruthenium in highly oxidising atmospheres, for example, following air ingress into the vessel.

The basic aerosol processes in an elementary fluid element are schematically shown in Figure 5.14. A distinction has to be made between the so-called internal and external processes. Internal are the processes that lead to changes of the aerosol characteristics due to processes that occur within a volume. There are two broad classes of internal processes: coagulation/agglomeration and gas-to-particle conversion. Coagulation/agglomeration occurs as particles collide due to their relative motion. Relative motion between aerosol particles is generated by a variety of mechanisms, such as Brownian motion, fluid turbulence, and fluid shear, or it is due to field forces (e.g., gravitational, electrostatic). When two particles collide, they may fuse together and lose their identity. In this case the process is called coagulation. Eventually, they may retain their identity and shape. In this case larger particles are built-up as a chain of primary particles. They are called agglomerates, and the process is called agglomeration.

Gas-to-particle conversion occurs as physical or chemical processes generate a supersaturated vapor in the gas phase. In severe accidents the basic physical process that leads to supersaturation is cooling. The metastable vapor state relaxes to equilibrium via two distinct routes: (1) the generation of new particles, called nuclei or embryos, via a process known as homogeneous nucleation; and (2) the growth of existing particles via a process known as condensation or heterogeneous nucleation.

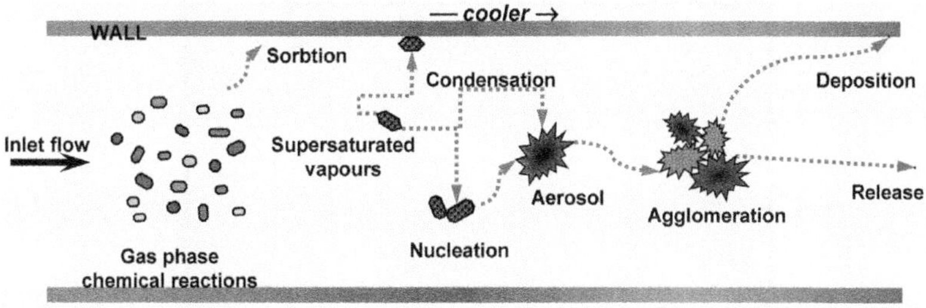

FIGURE 5.14 Schematic representation of aerosol processes (internal and external) within an elemental volume.

External are the processes that induce changes in the aerosol properties by transporting particles across the volume boundaries. The mechanisms driving the external properties are practically synonymous with the particle motion and deposition mechanisms, because their effect is to make particles migrate from the fluid element and deposit to an available surface. Several mechanisms drive particle motions, including gravitational settling, Brownian diffusion, inertial motion, and phoretic forces.

The following sections present a brief account on the physical basis and the basic theory for the internal aerosol processes and the particle motion mechanisms.

Besides describing the fundamental aerosol theory, a brief discussion of aerosol particle sizes is also in order. Indeed, two are the primary quantities of interest that characterize the source term: the chemical composition of the particles and the particle-size distribution. The fundamentals of describing aerosol size distribution are summarized in the following section. The summary is mostly based on the standard textbook of Hinds [43], where the interested reader is referred for more information.

5.4.3. Particle-size Distribution Fundamentals

Typically, in the RCS or the containment, the various nuclides reside in particles of 1 μm in diameter. The dot over the letter i is several hundreds of times larger than a 1 μm aerosol particle [43]. Still, it is important to realize that although we are dealing with "objects" that are extremely small, they should not be confused with gases or vapors. This is illustrated by Figure 5.15, which compares the relative size and spacing of an "air molecule" and a 0.1 μm aerosol particle (where λ denotes the mean free path, which is defined as the average distance traveled by a molecule between successive collisions; for air this is 0.066 μm at 1 atmosphere and 20°C). It is clear that aerosol particles are expected to behave in a fundamentally different way than gas molecules.

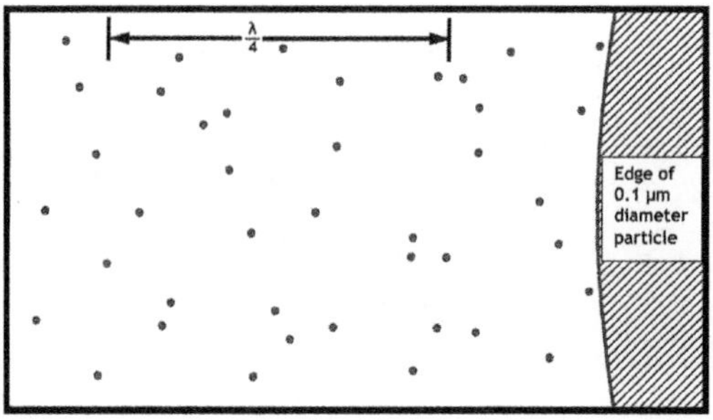

FIGURE 5.15 Illustration of relative sizes and spacing between "air molecules" and a 0.1 μm aerosol particle *(from [43] with permission from John Wiley & Sons Inc.).*

It is also important to realize that, unlike gases and vapors, particles are characterized by a size distribution. An aerosol in which all particles are of the same size is called monodisperse, whereas polydisperse is an aerosol characterized by a size distribution. Monodisperse aerosols do not occur naturally. They are man-made for special technological applications. Their generation is generally difficult and requires sophisticated technological methods. Naturally occurring aerosols, on the other hand, are polydisperse. In severe accidents, the aerosols in the RCS and the containment are of high polydispersity. The size distribution is broad—that is, the particle diameters span over several orders of magnitude, from a few nanometers to more than 10 μm. Such size distributions are necessarily described with statistical means. A common metric is the number distribution, defined as follows:

$$n_N(d_p) : n_N(d_p)dd_p \;=\; \text{number of particles per unit volume, say } 1\ \text{cm}^3$$
$$\text{of gas, having diameters in the range } d_p \textit{ to } d_p + dd_p.$$

The number distribution should not be confused with the number concentration. The number concentration, n, is the total number of particles in a unit volume of gas and is given by $n = \int_0^\infty n_N(d_p)dd$. The units of n are L^{-3} (number cm^{-3}), whereas the units of n_N (d_p) are L^{-4} (number cm^{-3} μm^{-1}).

The number distribution is not always the most appropriate metric to characterize an aerosol. Several aerosol properties depend on particle surface area, volume, or mass. For instance, the condensation rate of vapors onto particles depends on the aerosol surface area. The activity of a nuclear aerosol depends primarily on the mass of the aerosol rather than the number of particles. Accordingly, one defines, in a way similar as above, the distributions with respect to surface area, volume, or mass. These definitions are as follows.

- Surface area distribution $n_S(d_p)dd_p = n_S(d_p)$: $n_S(d_p)dd_p$ = surface area of particles per cm^3 of gas, having diameters in the range d_p to $d_p + dd_p$. The unit is L^{-2} ($\mu m^2\ cm^{-3}\ \mu m^{-1}$).
- Volume distribution $n_V(d_p)$: $n_V(d_p)dd_p$ = volume of particles per cm^3 of gas having diameters in the range d_p to $d_p + dd_p$. The unit is L^{-1} ($\mu m^3\ cm^{-3}$ μm^{-1}).
- Mass distribution $n_M(d_p)$: $n_M(d_p)dd_p$ = mass of particles per cm^3 of gas with diameters in the range d_p to $d_p + dd_p$. The unit is $M\ L^{-4}$ ($\mu g\ cm^{-3}$ μm^{-1}).

For the same aerosol the number, surface, volume, or mass distributions may be significantly different. To illustrate this, consider as an example the case of a basket containing nine grapes and one apple. In terms of number it consists of 90% grapes, whereas in terms of mass it consists of 90% apples. Caution is required in selecting a metric to describe an aerosol, depending on the property we need to characterize.

In severe accidents the source term size distribution is described with the help of a lognormal distribution. The lognormal distribution is the most commonly employed distribution in aerosol science and technology and is given by:

$$f = \frac{1}{\sqrt{2\pi} d_p \ln \sigma_g} \exp\left(- \frac{(\ln d_p - \ln D_{pg})^2}{2(\ln \sigma_g)^2} \right) \tag{15}$$

As can be seen, the lognormal distribution is described with two independent parameters: the median diameter D_{pg} and the geometric standard deviation. σ_g An example is given in Figure 5.16. As can be seen, the lognormal distribution has a skewed shape, with a long tail at the large sizes. The median diameter is the diameter for which one-half of the particles are smaller and one-half are larger; that is, the median diameter divides the distribution curve into equal areas and is always to the right of the highest point. The highest point corresponds to the most frequent size and is called mode of the distribution. An important property of the lognormal distribution is that it is self-preserving. If the number distribution n_N (d_p) is lognormal, the surface distribution and the volume distribution will also be lognormal, having the same σ_g as the parent distribution and median diameters easily calculated from the known median diameter. For example, the volume median diameter D_{pg}^V and the number median diameter D_{pg}^N are simply related by:

$$D_{pg}^V = D_{pg}^N \exp\left(3 \ln^2 \sigma_g\right) \tag{16}$$

These conversion equations, known as the Hatch-Choate equations [44], are the real power and utility of the lognormal distribution.

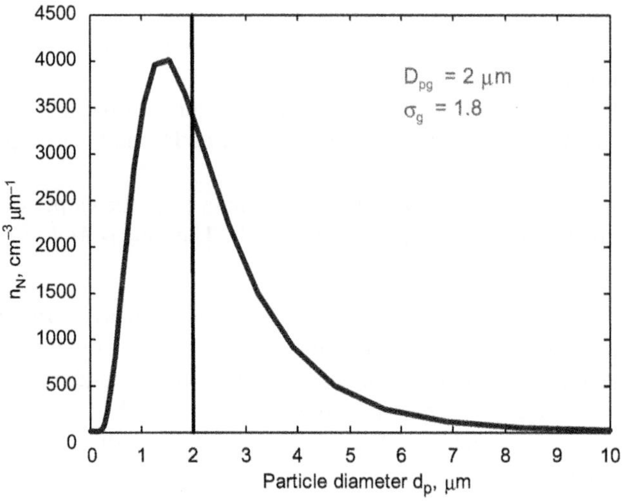

FIGURE 5.16 Example of lognormal number distribution having a median diameter of 2 μm and a geometric standard deviation of 1.8.

5.4.4. Synopsis of RCS Phenomena

Accident simulations with severe accident codes in PWRs indicate that during the in-vessel phase the temperature of the gas passes from 1500–3000 K in the core to 450–1000 K in the primary circuit. As a result, most FPs and structural materials emitted as vapors from the degraded core will be found under high supersaturation levels when they are transported through the reactor coolant system. The supersaturated vapors will contribute to the formation and growth of aerosols, via homogeneous and heterogeneous nucleation, respectively. Notable exceptions to this general trend are the noble gases (Xe, Kr), the highly volatile nuclides such as iodine, or ruthenium that may be transported through the RCS as vapors. In general, control rod materials, cladding materials, fuel, and stainless steel constituents form species of low volatility, which nucleate rapidly. More volatile FP species nucleate downstream in cooler regions of the circuit. It is reasonable to anticipate that aerosol formation may occur even before entering the RCS, in the cooler upper part of the core, especially for the low-volatility elements. There is experimental evidence from the Phébus FP tests that nucleation may indeed begin in the upper part of the core. Clearly, the condensation of the various elements onto particles or structural areas is controlled mainly by volatility. In this respect, the role of chemistry is very important. The conditions in the RCS favor the chemical interactions between different species residing in the gas phase as vapors, or between vapors and aerosol particles. The formed compounds may have physical properties, volatility in particular, significantly different than that of the initial elements.

Therefore, the role of chemistry is central in determining the transport of the emitted FPs and other materials out of the core and through the primary circuit.

Homogeneous nucleation leads to the generation within the gas flow of single-component nuclei. These nuclei, called embryos, are extremely small particles of the nanometer size range or smaller. The so-formed ultrafine particles serve as sites for condensation of the more volatile species, which still remain in the gas phase in vapor form. During this process a competition takes place between vapor condensation on preexisting particles and condensation on the pipe walls. The fraction of vapor that condenses on the aerosol particles is determined by a complicated balance between vapor transport to the particle surface and vapor transport through the background gas to the pipe walls. In general, the former route is more favored because the surface area of the aerosol particles available for vapor condensation is some orders of magnitude larger than the structural surface areas. This heterogeneous nucleation process makes the initially formed particles become larger and multicomponent. Besides that route, coagulation/agglomeration is another process that contributes to the increase of the aerosol size spectrum. In particular, the homogeneously nucleated primary particles are very mobile because of their small size (diffusivity is inversely proportional to particle size). The intense Brownian motion promotes coagulation rapidly, leading to the formation of larger particles.

The overall effect of the above described phenomena is that in the RCS the source term is in the form of multicomponent polydisperse particles, suspended by the steam flow. This general picture has been experimentally confirmed by the Phébus FP tests. The FPT0 and FPT1 tests have effectively shown (see the evaluation [45]) that the nuclear aerosol consists of multicomponent aerosol particles of well-mixed composition, containing all major released products, with the exception of iodine. The size distribution of the aerosol is broad, covering an extended range of particle sizes. The median diameter of the distribution, measured at different stations along the circuit, was found to increase, which is a clear indication of continuous particle growth. The aerosol has a mass median diameter of around 1 μm and a geometric standard deviation of more than 2. On the basis of Equation (15) of Section 5.4.3, it follows that the aerosol mass is distributed over the size range from 0.01 μm to 10 μm. For the conditions expected in the RCS, the Knudsen number (defined as the ratio of the molecular mean free path length to a representative physical length scale, for example, the radius of a body in a fluid) varies, respectively, between 10 and 10^{-3}. Therefore, the size distribution covers all the regimes, from the free molecular regime (high Knudsen number, $Kn \rightarrow \infty$) to the continuum regime (small Knudsen number, $Kn \rightarrow 0$), respectively.

During their transport in the RCS the various elements, whether they are separate or chemically recombined, are depositing on the walls of the pipework as vapors or aerosols. Deposition is highly beneficial because it attenuates the source term transmitted to the containment. Vapor deposition is mainly connected with wall condensation or, sometimes, surface absorption. Aerosol deposition, which is the most important, depends on particle-size and gas-flow conditions. The flow regime in the RCS depends on the steam-flow rate, the system geometry, and the thermal state of the flow at the core outlet, which in turn depends on the accident sequence, especially on the degree of depressurization and the exothermic reactions in the core. The flow regime in the RCS also depends on the diameters of the pipework and the prevailing gas temperatures along the circuit. Simulations with codes indicate that all flow regimes may occur in the RCS, ranging from the laminar through the transitional to the fully turbulent regime. In the Phébus FPT0 experiment, the gas flow generally remained laminar in the hot part of the circuit. At the steam generator inlet reached the transition regime and probably also became turbulent. In the cold part of the circuit the gas flow was turbulent.

Along the circuit, aerosol deposition is not uniform. Instead, there are areas of preferential deposition. Such areas are the water separators and steam dryers in BWRs, or the steam generator tubes in PWRs. In Phébus FP tests most circuit deposition was localized in the steam generator hot leg, with thermophoresis being the dominant removal mechanism (movement of particles relative to a fluid under the influence of a temperature gradient). An important effect that may take place in the RCS and critically influence the source term is aerosol resuspension or revaporization of the deposited matter. Aerosol resuspension is a complicated process that is not yet well understood. Many factors seem to play an important role, such as the gas velocity variation, the level of turbulence, the roughness of the surface, and the morphology and porosity of the deposits. Revaporization may arise as the RCS walls are continuously heated due to the decay heat of the deposited FPs. Currently, there is uncertainty as to the impacts of resuspension or revaporization on the source term transmitted to the containment. However, it may be stated that resuspension may be important in large LOCA scenarios where flow rates may be high, whereas revaporization may be important in severe accidents where coolant system surfaces may heat up as the transient progresses, leading to a late contribution to the source term.

Not only do the phenomena occurring in the RCS reduce the quantity of material released into the containment, but they also condition its physico-chemical form. The formation of highly volatile chemical species leading to vapor-phase contributions to the source into the containment is of particular concern. The importance of the different phenomena in conditioning the source from the RCS varies in terms of the accident sequence, for example, gravitational settling of aerosols is very minor for a hot-leg break sequence since horizontal surfaces seen by the transported aerosols are limited, much radioactive material remains in the vapor phase due to the relatively high temperatures between core and breach, and transport times through the RCS are short due to the short distance traveled before release into the containment and the high-flow rates. Following is a list of phenomena representing the principal phenomena arising beyond that of convection of gases and aerosols by the steam-hydrogen flow; those that are underlined are never negligible in the conditions of a NPP severe accident while the importance of the others may vary greatly. Detailed descriptions of the underlying physics and mathematical modeling of these physical mechanisms are presented in the subsequent sections of the present chapter.

- Chemical reactions of gases/vapors with: other gases/vapors (gas-phase chemistry); aerosols; and structural surfaces (including chemisorption).
- Homogeneous nucleation of vapors.
- Condensation of vapors on: aerosols (heterogeneous nucleation) and structural surfaces.
- Agglomeration of aerosols by: Brownian diffusion; sedimentation; and turbulence (shear and inertial effects).
- Deposition of aerosols due to: Brownian diffusion; thermophoresis (movement of particles relative to a fluid under the influence of a temperature gradient); diffusiophoresis (movement of particles relative to a fluid under the influence of a concentration gradient); electrophoresis (movement of particles under the influence of a spatially uniform electric field); sedimentation (gravitational settling); flow-geometry changes (inertial impaction); turbulence (inertial impaction due to eddies); and pool scrubbing.[12]
- Remobilization of deposits due to: revaporization (decay heating or hotter flow); and chemical reaction in the deposit and/or with the gas phase producing higher-volatility species; mechanical resuspension due to flow acceleration.

Despite the focus of this report being on aerosol behavior, brief treatment of chemical behavior is required. In effect, just as there is a requirement for a correct understanding of thermal-hydraulic, chemistry can also have a strong influence on the aerosols formed and the ultimate source term to the containment. A second aspect worth noting before dealing with the RCS phenomena is that there are a number of excellent sources of fundamental and background material on aerosol processes, including the following: [42], [43], [46], [47], [48], [49], [50], [51]. Hence, rather than recap this extensive material on aerosols here, the objective is to indicate how well nuclear-safety computer codes deal with significant phenomena.

Chemistry

In computer codes that handle chemistry, the reactions are resolved using an equilibrium approach. That is, codes use thermodynamic data in some form, usually Gibbs energy, to find the equilibrium as a function of the temperature

12. Note that pool scrubbing represents an ensemble of phenomena producing some of the phenomena already listed in this Introduction as well as specific effects, in particular jet impaction on the liquid interface and hydrodynamic and two-phase phenomena.

and the concentrations of reacting elements. The approach comprises minimization of the total Gibbs energy of the system since, according to the second law of thermodynamics, equilibrium is the state that maximizes the entropy and this minimizes Gibbs energy in accordance with the following definition,

$$\Delta G(T) = \Delta H(T) - T\Delta S(T) \tag{17}$$

where G is the Gibbs energy, H is enthalpy, S is entropy, and T is temperature.

The Gibbs energy is calculated from the specific heat capacity, which is defined as an empirically determined function of temperature, that is,

$$-T\left(\frac{\partial^2 G}{\partial T^2}\right)_P = C_P = a + bT + cT^2 + dT^{-2} \tag{18}$$

where P is pressure, C_p is the specific heat capacity at constant pressure for the particular species, and a, b, c, and d are empirical constants. This leads to compilation of code databases containing the values of the six empirical coefficients in the following typical relationship (the exact form may vary) for each chemical species covered:

$$G^\circ(T) - \Delta H^\circ_{298.15K} = A + BT + CT\ln T + DT^2 + ET^3 + FT^{-1} \tag{19}$$

The quality of the data in the code databases is the same as that in general thermodynamic codes, that is, quite variable and often involving estimated values (i.e., in the absence of measured data, analogies are drawn based on chemical, structural, and/or vibration-mode similarities). Furthermore, divergence between data series originating from different sources can be considerable. Hence, it can be said that for many species uncertainties associated with the data are large, even for some important FP systems such as I-Te, I-In, and Cs-Te. However, there seems to be little prospect at present of engaging new and extensive experimental work to produce better data. Nevertheless, with regard to the ASTEC code (its SOPHAEROS module is in charge of calculation of FPs and aerosol behavior in the circuits), there has been a special action to verify thoroughly the thermochemical data used and complete the species considered [53] where this code currently covers nearly 800 vapor species generated from 65 elements. (By comparison, VICTORIA, the only other code modeling chemistry and having today a significant user base, covers 22 elements generating nearly 300 species [54].)

Once the codes have calculated the vapor-phase composition, the vapors that are supersaturated constitute a driving force for homogeneous nucleation and condensation onto either aerosols and/or structural surfaces (as a function of surface areas, temperatures, and diffusion coefficients of the species). These processes are calculated either kinetically or at equilibrium (see below).

It is commonly accepted that the thermodynamic approach is an approximation valid only at high temperature—that is, where the characteristic time of the slowest chemical reactions is almost certainly much shorter than that of the transport of the species. In other words, at intermediate and lower temperatures (i.e., roughly speaking, accident sequences involving conditions cooler than those in the reactor-vessel and hot-leg), some reaction kinetics become comparable to or slower than convection rates where this leads to a nonequilibrium state as the flow moves downstream. Indeed, interpretation of the Phébus FP tests has led to the suggestion that slow reaction rates can explain the gaseous iodine species produced at low temperature in the circuit [55], [56]. However, it should be understood that data on reaction rates are generally scarce, and it is unlikely that sufficient rate data will ever be available on all possible reactions in severe accidents for development of a full kinetic model. An attempt is being made to look at reaction rates involving some of the iodine system (see the CHIP program in [57] where limited kinetics modeling for key species could perhaps be introduced into computer codes. So, while the thermodynamic approach is pragmatic, it probably encounters significant limitations when addressing accident scenarios involving secondary-side or cold-leg conditions. For this reason, both the ASTEC and VICTORIA codes allow the possibility of stopping the chemistry calculation (other phenomena continuing) below a user-determined temperature: a crude but interesting option equivalent to the opposite extreme of allowing equilibrium chemistry to continue whatever the temperature.

Lastly, it should be remembered that the full picture should include radiolysis since this may well produce significant quantities of radical and exotic species unpredicted by thermodynamic and/or kinetic models alone. Although radiolysis effects may have little impact in the core region where temperatures are so high that only simple atomic and radical species would exist, they may be important in cooler regions of the RCS where significant deposition occurs (high local dose rate). This aspect requires further research.

Homogeneous Nucleation of Vapors

Even in an unsaturated vapor, molecular clusters exist, but they are unstable. Once supersaturation is established for a vapor, it is ready to become aerosol or to condense on a structural surface. It can be broadly considered that homogeneous nucleation only occurs in conditions of extreme supersaturation in the absence of preexisting aerosols. This is because the high surface energy of the small particles formed by homogeneous nucleation constitutes a greater barrier than that due to condensation on existing surfaces (aerosol or structural). Nevertheless, it must be remembered that in the present context, the high radiation field reduces this resistance by providing abundant ionic nucleation sites. In general, therefore, if aerosols are not being produced by mechanical phenomena (e.g., bursting and spraying of control-rod alloy or a steam explosion in the reactor vessel), moving downstream from the point of release, the first aerosols are produced by homogeneous nucleation of the most refractory species. This may be, for example, silver vapor or, in a later stage of the core degradation process, uranium dioxide (or, more likely, uranium trioxide vapor noncongruently reverse-subliming to uranium dioxide). The molecular clusters, or embryo particles, that form then quickly agglomerate while being simultaneously the target of further condensation of the initial nucleating vapor as well as other supersaturating low-volatility species. The picture may well be more complicated since more than one species may supersaturate and, rather than unary nucleation, binary or even ternary nucleation will occur.

Beyond the vapor pressure itself, models require fundamental properties including chemical activities, surface tensions, and densities. In nuclear-safety computer codes, the phenomenon is simplified.

- First, the size of the initial clusters is not calculated. In the ASTEC and VICTORIA codes, the nucleated mass is simply consigned to the smallest size bin of the discretized aerosol-size distribution. While VICTORIA leaves this bin size as a user-defined parameter [54], it recommends checking the sensitivity of the calculation with respect to this parameter. ASTEC assigns a default radius of 10 Å (1 nm) to the smallest size bin which, when used by the homogeneous nucleation model, implies a critical cluster comprising around 100 simple molecules. Once such tiny particles have been formed, their agglomeration (Brownian diffusion) in most reactor accident situations is rapid since the nucleating mass is large and a high number concentration of nucleated particles results. This agglomeration reduces sensitivity to the exact value of the initial cluster size, though in circumstances where a weak source of vapors is condensing (say, late-phase revaporization of deposits), the value of this parameter is much more critical.
- Second, VICTORIA uses an equilibrium approach that restores supersaturated vapors to a saturated condition immediately following the thermodynamic resolution of the vapor phase. ASTEC calculates unary homogeneous nucleation for each supersaturated species as a kinetic process in competition, in the general case, with heterogeneous condensation (structural and aerosol surfaces). In this way, it can be expected that ASTEC allows a better representation of the physical state of a species (though it remains to explore this advantage in detail with respect to its impact on a reactor case). The unary nucleation model used by ASTEC is that of Girshick et al. [58], where this is used to provide only the mass rate of production of nucleated particles (but not, as should be understood from the above, the particle size).

The reader is referred to Section 5.4.4 for additional information on this phenomenon. In conclusion, it can be said that in this area work is lacking on model validation and assessment of the impact of nontrivial phenomenological simplifications on the results of reactor calculations. Furthermore, nucleation rates are extremely sensitive to surface tension values, a parameter that is not well known for many substances (especially at high temperature). It is recommended that the surface tensions used in computer codes modeling homogeneous nucleation be reviewed with respect to key refractory species, giving rise to the first aerosols in the RCS, for example, UO_2 (see [59]), Ag, etc...

Condensation of Vapors on Aerosols and Structural Surfaces

It should first be noted that the process of heterogeneous condensation onto existing aerosols and onto surfaces can be considered a critical phenomenon: in the accident context it is fundamental to the quantities of volatile species reaching the containment in aerosol form rather than being retained on RCS surfaces. The volatile species concerned—that is, those more volatile than the more refractory species creating the first aerosols—comprise the majority of the most important FPs radiologically (e.g., caesium iodide).

Heterogeneous condensation occurs when a vapor supersaturates and a surface (aerosol or structure) is available. The phenomenon is governed mainly by mass-transfer limitations where, in the RCS, such limitations usually constitute considerably lower resistance to condensation than that of homogeneous nucleation. It must also be remembered that while a vapor may be subsaturated in the bulk flow, it can be supersaturated with respect to a structural surface exhibiting a cooler temperature where (as computer codes assume) aerosols are at the same temperature as their surroundings—that is, their minuscule heat capacity and large surface-to-volume ratio mean that thermal equilibration with the surrounding gas is always fast. Hence, condensation on a wall can arise without competition from aerosol formation or condensation on aerosols. Once a vapor supersaturates in the bulk flow and aerosols are already present, condensation onto these aerosols is efficient since a small amount of aerosol (in terms of mass) represents a large surface area.[13]

Heterogeneous condensation onto surfaces is dealt with by traditional techniques: adoption of the heat transfer analogy and application of classic correlations as a function of thermal-hydraulic conditions. One oversimplification may be that the correlations used are for steady-state fully developed flows where, in fact, developing flow can induce higher condensation rates locally (see [56] for an illustration). The significance of this increased deposition with respect to a reactor analysis remains to be assessed.

Heterogeneous nucleation onto aerosols, including the significant negative feedback from warming of the particle due to latent-heat release by the condensing species, has been correctly understood for many years [60]. Nevertheless, three complications exist. One arises from an aerosol particle's considerable surface curvature which, combined with the inherent surface tension of the condensed species, induces an increase in effective vapor pressure at the liquid surface formed on the aerosol. This so-called Kelvin

13. To consider a representative situation, given an aerosol loading of the flow of 5 $g.m^{-3}$, assuming aerosols to be spherical with a uniform diameter of 1 μm and a material density of 3000 $kg.m^{-3}$, the aerosol surface area is 10 m^2 per cubic meter of the flow.

effect [61] can both reduce total condensation and considerably bias it toward larger particles in an aerosol population (such particles inducing lower surface curvature of the liquid phase). A second complication arises when the particle size is smaller, or of a similar order, than the mean free path between molecular collisions (i.e., large Knudsen number). If this is the case, consideration of the particle in a continuum regime breaks down and the condensation rate must be corrected ([62], [63]). A third complication arises from the probable heterogeneous nature of the aerosol surface affecting sites for condensation: this introduces considerably greater complexity into the condensation phenomenon [64].

Computer code treatment of these condensation phenomena diverges. For VICTORIA, for example, heterogeneous nucleation onto aerosols in the bulk gas is treated simply by reestablishing equilibrium (cf. the treatment applied to homogeneous nucleation activated in the absence of preexisting aerosols). In ASTEC the modeling is kinetic by treating the vapor diffusion process onto a stationary sphere (particle moving with the fluid, no relative motion) driven by the supersaturation. Account is taken of both the non-continuum situation [62] and the Mason effect (latent heat). On the other hand, the Kelvin effect (effect of curvature) is not considered in the current ASTEC V2 versions, though action will be taken to include it in future. Neither code considers heterogeneous particles, but the true nature of aerosols in the RCS during a real accident is open to some question (see Section 5.4.4).

In terms of recommendations, it is evident that this area of modeling lacks validation. This would allow assessment of the impact of the modeling simplifications where the VICTORIA approach could be considered conservative (leading to larger amounts of FPs reaching the containment as aerosol) while that of ASTEC is best-estimate. Nevertheless, increased retention (larger deposits) can lead to greater decay heating and, hence, greater revaporization of deposits where this is a very penalizing phenomenon. In this case, the VICTORIA approach would seem to be less conservative. Data should be sought to check the modeling.

Chemisorption of Vapors

Chemisorption of certain vapors, that is, their chemical reaction with structural materials, is a well-known phenomenon but little data is available on it. Some caesium and tellurium species react with metal alloys, and empirical chemisorption rates as a function of temperature have been derived for CsOH, CsI, Te (and by extension, to SnTe) for stainless steel and Inconel surfaces. There are also data for a handful of other species with respect to specific materials such as Zircaloy. However, it can be said that this is a poorly investigated area where a vast amount of experimental work would be required to provide rates for the species likely to be affected. In the first instance, it would be wise to proceed with review of the dominant vapor species in the RCS, followed by a search for relevant data for these species and assessment of the adequacy on coverage.

Agglomeration of Aerosols

Agglomeration occurs due to particle collisions arising from their differing velocities. Particle motion is induced by Brownian diffusion, sedimentation, and turbulence (shear and inertial effects) where other influences such as electrical forces and acoustic influences are less relevant in the present context (see the argument for limited charging of particles in the RCS under Section 5.4.4). Particles combine due to Van der Waals forces, changes in surface free energies, and/or chemical reactions where codes generally assume that the sticking efficiency is unity (i.e., colliding particles always stick together). In the RCS, it is the Brownian mechanism that is of most importance since, once embryo particles have formed, this phenomenon will rapidly lead to larger (and fewer) particles. Once the aerosols have grown to a greater size, the other agglomeration mechanisms (properly termed kinematic agglomeration) will come into play. However, the generally short residence times of aerosols and turbulent conditions in the RCS mean that sedimentary agglomeration is usually insignificant.

A particular point that should not be overlooked is that if agglomeration is a significant mechanism, then the numerical treatment of the aerosol population is critical in reproducing what the models intend. In severe accident codes, most commonly the so-called sectional method is employed. According to this method, the particle spectrum is discretized and divided into a number of sections (particle-size bins), thereby approximating the particle-size distribution by a histogram. The aerosol population is then solved for each section. If the particle size bins remain fixed then a high number (50 or more) of size bins is usually required to avoid significant spurious diffusion in the re-sizing scheme: a newly formed particle will not find a size in the discretized scheme that suits it perfectly, and fractioning is needed between two adjacent size classes. The coarser the discretization is, the worse the spurious diffusion.

An important uncertainty in this area is related to particle shape. It is common to associate two shape factors with an aerosol particle, one affecting its mobility (or dynamic) properties, the other its collision properties. Spheres are the most compact particle form possible, and so any deviation from this has some impact on the resistance to movement and the probability of colliding with another particle. Particles in the presence of high steam humidity tend to collapse to compact forms under the influence of water surface tension. However, generally RCS conditions are highly superheated, and compaction due to steam is likely to occur only near the breach for particular sequences producing saturated or near-saturated conditions in this region (cold-leg break or in the steam-generator tube in the case of a steam-generator tube rupture). Nevertheless, perhaps other condensing species are abundant enough to cause a compacting effect since there is evidence from representative experiments that particles are, in fact, fairly compact despite superheated steam conditions. This means that high values for the shape factors, such as for chain-like agglomerates, can probably be excluded. Nevertheless, the shape factors and their evaluation remain a significant uncertainty. Evaluation techniques are often empirical where, in relation to an arbitrary particle, there are no reliable analytical techniques for estimating the appropriate values for use in an agglomeration model. Nonetheless, it is desirable that review of the most representative experiments is undertaken with the objective of proposing more realistic values for the shape factors, these values becoming the default values (rather than unity as is now assumed) in nuclear-safety computer codes.

Brownian agglomeration is most significant for small particles where the free-molecular regime (Knudsen number $\gg$ 1) and transition regime (Knudsen number of the order of 1) must be considered. The mobility of small particles is very large, but this effect is tempered by the reduced target area that they present. Brownian agglomeration is most effective between very small and larger particles. In general, models derive from Brownian diffusion theory with correction factors for the free-molecular regime and nonspherical particles. The VICTORIA code uses a summation covering the free-molecular (the classic Brownian diffusion theory with the correction for the free-molecular regime proposed by Fuchs [62]) and continuum regimes. In contrast, ASTEC uses the Davies model [65] for the free-molecular and transition regimes and classic Brownian diffusion theory (including the two shape factors for nonspherical corrections) for the continuum regime.

Gravitational agglomeration is most clearly understood in terms of particle terminal velocities showing the phenomenon to be proportional to the difference in the velocities of the two particles and the sum of their projected areas (the target). Disparity arises in a factor termed the collision efficiency (see [66], [67]) where this constitutes a correction from the ideal situation in which the larger particle sweeps and collects with perfect efficiency all the smaller particles in its projected cylinder during free fall. The correction reduces the efficiency due to hydrodynamic effects where smaller particles tend to flow around the larger particles, allowing some to avoid collection. In the RCS, the limited impact of gravitational agglomeration means that exploration of the different efficiencies is not required here where, in any case, ASTEC and VICTORIA are in agreement in using the Pruppacher and Klett formulation [50].

Turbulent agglomeration arises due to the relative particle velocities induced by the shearing flow field and particle drift relative to the flow arising from inertial differences. This latter contribution is zero for particles of the same size, and turbulent agglomeration reaches a minimum in this case. The Saffman and Turner approach [68] is used in both the ASTEC and VICTORIA codes. An extremely important parameter of this model is the rate of energy dissipation per unit mass of the fluid due to turbulence. All codes refer to a correlation attributed to Laufer [69]:

$$\varepsilon = 0.03146 \cdot \frac{U^3}{\mathrm{Re}^{3/8} \cdot D} \tag{20}$$

where Re is the flow Reynolds number, D is the hydraulic diameter of the pipe, U is the mean flow velocity, and the rate of energy dissipation ε per unit mass of fluid is in $\mathrm{J \cdot kg \cdot s^{-1}}$. While from the point of view of dimensional arguments the above expression is satisfactory, its real origin is somewhat obscure and requires review. A second area of uncertainty arises from how to add together the two different turbulent contributions and how to add these to the other contributions to agglomeration. It has been recommended [66] that the turbulent shear, turbulent inertial, and sedimentation contributions be added in quadrature and then this combined contribution be added linearly to the Brownian contribution.

In ASTEC this is the approach used, which, in terms of agglomeration/coagulation kernels K (see Eq. (24) below), means:

$$K_{tot} = K_{Brown} + \left[K_{turb.shear} + K_{turb.inertia} + K_{sedim}\right]^{1/2} \tag{21}$$

where K_{tot}, K_{Brown}, $K_{turb.shear}$, $K_{turb.inertia}$, and K_{sedim} represent, respectively, the total agglomeration kernel, the agglomeration kernel due to Brownian diffusion, the agglomeration kernel due to turbulent shear, the agglomeration kernel due to turbulent inertia, and the agglomeration kernel due to sedimentation.

In VICTORIA only the turbulent terms are added in quadrature; then this is added linearly to the Brownian and gravitational contributions. The small contribution of the gravitational term in the RCS means that this divergence is slight where investigating and justifying the expression used for the turbulent-energy dissipation rate is of a higher priority.

Deposition of Aerosols

Aerosol deposition will occur in the RCS due to Brownian diffusion, thermophoresis, diffusiophoresis, electrophoresis, sedimentation (gravitational settling), inertial impaction (projection onto surfaces by flow-geometry changes and turbulent eddies), and pool scrubbing. Among these phenomena, not all are certain to be significant since they can be considered scenario-dependent. Brownian deposition will be significant if particles, once formed, remain small. Pool scrubbing phenomena (removal of aerosol particles suspended in gas bubbles rising in a liquid pool) may occur in the pressurizer (e.g., high-pressure sequence such as station blackout) or the steam-generator secondary side (steam-generator tube rupture). Probably only thermophoresis can be guaranteed to produce significant deposition whatever the accident sequence.

Deposition Due to Diffusion

Diffusional deposition is most relevant to the RCS zone where the first vapors nucleate and particles remain small, that is, within the reactor vessel. For laminar flows, use of the classic Brownian-diffusion approach (Fick's law) is only appropriate in the limit of stagnation. Hence, deposition from a laminar flow due to diffusion is often based on empirical models where that of Gormley and Kennedy [70] is used in ASTEC and VICTORIA (ASTEC does switch to the classic Brownian-diffusion model corrected for slip in the special case of a creeping flow). A key parameter in this model is the length over which diffusion occurs. In ASTEC, for example, a fixed value of 10^{-4} m is used. For turbulent flows, the phenomenology changes where the turbulence brings particles close enough to the wall for diffusion to become the dominant transport mechanism and lead to particle deposition; the codes VICTORIA and ASTEC concur, again, in the use of the same model, that of Davies [65]. With respect to these phenomena, it seems unsatisfactory that there is not a better analytical basis for the diffusion distance, the current situation being somewhat arbitrary; the discussion in [66] for the turbulent regime is of interest here. The situation could be saved by the fact that diffusional deposition, affecting above all highly submicron particles, is usually such a small contribution to overall deposition that the modeling approximations do not matter, though this should be checked for different accident sequences.

Deposition Due to Thermophoresis

Thermophoresis is a phenomenon that, unusually, does not depend strongly on particle size, though is larger for submicron particles. As a result of unbalanced collisions between gas molecules and aerosol particles, particles in a temperature gradient experience a force directed toward cooler temperatures (countered by the hydrodynamic drag force). Highly conducting (e.g., metallic) particles are less affected by thermophoresis than are particles of more insulating materials. For larger particles a thermal gradient may be established in the particle introducing considerable complexity.

Thermophoretic deposition in tube flow has been long studied in the context of many practical applications, with nuclear safety being one of those. A rigorous computational approach requires first that the gas-flow velocity and temperature fields be determined from CFD codes, or even, sometimes, from LES/DNS calculations. Then, the motion of individual aerosol particles is tracked by solving the particle equations of motion with the addition of the thermophoretic force acting on the particle. This method is called the Lagrangian particle tracking method and is usually computationally very demanding. For this reason, severe accident codes use simple one-dimensional pipe flow models, where thermophoretic deposition is accommodated by considering an additional deposition velocity (the thermophoretic deposition velocity). Recent investigations demonstrate that the simplified 1D approach is valid in all cases in laminar flow [71], [72]. In turbulent conditions, the approach is adequate for small gas-to-wall temperature differences, but for elevated temperature differences a multidimensional description may provide more accurate predictions [72].

Thermophoretic deposition velocity is the principal parameter required in nuclear safety codes. A number of expressions exist for calculating this parameter, though it is common to use the Brock formula with numerical coefficients provided by Talbot et al. [73] that allow good agreement with experimental data over the free-molecular, transition, and continuum regimes. For brevity, the modeling of thermophoretic deposition velocity is presented, together with the modeling of diffusiophoretic deposition velocity, in Section 5.7.

Deposition Due to Diffusiophoresis

Diffusiophoresis is important in RCS, especially in cold leg break accidents. In other situations, however, it is of secondary importance. Also, in the RCS, diffusiophoresis never occurs alone but must necessarily be accompanied by thermophoresis, so we actually observe thermo-diffusiophoresis. Diffusiophoresis will be described in Section 5.7.

Deposition Due to Electrophoresis

Deposition may be enhanced due to electrophoresis, a complex process dependent on aerosol decay activity, size, concentration, and confining geometry. Radioactive aerosols self-charge electrically where, broadly, due to the dominance of α decay (this event generally stripping out more electrons from the particle than the two positive charges it carries away) and β^- decay, a positive charge develops, though very small particles can develop a small negative charge. (See the relevant theoretical [74], [75] and experimental [76] work.) However, the theoretical work indicates that particle charging is very sensitive to two factors: in the RCS the small size of the particles should considerably limit their capacity to self-charge, while their high number concentration should reduce their self-charging by facilitating diffusive combination with gas phase ions. Then again, it must be acknowledged that, as predicted in [75], confinement of the aerosols, such as in the RCS, may aid charging by lowering ion concentrations. In summary, the theoretical work implies limited self-charging in these conditions except, perhaps, for regions of particularly confined flow.

At present, nuclear-safety computer codes do not take account of this effect, and, in the first instance, a proper assessment of its impact is required.

Deposition Due to Sedimentation

Particles settle under the influence of gravity where the settling (or terminal) velocity is broadly proportional to the square of the diameter. Gravitational settling is not particularly significant in the RCS due to the usually limited size of particles (small settling velocities) and their short residence time before release to the containment. The most common modeling approach is to assume "Stokesian" particles, namely, as particles move, the flow around them is in the Stokes regime ($Re_p << 1$, where Re_p is the particle Reynolds number). In this situation the drag force is given by the well-known Stokes law for a moving sphere in a fluid. An important point that differentiates the motion of an aerosol particle from the motion of a macroscopic sphere is the so-called slip effect. If the particle diameter is much larger than the mean free path of the molecules of the suspending gas, then the aerosol particle moves, indeed, like a macroscopic object and Stokes law may apply (this corresponds to the case of small Knudsen number, i.e., to the continuum regime). Instead, if particle diameter is comparable or smaller than the mean free path of the gas molecules, then the surrounding gas is no longer seen as a continuum (Knudsen number of order of unity, or large, corresponding, respectively, to the transitional regime and the free molecular regime). The Stokes law is no longer accurate, and the drag force needs to be reduced by a factor known as the Cunningham correction factor or the slip correction factor. Also, corrections must be introduced to the situation of particle motion outside the Stokes regime—that is, for larger particles where flow is no longer perfectly laminar and wake effects can be induced. RCS computer codes do not take this latter effect into account. As already stated, however, the settling mechanism most often has a limited impact that may well mean that modelling approximations has little effect on predictions of the source term to the containment.

Deposition Due to Inertial Impaction

Particles can be projected onto surfaces by flow-geometry changes where, due to their inertia, they deviate from the mean-flow direction to come into contact with confining walls or obstructions. Such geometries are common in the RCS, but most can be generally considered as bends, contractions, expansions, or combinations of these. The impaction phenomenon clearly comes into play as particle size increases where, in general, submicron particles are not concerned, being more influenced by other deposition mechanisms. Models for simple geometry changes do exist in the literature but have received little validation in reactor-relevant conditions, particularly highly turbulent flows. Furthermore, there are empirical models for these geometries as well as models for specific complex geometries encountered in reactors such as the components of the steam generator secondary side.

Computer codes use a mixture of models for pipe bends where there is general agreement on the use of the Cheng and Wang model [77] for laminar flow and the Pui et al. [78] model for turbulent flow. For contractions, the models used are that of [79] for abrupt narrowings and [80] for more gentle narrowings where these numerically based models were developed for laminar flows. Due to lack of an alternative, these models are also used in turbulent conditions, though the justification for this is slender and requires validation. Deposition in pipe expansions does occur but is not considered by codes, and relevant literature is scarce. Neither is deposition in tube bundles considered where this will occur to perhaps a limited extent in the reactor vessel where flow passes through upper core structures on its way to the hot leg as well as in the steam generator tube bundle in the case of a steam generator tube rupture. A model for this phenomenon exists for turbulent conditions [81] but does not appear to have been investigated for reactor applications. The VICTORIA code includes models for the separator of a steam generator secondary side that is based on simple considerations of centrifugal force and for dryers treating them as a succession of bends. Validation of models is necessary in this area, and it is hoped that the ARTIST program [82] will allow assessment of the tube bundle, separator, and dryer models.

In turbulent flow, eddies act to project particles toward surfaces where sufficient momentum is imparted to the particles that they can cross the boundary layer and come into contact with the wall. The phenomenon is highly sensitive to particle mass, or more strictly to the relaxation time of the particle. A number of theoretical models exist but with varying degrees of agreement with respect to data. Hence, more empirical models are often used, such as the model derived from the careful experiments of Liu and Agarwal [83]. Since few particles usually become quite large (supramicron) in RCS conditions, this phenomenon is not a dominant mechanism.

One phenomenon that is overlooked in nuclear safety computer codes associated with inertial impaction phenomena is that of particle bounce. The sticking efficiency of particles is not perfect, especially for the higher impact velocities. It is possible for impacting particles to bounce, to disintegrate, and/or to dislodge particles already on the surface. Significant data requirements are apparent since the models of particle sticking often involve parameters such as the Hamaker constant [84] (which reflects the strength of the van der Waals force between two particles, or between a particle and a substrate). What is certain is that for high-velocity flows in dry (superheated steam) conditions with hard (salt, ceramic, and metal) particles, bounce will occur. Integration of this phenomenon into computer codes is desirable.

Remobilization of Deposits

Revaporization of deposits due to decay heating or a hotter flow is a straightforward evaporation process involving mass transfer that is adequately modeled by existing techniques. Far more uncertain are the chemical reactions in the deposit and/or with the gas phase producing higher-volatility species, or even daughter products of decay chains producing more volatile species (e.g., $^{132}Te \rightarrow {}^{132}I$). Results of the Phébus FP tests, such as FPT2, provide evidence for chemical revaporization of caesium and other FPs [29]. This is potentially a very important phenomenon since it could lead to a relatively late source of FPs from the RCS into the containment. Work is generally lacking in this area, and assessment of the significance of the phenomenon and its consequences is required in order to judge the priority of further work.

Mechanical resuspension due to flow accelerations (typically a steam spike in the RCS due to core collapse) can be considered an unsatisfactory area for models despite integration of state-of-the-art modeling such as [85] into nuclear-safety computer codes. For safety needs in the current RCS context, the intervention of this phenomenon in the accident sequence is so close to the initial release event from the core that, in the containment, resuspended mass rapidly becomes indistinguishable from the initial source. In other words, it does not lead to any significantly higher suspended mass in the containment after, say, 24 hours. In the event of a containment bypass sequence, however, this phenomenon requires accurate evaluation. Further work in this area is required.

5.4.5. RCS Transport Modeling

The starting point here is a mixture of gases and vapors at a given temperature where this input is assumed along with thermal-hydraulic boundary conditions for a particular RCS geometry and accident scenario. It is worth noting, however, that there is a need to understand limitations with respect to thermal-hydraulic modeling since its influence can, beyond flow characterization (i.e., laminar or turbulent), be more subtle as in the correlations applied to reconstitute temperature gradients determining rates of mass transfer as in condensation on structures and thermophoresis.

The calculation of the source term necessitates the simulation of the aerosol formation, transport, and deposition in the RCS. The complicated interactions between transport, deposition, and size distribution changes can be described by a population balance equation, which in the aerosol literature is known as the General Dynamic Equation (GDE). The numerical solution of the GDE is one of the most essential steps in the various codes that simulate transport in the primary circuit or the containment (e.g., the SOPHAEROS module in ASTEC, or the containment code COCOSYS). In its general form, the GDE is three-dimensional in space. In RCS codes the GDE is used in a simplified one-dimensional form (1D pipe flow along the RCS), which can be written as follows:

$$\frac{\partial n}{\partial t} + \frac{\partial}{\partial x}(n u_g) + \frac{4}{d_h} V_{dep} n = \frac{\partial}{\partial x}\left(D \frac{\partial n}{\partial x}\right) + \left.\frac{\partial n}{\partial t}\right|_{g-p} + \left.\frac{\partial n}{\partial t}\right|_{\text{coag}} \tag{22}$$

In Eq. (22) the independent variables are position x (axial location along the RCS), time t, and a third variable that may describe particle size. As such, particle diameter d_p can be used. In practice, particle volume v is also used because volume is conserved during coagulation. The dependent variable is a multivariate size distribution function: $n = n(d_p; x, t)$, or equivalently $n = n(v; x, t)$.

On the left-hand-side, the first term is the accumulation term, the second is the convective term (u_g is gas velocity), and the third gives the deposition flux (drift flux) due to the various external processes (d_h is the hydraulic diameter of the duct). Therefore, appropriate correlations for deposition velocity V_{dep} are needed to account properly for effects like sedimentation, diffusional deposition, inertial impaction, and thermophoresis. A great deal of effort is invested by the code developers in seeking and validating candidate correlations. On the right-hand-side, the first term describes axial diffusion (not very important in most cases), and the next two terms describe the internal processes, namely, gas-to-particle conversion and coagulation. The terms giving the internal processes can be further written as following:

$$\left.\frac{\partial n}{\partial t}\right|_{\mathrm{g-p}} = J_{\mathrm{nuc}}\delta(d_p - d_p^*) - \frac{\partial}{\partial d_p}\left(n\frac{dd_p}{dt}\right) \tag{23}$$

$$\left.\frac{\partial n}{\partial t}\right|_{\mathrm{coag}} = \frac{1}{2}\int_0^v d\tilde{v}K(\tilde{v}, v-\tilde{v})n(\tilde{v})n(v-\tilde{v}) - n(v)\int_0^\infty d\tilde{v}K(v,\tilde{v})n(\tilde{v}) \tag{24}$$

In Eq. (23) the first term on the right-hand-side models homogeneous nucleation (δ is the Dirac function). Clearly, this requires knowledge of the nucleation rate J_{nuc} and of the critical diameter, d_p^*, which is the size of the homogeneously nucleated embryonic particles. These two important parameters are the object of the various nucleation theories. A nucleation theory referred to as the classical nucleation theory is mostly used in severe accident codes. Modeling of homogeneous nucleation generally involves thermodynamics and physical chemistry. Further details can be found in [42], [48], and [50]. The second term on the right-hand-side models particle growth by condensation (i.e., heterogeneous nucleation). Clearly, the important parameter to know is the derivative, dd_p/dt, known as the growth law. Knowledge of the growth law involves complicated thermodynamics, chemistry, and heat and mass transfer. In Section 5.7.2 we will address the fundamental case of particle growth due to steam condensation, which is the basic effect taking place in the containment.

Equation (23) describes coagulation/agglomeration. It is usually called the Smoluchowski equation, after Marian von Smoluchowski, who in 1916, presented a consistent mathematical treatment of particle coagulation [86]. The important parameter involved here is the coagulation kernel K, which is a bivariate function of the sizes of the colliding particles, $K = K(v_1, v_2)$, giving the collision frequency between particles of volume v_1 and v_2. The most important mechanisms that force particles to collide are Brownian motion, relative motion due to laminar shear or fluid turbulence, and gravitational settling. There exist expressions giving the collision frequency due to each individual mechanism. The overall kernel used in Eq. (23) involves some kind of addition of the individual kernels, as discussed earlier in Section 5.4.4.

It is obvious from the above introductory discussion that solution of the GDE is a challenging task. On one hand, it requires sophisticated numerical methods to cope with the complicated integro-differential form of the mathematical expressions. On the other hand, it requires valid correlations for the various physical parameters described above, such as the nucleation rate and the coagulation/agglomeration kernels. The physical behavior of these parameters is not generally known under severe accident conditions. For this reason, large-scale experiments have been organized, Phébus FP in particular, to confront experimental data with theoretical predictions.

5.4.6. Conclusions on Transport of Fission Products in the Reactor Cooling System: Remaining Modelling Uncertainties

Many areas are dealt with in a satisfactory manner. To improve modeling capabilities particular areas for further work are as follows:

- Chemistry where both verification of thermodynamic data for the relevant species and investigation of reaction kinetics in key systems must be pursued;
- Assessment of the impact of homogeneous agglomeration where the influence of the embryo-particle size in typical accident sequences should be checked as well as the influence of some simplifications such as reestablishing saturated conditions for the vapor(s) rather than using a kinetic model;
- Adequacy of data for chemisorption of the dominant vapor species in the RCS (to be identified);
- Assessment of typical particle-shape factors from prototypical experiments for use as default values in agglomeration models (among others);
- Assessment of the potential impact of electrophoresis (currently not modeled in nuclear-safety computer codes);
- Review of models for impaction due to flow-geometry changes;
- Review of mechanical resuspension modeling, which can be considered inadequate at present.

5.5. CONTAINMENT BYPASS

(S. Guentay, L. Herranz, and T. Lind)

5.5.1. Background

A containment bypass involves a direct release of radioactive material to the environment or into reactor buildings which may not retain the FPs effectively, by which the retention capability of the containment is bypassed. Examples include PWR steam generator tube ruptures (SGTRs) (spontaneous or induced tube ruptures), which allow radionuclides to be released through the secondary side of the steam generator into the environment due to failure of the steam generator safety valves, or interfacing systems LOCAs (IS-LOCA), which allow radionuclides to be released through a breach in a system outside the containment that interfaces with the reactor coolant system (e.g., in the auxiliary building [87]). As an example, failure of a check valve on ECCS piping would cause a small-break LOCA outside of the containment if the ECCS water tank is located outside of the containment (V-sequence). The FPs can therefore be transported out of the containment, if the accident progresses into a severe accident. A steam or feedwater line break outside of the containment and concurrent steam generator tube rupture will also allow FPs to bypass the containment.

Steam generators are key components for overall plant performance of PWRs. This fact has resulted in design and manufacturing improvements of these shell and tube heat exchangers as well as new modes of operation. Despite this, a number of SGTRs have been reported in several nuclear power plants [88].

The regulatory authorities in many countries consider spontaneous steam generator tube rupture accidents as design basis accidents for Western PWRs. Plants are designed to cope with such accidents, and no major consequences should be expected. However, a particular safety challenge arises from an SGTR in combination with other failures. For example, if the safety valves of the failed steam generator are stuck open, the result will be a loss of coolant that, eventually, will lead to core degradation and meltdown if the operator actions to cool down the reactor and stop the leak are not effective. Under these conditions, FPs and aerosols released from the reactor would bypass the containment. These accident scenarios are very unlikely, but, given the potential consequences of a direct path for FPs from the primary coolant system to the environment, they are estimated to be important risk contributors [89].

In addition to spontaneous SGTR, steam generator tube integrity may be challenged by high-temperature and high-pressure conditions during severe accidents. Consequently, they may have a potential to fail due to creep rupture or other flaws. In these sequences, it is important to estimate whether a steam generator tube ruptures first in the reactor coolant pressure boundary leading to containment bypass [90]. These sequences are called severe accident-induced SGTRs.

The potential retention within the secondary side of a failed steam generator during a SGTR severe accident sequence was seen as one of the largest uncertainties in the analyses reported in NUREG-1150 [89]. An expert elicitation panel [91] considered that little retention of radionuclides would occur in the reactor coolant piping and the failed steam generator. They estimated the overall transmission factor from the reactor to the environment to be higher than 75% for all radionuclides considered. Consequently, attenuation of this magnitude was attributed to retention in the primary coolant piping. Given the absence of a comprehensive database or specific model for the retention of radionuclides in the secondary side of the failed steam generator, Probabilistic Risk Assessments (PRAs) usually give no credit to any potential decontamination within the secondary side of a steam generator [92].

Nonetheless, it has been experimentally demonstrated that aerosol retention in the steam generator secondary side takes place and that it is highly dependent on the governing thermal-hydraulic conditions and the location and size of the break [92], [93]. In particular, the presence of water in the secondary side appears to be a key factor [94], since a substantial fraction of particles carried by gas might be scrubbed by the water. This is the reason supporting flooding of the failed steam generator as an accident management measure to minimize the release of FPs from a defective steam generator [92]. However, even if no water is present, gas interaction with internal structures (i.e., tubes, support plates, separator, etc.) result in FP and aerosol retention.

The EU-SGTR project of the fifth Framework Program of Euratom was the first European project (2000–2002) that generated a systematic understanding of possible mechanisms for retaining aerosol particles in tubes and the complex structures of the secondary side of a SG [94]. The International ARTIST project (AeRosol Trapping In a Steam GeneraTor) continued and extended those previous studies. Run and coordinated by PSI (2003–2007), the ARTIST project investigated aerosol retention in a Western design inverted U-tube SG [95]. A wide range of operational conditions of the components (dry, wet, and transition) and thermal-hydraulic and aerosol parameters were addressed. In addition, the droplet behavior in the separator and dryer was systematically investigated to address design basis accidents. An extensive interpretation of the data and modeling efforts has accompanied the experimental program. The investigation is presently being continued until the end of 2011within the frame of a second stage of the ARTIST project, called ARTIST2.

Beyond the SGTR and ARTIST projects, the SARNET (Network of Excellence for a Sustainable Integration of European Research on Severe Accident Phenomenology) project of the sixth Framework Program of Euratom (2004–2007) provided a further forum where additional efforts were coordinated to understand the complex aerosol phenomena and to model the relevant available data [96].

5.5.2. Phenomenology

This section highlights the phenomena that have a role in determining the source term within a vertical steam generator during severe accident SGTR sequences. Steam generator complexity, as well as broad variable ranges during SGTR sequences, makes it difficult to analytically determine the governing mechanisms for aerosol retention in the steam generator. However, trends and dependences of the retention on different parameters have been identified and are presented in this section. Further discussion may be found in [51].

The following discussion assumes that the source term enters essentially as particles into the secondary side of the steam generator. Phébus FP experiments [28] have shown that a significant fraction of some radionuclides, like iodine, could reach the steam generator as a vapor. Hence, vapor deposition might be the major retention mechanism in such a case. In addition, in case of a flooded steam generator, a fraction of iodine entering the secondary side might eventually become dissolved in the pool and undergo complex iodine chemistry reactions that have not been investigated under the relevant conditions so far. Therefore, these aspects are considered as open issues in the area of source term.

Dry Secondary Side

In-Tube Deposition

Aerosol retention in the SG may take place in various locations. The aerosol enters the SG through the broken tube before reaching the break. Flow out of the broken tube is choked and causes in-tube velocities on the order of several 100 m/s [92]. Under these conditions, inertial impaction, turbulent deposition, and particle resuspension determine the retention, so that the net efficiency of deposition is lowered by the resuspension effect. ARTIST experimental data show that aerosol retention in fact is dynamic, with high aerosol retention when aerosols are first introduced to the broken tube. The dynamic nature of coupled deposition/resuspension phenomena requires very detailed and simultaneous treatment of the flow dynamics, aerosol dynamics, and thermodynamics; therefore, analytical approach of this topic has not yet been attempted. The FP vapor deposition on the particles and on the tube inner surface as the temperature drops as a result of the expansion was not studied in the ARTIST projects.

In-Bundle Deposition

Three major locations for aerosol retention in the steam generator tube bundle should be distinguished by the flow behavior: (1) the vicinity of the tube break where a very high- velocity jet is being released from the break, (2) a support plate that obstructs the flow, and (3) a bundle far-field where flow is developed and has a relatively low velocity.

In the vicinity of the break, inlet jet velocities range from around 100 m/s to sonic levels (i.e., several 100 m/s). Investigation of aerosol retention on single tubes showed that collection efficiency is correlated as a function of the Stokes number under the conditions explored [81]. The Stokes number is a dimensionless number used to describe the curvilinear motion of a particle. It is the ratio of the stopping distance of a particle to a characteristic dimension of the obstacle causing the curvilinear motion. The Stokes number is defined as Stk:

$$\mathrm{Stk} = \frac{\rho_p d_p^2\, U C_s}{18\,\nu D} \tag{25}$$

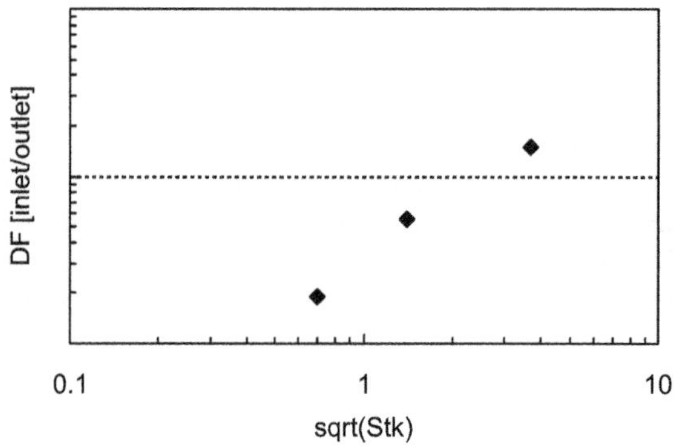

FIGURE 5.17 Decontamination factor in the vicinity of the break with spherical aerosol particles.

Here ρ_p is the particle density, d_p the particle diameter, U the flow velocity, C_s the slip correction factor, ν the dynamic gas viscosity, and D the characteristic dimension of the obstacle.

ARTIST tests showed that the aerosol particle collection efficiency of in the break vicinity was in fact correlated with the Stokes number in the case of spherical particles under similar flow conditions when the break was a symmetrical guillotine break (Figure 5.17). In this figure, the particle retention in the break vicinity is given as a decontamination factor DF, typically used to describe retention of radioactivity:

$$DF = \frac{MF_{IN}}{MF_{OUT}} \tag{26}$$

MF_{IN} is the mass flow of aerosol particles flowing into the retention stage and MF_{OUT} the mass flow of particles flowing out of the retention stage.[14]

The scenario is more complex in the actual case, and retention was found to be a function of several parameters: break type, particle type (spherical, agglomerate), particle stickiness, and the presence of steam. In addition, agglomerate particles were found to de-agglomerate in the vicinity of the break (Figure 5.18) [97]. Such a complexity makes it very unlikely to develop a mechanistic model for collection efficiency, as further discussed in [98]. Inertial impaction, turbulent deposition, and resuspension are the major contributors to the aerosol retention. Experiments carried out in the SGTR and ARTIST projects have shown a rather uneven distribution of aerosol deposits that seems consistent with the gas aerodynamics investigated [99]. The vapor deposition on existing particles and on surfaces was not studied.

14. Note: The decontamination factor can also be expressed in terms of the total mass accumulated over a given period.

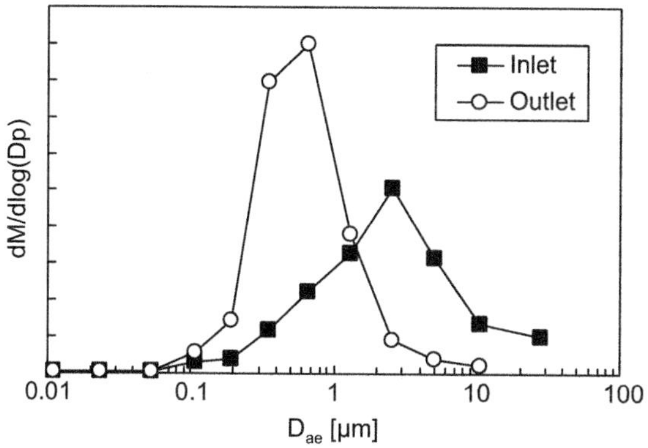

FIGURE 5.18 Particle-size distributions of agglomerate particles before (inlet) and after (outlet) the tube break as a function of particle aerodynamic diameter D_{ae}.

Away from the break, the flow will spread and move upward toward the support plate. As a result, the velocity will decrease considerably compared to the sonic speed at the break point. The flow exits the break stage through passages in the support plate. The passages are narrow, and some aerosol retention takes place at the support plate. However, compared to the retention in the break vicinity, the retention at the support plate is low.

Beyond the break vicinity, the flow will be mainly in the vertical direction, and little retention by inertial mechanisms on the tube structures is taking place at a gas mean velocity of the order of 0.2 m/s. ARTIST experiments showed that electrophoresis may increase aerosol retention in the bundle far field considerably. However, determination of the effect of electrophoresis under realistic reactor conditions is very difficult because the aerosol charging and, consequently, electrophoresis depend on the radioactive decay of the radionuclides and the geometry of the surrounding structures [75].

Table 5.3 compiles the nondimensional Stokes number estimated in the break stage, in the far field, and near the U-bend section under anticipated conditions [51]. Two values have been calculated in the break stage to consider the large variation of flow velocity U in that region. Three different particle sizes given as particle aerodynamic mass median diameter (AMMD) have been considered.

Deposition On Separators and Dryers

Separators and dryers, due to their complex geometry, produce very complex and changing flow patterns and velocities. Inertial impaction, interception, and turbulence- initiated particle agglomeration are the main processes that affect the aerosol removal. According to the ARTIST tests, aerosol deposition in these

TABLE 5.3 Typical Nondimensional Stokes Numbers at Different Locations of the Steam Generator

	Stokes number $\left(Stk = \frac{\rho_p d_p^2 U C_s}{18\mu D}\right)$			
	Break stage			U-bend section
AMMD μm	U = 300 m/s	U = 10 m/s	Far Field	
1	0.09	0.003	$1.5 \cdot 10^{-4}$	$6.2 \cdot 10^{-5}$
3	0.72	0.024	$2.5 \cdot 10^{-3}$	$4.8 \cdot 10^{-4}$
10	6.0	0.200	$9.6 \cdot 10^{-3}$	$4.0 \cdot 10^{-3}$

components is small, as anticipated based on the low gas velocity and small aerosol particle size.

Other Processes of Interest

The above discussion highlights the inertial and turbulent mechanisms as the dominating aerosol retention mechanisms in a dry steam generator. Nonetheless, other mechanisms need further consideration: (1) thermophoresis, driving the particles toward the colder surfaces, (2) particle de-agglomeration, which may take place at the tube break if the particles are agglomerates with relatively loose structures, resulting in particles that are much smaller than the original agglomerates and therefore, more difficult to retain, (3) resuspension at the high-velocity regions, and (4) recirculation regions, for example, between the outlet of the separator and the riser through the downcomer.

Flooded Secondary Side

For accident management purposes, water injection in the dry secondary side may be an option in order to reestablish heat removal and provide a pool where the incoming aerosols can be scrubbed. The POSEIDON [100] bare pool scrubbing experiments, for example, provided a database on the aerosol removal efficiency of hot pools showing that increasing the pool depth increased aerosol retention significantly.

In the SGTR and ARTIST experiments, a tube bundle was included in the experimental setup for flooded secondary side. It was seen that compared to bare pools, aerosol retention was much higher in the flooded tube bundle [101]. As expected, the flooded bundle also showed much higher aerosol retention than the dry bundle. In the flooded bundle, aerosol retention, even with very low submergence of the break, 0.3 m, was significant, with decontamination factors

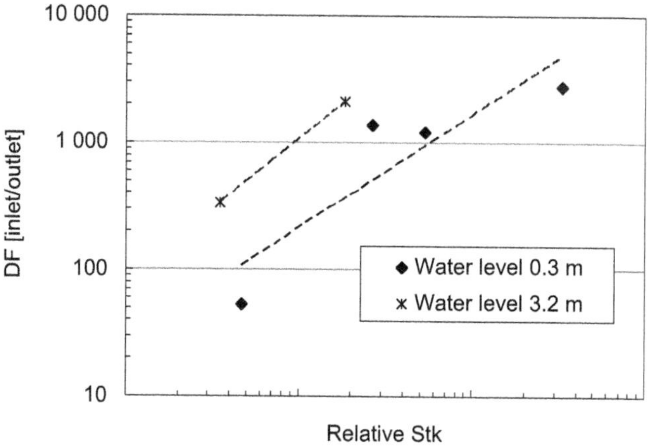

FIGURE 5.19 Aerosol decontamination factor in a steam generator bundle under flooded conditions.

depending on the test conditions in the range between 50 and almost 3,000. The particle retention increased with increasing particle inertia, that is, particle size and velocity (Figure 5.19).

Present pool scrubbing models are not capable of calculating the aerosol decontamination under flooded pool conditions [102], and they tend to underestimate the decontamination. Efforts are being made to develop models that could better describe the flow and aerosol dynamics under these extremely complex multiphase flow conditions.

5.5.3. Present Status of Investigations

As mentioned earlier, the only experimental program presently in progress is the International ARTIST2 project, which is a continuation of the preceding EU-SGTR and ARTIST projects. In the preceding text, a good number of references have been cited, so that most of the specifics can be found.

The present status can be synthesized through discussion of current analytical capabilities. In-tube deposition mechanisms seem to be well understood; however, the predictive capability weakens as one moves to high Reynolds number (as those expected in SGTR sequences) where resuspension comes into play. Even though a number of resuspension models exist in the literature (one of the best known being the so-called Rock'n Roll model [103]), it seems that none of them are capable of capturing scenarios with Reynolds numbers above 70,000 and for multilayer deposits. Efforts are being made to overcome this limitation in the frame of the ARTIST2 project.

Attempts to simulate aerosol retention in the break stage of a dry steam generator have been made based on the filter concept [98]. However, the extreme complexity of aerosol behavior through the tube bundle makes it very difficult to develop a mechanistic approach. However, some progress has been made by using a semi-empirical approach that considers simple models and correlations for both deposition mechanisms and resuspension, and an approximate expression of gas velocity based on 3D gas flow dynamics simulations [99]. Comparisons to data seem to indicate consistency of the predictions [104]. All these efforts are being conducted within the frame of the ARTIST2 project.

With regard to flooded scenarios, it has been shown that the traditional SPARC code approach [105] needs to be profoundly revised before being applied to the SGTR scenarios. Some models implemented in the code were validated under conditions drastically different from those of the SGTR. The presence of tubes in the water and the high velocities characterizing the jet injection make modeling of the two-phase flow extremely difficult under SGTR conditions.

In summary, despite considerable advances in this area, it seems that there is still a long way to go to accomplish a mechanistic aerosol retention model in the secondary side of the steam generator. New methods—for example, application of advanced modeling techniques like CFD—are being used to help in understanding the prevalent phenomena. Until then, the experimental data provided by ARTIST is helping us to better evaluate the source term from SGTR severe accidents.

5.6. EX-VESSEL FISSION PRODUCT RELEASE

(C. Spengler, M. Kissane, B.R. Sehgal, and M.K. Koch)

The evolution of aerosols in the containment is more problematic than that for aerosols in the RCS. Not only does a wide diversity of accident sequences exist with varying preconditioning of the source in the RCS before release to the containment, but also information for (more or less) prototypical particles is dependent on just one experimental program—Phébus FP (see, e.g., [106]), where information before this program was highly speculative (see [107]). In addition, the time scale of evolution of aerosols in the containment—about one day—leaves room for transformations due to radiolysis, oxidation, formation of (bi)carbonates, and the like, to occur.

Further complications arise in the containment from the (potential) occurrence of major secondary sources of aerosol material, that is, other than the direct source generated by a degrading core. These are, in particular, pressurized ejection of molten corium (high-pressure sequences), hydrogen deflagrations, and molten core–concrete interaction (MCCI).

Secondary sources of aerosol material in the containment can add significant masses of aerosol to that of the primary source. In the case of MCCI, significant diversification of suspended aerosol composition occurs due to the addition of a large amount of largely nonactive aerosol material in the size range of the existing aerosols. Addition of this source of aerosols would act to promote agglomeration and accelerate depletion of activity suspended in the containment atmosphere.

5.6.1. Phenomenology

In general, high-volatile FPs will be released in-vessel during the core degradation phase. Later aerosols may be generated ex-vessel by core–concrete interaction, with pool boiling and resuspension processes also leading to FP release. Aerosols usually with low volatility produced in the containment through the core–concrete interaction provide a long-term aerosol and FP source, after the in-vessel release phase also [108].

For aerosol generation four important mechanisms have been identified:

- Bursting of bubbles;
- Entrainment of melt in gas flow and subsequent break-up;
- Vapor condensation following vapor release from bubbles collapsing at the melt surface; and
- Condensation of vapors/volatile FPs released from the melt surface.

Here, the melt temperature also has a major influence on the quantity of aerosol and FP release. It is assumed that volatile FPs are released rapidly during the core–concrete interaction. For the less volatile components, the nonvolatile release fraction depends on the molten debris structure. As the melt reacts with concrete components and as melt temperature decreases, releases decline [108].

5.6.2. Experimental Investigations on Ex-vessel Fission Products/Aerosol Release

Early experimental results for the FP release under realistic conditions during the interaction between core melt and concrete (MCCI) were obtained in special experiments in the **SASCHA** facility at Forschungszentrum Karlsruhe (see also Section 5.3.1 and [11], [109]). In these special tests dedicated to MCCI conditions, a corium containing unirradiated UO_2 and some FP tracers was generated and heated within a siliceous concrete crucible. Release data for some elements as, for example, I, Ag, Te, and Mo were measured during approximately 12 minutes of core–concrete interactions. It was found that the fraction of iodine remaining in the melt after its heat-up was nearly completely released in the first three minutes of interactions. Based on this finding, a similar trend could be expected for high-volatile elements such as Kr, Xe, Br, Rb, Cs, and Cd. The elements Te and Ag with significant impact on source term and aerosol behavior were released up to a fraction of 40 and 60%, respectively, during the interaction. The release of the medium volatility FPs Mo and Ru was very low, despite the fact that their volatility in oxidic forms was expected to be much higher.

Further early tests of interest are the NSS series and the Transient Urania Reacting with Concrete (TURC) series, both performed at Sandia National Laboratory (SNL) and reviewed by Brockmann [110]. Neither of these series included sustained heating of the corium-simulating decay power, so the composition of the generated aerosols was potentially biased toward a higher-than-average fraction of the more volatile species. The results of the NSS series show that much of the generated aerosol is composed of nonradioactive materials, which do not depend strongly on composition of concrete type. However, the interaction of corium with concrete lasted only 1 to 2 minutes in these tests. In the TURC tests in one-dimensional concrete crucibles (with a concrete bottom and MgO sidewalls), iron-alumina based melts (TURC-1T) and urania-zirconia based melts (TURC-2 and -3) with FP mockups were used. Due to the transient character of the tests (i.e., without sustained heating), no significant concrete erosion was observed in tests 2 and 3. Both test series showed size distributions of the aerosols that are dominated by a single mode of about 1 μm mass median aerodynamic equivalent diameter. Some time-dependent data for aerosol mass release rates (e.g., for Te and La) and gas flow rates were obtained in TURC that could be used for model validation.

The large-scale Sustained Urania Reacting with Concrete (SURC) tests at SNL also addressed aerosol release as well as thermal-hydraulic phenomena in the cavity associated with prototypical core-melt materials in various types of concrete crucibles (limestone, basaltic), including sustained heating of the melts [111], [112]. Tests 1 and 2 addressed gas release, aerosol release and concrete ablation during sustained corium heating. Tests 3, 3a, and 4 investigated the additional effect of zirconium metal oxidation. In tests 1 and 2 the aerosol mass release rates were estimated in the range 1 to 10 g/min, mostly attributed to the concrete constituents Si, Na, K, and Ca. Aerosol release rates are roughly proportional to the gas release rates. It was further observed that the silicate concrete chemistry (SURC-2) retains more of the FP simulants in the melt than in the case of the limestone concrete chemistry (SURC-1). In tests 3 and 4 the aerosols comprised primarily Te, Na, Mn, K, and Fe oxides and were dominated by Te. The fractions of other simulated low-volatility FPs in the aerosol were generally very small (below 1% for Ce, U, and Ba, below 0.02% for others).

At Forschungszentrum Karlsruhe, a broad range of experiments was performed in the BETA-facility to investigate the interaction of iron-alumina-based thermite melts with concrete [113]. As in these experiments, the sustained heating was achieved by an induction technique, the power was released predominantly in the metal phase, which was layered in the crucible below the oxidic phase, because of its larger density. Thus, the results of these experiments are dominated by the interaction between the heated metal layer and the concrete. This is more prototypic of the MCCI's late stage, during which the oxidic layer, which throughout the concrete ablation process is continuously enriched by light concrete decomposition products, is already less dense than the metal layer. In the test series I, 19 experiments were performed at low and high volumetric power levels in axisymmetric cavities made from basaltic and limestone concretes. Aerosols were measured, but FPs were not simulated in the melt. There was a short, intense aerosol release during and immediately after the pouring of the melt into the crucible. After this initial time period of approximately 2 to 3 minutes the aerosol release decreases below 0.1 g/mol of the offgas flow. It was observed that in the case of limestone concrete a dense white aerosol consisting of CaO crystals with traces of Na and K was strongly released (>1.2 g/mol offgas) from the eroded concrete sidewalls above the melt, which are exposed to thermal radiation. In the test series II another 6 experiments were performed to investigate the additional influence of effects like the oxidation of zirconium or the behaviour of serpentinite concrete. In test series II, FP mockups were added to the melt. During Zr oxidation a high initial aerosol release was observed, which was substantially reduced after depletion of Zr. From the FPs added to the melt a substantial release of Te was measured, whereas the release of Ce, La, Sr, Ba and Mo is small and mostly below the detection limit. An electron microscopy analysis showed that the primary aerosol particles are spherical, with a typical diameter of 0.1…0.5 μm.

Finally, the richest source of information comes from the seven large-scale experiments of the international Advanced Containment Experiments (ACE) program on melt behavior and aerosol release during MCCI [114]. The experiments addressed four types of concrete (siliceous, limestone/sand, serpentine, and limestone) and a range of corium compositions (i.e., degree of oxidation) for both boiling water and pressurized water reactor scenarios. In the ACE tests, the concrete basemat had a square diameter (50 cm $\times$ 50 cm) and was 30 cm thick (see Figure 5.20 from [115]). The four vertical sidewalls of the test section containing the corium were either used for the simulation of internal decay heat (by the supply of direct current through the corium from tungsten electrodes) or designed as inert walls, so the MCCI investigated in the ACE tests has a 1-dim characteristic in axial direction. The zirconium metal was initially present in special layers embedded in the concrete basemat. The net electric power averaged approximately 100 kW during basemat ablation. The low-volatility FPs BaO, La_2O_3, SrO, CeO_2, MoO_2, SnTe, $ZrTe_2$, and Ru were present in the melt at concentration levels factor 2 or 4 higher than is typical of reactors to improve their detectability in the aerosols. The released aerosols contained mainly constituents of the concrete. Releases of tellurium and control rod materials—silver, indium and boron (from boron carbide)—were significant (Figure 5.21 from [116], Figure 5.22 from [115]). Fission product species other than Te contributed only <1% of the aerosol mass. For a given type of concrete, the release fractions of low-volatility FP elements decreased with decreasing concentration of metal in the melt. In the tests L1, L2, L6, and L7 with metal and limestone/sand or siliceous concrete, silicon compounds comprised 50% or more of the aerosol mass (Figure 5.22). It was found that the formation of barium and strontium silicates reduced the release for Ba and Sr. Based on these findings, it is concluded that precise thermodynamic data for some more silicates and zirconates (i.e., cerium, lanthanum) are necessary for adequate modeling. The significant contribution of large particles to the total uranium and zirconium releases indicated that mechanical processes are important for the release of these elements. Further, the condensed phase endothermic reaction between Zr and SiO_2 to produce SiO vapor was identified as important in predicting Si, SiO_2, and SiC aerosols (coming from SiO vapor followed by gas phase interactions). During ablation of the concrete, the aerosol composition remained fairly stable and particles were

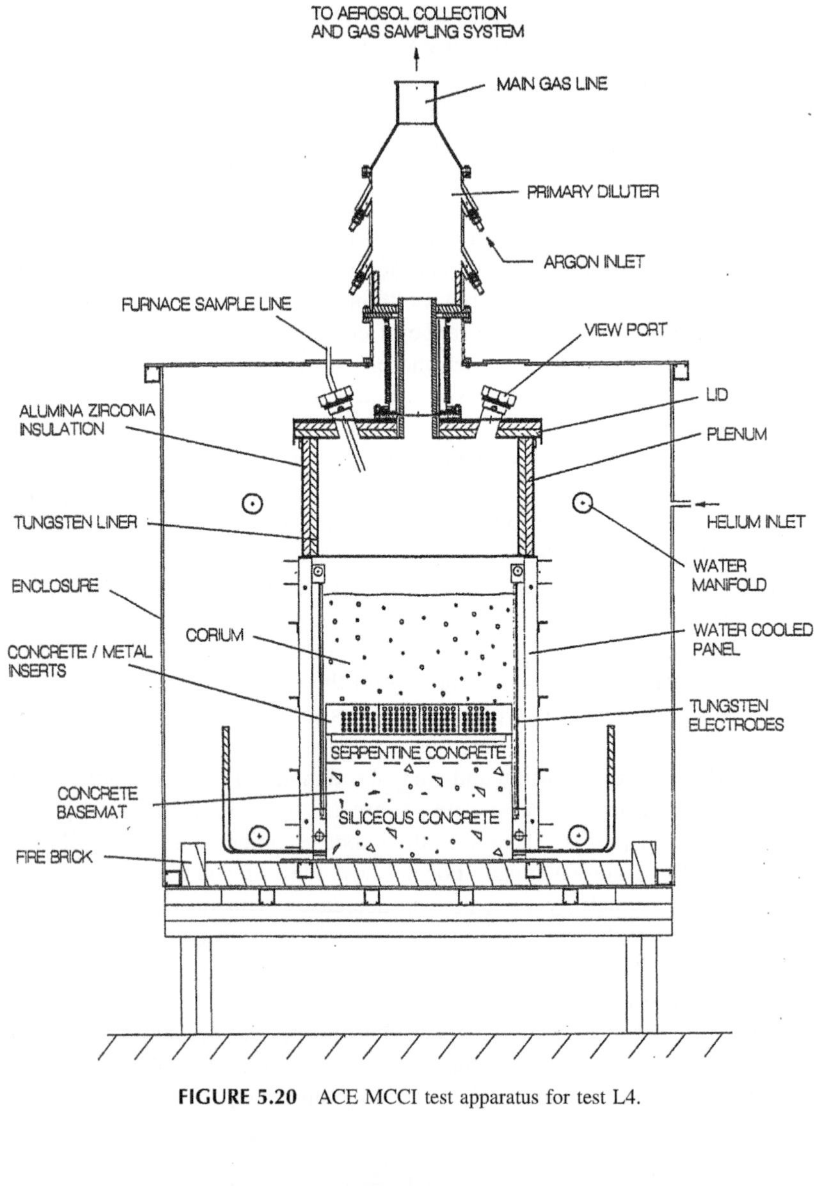

FIGURE 5.20 ACE MCCI test apparatus for test L4.

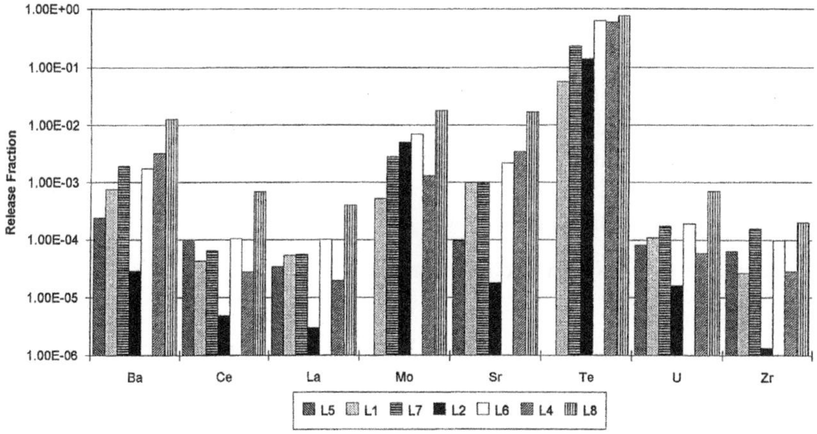

FIGURE 5.21 Total release of fission product elements and uranium in the ACE MCCI tests. *(from [116].)*

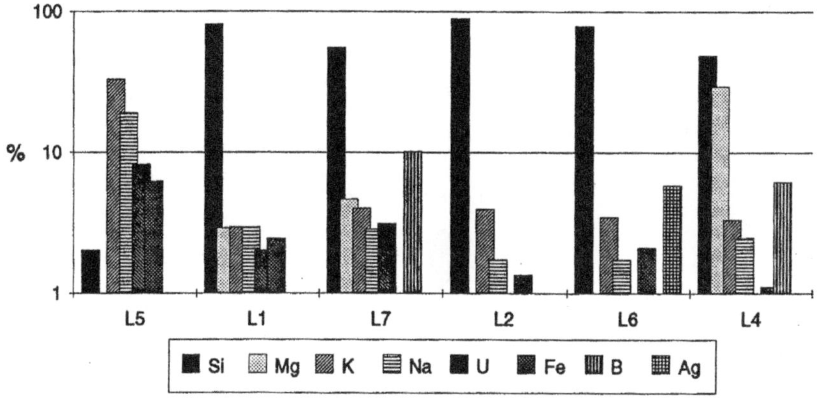

FIGURE 5.22 Mass fractions of aerosol elements in ACE MCCI tests. *(from [115].)*

compact but varied considerably in size, the majority being typically micron-sized (geometric diameter) but some with considerably larger sizes.

5.6.3. Models and Codes

The VANESA model [117] was developed under the sponsorship of the US-NRC and has been used in coupling with the MCCI code CORCON for the prediction of FP and aerosol release during MCCI. Both models are part of the US integral code MELCOR and of US containment code CONTAIN. VANESA provides mechanistic models for aerosol and FP release from molten core–concrete interaction. It considers releases coming from two processes: vaporization of substances into sparging gas bubbles and fragmentation of melt into small particles by the bursting of gas bubbles. Entrainment of melt into the gas flow is assumed to be small compared to the release by vaporization and therefore neglected. Inputs to the model are

- Chemical composition of the corium;
- Temperature of the corium;
- Chemical composition of the concrete;
- Mass flow rates of concrete decomposition products (slags, gases);
- Top surface area of the melt pool.

The output of VANESA encompasses:

- Aerosol mass flow rate;
- Chemical composition of the aerosol (FPs, concrete constituents, structural material);
- Particle-size distribution of released aerosol;
- Density of the released aerosol;
- Gas flow rate and chemical composition of the gas from MCCI.

A basic assumption in the code is the stratification of the melt into a bottom metal layer and a top oxidic layer (Figure 5.23, [117]). The first step in the VANESA analysis is to partition the material species in the melt between the oxide and the metal layer. The next step is to determine the shapes, sizes, and

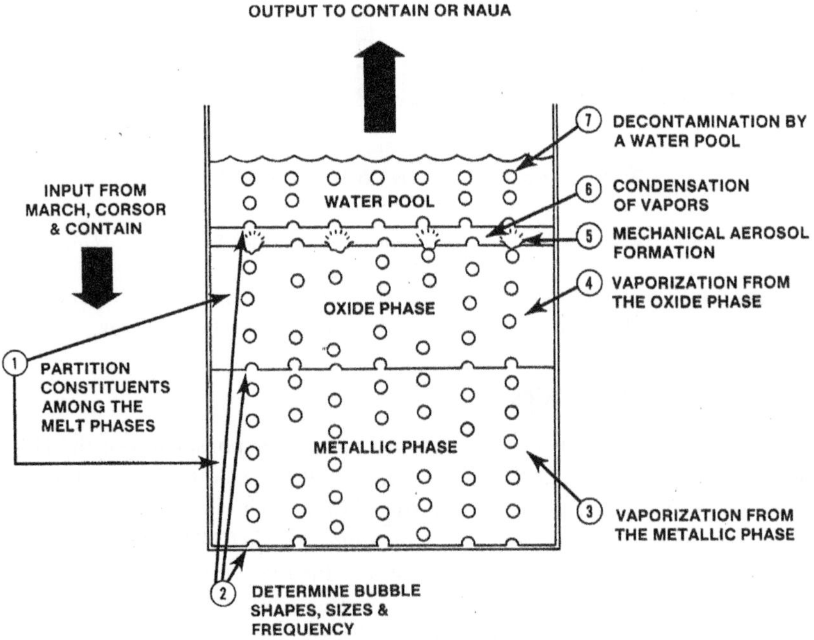

FIGURE 5.23 Schematic diagram of the calculation steps in the VANESA model.

frequency of bubbles sparging through the bottom layer (metal), that is, to determine the free surface for vaporization. For the next two steps, the calculation of vaporization in the metallic and the oxidic phase, both thermodynamic and kinetic considerations (separately for oxide and metal), are taken into account. The driving force (i.e., the pressure difference between the equilibrium and actual partial pressure exerted by that species) and the maximum extent of vaporisation are governed by thermodynamic considerations, whereas the kinetic considerations determine how the maximum extent is approached. For the calculation of vaporization, the equilibrium vapor pressure is calculated for selected species based on a ternary system for the element under consideration in combination with the elements oxygen and hydrogen. The assumption of ideal solution behavior was originally postulated [117] and means that the activities of the melt constituents are equal to their mole fractions—that is, that the activity coefficients are one. Oxide and metal phases were assumed not to equilibrate except that the chemical potentials of oxygen in the two phases were assumed to be the same. Later, some improvements were introduced in order to take into account the condensed phase chemistry of nonideal solutions [118]. When the gas bubbles reach the surface, they may burst and produce some amount of aerosol-sized droplets from the melt surface. Afterward the vapors may condense either by homogeneous nucleation or by deposition on surfaces such as the aerosol surfaces. The mechanical aerosol generation by bubble bursting in VANESA is based on available experimental data for droplet fragmentation due to bubble bursting in combination with the following assumptions:

- The size of aerosol particles is about 1 μm;
- A single bubble burst leads to approximately 2,000 aerosol particles;
- Gas bubbles are spherical and have a diameter of 2 cm.

Other approaches to model the FP release by vaporization rely on the application of conventional thermodynamic equilibrium models of an ideal or nonideal melt in contact with a gaseous phase. To calculate the thermodynamic equilibrium, a representation of the Gibbs energy of the total system as a function of temperature and composition is required. The equilibrium state is then obtained by minimizing the total Gibbs energy with respect to the composition under various conditions. The total Gibbs energy of the system is simply given by the algebraic sum of the Gibbs energies of the individual phases present at equilibrium under consideration of appropriate models for the excess Gibbs energy term in the case of solution phases.

The ELSA module of the ASTEC integral code for calculation of FPs has been presented in Section 5.3.3 [37]. The ELSA approach is also used for calculating FP release from the ex-vessel pool in combination with the MEDICIS module that calculates in ASTEC the interaction between molten core and concrete. In ELSA a chemical equilibrium in a well-mixed molten pool is assumed, and the release is referred to mass-transfer limited evaporation from the free surface of the pool. ELSA uses a database of Gibbs free-energy functions covering elements and their compounds via the MDB (Material Data Bank) tool in ASTEC in order to minimize the global Gibbs free energy of the system. This determines the species in the pool. The thermodynamic database for usage in ELSA has been extended for species of interest with regard to the ex-vessel scope. The release from the molten pool is calculated from the saturated partial pressure of species; it is obtained by equating the Gibbs free energy of species in the liquid and gas phases. The results have been compared to experimental data of the ACE program. Mechanical FP release is not considered.

The conclusion for all available approaches is that the release of FPs in the ex-vessel situation depends a great deal on the temperature of the core melt during the interaction and on the gas flow rate coming from the concrete decomposition. Since the uncertainties for the prediction of these quantities are still large, the precision of calculated releases—in view of the inherent additional uncertainties for thermodynamic modeling—is currently estimated to be at least one or two orders of magnitude.

5.7. FISSION PRODUCT TRANSPORT IN CONTAINMENT

(C. Housiadas and L. Herranz)

5.7.1. Phenomenology

The FPs arrive at the containment with the physical and chemical characteristics they have acquired during their transport in the RCS. Therefore, all FPs except noble gases and some free iodine and ruthenium arrive at the containment as aerosols (the latter when oxidizing conditions prevail in the core, i.e., in air ingress sequences). The various elements, either separately or chemically bound, are expected to be well mixed within the particles. The Phébus FP experimental observations support this assumption. It was found that the aerosol particles in the containment were a mixture of the various nuclides, and, moreover, the composition was independent of particle size and fairly homogeneous [46], [56], [106].

Severe accident simulations indicate that during the in-vessel phase of the accident the containment atmosphere is a steam-rich environment, at saturated conditions or close to saturation. The flow circulation in the containment atmosphere plays an important role because it controls the transport and distribution of steam, other noncondensable gases or vapors, and aerosols. Flow field calculations in conjunction with mass and heat transfer calculations determine the amount of steam that is condensed on the cool containment walls or other structural surfaces. Containment thermal hydraulics has been extensively investigated both experimentally and analytically, and in general a good level of understanding has been reached. A great deal of work was performed within the framework of the international experimental program on the ThAI

(Thermal Hydraulics, Aerosols and Iodine) facility in Germany [119] and in the earlier DEMONA experiments [120], discussed in [121].

The presence of high humidity in the containment induces an intense condensational growth of the aerosol particles. Steam condenses on the aerosol surface, and a fraction of the particle material can become dissolved. This process is enhanced by the presence of hygroscopic substances within the aerosol (e.g., CsOH). Although the containment saturation phase may be brief in time, the mechanism of hygroscopic condensational growth has very important implications for the evolution of the source term. Steam condensation significantly increases the size of the particles. The increase of particle size enhances the gravitational settling rate and therefore, aerosol depletion through deposition on surfaces. Also, coagulation/agglomeration is intensified as larger falling particles collide with smaller ones. Enhanced coagulation/agglomeration further increases the particle mass and size. Due to its importance, the mechanism of hygroscopic condensational growth is described in more detail in Section 5.7.2. Another mechanism contributing to enhanced deposition in the containment is diffusiophoresis, which is usually combined with thermophoresis. Diffusiophoresis is driven by the steam mass flux toward the condensing wall surfaces and the associated Stefan flow, whereas thermophoresis is driven by the steep temperature gradient between the containment atmosphere and the containment walls. These two mechanisms are the most important phoretic effects for the nuclear aerosols in the containment (and the RCS). Their modeling is presented in Section 5.7.2.

Computer code analyses and experimental observations effectively confirm that size distribution increases in the containment and that suspended aerosol mass is quickly removed. The dominant deposition mechanisms are gravitational settling and diffusiophoresis/thermophoresis. The overall trend is that larger particles are eliminated from the containment atmosphere, in which smaller particles remain airborne. The net result is a drastic decrease with time of the in-containment source term. The findings of the first Phébus test FPT0 are characteristic in this respect: during the core degradation phase and the following 20 minutes, 80% of the initially suspended mass in the containment was deposited on the containment walls. The mass median diameter of the airborne matter was initially equal to approximately 7 μm, very quickly decreased to roughly 4 μm, and finally was stabilized to 0.5 μm [55], [106]. Data are also available from the DEMONA experiments mentioned above, with and without steam atmospheres.

Iodine behavior in the containment is a critical parameter in the evolution of the source term. Iodine-bearing aerosol particles are dissolved in the sump water where iodide ions undergo a number of chemical reactions, especially with water radiolysis products. This promotes the formation of volatile inorganic and organic iodine that partitions with the gas phase. According to some Phébus FP experiments, if an oxidized chemical form of silver is present in sump water with a large excess as compared with iodine, the formation of nonsoluble silver iodide inhibits the formation of volatile iodine through radiolytically driven processes. Gaseous inorganic iodine interacts with surfaces through adsorption/desorption processes, painted surfaces being a much more efficient trap than steel ones. Part of the inorganic iodine trapped by painted surfaces is converted into organic iodine. Gaseous iodine reacts with air radiolysis products, with part of the inorganic being oxidized and part of the organic being destroyed. The balance between all formation and destruction phenomena, in aqueous and gas phases, determines the concentration of gaseous iodine in the containment atmosphere. Iodine behavior modeling is discussed in more detail in Section 5.7.2.

A very important aspect of the phenomenology in the containment is related to the behavior, eventually burn, of the accumulated hydrogen. This important part of phenomenology has been addressed separately in Chapters 2 and 3 of this book. Here we are limited to an analysis of FP behavior. Passive autocatalytic recombiners (PARs) are envisaged for converting hydrogen to water and thus lower the risk for hydrogen explosion. However, from the source term point of view, the operation of PARs may have undesirable side effects. Aerosol particles passing through the catalytic elements of PARs are heated up with significant evaporation of more volatile chemical species. This could aggravate the source term by converting easily filtered aerosol material residing in the containment into more troublesome vapors and gases. Experimental [122] and modeling [123] results based on the experimental program RECI demonstrated that potential exists for PARs to generate volatile forms of iodine, namely, molecular iodine, by thermocatalytic decomposition of metal-iodide aerosols.

Several safety-engineered systems are provided to lower the concentration of FPs in the containment atmosphere (mitigation measures). Efficient engineered systems for aerosol removal are sprays and suppression pools (in BWRs). They can also remove an important part of inorganic iodine from the atmosphere. Keeping the sump water alkaline is an efficient measure for preventing the formation of volatile iodine in the liquid phase. Filtered containment venting in case of containment overpressure can reduce release into the environment. A summary of the important mitigation processes used to achieve decontamination of the containment atmosphere is presented in Section 5.7.3.

5.7.2. Modeling of Basic Processes

Hygroscopic Condensational Growth

In the presence of highhumidity, aerosol particles grow. Condensation of water vapor is driven by the pressure difference between the ambient vapor pressure and the vapor pressure at the surface of the particle. Since latent heat is released at the particle surface during steam condensation, the treatment of particle growth requires consideration of the vapor mass transfer, as well as of the heat transfer between the particle and the ambient gas mixture. In general, growth processes are controlled either by gas phase transport or chemical reactions on the particle. The so-called growth law, that is, the derivative dd_p/dt that expresses the rate of change in particle size as a function of particle size and chemico-physical properties of the aerosol system (see Eq. (29)), is transport-limited in the former case and reaction-limited in the latter case.

Hygroscopic condensational growth is a typical process in which the growth law is transport-limited. The growth rate is obtained by solving the coupled equations for heat and mass transfer. At the beginning of 1970s, Mason [60] introduced an analytical approximation to the two coupled equations for the growth rate of a droplet by condensation. The Mason analysis can be summarized by the following expression

$$\frac{ddp}{dt} = \frac{4}{d_p}\left(\frac{S-1}{f_{\text{mass}} + f_{\text{heat}}}\right) \tag{27}$$

where $S = p_\infty / p_{\text{sat}}(T_\infty)$ is the saturation ratio, defined as the ratio between vapor pressure in the ambient gas, p_∞, and vapor saturation pressure over a flat surface, p_{sat}, at the gas temperature far away from the particle surface T_∞. The term f_{mass} represents the contribution associated with vapor diffusion through the gas to the particle surface and is given by:

$$f_{\text{mass}} = \frac{\rho_{\text{liq}}\, R_V T_\infty}{D p_{\text{sat}}(T_\infty)} \tag{28}$$

where ρ_{liq} is the density of the bulk liquid (water) and R_V is the water vapor gas constant. The term f_{heat} is the contribution to the growth rate due to heat conduction away from the particle through the gas and is given by

$$f_{\text{heat}} = \left(\frac{L_H}{R_V T_\infty} - 1\right)\frac{L_H\, \rho_{\text{liq}}}{T_\infty k_g} \tag{29}$$

where L_H is the latent heat of vaporization and k_g is the thermal conductivity of the gas.

Equation (27), known as Mason's equation, is based on equilibrium assumptions that may lead to significant inaccuracies when applied to small aerosol droplets. Three important thermodynamic and kinetic effects need to be taken into account:

- The so-called Kelvin effect, which is a curvature effect, arising from the fact that equilibrium vapor pressure P_{sat} over the highly curved particle surface may be significantly higher than that over a flat surface;
- The solute mass effect, which is due to the presence of dissolved molecules in the liquid droplet, causing a lowering of the vapor pressure with respect to vapor pressure over pure water;
- The so-called Fuchs effect, connected with the departure from the continuum regime for small particles.

The Kelvin and solute-mass effects may be incorporated in Mason's equation by modifying the droplet's equilibrium surface pressure, namely, by replacing the term $S-1$ in the numerator of Eq. (27) by

$$S - 1 - f_{\text{Kelvin}} + f_{\text{solute}} \tag{30}$$

$$f_{\text{Kelvin}} = \frac{4\sigma_1 \overline{v}_1}{R_u T_\infty d_p} \tag{31}$$

$$f_{\text{solute}} = \frac{6 n_2 \overline{v}_1}{\pi d_p^3} \tag{32}$$

Equations (30)–(32) are known as the Köhler equations [124]. In these equations R_u is the universal gas constant, σ_1 the surface tension of the solvent (water), n_2 the number of moles of the solute, and $\overline{v}_1$ the molar volume of the water in the solution, which, for a dilute solution may be approximated by the molar volume of pure water.

The Fuchs effects arising from departure from the continuum regime can be determined by multiplying the growth rate with a correction factor, known as the Fuchs and Sutugin interpolation formula [125]. This formula is given by:

$$f_{\text{FS}} = \frac{1 + 2\lambda/d_p}{1 + 3.42\lambda/d_p + 5.33(\lambda/d_p)^2} \tag{33}$$

With the above corrections, a modified Mason's equation can be inferred as follows:

$$\frac{dd_p}{dt} = \frac{4}{d_p}\left(\frac{S - 1 - f_{\text{Kelvin}} + f_{\text{solute}}}{f_{\text{mass}} + f_{\text{heat}}}\right) \cdot f_{\text{FS}} \tag{34}$$

Equation (34) is valid for determining the growth rate of the aerosol particles in the containment atmosphere covering the whole range from the continuum regime, through transitional, to the free-molecular regime.

Thermophoretic Deposition

Aerosol particles suspended in a nonisothermal gas tend to migrate from the hotter to the colder regions. The macroscopic result is a net transport in the direction of decreasing temperature, thus giving deposition on surfaces that are colder than the gas stream. The thermophoretic velocity V_{th} is expressed as a function of the local temperature gradient as follows

$$V_{th} = -K_{th}\frac{\nu_g}{T}\nabla T \tag{35}$$

In the preceding equation, T is gas temperature and K_{th} is a dimensionless parameter, the thermophoretic coefficient, which depends on properties of both the gas and the particle. Theoretical expressions for K_{th} have been rigorously derived only in the limits of continuum flow ($Kn \to 0$), the so-called Epstein [126] expression, or in the limit of free-molecular flow ($Kn \to \infty$), the so-called

Waldmann [127] expression. Talbot and co-workers [128], based on a theoretical expression suggested by Brock [129], proposed a fitting formula over the entire range $0 \leq Kn = 2\lambda/d_p \leq \infty$ that agrees to within 20% or better with most of the available experimental data. The Brock–Talbot expression reads

$$K_{th} = 2C_s \frac{C_c(k_g/k_p + 2C_t\lambda/d_p)}{(1 + 6C_m\lambda/d_p)(1 + 2k_g/k_p + 4C_t\lambda/d_p)} \quad (36)$$

and is the most widely used expression for calculating the thermophoretic coefficient. In Eq. (36) k_g/k_p is the ratio of gas to particle thermal conductivity, and C_c is the Cunningham correction factor. The constants C_s, C_t, C_m arise from noncontinuum effects at the particle interface: they are associated with the temperature and velocity boundary conditions (jump conditions) at the particle–gas interface. The recommended values are $C_s = 1.17$ (thermal creep coefficient), $C_t = 2.18$ (temperature jump coefficient), and $C_m = 1.14$ (velocity jump coefficient).

Figure 5.24 shows the thermophoretic coefficient K_{th} as a function of the Knudsen number (ratio of the mean free path of the gas molecules surrounding a particle to the particle radius) and parameterized by k_g/k_p. As can be seen, K_{th} has values around 0.5 and becomes a constant for very small particles (free-molecular regime). For large particles it depends both on particle size and on the gas-to-particle thermal conductivity ratio. This dependence is due to the nonnegligible temperature gradient established within a particle of appreciable size, which affects locally the temperature gradient of the gas.

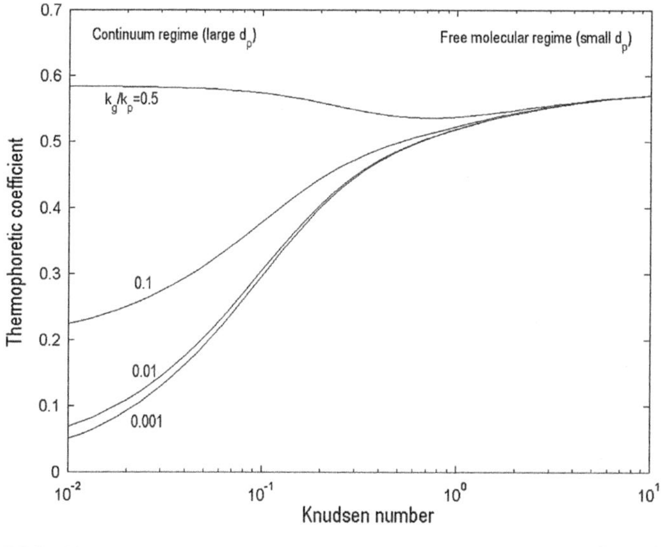

FIGURE 5.24 Thermophoretic coefficient K_{th} as function of Knudsen number, $Kn = 2\lambda/d_p$, for various gas-to-particle thermal conductivity ratios k_g/k_p.

Diffusiophoresis and Entrainment by Stefan Flow

Particles suspended in a nonequilibrium binary gas mixture, where the two gases interdiffuse, experience unequal momentum transfers by collisions with the molecules of the gas diffusing in one direction and the molecules of the other, oppositely diffusing gas. A net force is exerted on the particle in the direction of diffusion of the heavier gas molecule. The resulting motion, which is driven by the prevailing concentration gradient, is called diffusiophoresis. The above picture is further complicated when diffusion is concerned with the transfer of a condensable vapor through a background gas toward a condensing surface, as is precisely the case during wall condensation of steam in the containment. Under such circumstances, a secondary macroscopic flow is established within the gas, known as Stefan flow, which arises from a subtle mass balance between the diffusive flows of the gas and the water vapor. The direction of Stefan flow is always away from an evaporating surface or toward a condensing surface. Therefore, the motion of the particles is determined by the combined action of Stefan flow and diffusiophoresis. For such a combined motion in the mid-1060s Waldmann and Schmitt [130] provided the following theoretical expression for the particle drift velocity:

$$V_p = -\frac{\sqrt{\mathfrak{M}_1}}{\psi_1\sqrt{\mathfrak{M}_1} + \psi_2\sqrt{\mathfrak{M}_2}}\frac{\mathfrak{D}_{12}}{\psi_2}\nabla\psi_1 \tag{37}$$

where $\mathfrak{M}_1$, $\mathfrak{M}_2$ are the molecular weights of the condensable vapor (steam) and the background gas, ψ_1, ψ_2 the molar fraction of the vapor and the gas, and $\mathfrak{D}_{12}$ the binary gas diffusion coefficient. Note that the velocities obtained with Eq. (35) are independent of particle size. The predictions of Eq. (37) are in very good agreement with experimental data obtained with small nichrome particles ($0.02 \le d_p \le 0.2$ μm) in diffusing water vapor in air and helium. The simple model above is valid for small particles in the free-molecular regime only. The use of Eq. (35) for larger particles requires caution. A more complex and most generally valid model, covering the free- molecular and continuum particle flow regimes, is that of Loyalka [63].

For the case of water vapor in air, a practical approximate formula to use is:

$$V_p = -1.9 \times 10^{-4}\ (\mathrm{cm^2\ s^{-1}\ mb^{-1}})\ \frac{dp_1}{dx} \tag{38}$$

where dp_1/dx is the gradient of the partial pressure of the water vapor. The above correlation is within ±5% accurate with available measurements. If Eq. (37) is evaluated at standard conditions of pressure and temperature, a numerical expression identical to Eq. (38) is recovered (note that $(1/\psi_2)d\psi_1/dx$ can be written as $(1/p_2)dp_1/dx$ where p_2 is the partial pressure of the gas). Therefore, Eq. (38) has the appealing property of being both an empirical and a theoretical formula suitable for calculating the particle drift velocity due to combined diffusiophoresis and Stefan flow in water vapor–air systems.

Iodine Behavior

As stated earlier, most radioactive materials enter containment as particles and undergo aerosol physics processes such as settling. As a consequence, a substantial fraction ends up in the containment sump, where radionuclide decay builds up a radiation field that affects the behavior of some of them, particularly those containing iodine. The multiplicity of oxidation states of iodine together with its potential radiological impact make iodine chemistry an area of particular relevance in source term studies. Particular aspects of iodine behavior are summarised in the following sections, which rely heavily on the OECD/NEA State-of-the-Art Report on Iodine Chemistry [131] and on lecture notes for a SARNET course on severe accidents [132]. In addition, the Phébus FP project [133] should be highlighted as one of the most insightful projects regarding the chemical behavior of iodine under LWR severe accident conditions.

Water Radiolysis

Upon absorption of radiation, water decomposes and produces chemically reactive radicals and species $\cdot OH$, $\cdot H$, H^+, e^-_{aq}, H_2 and H_2O_2. These species would react with each other to reform water, a process during which additional species such as HO_2, O_2 and $\cdot O_2^-$ are also formed. This is illustrated in Figure 5.25.

The water radiolysis products will react with dissolved FPs. The important case of iodine is treated in the next two sub-sections.

Liquid Phase Chemistry of Iodine

The iodine aerosol particles injected into the containment are metal iodides such as caesium iodide. They are generally soluble, with the important exception of silver iodide AgI. Once dissolved, they produce iodide ions I^-. A number of reactions will take place, involving water radiolysis products as schematized in Figure 5.26. The number of existing reactions is too large to

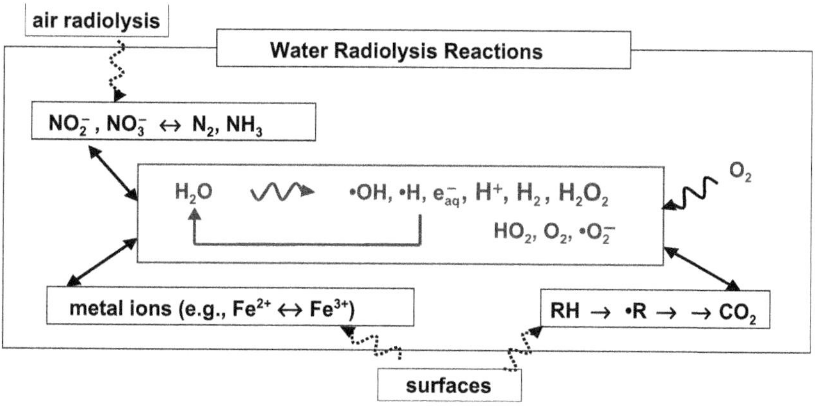

FIGURE 5.25 Water radiolysis reactions *(from [131])*.

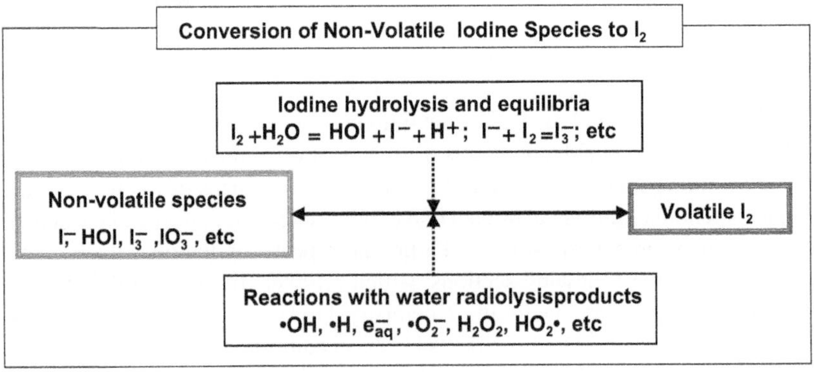

FIGURE 5.26 Conversion of nonvolatile iodine species to I_2 *(from [131])*.

give a comprehensive description of them in this course. What is to be remembered is that the net effect is the oxidation of iodide ions I^- into volatile molecular iodine I_2 that has then the potential to be released into the containment atmosphere.

The net formation of volatile I_2 depends on a number of different parameters, one of the most important being the pH of water: the lower the pH, the higher the I_2 production. This is illustrated in Figure 5.27 taken from parametric studies performed in the framework of the International Standard Problem no. 41 (ISP 41)[15] [134]. One can observe in that figure that the amount of volatile I_2 calculated by the codes listed on the right-hand side can vary by about one order of magnitude or more per pH unit.

The other main influential parameters of iodine volatility are temperature, irradiation, and dose rate: it decreases with temperature and increases with dose rate. The effect of the dose rate can be seen in Figure 5.28, again coming from ISP 41.

Organic materials are present in the sump coming from different sources such as paints and cables. Their radiolysis leads to the formation of organic radicals •R that will react with I_2 to form organic iodides RI. The simplest and the most volatile of these iodides is methyl-iodide CH_3I. Not only CH_3I can be formed, but also higher molecular weight species. As they are less volatile, they are generally not considered in safety evaluation studies.

Reactions of hydrolysis and radiolysis can convert organic iodides RI into iodide ions I^-. This is illustrated in Figure 5.29.

A large number of pressurized water reactors use silver as a neutron absorber in their control rods. When the control rod is heated up, silver is

15. International Standard Problems are exercises organized by the OECD Committee on the Safety of Nuclear Installations (OECD/CSNI) in which the results of code calculations are compared against a background of well-qualified experimental data.

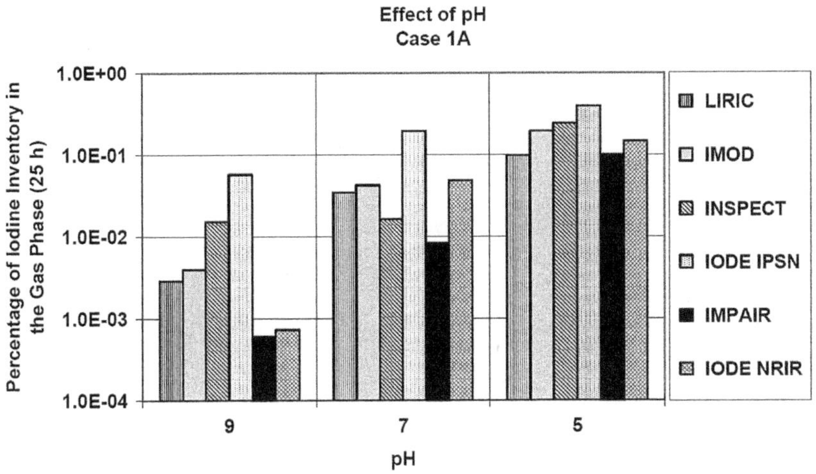

FIGURE 5.27 Effect of pH on production of volatile I_2 at 90°C and 1kGy/hr *(from [134])*.

vaporized, and some may be finally released into the containment and drained into the sump water. It may be present under two forms: metallic insoluble Ag or soluble oxide Ag_2O (the partial oxidation of silver particles takes place mainly in the containment). Both the metal and the oxide forms can react with iodine to form insoluble silver iodide AgI, the reaction with the oxide being more efficient. The formation of AgI consumes I_2 and iodide ions I^- that are no longer available to form volatile I_2 through radiolytically driven processes:

$$2Ag + I_2 \rightarrow 2AgI_{\downarrow}$$

$$Ag_2O + 2I^- + 2H^+ \rightarrow 2AgI_{\downarrow} + H_2O$$

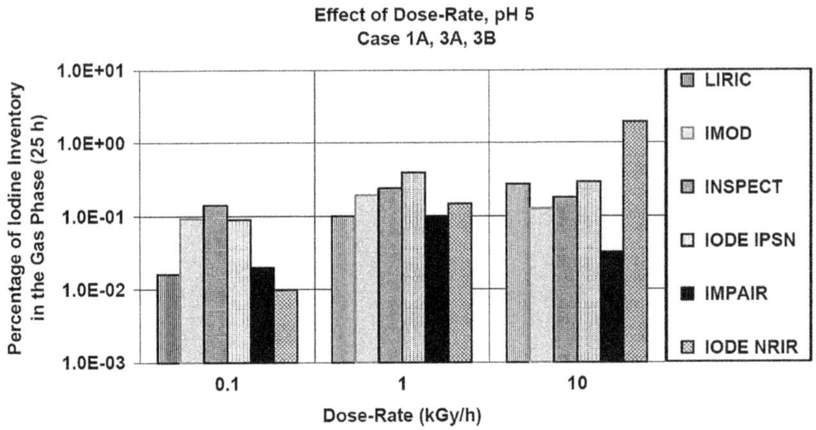

FIGURE 5.28 Effect of dose rate on iodine volatility at pH 5 and 90°C *(from [134])*.

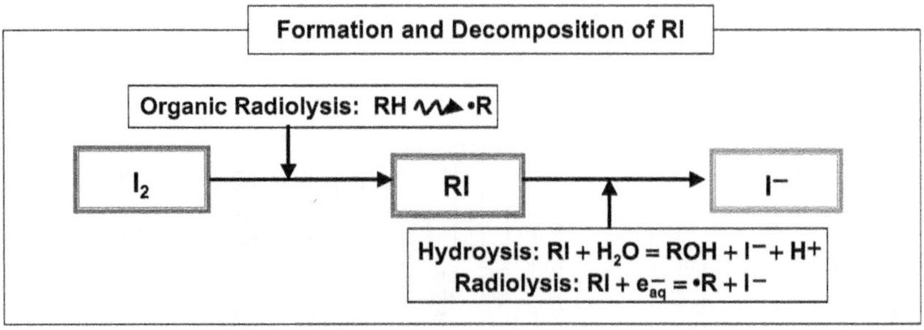

FIGURE 5.29 Formation and decomposition of organic iodides in liquid phase *(from [131])*.

It should be noted that silver must be in excess as regards iodine to be an efficient trap, as silver iodide AgI is not stable under radiation. The effect of silver on iodine volatility is illustrated in Figure 5.30, coming from ISP 41.

Note that there is one calculation showing no effect, just because at that time there was no model for Ag-I interaction in the code.

Gaseous iodine present as dissolved I_2 or RI in the sump water will partition to the gas phase. This partition depends on mass transfer coefficients in the liquid and the gas phase related to thermal-hydraulic conditions, while the ratio between the concentrations in gas and liquid phases at the gas–water interface obeys Henry's law.

Gas Phase Chemistry of Iodine

Gaseous iodine in the containment comes from two different sources: the RCS and the sump water as explained in previous subsections.

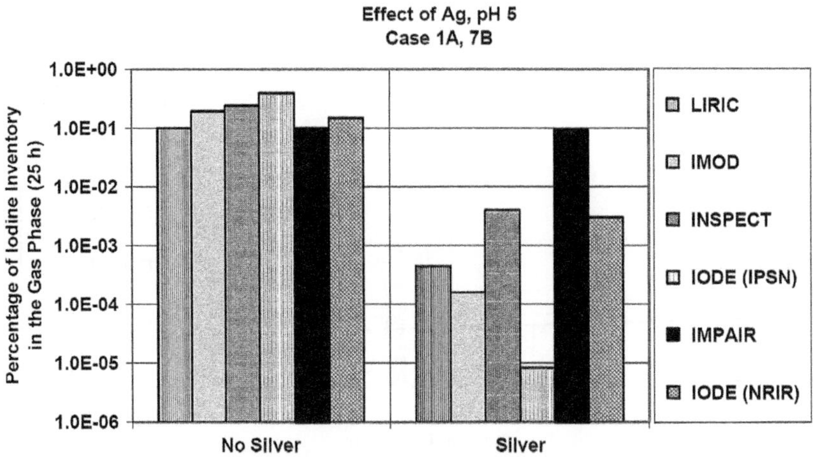

FIGURE 5.30 Effect of silver on iodine volatility in liquid phase *(from [134])*.

Inorganic iodine I_2 interacts with the containment's surfaces via adsorption and desorption mechanisms. Most studies have so far focused on organic-based painted and stainless steel surfaces. A first outcome of these studies is that I_2 can be much more efficiently trapped by painted surfaces than by stainless steel ones. The paints have a twofold action: they act not only as a sink for volatile inorganic iodine I_2, but also as a source of volatile organic iodides. Radiation plays a strong enhancing role in the latter process and has much more influence than temperature.

There are still large uncertainties in the quantification of these processes, and efforts are underway for reducing them, especially with the EPICUR experiments, part of the International Source Term Program, the results of which were analyzed in the SARNET network [96], [135].

The atmosphere of the containment is composed mainly of humid air that will, under radiation, form radiolysis products, including active radicals and molecules. Inorganic and organic iodine will react with these air radiolysis products to form new species. This is schematized in Figure 5.31. Inorganic iodine I_2 is oxidized in different iodine oxides species I_xO_y (e.g. I_2O_5, I_4O_9). These less volatile species should condense to form very fine particles that may either deposit on surfaces or be drained to the sump water to form dissolved iodate ions IO_3^-.

Organic iodides are decomposed to form inorganic species.

The quantification of the processes associated with air radiolysis products has made good progress mainly through the recent PARIS [136] project initiated by IRSN. It should be noted, however, that there are still large uncertainties about the fate of the new species formed, such as iodine oxides, which depend on thermal hydraulics in the containment and physics of fine aerosol particles.

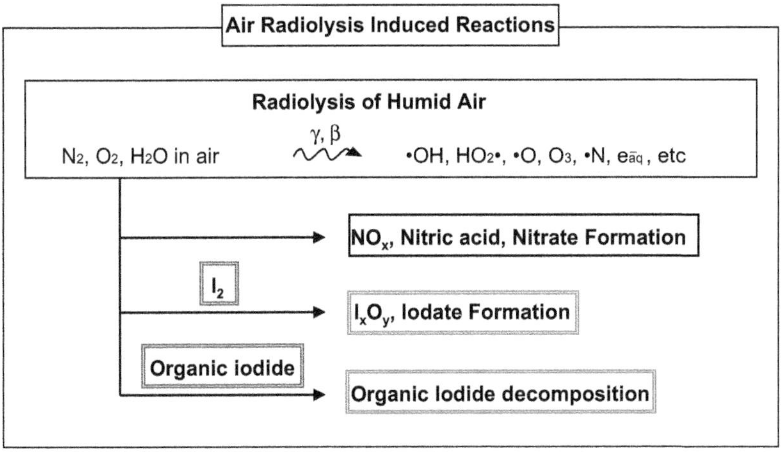

FIGURE 5.31 Air radiolysis-induced reactions *(from [131])*.

Aerosol Leaks through Cracks

In the course of an accident, one may expect the leak-tightness of the containment to deteriorate because of the pressurization effects. As a result, the mobile nuclides present in the containment as aerosols or vapors may be released to the environment by leakages through the containment cracks and/or the failed seals and penetrations. Currently, the assessment of the source term to the environment is generally based on the assumption that there is no retention of FPs in the leak paths from the containment. Therefore, the transmission of FPs from the containment is taken to occur with the general leakage rate. There is experimental and theoretical evidence of strong retention of aerosols in the leak paths, leading even to complete plugging of the leak flow. There is interest in knowing the impact of aerosol retention in leakage paths, to further reduce the uncertainty margins in the source term assessment and the off-site radiological consequences. Such improvements will give future NPPS additional importance, as most European safety authorities require that severe accidents be considered in the design.

Several small-scale laboratory studies on aerosol trapping in small tubes and capillaries have been performed in the past. In these early studies, the used tubes or capillaries had diameters ranging from a few micrometers to a few millimeters, and the employed pressure differences were up to several bars. In parallel, a number of theoretical analyses were performed. The most frequently cited model for evaluating plugging by the deposited aerosol was developed by Vaughan [137] and Morewitz [138], [139] in the 1980s. This model simply states that the total mass transported through a duct prior to complete plugging is proportional to the cube of the duct diameter:

$$m = k \cdot D^3 \tag{39}$$

The coefficient k was quantified as 30 ± 20 g/cm^3 through experimental data fittings over a wide range of duct diameters (20 μm to 26.5 cm) and pressure differences (0.3 kPa to 0.7 MPa), with a variety of aerosol materials. More recently, Powers [140] has reviewed available data and models relating to aerosol penetration of leak pathways, concluding that "the Morewitz-Vaughan correlation for leak pathway plugging does not provide a complete description of aerosol behaviour in leaks; in fact, the correlation is applicable for only a narrow portion of the range of aerosol concentrations and particle sizes of interest for the analysis of reactor accidents." The criteria of Agarwal and Liu [141], and of Davies [142], were preferred for assessing the suitability of leak paths for plugging.

Large-scale experiments have been performed in the United States (the Containment System Experiment program at Battelle Memorial Institute), using a one-fifth linear scale model of a typical 1000 MWe PWR. A series of aerosol leakage tests gave a decontamination factor 15 for iodine and 100 for caesium in dry conditions, and almost complete retention in wet conditions (i.e., when steam was condensing on the containment walls). Other large-scale experiments were performed in Japan by NUPEC on actual containment penetrations of a BWR plant using dry CsI aerosol particles: they indicated decontamination factors between 10 and 1,000.

In Europe large-scale tests on crack leakage have been performed in the MAEVA facility [143], which is a containment mockup. This facility (run by EDF) contains a cracked internal enclosure that is pressurized with air or an air–steam mixture up to 6.5 bars and an external enclosure that is divided into four parts in which global and local flow rates are measured. Also, experimental results from small-scale analytical tests are available [144]. These tests, performed by IRSN, deal with aerosol penetration through a shear-cracked concrete slab. Also, an experimental campaign has been conducted at the COLIMA facility of the PLINIUS Platform [145]. The COLIMA leak tests used using prototypical aerosols, produced from a piece of corium heated up to 2000–3000 K, whereas the cracked samples are made of representative limestone concrete [146], [147].

Advances in the mechanistic modeling of particle penetration through tubes and cracks have been made, with the view of developing methods suitable for large system codes like ASTEC. In particular, an Eulerian model has been developed to calculate deposition through cracks, as well as plug formation [148]. The model is based on the numerical solution of the aerosol transport equation, considered in one-dimensional form along the flow path. This model assumes that aerosol transport is almost steady state and that plug formation is uniform over the duct circumference. The particle deposition velocity is determined by considering Brownian diffusion and gravitational settling. In case of turbulent flow, the mechanisms of turbulent diffusion and eddy impaction are also included. Encouraging results have been obtained by comparing the predictions of this model with the measurements of [144].

5.7.3. Mitigation Measures

Mitigation stands for all processes that may lead to a decontamination of the containment atmosphere, these by reducing them the potential for significant release of radioactive material to the environment.

Natural processes act to remove aerosols from the atmosphere. They can be augmented by engineered systems and management measures to clean the atmosphere more rapidly. Engineered systems can also reduce the concentration of gaseous iodine in the atmosphere.

Natural Processes for Aerosols

The natural processes that remove aerosol particles from the containment atmosphere are deposition processes. The most efficient ones are gravitational settling, thermophoresis, and diffusiophoresis. The aerosol growth by coagulation with other particles or due to steam condensation or adsorption is also important as it increases their settling velocity.

Engineered Safety Systems for Aerosols

Containment spray is used for reducing pressure in the containment atmosphere by inducing condensation of steam. It can be used in two modes: direct or recirculating. In the direct mode, water from an outer tank is sprayed through nozzles in the upper part of the containment, generating a large number of cold droplets. In the recirculating mode water is taken from the sump, cooled down in a heat exchanger, and then sprayed.

The falling water droplets remove the aerosol particles by several mechanisms:

- Sweep-out of the particles that are unable to avoid the falling droplet;
- Interception of particles as they follow stream-lines around the droplets;
- Diffusion of particles to the droplet surfaces;
- Phoretic capture due to temperature gradients and steam condensation.

Overall, spray is very efficient in removing particles from the atmosphere. A typical decontamination factor (expressed as an inverse time constant for the airborne aerosol concentration decrease) is about 10 hr^{-1}, whereas it is about 0.5 hr^{-1} by natural processes without spray.

The spray efficiency depends on the particle-size distribution: fine and large particles are more efficiently removed than intermediate-size particles. Large particles are, for instance, more sensitive to sweep-out and interception, whereas small ones are more sensitive to diffusion.

In BWRs, the gases laden with aerosol particles can be directed into water pools through so-called quenchers. The quenchers are submerged pipes with many holes through which the gas will sparge the water pool, breaking the flow into bubbles. The aerosol particles are removed from the bubbles by several mechanisms:

- Condensation of supersaturated steam;
- Sedimentation within the bubbles;
- Diffusion of particles to the bubble's surface;
- Inertial impaction of particles on the bubble surface;
- Sweep-out of particles by bubble oscillations.

Mitigation of Iodine

A first mitigation measure is to prevent the formation of gaseous iodine. As explained in previous sections, gaseous iodine can be injected in the containment atmosphere either from the RCS or from the sump water. The second route of gaseous iodine formation can be largely prevented if the pH of the sump water is kept alkaline—a value of 9 or above 7 would be adequate—by safety engineered systems. Such measures are implemented in a number of power plants using soda NaOH, potassium hydroxide KOH or trisodium phosphate Na_3PO_4. For instance, in French nuclear power plants, soda is present in the tank used for spray. This means that the mitigation will be efficient if the spray is available.

Another way of mitigation is to remove the gaseous iodine from the containment atmosphere. Engineered systems such as spray or suppression pools can remove an important part of inorganic gaseous iodine I_2 from the atmosphere. Their efficiency depends on the pH of water, which should be alkaline or at least not acidic. They are generally considered as inefficient for the removal of organic iodine.

After many years of research and development, PSI invented a process that deals with rapid reduction of gaseous and nongaseous iodine species into nonvolatile iodide ions in aqueous solution and suppression of thermal and radiolytic oxidation to practically eliminate any gaseous iodine release from the solution.[16] The process uses a reducing agent as an additive and a co-additive to catalyze the reaction for the reduction and at the same time to bind iodide. The reducing agent is a nucleophilic substance, and the co-additive is a phase transfer catalyst and ion exchanger. The process can be applicable to many different reactor applications, such a wet containment venting filters.

Filtered Containment Venting

In some nuclear power plants with large dry containments, procedures are implemented to vent the containment in case of overpressure to avoid its failure and a large early release of radioactive materials. The released gases pass through filters that must have a large aerosol retention capacity to avoid their clogging. Their retention efficiency is generally much greater for aerosol particles than for gaseous iodine. As an example, for filtering systems used in French nuclear power plants (using sand-bed filters), the decontamination factors are about 1,000 for aerosols, 10 for inorganic iodine, and 1 for organic iodine.

REFERENCES

[1] L. Soffer, S.B. Burson, C.M. Ferrell, R.Y. Lee, J.N. Ridgely, Accident source terms for light-water nuclear power plants, Final Report, NUREG-1465 (1995).

[2] H. Bateman, The solution of a system of differential equations occurring in the theory of radioactive transformations, Proc. Cambridge Philos. Soc. 15 (1910) 423–427.

[3] A.G. Croff, ORIGEN2: a versatile computer code for calculating the nuclide compositions and characteristics of nuclear materials, Nuclear Technology 62 (1983) 335.

[4] A. Tsilanizara, C.M. Diop, B. Nimal, M. Detoc, L. Luneville, M. Chiron, T.D. Huynh, I. Bresard, M. Eid, J.C. Klein, DARWIN: an evolution code system for a large range of

16. Patent 20090127202 Fast Reduction of Iodine Species to Iodide 05-21-2009 (Switzerland).

applications, Proc. 9th International Conference on Radiation Shielding (ICRS-9) (October 17–22, 1999). Tsukuba, Japan.

[5] Y. Pontillon, P.P. Malgouyres, G. Ducros, G. Nicaise, R. Dubourg, M. Kissane, M. Baichi, Lessons learnt from VERCORS tests. Study of the active role played by UO_2-ZrO_2-FP interactions on irradiated fuel collapse temperature, J. Nuclear Materials 344 (2005) 265–273.

[6] M. Baichi, Contribution à l'étude du corium d'un réacteur nucléaire accidenté: aspects puissance résiduelle et thermodynamique des systèmes U-UO2 et UO2-ZrO2, Thèse INP Grenoble, France, 2001.

[7] H. Kleykamp, The chemical state of the fission products in oxide fuels, J. Nuclear Materials 131 (1985) 221–246.

[8] T.J. Haste, B. Adroguer, R.O. Gauntt, J.A. Martinez, L.J. Ott, J. Sugimoto, and K. Trambauer, In-vessel core degradation validation matrix, OECD/NEA/CSNI/R(95)21, OCDE/GD(96)14, January 1996.

[9] G. Ducros, P.P. Malgouyres, M. Kissane, D. Boulaud, M. Durin, Fission product release under severe accidental conditions; general presentation of the program and synthesis of VERCORS 1 to 6 results, Nucl. Eng. Des. 208 (2001) 191–203.

[10] Y. Pontillon, G. Ducros, P.P. Malgouyres, Behaviour of fission products under severe PWR accident conditions VERCORS experimental programme—Part 1: General description of the programme, Nucl. Eng. Des. 240 (2010) 1843–1852.

[11] B.J. Lewis, R. Dickson, F.C. Iglesias, G. Ducros, T. Kudo, Overview of experimental programs on core melt progression and fission product release behaviour, J. Nucl. Mater. 380 (2008) 126–143.

[12] H. Albrecht, V. Matschoss, H. Wild, Release of fission and activation products during light water reactor core meltdown, Nuclear Technology 46 (1979) 559–565.

[13] R.A. Lorenz, M.F. Osborne, A summary of ORNL Fission Product release tests with recommended release rates and diffusion coefficients, ORNL/TM-12801—NUREG/CR-6261, 1995.

[14] Z. Lui, D.S. Cox, R.S. Dickson, P. Elder, A summary of CRL fission product release measurements from UO2 samples during post-irradiation annealing (1983-1992), Report COG (1994) 92–377.

[15] J.P. Léveque, B. Andre, G. Ducros, G. Le Marois, G. Lhiaubet, The HEVA experimental program, Nuclear Technology 108 (1994) 33–44.

[16] Y. Pontillon, G. Ducros, Behaviour of fission products under severe PWR accident conditions VERCORS experimental programme—Part 2: Release and transport of fission gases and volatile fission products, Nucl. Eng. Des. 240 (2010) 1853–1866.

[17] Y. Pontillon, G. Ducros, Behaviour of fission products under severe PWR accident conditions VERCORS experimental programme—Part 3: Release of low-volatile fission products and actinides, Nucl. Eng. Des. 240 (2010) 1867–1881.

[18] A. Hidaka, Outcome of VEGA Program on Radionuclide Release from Irradiated Fuel under Severe Accident Conditions, Journal of Nuclear Science and Technology 48 (1) (2011) 85–102.

[19] A. Goryachev, et al., Techniques and first results of fission product release study in QUENCH tests with irradiated VVER rod simulators, Proc. 12th Int. QUENCH workshop (2006). FZ Karlsruhe, Germany.

[20] R.R. Hobbins, D.A. Petti, D.L. Hagrman, Fission product release from fuel under severe accident conditions, Nuclear Technology 101 (1993) 270–281.

[21] M.D. Allen, H.W. Stockman, K.O. Reil, A.J. Grimley, W.J. Camp, ACRR fission product release tests ST-1 and ST-2, Proc. Int. Conf. Thermal Reactor Safety vol. 5 (October 2–7, 1988). Avignon, France.

[22] D.A. Petti, Z.R. Martinson, R.R. Hobbins, D.J. Osetek, Results from the Power Burst Facility (PBF) Severe Fuel Damage (SFD) test 1-4: a simulated severe fuel damage accident with irradiated fuel rods and control rods, Nuclear Technology 94 (1991) 313–335.

[23] D.D. Lanning, N.J. Lombardo, D.E. Fitzsimmons, W.K. Hensley, F.E. Panisko, Coolant boilaway and damage progression program data report: Full-Length High Temperature (FLHT) Experiment 5, PNL-6540, Pacific Northwest Laboratories, April 1988.

[24] M.L. Carboneau, V.T. Berta, M.S. Modro, Experiment analysis and summary report for OECD LOFT Project Fission Product Experiment LP-FP-2, OECD LOFT-T-3806, Organization for Economic Cooperation and Development, June 1989.

[25] M. Schwarz, G. Hache, P. von der Hardt, Phebus FP: A severe accident research programme for current and advanced light water reactors, Nuc. Eng. Des. 187 (1999) 47–69.

[26] B. Clément, R. Zeyen, The PHEBUS Fission Product and Source Term International Programme, Proc. Int. Conf. Nuclear Energy for New Europe (2005). Bled, Slovenia.

[27] N. Hanniet-Girault, G. Repetto, FPT-0 Final Report, Document Phébus PF, IP/99/423, IRSN (1999).

[28] D. Jacquemain, S. Bourdon, A. De Bremaecker, M. Barrachin, FPT1 Final Report (Final version), IPSN/DRS/SA/PDF report SA1/00, IP00/479, France, December 2000.

[29] A.-C. Gregoire, P. March, F. Payot, G. Ritter, M. Zabiego, A. de Bremaecker, B. Biard, G.Gregoire, S. Schlutig, FPT2 Final Report, Document Phébus PF IP/08/579, IRSN (2008).

[30] F. Payot, T. Haste, B. Biard, F. Bot-Robin, J. Devoy, Y. Garnier, J. Guillot, C. Manenc, P. March, FPT3 Final Report, Document Phébus PF IP/10/587, IRSN (2010).

[31] P. Chapelot, A.-C. Gregoire, G. Gregoire, FPT4 Final Report, Document Phébus PF IP/04/553, IRSN (2004).

[32] M.R. Kuhlman, D.J. Lehmicke, R.O. Meyer, CORSOR user's manual, BMI-2122, NUREG/CR-4173, 1985.

[33] M. Ramamurthi, M.R. Kuhlman, Final Report on Refinement of CORSOR – An empirical in-vessel fission product release model, Battelle Columbus Laboratories (BMI) Report, US Nuclear Regulatory Commission, October 1990.

[34] A. Fick, On liquid diffusion, Phil. Mag. and Jour. Sci 10 (1855) 31–39.

[35] A.H. Booth, A method for calculating fission gas diffusion from U02 fuel and its application to the X-2-f loop test, AECL-496, Atomic Energy of Canada Limited, 1957.

[36] B. Andre, G. Ducros, J.P. Leveque, M.F. Osborne, R.A. Lorenz, D. Maro, Fission product release at severe LWR accident conditions: ORNL/CEA measurements versus calculations, Nuclear Technology 113 (1996) 23–49.

[37] W. Plumecocq, M.P. Kissane, H. Manenc, P. Giordano, Fission product release modelling in the ASTEC integral code: the status of the ELSA module, 8th Int. Conf. on CANDU fuel (2003). Honey Harbour, Ontario.

[38] G. Ducros, M.P. Ferroud-Plattet, M. Baichi, P.P. Malgouyres, C. Poletiko, H. Manenc, D. Boulaud, M. Durin, Fission product release on VERCORS HT1 experiment, Cooperative Severe Accident Partners Meeting (CSARP), Albuquerque, New Mexico, USA, May 1999.

[39] M.S. Veshchunov, V.D. Ozrin, V.E. Shestak, V.I. Tarasov, R. Dubourg, G. Nicaise, Development of the mechanistic code MFPR for modelling fission-product release from irradiated UO2 fuel, Nucl. Eng. Des. 236 (2006) 179–200.

[40] M.P. Kissane, N. Davidovich, R. Dubourg, C. Fiche, P.K. Mason, W. Plumecocq, Fission-product release and transport in severe-accident conditions: strategy and illustrations, Fuel Safety Research Specialists' Meeting, JAERI/Tokai Research Establishment (Japan), March 4–5, 2002.

[41] M.S. Veshchunov, V.D. Ozrin, V.E. Shestak, V.I. Tarasov, R. Dubourg, G. Nicaise, Modelling of defect structure evolution in irradiated UO2 fuel in the MFPR code, Paper 1085, Proceedings of the 2004 International Meeting on LWR Fuel Performance, Orlando, Florida (USA), September 19–22, 2004.

[42] Y. Drossinos, C. Housiadas, Aerosol Flows, in: C. Crowe (Ed.), The Multiphase Flow Handbook, CRC Press—Taylor & Francis, Boca Raton, FL, USA, 2005. Chapter 6.

[43] W.C. Hinds, Aerosol Technology: Properties, Behavior, and Measurement of airborne particles, second ed., John Wiley & Sons, Inc, New York, NY, USA, 1999.

[44] T. Hatch, S.P. Choate, Statistical description of the size properties of non-uniform particulate substances, J. Franklin Inst. 207 (1929) 369–387.

[45] M.P. Kissane, On the nature of aerosols produced during a severe accident of a water-cooled nuclear reactor, Nucl. Eng. Des. 238 (2008) 2792–2800.

[46] S.K. Friedlander, Smoke, dust and haze: fundamentals of aerosol dynamics, second ed., Oxford University Press, Oxford, UK, 2000.

[47] M.M.R. Williams, S.K. Loyalka, Aerosol science theory and practice: with special applications to the nuclear industry, Pergamon Press, New York, NY, USA, 1991.

[48] J.H. Seinfeld, S. Pandis, Atmospheric Chemistry and Physics: From Air Pollution to Climate Change, John Wiley & Sons, 1998. Chapters 7–12.

[49] Colbeck, Physical and Chemical Properties of Aerosols, Blackie Academic & Professional, Chapman & Hall, London, UK, 1998.

[50] H.R. Pruppacher, J.D. Klett, Microphysics of Clouds and Precipitation, second ed., Kluwer Academic Publishers, Heidelberg, Germany, London, UK, New York, NY, USA, 1997.

[51] H.J. Allelein, A. Auvinen, J. Ball, S. Güntay, L.E. Herranz, A. Hidaka, A.V. Jones, M. Kissane, D. Powers, G. Weber, State-of-the-art report on nuclear aerosols, OECD/CSNI report NEA/CSNI/R(2009)5, 2009.

[52] P.A. Baron, K. Willeke, Aerosol Measurement: Principles, Techniques, and Applications, John Wiley & Sons, Inc, New York, NY, USA, 2001.

[53] M.H. Kaye, M.P. Kissane, P.K. Mason, Progress in chemistry modelling for vapour and aerosol transport analyses, Int. J. Mater. Res. 2010/12 (2010) 1571–1578.

[54] N. Bixler, VICTORIA 2.0: a mechanistic model for radionuclide behavior in a nuclear reactor coolant system under severe accident conditions, USNRC, 1998. NUREG/CR-6131.

[55] L. Cantrel, E. Krausmann, Reaction kinetics of a fission product mixture in a steam-hydrogen carrier gas in the Phebus primary circuit, Nuclear Technology 144 (1) (2003) 1–15.

[56] M.P. Kissane, I. Drosik, Interpretation of fission-product behaviour in the Phébus FPT0 and FPT1 tests, Nucl. Eng. Des. 236 (2006) 1210–1223.

[57] B. Clément, Towards reducing the uncertainties on source term evaluations: an IRSN/CEA/EDF R&D programme, Proc. EUROSAFE Forum (November 8–9, 2004). Berlin (Germany), http://www.eurosafeforum.org/products/data/5/pe_90_24_1_2_07_source_term_evaluat_clement_031104.pdf

[58] S.L. Girshick, C.P. Chiu, P.H. McMurray, Time dependent aerosol models and homogeneous nucleation rates, Aerosol Sci. Tech. 13 (1990) 465–477.

[59] R.O.A. Hall, M.J. Mortimer, D.A. Mortimer, Surface energy measurements on UO2—a critical review, J. Nucl. Mater. 148 (1987) 237–256.

[60] B.J. Mason, The Physics of Clouds, second ed., Clarendon Press, Oxford, UK, 1971.

[61] E.R. Lewis, The effect of surface tension (Kelvin effect) on the equilibrium radius of a hygroscopic aqueous aerosol particle, J. Aerosol Science 37 (2006) 1605–1617.

[62] N.A. Fuchs, The Mechanics of Aerosols, Pergamon Press, Oxford, UK and Macmillan, New York, NY, USA, 1964.

[63] S.K. Loyalka, J.W. Park, Aerosol growth by condensation: a generalization of Mason's formula, J. Colloid Interface Sci. 125 (1988) 712–716.

[64] L.J. Willett, S.K. Loyalka, R.V. Tompson, Adsorption on heterogeneous regular surfaces, J. Colloid Interface Sci. 238 (2001) 296–309.

[65] C.N. Davies, Aerosol Science, Academic Press, New York, NY, USA, 1966.

[66] I.H. Dunbar, J. Fermandjian, Comparison of sodium aerosol codes, Commission of the European Commission report EUR 9172 (1984).

[67] R.L. Buckley, S.K. Loyalka, Implementation of a new model for gravitational collision cross-sections in nuclear aerosol codes, Nuclear Technology 109 (1995) 346–356.

[68] P.G. Saffman, J.S. Turner, On the collision of drops in turbulent clouds, J. Fluid Mech. 1 (1956) 16–30.

[69] J. Laufer, The structure of turbulence in fully developed pipe flow, National Advisory Committee Aeronaut, (NACA) Tech. Rep 1174 (1954).

[70] P.G. Gormley, M. Kennedy, Diffusion from a stream flowing through a cylindrical tube, Proc. Roy. Irish Academy, Sect. A 52 (1949) 163–169.

[71] R. Muñoz-Bueno, E. Hontañoćn, M.I. Rucandio, Deposition of fine aerosols in laminar tube flow at high temperature with large gas-to-wall temperature gradients, J. Aerosol Sci. 36 (4) (2005) 495–520.

[72] C. Housiadas, Y. Drossinos, Thermophoretic deposition in tube flow, Aerosol Sci. Tech. 39 (2005) 304–318.

[73] L. Talbot, R.K. Cheng, R.W. Schefer, D.R. Willis, Thermophoresis of particles in a heated boundary layer, J. Fluid Mech. 101 (1980) 737–758.

[74] C.F. Clement, R.G. Harrison, The charging of radioactive aerosols, J. Aerosol Sci. 23 (1992) 481–504.

[75] C.F. Clement, R.G. Harrison, Enhanced localized charging of radioactive aerosols, J. Aerosol Sci. 31 (2000) 363–378.

[76] F. Gendarmes, D. Boulaud, A. Renoux, Electrical charging of radioactive aerosols—comparison of the Clement-Harrison model with new experiments, J. Aerosol Sc. 32 (2001) 1437–1458.

[77] Y.S. Cheng, C.S. Wang, Motion of particles in bends of circular pipes, Atmospheric Environment 15 (1981) 301–306.

[78] D.Y.H. Pui, F. Romay-Novas, B.Y.H. Liu, Experimental study of particle deposition in bends of circular cross-section, Aerosol Sci. Tech. 7 (1987) 301–315.

[79] Y. Ye, D.Y.H. Pui, Particle deposition in a tube with an abrupt contraction, J. Aerosol Sci. 21 (1990) 29–40.

[80] D.R. Chen, D.Y.H. Pui, Numerical and experimental studies of particle deposition in a tube with a conical contraction—laminar regime, J. Aerosol Sci. 26 (4) (1995) 563–574.

[81] P.L. Douglas, S. Ilias, On the deposition of aerosol particles on cylinders in turbulent cross flow, J. Aerosol Sci. 19 (4) (1988) 451–462.

[82] S. Güntay, D. Suckow, A. Dehbi, R. Kapulla, ARTIST: introduction and first results, Nucl. Eng. Des. 231 (1) (2004) 109–120.

[83] B.Y. Liu, S.K. Agarwal, Experimental observation of aerosol in turbulent flow, J. Aerosol Sci. 5 (1974) 145–155.

[84] H.C. Hamaker, The London-Van der Waals attraction between spherical particles, Physica 4 (10) (1937) 1058–1072.

[85] L. Biasi, A. De los Reyes, M.W. Reeks, G.F. De Santi, Use of a simple model for the interpretation of experimental data on particle resuspension in turbulent flows, J. Aerosol Sci. 32 (2001) 1175–1200.

[86] M. Smoluchowski, Drei Vorträge über Diffusion, Brownsche Molekularbewegung und Koagulation von Kolloidteilchen, Phys. Z 17 (1916) 557–571 and 585–599.

[87] L.M.C. Dutton, S.H.M. Jones, J. Eyink, Plant assessments, identification of uncertainties in source term analysis, EUR 16502 EN, ISSN 1018–5593, 1995.

[88] P.E. Macdonald, V.N. Shah, L.W. Ward, P.G. Ellison, Steam Generator Tube Ruptures, NUREG/CR-6365 (April 1996).

[89] USNRC, Severe Accident Risks: An Assessment for Five U.S. Nuclear Power Plants, NUREG-1150, vol. 2 & 3, December 1990.

[90] Y. Liao, S. Guentay, Potential steam generator tube rupture in the presence of severe accident thermal challenge and tube flaws due to foreign object wear, Nucl. Eng. Des. 239 (2009) 1128–1135.

[91] USNRC, Severe Accident Risks: An Assessment for Five U.S. Nuclear Power Plants, NUREG-1150, App. C, December 1990.

[92] S. Güntay, A. Dehbi, D. Suckow, J. Birchley, Accident management issues within the ARTIST project, Proc. of a Workshop on Implementation of Severe Accident Management Measures, OECD/CSNI report NEA/CSNI/R(2001)20, November 2001.

[93] L.E. Herranz, F.J.S. Velasco, C. López del Prá, Aerosol retention near the tube breach during steam generator tube rupture sequences, Nuclear Technology 154 (1) (2006) 85–94.

[94] A. Auvinen, J.K. Jokiniemi, A. Lahde, T. Routamo, P. Lundstrom, H. Tuomisto, J. Dienstbier, S. Güntay, D. Suckow, A. Dehbi, M. Slootman, L.E. Herranz, V. Peyres, J. Polo, SG tube rupture (SGTR) scenarios, Nucl. Eng. Des. 235 (2005) 457–472.

[95] D. Suckow, A. Dehbi, T. Lind, R. Kapulla, S. Danner, S. Güntay, ARTIST international consortium project: facilities and preliminary results, Cooperative Severe Accident Research Program (CSARP), and MELCOR Code Assessment Program (MCAP) Technical Review Meetings, Albuquerque, New Mexico, USA, September 18–20, 2007.

[96] T.J. Haste, P. Giordano, L.E. Herranz, J.C. Micaelli, SARNET: Integrated severe accident research in Europe—safety issues in the source term area, International Conference on Advances in Nuclear Power Plants, ICAPP'06, Reno (USA), June 2006.

[97] T. Lind, Y. Ammar, A. Dehbi, S. Güntay, De-agglomeration mechanisms of TiO_2 aerosol agglomerates in PWR steam generator tube rupture conditions, Nucl. Eng. Des. 240 (2010) 2046–2053.

[98] L.E. Herranz, C. López del Prá, A. Dehbi, Major challenges to modeling aerosol retention near a tube breach during steam generator tube rupture sequences, Nuclear Technology 158 (2007) 83–93.

[99] C. López del Prá, F.J.S. Velasco, L.E. Herranz, Aerodynamics of a gas jet entering the secondary side of a vertical shell-and-tube heat exchanger: numerical analysis of anticipated severe accident SGTR conditions, Engineering Applications of Computational Fluid Mechanics 4 (1) (2010) 91–105.

[100] A. Dehbi, D. Suckow, S. Güntay, The effect of liquid temperature on pool scrubbing of aerosols, J. Aerosol Science 28 (Suppl. 1) (1997) S707–S708.

[101] T. Lind, A. Dehbi, S. Güntay, Aerosol retention in the flooded steam generator bundle during SGTR, Nucl. Eng. Des. 241 (2011) 357–365.

[102] L.E. Herranz, J. Fontanet, Assessment of ASTEC-CPA pool scrubbing models against POSEIDON-II and SGTR-ARTIST Data, International Conference on Advances in Nuclear Power Plants, ICAPP'09, Tokyo, Japan, May 10–14, 2009.

[103] M.W. Reeks, D. Hall, Kinetic models for particle resuspension in turbulent flows: theory and measurement, J. Aerosol Science 32 (2001) 1–31.

[104] C. López del Prá, L.E. Herranz, Modeling aerosol retention in the break stage of a failed SG during a core meltdown sequence, International Aerosol Conference 2010 (IAC2010) Helsinki (Finland), August 29–September 3, 2010.

[105] P.C. Owczarski, K.W. Burk, SPARC-90: a code for calculating fission product capture in suppression pools, NUREG/CR-5766, PNL-7223 (1991).

[106] B. Clément, N. Hanniet-Girault, G. Repetto, D. Jacquemain, A.V. Jones, M.P. Kissane, P. von der Hardt, LWR severe accident simulation: synthesis of the results and interpretation of the first Phebus FP experiment FPT0, Nucl. Eng. Des. 226 (2003) 5–83.

[107] J. Jokieniemi, I. Dunbar, M. Kissane, I. Ketchell, J. Gauvain, J.-P. L'Heriteau, E. Schrödl, Physical and chemical characteristics of aerosols in the containment, OECD/NEA/CSNI/R(93)7 (1993).

[108] J.N. Lillington, Light Water Reactor Safety, Elsevier, Amsterdam, New York, Tokyo, 1995. ISBN 0 444 89741 0.

[109] H. Albrecht, Results of the SASCHA program on fission product release under core melting conditions, High Temperature Science 24 (3) (1987) 123.

[110] J.E. Brockmann, Ex-vessel releases: aerosol source terms in reactor accidents, Progress in Nuclear Energy 19 (1) (1987) 7–68.

[111] S.B. Burson, D. Bradley, J. Brockmann, E. Copus, D. Powers, G. Greene, C. Alexander, United States Nuclear Regulatory Commission research program on molten core debris interactions in the reactor cavity, Nucl. Eng. Des. 115 (2–3) (1989) 305–313.

[112] E.R. Copus, Sustained uranium dioxide/concrete interaction tests: the SURC test series, OECD/CSNI NEA Specialist Meeting on Molten Core Debris-Concrete Interactions, OECD/CSNI Report NEA/CSNI/R(92)10, Forschungszentrum Karlsruhe (April 1992).

[113] H. Alsmeyer, Review of experiments on dry corium concrete interaction, European Commission—Nuclear Science and Technology: Molten Corium/Concrete Interaction and Corium Coolability—A State of the Art Report, EUR 16649 EN (1995) 37–82.

[114] J.K. Fink, D.H. Thompson, B.W. Spencer, B.R. Sehgal, Aerosol and melt chemistry in the ACE molten core-concrete interaction experiments, High Temperature and Materials Science 33 (1) (1995) 51–76.

[115] J.K. Fink, D.H. Thompson, B.W. Spencer, B.R. Sehgal, Aerosols released during large-scale integral MCCI tests in the ACE program, OECD/CSNI NEA Specialist Meeting on Molten Core Debris-Concrete Interactions, OECD/CSNI Report NEA/CSNI/R(92)10, Forschungszentrum Karlsruhe (April 1992).

[116] J.K. Fink, D.H. Thompson, Compilation, analysis and interpretation of ACE Phase C and MACE experimental data: volume II—aerosol results, Electric Power Research Institute (EPRI), Palo Alto, California, USA, 1997. ACEX TR-C-14.

[117] D.A. Powers, J.E. Brockmann, A.W. Shiver, VANESA: A mechanistic model of radionuclide release and aerosol generation during core debris interactions with concrete, Sandia National Laboratories (1986). Prepared for US-NRC, NUREG/CR-4308.

[118] D.A. Powers, Non-ideal solution modeling for predicting chemical phenomena during core debris interactions with concrete, Proc. OECD/CSNI NEA Specialist Meeting on Molten Core Debris-Concrete Interactions (April 1992). OECD/CSNI Report NEA/CSNI/R(92)10, Forschungszentrum Karlsruhe.

[119] M. Sonnenkalb, G. Poss, The international test programme in the ThAI facility and its use for code validation, Proc. EUROSAFE meeting, Brussels, November 2–3, 2009.

[120] J.-O. Liljenzin, J. Collen, W. Schöck, F. Rahn, Report from the MARVIKEN-V/DEMONA/LACE workshop: Proceedings of the workshop on aerosol behaviour and thermal–hydraulics in the containment, OECD CSNI Report 176 (1990).

[121] K. Fischer, T. Kanzleiter, Experiments and computational models for aerosol behaviour in the containment, Nucl. Eng. Des. 191 (1999) 53–67.

[122] J.C. Sabroux, F. Deschamps, Iodine chemistry in hydrogen recombiners, Eurosafe, Brussels, Belgium, 2005. http://www.eurosafe-forum.org/forums/eurosafe_2005.html. November 7–8, 2005.

[123] M.P. Kissane, D. Mitrakos, C. Housiadas, J.C. Sabroux, Investigation of thermo-catalytic decomposition of metal-iodide aerosols due to passage through hydrogen recombiners, Nucl. Eng. Des. 239 (2009) 3003–3013.

[124] H. Köhler, The nucleus in and the growth of hygroscopic droplets, Trans. Faraday Soc 32 (1936) 1151–1161.

[125] N.A. Fuchs, A.G. Sutugin, in: G.M. Hidy, J.R. Brock (Eds.), Topics in current aerosol research (Part 2), Pergamon, New York, 1971.

[126] P.S. Epstein, On the resistance to motion experienced by spheres in their motion through gases, Phys. Rev. 23 (1924) 710.

[127] L. Waldmann, Über die Kraft eines inhomogenen Gases auf kleine suspendierte Kugeln, Z. Naturforsch. A 14a (1959) 589.

[128] L. Talbot, R.K. Cheng, R.W. Schefer, D.R.Willis, Thermophoresis of particles in a heated boundary layer, J. Fluid Mech. 101 (1980) 737–758.

[129] J.R. Brock, On the theory of thermal forces acting on aerosol particles, J. Colloid Sci. 17 (1962) 768.

[130] L. Waldmann, K.H. Schmitt, Thermophoresis and diffusiophoresis of aerosols, in: C.N. Davies, Aerosol Science, Academic Press, London, 1966.

[131] B. Clément, L. Cantrel, G. Ducros, F. Funke, L. Herranz, A. Rydl, G. Weber, C. Wren, State of the art report on iodine chemistry, OECD/CSNI Report NEA/CSNI/R 1 (2007) 2007.

[132] B.R. Sehgal P. Piluso (Eds), K. Trambauer, B. Adroguer, F. Fichot, C. Müller, L. Meyer, W. Breitung, D. Magallon, C. Journeau, H. Alsmeyer, C. Housiadas, B. Clément, M.L. Ang, B. Chaumont, I. Ivanov, S. Marguet, J.·P. Van Dorsselaere, J. Fleurot, P. Giordano, M. Cranga, SARNET lecture notes on nuclear reactor severe accident phenomenology, CEA-R-6194, ISSN 0429–3460, September 2008.

[133] L.E. Herranz, B. Clément, In-containment source term: Key insights gained from a comparison between the PHEBUS-FP programme and the US-NRC NUREG-1465 revised source term, Progress in Nuclear Energy 52 (2010) 481–486.

[134] J. Ball, G. Glowa, J. Wren, F. Ewig, S. Dickinson, Y. Billarand, L. Cantrel, A. Rydl, J. Royen, International Standard Problem (ISP) n°41, Follow-up exercise – containment iodine computer code exercise (parametric studies), OECD/CSNI report NEA/CSNI/R 17 (2001) 2001.

[135] J.-P. Van Dorsselaere, A. Auvinen, D. Beraha, P. Chatelard, C. Journeau, I. Kljenak, S. Paci, Th.W. Tromm, A. Miassoedov, R. Zeyen, Some outcomes of the SARNET network on severe accidents at mid-term of the FP7 project, International Conference on Advances in Nuclear Power Plants, ICAPP'11, Nice (France), May 2011.

[136] L. Bosland, F. Funke, N. Girault, G. Langrock, PARIS project: Radiolytic oxidation of molecular iodine in containment during a nuclear reactor severe accident Part 1. Formation and destruction of air radiolysis products—Experimental results and modelling, Nucl. Eng. Des. 238 (2008) 3542–3550.

[137] E.U. Vaughan, Simple model of plugging of ducts by aerosol deposits, Trans. American Nuclear Society 22 (1978) 507.

[138] H.A. Morewitz, Leakage of aerosol from containment buildings, Health Physics 42 (1982) 195–207.

[139] H.A. Morewitz, R.P. Johnson, C.T. Nelson, E.U. Vaughan, C.A. Guderjahn, R.K. Hilliard, J.D. McCormack, A.K. Postma, Attenuation of airborne debris from liquid-metal fast breeder reactor accidents, Nuclear Technology 46 (1979) 332–339.

[140] D.A. Powers, Aerosol penetration of leak pathways—an examination of the available data and models, Sandia Report SAND2009-1701 (April 2009).

[141] J.K. Agarwal, B.Y.H. Liu, A criterion for accurate sampling in calm air, American Industrial Hygiene Association Journal 41 (1980) 191–197.

[142] C.N. Davies, The entry of aerosol into sampling tubes and heads, British J. Applied Physics, Journal of Physics D 1 (1968) 921–932.

[143] L. Granger, P. Labbe et al., A mock-up near Civaux nuclear power plant for containment evaluation under severe accident—the CESA Project, Proc. of the FISA-97 Symposium on EU Research on Severe Accidents. Luxembourg, EUR 18258 EN, 293–302, 1998.

[144] T. Gelain, J. Vendel, Research works on contamination transfers through cracked concrete walls, Nucl. Eng. Des. 238 (2008) 1159–1165.

[145] A. Miassoedov, T. Jordan, L. Meyer, M. Steinbrück, W. Tromm, C. Journeau, J.M. Ruggieri, P. Piluso, L. Ferry, P. Fouquart, N. Cassiaut-Louis, LACOMECO and PLINIUS experimental platforms at KIT and CEA, 4th European Review Meeting on Severe Accident Research (ERMSAR-2010), May 11–12, 2010.

[146] F. Parozzi, E. Caracciolo, L.E. Herranz, C. Housiadas, D. Mitrakos, C. Journeau, P. Piluso, Investigation on aerosol leaks through containment cracks in nuclear severe accidents using prototypic materials, European Aerosol Conference, Thessaloniki, Greece, August 24–29, 2008.

[147] F. Parozzi, S. Chatzidakis, T. Gelain, L.E. Herranz, E. Hinis, C. Housiadas, C. Journeau, E. Malgarida, G. Nahas, P. Piluso, W. Plumecocq, J. Vendel, Investigation on aerosol transport in containment cracks, International Conference on Nuclear Energy for New Europe, Bled, Slovenia, September 5–8, 2005.

[148] D. Mitrakos, S. Chatzidakis, E.P. Hinis, L.E. Herranz, F. Parozzi, C. Housiadas, A simple mechanistic model for particle penetration and plugging in tubes and cracks, Nucl. Eng. Des. 238 (2008) 3370–3378.

[149] J.H. Vincent, Aerosol Science for Industrial Hygienists, Elsevier, New York, NY, USA, 1995.

[150] G.M. Hidy, J.R. Brock, The Dynamics of Aerocolloidal Systems, Pergamon Press, 1970.

[151] S.K. Loyalka, Velocity slip coefficient and diffusion slip velocity for a multicomponent gas mixture, Physics of Fluids 14 (12) (1971) 2599–2604.

[152] L.Z. Waldmann, On the motion of spherical particles in non-homogeneous gases, Rarefied Gas Dynamics, Academic Press, New York, 1961.

[153] G. Choppin, J. Rydberg, J.-O. Liljenzin, Radiochemistry and Nuclear Chemistry, third ed., Butterworth-Heinemann, New York, NY, USA, 2001.

[154] H. Albrecht, Release of fission and activation products at LWR core melt: Final Report of the SASCHA program, Karlsruhe Institute of Technology Report KfK 4264, June 1987.

[155] M.F. Osborne, R. A. Lorenz, J.L. Collins, C. S. Webster, Data summary report for fission product release test HI-1, NUREG/CR-2928, ORNL/TM-8500, December 1982.

[156] M.F. Osborne, R. A. Lorenz, J.L. Collins, J. R. Travis, C. S. Webster, T. Nakamura, Data summary report for fission product release test VI-4, NUREG/CR-5481, ORNL/TM-11400, January 1991.

[157] A. Hidaka, T. Nakamura, T. Kudo, R. Hayashida, J. Nakamura, T. Otomo, H. Uetsuka, Current status of VEGA program and a preliminary test with cesium iodide. Proc. SARJ Meeting, Tokyo, 1999, JAERI-Conf 2000-015 paper 9.1.

[illegible] M[illegible] R.[illegible] [illegible] V.K. Dhir, J.D. McCormack, [illegible] [illegible] Technology [illegible]

[illegible] Sandia Report SAND2009-[illegible] April 2009.

[illegible] American Industrial Hygiene Association Journal [illegible]

[illegible] The theory of [illegible] surfaces [illegible] Applied Physics [illegible]

[illegible] power plant [illegible] Energy [illegible] 1986.

[illegible]

[illegible]

[illegible]

[illegible] Nucl. Eng. [illegible]

[illegible]

[illegible] Academic Press, New York, 1961.

[illegible]

[illegible]

[illegible] December 199[illegible].

[illegible] January 1999.

[illegible] Meeting, [illegible]

Chapter 6

Severe Accident Management

Chapter Outline

Nuclear Safety in Light Water Reactors. DOI: 10.1016/B978-0-12-388446-6.00006-X

6.1. SEVERE ACCIDENT MANAGEMENT GUIDELINES (SAMG)*

(G. Vayssier)

6.1.1. Introduction

Emergency operating procedures (EOPs) are in place in nuclear power plants to guide the operators, should incidents and accidents occur despite preventative measures. The objective is to reach a final stable state of the core and other sources of fission products, within prescribed limits. In most cases, EOPs also cover events that are outside the design basis, such as Anticipated Transient without Scram (ATWS), Station Blackout (SBO), and loss of all secondary feedwater (LOFW, for PWRs).

Due to multiple failures, however, potential design errors, unforeseen internal or external events, and/or operator errors, EOPs may be not effective and so severe core damage may occur; the event is accordingly called a severe accident. In order to mitigate the consequences of such events, severe accident management guidelines (SAMG) have been developed. The prime objective of SAMG is to protect remaining (if any) fission product boundaries and to limit actual or possible releases. SAMG's prime objective is therefore not to protect the core (as is the case with EOPs): core integrity is lost, and most probably beyond repair. Core and debris cooling remains relevant to limit fission product release and to provide an ultimate heat sink for the decay heat.

SAMGs provide systematic guidance to be followed during the entire course of a severe accident and specifies appropriate actions to be taken at each phase, with the intention of either interrupting the course of the accident or, if this is not possible, reaching a controlled and stable final state, thereby minimizing the radioactive releases.

In the United States, the different Owners Groups had decided to set up a SAMG system and had committed themselves to have it implemented

---

* A list of acronyms used in this section is presented at the end of the chapter.

nationwide by the end of 1998. Also in many other countries, guidance and/or procedures with similar objectives have been developed. Many were based on the U.S. approach, but many others developed their own product.

It should be noted that SAMG is an aid to the operator and other plant staff to mitigate the possible consequences of core damage events. As most plants have not been designed to control such events in their design basis, the mere application of SAMG may not prevent large releases. But it can delay or reduce those releases so as to give the emergency preparedness organization more time to execute protective actions for the population in the vicinity of the plant. By these actions, casualties in the civil population can be largely or wholly prevented, yet the large socio-economic and environmental consequences of large releases, as we have seen in the Chernobyl and Fukushima accidents, may not be fully mitigated. These can only be avoided by deliberate design measures against severe accidents, such as are available in some advanced reactor designs, commonly known as Generation III reactors.

Note also that SAMGs make use of the availability of AC electric power and cooling water; it is at present largely unable to mitigate events where these two are lost for an extended time. DC power is assumed to be available during all or most of the time, as it is needed to read instruments and to operate components such as valves. However, it is still possible to mitigate the event upon loss of all AC, DC, and cooling water, as will be mentioned later.

6.1.2. Objectives and Scope

This Section deals with the SAMG situation in various countries, with emphasis on developments in the United States, as they preceded developments in most other countries, which in many cases decided to follow the U.S. approaches. The Section describes in a limited way the guidance and guidelines that have been developed, the tools that are necessary for their execution, such as the instrumentation and other operator support, the associated organisational infrastructure needed for proper execution, and the interface with the emergency preparedness organisation.

6.1.3. Development and Implementation of SAMG in the United States

Background and History

After the accident at Three Mile Island, the USNRC identified severe accidents as an "unresolved safety issue"—that is, an issue that required specific attention and industry and regulatory action to be resolved, as already indicated in Chapter 1. The NRC announced rulemaking on the subject and defined its policy in SECY 88-147 [1]. A Generic Letter 88-20 [2] was sent out to the licensees requesting a systematic search—an individual plant examination (IPE)—for severe accident vulnerabilities.

In the SECY 88-147 the USNRC described its integration plan for closure of severe accident issues.

Four areas were defined where licensee actions were required:

- individual plant examination (IPE),
- containment performance improvements (CPI),
- individual plant examination of external events (IPEEE),
- accident management.

Accident management was considered to be an essential element of this plan. It encompasses[1] all those actions taken during the course of an accident by the plant operating and technical staff to:

- prevent the accident from progressing to core damage,
- terminate core damage progression once it begins,
- maintain the capability of the containment as long as possible, and
- minimize on-site and off-site releases and their effects.

The latter three actions constitute a subset of accident management (A/M) referred to as severe accident management (SAM), or severe accident mitigation.

The USNRC worked on an Accident Management Program that would require the licensees to develop a framework for preparing and implementing severe accident operating procedures, and for training operators and managers in these procedures. The main framework elements were:

- developing technically sound strategies and ensuring that procedures and guidance were in place to implement these strategies;
- assuring that instrumentation and equipment calling for diagnosis and response be identified and that their availability and capabilities be assessed and, where needed, incremental improvements to relevant systems assessed;
- assuring that well-established, clear lines of communication and authority exist, as well as assigned responsibility for key decisions, and authority for procedural overrides and ad hoc modifications;
- assuring that operators, plant technical staff, and plant management are well trained in the procedures and guidance; and
- providing a technical basis for assessing the effectiveness of the severe accident management strategies and capabilities.

---

1. Definitions according to NEI 91-04, rev.1 [3]. Note also that slightly different definitions are in use, for example, those by the IAEA in NS-R-1 [4].

Hardware changes or other plant modifications designed to reduce the frequency of severe accidents were *not* a central aim of this program, although minor modifications might be identified in developing the A/M plan. Note that this position was not followed in most European countries, as already discussed in Chapter 1. European countries focused on hardware changes, with — initially — limited development of SAMG.

The Electric Power Research Institute (EPRI) started to develop the technical basis for A/M strategies, which was documented in the Technical Basis Report, known as the TBR [5], and which encompasses the physics and phenomenology of severe accidents, as described in other chapters of this book. The TBR also describes the potential countermeasures available to plant operators and supporting technical staff, and *analyses their potential effects during a variety of plant damage states*[2]. From this information, the Owners Groups set up generic SAMG.

During 1997 and 1998, they transferred their strategies and methods to the individual licensees, from which these developed their plant specific procedures/guidelines. As stated above, the implementation was completed by the end of 1998. European plants mostly started this process after that time. At the time of this writing (June 2011), most plants in Europe, Korea, and Japan have also completed their SAMG; others (e.g., Ukraine, Russia, and China) are developing SAMG for their nuclear power plants.

After the September 11, 2001 events, both the USNRC and the U.S. Department of Homeland Security (DHS) conducted assessments of all operating reactors in the United States. These assessments resulted in plant modifications, additional procedures (called Extensive Damage Mitigation Guidelines, or EDMGs), and additional coping capabilities outside the plant. Note that the EDMGs also span the *preventive* domain (i.e., before core damage).

Major Elements of Implementation

The following are the main elements of the SAMG implementation.

Strategies and Guidance

Guidance should be developed to help the emergency response organization (ERO) personnel in assessing plant damage, planning and prioritising response actions, and implementing strategies that delineate actions inside and outside the control room. Strategies and guidance should be interfaced with the EOPs and the emergency plan.

The guidance should include an approach to evaluate plant conditions and challenges to plant safety functions; operational and phenomenological conditions that may influence the decision to implement a certain strategy;

2. This latter process is a unique development of those days; it took almost 20 years before utilities started to analyze the effect of SAMG actions with the PSAs [6] [7].

a basis for prioritising and selecting appropriate strategies; and the means to evaluate the effectiveness of selected actions.

The strategies should make maximum use of existing plant capabilities, including other than typical safety-related systems and other than usual alignments. Critical resources and procedures to implement strategies should be identified.

Assignment of Responsibilities and Training

Three main categories of personnel are identified and should be trained specifically for their assignments:

- Evaluators: the people responsible for assessing plant symptoms in order to determine plant damage and potential strategies to be followed;
- Decision makers: the ERO personnel designated to assess and select the strategies that should be implemented. Usually, these are high-level managers;
- Implementers: the people who actually execute the actions that decision makers decide upon—normally the control room staff and other operational/ technical staff.

The areas of emphasis and the level of detail of training for the staff involved should be commensurate with the different functions. An implementer may have training that differs from that for an evaluator or a decision maker.

Usually, a specific location is available with appropriate equipment to execute the tasks by the evaluators and decision makers, called the Technical Support Center (TSC).

Technical Support, Computational Aids

Technical support in the form of computational aids (CAs) should be available to estimate, as appropriate, key plant parameters and plant response relative to accident management decisions. The CAs should be easy to use and need not to be computer based.

Instrument Availability and Survivability

The availability and survivability of information sources (instrumentation) should be considered. As measurements may not be reliable because of harsh environmental conditions or because they are outside the instrument design range, multiple indications for needed parameters are advisable. Indirect means for providing the necessary information should also be considered. Expected deviations from instrument readings as a consequence of differing environmental conditions should also be investigated and documented (e.g., the SG-level measurement depends on the containment pressure).

Decision Making Responsibility; Interface with the Plant Emergency Plan (EP)

Utilities should ensure that responsibilities for authorising and implementing accident management strategies are delineated as part of the EP.

Utility Self-assessment

Finally, a self-assessment should be set up to ensure the feasibility and usefulness of the capability of the severe accident response. After the SAMG has been created, an initial evaluation should be performed, followed by periodic table-tops and/or mini-drills, to ensure that ERO personnel are familiar with the use of SAMG and the delineation of the responsibilities inside the ERO during the use of SAMG.

Screening Criteria

In the U.S. program, screening criteria have been developed that basically state that for sequences with a relatively large core damage frequency (CDF) or containment bypass frequency, measures should be taken (i.e., administrative, procedural, or hardware modification), which are directed mainly at reducing the likelihood of the source of the accident sequence initiator, whereas for sequences with a relatively small CDF or containment bypass frequency, such measures need not be taken, but SAMG should be in place. Below the 10^{-6}/reactor-year for CDF and below the 10^{-7}/reactor-year for bypass frequency, no actions are required (i.e., also no SAMG is required). In practice, however, this lower limit has not been used to exclude the development of SAMG, as long as meaningful SAMG can be developed for events in this lower frequency range.

Fundamentals of the SAMG Approach

Strategies and Guidelines

As was discussed earlier, EPRI developed a technical basis for selecting and determining potential countermeasures and documented this basis in the Technical Basis Report (TBR) [5] [6]. The TBR provides various plant damage conditions to describe a severe accident progression, along with their anticipated symptoms and related phenomena. Principal potential countermeasures, so-called candidate high-level actions (CHLAs), are identified. The actual development of severe accident management guidance was not attempted; this was left to the Owners Groups and the individual utilities.

The different damage conditions of the core were in the TBR captured in three major core damage states (formally called damage condition descriptors):

- Oxidised fuel but core intact, acronym OX;
- Badly damaged core (wholly or partly relocated to the lower plenum), acronym BD;
- Significant core debris ex-vessel, acronym EX.

A similar concept was followed for the containment. Here also three damage states were defined:

- Containment closed and cooled (i.e., intact, not challenged), acronym CC;
- Containment closed but challenged (e.g., by hydrogen combustion), acronym CH;
- Containment impaired but building intact, acronym I;
- Containment bypassed, that is, RCS failure outside containment, acronym B.

Fifteen CHLAs were defined (see Table 6.1), and the TBR *investigated the response of each of the core and containment damage states to each one of the CHLAs*[3]. Later, three additional actions were considered, being external cooling of the RPV (the in-vessel melt retention [IVMR] mechanism discussed in Sections 6.2 and 6.3), steam inerting of the containment and in-vessel cooling. It was recommended that these should also be considered when developing plant-specific guidance.

Note that the TBR goes beyond the scope of most PSAs, as the impact of countermeasures is usually not treated in a PSA. Such studies were performed only recently [7] [8].

Nature of Guidance: Probabilistic, Mechanistic

As stated earlier, the TBR did not direct its evaluation to certain events or scenarios (i.e., the initiating event plus its evolution), but was entirely phenomena-oriented. Core and containment damage conditions were described regardless of their probability. This principle was repeated among others by the BWROG in [9], where they stated: "Any mechanistically possible plant conditions for which operational guidance can be developed are addressed as appropriate to minimize the impact on public health and safety, irrespective of the probability of occurrence. These conditions included multiple equipment failures and operator errors." This principle of SAMG development has also been included in IAEA's NS-G-2.15 [10].

Responsibility for Evaluation and Actions

In the United States, as in many countries, training of licensed operators focuses on the prevention of accidents and handling of accidents within the plant's design basis.[4] Therefore, once it was decided to cover also the area of core damage and possible containment damage in guidelines and procedures, the responsibility for handling the plant in this regime was brought partly or

3. As observed before (in footnote 2), this was a unique feature of the TBR, only many years later followed by the analysis of such measures in some PSAs.

4. Note that this may also involve accidents beyond the licensed design basis, such as ATWS, SBO. Many plants are designed to mitigate such events, but their consequences may go beyond prescribed regulatory limits for design-basis accidents (DBAs). They are therefore called beyond design-basis accidents (BDBAs).

TABLE 6.1 Candidate High-Level Actions (CHLAs) and Special Considerations

	Candidate High-Level Actions
1.	Inject into (makeup to) RPV/RCS
2.	Depressurise the RPV/RCS
3.	Spray within the RPV (BWR)
4.	Restart RCPs (PWR)
5.	Depressurise steam generators (PWR)
6.	Inject into (feed) the steam generators (PWR)
7.	Spray into containment
8.	Inject into containment
9.	Operate fan coolers
10.	Operate recombiners
11.	Operate igniters
12.	Inert containment with noncondensables (BWR)
13.	Vent containment
14.	Spray secondary containment
15.	Flood secondary containment
	Special Considerations
1.	External cooling of RPV/RCS
2.	Steam inerting of the containment

totally outside the control room. Another reason to do this is the complexity of a severe accident, which requires additional support to mitigate its consequences. This support may go far beyond the capabilities and authorisation of the control room staff (e.g., timing of containment venting may depend on activities needed to evacuate a nearby school).

It should be noted that some plants in Europe deviate from this path. They argue that one who is in the middle of an accident should not transfer authority to take actions to another body; it would be too confusing for those handling the event. In principle, both options have their merits. The most important consideration here is that plants consider this problem thoroughly and select the option that would be most suitable for them.

The proper handling of SAMG requires the multiple functions of evaluating, decision making, and implementing. Often, evaluation and decision making are in the TSC, and implementation is by the control room staff and, if applicable, by operators for local manual actions.

Interface with EOP Space

For severe accidents to occur, certain EOP actions must fail. Therefore, EOPs should contain exit conditions that also function as entry conditions into the SAMG space. These exit conditions need to be defined and implemented in the applicable EOPs. An example is the threshold for the core temperature in the Westinghouse Owners Group SAMG: if the core exit temperature exceeds 650°C *and* all recovery actions have failed, the EOPs are exited and SAMG space is entered. Actions under EOPs that still would be useful under SAMG are included in SAMG.

In other applications of SAMG, EOPs are not closed, but are checked for inconsistency with SAMG, where the latter has priority.

Reliability of Instrumentation

During a severe accident, many instruments are subject to harsh environmental conditions for which they have not been designed. However, as the basic philosophy is to go through the event with the plant hardware as it has been built, no (major) modification to instrumentation is foreseen.

Nevertheless, all instrumentation is scrutinized for its design-basis conditions, so that it is known beforehand what instruments can be relied upon. In some SAMG approaches, potential deviations from instruments due to their environment are systematically investigated and quantified. As already mentioned, an example is the reading of the SG level, which changes as the containment pressure increases.

Where data are lacking, an attempt is made to obtain the necessary information from other sources. Primary pressure, for example, can be measured by appropriate pressure sensors, but if they have failed, one could try to infer the primary pressure from the ECCS suction line pressure measurement. Such guidance is thought about in advance, so that alternate sources of data are available when called upon during the event. Throughout a severe accident, as discussed, multiple sensors are read (where these are available), to obtain the best possible measurements.

Computational Aids, Computerized Support

For some plant parameters, quantitative information is useful during the event. For example, is the containment atmosphere flammable with the amount of hydrogen that is observed (or estimated from other sources)? Or will the containment atmosphere be flammable after we have vented the containment? Or can we still blow down the RPV when the suppression pool temperature has gone up to a certain temperature? Such information can be precalculated and will then be available in the form of graphs, tables, etc., often called computational aids (CAs).

The TSC generally avoids using computers during an event, as errors are easily made and also computers themselves may fail, running out of power, and so on.

Repair Priorities

A severe accident should not occur unless multiple component and/or system failures occur. Hence, one of the important tasks is to restore important systems to service. As not all systems can be repaired at the same time, priorities have to be set. In addition, some systems depend on others to function properly. Some plants have developed guidance for the TSC in this respect. For all systems, the minimum requirements needed to function are documented, as well as the systems on which they depend. Also, the radiation levels in important equipment rooms are monitored (or even precalculated), so that repair possibilities (but also local actions such as manual operation of a valve) can be deduced.

Interface with the Emergency Plan

All plants have an emergency plan (EP), which activates an emergency response organisation (ERO) on certain predefined criteria. The ERO coordinates all actions necessary to cope with the event and to communicate with the public authorities. The TSC is, therefore, set up as part of the ERO. The ERO is the proper vehicle needed to organise component and system repairs, once the TSC decides upon such actions. Where decisions involve actions by plant personnel outside the TSC, decisions are coordinated with the ERO director (who is often also the decision maker in the TSC).

Verification and Validation (V&V) of SAM guidelines

Once strategies have been defined and guidelines have been developed and transformed into plant-specific situations, a process of verification and validation is started. Verification means a process to establish that the guidelines are technically correct — that is, the correlations used are applicable and numerical results are reliable. Validation means that the guidelines and procedures can be executed by trained personnel. Verification and validation (V&V) is usually a complex process that may include use of the plant simulator to actually see whether preplanned actions are executable. Where the simulator does not hold any longer, usually after core damage has occurred, precalculated scenarios and templates are used, which have been designed with the help of sophisticated severe accident computer codes (e.g., MAAP, MELCOR, ASTEC).

Training and Templates

As stated earlier, most SAM guidance is set up to be used by available plant personnel. The training must provide for the information needed to carry out the duties of evaluator, decision maker, and implementer. The training should not transform these people into severe accident specialists. Hence, an amount of training is envisaged that is limited to a few days or hours, depending on the nature of the function.

Parts of the training are exercises and demonstrations, where an actual event is played. For the initial part, that is, the part without core damage, a simulator is the proper tool, as already indicated. And the later parts, for which usually no simulator exists, are precalculated with advanced codes such as MAAP, MELCOR, or ASTEC. The templates are set up dynamically; that is, they have sufficient latitude — up to a certain extent — to follow the actions that the TSC has decided on. The templates are configured so as to encompass most or all SAMG actions.

Apart from the plant ERO, often a link is available to the vendor for support in highly complex domains and/or to a national crisis center, with experts from government, research, and industry.

6.1.4. Future Developments

This section includes contributions from [17] and [18].

As was discussed previously, the basis for the SAMG is a variety of sources: the TBR, PSAs, research programs, and so forth. This basis is not static; rather, it is expanding continuously. Therefore, plants that implement SAMG should also have a mechanism to process new knowledge and new information. As SAMGs are fully institutionalised in the United States—that is, the program is carried out under a general plant commitment, as was expressed in the industry position, and also resources in manpower and budget are allocated—a basis is present to acknowledge and process future developments. It appears, however, that the technical basis of the generic SAMG programs, which was based on the TBR [5], may need to be updated in view of the new insights. In Europe, the review of the SAMG of a plant is mostly part of the 10-year periodic safety assessment and, through this mechanism, would be periodically compared to the latest status in research and development.

New insights that were already developed during the SAMG implementation effort concern external cooling of the RPV, direct containment heating, ex-vessel coolability, and in-vessel cooling of debris. At the time of this writing, uncertainties still exist in these and a number of other areas, such as ex-vessel steam explosion, hydrogen combustion, and core–concrete interaction. Reference is made to the SARNET determination of areas that warrant further research [10]. Implementation of SAMG programs should include these new insights, as well as take note of the concerns and uncertainties still present.

Many plants still do not possess SAMG for shutdown states, so-called SSAMG, which includes "head off" and "containment open" and also considers the spent fuel pool. There is an urgent need to fill this gap in SAMG approaches.

A typical matter in many SAMG decisions is the assessment of potential negative effects of proposed actions. The present SAMG approaches, however, do not provide for proper tools to estimate these negative effects; neither is the training of TSC staff always sufficient to fulfill this duty. These matters should be improved.

In some new reactor designs, features have been included that are specifically designed to cope with severe accidents, such as the IVMR strategy or a core catcher. Existing SAMG approaches have been/are being modified to benefit from these additional features.

We can learn various lessons from the Fukushima accident in March 2011, lessons that would, most probably, impact the SAMG as well as the EOP domain. The accident was characterized by a long-term loss of AC and DC power, as well as cooling water. The anticipated effects on SAMG (and, partly, EOPs) are as follows:

- Longer "coping times" for accidents that include loss of basic supplies (AC and/or DC, cooling water).
- Methods to use SAMG (and EOPs) under minimum availability or temporary loss of instrumentation and control power, which may include the use of methods to read I&C without station battery power. This may be possible at local positions, powering I&C with carry-on DC.
- Methods to locally operate valves, which may require additional equipment to do so, plus vigilance about the locations that must be entered for such actions.
- Methods to hook on DC, AC, and cooling water from mobile sources. Many plants already are able to feed steam generators from, for example, the fire brigade trucks. Such cooling provisions could be expanded to include directly the core[5] and other storage of highly radioactive material (e.g., the spent fuel pool) and may include special storage on- or off-site for the necessary equipment; consideration should be given to using seawater, dirty water, and unborated water.
- Standard use of borated water to cool spent fuel pools and maintain subcriticality to be subject to additional considerations, as emergency cooling water (possibly from a river or the sea) may not be borated.
- More explicit attention to the availability of an ultimate heat sink, which is often more implicitly considered.
- Have plants develop SAMG for other large sources of radioactive material, including spent fuel pools (*note:* some plants have this already as part of the shutdown SAMG, the SSAMG). This may include hardware changes to provide coolant and to control any hydrogen that may arise. It may also include extension of SAMG to areas where containment atmosphere gases may leak through cracks and bypasses such as SGTR or the interfacing system LOCA (ISLOCA). An example is hydrogen control in such areas.

5. This option is already available at some plants.

- More time perhaps needed for executing the methods and actions described, which should be reflected in the guidelines.

Future SAMGs will therefore most probably include measures and methods to deal with extensive damage on the NPP site, possibly caused by a variety of external events with high damage potential, including explosions and large fires (as in the Extensive Damage Management Guidelines, Section 6.2.3). As discussed, this will also be part of the preventive domain. As these operations are complex, the TSC's guidance/decision making in both the EOP and SAMG domain may need expansion, as well as training of the TSC to fulfill these complex tasks.

At present, a variety of SAMG approaches exist, each having its strengths and weaknesses. It is worthwhile to strive for harmonisation in an approach that combines the strengths, avoids the weaknesses, and possibly uses a uniform format. This will increase the possibilities for conducting peer review and for sharing plant experiences in exercises and drills.

6.1.5. Regulatory Position

In most countries, there is no formal regulatory approval of the SAMG, as it is an area outside the licensed design basis.

The absence of formal regulation has certain drawbacks: the SAMG programs are not assessed and inspected against a certain set of predefined criteria. This is more or less inherent in an area that is beyond the licensed design basis. Therefore, the value of a SAMG-program must primarily be assessed by the utility itself.

European regulators have involved themselves more directly, sometimes defining criteria that must be met (as in Sweden and Finland).

The IAEA has developed requirements and guidance in this area, notably in NS-R-1 and NS-G-2.15 [4] [10].

6.1.6. Examples of SAMG Approaches

In subsequent sections, practical applications of SAMG approaches will be presented. First, the Westinghouse Owners Group SAMG approach will be treated, as this is the most widely used one. Other approaches are in use as well, equivalent in objectives and quality. They will be briefly mentioned afterward, with some detail for the BWROG SAMG.

Approach by the Westinghouse Owners Group (WOG SAMG)

The basic vehicle of the WOG SAMG[6] consists of two interlinked decision trees. One directs plant personnel to take actions to restore the plant to a stable,

6. The WOG has been expanded in the meantime to a PWR Owners Group, PWROG. In this section, however, we will keep the acronym WOG SAMG, as this abbreviation is most widely known.

controlled state; this is the diagnostic flow chart (DFC). Actions are decided upon after consideration of both positive and negative potential consequences and generally are taken only if the positive outweighs the negative. The guidelines are shaped this way; that is, they contain this deliberation process. If, for example, the containment pressure is increasing, it is considered to start the containment spray. But this will be done only if there is no risk for hydrogen explosion, because the spray will de-inert the containment atmosphere, or if this risk is deemed to be acceptable.

The other decision tree directs plant personnel to take action if an immediate challenge to a fission product (FP) barrier is observed; this is the Severe Challenge Status Tree (SCST). As the challenge is immediate, response must be immediate and does not leave (much) time for consideration of potential negative consequences. The corresponding guidelines, therefore, do not contain the process of deliberation about potential negative consequences. For example, if the containment failure pressure is about to be reached, the containment vent is opened, regardless of the potential negative consequences of such an action.

Both logic trees are monitored and followed in parallel, where the SCST has priority. In most exercises, actions are considered and decided upon primarily using the DFC. If, while executing an action that belongs to the DFC, a challenge to a FP boundary is observed, priority shifts to the SCST.

The actions considered in the DFC are defined in the severe accident guidelines (SAGs), the ones in the SCST are defined in the severe challenge guidelines (SCGs).

Approaches by Other Owners Groups and by Other Utilities and Countries

Apart from the WOG SAMG, the Owners Groups of Combustion Engineering (CEOG) and Babcock & Wilcox (B&WOG) reactors have also developed SAMG for their groups of plants. Also the Boiling Water Reactors Owners Group (BWROG) has developed its SAMG.

The CEOG and B&WOG approaches are quite different from the WOG SAMG, as they require a certain diagnosis of the plant damage state (**not** of the scenario, as already explained). For example, recognition of vessel failure has an impact on the type of countermeasures initiated. Guidance is also available should this insight get lost during the evolution of the accident.

Basically, the CEOG and B&WOG SAMG methods depart directly from the CHLAs from the EPRI Technical Basis Report. That is, they link their actions to observed plant damage states and shape their guidelines accordingly (which differs from the WOG SAMG, which does not ask about the plant damage states).

The BWROG SAMG is a further development of their EOPs,[7] which already went quite far in the severe accident domain. It has three severe

7. In BWROG language usually called Emergency Procedure Guidelines (EPGs).

accident guidelines (SAGs): the first one is a guideline for integrated flooding of the RPV and the containment, the second one is a radioactivity release control guideline, and the third is a hydrogen control guideline.

Also, the BWROG SAMG approach is symptomatic; that is, it departs from the observation of symptoms, and not from recognition of scenarios. But, like the CEOG and B&WOG methods, it requires recognition of RPV failure, for which a "vessel failure recognition" signature has been developed (i.e., a combination of parameters and trends of parameters whose simultaneous appearance is indicative for vessel failure).

Also in view of the accident at the Fukushima nuclear power plant, a short overview of these latter guidelines is presented (although it is not claimed that these are implemented at the Japanese BWRs).

Principles of the BWROG SAMG

The main vehicle for coping with transients and accidents at BWRs is their set of Emergency Procedure Guidelines (EPGs), Rev. 4, which uses a symptom-based approach. These EPGs address a very broad spectrum of accidents, which, unlike the PWR case, originally included several severe accident phenomena.

As stated in [9], the EPGs still were not a full package of SAMG, it was decided to develop separate SAGs, defining a transition from the EPGs to the SAGs—being the onset of core damage—and restrict the EPGs to strategies prior to the transition. The shift in responsibility from the main control room to the TSC (respectively ERO) should also occur at the transition point. As a consequence, considerable changes to the EPGs had to be made.

In developing the SAGs, the bulk of severe accident knowledge was reviewed for relevance and applicability for BWRs, as was the effect of the candidate high-level actions (CHLAs).

The development of the SAGs and the construction of the interface EPGs-SAGs, plus the new insights in EPG-related phenomena, led to numerous changes in the EPGs.

It was decided to structure the injection and set priorities in the SAGs, dependent on the amount of water that was available and the damage condition of the plant. Priorities for injection were set as follows and in sequence:

1. try to flood the core;
2. try to cool the debris;
3. try to keep the debris in the reactor pressure vessel (RPV);
4. try to keep the containment intact;

where a higher number means a lower priority.

For all situations, the following questions arise: should we inject, or should we spray, and/or should we vent? These questions are treated systematically in the SAGs, specifically in SAG-1.

Hence, the basis of SAG-1 is that there is no, little, sufficient, or much water available for injection, and no, little, or much damage to the reactor cooling system (RCS). This creates unique situations that warrant different follow-up measures. As a result, a flooding strategy decision loop had been made, which is basically connected to the amount of water available for injection.

The SAGs, together with the modified EPGs, form an integrated set of symptomatic instructions that attempt to cover all possible accident sequences. They address a spectrum of conditions, both less and more severe than the plant design basis. As with the WOG SAMG, any mechanistically possible plant condition for which generic guidance can be developed is addressed, irrespective of the probability of occurrence.

The BWROG developed specific guidance to strengthen the TSC [12]. The enhancements include the following:

- current and forecast values of the EPG and SAG control parameters;
- alternate means of determining EPG/SAG limits;
- continuing assessment of plant status or condition;
- evaluation of continued availability of, or time required to restore and repair plant systems, needed to implement the EPGs and plant specific SAMG (PSAMG);
- developing priorities for system restoration; and
- timing for EPG or PSAMG actions.

These guidelines are called technical support guidelines (TSGs).

A concise but complete description of the BWROG philosophy and SAMG development and implementation can be found in [9] and [12]. Other information in this section is based on [14] [15]. Adaptations of the SAMG for a newer design (ESBWR) are described in [16].

European Approaches

In Europe, developments have been somewhat different. As previously mentioned, the regulators in some countries demanded hardware changes, such as filtered containment vents, passive autocatalytic recombiners (PARs), specific systems to flood the drywell (some BWRs) or the cavity (some PWRs), and upgraded instrumentation.

Severe accident management guidelines have been developed in different formats. Some followed the detailed U.S. approaches, described earlier; this was done notably in countries that follow (largely) the regulatory guidelines of the vendor country (e.g., Spain, Slovenia, and Belgium), but also in the Netherlands and in countries operating VVERs (Czech Republic, Slovak Republic, Bulgaria, Hungary). Most of these countries follow the WOG SAMG. Others have developed more general guidelines or manuals to be used during a severe accident. This latter approach has been followed in Sweden.

The UK developed some form of SAMG for Sizewell B at the design stage and compared it later to the WOG SAMG elements.

Some countries (notably France) developed a well-equipped central crisis center in Paris, capable of supporting any stricken plant with the necessary recommendations, compiled by a large group of experts.

Again, others (notably Finland for Loviisa) developed their own technical basis and from there derived the applicable SAM guidance. In some German and Swiss plants, a derivative from the CEOG approach is being used, that is, SAMG that derives from recognition of the plant damage states.

Outside Europe, notably in Korea and South Africa, use was made of the WOG SAMG, also for non-Westinghouse plants.

For the EPR™, a specific development has been made, which has some elements in common with the WOG SAMG, but more specific guidance is given to assure, for example, subcriticality and to have an ultimate heat sink available. It is called OSSA (Operation Strategies for Severe Accidents). It focuses on three safety functions: the first one is to control releases, the second one is to keep the containment and other fission product barriers intact, and the third is to remove decay heat. These safety functions can have four different states. A diagnostic is present to identify the status of each safety function and the presence and severity of their challenges [70].

Only a number of plants have so far developed SAMG for the shutdown stage, the SSAMG, for example, plants in France and Finland, Koeberg (South Africa), and the Swiss plants and VVERs in Central Europe. SSAMG was not developed by the U.S. Owners Groups.

A more detailed overview of approaches in Europe is given in [13]. Note that this document dates from 2000 and does not capture developments since. The general picture, however, has not changed much.

6.1.7. Concluding Remarks on SAMG

As has been seen, the setup of SAMG at a plant is a very structured endeavor, leading to a number of considerations and actions in a predefined sequence. The main goal—to take deliberate actions during the turmoil of a severe accident—seems to have been achieved. Actions derive mainly from plant parameters, and not from insights in the underlying scenario, thereby avoiding the need for plant people to have an understanding of the cause of the accident or sophisticated knowledge about severe accident phenomenology.

Actions are considered for their possible negative outcome before any decision is taken. To be able to estimate such negative consequences requires, however, substantial knowledge, however; this knowledge must be provided by the TSC, which requires considerable training.

For some of these consequences, for example, possible steam explosions in the cavity, research is ongoing, and the outcome cannot be predicted at this time. Therefore, the SAMG developer must follow international progress, for which a mechanism is foreseen in each SAMG program.

With the exception of some new stations, plants have not been designed against severe accidents. Therefore, applying SAMG is no guarantee that the severe accident can be fully mitigated. However, it can be anticipated that application of SAMG in a severe accident will reduce the potential consequences. As such, application of SAMG should reduce risk.

It is recommended that the amount of risk reduction by SAMG be investigated using probabilistic safety analysis (PSA). Although such a "test" has inherent limitations, as the accident evolution cannot fully be predicted, neither the capability of a plant to restore damaged equipment or to bring it back to service.

It is anticipated that the lessons from the Fukushima accident will further enhance SAMG features, notably for prolonged absence of DC, AC, and cooling water.

For a number of newer plants, with core catchers and other mitigative devices, SAMG has not yet been fully developed[8]. The main principle—to develop SAMG for all conditions mechanistically possible—should remain, including failure of the mitigative devices, as long as such guidance can be written. As Edward Frederick, operator of TMI-2, said: "No operator should ever be placed in a situation that has not been pre-analysed by an engineer."

6.2. TECHNIQUES APPLIED IN SEVERE ACCIDENT MANAGEMENT GUIDELINES

As discussed in Section 6.1, a variety of actions can be taken to mitigate the consequences of a severe accident. An overview is presented in Section 6.1, Table 6.1, denoted as candidate high-level actions (CHLAs). The present section describes the various CHLAs, together with their potential effect on the evolution of the severe accident. In principle, each CHLA should be considered for each plant damage state, which is a highly plant-specific assessment. This, however, has not been attempted in this section; detailed information is provided in [5] and [6].

6.2.1. Inject into the RPV/RCS

Water injection into the RPV/RCS seems a most obvious way to mitigate a severe accident. However, the focus here is on injection after core damage has occurred. And core damage has occurred because water injection was not available for a prolonged period of time. During that time, fuel cladding may have been oxidized (plant damage state: oxidized, OX), fuel may have relocated to the lower plenum (plant damage state BD), ultimately resulting in a melt-through of the RPV (plant damage state: EX). Water injection during these various phases has different consequences.

---

8. With the exception of the EPR™, for which the OSSA has been developed, and the AP1000, which uses an approach derived from the WOG SAMG.

The heat that must be removed is stored heat, decay heat, and heat from the metal-water reaction. In principle, two mechanisms are available:

- Supply so much water to the core that the temperature will raise to the local saturation value, denoted as W-sat.
- Supply so much water that the entire volume injected evaporates, denoted as W-vap.

In the second case, one could also take credit for an amount of superheat of the steam; that is, even less water could be sufficient for cooling. However, if not all metals are yet oxidized, there is a risk that the superheated steam will start (again, possibly) to oxidize the remaining metals and, by this process, again creates—temporarily—a large source of heat. It is generally believed that injection of an amount of water equal to W-sat will not generate much hydrogen, whereas injection of W-vap may already generate a substantial amount of hydrogen. Apart from this, the core will be further damaged, and the state BD may be reached more quickly.

If borated water is used for injection of W-vap, boron salts will precipitate, decreasing the porosity of the core and increasing the amount of damage to the core, that is, to the BD state. Similar concerns hold for dirty water, seawater, and so on.

The injected water may not be able to reach all parts of the core; that is, it may not be able to prevent the formation of a corium pool, as it has been seen in the TMI-2 event. Consequently, injection alone may not be sufficient to prevent vessel melt-through (although the TMI-2 vessel remained intact). The Korean SONATA project has investigated this issue further [19].

Another problem that may arise from water injection is possible recriticality of the core, notably in the time frame that the control rods have melted away, but the stack of fuel elements is still largely intact, as a consequence of the injection of water that is not borated, or at least adequately borated. In principle, power could increase to the amount that corresponds to evaporation of all water injected. Power peaks are possible if the feedback mechanisms of reactivity addition are not fast enough.

Injection of water depends on the possibility of depressurizing the RCS for low-pressure injection sources (see Section 6.2.2). Note that in the SAMG domain, water is injected from all available sources, which may include potable water, groundwater, fire-extinguishing system water, water from a fire brigade truck, water from lakes, and, ultimately, water from the sea.

Water addition to an overheated core may lead to increased RCS pressure and, hence, to an increased risk for SG tube creep rupture. Such a rupture is a containment bypass and may result in very severe off-site consequences, already early in the accident evolution.

Should water be injected after vessel melt-through (possibly also via sprays), the containment may be subject to a pressure peak. Some studies indicate that this pressure peak can be quite high and could possibly endanger containment integrity [20]. Quenching of debris may also result in additional hydrogen generation.

If the intent is to flood the remaining debris in the core by having the water flowing into the vessel through the melt-through hole, the vessel may need to be vented. This is notably the case for the BWR vessel.

Finally, if water is injected from outside sources, it will gradually increase containment pressure and may lead to an earlier need for containment depressurization with venting. Where such depressurization is unfiltered, large releases will occur.

As a result, the injection of water after core damage may limit further core damage, but may also further degrade the core configuration. It may result in temporary recriticality of the core. It may also have large consequences for the amount of hydrogen in the containment and produce SG tube creep rupture, leading to a need to vent the containment earlier.

A positive effect is that volatile fission products will be scrubbed by the overlaying water.

6.2.2. Depressurize the RPV/RCS

In the EOP-domain, the depressurization of the RPV/RCS is a method that makes low-pressure injection sources available to the RCS. This includes an ultimate cooling method, called "feed and bleed" (or "bleed and feed"). It consists of starting the ECCS and letting it pump through the SVs (Safety Valves) or PORVs (Pressurizer Operated Relief Valves) (feed and bleed), or opening the RCS PORVs, then starting the ECCS and letting it pump the water through the PORVs (bleed and feed). As the SVs/PORVs usually are designed for steam flow only, they may need to be replaced by valves that can withstand single-phase water flow and two-phase flow (it has been done at many plants).

Depressurization—through the PORVs—is also used in some EOP-approaches where it is concluded that the core is already partly uncovered, to flash the remaining water through the core and thereby gain time to restore other ways of injection.

In the SAMG domain, depressurization is used to (again) try to make low-pressure injection sources available to the core, as discussed in Section 6.2.1. The prime objective, however, is to protect fission product boundaries. Two processes are challenging the containment integrity: direct containment heating (DCH) and SG tube creep rupture. A remote additional containment failure mechanism is "rocketing" the RPV through the containment, which is possible if the lower head vessel weld would unzip under the attack by corium (which is considered to be unlikely).

The two main depressurization techniques (PWR) are:

- Depressurizing the secondary system, for example, through steam dump, preferably from an intact SG;
- Depressurizing the primary system, for example, through opening the PORVs.

For the BWR, depressurization is done by manual operation of the automatic depressurization system (ADS).

By depressurization through the secondary system, the primary inventory is kept. It requires, however, a sustained thermodynamic coupling between the primary and secondary systems. Later in the severe accident evolution, the primary and secondary circuits may be decoupled, which initiates during BD (and is complete during EX). Secondary depressurization may then no longer be effective, and depressurization must be done by opening the PORVs (or other suitable relief valves, possibly even designed for this purpose).

Depressurization through the PORVs releases RCS inventory to the containment. As depressurization will also flash steam through the core, it may generate additional hydrogen, which then also is released to the containment. The containment will also be heavily loaded by fission products. However, depressurization can also be used to deliberately bring steam into the containment, which may help to inert the containment against hydrogen combustion (see Section 6.2.17).

Depressurization increases the risk for a steam explosion in the RPV, when the hot corium will contact remaining water. It is generally believed, however, that a damaging steam explosion in the RPV is an unlikely event and can be considered to be a residual risk (i.e., a risk that is considered to be so small that no mitigating measures need to be developed). The damage mechanism considered was a steam explosion that would rupture the RPV-lid and propel it upward through the containment wall.

Depressurization may also lead to the interruption of any remaining natural circulation flow in the core, which is favorable for the hot legs and surge line; that is, less heat will be transferred to these components.

No PORV is qualified to sustain the high temperatures that occur during RCS depressurization in a core-melt accident. As some of the PORVs have plastic parts, they have a greater risk of not functioning adequately. A compromise would be to select valves that do not have parts that are typically vulnerable to elevated temperatures (i.e., have no plastic parts).

It should also be noted that some of the PORVs need the RCS pressure to open and stay open. This could mean that they may not stay open at low pressure. This consequence should be considered, and possibly the valves (or parts thereof) should be replaced.

Not depressurizing the RCS will keep it at high pressure, possibly with a prolonged popping of SVs/PORVs, as a way to relieve overpressure. This may result in a failure to operate the valves later for a deliberate depressurization.

Although SAMG usually recommends opening all PORVs, it may be beneficial to open only one or two valves. This may result in a much longer time to vessel failure and, hence, in increased probability of restoring important systems, such as emergency diesels, to service [21].

6.2.3. Spray within the RPV (BWR)

This CHLA is similar to "Inject into the RCS" (Section 6.2.1). As the spray will also reduce pressure, power oscillations may occur if the reactor has not yet been scrammed. Spray water could also contribute to hydrogen generation. A beneficial effect is the scrubbing of volatile fission products.

6.2.4. Restart of RCPs (PWR)

In the EOP domain, RCPs are usually stopped because their continued running could increase the loss of inventory from the RCS through the break. Restart of RCPs is an option meant to sweep water through the core that has been collected/remained in the loop seals. Such cooling may prolong the time to core damage substantially (hours).

In the SAMG domain, the RCS is considered to be more voided. RCP restart could enhance steam flow through the core and, hence, contribute to cooling if SGs are still cooled. However, should there still be a considerable amount of water in loop seals, the RCP restart could substantially increase the RCS pressure and thus increase the probability for SG tube creep rupture in the unwetted area of the tubes. Some SAMG, therefore, totally abstain from this CHLA. Other SAMGs include a warning.

If there is a bypass of the containment, the RCP restart will increase the release of fission products through the bypass opening.

Finally, RCP restart could also increase the hydrogen generation.

6.2.5. Depressurize the Steam Generators

As discussed in Section 6.2.2, depressurization of the SGs also produces a depressurization of the RCS, assuming the primary and secondary circuits are still thermodynamically coupled (i.e., heat transfer between both circuits occurs). There is also the risk, however, that the pressure difference between the RCS and the SG tubes increases for some time, and then the risk for SG tube creep rupture may increase. For example, if a considerable amount of hydrogen and/or other noncondensable gases has accumulated at the SG tubes' hot side, the heat transfer is limited and RCS pressure may not decrease much.

Under ATWS, the cooling of the RCS may cause recriticality.

If a SGTR must be feared, it may be wise to keep a slight overpressure from the SG secondary side to the primary side.

Depressurization is carried out using the atmospheric dump valves, or the SG blow down path (which is a slow path, if it functions at all).

6.2.6. Inject into (Feed) the Steam Generators (PWR)

This action is beneficial in almost all plant conditions. The injection would provide a path for cooling the RCS (assuming the primary and secondary systems are still thermodynamically coupled), for depressurization of the RCS, and for providing scrubbing for volatile fission products in case of a SGTR. It would also offer protection against SG tube creep rupture. As this is a major threat, some SAMGs start even with this action (WOG SAMG).

A risk is associated with filling an empty SG, as it may cause thermal shock on the tube sheet and the tubes. Therefore, SAMGs often contain a restriction on the amount of water to be injected initially into an empty SG.

6.2.7. Spray into the Containment

Spraying into the containment can have various effects:

- It cools and depressurizes the containment atmosphere;
- It washes out fission products;
- Water may collect in areas where it may cool the RPV from outside or cool debris (is design specific, for example, drywell spray in BWR, which also protects the liner in Mark I containment);
- And it may de-inert the containment atmosphere.

The last item is a negative one: the spray may promote hydrogen combustion.

With regard to this last matter, SAMGs often contain so-called computational aids that help the TSC estimate the risk for hydrogen combustion when using the containment spray.

In case of a non-leak-tight containment or a containment failure, the spray is important to limit the releases.

Not all plants have a containment spray, notably not PWRs of German (AREVA) design.

Spray water often contains pH additives that provide better control of fission products, but that may damage components, notably by inadvertent operation during normal plant operation.

If the spray already has been used before RPV melt-through (e.g., an action in the BWROG SAMG) and if water has reached the cavity, the debris will fall into a water pool and thereby cause a sharp pressure spike in the containment. Although it is expected that the spike in absolute terms is smaller than the pressure peak caused by LBLOCA (which is the design basis for the containment), it may occur at elevated containment pressure and, thereby, cause a risk for containment failure.

If the spray has been used after venting, containment subatmospheric pressure may ultimately occur and cause a containment failure, as the negative design pressure is often quite small (e.g., 0.3 bar). This will lead to inflow of oxygen, which will promote hydrogen combustion.

Often, the containment is protected by vacuum breakers. But if they open, they will also lead to oxygen inflow and, hence, possibly to hydrogen combustion.

Note: Some designs use a spray to cool the (steel) containment from outside, which can be an effective measure to cool and depressurize the containment.

6.2.8. Inject into the Containment

Water can be added to the containment in three time frames:

- Before the corium is ex-vessel;
- After the corium has been released from the vessel and forms a pool of molten material;
- After the pool of corium has been solidified.

In the first time frame, there is a risk for a high-pressure spike in the containment, possibly even a steam explosion, that may damage the cavity walls. Whether such a damaging steam explosion may occur is part of an project [22]. Ex-vessel steam explosions have been observed in Korean tests at KAERI. It should be noted that spontaneous steam explosions (i.e. not triggered) have never been observed in earlier test facilities for prototypical materials (KROTOS, FARO).

If sufficient water is supplied before vessel melt-through, RPV failure may be prevented (in-vessel melt retention, IVMR, through external cooling). This effect has been proven for smaller reactors (e.g., Loviisa in Finland and AP600 Westinghouse design), and possibly also holds for higher power levels (e.g., the USNRC has accepted the cooling method for the AP1000). It has been claimed to be effective even for higher power levels, in conjunction with RPV internal flooding (Korean APR1400). Other designers provide for an external core catcher at power levels of 1000 MWe and above (e.g., EPR™, AREVA design; BiMAC, GEH design; new Russian VVER-designs).

The external cooling mechanism is likely to fail if the vessel skirt (e.g., some GEH earlier BWR-designs) is not vented, as the skirt is then an air trap that prevents water from reaching the vessel wall at the elevation of the skirt.

Even if injection is not able to prevent RPV failure, it will slow down boil-off of reactor inventory, thus delaying the time of RPV failure. Gaining time is important because it may be able to restore failed equipment back to service before RPV failure. For example, if RPV failure can be delayed to 13 hours in an SBO-initiated severe accident, it has been calculated that there is a 98% probability that a diesel can be brought back to service in that time frame [21].

If no core catcher is provided but sufficient cavity surface is available, designers claim that cooling by a water layer from above will stop the interaction of the corium with the underlying concrete. EPRI has indicated that this requires a cavity surface higher than 0.02 m^2/ MWth. It should be noted that the USNRC has neither rejected nor endorsed this value [23].

It should also be observed that injecting from external sources will increase containment pressure. Therefore, if a containment bypass or leakage occurs, such injection should be minimized.

In principle, however, injection must be done up to top of active fuel (TAF) in order to make sure that all rests of fuel, including the fuel that has been left in the RPV, are flooded by water.

During such flooding, important functions may get lost, such as the pressure suppression function of the suppression pool (BWRs), the capability to relieve pressure through the PORVs (BWRs), instrumentation, containment vent paths (loss of wetwell venting for GEH BWR Mark I and II), fan coolers (if their suction or discharge is covered by water—see also Section 6.2.9). If the required pH may not be obtained, damage to equipment for heat removal may occur, as well as reduced retention of fission products. External flooding may also result in a need to (earlier) vent the containment.

In addition, the injected water will absorb the corium decay heat to a large extent; thus a mechanism must be provided to remove this heat to an ultimate heat sink. If no closed cooling circuit is available, water itself must be removed and the capability must be provided to store and process this large amount of heated and highly radioactive water for prolonged time.

The long-term containment injection to the TAF level is, therefore, a delicate process that must be adequately considered and guided by the TSC. Note that injecting to TAF may also create increased static loading on the lower containment parts/basemat foundation.

The mechanism of IVMR and the various designs of a core catcher are described in Sections 6.3 and 6.4. IVMR was part of a joint OECD/SARNET workshop [24].

6.2.9. Operate Fan Coolers

Fan coolers are capable of removing a small percentage of nominal reactor power if there is a high steam concentration in the containment atmosphere. In this way, fan coolers could provide an ultimate heat sink for the decay heat, as discussed in Section 6.2.8. Note that this assumes that there is a relatively high steam concentration in the atmosphere and that other gases such as hydrogen, CO, and CO_2 may reduce the cooling effectiveness.

In removing heat from the containment atmosphere, they could also de-inert the atmosphere and thereby provoke hydrogen combustion. Their operation will homogenize the containment atmosphere, which will reduce the risk of hydrogen pockets. As they operate with electric motors, they can provide sparks for hydrogen ignition, if such ignition is required to remove the hydrogen (although it is presumed that they did not provoke the hydrogen burn that was observed at TMI-2).

Fan coolers may be damaged from deposition of aerosols, arising from core–concrete interaction, and they may fail due to flooding of the suction or discharge (see Section 6.2.8). Certain types of fan coolers are made of aluminium and hence, if heated, could contribute to the hydrogen source in the containment.

6.2.10. Operate Recombiners

In this section, only the passive autocatalytic recombiner (PAR) is discussed.

PARs are usually small steel boxes with plates coated with a catalytic material. There are open ends at the bottom and the top, so that gases can easily circulate through them.

PARs recombine the hydrogen with the oxygen already at a low (~2%) hydrogen, also in the presence of steam. They start automatically in the presence of hydrogen and oxygen. PARs have been qualified in hostile environments, including aerosols. As they draw gas in on the lower side and exhaust on the upper side, they contribute to hydrogen mixing.

Their exhausts can be quite hot and, for hydrogen concentrations above about 10%, may act as igniters. It is therefore essential that the hydrogen concentration in the containment does not reach that level. Some countries even want to restrict the maximum hydrogen concentration to 8%. PARs can also have a relatively low concentration of catalytic substance on the surface or other techniques to exclude overheating of the surface [25].

PARs operate relatively slowly. Hence, it may happen that the containment atmosphere is still flammable in the early hours of the severe accident. Note that the presence of steam prevents combustion only if the steam content is quite high, around 65% by volume (for detailed information see the well-known Shapiro diagram).

On the other side, steam has a high heat capacity and, thereby, opposes flame acceleration, so that the hydrogen combustion in the presence of steam should have a relatively mild character. Even combustion can be incomplete—that is, not all hydrogen will burn, which will result in even lower combustion loads.

Where the containment atmosphere is still flammable during some time, it is either strived for to eliminate all potential ignition sources during that time, or deliberate ignition is provoked by igniters (see Section 6.2.11). In the latter option, it is assumed that ignition close to the flammability limit results only in mild deflagration, even more under the presence of steam, as discussed.

The number of PARs varies. For only DBA recombination, about 2 PARs should be sufficient. For severe accidents, about 20 igniters are installed, up to about 50. The larger number is associated with smaller recombiners. They are distributed in the containment.

PARs have been extensively studied and documented, and much of this information is accessible through search engines.

6.2.11. Operate Igniters

Above about 4% hydrogen concentration, the containment atmosphere is flammable, also in the presence of steam. From the Shapiro diagram, it can be seen that the atmosphere remains flammable up to about 65% steam concentration.

The earliest concept of hydrogen mitigation was, therefore, ignition. At concentrations up to about 8%, the combustion is a mild deflagration. The pressure that results can be calculated by the heat input from the combustion. It is called adiabatic isochoric complete combustion (AICC). At higher concentrations flames may accelerate, and, at even higher concentrations, flames may accelerate to detonation: deflagration to detonation transition (DDT). Spontaneous detonation occurs at even higher hydrogen concentrations, but these are not easily obtained in most NPP containments.

Criteria have been developed to investigate the possibility of flame acceleration, via the sigma criterion, and DDT, via the lambda criterion. As the combustion loads from accelerated flames and from DDT can be very high, the mitigation strategy is to ignite the hydrogen long before these concentrations have been achieved.

Ignition can be achieved by two types of igniters: glow plug igniters and spark plug igniters. The glow plus igniters use AC and, hence, must be coupled to some AC source. If plant AC is not available, a solution is through mobile generators. Spark plug igniters have a battery and, hence, do not require a separate power supply. The glow plug igniter is a permanent source for ignition, as it is constantly glowing after having been put into operation. The spark plug igniter produces a spark every few seconds and does not need outside activation, it activates itself.

Apart from these designed features, one could also use other techniques to ignite hydrogen, for example, by cycling valves or operating other equipment that may generate a spark. In older times, even ring phones were used in the SAMG: they had equipment inside that could produce a spark.

An inherent risk of the igniter is that it does not function in a steam-inert atmosphere. If such an atmosphere has been created during the severe accident evolution, the ignition will not work and hydrogen can accumulate. If later on steam is condensed, either by spray operation or by natural processes, ignition will occur at elevated concentrations of hydrogen, and the sigma and/or lambda criterion may be violated, causing high combustion loads on the containment.

PWRs with a very large and dry containment may not need any hydrogen mitigation feature: hydrogen concentration will be low, and any ignition, spontaneous or provoked, will not jeopardize the containment integrity. Already in the early days of research on hydrogen combustion in containments, a qualitative analysis has been made on the appearance of possible dangerous local concentrations with only a very few places where hydrogen could accumulate [26].

PWRs with the smaller ice condenser containment and BWR Mark III containments have been equipped with igniters. The ice condenser containment outside the United States has both igniters and PARs, possibly in future with PARs only. The Mark III igniters are located above the suppression pool surface and burn the hydrogen when it escapes through the pool surface. This may even result in standing flames.

Note also that other gases are created: CO and CO_2 from core–concrete interaction. CO will also be burned by igniters, and CO_2 may contribute to inertization of the containment atmosphere.

Hydrogen can also be generated in the spent fuel pool, which is outside the containment in a number of plants. Hydrogen can also leak to surrounding compartments if the containment has a leak or rupture, or is bypassed. To the knowledge of the author, no plant is equipped with igniters (or other hydrogen mitigation devices) capable of mitigating the risk of hydrogen accumulation and combustion outside containment. Hydrogen combustion outside the containment will generate a negative pressure pulse on the containment, where the containment usually has been designed only for small negative pressure differences, as discussed before, so that such a burn may jeopardize (any remaining) containment integrity.

6.2.12. Inert Containment with Noncondensables

Smaller containments, notably the GEH BWR Mark I and Mark II designs, as well as early AREVA BWR-designs, are inerted with nitrogen gas that prevents all combustion. During start-up, the containment may not yet have been fully inerted, so that a severe accident during that phase may require hydrogen mitigation techniques. If the inertization is initiated upon observation of a severe accident, core damage and hydrogen generation can occur before the inertization has been completed.

If the containment has been vented and later on steam has been condensed, air may be drawn inside the containment, thereby possibly de-inerting the containment.

In non-nuclear applications, inerting is also considered a mitigation measure during the evolution of the accident, for example, a large fire. Such systems consist of large volumes of CO_2 or nitrogen gas, which are injected into the compartments hit by the accident (fire). Such techniques are not used in NPP containments, as the containment pressure will rise to about 3 bars, still to be augmented with pressure caused by steam, hydrogen, CO, and CO_2.

6.2.13. Vent Containment

Due to the production of steam and gases (H_2, CO, CO_2), the containment pressure may rise to a level where containment failure may occur. If the corium decay heat is also rejected to the containment, the pressure rise may even be larger.

In those cases, the containment should be vented, preferably through a path that reduces the release of fission products, such as through pools or through a dedicated filter. A variety of filters exist: pool scrubbers, sand bed filters, and venturi scrubbers. A large amount of literature is available in which these filters are described, accessible through search engines.

If during the pressure increase of the containment a hydrogen burn is also possible, the selection of the pressure level at which to vent the containment should include the potential pressure rise due to combustion—that is, it should already be vented below an ultimate pressure-bearing capability.

Venting the containment may also be needed for other reasons, including making it possible to inject by low-pressure water systems. If a bypass or a leak is present, venting the containment will reduce the release of fission products along uncontrolled pathways.

Venting can also reduce the hydrogen mass inside the containment (it does not change the hydrogen concentration, hence, not the flammability limits).

After venting, it should be realized that the containment may reach subatmospheric pressure upon condensation of steam (as already discussed before), for which proper SAMG should be in place.

6.2.14. Spray Secondary Containment

Accident sequences that involve a containment leak or bypass may affect the secondary containment. Release will include steam, fission products, and hydrogen. The secondary spray system, if available (it could be the fire extinguishing system), can wash out fission products. A potential negative consequence may be that the spray water may affect equipment that is used to mitigate the accident (e.g., ECCS) or may be used in the longer term to recover from the accident.

6.2.15. Flood Secondary Containment

Remarks as in Section 6.2.14 apply. Typically, such flooding is meant to cover the leak with water in order to scrub escaping fission products. The technique is only practical at low-elevation leaks/bypasses.

6.2.16. External Cooling of the RPV

This CHLA refers to the IVMR, discussed in Section 6.2.8.

6.2.17. Steam Inerting of the Containment

As discussed in Section 6.2.12, noncondensable gases can be used to inert the containment. The containment can also already be inert due to the steam that escapes from the RCS. If an inert condition is desirable but injection of noncondensable gases is not possible or not feasible, one could use auxiliary boilers to inject steam into the containment (some plants do this). A drastic but feasible action could be the opening of the primary PORVs, as this will release steam into the containment. This action would, however, also result in a release of hydrogen to the containment, if the core degradation has already proceeded substantially.

6.2.18. Conclusions on the Severe Accident Management Techniques

A variety of severe accident management techniques have been described in this section. As discussed, techniques can mitigate the consequences of severe accidents. However, some techniques still have uncertainties, or their potential success depends on the proper moment to initiate actions. Some even can have negative effects if they are not implemented properly (an example is a spray system that de-inerts the containment atmosphere and provokes a severe hydrogen combustion). They should, therefore, always be embedded in a well-defined series of severe accident management guidelines and developed on a plant-unique basis, as is described in Section 6.1. The following sections describe two examples of SAMG with more details: respectively, in-vessel melt retention and ex-vessel corium retention.

6.3. IN-VESSEL MELT RETENTION AS A SEVERE ACCIDENT MANAGEMENT STRATEGY

(B.R. Sehgal and Nam Dinh)

6.3.1. Introduction

A postulated severe accident in an LWR assumes the occurrence of a meltdown of the core, breaching the first barrier of the clad to release the radioactive fission products. The accident proceeds further, and the molten corium moves to the bottom head of the vessel. The bottom head will fail if the corium melt remains uncooled, thereby failing the second barrier to the release of radioactivity to the environment. The corium melt released to the containment may fail the containment in a short time if some energetic reactions, for example, hydrogen burn (explosion), steam explosion, or direct containment heating occurs. If such energetic interactions do not occur, or are managed not to occur, the containment could fail later (by several hours or a few days) due to the attack of the core melt on the concrete, which would release noncondensible gases pressurizing the containment and possibly cause the melt-through of the basemat. The containment structural failure or the basemat melt-through constitutes the failure of the third and last barrier to the release of radioactivity to the environment.

Severe accident management (SAM) consists of actions (measures) that would prevent the failure of barriers 1 to 3. The first aim of SAM is to prevent damage to the clad on the uranium fuel pins. If that is not possible due to the lack of timely injection water to the vessel, the second aim of SAM becomes the prevention of the failure of the bottom head of the vessel. If that aim is not achieved due to either the inability to inject water to the vessel to cool and quench the melt pool in the lower head; the next aim becomes prevention of the failure of the containment and/or the basemat melt-through so that there is no significant release of radioactivity to the environment. That event—early or late failure of the containment—should not occur as per the European utility requirements for new plants. The existing plants have also agreed to install equipment for SAM actions that would prevent an early failure of the containment and that would reduce the conditional probability of the late failure of that containment to 10% and prevent the failure for at least 24 hours.

One key to the SAM actions preventing failure of clad, vessel, or containment is coolability, be it of the hot fuel rods, of the melt pool that could form in the original core geomentry, or in the bottom head or on the floor of the basemat. The corium melt temperature is >2000°C, which is greater than the melting temperature of the vessel bottom head steel and the melting temperature of the concrete basemat. Thus, after the failure to prevent damage to the core in its original confines, the SAM actions are directed primarily toward the coolability of the melt pool in the bottom head or on the containment floor when it is interacting with the concrete in the basemat. Such SAM actions, if successful, may be termed as stabilization and termination of the accident, without preventing a release of radioactivity to the environment.

The stabilization and termination of the accident, if it is successful with the coolability of the core melt in the bottom head, is called in-vessel melt retention (IVMR) and the same, if successful with the coolability of the melt on the concrete basemat, is termed ex-vessel melt retention (also called core catcher).

6.3.2. IVMR SAM Strategy

The idea of the retention of core melt inside the vessel as a severe accident management (SAM) strategy was conceived during the deliberations on the safety upgrade of the VVER-440 plant, Loviisa, in Finland. The Loviisa plant has an ice-condenser containment of relatively low-pressure rating, quite vulnerable to melt attack on the basemat. In the large LOCA scenario, the vessel cavity would be full of water due to the melting of ice. Thus, out of necessity, opportunity, and vision, investigations of whether the heat removed by water from the outer surface of the vessel would be greater than the heat flux imposed by the melt in the lower head were started jointly by IVO (currently called Fortum) and by Professor Theofanous at the University of California at Santa Barbara. The case for in-vessel melt retention (IVMR), as the SAM strategy for Loviisa, was submitted to STUK, the Finnish Regulatory Authority [27], which after due deliberation approved this SAM strategy for Loviisa [28].

The IVMR strategy was next proposed for AP600, the Gen.III+ plant designed by Westinghouse. The technology and methodology developed by Theofanous and colleagues [29] was extended to that required for the case to be made for AP-600. The AP600 has been certified by the United States Nuclear Regulatory Commission.

More recently, Westinghouse has incorporated the IVMR SAM strategy into the AP1000 design. This required further development of the technology and methodology, in particular, for enhancing the critical heat flux (CHF) on the outer surface of the wall of the vessel lower head in order to obtain a sufficient margin of safety to cover the uncertainties. Figure 6.1 depicts the configuration of the vessel with a melt pool in the bottom head and the cooling of the outer wall of the vessel with two-phase flow along the vessel.

The IVMR SAM strategy was also incorporated into the SWR-1000 (now KERENA) designed by Siemens, prior to its merger with Framatome (AREVA NP). The BWR vessel is much larger in diameter than a PWR vessel, and the BWR contains much greater amounts of steel and zirconium. Clearly, as we shall see, these parameters are helpful for the case of IVMR SAM strategy. However, the BWR also has a large number of penetrations in the vessel bottom, which will have to withstand any direct attack of melt, without failure.

The Korea APR-1400 also employs the IVMR SAM strategy. For a 1400 MWe plant, the margin to even the enhanced CHF on the vessel external surface is either very small or non-existent. However, an additional safeguard employed by the Korean plant is to add water to the top of the melt pool; with a separate circuit of water. The water addition would substantially reduce the focused heat flux from the metal layer on the vessel and thus preserve a margin to the CHF.

Clearly, the IVMR SAM strategy is, inherently, very appealing because its success ensures that the PWR and BWR containments designed for a large

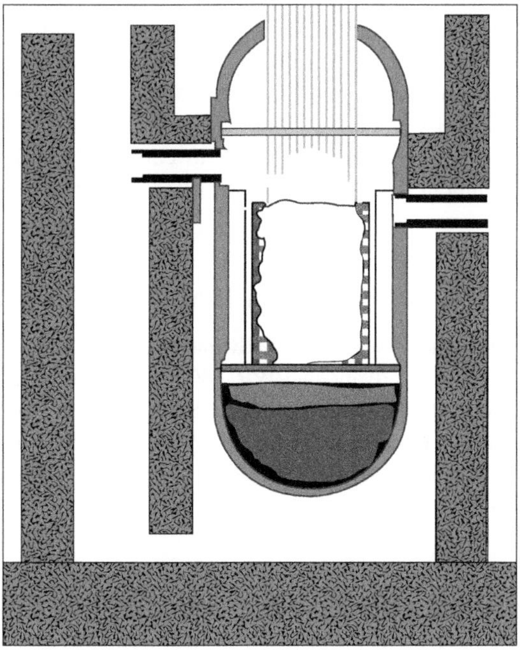

FIGURE 6.1 In-vessel melt retention (IVMR).

LOCA cannot be loaded to catastrophic failure, either soon after vessel failure (e.g., by direct containment heating [DCH]) or later, by corium–concrete interactions (CCI) or basemat melt-through. It is possible to assure the public that a large release of radioactive fission products will not occur in a postulated severe accident at a PWR plant.

6.3.3. The Requirements on the IVMR

The IVMR required the development of new phenomenology. Specifically, it required the description of how the melt relocates from the core region to the lower head and the determination of the probability (or possibility) of vessel failure due to (1) the melt jet attack on a particular location in the vessel and (2) a steam explosion generated by the entry of melt into the water contained in the lower head. It also required the determination of the melt-pool composition, configuration (e.g., stratification, etc.), and description of the melt-pool convection process, in order to determine the magnitude and polar angle distribution of the heat flux, imposed by the assumed melt mass in the lower head, on the vessel wall. Simultaneously, it required the determination of the magnitude and the polar angle distribution of the heat removal (CHF) obtained with the two-phase (water-steam bubbles) flow around the 3-D vessel. For this purpose, it was necessary to ensure that there would be a two-phase natural circulation flow around the vessel to provide flow boiling instead of pool boiling, whence the CHF of 1 MW/m^2 is increased substantially. The ultimate requirement for a particular reactor design is to show that the heat flux imposed by the melt resident in the lower head is less than the CHF of heat removal at all polar angle locations of the vessel wall, for all probable melt configurations (e.g., stratifications) with a sufficient margin to cover uncertainties. If such a case could be made, the IVMR SAM strategy for that particular reactor design would be acceptable and there would be no need to consider vessel failure and to analyze the containment response to a melt discharge into the containment.

6.3.4. Development of the Phenomenology for IVMR

The dominant work for the phenomenology of the IVMR has been performed for Loviisa, AP-600, AP-1000, Korean advanced PWR-1400, which are all PWRs and for the 1000 MWe BWR. The Russians have also performed work of a similar nature for a 640 MWe PWR in St. Petersburg. We shall consider primarily the R&D performed for the Loviisa, AP-600 and the AP-1000 plants in the following parts of the text.

Melt Relocation

The PWR core meltdown and melt-relocation (to the bottom head) scenarios follow what occurred in the TMI-2 accident (see Figure 6.2). There can be other

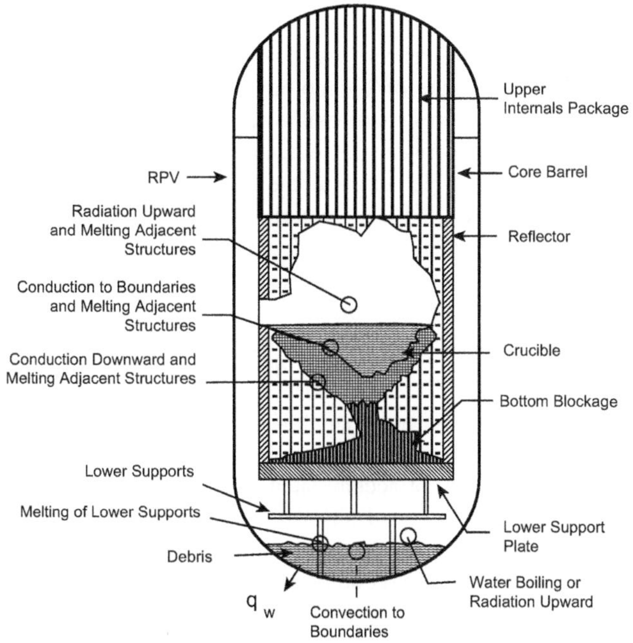

FIGURE 6.2 Material configuration and heat transfer processes at an intermediate state of melt progression, after initial major relocation event and before final steady state [29].

variations: for example, the melt comes through the core lower plate, instead of through the downcomer space as occurred at TMI-2. There may be some dispute about how the melt relocates to the bottom head; for example, melt could arrive in the bottom head in the form of several concurrent jets. Whatever the form of the melt entry into the bottom head, the fact remains that the bottom head contains a lot of water and the interaction of melt and water should lead to fragmentation of a large quantity of melt.

If the melt relocates through the downcomer, as it did at TMI-2, there is a hazard of melt jet attack on the vessel wall, as well as that of a steam explosion in the bottom head. Both of these events were investigated in turn, and the evaluation of these hazards required performance of experiments and development of models.

The melt relocation from the core region to the bottom head begins with failure of crust of the melt pool in the core region and the ablation of the core radial barrel cylinder or the core plate. The hole ablation as the melt flows increases the jet diameter and the rate of the discharge of the melt to the bottom head. The melt jet will fragment in the water of the bottom head, unless it is of 10 to 15 cm diameter, since according to Saito's correlation [30] the break-up length should be 10 to 15 times the jet diameter. It is possible that with a large jet there may be some ablation of the vessel wall.

Melt Jet Attack on Vessel Wall

The melt jet attack on the vessel wall was investigated through an experimental program at the Royal Institute of Technology (KTH). Experiments were conducted with high-temperature jets of oxidic stimulant materials impinging on lead plates, so that the ΔT between the jet temperature and the melting point of the plate was close to that which exists in the prototypic situation of corium jet impinging on the steel wall. Later, experiments were also performed with simulant metallic melt materials impinging on other metal plates. In these experiments, the heat transfer rates imparted to the plates were measured as a function of time and the ablation profile.

An evaluation [31] was made of the prototypic oxidic and metallic melts with and without water in the lower head. The findings were as follows:

- Without water in the lower head; small size (<5 cm), long-duration jets are more damaging to the vessel wall than large-diameter short-duration jets to deliver a certain quantity of accumulated melt in the core region.
- With water in the bottom head jets will break up, and the probability of vessel wall failure due to melt jet attack is very low indeed.
- It is physically unreasonable (conditional probability $<10^{-3}$) to expect that the 15 cm thick wall will fully ablate to fail the vessel, during the delivery of the melt from the core region to the bottom head.

Steam Explosion in the Bottom Head

The melt jet entry into the saturated water could cause a steam explosion. The question that arises is whether the resulting dynamic loads are large enough to fail the bottom head locally. If such failure occurs, the whole concept of IVMR becomes moot. We must note here that the traditional impact load that was the focus of attention for the steam explosion in the bottom head is that which could fail the bolts in the upper head, making it airborne and hit the containment wall with sufficient force to fail the containment locally. However, the focus of the investigation here is on determining whether the pressure and impulse generated will fail the bottom head locally. We should note here that the AP-600 and AP-1000 vessels do not have penetrations in the bottom head.

Professor Theofanous [32] performed a study of the steam explosion loads on the bottom head with the codes ESPROSE and PM/ALPHA, which had an assumed rate of melt delivery of 200 kg/sec. The steam explosion was modeled, and the impulse load history was determined. This history was input to a structural calculation with the ABAQUS code [33]. The results of the ABAQUS code were compared to the fragility curve for the reactor vessel to determine the margin to failure. In addition, sensitivity and uncertainty calculations were performed with a variation of parameters to ascertain that indeed the conditional probability of failure of the vessel due to a steam explosion with quite a large rate of melt delivery would be $<10^{-3}$, that is, physically unreasonable.

Melt-pool Formation in the Bottom Head

The melt delivery to the water in the lower head leads to partial fragmentation and quenching. There is a considerable amount of water in the lower head, which will eventually (in 1 to 4 hours) evaporate. The partially fragmented and cooled corium debris will dry and heat up to form a melt pool in the lower head. During this process, it will also melt the steel structures in the lower head and even the lower core support plate if it is submerged in the debris.

The composition and configuration of the melt pool as it is formed are of concern. In particular, we need to know

- What is the content of zirconium in the melt?
- How much steel has been melted, and does it stratify?
- What are the possible chemical reactions among the constituents with the availability of steam, and how do they affect the physical configuration and stratification of the pool?
- How do the possible stratified configurations affect the thermal loading on the vessel wall?

This is a complicated and still not fully resolved story. It requires research—and the RASPLAV and the MASCA Projects are actors in this play.

The Magnitude of the Melt Pool

The magnitude of the melt in the lower head will change as a function of time. A key assumption needed is whether one should analyze all the developing melt-pool configurations, stratifications, and magnitudes of melting of individual components, for example, steel. Professor Theofanous, in his analysis of the AP-600, assumed a steady-state configuration that would provide the largest (bounding) heat source in the lower head [29]. This bounding loading would be a complete fill-up of the lower head with the oxidic and metallic melts. This implies that practically the whole core has melted and filled up the lower head. This assumed loading would produce the largest thermal load on the vessel wall. It was established that a partially full lower head would exert less thermal load on the vessel wall (see Figure 6.3).

This steady-state, extreme thermal loading configuration became the basis for the analyses Theofanous performed for Loviisa, AP-600 and AP-1000 reactor designs.

The potential for pool stratification due to the presence of a large amount of steel in the bottom head was recognized in the early phase of the studies. Steel, being lighter and immiscible, would separate from the oxide melt and ride as a layer on top of the heat-generating oxide melt. Thus, the configuration consisted of:

- an oxidic pool containing almost the whole core and the oxidized Zr,
- an upper layer of unoxidized Zr and all the steel that could melt and join the pool, as indicated in Figure 6.4 for AP-600.

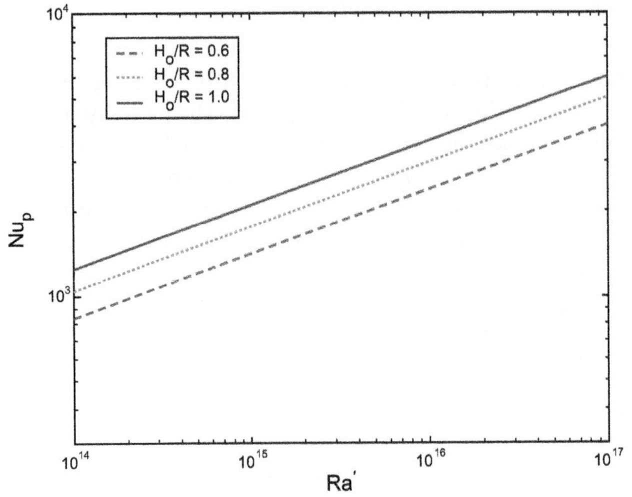

FIGURE 6.3 The oxide pool Nusselt number, as a function of the Rayleigh number and the "fill" fraction, Ho/R [34].

Melt-pool Convection Thermal Loads

The oxidic melt pool—generating several MW of heat, topped by a metal layer, possibly generating some heat (due to metallic fission products), cooled from the vessel exterior by a flowing water–steam mixture and radiating heat away to

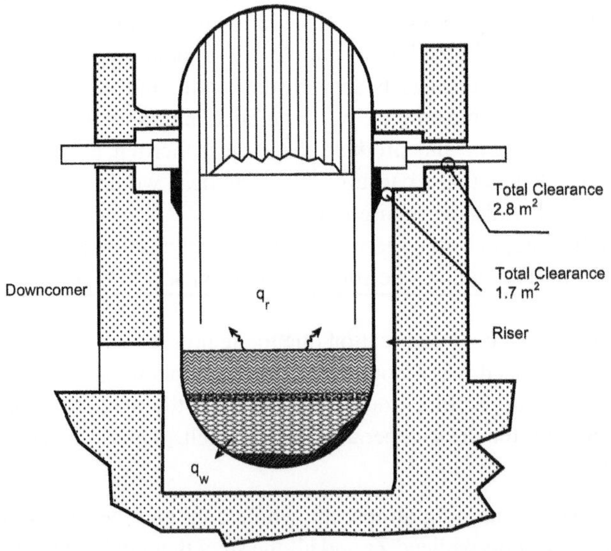

FIGURE 6.4 Configuration of materials in the AP-600 lower head [29].

the vessel components above—became the model for the phenomenology development (see Figures 6.4 and 6.5).

The oxidic pool will be surrounded by crust all around, but inside it will perform natural circulation motion. The characteristics of such motions were studied by Steinberner and Reineke in Germany [35] and Kulacki et al. in the United States [36]. It was found that the character of the convection is as follows:

- Stationary stratified region near the bottom, where heat transfer is conduction-dominated;
- Highly turbulent region above the stationary region in which plumes of fluid rise to bring heat to the top. The plumes cool down and return to the lower part of the turbulent region;
- Convective flow along the spherical boundaries of the pool from the top to the bottom, replenishing the fluid in the stationary and the highly turbulent regions of the pool.

The most remarkable fact about melt-pool convection is that the total upward and downward heat transfer correlates almost to a single non-dimensionless number: the internal Rayleigh number, which is related to the Grashof number and the Prandtl number. It incorporates physical dimension to the power 5. Thus, for prototypic vessel size, the values are 10^{16} to 10^{17}. The upward and downward Nusselt numbers had been measured for Ra numbers up to ~10^{12}, but a correlation was needed for higher values.

For the work on the Loviisa reactor, an experimental facility named COPO was built, which employed a slice of ½ scale employing salted water heated electrically. Besides measuring the upward and downward heat fluxes, measurements were also made for the variation of heat transfer as a function of the polar angle. Clearly, as the heat rises, greater heat is transferred to the vessel

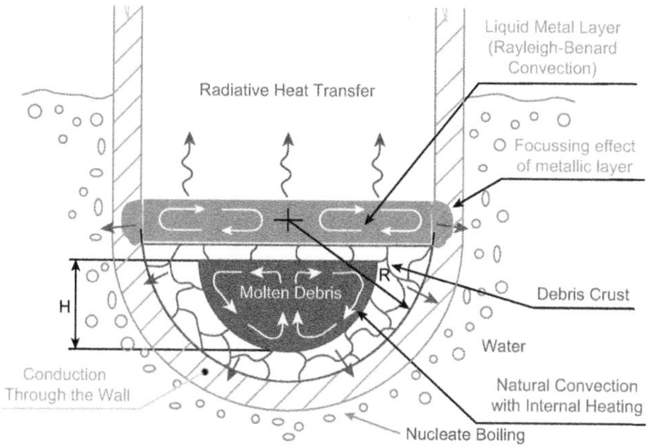

FIGURE 6.5 Phenomena of in-vessel melt retention.

wall at higher polar angles. The maximum temperature is close to the top corners of the oxidic pool.

The COPO facility [37] employed a torospherical lower head as found in all VVERs. Professor Theofanous built a facility named ACOPO [38], which was a full-scale hemisphere. Instead of heating salted water, he employed hot water, which was cooled down. The cool-down technique was found to be almost equivalent to the fluid-heating technique. The ACOPO facility is shown in Figures 6.6 and 6.7.

The oxidic melt-pool convection was effectively modeled by the correlations obtained in simulant material facilities: COPO, ACOPO and some others. Further work on melt-pool convection was performed at the Royal Institute of Technology (KTH) with the SIMECO facility and at the Kurchatov Institute (KI) with molten salt pools, which formed crust boundaries matching the prototypic situation. Later, KI performed RASPLAV experiments employing UO_2-ZrO_2-Zr mixture melts. The data obtained in the SIMECO [39] and RASPLAV [28] facilities were essentially similar to the COPO and ACOPO facilities.

Thus, with the correlations, the heat split between the top and the bottom could be obtained. More heat is transferred to the top than to the bottom. The upper corners experience large heat flux.

Figures 6.8–6.10 provide some of the data that were measured in various facilities for Nusselt number versus Rayleigh number and its variation versus the angle along the vessel bottom head, with zero being at the bottom pole position.

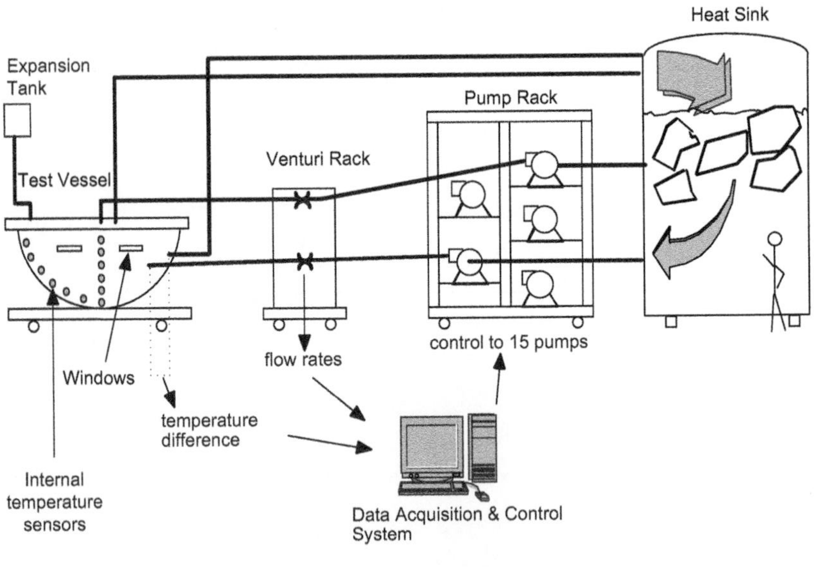

FIGURE 6.6 Scheme of the ACOPO facility [38].

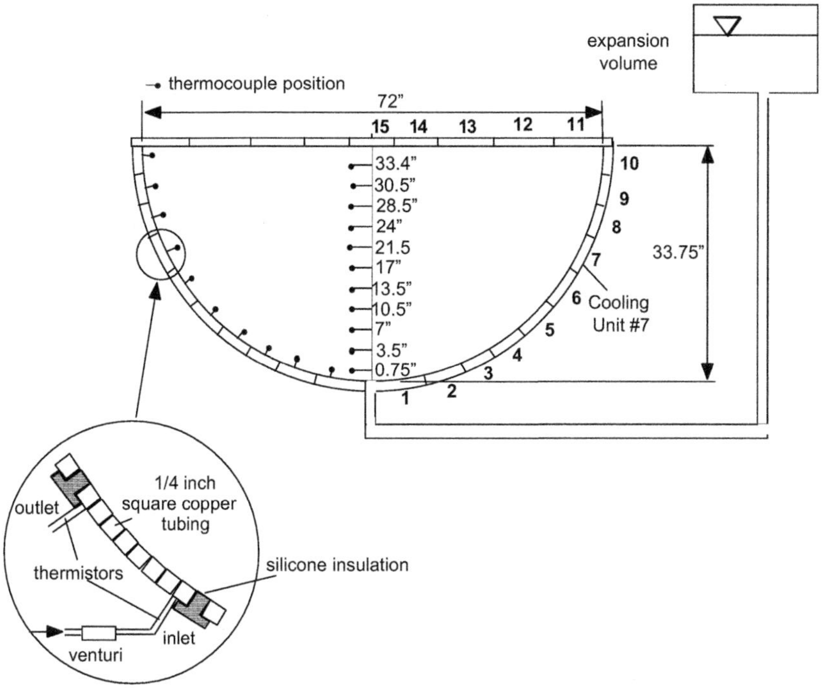

FIGURE 6.7 The ACOPO facility [38].

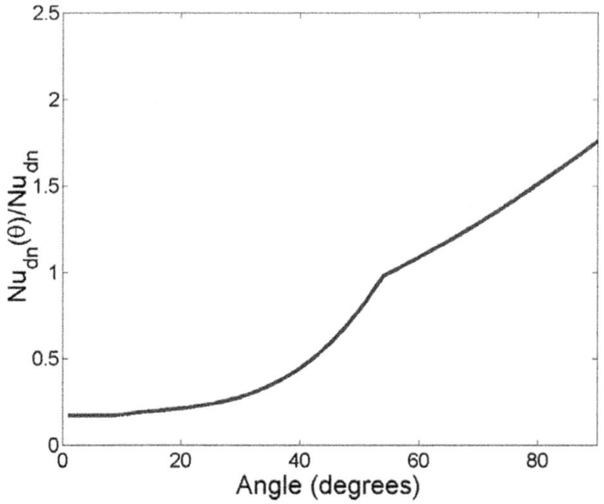

FIGURE 6.8 The heat flux distribution on the lower boundary of a naturally convecting hemispherical pool [29].

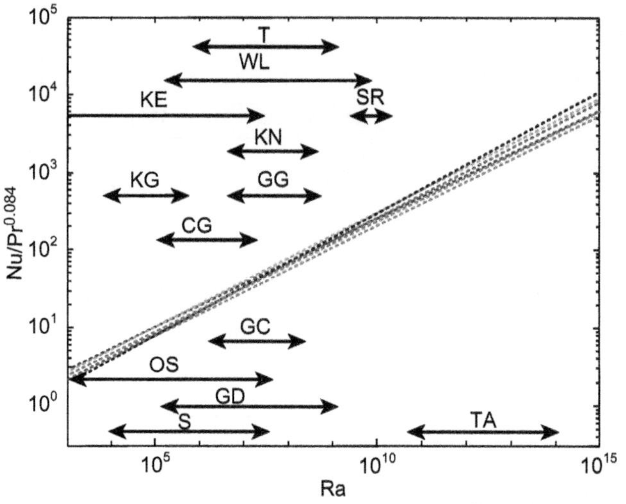

FIGURE 6.9 Nusselt number dependence on external Rayleigh number. Abbreviations: CG—Chu and Goldstein, GC—Goldstein and Chu, GD—Globe and Dropkin, GG—Garon and Goldstein, KE—Kulacki and Emara, KG—Kulacki and Goldstein, KN—Kulacki and Nagle, OS—O'Toole and Silveston, S—Silveston, SR—Steinberner and Reineke, T—Threlfall, TA—Theofanous and Angelini [34].

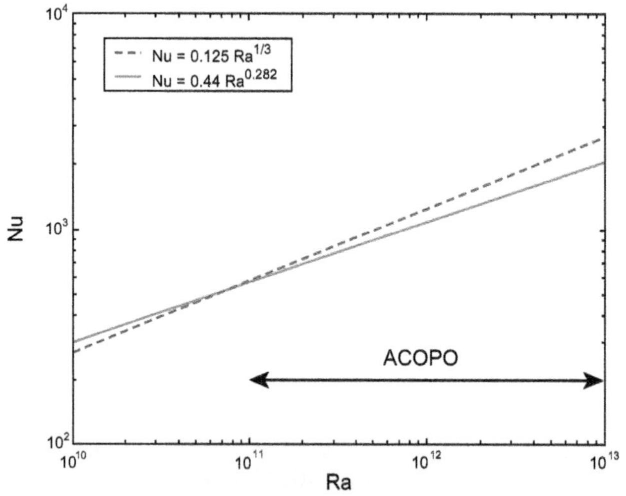

FIGURE 6.10 ACOPO heat flux at the pool upper corner (the dotted line is Churchill and Chu's Correlation [34]).

Metal Layer Induced Heat Flux Focusing Effect

The metal melt layer, riding on top of the crusted upper boundary of the heat-generating oxidic melt pool, receives the heat sent to its lower boundary. It performs the classic Rayleigh-Benard convection that transfers heat to its edges, which are touching the part of the vessel wall above the crust of the oxidic pool. This is the well-known heat flux focusing effect. Thus, whatever heat is not radiated out from the upper surface of the crusted metal layer can become incident on the area of the vessel wall touching the metal melt layer. Since the heat lost by radiation from the metal layer is an important element in the heat balance for the metal layer; the emissivity value becomes an important factor. Another is the temperature of the vessel internals above the metal layer. These internals would heat up and reach temperatures close to the steel melting point, whence the radiative heat flux would decrease and more heat would be transferred to the vessel wall.

Clearly, the heat flux on the vessel wall will be determined by the area of the metal layer in contact with the vessel. This area is directly proportional to the thickness of the metal layer or to the mass of the steel melted. Thin metal layers may produce quite large heat flux, which could be greater than the CHF outside. This will lead to heat-up and melting of the vessel from the inside, which may increase the thickness of the metal layer. However, this requires that the vessel steel melt mix with the already existing metal layer. A thick metal layer may not have heat flux greater than the CHF outside. Thus, the success of the IVMR SAM strategy depends on the amount of steel that melts during the melt accumulation process in the lower head. Clearly, there may be some chance of sustaining a thin metal layer for a short duration even when the bounding configuration has a thick metal layer. Such possibilities should be investigated.

The focusing effect can be much reduced if the upper face of the metal layer is cooled on top. This is an accident management requirement that calls for special handling. The Korea APR-1400 intends to establish a special water-addition circuit to add water on top of the metal layer. Of course, it is not easy to add water that will reach the lower head and stay on top of the metal layer. If the metal layer has no crust, there is also danger of a stratified steam explosion.

Heat Removal Capability at Vessel External Surface

The IVMR strategy requires the flooding of the containment cavity around the vessel. In general, the water level is established up to the hot legs, so that the vessel wall is kept cool above the melt pool, in spite of the heat flux that is imposed by the hot melt pool radiating into the vessel. It is necessary to establish a heat transport cycle that takes heat away from the vessel wall. This is done by providing a clear passage way for the steam rising from the vessel external surface and a condenser to condense the steam and return the water to the water pool from which the water employed for cooling the vessel was drawn. A two-phase natural convection flow circuit helps to increase the CHF much above its value for pool boiling.

The IVMR requirement is that the heat removal heat flux on the vessel wall surface should be greater than the heat flux imposed on the inner surface of the vessel wall by the convecting melt pool. The critical polar locations on the vessel are:

- near the top of the oxidic pool.
- facing the metal layer.

The critical heat flux that can be removed on the outer surface of the vessel was measured in the SULTAN facility [41], where long-heated plates were inclined at different angles. More importantly, CHF as a function of polar angle was measured by Professor Theofanous and colleagues at the University of California at Santa Barbara in the ULPU facility (see Figures 6.11 and 6.12). This is a full vertical-scale facility that employed a hemispherical-slice copper-block, which is heated by electrical heaters to provide the heat flux measured in the ACOPO facility as a function of the

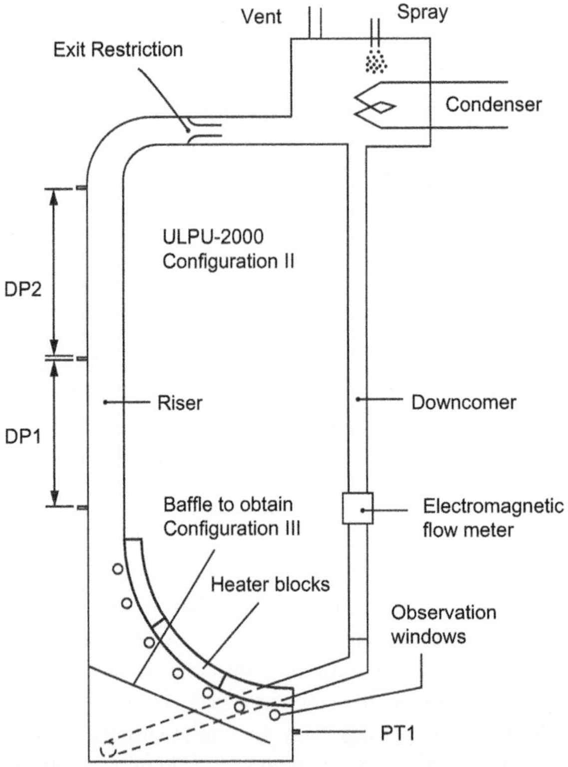

FIGURE 6.11 Scheme of the ULPU facility: Configuration II [34].

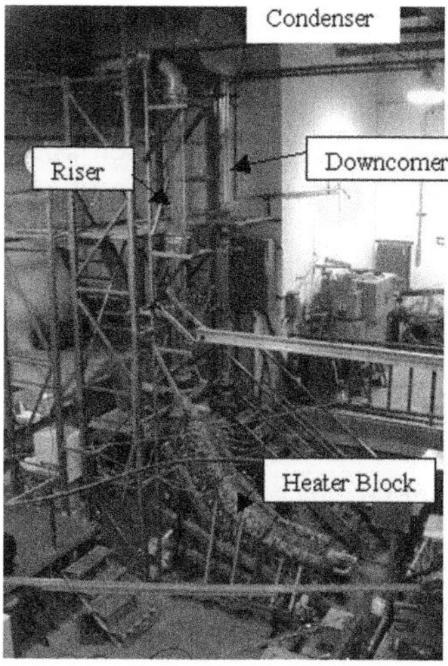

FIGURE 6.12 The ULPU facility [34].

polar angle. A temperature transient signifying the response of a thermocouple at a location about to experience CHF is shown in Figure 6.13. The experimental procedure consists of increasing the power input to the test section to increase the imposed heat flux until CHF conditions are experienced at a location. The input power is decreased immediately after the temperature trend is evident.

The ULPU facility establishes the two-phase flow (water–steam) natural circulation circuit outside the heated vessel wall. The pressure drop and the

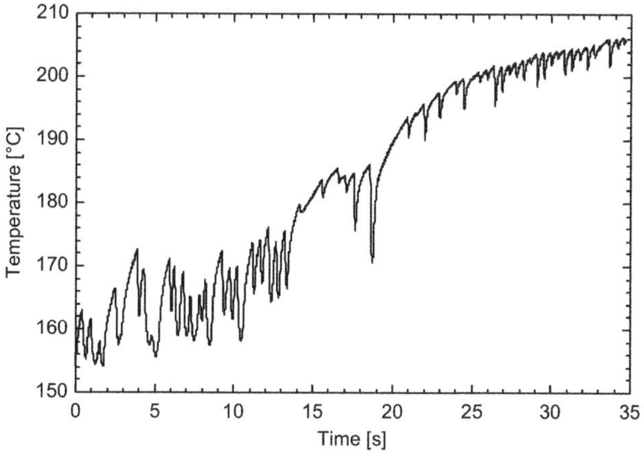

FIGURE 6.13 A temperature transient (local microthermocouple response) associated with boiling crisis [34].

cooling coil are sized as they would be in the AP-600 and AP-1000 flow circuits. Thus, the scaling, except for the slice nature of the facility, is exact.

It is another remarkable fact that the measured CHF increases with polar angle almost in the same fashion as does the incident heat flux from melt-pool convection (see Figure 6.14). If it were a different behavior, IVMR could not succeed. The increase in CHF with polar angle is due to the ease of bubble departure and the higher velocity of flow as the angle increases.

The measurements made in ULPU-2000 without any shaping of the two-phase flow showed the maximum value of CHF to be ~1.5MW/m^2 (see Figure 6.14). This was sufficient for the AP-600 case, as we shall see presently. However, later when an IVMR case had to be made for AP-1000, which has 66% higher power, without the commensurate increase in vessel surface area, increasing the value of CHF became a necessity.

Increase in CHF was obtained by:

- making changes in the flow circuit to decrease the pressure drop,
- shaping the flow circuit along the hemispherical head with a baffle; in order to increase the flow velocity.

New configurations of ULPU were designed called ULPU-configuration IV (Figure 6.15) in which the maximum value of CHF increased to ~1.8MW/m^2 (Figure 6.16). This value of CHF did not provide sufficient margin. Additional research was conducted, and later the ULPU-V configuration was developed with detailed shaping of flow (Figure 6.17) that provided a further increase in maximum value of CHF (Figure 6.18). In addition to flow shaping, surface effects were studied. It was found that the paint, which is generally used on the

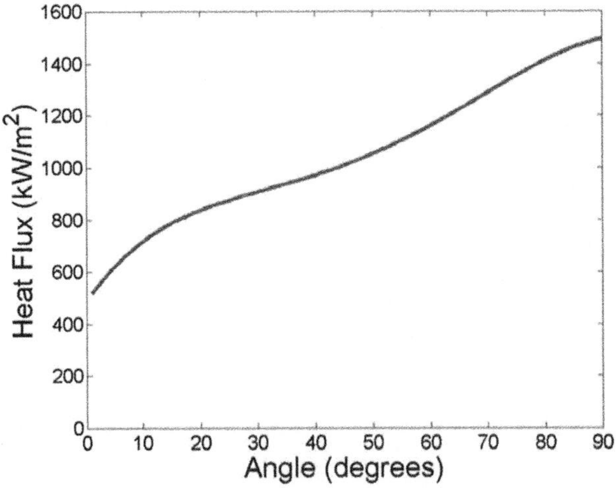

FIGURE 6.14 Critical heat flux as a function of angular position on a large-scale hemispherical surface [34].

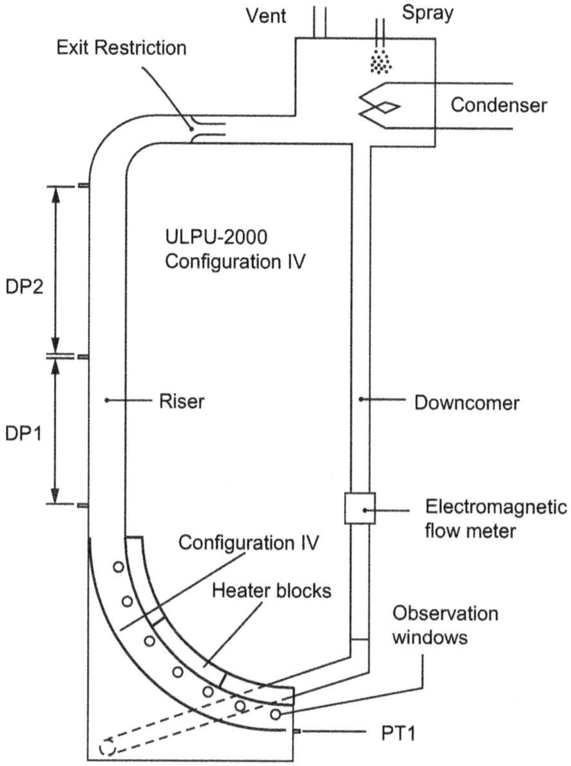

FIGURE 6.15 Scheme of the ULPU facility: Configuration IV [34].

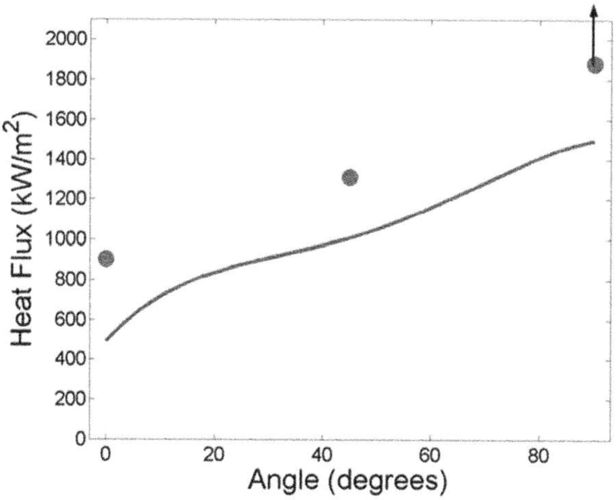

FIGURE 6.16 New configuration IV CHF results (data points) [34].

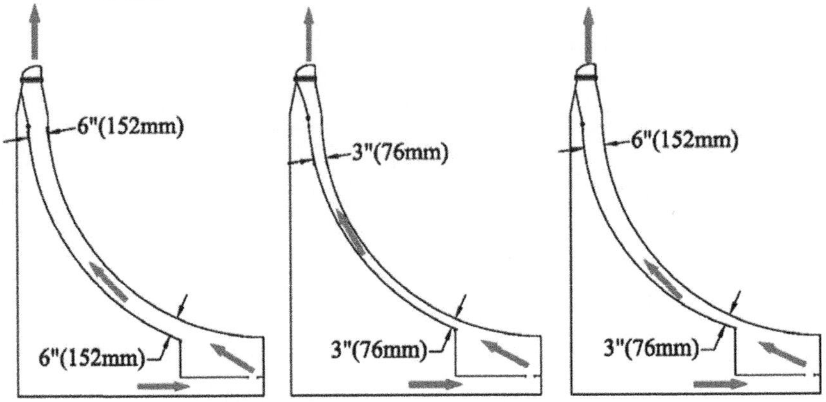

FIGURE 6.17 ULPU-V: Three baffle configurations [42].

manufactured vessel, was not conducive to increase of CHF. A sand-blasted clean surface was found to be more effective. Special chemical treatment of water was also devised to increase CHF some more. The ULPU-V configuration produced CHF values of ~$2 MW/m^2$ (see Figure 6.18), which is a substantial increase. As a result, the case for AP-1000 became feasible (i.e., with reasonable margin to cover uncertainties). ULPU-V constructed the complete natural circulation path of the AP-1000 design as a 1/84-slice and matched resistance (flow area and geometry) as specified in the design.

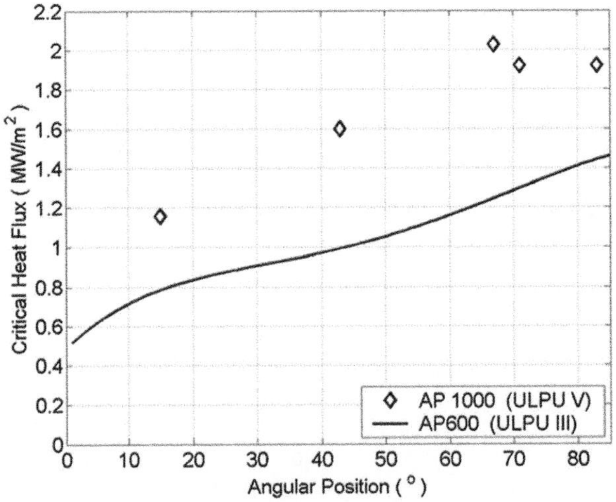

FIGURE 6.18 ULPU-V reference data for AP-1000 IVMR conditions [42].

6.3.5. The IVMR Case for AP-600

It should be clear by now that the metal layer focusing effect is the Achilles heel for making a case for the IVMR. The focused heat flux can be within the bounds of the CHF only if it can be shown that there will be a large mass of steel melted in the bottom head and that all of that will be in the metal layer at the top of the heat-generating oxidic melt pool. This ensures that the area of the metal layer touching the vessel is large enough to keep the imposed heat flux below the value of CHF of ~1.5 MW/m^2.

This requirement is met by AP-600, since the core bottom plate, which has many tons of steel, hangs low into the bottom head. It would be buried into the fragmented corium debris and on melting of corium it would also melt. This condition is also met by the bottom head in the Loviisa reactors, since the head is torospherical and of shallow depth. Clearly, it depends on the vessel design and in particular, on the design of the bottom head and its steel content.

As mentioned earlier, the radiative heat rejected by the metal layer is a factor in evaluating the focused heat flux. In this context the emissivity factor will be the variable, since the temperature of the metal layer will be close to the melting point of steel and Zr metal alloy. The coolability region with in-vessel retention for an AP-600 type reactor is shown in Figure 6.19. In this figure, ε, the value of the emissivity of the steel layer, varied between 0.45 and 0.8, and the CHF value for the outside cooling is taken as ~1 MW/m^2 for pool boiling and ~1.5 MW/ m^2 for the natural circulation boiling on the outer wall of the vessel. Coolable and not-coolable regions can be identified with the parameter of quantity of steel and Zr on (or the thickness of) the metal layer. The AP-600

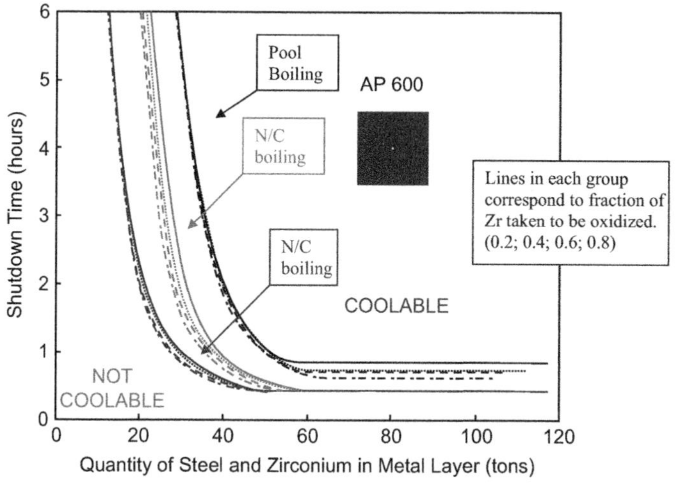

FIGURE 6.19 The coolability region of an AP-600 reactor for different cooling options and metal layer emissivity [34].

position in this graph is identified along with the uncertainties in the analysis. It is seen that the bounding configuration AP-600 is coolable with the accident management strategy of in-vessel melt retention (IVMR).

6.3.6. The IVMR Case for AP-1000

The AP-1000 case, in terms of the focusing effect, is similar to that for AP-600—that is, a large amount of steel will melt for the bounding configuration and will not provide an extremely large focusing effect.

The maximum thermal load for AP-1000 was estimated to be 1.3MW/m^2, which is too close for comfort to the maximum CHF value of 1.5MW/m^2 as measured in ULPU-2000, that is, without any special shaping of the flow.

The AP-1000 case, with sufficient margin to CHF, could only be made after establishing that the CHF could be increased to 1.9 to 2MW/m^2, with baffling to shape the two-phase flow field on the vessel external surface, removing the paint on the surface, reducing the pressure drop in the flow circuit, and so on.

The AP-600 design has been certified by USNRC. The AP-1000 design is currently undergoing the USNRC certification process. This is a success story.

6.3.7. Conclusions on In-Vessel-Melt-Retention

The IVMR case established for the Loviisa VVER 440 and the AP-600 appears to have significant margin between the heat flux imposed by the melt pool convection on the vessel wall and the CHF for heat removal on the outside surface of the vessel wall. No chemical reactions were included in that assessment.

Similarly, the IVMR case was established for AP-1000, for which it was necessary to increase the value of the CHF for cooling the vessel from outside. This was made possible by employing baffling to increase the velocity of the two-phase flow on the vessel outside surface. This provided the needed margin for CHF above the heat flux incident on the vessel wall from the melt pool.

We believe that the case of IVMR as a SAM strategy for the high- and the very high-power reactors can be secured only through further analytical and experimental efforts on the effects of chemical reactions between the different components of the melt. In this context, the results of the MASCA experiments [43] would have to be considered. In general, it may be necessary to describe the coupled thermal-hydraulic and chemical scenarios that may be occurring in the melt pool.

Validation of the methodology may need further experimental efforts with prototypic materials. The experiments have to consider the scale effects in their design.

6.4. EX-VESSEL CORIUM RETENTION CONCEPT

(C.Journeau and M.Fischer)

6.4.1. Introduction

Two PWR designs currently rely on ex-vessel retention for the management of hypothetical severe accidents: the EPR™ and the VVER 1000-91/99. In these designs, it is considered that in-vessel retention cannot be proven for large power reactors in all severe accident scenarios. As a result, dedicated core catchers have been designed that can gather the corium and cool it safely.

Ex-vessel retention is also considered in a flooded vessel cavity for Nordic BWRs [44] [45]. In these reactors, it is expected that after a vessel melt-through, the corium will be fragmented in the flooded cavity into a coolable debris bed.

This section concentrates on the core-catcher concepts for the EPR™ and VVER1000.

6.4.2. EPR™

The European Pressurized Water Reactor (EPR™) is a 4-loop reactor with a power level of about 1600 MW_{el}, developed by Framatome France and Siemens (Germany), which recently merged into AREVA-NP. Being an evolutionary type of reactor, the EPR™ relies on the proven designs and technologies implemented in its predecessor plants, the French N4 and the German Konvoi PWR. The EPR™ has inherited its general safety philosophy from these two designs.

The EPR™ introduces improvements on all levels of the defense-in-depth approach to further decrease the probability of the occurrence of a severe accident (SA) with core melting. Beyond this, it involves measures at the design stage that help to drastically limit the consequences of a SA in the extremely unlikely case that such an event should occur. The target is to preserve containment integrity in the short and long term and to avoid the need for evacuating the surrounding population, except in the immediate vicinity of the NPP site and avoid the complementary major restrictions with regard to the consumption of locally grown food [46].

Implementation of the corresponding design measures is complemented by a probabilistic risk assessment (PRA) to justify the coherency of the choice of measures and to demonstrate that the achieved level of the residual risk is sufficiently low.

According to the classification in WASH-1400 (USNRC, 1975), the following challenges and related containment failure modes must be addressed in a severe accident:

- α-mode: impact loads from missiles generated by energetic events, for example, an in-vessel steam explosion,

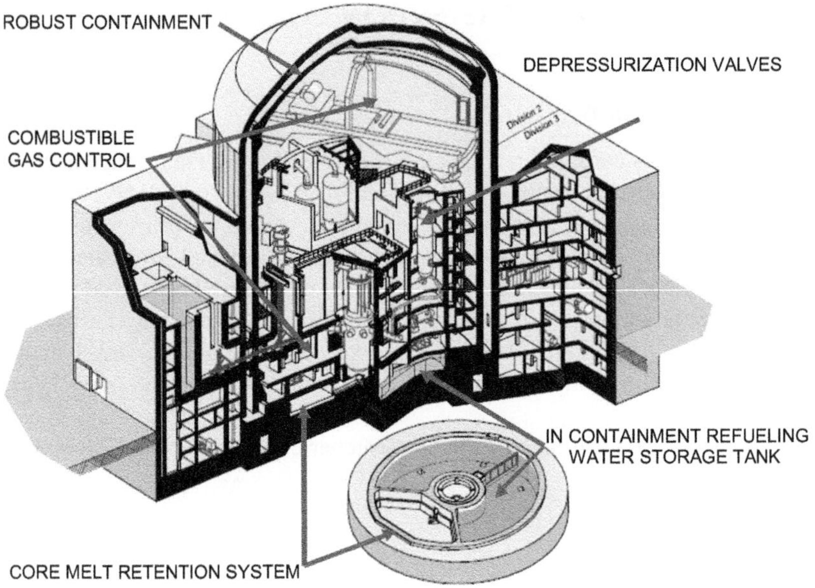

FIGURE 6.20 EPR™ features related to the mitigation of a severe accident.

- β-mode: containment bypass via tubes and penetrations,
- γ-mode: pressure loads caused by the combustion of burnable gas released into the containment atmosphere,
- δ-mode: overpressure failure following continued gas and steam release into the containment atmosphere,
- ε-mode: basemat ablation by molten corium–concrete interaction.

With the exception of containment bypass, which is addressed through preventive features and demonstrated in the Level-2 PRA, all these containment failure modes are avoided and mitigated in the EPR™ by the dedicated design provisions, as shown in Figure 6.20.

High-pressure Core-melt Sequences

To prevent steam generator tube rupture with potential containment bypass, as well as RPV failure at high primary pressure with potential missile generation and direct containment heating (DCH), the EPR™ design includes two trains of primary depressurization system (PDS) valves (see Figure 6.21). As a supplement to the existing feed-and-bleed (F&B) valves, they provide an additional total discharge capacity of 900 Mg/h at reactor coolant system design pressure.

The opening of either one F&B-valve or PDS-valve is sufficient to reduce the pressure in the primary circuit down to a few bars at RPV melt-through.

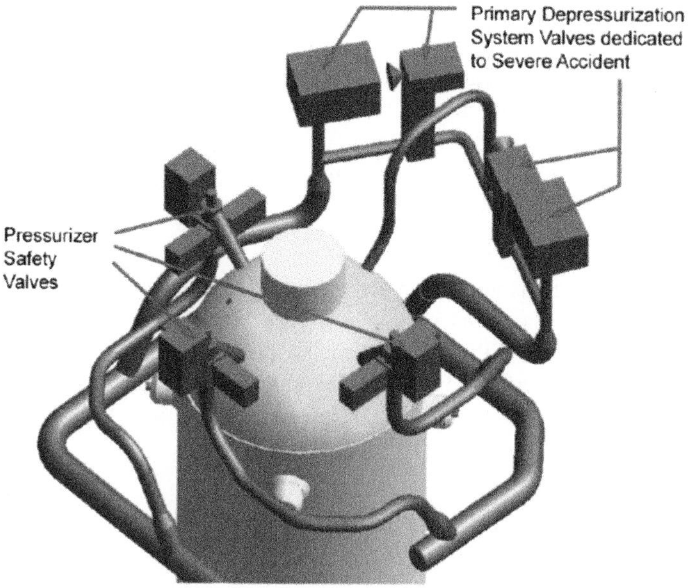

FIGURE 6.21 Two trains (each with two valves in series) of dedicated primary depressurization system (PDS) valves on top of the pressurizer head.

Even in case of a postulated late reflood of the core, at the most unfavorable moment, the one open valve will keep the primary pressure below 20 bar at vessel failure.

The dedicated valves will be activated manually, at the latest, when the core outlet temperature exceeds the predefined value of 650°C. In this way, a significant heat-up of the SG tubes is avoided.

The I&C[9] that controls the two PDS valves in series for each train is associated with different divisions to prevent uncontrolled opening as a consequence of internal hazard or equipment failure in one division. To achieve redundancy, the two lines are also activated from different divisions. The power supply for the valves is backed up by SBO[10] diesels and by dedicated long duration (12 h or more) severe accident batteries that address a total loss of on-site electrical systems and off-site electrical supply, which could result in the core-melt accident.

The valves feed discharge lines that lead from the pressurizer head to a common quench tank in the lower equipment rooms. This tank provides sufficient water inventory to condense the steam released through safety valves in case of RCS overpressurization events.

9. Instrumentation and Control

10. Station Black-Out

The opening of a PDS valve will result in the opening of the quench tank's rupture disk, after which the steam is released through connected pipes into the gas space of the lower steam generator compartments. For the EPR™, such direct release into the containment atmosphere was preferred to a discharge into the internal refueling water storage tank (IRWST) because of the advantageous effect of the steam for the mixing and dilution of hydrogen.

Hydrogen Deflagration/Detonation

In addition to the steam inertization and the mixing achieved by discharging the RCS inventory into the lower equipments room, the combustible gas control system (CGCS) of the EPR™ relies on the large free containment volume of about 80,000 m^3, good overall convective mixing achieved by "open" compartments, and ultimately on a large number of passive autocatalytic recombiners (PARs) (see Figure 6.22).

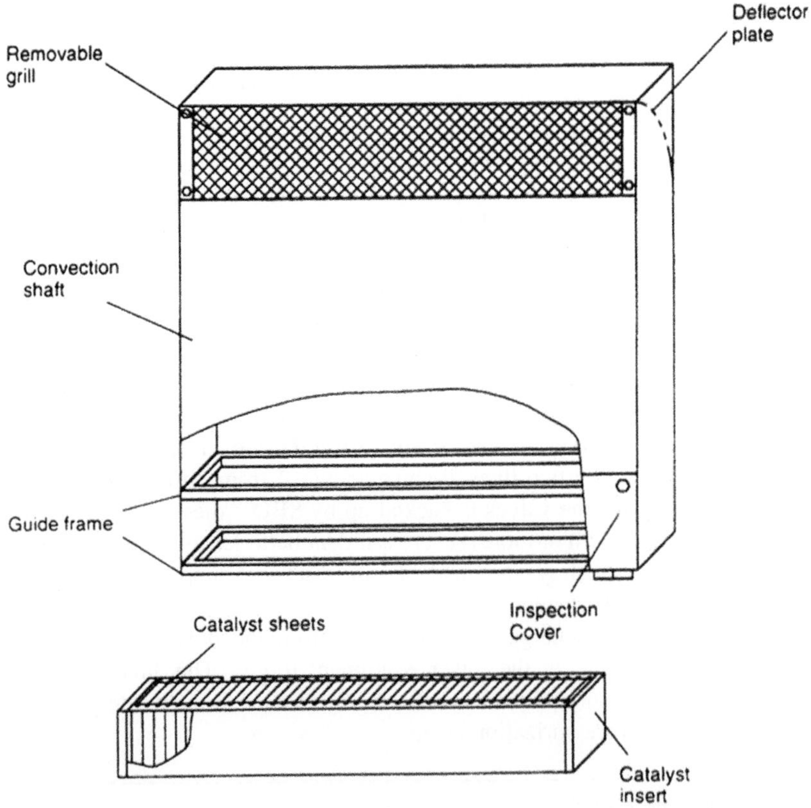

FIGURE 6.22 Large passive autocatalytic recombiner (PAR) used in the EPR™.

PARs were preferred to igniters because they also operate in steam-inerted atmosphere and for H_2 concentrations below 4 vol%. Their use is also not sensitive to an activation of sprays at the moment of high hydrogen and steam concentration nor to the availability of electric power. In the EPR™ containment, the PARs are positioned in a way that they enhance the overall convection within the containment and thus promote homogeneous atmospheric conditions and avoid high local concentrations.

The transfer of the two-room containment, with inner equipment rooms surrounded by accessible operational compartments, into a well-mixed one-room containment, is achieved by rupture- and convection-foils atop the steam generator compartments, in combination with mixing dampers that open large overflow cross sections between equipment and operational rooms above the IRWST passively at certain pressure and/or temperature differences, when a limiting absolute pressure is exceeded, or the electric power supply is lost. Alternatively the mixing dampers can be opened actively by the operator.

All components of the CGCS survive severe accident-relevant ambient conditions. The recombiners are further resistant against the gas and aerosol species released into the containment atmosphere during core melt-down.

For the EPR™ it is demonstrated that, by these measures, the maximum average H_2 concentration in the containment will be kept below 10 vol%, in any case and at any moment, based on the assumption of 100% clad oxidation, and considering realistic rates of in-vessel and ex-vessel burnable gas production. In addition, the de-inerting effect of the containment spray is taken into account. The compliance with the H_2 reduction targets is checked through analysis for bounding scenarios under conservative assumptions.

The justification is based on a staged approach. First, calculations with integral codes are performed for a large number of different accident scenarios, giving their specific hydrogen release histories. Based on these calculations, a subset of scenarios is chosen that could lead to rapid hydrogen deflagration. Criteria are, for example, high released hydrogen mass, high release rate, or a low steam content of the atmosphere.

The next step is the analysis of this subset with a lumped parameter code: to analyze and demonstrate compliance with global goals (e.g., global hydrogen concentration) and long-term behavior (temperature development, hydrogen depletion by recombiners, etc.) and to identify scenarios and periods of time, for which further analysis with CFD codes is necessary (e.g., due to large atmospheric inhomogeneities or high "local" hydrogen concentrations). These periods are then investigated with a CFD code. Results are, for example, local gas and temperature distributions as well as a measure for the risk of flame acceleration, the so-called sigma-criterion. Flame acceleration is a prerequisite for a deflagration-to-detonation transition, with possibly high mechanical loads on the containment and its internals.

For situations, where flame acceleration cannot be excluded, a specialized CFD code is used to calculate the combustion of the hydrogen present in the atmosphere. The ignition is assumed to occur at the most unfavorable moment and location. This calculation gives the histories of maximum flame speed and absolute pressure and pressure differences due to the combustion. Finally, if these pressures are not benign, an analysis of the structural response would have to be performed to demonstrate that the reactor building withstands these loads.

Steam Explosion

With respect to in-vessel steam explosion, the EPR™ involves no specific design measures. Instead, the occurrence of events that could lead to early containment failure is practically eliminated, based on the following procedure. First, the selected relevant scenarios, boundary conditions, and premixing masses of corium and water are evaluated. Based on these evaluations, the maximum fraction that can be converted into mechanical energy by a steam explosion is assessed. Finally, the final kinetic energy of the moving slug and the resulting behavior of the vessel head under impact are assessed and judged. For the EPR™, it was found that all obtained loads remain far below the failure limit of the RPV and its upper head.

As the EPR™ uses an ex-vessel melt-retention concept, steam explosions could potentially also occur during and after the release of the melt from the RPV. The interaction of melt and water during these events is avoided by ensuring a dry pit and spreading area.

For the pit, this is achieved by the exclusion/elimination of RPV nozzle failure, a leak-tight seal around the upper RPV, specific designs of the ex-core instrumentation, and ventilation tubes, as well as by the avoidance of water spillage into the pit from outside pipe breaks and the exclusion of condensation after internal hazards.

The spreading area is completely decoupled from any possible water inflow and is connected with the containment only by a vertical steam exhaust channel, the entrance of which is shielded against water ingress after pipe breaks. Only limited condensation of steam is possible. The final flooding of the melt will be performed in a controlled way and so that critical conditions are avoided.

Basemat Protection

Due to the high projected power rating and the resulting narrow margins, the concept of in-vessel melt retention by outside vessel cooling is not applied for the EPR™. Instead, an ex-vessel strategy is followed for avoiding the interaction between the molten core and the structural concrete with the ultimate risk of basemat penetration.

Instead, an ex-vessel strategy is used to avoid the attack of the molten core on the structural concrete and the penetration of basemat and embedded liner. The key element of the EPR™ core-melt stabilization system (CMSS) is the spreading of the melt on a large outside-cooled metallic structure—the core-catcher, which is located lateral to the reactor pit. The increase in surface-to-volume ratio related with the spreading process allows an effective quenching and long-term cooling of the melt. One-step spreading into the core-catcher is promoted by a preceding phase of temporary melt retention in the pit, which is aimed at collecting and conditioning the core melt before spreading [47].

Melt accumulation is achieved through a sacrificial concrete layer in the lower pit, backed-up by a cylindrical shielding consisting of refractory material. The length of the retention period is determined by the time needed to ablate the sacrificial layer and to thermally destroy an embedded melt plug at the pit bottom, which opens the way to the melt-discharge channel and the core-catcher (see Figure 6.23).

The slow destruction of the sacrificial concrete layer in the pit results in an effective accumulation of the melt, while the admixture of concrete decomposition products narrow the spectrum of melt states, as the melts properties become dominated by the rather well-defined concrete fraction. This fraction reduces the mixture solidus temperature and keeps the mixture fluid much longer so that it can spread easily. This makes the spreading process and all subsequent measures independent of the inherent uncertainties associated with in-vessel melt pool formation and RPV failure [48] [49].

The separation of functions between pit (accumulation) and spreading compartment (cooling) allows a robust design of the pit, which during RPV failure is exposed to potentially high thermal and mechanical loads, including melt jets and the mechanical impact of the detached lower head. The latter is absorbed by dedicated concrete structures, situated around the center of the pit.

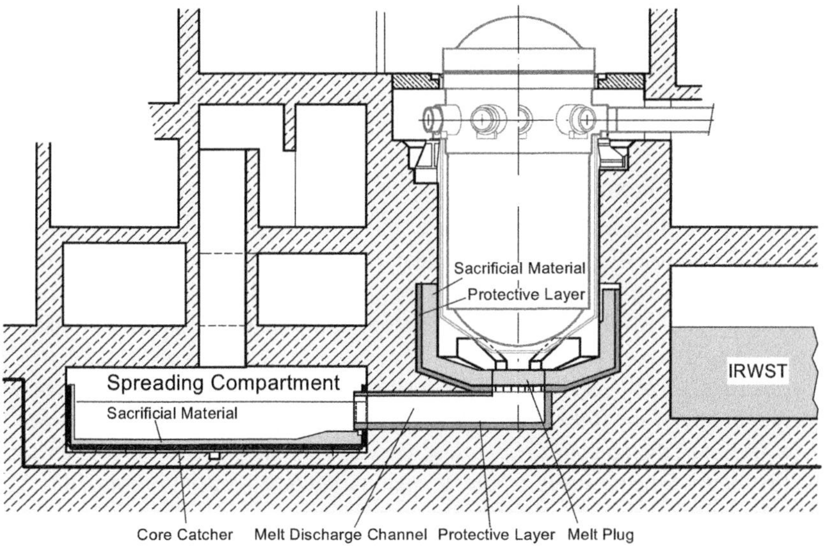

FIGURE 6.23 Components of the EPR™ core-melt stabilization system.

The core catcher, into which the melt finally relocates and which provides the long-term cooling function, is well protected against these loads by its remote and isolated location.

The available experiments and computer simulations [50] show that, for an expected accumulated melt mass of ~400 Mg (including metal and sacrificial concrete) spreading will be complete after less than 5 min and result in a complete coverage of the core catcher area. As the spreading compartment is a dead-end room, isolated from the rest of the containment, there is no prior inflow of water from sprays, leaks, or other kinds of spillage. As a consequence, the melt spreads under dry conditions.

The arrival of the melt in the core catcher triggers the opening of passive, spring-loaded valves that initiate the gravity-driven overflow of water from the IRWST. The water first fills the space below and around the core catcher and finally pours onto the surface of the melt from parts of the circumference. For the chosen initial flooding rate of about 100 kg/s, the fill-up process will take more than 5 min. Water overflow continues until the hydrostatic pressure levels between spreading room and IRWST are balanced.

The bottom and sides of the core catcher consist of flexible connected cast-iron elements with integrated open cooling channels in which the heat from the melt is transfered to the coolant. Experiments in a full-scale 5 m long channel have demonstrated that the design is capable of transferring much more than the maximum possible downward heat flux to the water [49]. This excellent performance is based on the fact that the heat can enter the water through both the horizontal and vertical surfaces of the cooling channels. Thanks to this capability, the formation of an insulating steam layer in the upper part of the channel does not result in a local dry-out.

As the steady availability of coolant is guaranteed through the open connection with the IRWST, permanent passive cooling of the melt is ensured. No pumps or external water supplies are needed.

Containment Heat Removal

The transport of the residual heat out of the EPR™ containment is performed by a dedicated active containment heat-removal system (CHRS). It draws water from the IRWST, feeds it though a closed-circuit external heat exchanger and re-injects it into the containment, thus avoiding outside dispersal of activity.

The EPR™ CHRS has two fully independent trains. Both can independently operate in one of the following modes. In the first mode, the recirculated water is sprayed into the containment atmosphere. This is the preferred mode during the first days of the accident, when the main target is to reduce the containment pressure and to wash out airborne fission products.

In the long-term, the CHRS can be used to feed water into spreading compartment (Figure 6.24). In this mode, the water in the cooling channels and

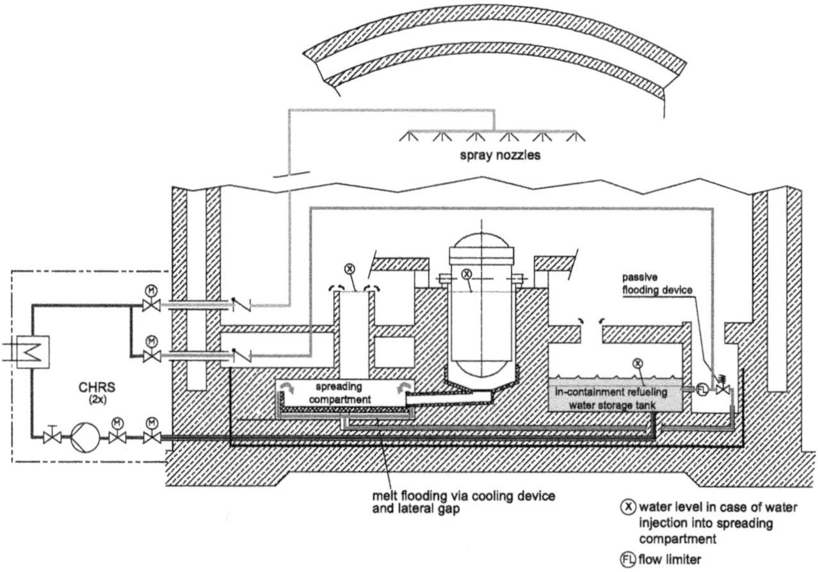

FIGURE 6.24 Scheme of the EPR™ severe accident containment heat-removal system.

atop the spread melt becomes sub-cooled. Decay heat is then removed from the melt by single-phase flow, instead by steam release into the containment atmosphere. This way ambient containment pressure can be reached without activation of the containment venting system.

6.4.3. VVER-1000

Two VVER 1000-91/99 units have been constructed in Tian Wan (China) and planned for construction at KudanKulam in India [54]. The TianWan reactors have the first ex-vessel core catcher ever built (Figure 6.25).

The main practical targets during severe accidents management at VVER plants are the following:

- Core-melt localization (in-vessel localization for VVER of medium capacity or ex-vessel localization for VVER of high capacity);
- Hydrogen safety;
- Long-term containment failure prevention;
- Secondary containment tightening;

Technically, the main design differences of VVER-91/99 from the reference VVER1000 are the lifetime of the main equipment prolonged to 40 years, the extra safety train with 100% capacity, the double concrete containment with spray system, and a set of severe accident management engineered features, including passive hydrogen recombiners and a core catcher.

FIGURE 6.25 Photograph of the Tian Wan 1 core catcher before insertion of the sacrificial material (courtesy of SpBAEP).

The core catcher is intended for reception, localization, and cool-down of the core melt during severe accidents connected with core degradation and reactor pressure vessel failure in order to reduce the radiation consequences of such accidents down to the safe level.

The following factors were also taken into account while choosing the concept [51] for the VVER-91/99 project:

- Experience from the R&D for the in-vessel core-melt retention concept for VVER-640 design;
- Availability of a free volume in under-reactor concrete cavity in the VVER-91/99 design, even if it is smaller than the EPR™ spreading area;
- Availability of large water inventory inside the containment.

Taking into account these factors, a crucible core-catcher concept [52] [53] was designed that combines elements of in-vessel core-melt retention (passive water cooling of metal boundaries of the core-melt localization area) and elements of conditioning the corium physico-chemical properties (thanks to the application of sacrificial material).

Figure 6.26 shows a large vessel that will catch the core melt in case of vessel failure. This vessel is filled with a sacrificial material that will totally oxidize the metallic melt and prevent any focusing effect. This core catcher will be externally cooled by water flow.

During a postulated severe accident, the molten corium relocates to the core catcher through the central hole in the lower plate and interacts with the sacrificial

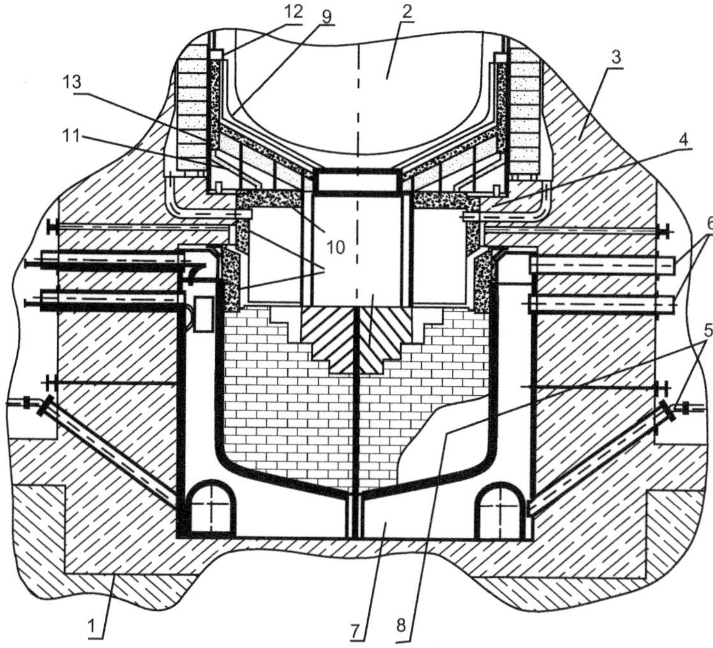

FIGURE 6.26 Scheme of the VVER 1000 core catcher—1: Containment, 2: Reactor, 3: Concrete cavity, 4: Cantilever, 5: Coolant supply, 6: Coolant outlet, 7: Ring heat exchanger, 8: Core catcher, 9: Protection, 10: Heat insulator, 11: Air-cooling channel, 12: Heat insulator, 13: Lower plate.

material located in the basket. Heat transfer from the melt is carried out through a heat exchange wall to cooling water circulating inside a heat exchanger.

The water is fed to the heat exchanger through supply channels. Generated steam is removed through steam discharge channels to the containment atmosphere, and the excess water overflows through the channels to the containment sump. Heat-insulating panels are used for transient protection of building constructions and support elements of the core catcher against thermal radiation from the corium. To maintain long-term protection, steam water cooling is used; and at a specified time after melt relocation to the core catcher, water is supplied onto the corium surface.

The water is supplied in a section heat exchanger and onto the melt surface from the reactor internals inspection shaft and fuel pool with the help of specially designed system (Figure 6.27). The water is supplied to the core catcher after the operator opens the electrical-driven gate valves. Water inventory contained in the reactor internals inspection shaft and fuel pool is sufficient to ensure melt cooling during 24 hours after the beginning of core degradation.

The key feature of the VVER-91/99 core catcher is the sacrificial material: an equimolar mixture of Fe_2O_3, Al_2O_3 (plus 1% gadolinium oxide) [54] and

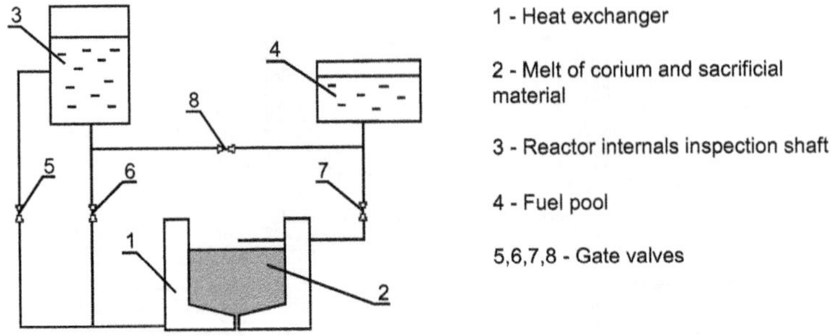

FIGURE 6.27 System of emergency use of water from a reactor internals inspection shaft and fuel pool.

sacrificial steel. Oxide sacrificial material is made as ceramic triangular bricks assembled in hexagonal steel cartridges. The cartridges are located in the core-catcher basket (see Figure 6.28).

The sacrificial material provides the following functions:

- Reduction of core-melt temperature due to integral endothermic effect during interaction of sacrificial material with the corium melt;
- Increase of heat transfer surface between corium and cooling water in the heat exchanger, reduction of heat flux at its walls, and increase of the CHF margin;
- Inversion of the oxidic and steel layers in the molten pool;
- Decrease of the corium chemical activity due to oxidation of its components by sacrificial material;

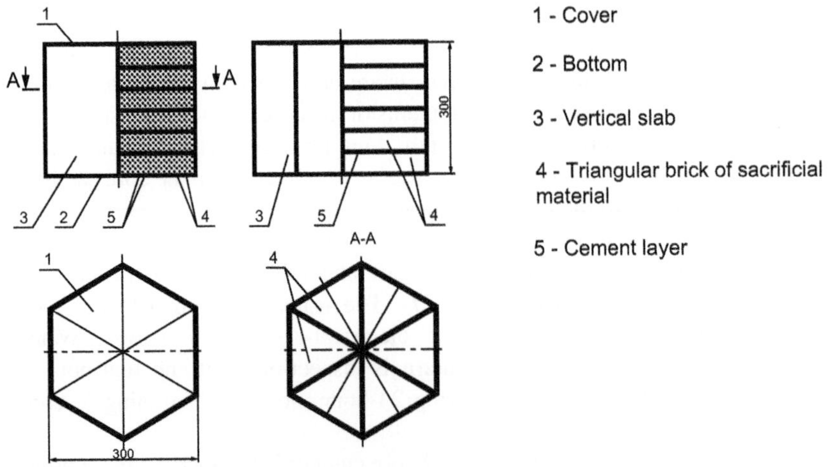

FIGURE 6.28 Sacrificial material cartridge.

- Minimization of hydrogen generation after oxidation of metal zirconium contained in the melt during initial stage of corium interaction with ceramic sacrificial material;
- Guarantee of core-melt subcriticality.

After inversion, water can be supplied onto the oxidic melt surface without risk of steam explosions and hydrogen generation from steam–metal reactions.

The severe accident management with the help of a core catcher is carried out according to the following simplified process:

- Passive steady heat transfer from the melt to the cooling water; water boils in the section heat exchanger, steam is discharged via the steam channels to the containment volume.
- Recirculation of sump water along with its cleaning and cooling is provided after power supply recovery. Also, water inventory in the reactor internals inspection shaft and fuel pool is replenished.
- At the melt cooling and subsequent crystallization stage, the steady decrease of corium temperature is achieved. According to the estimations carried out, complete bulk corium crystallization occurs approximately after one year from the moment of relocation to the core catcher.

It can be emphasized that the core-catcher operation is based mainly on passive principles. The only active devices are electro-driven valves installed at the pipelines connecting the reactor internals inspection shaft and the fuel pool with the core catcher.

The importance of the core catcher in the general safety concept for VVER-91/99 and the novelty of many design solutions have led to raised requirements for justification of the design solutions [56] [57] [58] [59]. All activities for justification can be generally divided into the following basic directions:

- Experimental investigation of local processes;
- Calculation model of core-catcher operation;
- Calculation investigation for justification of separate core-catcher elements;
- Development and investigation of new materials.

6.4.4. Conclusions on Ex-vessel Retention

In order to guarantee reliable ex-vessel retention, it is necessary to provide cooling of the corium from below and from above. Several designs have been proposed. They have in common the use of some sacrificial material that ensures a delay between vessel melt-through and the start of water cooling.

The VVER-1000 core catchers in TianWan (P.R. China) were the first ones installed under operating reactors. The EPR™ reactor is under construction in Olkiluoto (Finland) and will soon be a second operating example of ex-vessel retention.

ACKNOWLEDGMENTS

The description of the VVER core catchers was largely inspired by EUROCOURSE-2003 "CORIUM" lecture notes authored by S.V. Svetlov, V.V. Bezlepkin, Yu.G. Leontiev (AtomEnergoProekt, St. Petersburg).

List of Acronyms

A/M	Accident Management	
ATWS	Anticipated Transient Without Scram	
B	Bypassed	containment damage state
B&WOG	Babcock & Wilcox Owners Group	
BD	Badly Damaged	core damage state, core relocated to the bottom of the RPV
BDBA	Beyond Design Basis Accident	
BWROG	Boiling Water Reactor Owners Group	
CC	(Containment) Closed and Cooled	containment damage state (containment intact)
CDF	Core Damage Frequency	
CEOG	Combustion Engineering Owners Group	
CFD	Computational Fluid Dynamics	
CGCS	Containment Gas Control System	
CH	(Containment) Challenged	containment damage state
CHF	Critical Heat Flux	
CHLA	Candidate High-Level Action	
CHRS	Containment Heat Removal System	
CMSS	Core Melt Stabilization System	
CPI	Containment Performance Improvements (program)	

List of Acronyms—cont'd

DFC	Diagnostic Flow Chart	WOG SAMG logic diagram
EDMG	Extensive Damage Mitigation Guideline	
EOP	Emergency Operating Procedure	
EP	Emergency Plan	
EPG	Emergency Procedure Guideline	used in BWROG, equals EOP
EPR™	European Pressurized Water Reactor	
EPRI	Electric Power Research Institute	
ERMSAR	European Review Meeting on Severe Accident Research	www.sar-net.eu
ERO	Emergency Response Organisation	
ESBWR	Economic Simplified Boiling Water Reactor	
EX	Ex-Vessel	core damage state
FP	Fission Product	
FPL	Florida Power and Light	U.S. utility
GRS	Gesellschaft für Reaktorsicherheit (Reactor Safety Company)	main German Technical Support Organization
I	Impaired	containment damage state
I&C	Instrumentation and Control	
IPE	Individual Plant Examination	simplified PSA
IPEEE	IPE for External Events	
IRWST	In Reactor Water Storage tank	
ISAMM	Implementation of Severe Accident Management Measures	OECD workshops
IVMR	In-Vessel Melt Retention	core/debris cooling by external flooding of the RPV lower head

(*Continued*)

List of Acronyms—cont'd

LOFW	Loss Of Feed Water	
KIT	Karlsruhe institute of Technology, formerly FZK	
NEI	Nuclear Energy Institute	represents U.S. NPPs to the NRC
NPP	Nuclear Power Plant	
OSSA	Operational Strategies for Severe Accidents	SAMG for EPR™ (and other)
OX	Oxidised Fuel	core damage state
PAR	Passive Autocatalytic Recombiner	
PSAMG	Plant-Specific SAMG	
PWROG	Pressurized Water Reactor Owners Group	
SAM	Severe Accident Management	
SAMG	Severe Accident Management Guidance *or* Guideline(s)	
SCG	Severe Challenge Guideline	
SCST	Severe Challenge Status Tree	WOG SAMG logic diagram
SGTR	Steam Generator Tube Rupture	
SSAMG	Shutdown SAMG	SAMG for shutdown stages
TBR	Technical Basis Report	EPRI manual for SAMG development
TSC	Technical Support Centre	or Technical Support Group
TSG	Technical Support Guideline	guideline used by TSC to support execution of the SAMG
VVER	Water-Water Energetic Reactor	Russian designed PWR with horizontal steam generators
WOG	Westinghouse Owners Group	

REFERENCES

[1] Integration Plan for Closure of Severe Accident Issues, USNRC, SECY 88—147, May 1988.
[2] Individual Plant Examination for Severe Accident Vulnerabilities, USNRC, Generic Letter 88–20, 1988.
[3] Severe Accident Issue Closure Guidelines, Nuclear Energy Institute, NEI 91-04, Revision 1, December 1994.
[4] Safety of Nuclear Power Plants: Design, Requirements, IAEA, NS-R-1, Vienna, 2000.
[5] R.E. Henry et al., Severe Accident Management Guidance Technical Basis Report (TBR); vol. I: Candidate High-Level Actions and Their Effects; vol. II: The Physics of Accident Progression, EPRI-TR 101869, vols.1 and 2, December 1992.
[6] B. Chexal, et al., Update on the Technical Basis for the Severe Accident Management Guidelines, Electric Power Research Institute, OECD Specialist Meeting on Severe Accident Management Implementation, Niantic, CT, USA, 12–14, June 1995.
[7] B. Lutz, M. Lucci, B. Henry, K. Kiper, presented by G. Vayssier, Seabrook Station Level 2 PRA Update to Include Accident Management, ANS Summer Meeting / ICAPP '06, June 8, 2006.
[8] Various studies, a.o. by GRS Germany, Paks Hungary, Goesgen Switzerland, OECD/NEA Workshop on Implementation of Severe Accident Management Measures (ISAMM), Böttstein, Switzerland, 2009. http://sacre.web.psi.ch/ISAMM2009/isamm09-workshop.html
[9] BWR Owners' Group Accident Management Guidelines Overview Document, Rev. 1, BWROG Accident Management Working Group, May 30, 1996.
[10] Severe Accident Management Programmes for Nuclear Power Plants, IAEA Safety Guide, NS-G-2.15, 2009.
[11] SARNET ERMSAR meetings in Nesseber, Bulgaria, 2008 and Bologna, Italy 2010. www.sar-net.eu.
[12] E. Burns, J. Gabor, Technical Support Guidelines, Duane Arnold Energy Center, ERIN Engineering, December 1997.
[13] G. Vayssier, Severe Accident Management Implementation and Expertise (SAMIME), Final Report, Contract FI4S-CT98-0052, G.L.C.M. Vayssier, Nuclear Safety Consultancy (NSC), December 2000.
[14] Bwrog SAMG pilot demonstration at Duane Arnold Energy Center, guided by R.M. Harter et al., IES Utilities, February 1998.
[15] Ken L. Ross, Training course on BWROG EPG/SAG rev. 2, KLR Services, Philadephia, PA, 2002.
[16] Letter of GE Hitachi to USNRC re ESBWR design certification, MN 09-074, January 31, 2009.
[17] R.A. Hill, Personal communication, ERIN Engineering and Research, Campbell, CA, USA, May 2011.
[18] R.J. Lutz Jr, Personal communication, Westinghouse, Pittsburgh, PA, USA, May 2011.
[19] J.M. Kim, et al., Experimental study on Inherent In-vessel cooling Mechanism during a Severe Accident, KAERI, Taejon, Korea, 1999.
[20] R.J. Lutz Jr, et al., Seabrook Station Level 2 PRA Update to include Accident Management, ANS Summer Meeting, ICAPP '06, June 2006.
[21] J.H. Song, et al., Improved Molten Core Cooling Strategy in a Severe Accident Management Guideline, ISAMM 2009, Böttstein, Switzerland, October 2009.
[22] P. Piluso, et al., OECD SERENA: A Fuel Coolant Interaction Programme (FCI) devoted to reactor case, ISAMM 2009, Böttstein, Switzerland, October 2009.
[23] USNRC Policy Note, SECY 93-087, 1993.

[24] Joint OECD/SARNET workshop on In-vessel Coolability, NEA Headquarters, Paris, France, October 2009.

[25] I.M. Trasdorf, Research and Development of Catalytic Elements for Innovative Hydrogen Recombiners (in German), Jülich Nuclear Research Centre (KFA Jülich), Energy Technology, 2005. Report No. 36.

[26] M.P. Sherman, M. Berman, The Possibility of Local Detonation During Degraded Core Accidents in the Bellefonte Nuclear Plant, NUREG/CR-4803, SAND86-1180, Sandia National Laboratories, 1987.

[27] O. Kymäläinen, H. Tuomisto, T.G. Theofanous, In-Vessel Retention of Corium at the Loviisa Plant, Journal of Nuclear Engineering and Design 169 (1997) 109–130.

[28] H. Tuomisto, O. Kymäläinen, Confirming the In-Vessel Retention for the Loviisa Plant, OECD/NEA RASPLAV Seminar, Munich, Germany, November 14–15, 2000.

[29] T.G. Theofanous, C. Liu, S. Additon, S. Angelini, O. Kymäläinen, T. Salmassi, In-Vessel Coolability and Retention of a Core Melt, DOE/ID-10460, vols. 1 and 2, October 1996.

[30] M. Saito, K. Sato, S. Imahori, Experimental Study on Penetration Behaviors of Water Jet into Freon-11 and Liquid Nitrogen, Proceedings of ANS National Heat Transfer Conference, Houston, TX, July 24–27, 1988, 173–183.

[31] T.N. Dinh, W.G. Dong, J.A. Green, R.R. Nourgaliev, B.R. Sehgal, Melt Jet Attack of Reactor Vessel Lower Plenum: Phenomena and Prediction Method, Proceedings of the Eighth International Topical Meeting on "Nuclear Reactor Thermal Hydraulics - NURETH-8, Kyoto, Japan, 2 September 1997. 612–619.

[32] T.G. Theofanous, The Study of Steam Explosions in Nuclear Systems, Journal of Nuclear Engineering and Design 155 (1995) 1–26.

[33] T.G. Theofanous, W.W. Yuen, S. Angelini, J.J. Sienicki, K. Freeman, X. Chen, T. Salmassi, Lower Head Integrity under Steam Explosion Loads, Journal of Nuclear Engineering and Design 189 (1999) 7–57.

[34] T.G. Theofanous, T. Dinh, In-Vessel Coolability and Melt Retention, CRSS-02/01, Center for Reactor Safety Research, Santa-Barbara, CA, January 2002.

[35] U. Steinberner, H.H. Reineke, Turbulent Buoyancy Convection Heat Transfer with Internal Heat Sources, Proceedings of the 6th Int. Heat Transfer Conference 2 (1978) 305–310. Toronto, Canada.

[36] F.A. Kulacki, R.J. Goldstein, Thermal Convection in a Horizontal Fluid Layer with Uniform Volumetric Energy Sources, J. Fluid Mech. 55 (part 2) (1972) 271–287.

[37] O. Kymäläinen, H. Tuomisto, O. Hongisto, T.G. Theofanous, Heat Flux Distribution from a Volumetrically Heated Pool with High Rayleigh Number, J. Nuclear Engineering and Design 149 (1994) 401–408.

[38] T.G. Theofanous, M. Maguire, S. Angelini, T. Salmassi, The First Results from the ACOPO Experiments, Journal of Nuclear Engineering and Design 169 (1997) 49–57.

[39] G. Kolb, S.A. Theeerthan, B.R. Sehgal, Natural Convection in Stable Stratified Layers with Volumetric Heat Generation in the Lower Layer, Proceeding of the 34th National Heat Transfer Conference, Pittsburgh, USA, 2000.

[40] V. Asmolov, RASPLAV Project: Major Activities and Results, OECD/NRA RASPLAV Seminar, Munich, Germany, November 14–15, 2000.

[41] S. Rougé, SULTAN Test Facility for Large-Scale Vessel Coolability in Natural Convection at Low Pressure, Journal of Nuclear Engineering and Design 169 (1–3) (1997) 185–195.

[42] T.N. Dinh, J.P. Tu, T. Salmassi, T.G. Theofanous, Limits of Coolability in the AP-1000-Related ULPU-2400 Configuration V Facility, The 10th International Topical Meeting on Nuclear Reactor Thermal Hydraulics (NURETH-10), Seoul, Korea, October 5-9, 2003.

[43] V. Asmolov, D. Tsurikov, MASCA Project: Major Activities and Results, OECD/NEA MASCA Seminar, Aix-en-Provence, France, June 10–11, 2004.

[44] C.C. Chu, J.J. Sienicki, B.W. Spencer, W. Frid, G. Löwenhielm, Ex-vessel melt-coolant interactions in deep water pool: studies and accident management for Swedish BWRs, Progr. Nucl. Energ. 155 (1995) 159–213.

[45] I. Lindholm, et al., Dryout heat flux experiments with deep heterogeneous particle beds, Nucl. Eng. Des. 236 (2006) 2060–2074.

[46] F. Bouteille, G. Azarian, D. Bittermann, J. Brauns, J. Eyink, The EPR™ overall approach for severe accident mitigation, Proc. of ICONE-13, Beijing, China, 2005. Paper #50018.

[47] M. Fischer, The core melt stabilization concept of the EPR™ and its experimental validation, Proc. of ICONE 14, Miami FL, USA, 2006. Paper #89088.

[48] M. Nie, M. Fischer, Use of molten core concrete interactions in the melt stabilization strategy of the EPR™, Proc. of ICAPP 2006, Reno, NV USA, 2006. Paper #6330.

[49] M. Fischer, O. Herbst, H. Schmidt, Demonstration of the heat removing capabilities of the EPR™ core catcher, Proc. of the 3rd International symposium on two-phase flow modeling and experimentation, Pisa, Italy, September 22–24, 2004.

[50] M. Nie, Temporary Melt Retention in the Reactor Pit of the European Pressurized Water Reactor (EPR™), PhD Thesis, University of Stuttgart, Germany, 2005.

[51] I.V. Kukhtevich, V.V. Bezlepkin, V.S. Granovskii, V.G. Asmolov, S.V. Bechta, A.S. Sidorov, V.M. Berkovich, V.F. Strizhov, M.C. Khua, M.F. Rogov, V.P. Novak, The concept of localization of the corium melt in the ex-vessel stage of a severe accident at a nuclear power station with VVER-1000 reactor, Therm. Eng 48 (2001) 699–706.

[52] A.A. Sidorov, A.B. Nedorezov, M.F. Rogov, V.P. Novak, I.V. Kukhtevich, V.V. Bezlepkin, V.B. Khabinskii, V.V. Granovskii, S.V. Bechta, The Device for Core Melt Localization at the Tyan'van Nuclear Power Station with a VVER-1000 Reactor, Therm. Eng. 48 (2001) 707–712.

[53] S.V. Svetlov, V.V. Bezlepkin, I.V. Kukhtevich, S.V. Bechta, V.S. Granovsky, V.B. Khabensky, V.G. Asmolov, V.B. Proklov, A.S. Sidorov, A.B. Nedoresov, V.F. Strizhov, V.V. Gusarov, Yu.P. Udalov, Core Catcher for TianWan NPP with VVER-100 reactor. Concept, Design and Justification, Proc. 11th Int. Conf. Nucl. Eng. (2003). Tokyo, Japan, Paper ICONE11-36102.

[54] S.K. Agrawal, A. Chauhan, A. Mishra, The VVERs at KudanKulam, Nucl. Sci. Des. 236 (2006) 812–835.

[55] V.V. Bezlepkin, I.V. Kukhtevich, Yu.G. Leont'ev, S.V. Svetlov, The concept of overcoming severe accidents at Nuclear Power Stations with VVER reactors, Therm. Eng. 51 (2004) 115–123.

[56] V.V. Gusarov, V.I. Al'myashev, S.V. Bechta, V.B. Khabenskii, Yu.P. Udalov, V.S. Ranovski, Sacrificial Materials for Safety Systems of Nuclear Power Stations: A New Class of Functional Materials, Thermal Engineering 48 (2001) 721–724.

[57] Yu.B. Petrov, D.B. Lopukh, S.V. Bechta, A.Yu. Pechenkov, et al., Corrosive capacity of superheated corium melt, Advanced materials(No. 3) (1996) 374–378.

[58] D. Lopukh, et al., New Experimental Results on the Interaction of Molten Corium with Core Catcher Material, Proc. of ICONE 8th International Conference on Nuclear Engineering, Baltimore, USA, 2000. ICONE-8179.

[59] V.G. Asmolov, V.N. Zagryazkin, I.F. Isaev, I.M. Semenov, V.Yu. Vishnevskii, S.V. Bechta, V.S. Granovskii, V.V. Gusarov, Yu.P. Udalov, Choice of buffer material for the containment trap for VVER-100 core melt, Atom. Energ. 92 (2001) 5–14.

[60] H. Alsmeyer, W. Tromm, Experiments for a core catcher concept based on fragmentation, Proceedings of the ARS'94—International Topical Meeting on Advanced Reactor Safety, Pittsburgh, PA, USA, 1994.

[61] H. Alsmeyer, W. Tromm, The COMET concept for cooling core melts: evaluation of the experimental studies and use in the EPR™, Report FZKA 6186 (1999).

[62] H. Alsmeyer et al., Corium cooling by bottom flooding: results of the COMET investigations, Proceedings of the OECD Workshop on Ex-vessel Debris Coolability Karlsruhe, Wissenschaftliche Berichte FZKA-6475 (2000) 345–355.

[63] W. Tromm, et al., Ex-vessel corium cooling by passive water addition through porous concrete, Proceedings of ICONE 9, Nice, France, 2001.

[64] W. Widmann, M. Bürger, G. Lohnert, H. Alsmeyer, W. Tromm, Experimental and theoretical investigations on the COMET concept for ex-vessel core melt retention, Nucl. Eng. Des. 236 (2000) 2304–2327.

[65] D. Paladino, S.A. Theertan, B.R. Seghal, DECOBI: investigation of melt coolability with bottom coolant injection, Progr. Nucl. Ener. 40 (2002) 161–206.

[66] M.T. Farmer, B.W. Spencer, D.J. Kilsdonk, R.W. Aeschlimann, Quick Look Data Report for COMET Test U2, Argonne National Laboratory Report ANL-RE/98–1, 1998.

[67] C. Journeau, H. Alsmeyer, Validation of the COMET Bottom-Flooding Core catcher with Prototypic corium, Int. Congr. Advances nuclear Power plants (ICAPP06), Reno, NV, 2006.

[68] W. Schmid, Influence of multidimensionality and interfacial friction on the coolability of fragmented corium, Dissertation, IKE 2-149, University of Stuttgart, 2004.

[69] M. Bürger, M. Buck, W. Schmidt, W. Widmann, Validation and application of the WABE code: investigations of constitutive laws and 2D effects on debris coolability, Nucl. Eng. Des. 236 (2006) 2164–2188.

[70] OSSA – "An Optimized Approach to Severe Accident Management: EPR Application", E. Sauvage, R. Prior, S. Mazurkiewicz, K. Coffey, Areva France, presented at the ANS Winter Meeting, Reno, Nevada, USA, June 2006

Chapter 7

Environmental Consequences and Management of a Severe Accident

H. Schnadt and I. Ivanov

Chapter Outline

7.1. INTRODUCTION

An accidental release of radioactive substances from a nuclear plant has a great potential to affect the health of people living in the surroundings of the plant. The radiation doses to the population depend on a large number of influencing factors: the release strength, its isotopic composition, the release characteristics (e.g. release height, thermal energy, momentum), meteorological conditions during dispersion (wind speed and direction, turbulent status of the atmosphere = stability), distance of dwellings from

Nuclear Safety in Light Water Reactors. DOI: 10.1016/B978-0-12-388446-6.00007-1

the plant, behaviour of people (dietary habits, residence times in and outside of buildings), efficiency of countermeasures to protect people, just to name a few.

There are two major aims of consequence analyses:

- On the one hand one would like to assess the risk for different end-points of a release: end-points are for instance the number of fatal health effects, the appearance of cancer and leukaemia, the size of areas in which countermeasures would probably be implemented such as sheltering, evacuation, resettlement, restrictions of the food production etc. Such questions are object of probabilistic safety assessments (PSAs) level 3.
- On the other hand one needs decision support in the case of an actual release concerning countermeasures for lowering the possible negative health effects as well as social and economic consequences. Generally, the criteria for the implementation of countermeasures are expressed in dose reference values. Therefore one needs to predict radiation doses.

7.2. BASIC PHENOMENA

The following description is intended to provide an overview of the most important phenomena that determine the pathways of radioactive substances from an accidental release to the environment and the people living there.

However, due to their complexity, the phenomena that describe the flow of radioactivity from deposition to food of man are not discussed here. A rough overview is given in [1].

If we have a look on atmospheric transport of radioactive particles and gases stemming from accidental releases of nuclear plants we should have an idea about the characteristics of the layer into which the release is injected.

The earth is the lower boundary of the atmosphere which is called the troposhere and which has a height of about 11 to 15 km. The troposphere can be divided into the so called (planetary) boundary layer with a height of about 1 to 3 km above ground and a remainder which is called the free atmosphere. The boundary layer can be defined as that part of the troposphere that is directly influenced by the presence of the earth's surface, and responds to surface forces with a timescale of one hour or less. These forces include frictional drag, evaporation, transpiration, and heat transfer. Turbulence is next to advection the most important transport process in the boundary layer. The lower part of the planetary boundary layer, where temperature decreases with altitude, is often called mixing layer and is capped by the inversion layer where temperature increases with altitude. This is especially true when solar radiation warms up the earth's surface. If solar radiation is low or absent, for instance at night, the outgoing radiation will cool the earth's surface and this will lead to temperatures which decrease with altitude.

In most cases accidental releases resulting from loss of coolant and subsequent melting of the core will take place in the first hundred meters above the ground. Releases into higher parts of the atmosphere are possible if the release has a high thermal energy and/or an upwards directed mechanical momentum. The Chernobyl accident was of this type due to a massive power excursion resulting in an explosion which destroyed the reactor core. In this special case, radioactive particles and gases were transported to heights of more than 1 kilometer. In such a case special considerations and models are needed.

7.2.1. Atmospheric Dispersion

To help understand the routes by which radioactive material released in an accident can interact with the environment, it is useful to look at the processes involved in the transport and movement of materials through the atmosphere. Radioactivity released to atmosphere will behave much as any other chemical pollutant except that it undergoes radioactive decay. The radionuclides released may decay into a stable nuclide or into another radioactive progeny. The material is blown downwind (advection), mixes with the surrounding air (turbulent diffusion), becoming more diluted. Depending on its physical and chemical characteristics it may be deposited onto surfaces thereby contaminating them.

The main factors which influence the transport of pollutants downwind are wind speed, wind direction, and atmospheric stability (or the turbulence in the atmosphere). These factors dictate the dispersion, direction, horizontal and vertical spread of the plume. The degree of spread of the plume will affect the air concentrations; a well-dispersed plume will mean lower concentrations but a much wider area will be affected. The air concentration is approximately inversely proportional to the wind speed, i.e. the greater the wind speed the lower the concentration.

Wind

The most important informations in an actual release situation are the wind direction and wind speed. They determine the transport of particles and gases within air parcels with the wind characterized by wind velocity and wind direction (advection).

In the planetary boundary layer the wind increases with height and thus a wind profile is formed in the layer. The increase of wind speed is determined by frictional forces and turbulence. Figure 7.1 shows the wind profile functions for three turbulence situations. The diagram can be used to assess the wind speed u_z in a height z relative to a wind speed u_{10} measured at a height of 10 m.

The wind profile functions also depend on the roughness length z_0 which accounts for the effect of the roughness of a surface on the wind flow. Its value is

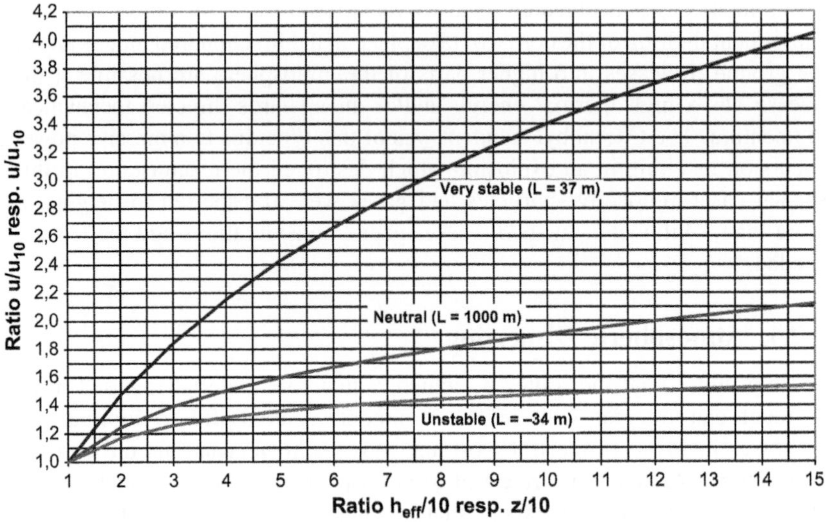

FIGURE 7.1 Wind profile functions for three turbulence situations (characterized by the Obukhov Length (cf. next section on turbulence) and a roughness length of 0.5 m).

between 1/10 and 1/30 of the average height of the roughness elements on the ground. Over smooth, open water one would expect a value of around 0.0002 m, while over flat, open grassland $z_0 \approx 0.03$ m, cropland $\approx 0.1-0.25$ m, and brush or forest $\approx 0.5-1.0$ m. Values above 1 m are rare and indicate very rough terrain.

Turbulence

Turbulent flow occurs at high Reynolds numbers and is dominated by inertial forces. It produces random eddies, vortices, and other flow fluctations. Turbulent diffusion is much more effective (5 to 6 orders of magnitude) than Fickian diffusion.

Typically, the turbulent state of the atmosphere is divided into categories. The best known scheme is that of Pasquill, which has six categories ranging from very unstable over neutral to stable [2]. A rough categorization can be done on the basis of synoptical observations according to Table 7.1 [3].

Turbulence can also be characterised by the (Monin-) Obukhov length L, which is a continuous variable. L describes the proportion between the mechanical production of the turbulent kinetic energy (TKE) and the buoyant production of the TKE. During the day L is the height at which the buoyant production of turbulence kinetic energy (TKE) is equal to that produced by the shearing action of the wind (shear production of TKE). The relationship between the Obukhov-length L and diffusion categories depends on the roughness length. In [4] a formula is presented which

TABLE 7.1 Relation of Turbulent Types to Weather Conditions [3]

	Day			Night	
	Day time insolation[1]			Outgoing radiation	
				Weak	Strong
				Cloudiness*	
Surface wind speed $m \cdot s^{-1}$	Strong	Moderate	Slight	$\geq 4/8$	$\leq 3/8$
$\overline{u} < 2$	A	A–B	B	(E–F)	(F)
$2 \leq \overline{u} < 3$	A–B	B	C	E	E
$3 \leq \overline{u} < 5$	B	B–C	C	D	E
$5 \leq \overline{u} < 6$	C	C–D	D	D	D
$\overline{u} \geq 6$	C	D	D	D	D

A – Extremely unstable conditions
B – Moderately unstable conditions
C – Slightly unstable conditions
D – Neutral conditions[a]
E – Slightly stable conditions
F – Stable or very stable turbulence conditions. In [3] category F is described as "moderately stable". However, most modern adaptions of this category designate it as described above.
[1] *Insolation = ingoing solar radiation*
**The degree of cloudiness is defined as that fraction of the sky above the local apparent horizon which is covered by clouds.*
[a] *Applicable to heavy overcast- day or night*

calculates the inverse Obukhov-length for the Pasquill diffusion categories.

Flow Around Buildings

In many cases of an accidental situation in a nuclear plant the release will take place anywhere out of the reactor building . The prediction of the flow around the building (cf. Figure 7.2) is very difficult as it depends on many factors: for instance the flow will vary with the form of the building, with wind direction, wind speed, and atmospheric stability. The most appropriate method is to investigate the flow using a model of the building in a wind tunnel. However, it is quite difficult to simulate a stable atmosphere in a wind tunnel. From a practical point of view, in most cases it is not necessary for decision makers to have detailed knowledge of this phenomon because its influence is restricted to distances which are very close to the plant.

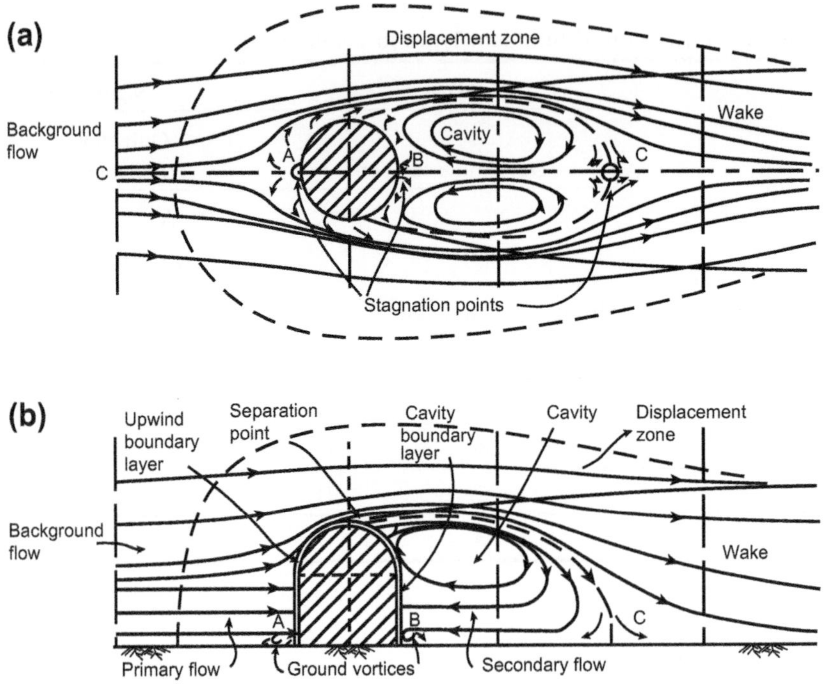

FIGURE 7.2 Flow around a rounded building. (a) Flow I a horizontal plane near the ground, (b) Flow in the longitudinal center plane *(from Halitzky, Golden, Halpern, and Wu, 1963, cited after [3])*.

Mixing Layer

The mixing layer is often limited at the upper edge by a temperature inversion (the temperature does not decrease with altitude as usual but increases). The inversion then acts as a barrier in the atmospheric exchange, and can effectivly limit the transport of atmospheric gases and particles from the near-surface layer into the free troposphere.

The height of the mixing layer depends on the turbulence state of the atmosphere (stability, wind speed) and varies with it. In the present operational model of the German Weather Service its calculation is part of the turbulence parametrisation of the weather forecast system [5].

Very simple models just use a table with mixing heights in dependence on the stability category (cf. Table 7.2).

It should be mentioned that low mixing layer heights may lead to higher concentrations if the release takes place beyond the inversion layer. Under unstable and neutral conditions real mixing layer heights can be considerably higher. The values for unstable and neutral conditions in the Table 7.2 are a rather conservative estimate. On the other side, the values for categories E and

TABLE 7.2 Mixing Heights and Diffusion Categories

Diffusion category	A	B	C	D	E	F
Mixing layer height	1600 m	1100 m	800 m	560 m	320 m	200 m

(after [6], cited in [1])

F may be meaningless as in these situations the vertical dispersion is very low. Since wind speed and turbulence often are subject to a diurnal change, also the mixing layer height changes with the same periodicity. Typical is the formation of a ground (height some meters) inversion in the early morning due to cooling of the earth by outgoing radiation. Later, when the ground warms up, the mixing layer height increases with sunrise. If the release is injected above the inversion the pollutants will not reach the ground until the inversion disappears (e.g. due to the increase of turbulence by wind and/or by insolation).

Plume Rise

If the release is characterised by a mechanical momentum and/or by thermal energy, this will lead to a plume, or alternatively, cloud rise. The gain in height ΔH (effective release height H minus stack height h_s; cf. Figure 7.3), which normally leads to lower concentrations of the pollutant at the ground, depends on the turbulence state (diffusion category). The neglection of these effects provides more conservative estimates of the

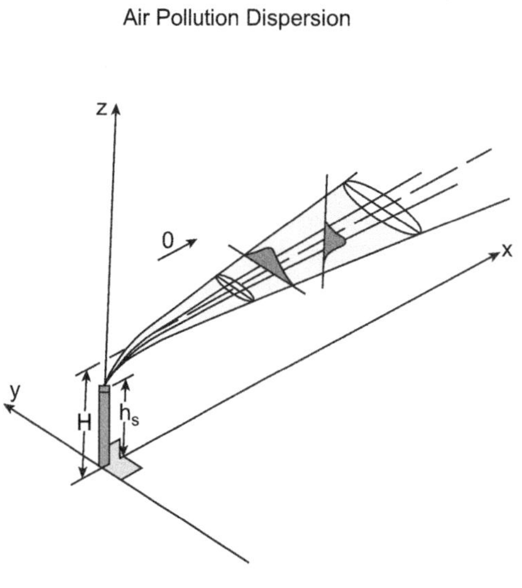

FIGURE 7.3 Spread of the plume by the air pollution dispersion.

concentration or dose. More information about the effect can be found in [7].

Deposition

All the sentences below in this sub-Section are taken from [1]. "Radioactive material, with the exception of radioisotopes of noble gases, can be deposited onto the ground, which leads to a further depletion of the cloud. The extent to which this occurs depends on the physical and chemical nature of the radioactive material released, such as whether it is released in a gaseous or particulate form. It also depends on the nature of the underlying surface, which, for example, might be concrete, tarmac, grass, soil or forest. Further it depends on the micrometeorology and a host of other factors.

There are two main routes by which material can be deposited onto the ground: dry deposition and wet deposition. Dry deposition includes the processes of adsorption, impaction, or sedimentation of particles and gases to the ground. The physico-chemical properties of the released radioactive material will strongly influence the deposition characteristics. The rate at which material is deposited onto the ground via the dry deposition pathway is known as the deposition velocity. It is the ratio between the time integrated air concentration ($Bq \cdot s \cdot m^{-3}$) and the amount of material deposited on the ground ($Bq\ m^{-2}$). The deposition velocity varies from zero for radioisotopes of noble gases up to the order of 10^{-4} to 10^{-3} $m \cdot s^{-1}$ for radionuclides release in particulate form. The deposition ratio or velocity may be as high as 10^{-2} $m \cdot s^{-1}$ for reactive vapours, such as inorganic forms of iodine.

Radionuclides may also be deposited onto the ground if it rains within or above the airborne plume. The process, known as washout, is more complex than dry deposition and may lead to concentrations of radionuclides on the ground being more variable and perhaps 10-100 times greater than what might arise from dry deposition alone. Moderate rain will often remove more in one hour than dry deposition will over two or more days. However, this excludes non-reactive gases that are not readily removed from the atmosphere. Snow and fog instead of rain can significantly alter the rates at which wet deposition occurs. The temporal and spatial variation associated with rain can often lead to localised areas of high deposition".

Depletion

Dry and wet deposition cause a reduction of the activity in the cloud. For dry deposition and wind speeds more than about 2 or 3 m s-1 the depletion is low. For distances up to about 20 to 30 km this effect can be neglected. For low wind speeds and greater distances, where actions as the blockage of the thyroid or the protection of the food chain may be of interest, the depletion will be an important issue, especially in the case of rain. The activity concentration in the cloud and the contamination of the ground will be considerably lower.

Resuspension

Radionuclides deposited on the ground may be resuspended into the atmosphere and subsequently inhaled. This exposure pathway is generally considered to be of much lower significance than the inhalation dose from the original passing cloud. Only a small proportion of the radionuclides in the initial plume is deposited at any one location. Only a small proportion is subsequently resuspended. Resuspension occurs because of natural disturbances such as wind or by human activity such as traffic, agricultural activities, and decontamination measures. Resuspension rates decrease with time [8].

Resuspension may be important for radionuclide depositions with a significant actinide component. When mostly gamma emitting radionuclides have been deposited the external irradiation will cause a larger dose contribution than resuspension.

Accidental Releases into Surface Water

Sources of radioactive releases could be circuit leaks, the spent fuel storage facility, equipment decontamination facilities, installations for iodine-exchange filter regeneration and purification, special laundries for means of protection, sanitary gates, radio-chemical laboratories, etc.

However, the accident of the Fukushima-Daiichi-plant in March 2011 showed that under accidental conditions large amounts of contaminated water, which in Fukushima was needed for cooling the damaged reactor cores and spent fuel storages, can be released into the sea or rivers. The dispersion of radioactivity in the water body and the sedimentation of the radionuclides are subject to specialised hydrological models. These are not discussed further here.

7.2.2. Models

Several types of models are available, which simulate the dispersion of pollutants in the atmosphere (e.g. radioactive gases and particulates, toxic substances, and volcano ashes). The mostly used types are (taken from [9]):

- Gaussian model — the Gaussian model is the most commonly used model type. It assumes that the air pollutant dispersion has a Gaussian distribution, meaning that the pollutant distribution has a normal probability distribution.
- Lagrangian model — the Lagrangian dispersion model mathematically follows pollution plume parcels (also called particles) as the parcels move in the atmosphere, and they model the motion of the parcels as a random walk process. The Lagrangian model then calculates the air pollution dispersion by computing the statistics of the trajectories of a large number of the pollution plume parcels.
- Eulerian model — the Eulerian dispersion model is similar to the Lagrangian model in that it also tracks the movement of a large number

of pollution plume parcels as they move from their initial location. The most important difference between both models is that the Eulerian model uses a fixed three-dimensional Cartesian grid as a frame of reference rather than a moving frame of reference. Eularian models are especially suited for the dispersion simulation of many sources which can be plane sources or volume sources. As they require very much computational power, they are seldom used in the field of nuclear accident consequence studies and therefore not further discussed here.

Gaussian Models

Gaussian Plume Model

At present, most of the assessments of a nuclear power plant (NPP) impact on the environment use the so-called Gaussian Plume Model.

The estimation of the ground-level concentration of the pollution is based on the equation:

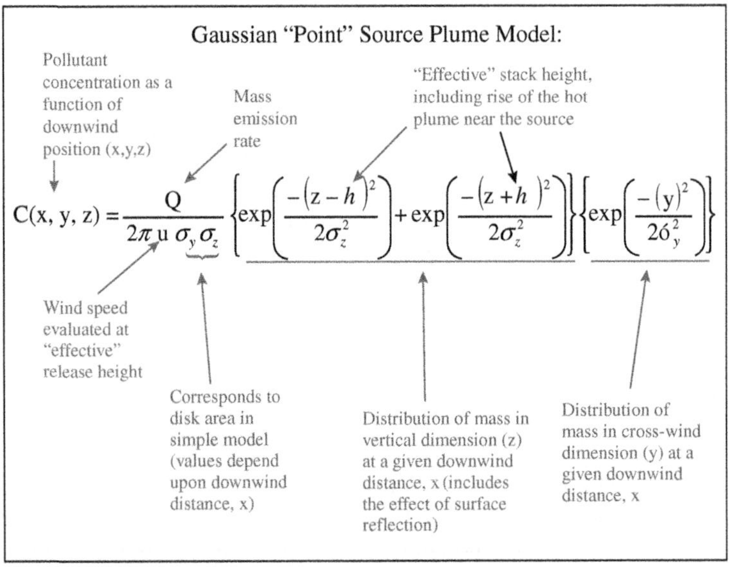

The spread of the plume by the air pollution dispersion is characterised by the standard deviations of the plume concentration in the Y and Z directions (cf. Figure 7.3).

Gaussian Puff Model

In a Gaussian Puff model the release is divided in a series of single puffs which bear the amount of pollutant released in a time interval (cf. Figure 7.4). The

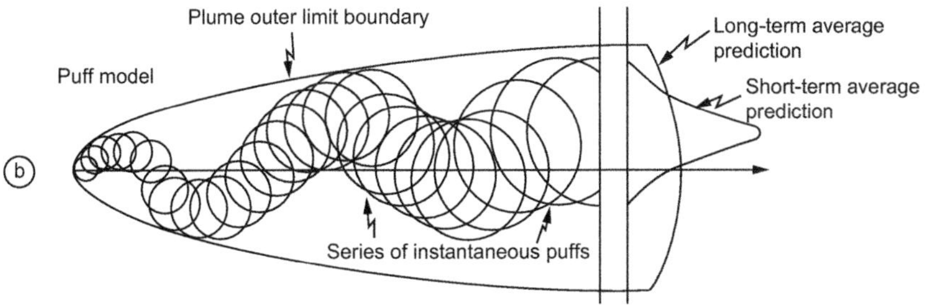

FIGURE 7.4 Horizontal sight on the series of puffs in a Gaussian puff model *(from [10])*.

time intervals may vary between some seconds and some minutes. They follow the advection according to the prevailing wind speed and wind direction while they are growing due to turbulent diffusion.

The contribution of single puffs to the concentration is calculated by the formula:

$$C_i(x, y, z) = \frac{Q_i}{(2\pi)^{3/2}\sigma_x\sigma_y\sigma_z} \cdot e^{-\frac{(x-ut)^2}{2\sigma_x^2}} e^{-\frac{y^2}{2\sigma_y^2}} \cdot \left\{ e^{-\frac{(z-H)^2}{2\cdot\sigma_z^2}} + e^{-\frac{(z+H)^2}{2\cdot\sigma_z^2}} \right\},$$

where

$C_i(x, y, z)$ contribution of a puff i to the concentration (e.g. Bq m^{-3}),
Q_i amount of a pollutant assigned to a puff i (e.g. Bq),
H release height (m) (assuming a reflection on the ground; more sophisticated approaches also consider reflections at the inversion layer),
x, y, z coordinates in a puff-following coordinate system (m),
σ_j diffusion parameter ($j = x, y, z$) (m) (normally it is assumed that the horizontal diffusion parameters σ_x and σ_y are equal: $\sigma_x = \sigma_y$),
u wind speed ($m \cdot s^{-1}$),
t travel time (s).

The big advantage of a Gaussian puff model over the plume model is that it can represent the temporal and spatial changes of wind and turbulence in a horizontal wind field. However, the vertical advective air motion cannot be simulated. The computational effort is considerably larger than that for the plume model. This is especially true if one has to compute the external dose from the cloud (cf. Section 7.3).

Diffusion Parameters of Gaussian Models

The standard deviations of the assumed Gaussian distribution σ_y, and σ_z are called the diffusion parameters. The selection of appropriate diffusion parameter sets is important as it determines the model output. There are very many diffusion parameter sets available around the world. As mentioned before, the most renowned example are the Pasquill-parameters ([2], cited after [3]). They were derived from prairie grass experiments and are therefore best suited for sites with low roughness. There are many other sets which take into account emission height, diffusion category, and roughness.

Frequently, the diffusion parameters are expressed in dependence on the travel distance of the cloud using a formula like

$$\sigma_{i,c} = p_{i,c} \cdot x^{q_{i,c}},$$

where

x distance from the source (m),
i index for direction x or y,
c index for diffusion category,
$P_{i,c}$, $q_{i,c}$ empirical constants.

Some formulations use a travel time dependency (e.g. the so called Doury parameters which are used in France [11]).

7.2.3. Lagrangian Particle Dispersion Model

Lagrangian Particle Dispersion Models (LPDM) are especially appropriate for the simulation of the dispersion of material emitted from point sources. The emissions are represented by a large amount of mass particles (e.g. 100,000), and the model simulates the transport and dispersion of these particles in the atmosphere. The concentration distribution of the radioactive material is calculated by the determination of the particle densities on regular grids.

Lagrangian Particle Dispersion models are available for all scales (small scale: up to about 10 to 20 km, mesoscale: up to about 50 to 100 km and large scale: more than about 100 km). Prominent examples in Europe are LPDM developed and operated by the German Weather Service (DWD) [12], and FLEXPART operated by the Austrian Weather Service (ZAMG) and a large number of research institutes ([13], [14]).

The long range dispersion simulations are normally based on the output data of the numerical weather prediction models. For the calculation of the particle trajectories the 3-dimensional wind and turbulence information of the atmosphere is used. Diffusion is simulated by a "Monte Carlo" technique. Radioactive decay as well as dry and wet deposition (at the ground) cause a reduction of the radioactivity esp. the mass of the particles. Dry deposition parameterisation follows a deposition velocity concept. Wet deposition and radioactive decay are evaluated using isotope-specific scavenging and decay coefficients. Additionally, a vertical mixing scheme for deep convection processes is included. Starting from selected receptor points, the models (e.g. LPDM) can be employed also in a backward mode to determine unknown source positions.

7.2.4. Trajectories

Another important service of some weather services is the generation of trajectories. These are based on calculational fields of the weather prediction models [15]. For two months after the Fukushima accident in March 2011 the German weather service generated daily prognostic trajectories with forecasts up to 72 h. The system is operational and will be activated on demand. It is able to produce backward trajectors in order to locate the origin of a potential release. A similar system – operated by the U.S. National Oceanic and Atmospheric Administration (NOAA) – is available on the web ([16], [17]) and is free to be used by everyone.

7.3. EXPOSURE PATHWAYS

There are six processes that account for most of the ways in which people can accumulate a radiation dose after an accidental release of radioactive material to the atmosphere:

- Exposure to external irradiation from the passing cloud (cloudshine)
- Exposure to external irradiation from deposited radionuclides (groundshine)
- Exposure to external irradiation caused by the contamination of skin and clothes
- Internal irradiation by radionuclides incorporated by the inhalation of contaminated air from the passing cloud
- Internal irradiation after inhalation of resuspended radionuclides, which were once deposited on the ground
- Internal irradiation by radionuclides incorporated with contaminated foodstuffs (ingestion)

Their importance depends on the composition of the release, the chemical and physical properties of the released radionuclides, deposition characteristics and exposure duration. Wet deposition, especially in the case of high rain intensity, leads to high doses due to groundshine and ingestion. Cloudshine as well as inhalation play only a role during the passing of the cloud. Cloudshine has a minor importance compared to inhalation and groundshine. However, if the release is dominated by noble gases (three or four orders of magnitude higher activity than iodine or particulates), then cloudshine will cause most of the dose. This could be the case after venting with intact filters. During the passage of the cloud, inhaled particles and gases can remain in the body for a longer period and thus cause a long lasting contribution to the dose. The importance of groundshine rises with exposure duration. After some days groundshine will be the predominant pathway. Resuspension and skin contamination will play a minor role. However, these exposure pathways are to be regarded for emergency workers on the scene and people who have visited a contaminated area.

Table 7.3 shows the relative importances of various exposure pathways, which are calculated for core melt scenarios of German PSAs (own calculations).

In the long run the ingestion pathway – if not interrupted by countermeasures - will play the major role followed by ground deposition. Inhalation due to resuspension of radioactive particles earlier deposited on the ground may be of importance for persons living or working in a contaminated environment in the late phase of an accident.

The calculation of the cloudshine is complicated because of the range of gamma radiation (depending on the gamma energy up to about some 100 m). Ideally, for each receptor point one has to sum up the dose contributions over the cloud which generally does not have a uniform activity distribution. Only far away from the release point a uniform distribution of activity in the mixing layer can be assumed. There the semi-infinite cloud model can be used.

There are several approaches to the calculation in dependence on the used model.

A simple model for a plume first calculates the dose according to the semi-infinite model. Then correction factors are applied for receptor points near to the

TABLE 7.3 Relative Importances of Exposure Pathways for Some Release Scenarios; Projected Effective Doses to Adults

Exposure pathway	Early release due to an 80 mm diameter hole (FK3)*	Loss of containement function due to overpressure, no filters (FK5)*	Filtered venting (FKI)*
Dry deposition			
Cloudshine	0.7%	4.2%	75.7%
Groundshine[a]	7.4%	14.1%	4.6%
Inhalation	91.8%	81.7%	19.7%
Total	100.0%	100.0%	100.0%
Wet deposition (rain intensity 2 mm per hour)			
Cloudshine	0.6%	3.6%	61.9%
Groundshine[a]	45.4%	44.0%	22.0%
Inhalation	54.1%	52.4%	16.1%
Total	100.0%	100.0%	100.0%

**The scenarios FK3 and FK5 are taken from the German Risk study [18]; FKI is taken from a German plant specific PSA Level 2 [19].*
[a]Integration time : 7 days.

release point. More sophisticated methods are based on line sources and plane sources and are analogous to the point source model. These models are used, along with the point source model, until the plume grows to a sufficient size so that the assumptions associated with the semi-infinite cloud model are met.

For Gaussian puff models special calculational methods exist which take into account the extension of the puffs. More detailed information on the approaches to this problem can be found in [20] and [21].

The calculation of cloudshine in Lagrangian particle dispersion models is theoretically quite simple as each particle can be regarded as a point source. However, as a large amount of particles is simulated, this approach cannot be realised for practical reasons. At present, calculational methods are under development which are able to solve this problem.

7.4. EMERGENCY PLANNING

7.4.1. Countermeasures

The main options for preventing and limiting exposures are generally the following ones:

Evacuation: The best strategy for preventing serious exposures, if feasible, is to evacuate people from the area before the radioactive materials arrive.

Sheltering: Placing barriers between the radioactive materials and people is effective for some releases. The most commonly available and suitable barrier is a building, the walls and roof of which attenuate to some extent the gamma radiation. The heavier the construction, the more effective the shielding; basements are particularly advantageous locations.

Respiratory protection: Breathing through any of a variety of materials – facemasks, tissues, towels, or other cloth – offers significant protection against the inhalation of particles.

Relocation: If large amounts of radioactivity persist in the area, sheltering is not a sufficiently protective measure, and people must be moved from the area until it is decontaminated.

Potassium iodide (KI) blockage: Iodine uptake by the body can be blocked by the ingestion of stable iodine prior to, or immediately after, exposure. If taken properly, potassium iodide will help reduce the dose of radiation to the thyroid gland from radioactive iodine, and reduce the risk of thyroid cancer. Detailed information can be found in [22] and [23].

Decontamination of people: Apart from removing people from the vicinity of radioactivity or using barriers, it is, in some situations, desirable to remove radioactive materials from the immediate vicinity of people. Decontamination includes removing contaminated clothing and washing off external contamination.

Decontamination of land and buildings: This is not generally considered an emergency response; however, it is important to remember that the significant off-site economic costs of a major accident will be for attempted decontamination and for property that is unusable because it cannot be sufficiently decontaminated. This item includes a lot of single measures, which are described in detail in [24].

Protection of the food chain: Ingestion of contaminated food and water can account for nearly half of the aggregate population's exposure to radioactivity. Food-chain interventions are thus crucial to emergency response efforts directed toward delayed health effects. This item includes a lot of single measures, too, which are described in detail in [25].

Medical treatment: Finally, there is a need for medical efforts to alleviate consequences. Medical care entails screening and follow-up capabilities and the possibility of deploying a significant medical infrastructure. More information on this item can be found in ([26], [27], [28]).

7.4.2. Radiological Basis of Emergency Planning and Preparedness

Health Effects

Adverse health effects as end points of a radiation exposure are divided into deterministic effects and stochastic effects [29].

- Deterministic effects occur if, as a result of the high deposition of energy, functionally significant numbers of cells are damaged or die or if re-generation is not possible or only after a considerable delay. Such effects may be temporary or permanent. They are subject to threshold doses which vary for different tissues, organs and individuals. Above the threshold dose the extent of the damage depends on the dose, but the probability of occurrence is 100 percent.
- Stochastic effects are observed at relatively low doses and include cancer induction and hereditary effects in the offspring. Injury occurs at a cellular level with modifications of the genome. The afflicted cell may be capable of dividing and eventually manifests its damage as cancer in the exposed person or as hereditary damage in his progeny. Such damage is called stochastic because it is not predictable for a given person. It can occur even after small doses with a certain, correspondingly small probability without the threshold dose having been exceeded. The probability of stochastic effects rises with dose.

The older literature concerning severe accident consequences adresses deterministic effects as early health effects and stochastic effects as late effects [30].

The new concept of ICRP with respect to nuclear and radiological emergencies

The International Commission on Radiological Protection (ICRP) is the leading organisation for radiation protection concepts in general and for the conceptual framework of nuclear and radiological emergencies in particular. The ICRP recommendations on protection strategies and their radiological grounds form the basis of international and national regulations for emergency planning and preparedness.

In the publications No. 103 [31] and No. 109 [32], the ICRP – in the following referred to as the Commission - revised its concept for radiological emergencies. Thus, their previous concept, which is described in the publications No. 60 [29] and No. 63 [33], was changed in some parts. The main changes in relation to emergencies are:

The previous process-oriented division of radiation protection into "practices" and "intervention" is altered in favour of a situation oriented division into planned exposure situations, emergency exposure situations, and existing exposure situations. The Commission will continue to use the term "practice" for an activity that causes an increase in exposure to radiation or its risk and where the radiation source can be controlled. "Intervention" previously described situations where protective actions were needed to reduce exposures. The Commission now defines [31]:

- "Planned exposure situations are situations involving the planned introduction and operation of sources (this type of exposure situation includes situations that were previously categorised as practices).
- Emergency exposure situations are unexpected situations such as those that may occur during the operation of a planned situation, or from a malicious act, requiring urgent attention.
- Existing exposure situations are exposure situations that already exist when a decision on control has to be taken, such as those caused by natural background radiation."

The radiation protection principles justification, optimisation, and application of dose limits and dose constraints (the latter for planned exposure situations) are not altered. In emergency situations, dose limits and dose constraints are not applicable. There is an additional principle to be applied for emergeny exposure situtions: the requirement to protect against severe deterministic injury. Situations in which the dose thresholds for severe deterministic injuries could be exceeded should always require actions. As a basis for decisions whether in emergency exposure situations as well as in existing exposure situations protective measures should be implemented, the Commission recommends the term "reference level". In an actual emergency situation, the reference level should serve as a benchmark for the success of the implemented measures.

The reference level should be set as a residual dose. The ICRP gives the following explanation (from [32]):

Residual Dose:

"The total residual dose is the total effective dose from the emergency exposure situation that is expected to remain after the implementation of the full strategy of protective measures. The total residual dose should be calculated as realistically as possible. Strictly, since emergency plans are developed to protect population groups, not specific individuals, the residual dose is derived as the dose to each of a set of "representative persons". Guidance on characterising the representative person is provided in Publication 101 [34]: "The total residual dose may be calculated over the period of time that a protective measure is active (in order to determine the dose averted by the action for comparison with appropriate intervention levels, or it may be calculated, taking account of the effectiveness of planned protective measures, for a year (or the duration of exposures, whichever is shorter) for comparison with the appropriate reference level. Where the implementation of additional protective measures is being considered during the actual response, the residual dose calculated for comparison with the reference level should include doses already received and committed, as well as those expected to be received in the future."

In the past, the principle of optimisation should be applied to single protective measures, and it was the averted dose that should be used as optimisation criterion. Now the optimisation shall be applied to a situation adapted overall strategy, which includes all exposure pathways and all appropriate measures to reduce exposure. The aim is that these overall strategies, as far as possible, will be formulated and optimised already in the planning phase. However, the averted dose remains an input to the development of strategies.

Projected Dose:
"Projected dose is the individual effective (or equivalent) dose that is expected to occur as a result of the emergency exposure situation, should no protective measures be employed. Projected doses should be calculated to representative persons. ... Projected doses may be used in several ways, within emergency planning:

- to give an initial indication of the scale of response planning required, by comparing them with the appropriate reference level(s)
- to determine the dominant exposure pathways and the likely time evolution of doses, for informing the emergency planning process with respect to the type and urgency of protection measures required
- to compare with threshold doses for severe deterministic injuries."

Averted Dose:
"The averted dose is the dose to the appropriate representative person (usually expressed as effective dose or equivalent dose) that is expected to be averted by the implementation of a protective measure or combination of protective measures."

The dosimetric quantities which are used for different issues are shown in Figure 7.4.

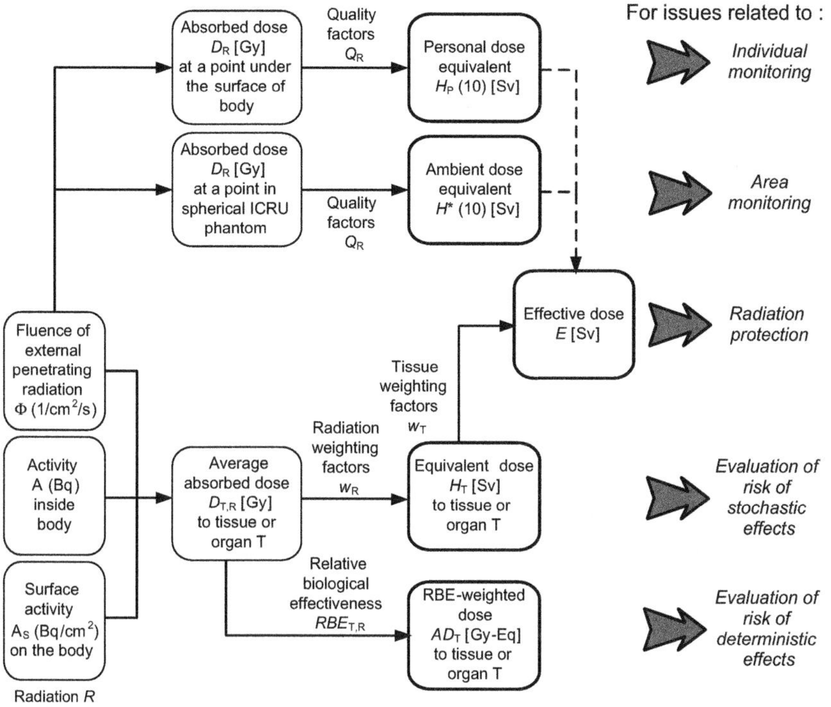

FIGURE 7.4 Basic dosimetric quantities and their application *(from [35])*

Reference Levels

With reference to emergency exposure situations, the ICRP recommends to set reference levels in two bands [31]:

"The band from 20 mSv to 100 mSv (residual dose within 1 year) applies for situations, where actions taken to reduce exposures would be disruptive. Action taken to reduce exposures in a radiological emergency is the main example of this type of situation. The Commission considers that a dose rising towards 100 mSv will almost always justify protective action. In addition, situations in which the dose threshold for deterministic effects in relevant organs or tissues could be exceeded should always require action." Severe nuclear accidents would regularly fall into this band.

The band between 1 mSv to 20 mSv applies for minor accidental events or to post-accident situations.

"At doses higher than 100 mSv, there is an increased likelihood of deterministic effects and a significant risk of cancer. For these reasons, the Commission considers that the maximum value for a reference level is 100 mSv incurred either acutely or in a year. Exposures above 100 mSv incurred either acutely or in a year would be justified only under extreme circumstances, either because the exposure is unavoidable or in exceptional situations such as the saving of life or the prevention of a serious disaster."

Generic criteria according to this concept have been published in [35]. Table 7.5 shows generic criteria for protective actions in emergency exposure situations to reduce stochastic effects as recommended by IAEA [35]. The cited document also contains a table for generic criteria for acute doses for which protective actions are expected to be taken under any circumstances to avoid or to minimize severe deterministic effects.

Following the recommendations of ICRP 60 [29] and ICRP 63 [33] many states have formulated reference levels for single countermeasures, frequently

TABLE 7.5 Generic Criteria for Protective Actions and Other Reponse Actions in Emergency Exposure Situations to Reduce the Risk of Stochastic Effects (from [35])

Generic criteria		Examples of protective actions and other response actions
Projected dose that exceeds the following generic criteria: Take urgent protective actions and other response actions		
$H_{Thyroid}$	50 mSv in the first 7 days	Iodine thyroid blocking
E H_{Fetus}	100 mSv in the first 7 days 100 mSv in the first 7 days	Sheltering; evacuation; decontamination; restriction of consumption of food, milk and water; contamination control; public reassurance
Projected dose that exceeds the following generic criteria: Take protective actions and other response actions early in the response		
E H_{Fetus}	100 mSv per annum 100 mSv for the full period of in utero development	Temporary relocation; decontamination; replacement of food, milk and water; public reassurance
Dose that has been received and that exceeds the following generic criteria: Take longer term medical actions to detect and to effectively treat radiation induced health effects		
E	100 mSv in a month	Screening based on equivalent doses to specific radiosensitive organs (as a basis for medical follow-up), counselling
H_{Fetus}	100 mSv for the full period of in utero development	Counselling to allow informed decisions to be made in individual circumstances

Note: *HT* — equivalent dose in an organ or tissue *T*; *E* — effective dose.

in terms of averted dose. Other states have decided to use reference levels in terms of projected dose. There is also a wide spectrum of integration times, which are needed to calculate groundshine doses. An overview for the different settings of reference levels in European states including the proposals of international organisations can be found in [23] and [36].

It is expected that the new recommendations with respect to reference levels will be introduced in the next years in most states as the new revision of the IAEA Basic Safety Standards (to be published in 2012) as well the European directive laying down basic safety standards for the protection of the health of workers and the general public against the dangers arising from ionising radiation ([37], under revision) will adopt the ICRP recommendations.

ICRP [32] and IAEA [35] recommend the use of Operational Intervention Levels (OIL) as triggers for the implementation of actions. These are indicators which can be measured (e.g. ambient dose rate, specific activities, contamination).

7.4.3. Zoning

In general, Emergency Planning Zones (EPZs) around each NPP are defined on a deterministic basis. They help to plan a strategy for protective actions during an emergency. The exact size and shape of each EPZ is the result of detailed planning which includes consideration of the specific conditions at each site, geographical features of the area and demographic information. Predetermined protection action plans are in place for EPZs and are designed to avoid or reduce dose from potential ingestion of radioactive materials. These actions include such action as sheltering, evacuation, using of stabile iodine tablets in the short term, food bans, relocation and decontamination in the longer term. There are various differences in European Union (EU) Member States in the way how emergency plans have been drawn up and how EPZs have been defined. Usually simplified deterministic approaches are used; in a few cases some probabilistic information from Probabilistic Safety Assessment (PSA) is being used as well. With the advent of improved understanding and characterisation of NPP severe accidents, emergency off-site actions should be planned and organized in a way so that the whole spectrum of their possible realisations and related probabilities, within the limits of practicality, is taken into account. A deterministic approach, as a basis, coupled with both PSA methodology and results, could play a significant role in the development of relevant policies by utilities and regulatory bodies.

A brief summary of the IAEA guidance on EPZ, actions and intervention criteria will be presented on the basis of the information from IAEA document EPR-METHOD 2003 [38] (under revision since 2011) and IAEA Safety Standards Series GS-R-2 [39a]. This provides an international perspective on the pertinent issues related to the emergency zoning requirements.

The IAEA Requirements are specified for 5 threat categories. The threat category I applies to facilities, such as NPPs, for which on-site events, including very low probability events, could give rise to severe off-site deterministic effects. The on-site events involve an atmospheric or liquid release of radioactive material or external exposure that originates from a location on the site.

The IAEA Requirements establish numerous prescriptions related to generic areas, both on-site and off-site. In addition, requirements for two off-site emergency zones are established: the Precautionary Action Zone (PAZ) and Urgent Protective Action Planning Zone (UPZ). Facilities in threat categories I and II (e.g. some type of research reactors) warrant extensive on-site and off-site emergency preparedness arrangements. In addition, threat category V is considered further, as it applies for activities not normally involving sources of ionizing radiation, but with products with a significant likelihood of becoming contaminated as a result of events at facilities in threat categories I or II, including facilities in neighbouring countries.

On-site area: this is the area under control of the operator or the first responder[2]. It surrounds the facility within the security perimeter (fence) and is under immediate control of the facility operator.

Off-site area: this is the area beyond that which is under the control of the facility operator or first responders.

The IAEA Requirements require that, for facilities in threat category I or II, arrangements shall be made for effectively making and implementing decisions on urgent protective actions to be taken off-site within (cf. Figure 7.6):

a) the PAZ for facilities in threat category I, within which arrangements shall be made with the goal of taking precautionary urgent protective action, before a release of radioactive material occurs or shortly after a release of radioactive material begins, on the basis of conditions at the facility (such as the emergency classification) in order to substantially reduce the risk of severe deterministic effects.

b) the UPZ, for facilities in threat categories I or II, within which arrangements shall be made for urgent protective action to be taken promptly, in accordance either with international or national standards, in order to avert off-site doses.

The PAZ and UPZ should be areas around the facility roughly circular in shape. Their boundaries should be defined, where appropriate, by local landmarks (e.g. roads or rivers) to allow their easy identification during a response (cf. Figure 7.6). It is important to note that the zones should not stop at national borders.

In addition to PAZ and UPZ there is also a longer-term protective action zone, which is called the Food Restriction Planning Zone (FRPZ). This is the

---

2. It is the first member of an emergency service to respond at the scene of an emergency.

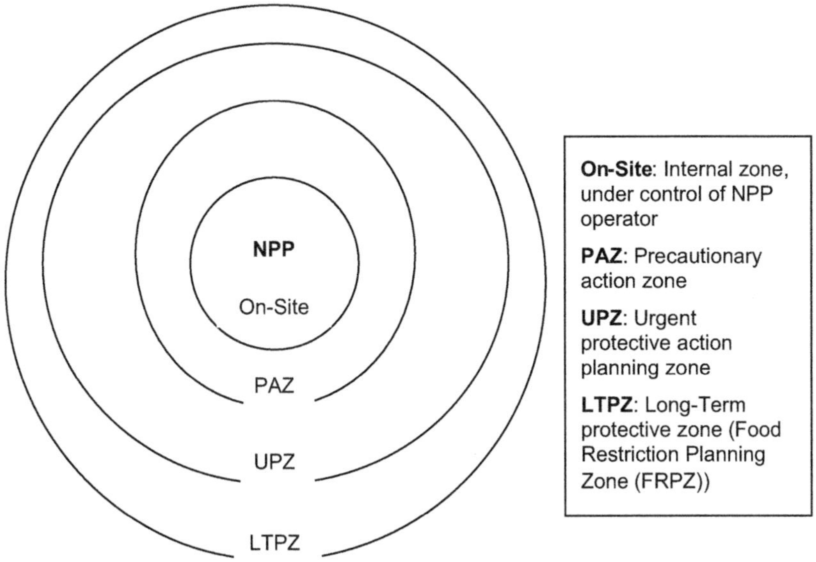

FIGURE 7.6 NPP emergency planning zones.

area around the facility where preparations for effective implementation of protective actions to reduce the long term dose, i.e. the risk of stochastic health effects from deposition and ingestion of locally grown food should be developed in advance. The FRPZ will of course include the PAZ and the UPZ, and extend to a larger radius. On the bases of severe accident studies, the US NRC/US FEMA has, for instance, adopted a radius of 80 km (50 miles) for the FRPZ (in [39] called "EPZ-ingestion"). It could, however, be much larger, up to several 100 km.

Concerning urgent protective action, it is the "action in the event of an emergency which must be taken promptly (normally within hours) in order to be effective, and the effectiveness of which will be markedly reduced if it is delayed. The most commonly considered urgent protective actions in nuclear or radiological emergency are evacuation, decontamination of individuals, sheltering, respiratory protection, iodine blockage and restriction of the consumption of potentially contaminated foodstuffs". The urgent protective actions are, effectively, the radiological exposure protective options, which represent the consequence mitigation part.

It should be mentioned that in some countries the term "emergency planning zone (EPZ)" is used instead of the above term UPZ. The differences in basic terminology have historical reasons, since the traditional terms were usually based on US NRC NUREG documents, particularly NUREG-0654 [39], which is still in force.

TABLE 7.6 Suggested Emergency Zones and Radius Sizes for NPPs

Facilities	PAZ radius	UPZ radius	FRPZ radius
Reactors > 1000 MW(th)	3–5 km	25 km	300 km
Reactors > 100–1000 MW(th)	0.5–3 km	5–25 km	50–300 km

(from [38]).

For threat category I, i.e. for NPPs, IAEA document gives the following suggestions for the approximate radius of the emergency zones and food restriction planning radii (cf. Table 7.6):

The radii were selected on the basis of calculations performed by using the RASCAL 3.0 computer code (cf. section 7.5.1). The calculations assumed average meteorological conditions, no rain, ground level release, 48 hours of exposure to ground shine, and calculates the centerline dose to a person being outside for 48 hours.

The suggested **sizes for the PAZ** are based on expert judgment under the following constraints:

1) Urgent protective actions taken before or shortly after a release within this radius will prevent doses above the early death thresholds for the vast majority of severe emergencies postulated for these facilities.
2) Urgent protective actions taken before or shortly after a release within this radius will avert doses above the urgent protective action Generic Intervention Level[3] (GIL) for the majority of emergencies postulated for the facility.
3) Dose rates that could have been fatal within a few hours were observed at these distances during the Chernobyl accident.
4) The maximum reasonable radius for the PAZ is assumed to be 5 km because:
- apart from the most severe emergencies, this is the limit to which early deaths are postulated;
- this radius provides a factor of ~10 reduction of dose as compared to the dose on the site;
- it is very unlikely that urgent protective actions will be warranted at a significant distance beyond this radial distance;
- 5 km is considered the practical limit of the distance to which substantial sheltering or evacuation can be promptly implemented before or shortly after a release; and
- implementing precautionary urgent protective actions to a larger radius may reduce the effectiveness of the action for the people near the site, who are at the greatest risk.

---

3. The level of avertable dose at which a specific protection action is taken in an emergency or situation of chronic exposure

1) The suggested **sizes for the UPZ** are also based on expert judgment by taking into account the following factors [38].
2) These are the distances to which studies suggest that monitoring to locate and evacuate hot spots (deposition) within hours/days may be warranted in order to significantly reduce the risk of early deaths.
3) At these radial distances there is a factor of reduction of around 10 in concentration (and thus risk) from a release as compared to the concentration at the PAZ boundary.
4) This distance provides a substantial base for expansion of response efforts.
5) 25 km is assumed to be the practical limit for the radial distance within which to conduct monitoring and implement appropriate urgent protective actions within a few hours or days. Attempting to conduct initial monitoring to a larger radius may reduce the effectiveness of the protective actions for the people near the site, being at largest risk. For average meteorological (dilution) conditions, beyond this radius, for most postulated severe emergencies, the total effective dose for an individual would not exceed the urgent protective action GILs for evacuation.

As far as the **FRPZ** is concerned, in general, protective actions such as relocation, food restriction, and agricultural countermeasures will be based on expert judgement by taking into account the following factors:

- Detectable excess stochastic effects (cancers) are very unlikely beyond this distance.
- Detailed planning within this distance provides a substantial basis for expansion of response efforts.
- Food restrictions were warranted to about 300 km following the Chernobyl accident in order to prevent detectable excess thyroid cancers among children.

A collection of zone settings in different states can be found in [36], [40], [41].

7.5. TOOLS FOR THE ASSESSMENT OF SEVERE ACCIDENT CONSEQUENCES

7.5.1. RASCAL

There are a great number of atmospheric dispersion models. Ref. [42] lists some hundreds. Most of them are dedicated to air pollution assessments with the end points of concentration or time integrated concentration of pollutants. Some models include the simulation of dry and wet deposition. Many programs are able to calculate radiation doses.

The program RASCAL (Radiological Assessment System for Consequence AnaLysis) [43] is a modern example of a tool intended to make dose projections after an actual accidental release of radionuclides into the atmosphere. It is used by US NRC's emergency operations center. The current version is RASCAL 3.0.5 which is distributed by OECD/NEA. The International Radiological Assessment System (InterRAS) developed on behalf of IAEA and described in [44] is based on an earlier version of RASCAL (V 2.1).

RASCAL 5.0.3 includes a quite sophisticated module for the determination of the source term.

RASCAL integrates a Gaussian plume dispersion model as well as a Gaussian puff dispersion model. The plume model is intended for shorter distances and uses a uniform wind field with respect to wind speed and wind direction. The puff model can handle changes of wind direction and wind speed. Included is a database describing the topography as well as meteorological observation stations around U.S. nuclear sites. The information can be used to calculate a wind field for the puff model. Additionally, RASCAL 3.0.5 has special procedures for calm winds which usually cannot be handled by Gaussian models.

Dose calculations are made for the most relevant pathways inhalation, external dose from the cloud and from the ground resulting in effective doses and organ doses to the thyroid, to the lungs and to the bone marrow. These are the doses which are to be compared with emergency reference levels in order to decide on early countermeasures.

A workbook is available [45] which is intended for the guided education of RASCAL Users.

7.5.2. Probabilistic Consequence Models

Introduction: Overview of PSA Level 3

PSA provides a comprehensive, structured approach to identify failure scenarios and derive numerical estimates of the risks to both the plant staff and the public at large. PSA provides a systematic approach to determining whether safety systems are adequate, whether the plant design is balanced, whether the defence in depth requirement has been realized and the risk is as low as reasonably achievable. These are characteristics of the probabilistic approach, which distinguish it from the deterministic approach. To date, PSAs have been performed for more than 200 NPPs worldwide and are under various stages of development for most of the remaining NPPs. All of them have been done at Level 1 to provide an estimate of core damage frequencies (CDF) for initiating events occurring during full/low power and shutdown operation conditions and in recent years many plant-specific Level 2 PSAs have also been carried out. However, only few PSAs Level 3 are available (e.g. [47]). For instance, no complete PSA from Level 1 up to Level 3 has been produced in Germany since 1979 when the results of the German Risk Study A were published [18].

The results of a PSA Level 2 are release categories, which are normally characterised by the associated probability, magnitude of radionuclides, isotopic composition, release timing, physical and chemical characteristics of the release, release heights (out of the stack, out of the building, ground release), heat content of the release (plume), etc.

These data describe the source term, which is part of the input of a PSA Level 3.

Database

A rough overview and characterisation of other input data follows.

- **Meteorological data**
  These data describe meteorological conditions prevailing at the sites under consideration, i.e. sequences of sets of wind fields (wind speed, wind directions, turbulence states, rain, snow) as a series of subsequent time intervals.
- **Population data**
  The population data describe the number of people at their dwellings in the possibly affected regions. Additionally, information on the diurnal and saisonal variation of their visits can be of interest. Usually the population distribution is projected onto a calculational grid.
- **Agricultural production data**
  The agricultural data describe the seasonal states of the crops (e.g. vegetables, grain) and of the milk and meat production. These data are also projected onto the calculational grid. Additionally, food distribution data can be used to assess which part the food production is consumed at its origin and which part of the food production is possibly exported.
- **Land data**
  These data comprise information on the infrastructure of the potentially affected areas, e.g. house types (shielding!).
- **Countermeasures**
  The data sets describe the emergency preparedness situation at the potentially affected sites. The data will include triggers for the implementation of short-term (e.g. sheltering, evacuation, intake of stable iodine) and long-term (e.g., relocation, land decontamination, food bans) as well as planning zones. Further on, the time needed for the implementation of countermeasures should be included.
- **Economic data**
  These data include costs e.g., for the temporary or permanent loss of industrial, and agricultural production sites, the cost of the various countermeasures, the compensations for the affected people including the cost for medical treatments.
- **Radioecological data**
  Radioecological data are data which are used to perform atmospheric dispersion calculations as well as data which are used for the calculation of food contamination (e.g., transfer factors for the different food pathways.

- **Radiological data**
  This data set consists of all the radiological data needed for dose and health effects calculation (e.g., isotopic data, dose coefficients, risk coefficients).

Course of Calculations

The rough course of calculational steps in a PSA Level 3 is:

- Atmospheric dispersion: release of the plume/cloud, direction and dispersion of the plume/cloud, surface deposition

For each grid element:

- Calculation of projected doses: cloudshine, inhalation, groundshine with different integration times, ingestion, (skin contamination, resuspension)
- Simulation of countermeasures in order to get residual doses
- Calculation of health effects: deterministic and stochastic (early and late) on the basis of the calculated residual doses
- Calculation of economic consequences

These calculations are done for a sufficient large number of different weather conditions, which could be selected from the meteorological sequences by sampling techniques. The data obtained can then be processed to display the results in the form of distributions and statistical parameters (e.g. mean values, confidence intervals).

Program Packages

It can be seen from the amount of data and the course of necessary calculations that sophisticated computer program packages are necessary. There are only a few packages of programs available around the world for this purpose.

One of these packages is the COSYMA- code which has been developed in the framework of the European MARIA project in the eighties and nineties of the last century ([6], [48], [49]). Another program package is MACCS2 ([50], [51]), which is described below.

MACCS2

The MELCOR Accident Consequence Code System (MACCS), distributed by U.S. government code centers since 1990, has been developed to evaluate the impacts of severe accidents at nuclear power plants on the surrounding public.

The principal phenomena considered are atmospheric transport and deposition under time-variant meteorology, short- and long-term mitigative actions and exposure pathways, deterministic and stochastic health effects, and economic costs. No other U.S. code that presently is publicly available offers all of these capabilities. Sometimes the MELCOR has been selected as the preferred code, for instance of the Swiss nuclear industry and of the Paul

Scherrer Institute (PSI Villigen, Switzerland) for severe accident analysis, on account of its integrated systems-level approach and validation against experiments and more detailed codes, while MACCS is commonly used by safety authorities for the independent assessment of off-site consequences, in particular health effects [30].

The MACCS2 package represents a major enhancement of its predecessor MACCS, with three primary enhancements: (1) a more flexible emergency-response model, (2) an expanded library of radionuclides, and (3) a semi-dynamic food-chain model.

The structure of MACCS2 is based on that of CRAC2 [52], as descended from the *Reactor Safety Study* [53]. MACCS2 is used to estimate the radiological doses, health effects, and economic consequences that could result from postulated accidental releases of radioactive materials to the atmosphere.

If contamination levels exceed a user-specified criterion, mitigative actions can be triggered to limit radiation exposures. If mitigative actions are triggered, the economic costs of these actions are calculated and can be reported.

The accident phases (early phase, intermediate phase, late phase according to U.S. EPA's Protective Action Guides [54]) determine which exposure pathways are considered and which dosimetric quantities and effects are calculated.

The original design of MACCS incorporated five separate FORTRAN programs executed in sequence, named ATMOS, EARLY, CHRONC, MERGER, and SUMMER. Functions of the former MERGER and SUMMER are now performed by the OUTPUT module. Results generated by ATMOS, EARLY, and CHRONC are written to binary files, which are then processed by OUTPUT in order to generate complementary cumulative distribution functions (CCDFs). Figure 7.7 gives an overview of the system.

Two types of doses are calculated by the code. They are referred to as "acute" and "lifetime doses" in this report. Acute doses are calculated for the sole purpose of estimating the "deterministic" health effects that can result from high doses delivered at high dose rates. Such conditions may occur in the

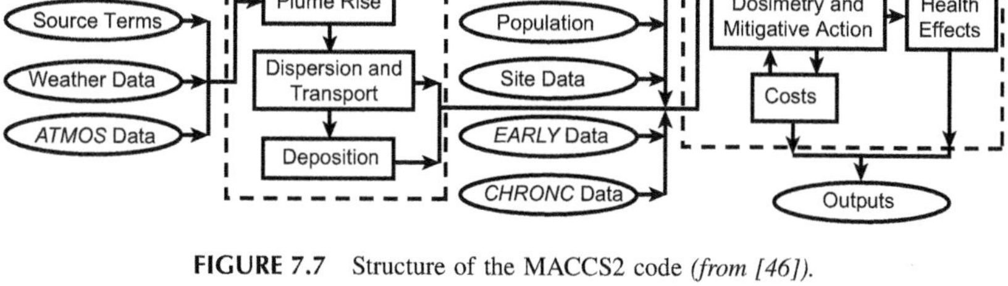

FIGURE 7.7 Structure of the MACCS2 code *(from [46])*.

immediate vicinity of a nuclear power plant following hypothetical severe accidents where containment failure has been assumed to occur.

Lifetime doses are the conventional measure of detriment used for radiological protection. These are the 50-year dose commitments to either specific tissues (e.g., red marrow and lungs), or a weighted sum of tissue doses defined by the International Commission on Radiological Protection (ICRP) [29] and referred to as "effective dose". Lifetime doses may be used to calculate the stochastic health effect risk resulting from exposure to radiation. MACCS2 uses the calculated lifetime dose in cancer risk calculations.

All the MACCS2 calculations are stored on the basis of a polar-coordinate spatial grid with a treatment that differs somewhat between (1) calculations of the emergency phase and (2) intermediate and long-term phases, as described later. The region potentially affected by a release is represented with an (r, θ) grid system centered on the location of the release.

Since the emergency-phase calculations utilise dose-response models for early fatalities and early injuries that can be highly nonlinear (again, as specified by the user), these calculations are performed on a finer grid basis than the calculations of the intermediate and long-term phases. MACCS2 calculations are performed in three phases: (1) input processing and validation, (2) phenomenological modeling, and (3) output processing.

Implementation of the Evans, Moeller, and Cooper [30] early health effects models requires a calculation method that takes account of dose protraction for radioactive material inhaled and retained in the respiratory system. MACCS2 applies dose reduction factors to protracted doses that contribute to early health effects. Dose reduction factors are derived from median lethal dose (LD_{50}) values that apply to a sequential set of time periods of fixed length. In addition, for the calculation of early fatalities and injuries in MACCS2, a new measure of dose is defined in order to reduce the computational demands of the calculations, which is referred to as effective acute dose.

An example for the probability for fatal outcomes is shown in Figure 7.8.

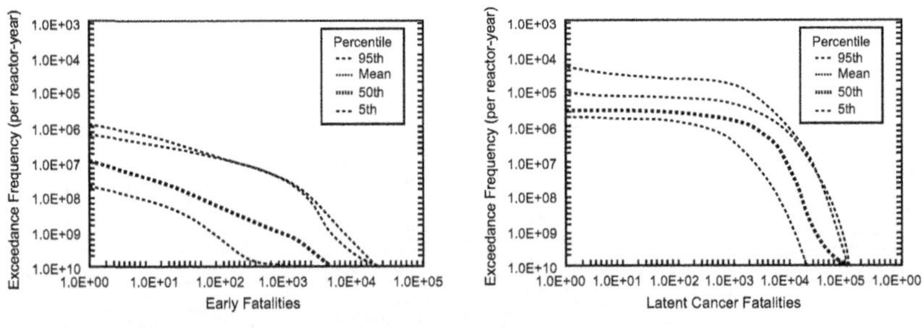

FIGURE 7.8 Example for complementary cumulative distribution functions (CCDFs) or exceedance plots: Probabilities for fatal outcomes *(from [47])*.

7.5.3. The RODOS System for Nuclear Emergency Management

As a project in the European Commission's 3rd, 4th and 5th Framework Programmes, the RODOS - Realtime Online Decision Support System-for nuclear emergency management [55] was established to respond to needs defined after the Chernobyl accident on 26 April 1986 and also to some extent after the tragic events in the U.S. in September 2001.

The Chernobyl accident had a profound effect on emergency preparedness and post-accident management worldwide and, in particular, in Europe. Deficiencies in emergency preparedness at both national and international levels (e.g., in world food trade) led to many problems of practical and political nature. Many lessons have been learnt, and considerable resources have since been allocated to improve emergency preparedness and post-accident management capabilities. Improvements have been made at national, regional, and international levels. They have been diverse in nature. However, more needs to be done and continuous effort is needed to ensure a timely and effective response to any future accident.

Emergency management, more generally, has received increased attention following the events in the U.S. in September 2001. Attacks with radiological dispersal devices (RDD), which spread radioactive material by aerosolising or dissolution in water reservoirs are currently under intense public, political, and scientific discussion (see e.g. [56]).

A number of requirements emerge from these considerations, such as:

- the need for a more coherent and harmonised response in Europe and in the different stages of an accident (in particular, to limit the loss of public confidence in the measures taken by the authorities for their protection);
- the exchange of information and data in an emergency to enable neighbouring countries to take more timely and effective actions;
- the necessity to make better use of limited technical resources and avoid duplication.

The RODOS project was established to respond to these needs and was launched in 1989 with funds from the European Commission and also with significant additional support by many national R&D programmes, research institutions, and industrial collaborators. In particular, the German Federal Ministry for the Environment, Nature Conservation, and Nuclear Safety (BMU) financially contributed to the project with a special focus on early emergency response. Up to 40 institutes from some 20 countries in the European Union, Central and Eastern Europe, and the Former Soviet Union were actively involved in the project [55].

As a result of these collaborative actions, a comprehensive decision support system (RODOS) has been developed which can be applied within Europe. It can be used in national or regional nuclear emergency centres, providing coherent support at all stages of an accident (i.e. before, during and after a release), including the long term management and restoration of contaminated areas. The system is able to support decisions about the introduction of a wide range of potentially useful countermeasures (e.g., sheltering and evacuation of people, distribution of iodine tablets, food restrictions, agricultural countermeasures, relocation, decontamination, restoration, etc.) mitigating the consequences of an accident with respect to health, the environment, and the economy. It can be applied to accidental releases into the atmosphere and into various aquatic environments. Appropriate interfaces exist with local and national radiological monitoring data, meteorological measurements and forecasts, and for adaptation to local, regional and national conditions in Europe.

The current version of the system is named PV 7.0. It has been, or is being, installed in national emergency centres of the following European countries for operational or pre-operational use, such as Germany, Finland, Spain, Portugal, Austria, the Netherlands, Poland, Hungary, Slovakia, Ukraine, Slovenia, Bulgaria, Romania, Russia and the Czech Republic. Installation is foreseen or under consideration in Switzerland, Greece and for the second time in Ukraine within the next few years.

The RODOS system has levels of decision support shown in Table 7.6.

In order to improve the usability of the system, to reduce the maintenance effort, to easier adapt to national requirements, and to facilitate the

TABLE 7.6 Levels of Decision Supports in RODOS

Level 0
Acquisition and quality checking of radiological data and their presentation, directly or with minimal evaluation, to the end-users, along with geographical and demographic information.
Level 1
Analysis and prediction of the current and future radiological situation (i.e., the distribution over space and time in the absence of countermeasures) based upon monitoring data, meteorological data and models, including information on the radioactive material released to the environment.
Level 2
Simulation of potential countermeasures (e.g., sheltering, evacuation, issue of iodine tablets, relocation, decontamination and food-bans, restoration), in particular, determination of their feasibility and quantification of their benefits and disadvantages.
Level 3
Evaluation and ranking of alternative countermeasure strategies by balancing their respective benefits and disadvantages (i.e. costs, residual dose, reducation of stress and anxiety, socio-psychological aspects, political acceptability, etc.) taking account of the judgements and preferences of decision makers.

implementation of new models, RODOS has been re-engineered in the last years [57]. Being a Java based product, the new RODOS version named JRodos introduces a cross-platform solution capable to run on most operation systems, including Windows and the main UNIX derivates. The re-engineered system preserves computational models from RODOS v6.0, adding a powerful GIS support and applying modern database technologies with flexible configuration possibilities. JRodos import functionality includes the support of EURODEP measurement files and GRIB 1 meteo data. The system operates successfully on Windows (NT based versions from Windows 2000 to Windows 7) and UNIX machines (OpenSuSe, Ubuntu, Solaris), both x86 and x64. Any RODOS further development will be realized in the JRodos version aiming to have this as the main operational version for the next 10 years.

The documents developed within the RODOS project contribute to a more comprehensive and uniform knowledge base for decision-making in nuclear or radiological accidents in Europe.

ACKNOWLEDGMENTS

The authors of Chapter 7 wish to acknowledge Mr. Kubanyi (European Commission - DG TREN) and Mr. Kirschteiger (European Commission - JRC/IET) for their support to the writing of this Chapter.

REFERENCES

[1] A. Sohier (Ed), A European Manual for 'Off-site Emergency Planning and Response to Nuclear Accidents', SCK·CEN Report R-3594, December 2002.
[2] F. Pasquill, The estimation of the dispersion of windborne material, The Meteorological Magazine 90 (1063) (1961) 33–49.
[3] D.H. Slade (Ed.), Meteorology and atomic energy, US AEC Report TID-24190, Washington DC, 1968.
[4] J.H. Seinfeld, Atmospheric Chemistry and Physics of Air Pollution, J. Wiley & Sons, New York, 1986.
[5] http://www.dwd.de; search for "Mixing heights".
[6] J.A. Jones, P.A. Mansfield, S.M. Haywood, I. Hasemann, C. Steinhauer, J. Ehrhardt, D. Faude, PC COSYMA (Version 2): an accident consequence assessment package for use on a PC, EUR Report 16239 (1995).
[7] http://en.wikipedia.org/wiki/Atmospheric_dispersion_modeling
[8] C. Walsh, Calculation of Resuspension Doses for Emergency Response, NRPB-W1, NRPB, 2002.
[9] http://en.wikipedia.org/wiki/Air_pollution_dispersion_terminology
[10] T. Mikkelsen, S.E. Larsen, S. Thykier-Nielsen, Description of the Risø Puff Diffusion Model, Nucl. Techn. 67 (1984) 56–65.
[11] A. Doury, Pratiques françaises en matière d'évaluation quantitative de la pollution atmosphérique potentielle liée aux activités nucléaires, CEC-Séminaire sur les Rejets Radioactifs et

leur Dispersion dans l'Atmosphère à la suite d'un Accident Hypothétique de Réacteur, Actes des Conférences, Risø I (1980) 403.
[12] http://www.dwd.de; search for "LPDM".
[13] A. Stohl, C. Forster, A. Frank, P. Seibert, G. Wotawa, Technical note: The Lagrangian particle dispersion model FLEXPART. version 6.2, Atmos. Chem. Phys. 5 (2005) 2461–2474, http://www.atmos-chem-phys.net/5/2461/2005/acp-5-2461-2005.pdf.
[14] S. Eckhardt, A. Stohl, H. Sodemann, A. Frank, P. Seibert, G. Wotawa, The Lagrangian particle dispersion model FLEXPART. version 8.0, http://zardoz.nilu.no/~andreas/flexpart/flexpart8.pdf.
[15] http://www.dwd.de; search for "Trajectories"
[16] R.R. Draxler, HYSPLIT (HYbrid Single-Particle Lagrangian Integrated Trajectory) Model access via NOAA ARL READY Website (http://ready.arl.noaa.gov/HYSPLIT.php) Silver Spring, MD (U.S.): NOAA Air Resources Laboratory, 2011.
[17] G.D. Rolph, Real-time Environmental Applications and Display System (READY) Website (http://ready.arl.noaa.gov), Silver Spring, MD (U.S.): NOAA Air Resources Laboratory, 2011.
[18] German Risk Study on Nuclear Power Plants, Cologne (Germany): Verlag TUV Rheinland, 1979 (Translation of Main Report, EPRI Report NP-1804-SR, April 1981).
[19] K. Köberlein, (Red.), Bewertung des Unfallrisikos fortschrittlicher Druckwasserreaktoren in Deutschland, GRS-175, Köln: Gesellschaft für Anlagen- und Reaktorsicherheit, 2001. (in German).
[20] S.A. McGuire, J.V. Ramsdell, Jr, G.F. Athey: RASCAL 3.0.5: Description of Models and Methods, NUREG-1887, U.S. NRC, August 2007.
[21] S. Thykier-Nielsen, S. Deme, E. Láng: Calculation Method for Gamma Dose Rates From Gaussian Puffs. Risø-R-775(EN), Risø National Laboratory, Roskilde (Denmark), June 1995.
[22] Distribution and Administration of Potassium Iodide in the Event of a Nuclear Incident, National Academic Press, Washington DC, 2004.
[23] J.R. Jourdain, K. Herviou, R. Bertrand, M. Clemente, A Petry: Medical Effectiveness of Iodine Prophylaxis in a Nuclear Reactor Emergency Situation and Overview of European Practices, RISKAUDIT Report No. 1337, Fontenay-aux-Roses: Riskaudit. http://ec.europa.eu/energy/nuclear/radiation_protection/doc/reports/2010_stable_iodine_report.pdf, January 2010.
[24] J. Brown, K. Mortimer, K. Andersson, T. Duranova, A. Mrskova, R. Hänninen, T. Ikäheimonen, G. Kirchner, V. Bertsch, F. Gallay, N. Reales, D. Hammond, P. Kwakman: Generic Handbook for Assisting in the Management of Contaminated Inhabited Areas in Europe Following a Radiological Emergency, EURANOS(CAT1)-TN(07)-02, May 2007.
[25] A.F. Nisbet, A.L. Jones, C. Turcanu, J. Camps, K.G. Andersson, R. Hänninen, A. Rantavaara, D. Solatie, E. Kostiainen, T. Jullien, V. Pupin, H. Ollagnon, C. Papachristodoulou, K. Ioannides, D. Oughton: Generic handbook for assisting in the management of contaminated food production systems in Europe following a radiological emergency, EURANOS(CAT1)-TN(09)-01, July 2009.
[26] Generic Procedures For Medical Response During a Nuclear or Radiological Emergency, EPR-MEDICAL, IAEA, Vienna 2005.
[27] C. Rojas-Palma, A. Liland, A. Næss Jerstad, G. Etherington, M. del Rosario Pérez, T. Rahola, K. Smith (Eds.): TMT Handbook, Triage, Monitoring and Treatment of people exposed to ionising radiation following a malevolent act, NRPA, Østerås (Norway), 2009, http://www.tmthandbook.org.
[28] Management of Persons Contaminated with Radionuclides, Scientific and Technical Bases, NCRP REPORT No. 161, December 2008.

[29] ICRP, Recommendations of the International Commission on Radiological Protection; Publication 60, Ann. ICRP 21 (1–3) (1991).

[30] J.S. Evans, S. Abrahamson, M.A. Bender, B.B. Boecker, E.S. Gilberts, B.R. Scott, S.S. Yaniv, Health Effects Models for Nuclear Power Plant Accident Consequence Analysis, Part I: Introduction, Integration, and Summary. NUREG/CR-4214, Rev. 2, Part I, ITRI-141, U.S. Nuclear Regulatory Commission, Office of Nuclear Regulatory Research, Division of Regulatory Applications, Washington DC, 1993.

[31] ICRP, The 2007 Recommendations of the International Commission on Radiological Protection. Publication 103, Ann. ICRP 37 (2–4) (2008).

[32] ICRP: Application of the Commission's Recommendations for the Protection of People in Emergency Exposure Situations, Publication 109, Annals of the ICRP,

[33] ICRP, Principles for intervention for protection of the public in a radiological emergency. Publication 63, Ann. ICRP 22 (4) (1991).

[34] ICRP, Assessing dose of the representative person for the purpose of radiation protection of the public. Publication 101, Ann. ICRP 36 (2) (2006).

[35] Criteria for Use in Preparedness and Response for a nuclear or Radiological Emergency, IAEA Safety Standards Series No. GSG-2, Vienna, 2011, http://www-pub.iaea.org/MTCD/publications/PDF/Pub1467_web.pdf

[36] Internationaler Vergleich der Modelle und Parameter zur Entscheidungsbegründung in Notfallsituationen, Cologne (Germany): TUV Rheinland, 2003 (in German). Part II of the report: Comparison of reference levels is available in English. Bonn: Federal Ministry for Environment, Nature Protection, and Reactor Safety, 2005.

[37] Council Directive 96/29/Euratom of 13 May 1996 laying down basic safety standards for the protection of the health of workers and the general public against the dangers arising from ionising radiation, Official Journal of the European Communities 29.6 (1996). No. L 159.

[38] IAEA, "Method for the Developing Arrangements for Response to a Nuclear or Radiological Emergency", EPR-METHOD (2003), ISBN 92-0-111503-2, IAEA, Vienna, 2003, http://www-pub.iaea.org/MTCD/publications/PDF /Method2003_web.pdf.

[39] Criteria for Preparation and Evaluation of Radiological Emergency Response Plans and Preparedness in Support of NPPs, NUREG-0654, USNRC, Nov. 1980, http://www.nrc. gov/reading-rm/doc-collections/nuregs/staff/sr0654/.

[39a] Preparedness and Response for a Nuclear or Radiological Emergency, IAEA Safety Standrds Serie No. GS-R-2, Vienna, 2002.

[40] I. Ivanov, et al., "EIA report of Kozloduy NPP", MoEW, Sofia, 1999/2000.

[41] I. Ivanov, et al., "EIA Report of Investment Proposal for Belene NPP Construction", MoEW, Sofia, 2004.

[42] http://en.wikipedia.org/wiki/List_of_atmospheric_dispersion_models

[43] "RASCAL 3.0, Description of Model and Methods". NUREG-1741, USNRC, 2001.

[44] Generic assessment procedures for determining protective actions during a reactor accident. IAEA-TECDOC-955, IAEA, Vienna, 1997.

[45] S.A. McGuire, J.V. Ramsdell Jr., G.F. Athey, RASCAL 3.0.5 Workbook. NUREG-1889, U.S. Nuclear Regulatory Commission, Washington, Office of Nuclear Security and Incident Response, Washington, DC, September 2007.

[46] F.E. Haskin, F.T. Harper, L.H.J. Goossens, B.C.P. Kraan, Probalistic Accident Consequence Uncertainty Analysis, Early Health Effects Uncertainty Assessment, Appendices, NUREG/CR-6545, EUR 16775, SAND07-2689 Vol. 2, Luxembourg: Office for Publications of the European Communities, 1997.

[47] "Severe Accident Risk: An Assessment for five US NPPs". NUREG-1150, USNRC, 1990, http://www.nrc.gov/reading-rm/doc-collections/nuregs/staff/sr1150/, 1990.
[48] KfK and NRPB, COSYMA - a new program package for accident consequence assessment. EUR 13028, Commission of the European Communities, Brussels, 1991.
[49] J.A. Jones, J. Ehrhardt, L.H.J. Goossens, R.M. Cooke, F. Fischer, I. Hasemann, B.C.P. Kraan, Probabilistic Accident Consequence Uncertainty Assessment Using COSYMA Uncertainty from the Early and Late Health Effects Module, EUR-18824, FZKA-6310 (2001).
[50] MACCS2 Computer Code Application Guidance for Documented Safety Analysis, Final Report, DOE-EH-4.2.1.4-MACCS2-Code Guidance, U.S. Department of Energy, Office of Environment, Safety and Health, Washington DC, June 2004.
[51] "Code Manual for MACCS2: Volume 1, User's Guide", NUREG/CR-6613, SAND97–0594. Division of Systems Technology Office of Technical and Environmental Support U.S. Nuclear Regulatory Commission U.S. Department of Energy, Washington, 1998.
[52] L.T. Ritchie et al., CRAC2 Model Description, NUREG/CR-2552, SAND82-0342, 1984.
[53] Reactor Safety Study: An Assessment of Accident Risks in U.S. Commercial Nuclear Power Plants, NUREG-75/014 (WASH-1400), U.S. NRC, 1975.
[54] Manual of Protective Action Guides and Protective Actions for Nuclear Incidents, EPA-400-R-92-001, U.S. EPA, 1992. http://www.epa.gov/radiation/docs/er/400-r-92-001.pdf.
[55] RODOS System Version PV6.0, EURATOM (2002–2006), EURANOS project, FI6R-CT-2004-508843.
[56] ICRP, Protecting people against radiation exposure in the event of a radiological attack. Publication 96, Ann. ICRP 35 (1) (2005).
[57] I. Ievdin, D. Trybushny, M. Zheleznyak, W. Raskob, RODOS re-engineering: aims and implementation details, Published online: 16 September 2010, DOI: 10.1051/radiopro/2010024, Radioprotection, vol. 45, (Suppl. 5) 2010.

Chapter 8

Integral Codes for Severe Accident Analyses

J.P. Van Dorsselaere, J.S. Lamy, A. Schumm and J. Birchley

Chapter Outline

8.1. INTRODUCTION

The TMI2 accident in 1979 accelerated the development of computer codes for simulation of severe accidents (SA) scenarios, first in the United States and then progressively in the 1980s in Europe and Japan. Three classes of SA codes can be defined, depending on their scope of application: integral codes, detailed codes, and dedicated codes.

- **Integral codes** (also called engineering-level codes): These codes simulate the overall nuclear power plant (NPP) response, that is, the response of the reactor coolant system, the containment, and, most important, the source

Nuclear Safety in Light Water Reactors. DOI: 10.1016/B978-0-12-388446-6.00008-3

term to the environment, using "integrated" models for a self-consistent analysis of the accident. They include a combination of phenomenological and parametric models for the simulation of the relevant phenomena. They must be (relatively) fast-running to enable a sufficient number of simulations of different scenarios to be performed, accompanied by uncertainty studies. These codes are not designed to perform best-estimate simulations, but rather to allow the user to bound important processes or phenomena, often using user-defined parameters. They are generally used to support Probabilistic Safety Assessment level 2 (PSA2) analyses, for a good estimate of risks for SA scenarios, and for the development and validation of accident management programmes. The principal internationally used codes today are MAAP (developed by Fauske & Associates Inc., USA), MELCOR (developed by Sandia National Laboratories, USA, under USNRC sponsorship), and ASTEC (jointly developed by IRSN, France, and the GRS, Germany).

- **Detailed codes** (also called mechanistic codes): They are characterized by best-estimate phenomenological models, consistent with the state of the art, to enable, as far as possible, an accurate simulation of the behavior of an NPP in the event of SA. In order to better illustrate the differences with the approach of integral codes, in most cases, a numerical solution is found for integral-differential equations, while in integral codes some correlations may be used. The basic requirements are that the modeling uncertainties are comparable with the uncertainties on the experimental data used to validate the code and that user-defined parameters are only necessary for phenomena that are not understood due to insufficient experimental data (including space scaling problems). The main advantages of these codes are to give a more detailed insight into the progress of a SA and to design and optimize mitigation measures. They can also be used for benchmarking the integral codes or to help to derive simplified models for implantation in integral codes. Due to the high computation time, they simulate only a part of the plant, for example, the reactor cooling system (RCS) or the containment. Their computation time depends on the scope of the application and on the level of space or time discretization, but it can span over days and weeks. The principal internationally used codes today are: for the RCS behavior and core degradation, ATHLET-CD (GRS, Germany), SCDAP/RELAP5 (INL, USA), RELAP/SCDAPSIM (ISS, USA) and ICARE/CATHARE (IRSN); and for containment CONTAIN (USA) and COCOSYS (GRS).
- **Dedicated codes**: These codes aim at simulating a single phenomenon. They have become important in the context of the requirements that regulatory authorities take into account SAs in the design of new NPPs and that the uncertainties of risk-relevant phenomena be reduced. Depending on their objective, they may be simple and consequently fast-running, or they have to be very complex, with the drawback of large calculation time. Typical issues for which

dedicated codes are required include: steam explosion and melt dispersal (e.g., MC3D at IRSN) and structure mechanics (e.g., CAST3M by CEA in France, or ABAQUS in the United States). This family of codes includes the CFD (Computational Fluid Dynamics) codes that solve Navier-Stokes thermal-hydraulics equations in 3D geometry, such as GASFLOW in KIT (Germany), TONUS in IRSN, CFX as commercial tool, and so on.

8.2. PROCESS OF CODE ASSESSMENT

The VASA project in the fourth Framework Program (FwP) of the European Commission [1] analyzed the assessment process of SA codes in some detail. Note that this general approach as described below is valid for other physical domains as well.

After implementation of physical models in the code with the adequate numerical scheme, a first stage, often called "verification", consists, for example, in checking the consistency between code specifications and coding; checking the accuracy of code solutions of a differential equation; verifying the conservation of mass, energy, and momentum; checking the stability of numerical results on diverse computer types.

The second stage of code assessment, often called "validation", aims at demonstrating that the correct physics is modelled, mainly by comparing the code results with results of experimental programs. This implies definition of a validation matrix covering diverse experimental conditions and scales, which leads to distinguishing three types of experiments:

- *Separate-Effects Tests (SET)* investigate a single phenomenon and provide data for development of a model that describes the effect and that is to be integrated as subroutine(s) in a code. The corresponding facilities are typically single-purpose, small-scale channels or boxes equipped with specialized, sophisticated, high-accuracy instrumentation.
- *Coupled-Effect Tests (CET)* investigate the coupling of two or more phenomena previously explored in SETs, and provide data for the appropriate integration of the corresponding models into a code. The corresponding facilities are typically of small to intermediate-scale, using a test vessel with comprehensive instrumentation adapted to the effects to be investigated.
- *Integral Experiments* represent all or part of a reactor accident scenario. They examine the multiple phenomena previously studied in SETs and/or CETs, and their mutual interactions. The data are needed to confirm that the modeling of the relevant phenomena and of their interactions is adequate, and that no important phenomenon has been neglected. The corresponding facilities are typically intermediate to large-scale models of a reactor part (core, circuits, containment).

It is mandatory that, during the validation process, the users do not vary the code input data until a reasonable fit to experimental results has been achieved.

The user must choose a single set of code options or parameters that allow achievement of a reasonable agreement for a large number of experiments covering a broad scope of conditions.

The final stage of code assessment consists in verifying the code capability to adequately simulate real SA scenarios at full scale. This stage includes several types of work:

- Benchmarking the code results of plant applications against results of other codes, either integral codes or detailed ones. The International Standard Problems (ISPs), organized in the framework of the OECD/NEA Committee on the Safety of Nuclear Installations (CSNI), provide a particularly valuable source of information since the experiments are well documented and extensive code-to-code and code-to-data comparisons were performed [2]. Over the last 35 years, 50 ISPs were organized. They are comparative exercises in which predictions of different computer codes for a given physical problem are compared with each other and with the results of a carefully controlled experimental study. The main goal of ISP exercises is to increase confidence in the validity and accuracy of tools that are used in assessing the safety of nuclear installations. They are often performed in two steps: first, a "blind" step where only the experimental conditions are given to the users, and second, an "open" step where the results of the experiment are made available to the participants before performing the calculations.
- Applying the code to real plant SA scenarios, which are very scarce except for the TMI-2 and Chernobyl accidents.
- Performing on one hand sensitivity analyses by making varying code parameters, in order to show the consistency and reliability of code results, including analysis of the influence of nodalization and of numerical time-steps; and on the other hand performing analyses of uncertainties (model parameters, extrapolation in scale and materials, etc.) [3].

Besides the inadequacies in the modeling of SA phenomena, users have to be well aware of the validation status of codes and must take into account their limitations when performing plant studies. Severe accident codes are difficult to handle, and their validation is not complete. They should not be used as "black boxes"; that is, their results should be interpreted according to the goal of the study for which they are used. There is a "user effect", as it appears clearly in ISPs where several teams apply the same codes with users' preferred choices for code inputs. Extensive training of new users should be mandatory, and efficient quality assurance procedures for reactor studies should be employed, including review of the results by experienced experts not directly involved in the work. Finally, users should not automatically trust the results of their calculation, but should instead make a critical analysis of their consistency and reliability. The importance of the quality of code documentation must also be underlined, in particular the users' guidelines.

For several of the codes, a formal and independent peer review of the models has also been an important part of the assessment of these codes, resulting in detailed in-depth assessment of the models (as, for example, the MELCOR Peer Review [4]).

8.3. DESCRIPTION OF THE MAIN SEVERE ACCIDENT INTEGRAL CODES

8.3.1. History

In the 1980s, after the TMI-2 accident, two integral codes were developed in parallel in the United States for PWR and BWR: MAAP as a commercial code, owned by EPRI and mainly used by the nuclear industry; and MELCOR, owned by the USNRC (the latter replaced STCP, the former system of codes that was developed by the USNRC in the 1970s [5]). In the 1990s, Japan and France started the development of similar codes, respectively, THALES (by JAERI [6]) and ESCADRE (by IRSN [7]), the latter only for PWR (including VVER). During this period, the two U.S. codes were released and used worldwide.

Then in the mid-1990s, IRSN in France decided to jointly develop with its counterpart German GRS a new integral code, ASTEC, on the basis of the best codes available in both organizations. The THALES project stopped in Japan, and instead NUPEC started the SAMPSON project. More recently, Russian organizations have launched the development of the SOCRAT system of codes.

Today five systems of codes exist worldwide: MAAP, MELCOR, and ASTEC, which are widely distributed and used, along with SOCRAT and SAMPSON. All these codes except for SAMPSON are still under active development. In the interest of completeness, one can note the existence of the ISAAC code, which is being developed by KAERI (Korea) for CANDU heavy water reactors.

8.3.2. Main Features

The purpose of integral codes is usually defined as yielding an overall analysis of SA with reasonably accurate estimates of the timing of the events and the consequences for the plant, or even for the environment. But, most important, the integral codes should provide sufficiently accurate estimates of the source term and should allow performance of PSA Level 2 studies for estimates of risks for SA scenarios.Their application to real accident sequences requires that the models capture the effects of essential physical phenomena: they cover almost all relevant SA phenomena (that were described in detail in the previous chapters), except in general steam explosion and containment structural mechanics. The codes must also simulate all the engineered safety systems (such as sprays and fan coolers) and the operator procedures that play a role in the SA scenario.

As regards the calculation time, it must be reduced as much as possible. A common agreement in European organizations on the fast-running aspect [1] led to the conclusion that computing time should be close to real time on workstations or PCs. Other requirements concern a robust numerical scheme and user-friendly pre- and post-processing tools (respectively for preparation of the inputs and analysis of the results).

Among the existing integral codes today, MELCOR and ASTEC have adopted the same philosophy as regards the complexity and flexibility of modeling. On the other hand, the priority of the MAAP code development is to obtain a very rapid and robust code, validated on the most relevant phenomena but with limited flexibility for plant nodalization and for selection of models: it is especially suited for large-scale PSA level 2 applications.

Common Features

The MELCOR, MAAP, and ASTEC codes integrate a large number of phenomena into a comprehensive plant simulation (reactor coolant system, secondary system, engineered safeguard system, containment, and auxiliary buildings). This allows calculating the surface/volume contaminations in the reactor buildings or the release of fission products (FP) to the environment.

They must be able to model in a coupled way all the SA interactive physical phenomena and transient processes. An important example is the close coupling between the primary system and the containment: the thermodynamic conditions in the reactor building are a function of the thermal-hydraulic evolution of the primary system and of the boundary conditions at the break. It allows, for example, taking into account automatic actions on the primary system when reaching a pressure threshold in the containment, or containment feedbacks on the primary system—for example, for simulation of shutdown conditions with an open vessel.

They are nodal codes; that is, the equations of conservation of mass and internal energy are solved for each control volume. The RCS (except MAAP4 where the number is fixed to 14) and the containment can be represented by any number of volumes: in the containment—in general between 10 and 50 for a usual plant application; and in the core that is represented in a two-dimensional axisymmetric geometry, generally 4 to 6 radial rings or meshes and 10 to 20 axial meshes.

These large codes—ca. 400,000 to 500,000 programming instructions and 1,000 to 1,500 subroutines—are organized in modules that correspond to parts of the reactor or to a set of phenomena and are linked through an executive module that synchronizes the time-step evolution, manages the data transfer, and maintains conservation of mass and energy.

Originally, the integral codes were conceived with as simple and as fast-running models as possible, with a substantial user flexibility in order to allow managing uncertainties through many sensitivity studies. Recently, the rapid increase of the computer performance enabled the substantial replacement of parametric models by semi-mechanistic (phenomenological) or mechanistic models, in a degree that depends on the codes.

Most of the models allow alternative options and a range of model parameters to enable bounding calculations and to address the many uncertainties.

The treatment of physical processes is complemented by special component models, for example, for containment sprays, fan coolers, and igniters.

The modeling of the in-vessel phases of the scenario presents a peculiarity with respect to other fields of physics: differently from design basis accidents, the geometry is continuously changing with time, with, for instance, the successive disappearances of materials or constituents due to melting, relocation, vaporization, etc. This is a real challenge for the numerical resolution of the physical equations. The corresponding in-vessel modules thus simulate the various stages of core melting, corium movement and interactions with vessel steel in the lower head, and corium ejection into the cavity: they must track the mass composition of materials as they melt, interact with each other, and are transported. They also track the masses and energy sources of radionuclides following fission product release and transport.

The containment phenomena are considered in four areas: thermal-hydraulic response, fission product and aerosol behavior, corium–concrete interaction, and iodine chemistry. The thermal-hydraulic response depends on steam condensation on heat-conducting structures and on the operation of engineered safety systems (condensers, fan coolers, recombiners, etc.). The corium–concrete interaction adds molten materials to corium and releases noncondensable and combustible gases. As the gases bubble through the corium pool, they entrain aerosols as well as some of the previously unreleased fission products, thus transporting them to the containment atmosphere. Their distribution is determined by the transport from the RCS break and the cavity to the containment, the flows between compartments, and the aerosol removal by filtering, deposition, and pool scrubbing. An important additional contributor to the source term is the gaseous iodine formed by chemical and radiochemical reactions in the containment sump, atmosphere, or on the painted walls.

The containment models thus provide the mechanical, energetic and fission product loads. The source term then depends on the capacity of the containment structure and engineered fission product removal features to accommodate those loads.

A crucial part of the codes is the user's definition of the accident scenario. The code must be very flexible and user-friendly to reproduce any possible scenario and operating procedure. This may involve user-defined functions of some code variables or parameters. It also implies simulating all main safety systems in the reactor, as well as the severe accident management (SAM) procedures:

- In RCS: depressurization systems, water injection inside the vessel on the core,

- In containment: passive autocatalytic recombiners (PARs), fan coolers, filters, spray, pressure suppression systems (as in VVER-440), ice condensers, igniter systems, and so on.

The experience shows that a coarse node representation of the hydraulic systems and the use of default models for physical processes provide a coherent account of the integral behavior despite a simplified treatment at component level, which fosters confidence in the suitability of these integral codes for most plant applications.

The main difference between these codes concerning their modeling approach lies in their modeling of thermal-hydraulics. It is detailed in the following discussion.

Approach for Thermalhydraulics

The thermal-hydraulic models are pivotal because they furnish the thermal and fluid conditions for all process and component models—degradation, fission product transport, corium–concrete, and so forth. But the levels of detail and completeness vary from one code to another:

- MELCOR: its thermal-hydraulic models are common for the whole code, and thus are used for both RCS and containment. The discretized hydrodynamic equations are formulated in a form that conserves mass and energy to a high level of accuracy. The numerical solution method is based on a semi-implicit solution of the equations, and the timestep is therefore subject to the material Courant limit. Sources, sinks, and other modifiers are calculated by closure relations and additional auxiliary models that are generally nonlinear.
- MAAP4: the momentum balances are considered as quasi-steady, which reduces them to algebraic expressions (Bernoulli-like equations). Therefore there are no differential equations for conservation of momentum, neither for the RCS nor for the containment.
- ASTEC: the RCS two-phase thermalhydraulics uses a five-equation numerical approach with a fully implicit solution. For containment thermalhydraulics, the mass transfer between nodes is described separately for gas and liquid flow by different momentum equations (unsteady, incompressible), taking into account height differences of the node centers. The mass flow rates are calculated without slip according to the composition of the source node. Furthermore mass transfer by diffusion is considered. The diffusion flow rate is calculated separately in a quasi-steady formulation for all gas components.

What is common and in the spirit of the coarse noding approach is that, especially for containment, fairly large regions are often represented by one hydraulic cell. Each cell normally comprises a pool and an atmosphere region that are normally considered to be in thermal nonequilibrium with each other, but each region may be two-phase and at saturation. The mass of each noncondensable gas species is also tracked in the atmosphere regions. Heat-conducting structures provide the physical and thermal boundary for the hydraulic system; these can ablate or undergo failure through thermomechanical loads. The codes also capture the opening of flow paths due to structural failure—for example, during degradation of a BWR channel box, RCS depressurization following surge line failure at high pressure, and ejection of core debris into the cavity.

Numerical Aspects

Some of the models can impose additional timestep reduction, while others may require subdivision of the overall timestep into subcycles. This fairly simple strategy does not necessarily optimize running speed but has the advantages of robustness and flexibility.

Running times depend on the codes. For MELCOR and ASTEC, typical plant calculations usually need several hours, sometimes days; the computing time is typically comparable (though often smaller) to real time on a modern PC, thanks to the use of coarse nodalization. For MAAP4, the running time is much smaller: about 2 hours to simulate 24 hours of real time for a sequence with core degradation. For all codes, this running time, besides the level of details of nodalization, also depends on the selected models since very detailed models might be an order of magnitude slower. Numerical difficulties are most commonly associated with events involving significant changes in geometry. These frequently result in timestep reductions and, in the worst cases, in failure to converge to a solution. Normally, problems of this kind are detected by the codes, so the calculation terminates cleanly; besides, the user can invoke various options to overcome numerical difficulties.

For many years, the codes ran on UNIX platforms for use in workstations. More recently, changes in the prevailing computing environment resulted in a move toward use of Windows and Linux Operating Systems on PCs.

Pre- and Postprocessing

Tools have been developed since the origin of the codes in order to help the user with code applications, both for preparation of input decks and for analysis of the calculation results.

For MELCOR, the commercially available simulator MELSIM provides a detailed interactive interface for management of input decks and calculations, online visualization, user intervention, replay, and post-rocessing. MELSIM is essentially an extension of MELCOR, where the base code has been coupled with the interface modules to provide an integrated package. For ASTEC and MAAP4, similar tools are available today. These tools are becoming continuously more and more user-friendly for all these codes.

Tools for evaluation of the propagation of uncertainties have also been coupled to these integral codes in the recent years.

8.3.3. ASTEC Code

The French IRSN and its German counterpart, the GRS, have jointly developed the ASTEC integral code since 1996: Accident Source Term Evaluation Code [8]. This code plays a leading role in the SARNET European network of excellence in the sixth and seventh FwP of the European Commission [10]. ASTEC integrates in the form of models all the knowledge generated by the network, and 28 network partners perform the assessment of the successive versions. One SARNET goal is to progressively make ASTEC the European reference code.

ASTEC is released to about 40 organizations in 20 countries, most of them in Europe, but also some in Russia and Asia. It can be applied to different types of PWR (French 900–1300 MWe, Konvoi 1300, Westinghouse 1000, etc.) and VVER-440 and 1000 NPPs. It simulates all stages of a severe accident for reactor operation or shutdown states. All types of SA scenarios can be simulated (e.g., SBO, all LOCAs, SGTR..), as well as most safety systems such as in-vessel early water injection, containment recombiners, spray or venting, or pressure suppression pools in VVER-440. Figure 8.1 illustrates the different code modules and the management of their coupling.

A strong feature of ASTEC is the high-quality modeling of fission product behavior, where the lessons learned from the Phébus FP experimental program and from many years of other analytical experiments around the world have been taken into account.

In the last years, ASTEC has been used intensively for the PSA2 IRSN studies on a French 1300 MWe PWR and for evaluation studies of the source term on

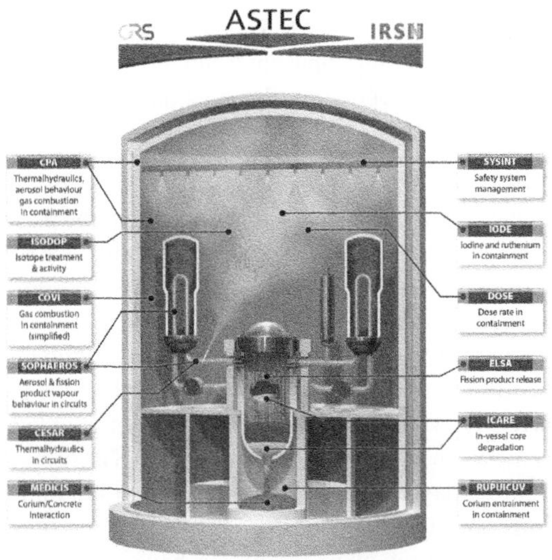

FIGURE 8.1 ASTEC V2 structural diagram.

French reactors. This work involved analysis of a considerable number of scenarios, assuming different initiating events and the operation of all safety systems. In addition, benchmarks were performed with calculations using the French thermal-hydraulic code CATHARE and with the ICARE/CATHARE IRSN mechanistic code. GRS has also performed several benchmarks on Konvoi-1300 plant applications against previous results from MELCOR calculations.

The international promotion of the code started in the EVITA European project (fifth FwP) that was devoted to assessment of the code by 15 organizations [11]. Within the SARNET context, this work was intensified so that 28 partners are performing validation tasks and benchmarking the newest versions of the code against other codes on various plant sequences in PWR 900, Konvoi 1300, Westinghouse 1000, VVER-440, and VVER-1000 reactors.

In the 1990s, a very intensive validation was performed on the codes from which ASTEC was developed (ESCADRE at IRSN, RALOC and FIPLOC at GRS), resulting in a solid and reliable reference basis. Taken all together, the code versions have been applied to a total of more than 160 experiments, including reference experiments such as ISP.

Figure 8.2 illustrates the validation of the ASTEC/IODE module on the Phébus FPT1 integral experiment (for more details on the experiment, see Section 8.4.2). Iodine is a highly reactive fission product due to its high level of oxidation, and evaluation of the iodine source term implies accounting for many physico-chemical reactions in the reactor cooling system and the containment. The module describes the behavior of all iodine chemical forms

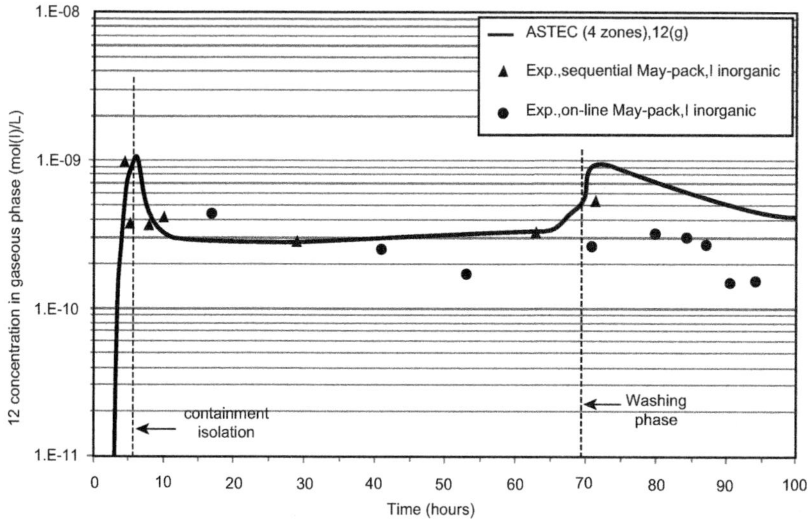

FIGURE 8.2 ASTEC validation on the Phébus FPT1 experiment: inorganic iodine evolution in the gaseous phase of the containment (circles: sequential measurement, triangles: online measurement, line: ASTEC computation).

in the containment (plus Ruthenium chemical forms) and it includes around 45 different physico-chemical reactions in gas and liquid phases and due to interaction of gaseous iodine with containment surfaces. Models for each of those reactions have been validated using analytical experiments, whereas the IODE module has been validated on all Phébus integral experiments. Computation of the evolution of inorganic iodine in the gaseous phase of the containment during the FPT1 experiment is compared in Figure 8.2 to both online and sequential experimental measurements, which shows a good agreement.

A new series V2 of the ASTEC code was released in mid-2009 by IRSN-GRS. The first version V2.0 is applicable to EPR, in particular modeling the core catcher. It includes the advanced models of the IRSN ICARE2 mechanistic core degradation code, in particular models of 2D magma-debris relocation within the core and vessel lower plenum. This allows taking full benefit of the ICARE2 validation work over more than 15 years [12]. Other new important models are vessel external cooling for in-vessel melt retention; chemical kinetics in the RCS gas phase in order to reproduce the Phébus FP experimental findings on the existence of gaseous iodine at RCS break; and behavior in RCS and containment of ruthenium species released after air ingress into the vessel following its failure.

In the next years, the code will continuously integrate models derived from interpretation of the current experimental programs, in particular the International Source Term Program (ISTP) at IRSN: CHIP for iodine chemistry in the RCS, and EPICUR for iodine chemistry in the containment; and effect of high fuel burn-up and of MOX fuel on core degradation and fission product release (VERDON at CEA). Other important model improvements will concern heat flux spatial distribution in the corium pool during molten-corium-concrete-interaction (MCCI) (VULCANO at CEA, OECD-CCI at ANL); and debris bed reflooding, based on the PEARL experiments in preparation in IRSN.

Work on adaptation of ASTEC models to BWR and CANDU-type reactors is ongoing. Most models currently in ASTEC have been shown to be applicable, except for core degradation where the efforts are currently focused.

ASTEC will keep its role as the repository of knowledge generated in the SARNET network through development of models on remaining key safety issues. In parallel, investigations have started to couple the code with dynamic PSA2 tools. Efforts are also focusing on acceleration of the code calculations, for instance, by parallelization of the programming, for its use in emergency response tools.

8.3.4. MAAP Code

The Modular Accident Analysis Program (MAAP) code was originally developed in the early 1980s by Fauske and Associates (FAI) as part of the

Industry Degraded Core Rulemaking (IDCOR) program, which was funded by EPRI on behalf of about 60 American utilities and other organizations. The overall goal of this program was to address the needs for a reliable tool for use in performing physical studies and in providing the SA consequences for conducting PRA/PSA studies.

Today EPRI is the owner of the MAAP code while FAI is still performing the code developments. Many nuclear power plant operators and designers have acquired a MAAP license and use the code for their safety studies. The MAAP Users Group gathers approximatively 50 organizations.

The current version of MAAP (version 4) can simulate the response of light water reactor power plants, both current designs and advanced light water reactor (ALWR) designs, during severe accident sequences, whatever the initial state of the installation is: full power or shutdown conditions [13]. There are two parallel versions of MAAP4, one for BWRs and one for PWRs (specific versions for VVERs and CANDUs also exist). The MAAP4 functional modeling is particularly adapted to investigate the impact of operator actions taken as part of accident management on the progression of sequences. MAAP4 can be used for the PSA level 2 studies supporting the elaboration of SAM procedures and for the design of mitigation systems.

FAI released a new version of MAAP4 (MAAP4.0.7) in January 2008. It includes the following models developed to support EPR design:

- Modeling of boron carbide thermal-chemistry (degradation, oxidation, and melting of B_4C),
- Modeling of core heavy reflector, including thermal attack, penetration by molten material, and corium draining down the side of the core barrel,
- Improvement of energy transfer from core debris to vessel and concrete during accumulation of corium in the reactor pit,
- Discharge of corium in the spreading compartment after breach of the melt plug,
- Diversity of floor and wall configurations in the reactor pit and spreading area, including concrete composition and heat removal from the bottom of the structural heat sinks,
- Modeling of AREVA NP PARs.
- Improvement of the pressurizer relief tank rupture and pressurizer breaks.

In practice, the physical validation of the MAAP4 code is a continuous process fed by the new experiments. Thus, FAI and EDF[1] regularly contribute to increasing the MAAP4 matrix of validation, in particular through taking part in OECD programs (ISP) and European projects (SARNET). They have applied the code to many experiments, most of which are listed in Table 8.1.

---

1. Since it acquired the code, EDF has developed its own know-how in developing and validating MAAP4 in addition to FAI activity. Since 1996, EDF has carried this logic of development further by producing EDF versions of MAAP that contain specific additional and improved modeling.

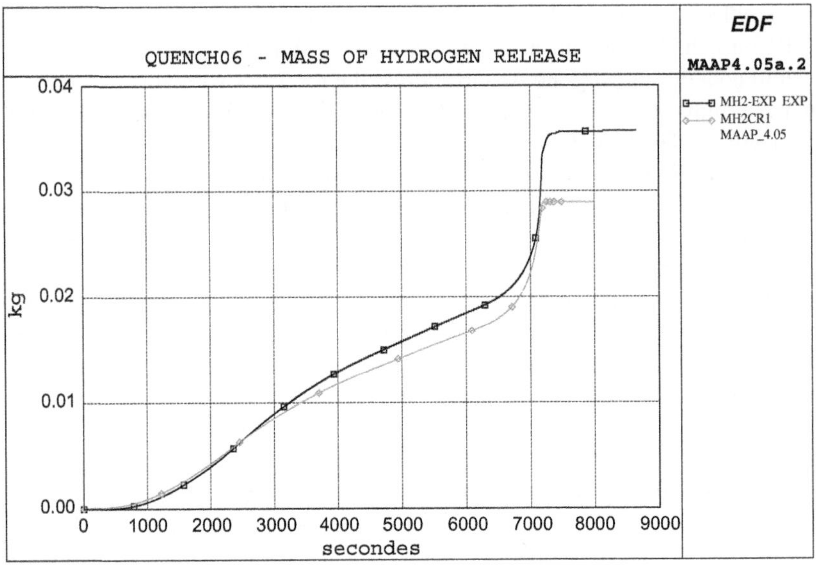

FIGURE 8.3 MAAP validation on the QUENCH06 reflood test—total mass of hydrogen released.

Figure 8.3 illustrates the validation on the QUENCH-06 experiment (KIT) that investigated the behavior of pre-oxidized LWR fuel rods on cooling down with water injection from the bottom. The test bundle is made up of 21 fuel rod simulators with a length of approximately 2,5 m. Twenty fuel rod simulators are heated over a length of 1 m, and one unheated fuel rod simulator is located in the center of the test bundle. The Zircaloy-4 rod cladding is identical to that used in LWRs. Heating is carried out electrically using tungsten heating elements that are installed in the center of the rods and that are surrounded by annular ZrO_2 pellets. In the heat-up phase, flows of superheated steam and argon enter the test bundle at the bottom. After control of the electrical power to reach a desired oxide layer thickness of about 200 μm, the power increase led to a transient heat-up rate of 0.32 K/s, which resulted in a significant increase of hydrogen release. The quench phase was initiated by injecting water at the bottom of the test section. The total measured hydrogen production resulted in around 35 g, with a maximum H_2 release rate of 0.24 g/s. Of the 35 g, about 2 g were estimated for the quenching phase. Figure 8.3 shows that MAAP reproduces the time evolution of hydrogen production, but with a 16% underestimation of the total mass. The final state of the bundle shows essentially an intact bundle geometry, whereas considerable cracking and some fragmentation of clad and ZrO_2 pellets occurred during the phase of water quenching.

Code-to-code comparisons are also performed. On the RCS thermal-hydraulics, MAAP4 results were confronted with results of mechanistic codes such as RELAP or RETRAN. The comparison was done on the basis of a large variety of sequences (LOCA, SBO, SGTR transients, etc.). A series of experiments (Westinghouse loop) made it possible to validate the modeling chosen to represent natural convection during high-pressure scenarios.

Further improvements to come will be released in a new major version of the code called MAAP5. These improvements are as follows:

- Significant increase of the maximum number of nodes in the core, lower head, and containment.
- Advanced containment modeling: gas flow induced by the substantial momentum of the jet discharging from a RCS break, which results in entrainment of condensate from wall surfaces that is recirculated as water droplets into the gas space.
- Advanced RCS model: individualization of the behavior of each coolant loop for asymmetric transients (such as blow-down of only one steam generator loop); a fully posed momentum equation that accommodates flow reversal in the event of a large transient such as a large break or an asymmetric trip of reactor coolant pumps; a mechanistic phase separation model in each node; extended nodalization (quadrant nodes) of flows in the vessel to account for pressure-driven flows and turbulent mixing.
- Miscellaneous new models: for neutronics, kinetics model (0-D point and 1-D space-time); shutdown modeling capabilities (mid-loop operation, reactor head open); break in the pressurizer, radiation from corium pool in cavity to reactor vessel wall, new Zr oxidation correlations, modeling of hybrid (Ag/In/Cd and B_4C) control rods, oxidation and degradation of B_4C control rods, new FP release correlations and improvements in iodine chemistry model; improvements of PARs modeling.

These evolutions are extensively benchmarked against a large variety of small-scale separate-effects tests and large-scale integral tests. This level of involvement makes it possible for the MAAP code to be a durable code for SA studies.

8.3.5. MELCOR Code

MELCOR has been developed at Sandia National Labs (SNL) since 1982 [14]. It was first released in the United States in 1986 and internationally in 1989. Several versions have been released to the user community during the ensuing years. The development has been almost entirely funded by the United States Nuclear Regulatory Commission (USNRC). The development is also strongly influenced by the participation of many international partners through the USNRC's Severe Accident Cooperative Research Program (CSARP). MELCOR is currently used by approximately 100 institutes in about 25 countries. It is designed primarily for PWRs and BWRs, but it has also been applied to other plant types and to nonreactor situations (e.g., spent fuel pools, active handling areas).

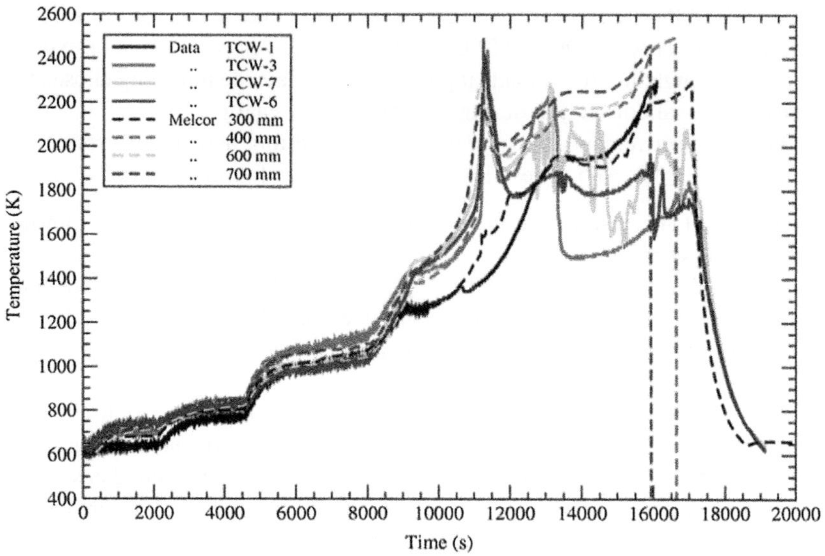

FIGURE 8.4 MELCOR 1.8.5 comparison with Phébus FPT1 bundle temperatures.

The current production version is identified with the series 1.8.6 released in 2006 [15][16][17]. Some of the component and subsystem modules were derived from existing codes already in use (e.g., CONTAIN, VANESA). Concurrently with these model improvements, the entire code has been rewritten in FORTRAN 95 in order to provide a more structured, more readily maintainable code, to benefit from modern compiler software, and to meet current standards of software engineering. This is designated as MELCOR 2, and version 2.1 was released in September 2009.

MELCOR validation has been the subject of numerous studies at the process, component, subsystem, and whole-plant levels; thus, they cover the full range of phenomena, most of which are listed in Table 8.1. Figure 8.4 shows the application to the Phébus FPT1 experiment (for more details on the experiment, see Section 8.4.2). MELCOR correctly reproduced the evolution of bundle temperatures (here four thermocouples are selected). The discrepancies on the maximum rod temperatures are about 100 to 200 K: underestimation on the first peak after the oxidation runaway, and overestimation in the next phase of bundle degradation.

The latest version 1.8.6 includes models for silver release and in-vessel melt pool evolution. The latter capability makes it possible to perform realistic estimates of lower head failure and to address issues such as the focusing effect of metallic layers on heat transfer from corium in the lower head to the vessel walls and in-vessel retention by external cooling of the vessel. Treatment of the vessel structures is extended to provide a more complete and consistent account of in-vessel behavior. Further generalization of the core material models may be necessary to capture intercomponent interactions, as highlighted by Phébus FPT3, which showed a strong effect of the B_4C control rod on degradation of the nearby fuel rods. The treatment of B_4C oxidation of gas-phase chemistry can furnish the conditions for formation of organic iodine in the circuit. An improved model is under development for Zircaloy cladding oxidation in air or steam-air mixtures, which is essential for its impact on the volatility of certain fission products. There is a strong need to improve the models for iodine chemistry in the containment.

Applications initially concentrated on broad source term evaluation and PSA support, but have since broadened into new areas such as accident management, mitigation, spent fuel issues, and operator training. The original simple treatments that were adequate for scoping and bounding calculations are being superseded by more physically based models. Among the questions being addressed now are whether to use igniters or passive autocatalytic recombiners for hydrogen control, and whether to inject available water within the vessel or flood the cavity. New applications include simulation of accidents in glove box facilities, reactor auxiliary buildings, nuclear warhead disassembly, and assessments of fires in nonnuclear buildings.

The wider range of application requires a greater level of detail and fidelity in representing the physical processes. Models with more detailed noding are now being used to resolve certain issues such as the effect of natural circulation in the vessel on core damage progression, and the effect of countercurrent flow in the hot legs and steam generator on pipe rupture. These extensions of the usage go hand in hand with the model developments identified previously. Related to this trend, MELCOR is being used to support the knowledge base through experiment planning and analysis, thus integrating it more closely into the learning cycle.

8.3.6. Other Integral Codes

SOCRAT

The SOCRAT system of codes is being developed by several Russian organizations (IBRAE, Kurchatov Institute, OPPE, etc.) for simulation of all stages of design and beyond design accidents, including failure of the pressure vessel and melt localization device [18][19].

The thermal behavior of the reactor systems and structural response is simulated by the set of the following codes: RATEG for thermalhydraulics in the primary system; SVECHA for core damage phenomena; HEFEST for late-phase core degradation; VAPEX for interaction of molten fuel with the coolant; RASPLAV and CONV for MCCI and corium interaction with the sacrificial materials in the core catcher; and KUPOL for thermalhydraulics in the containment.

Decay heat and radiological consequences of accidents are simulated by the set of the following codes: BONUS for inventories of FP and calculation of the decay heat; MFPR (developed in close cooperation with IRSN) for release of FP from the fuel; PROFIT for transport of FP through the primary system; and NOSTRADAMUS for migration of FP through the environment and impact on the population.

All constituent codes interact with each other using special interface routines that ensure feedback from one module to another. The system also contains a database on the thermo-physical properties of different materials.

Some of the codes were used for VVER plant applications, for instance, for Tian-Wan (China) and Kudamkulam (India) NPPs, in particular the justification of the system for hydrogen safety, analysis of core degradation, and relocation of the molten core to the core catcher used for the stabilization and termination of the accident.

SAMPSON

In the 1990s, the Nuclear Power Engineering Corporation (NUPEC, Japan) developed the severe accident analysis code SAMPSON in the framework of the IMPACT global R&D Japanese program [20]. The code is now jointly released by NUPEC and ISS (USA) [21].

The code initially aimed at being a system of mechanistic codes. It consists of 11 analysis modules to analyze the different phenomena (fuel rod heat-up, FP behavior, MCCI) and a control module that manages the analysis modules according to the progression of the accident. Recently, the merging of the SAMPSON and RELAP/SCDAPSIM codes started: the RELAP part of RELAP/SCDAPSIM was introduced into the in-vessel thermal-hydraulic module of SAMPSON, and the MATSIM part is used in the SAMPSON modules for calculation of physical properties.

SAMPSON was validated on several experiments, including calculations of OECD ISPs such as OECD ISP-45 (QUENCH-06 on reflood of degraded fuel rods) and ISP-46 (Phébus FPT1: see Section 8.4.2). The code has also been applied to various severe accident sequences in typical PWR and BWR plants.

8.4. VALIDATION OF THE INTEGRAL CODES

8.4.1. Validation Matrices

Based on the existence of different types of experiment, validation matrices have been built for the integral codes. Their objective is to link to each important phenomenon at least one experiment, or better a set of experiments at different scales, using simulant and/or prototypic materials. A basis for these matrices may be either the ISP experiments or the validation matrices that were built in the OECD/CSNI frame for core degradation [22] or containment thermalhydraulics [23].

In such matrices, the material effect must be taken into account because the properties of materials have a strong influence on the SA events: experiments with simulant materials can be selected, but experiments with prototypical materials, in particular for the corium behavior, should also be selected.

Scaling is also an important aspect of the validation process. Indeed, full-scale experiments cannot be performed for all the postulated accident conditions. Even at small scale, experiments employing prototypical reactor materials are expensive. The boundary conditions of the small-scale experiments are not typical for plant conditions (thermal-hydraulic conditions, heating methods). If the experiments performed are not scaled properly, the validation efforts would become meaningless.

TMI-2 was an accident at full scale, providing insights into the behavior of a plant during an SA, particularly during the late stage of the accident during and following the formation of a large ceramic molten pool. However, the available measurements from TMI-2 were insufficient to provide information about progression of the accident. An extensive analysis with the support of codes was necessary to get a good understanding of what happened during the accident.

The validation matrices that were adopted for the three main integral codes share a large common part: Table 8.1 summarizes these main experimental programs.

TABLE 8.1 Main Experimental Programs Used for Validation of Integral Codes

Physical process	Program name	Organization (country)
Integral tests	TMI-2 accident	–
	LOFT-LP- FP2	INEL (USA)
	Phébus FP (incl. FPT1/ ISP46)	IRSN (France)
RCS thermalhydraulics	BETHSY (ISP38)	CEA (France)
Core degradation	CORA (7, 12, 13/ISP31....)	KIT (Germany)
	QUENCH (01, 04, 06/ ISP45, 07....)	KIT
Fission product release	ORNL (VI-2 to VI-7)	ORNL (USA)
	VERCORS (1 to 6)	CEA
Aerosol transport in the RCS and in the containment	FALCON	AEAT (UK)
	VERCORS (HT1 to HT3)	CEA
	LACE	INEL
	KAEVER	Battelle (Germany)

(Continued)

TABLE 8.1 **Main Experimental Programs Used for Validation of Integral Codes—cont'd**

Physical process	Program name	Organization (country)
Vessel mechanical failure	LHF-OLHF	SNL (USA)
	FOREVER	KTH (Sweden)
Heat transfer in corium molten pools	COPO	Fortum (Finland)
	ULPU	USCB (USA)
	BALI	CEA
Fragmentation of corium in water	FARO (L06, L08, L11, L14...)	JRC Ispra
Corium entrainment during Direct Containment Heating (DCH)	ANL (U1B...)	ANL (USA)
	Surtsey (IET-1, IET-8B...)	SNL
	DISCO (C, H)	KIT
Molten corium–concrete interaction	BETA (v5.1, v5.2, v6.1...)	KIT
	SWICSS	SNL
	OECD-CCI (1, 2..., 6)	ANL
	ACE (L2, L5...), MACE	ANL
Iodine chemistry in containment	ACE/RTF	AECL (Canada)
	CAIMAN	CEA
Containment thermal-hydraulics	NUPEC (M4.3, M7.1....)	NUPEC (Japan)
	VANAM-M3 (ISP37...)	Battelle (Germany)
	TOSQAN (ISP47...)	IRSN
	MISTRA (ISP47...)	CEA
Hydrogen combustion in containment	HDR (E12.3.2...)	Battelle
	RUT	RRC-KI (Russia)

8.4.2. An Example of a Validation Exercise: ISP46

The ISP46 was devoted in 2002–2003 to a code comparison exercise based on the Phébus FPT1 integral experiment [24] (see Chapter 5 for more details on the overall Phébus FP program). It was a very important exercise because it aimed at assessing the capability of computer codes to reproduce an integral simulation of the SA physical processes, with prototypical materials and at a large scale (e.g., a large height of the bundle). It was conducted in four phases, with all relevant experimental results being available to the participants:

- Fuel degradation, hydrogen production, fission product, and structural material release;
- Fission product and aerosol transport in RCS;
- Thermal-hydraulics and aerosol physics in containment;
- Iodine chemistry in containment.

The emphasis was on integral calculations involving all phases. The aim was not to carry out interpretation work, but to use the codes as in plant studies—that is, with standard models/options as far as possible, representing the facility in a similar level of details. This constituted the mandatory "base case" calculation, typically: for the bundle, 11 axial nodes and three to five radial rings, with normally one or two thermal-hydraulic flow channels; for the circuit, 11 nodes; for the containment, one node for the main volume and one for the sump.

Fifteen different codes were used by 33 organizations from 23 countries, including ASTEC, MAAP, MELCOR, and SAMPSON. Nine integral calculations were performed: four with ASTEC, one with MAAP, three with MELCOR, and one with SAMPSON.

The FPT1 test bundle included 18 PWR fuel rods ($\sim$6.85% initial enrichment in U^{235}) previously irradiated to a mean burn-up of 23.4 GWd/tU, two instrumented fresh fuel rods ($\sim$3.5% enrichment in U^{235}), and one silver-indium-cadmium (AIC) control rod. The bundle was pre-irradiated for $\sim$7 days with a mean bundle power of $\sim$205 kW in the Phébus facility before the experimental phase of the test itself in order to generate short-lived fission products in the fuel. After the preconditioning phase, a period of 36 hours was necessary to establish the boundary conditions of the experimental circuits, in particular for drying the bundle using neutral gas. The experiment itself then began by injecting steam into the bundle and gradually increasing the core nuclear power.

The following phenomena/parameters were in general well simulated by the codes:

- Bundle: thermal response (given the adjustment of input nuclear power and shroud thermal properties within experimental uncertainties), hydrogen production (Figure 8.5), bundle final state material distribution (given the suitable reduction of the bulk fuel relocation temperature from the ceramic value), total release of volatile fission products (Figure 8.6);
- Circuit: total retention of fission products and structural materials;
- Containment: thermal-hydraulic behavior (as average gas temperature, pressure, relative humidity, and condensation rate) and aerosol depletion rates;
- Chemistry: Ag/I reaction in the liquid phase.

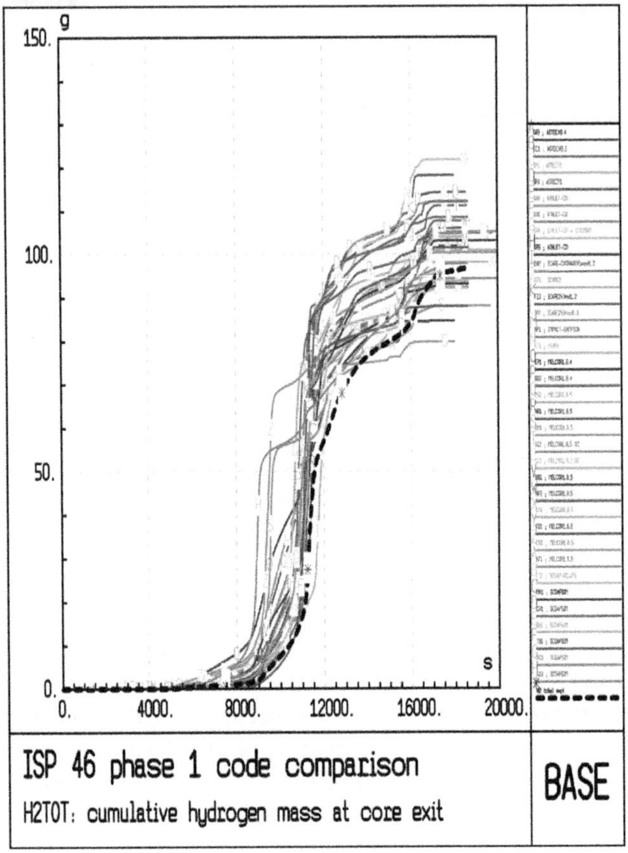

FIGURE 8.5 FPT1 bundle total mass of hydrogen released (experiment = dotted black line).

The following phenomena/parameters were reasonably well simulated, but some modeling improvement was judged as desirable:

- Bundle: outlet coolant temperatures (overprediction), time dependence of volatile FP release (generally too fast a release at low temperatures);
- Circuit: distribution of deposition in the circuit (underestimation in the upper plenum where vapor condensation and thermophoresis are the dominant mechanisms, overestimation in the steam generator hot leg where the mechanisms are thermophoresis for all elements and vapor condensation for I and Cd), noting that a too coarse noding leads to underestimation of deposition;
- Containment: relative importance of the two main depletion processes (diffusiophoresis and gravitational settling) (Figure 8.7).

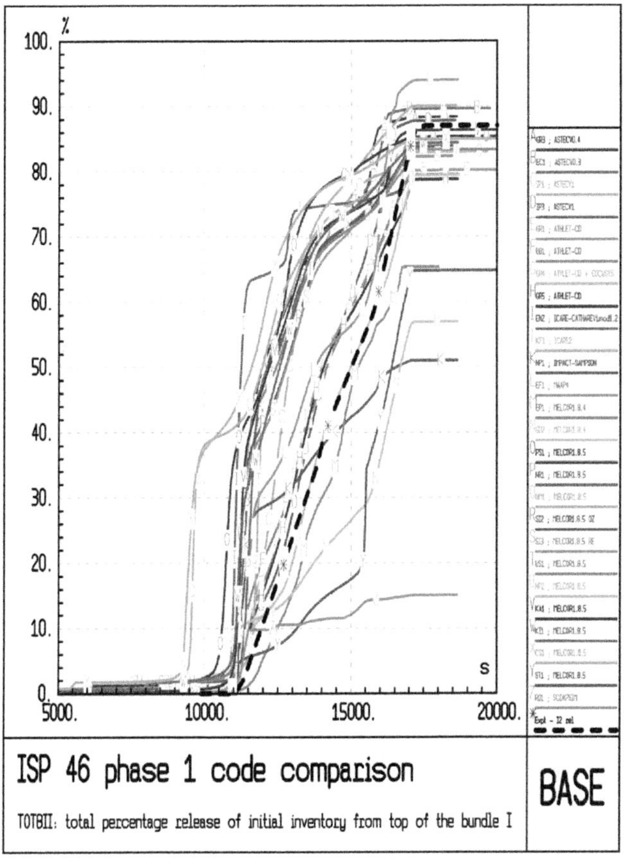

FIGURE 8.6 FPT1 iodine fraction released from the fuel (experiment = dotted black line).

The following phenomena/parameters were not well simulated, and substantial model development was judged as necessary:

- Bundle: release of medium and low-volatile FP and of structural materials (Ag/In/Cd from the control rod, tin from the Zircaloy cladding);
- Circuit: iodine speciation and physical form;
- Chemistry: gas phase reactions (Figure 8.8), organic iodine reactions, including production and destruction through radiolytic processes.

As a conclusion of this exercise, the results of integral codes were examined with respect to the "key signatures" of plant sequence calculations, namely, the core final state (relevant to in-vessel corium retention) and the fission product source term. Good agreement for the bundle final state could be obtained with suitable reduction of bulk fuel relocation temperature, but the need of

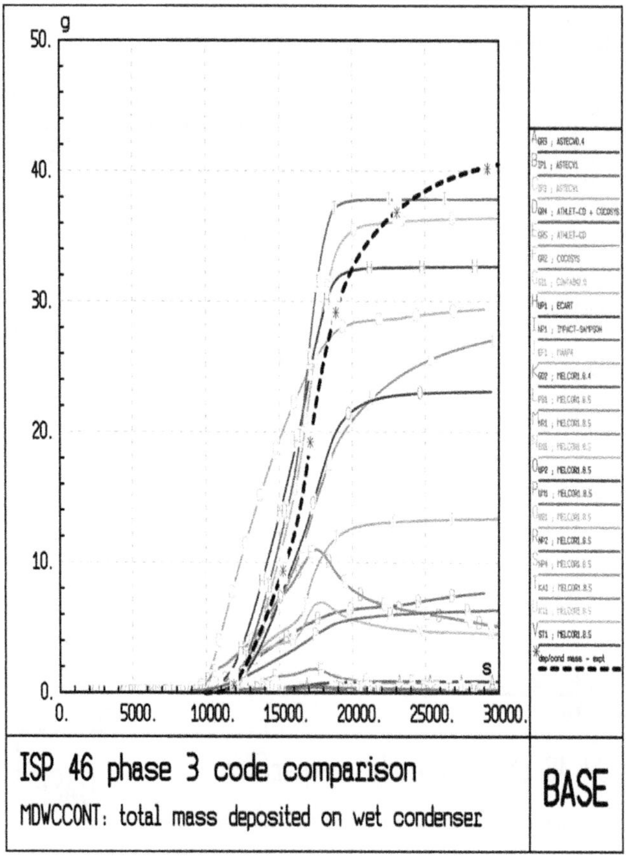

FIGURE 8.7 FPT1 mass of aerosols deposited on the wet condensers (experiment = dotted black line).

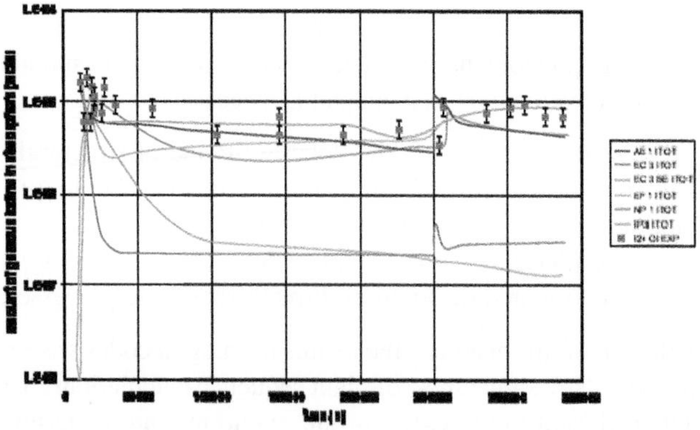

FIGURE 8.8 Iodine mass in the FPT1 containment gaseous phase (experimental results indicated by square dots).

sensitivity calculations on relocation temperature was underlined for plant studies. Further studies were recommended on control rod failure (influence of control rod materials) and the fuel rod oxide shell breach criterion (first movement of U/Zr/O melt). The good prediction of hydrogen production, generally near the upper bound of the experimental uncertainty range, was an important conclusion relevant to safety.

With regard to the source term, the accuracy of containment calculations with integral codes is highly sensitive to the results of previous stages through the propagation of uncertainties: for instance, kinetics and the final amount of release of FP and structural materials (Ag, In, Cd and Sn) from the bundle, or temperatures at the entrance to the circuit, which strongly influence the deposition pattern. The iodine speciation and physical form in the circuit were poorly predicted: no code reproduced the observed gaseous iodine fraction in the RCS.

There was considerable scatter among the results obtained from each code by different users, that is, the "user effect," as illustrated by Figures 8.6 and 8.7.

Several years later, it can be observed that most code models for the above poorly-simulated phenomena have been significantly improved, mainly concerning those for FP release from the fuel and iodine chemistry in the containment. A similar exercise is currently being performed on the Phébus FPT3 integral experiment (which involves a boron carbide control rod rather than Ag/In/Cd as in FPT1) in the SARNET frame. It should indicate the recent progress of code models, in particular in the ASTEC code where considerable efforts were focused on FP behavior.

8.4.3. Benchmarks on Plant Applications

Such benchmarks have been performed in the SARNET frame between 2004 and 2011 on diverse NPP types (PWR, VVER-440, and VVER-1000), mainly through comparisons of MAAP or MELCOR with the ASTEC code [25]. These code-to-code comparisons showed that the trends and orders of magnitude of results obtained with the three main integral codes are similar for in-vessel and containment phenomena. This work is continuing in this SARNET frame, using the new ASTEC V2 series of codes, focusing on the evolution of fission products in the RCS and the containment.

A useful illustration of such benchmarks is the OECD/NEA/CSNI exercise that was performed in 2008–2009 in order to determine the ability of current codes to predict core degradation in nuclear reactors [26]. The TMI-2 reactor was selected using a postulated core degradation scenario, specified with simple initial and boundary conditions in order to minimize the influence of the uncertainty of these conditions (especially the make-up and let-down coolant flows that were difficult to derive from the accident).

The following codes were used by eight organizations: ASTEC V1.3, MELCOR 1.8.5 and 1.8.6, MAAP 4.03, ATHLET-CD Mod 2.1A, ICARE/CATHARE V2.1.

The initial plant state corresponds to that of the standard TMI-2 accident sequence. The accident is initiated by a small break with a size of 0.001 m^2 located at 4 m along the hot leg A, and followed by a stopping of primary pumps when the mass of water contained in the primary system becomes lower than 85 tons. The high-pressure injection operation is delayed until 5000 s after the stopping of primary pumps. Another assumption is that no pilot-operated relief valve failure and no let-down flow occur. This scenario leads to a significant degradation and core melting before the reflooding of the core.

The steady state was well predicted by all codes with very small differences. The sequence was divided into three phases: initial thermal-hydraulic phase, core degradation phase, and reflooding phase.

The first thermal-hydraulic phase lasted up to the stopping of main pumps at about 5000 s. when the mass of water contained in the primary system became less than 85 tons. It is governed by the mass flow rate at the break and the heat transfer to the secondary circuit. Figure 8.9 shows a very good agreement of the results of mass flow rate for most of the codes. Discharge from the break can be divided in three parts: liquid discharge up to saturated conditions at about 200 s, two-phase discharge until about 4500 or 5000 s when the main pumps are stopped, and steam discharge up to the start of the reflooding phase at about 10,000 s.

The heat transfers with secondary side, the behavior of the steam generators, the core temperatures, and the pressurizer level show a very good agreement among almost all codes. When the pumps stop, the mass

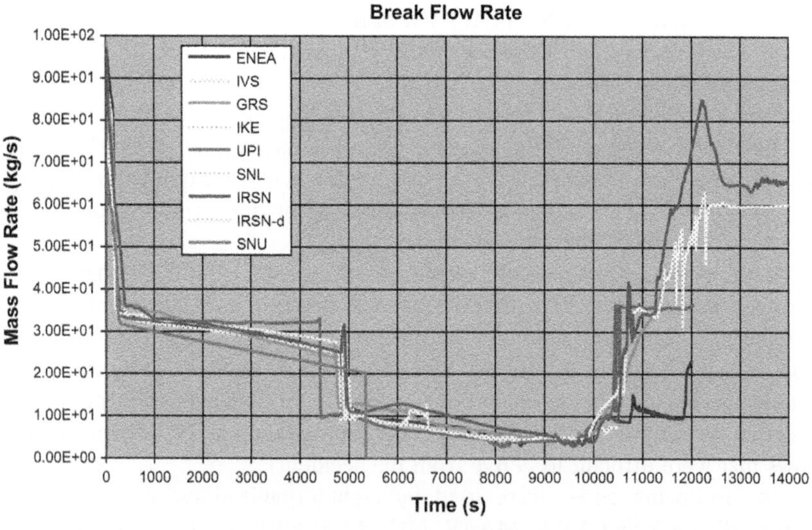

FIGURE 8.9 Break flow rate calculated in the TMI2-like exercise.

flow rate at the break is significantly reduced due to the separation of steam and liquid, and the primary mass decreases more slowly. Heat transfer to the secondary side is strongly reduced, and the evolution of the primary pressure becomes independent of the secondary pressure. This phase is characterized by a progressive dry-out and uncovery of the core, followed by the increase of temperature of the rods and the oxidation of the claddings.

After the pumps stop, water in the loops is drained down into the vessel, which leads to an increase of water volume in the vessel. All codes predict this rapid increase of the collapsed level. Then a progressive decrease of the water level is predicted, corresponding to the dry-out and core uncovery. Despite differences in the time of beginning of core uncovery, the agreement is good, in particular on the prediction of the core dry-out.

All codes show a slow decrease of the mass flow rate at the break, from approximately 19 kg/s to a rather stable value of approximately 4 kg/s before reflooding (Figure 8.9).

The evolution of core temperatures predicted by participants showed several differences. The main discrepancy is the time of beginning of heat-up, which varies from 500 s to 1800 s after pump stop, with a variance of 40%. This is partly due to the discrepancies on timing of core uncovery, but also to differences in the melt relocation models or cladding failure criteria that lead to a different propagation of the "hot" oxidation front.

The oxidation runaway is predicted with a good agreement between 7500 s and 8000 s, as well as the increase of temperature due to temperature escalation. The predicted temperature at the end of escalation is approximately 2300 K, which is related to the cladding failure criteria. The cladding failure and first melting relocation are also predicted at approximately the same time for most codes. In the next heat-up phase, the main difference between results is the maximum temperature, which is closely related to the choice of the melting temperature of UO_2.

The predicted rate of hydrogen generation is very similar for most results (approximately 0.6 kg/s in average), and the main differences are the times of maximum production (Figure 8.10) (linked to the differences in the core temperature curves).

The time evolution of the total mass of molten materials shows some differences but the agreement is rather good on the final mass (Figure 8.11). However, the predicted mass of molten metals shows a smaller variation for metals due to the better knowledge of their melting temperatures than for oxides.

The states of the core predicted at the beginning of the oxidation runaway appear to be rather consistent. The core temperature distributions are similar: a cold lower part and a hot upper part that is uncovered by water. When corium relocation starts, some differences appear, but just before the reflooding all calculations predict a very compact region of accumulated materials at

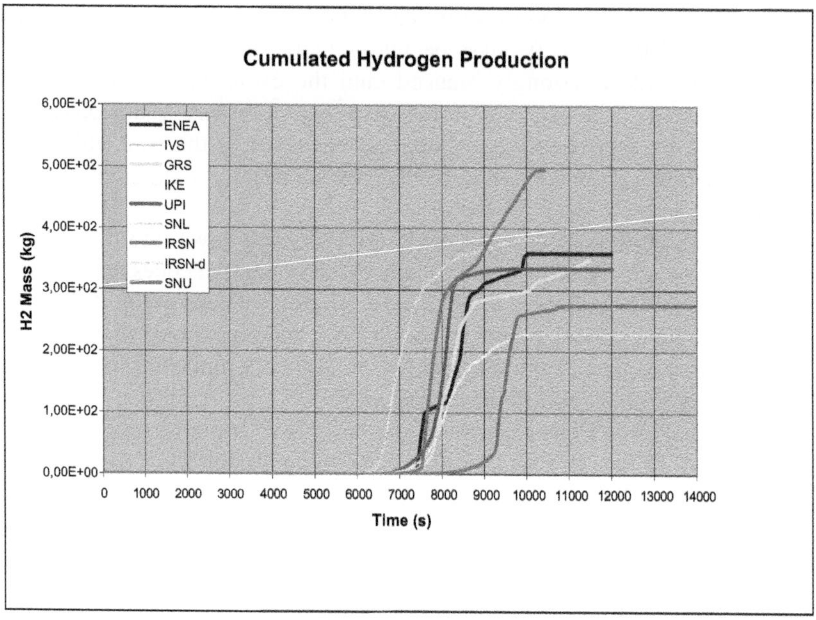

FIGURE 8.10 Cumulated in-vessel hydrogen production calculated in the TMI2-like exercise.

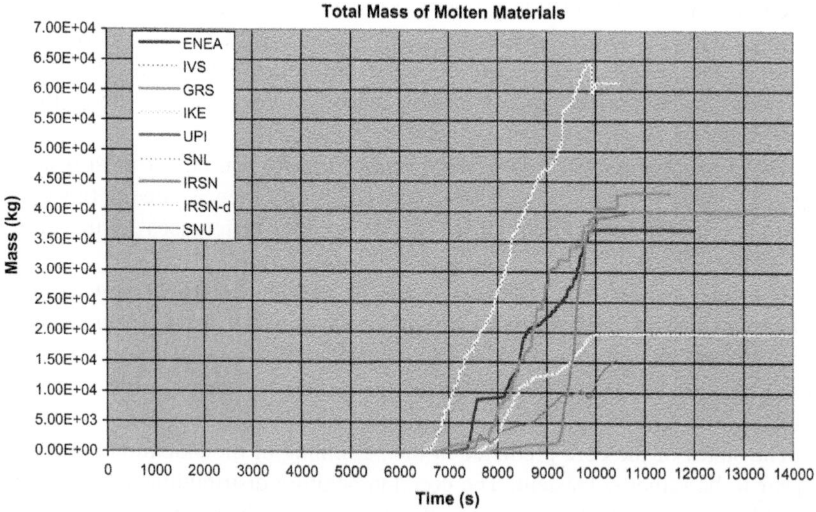

FIGURE 8.11 Evolution of the total mass of molten materials calculated in the TMI2-like exercise.

approximately 1 m elevation, which can be interpreted as a "crust" or "crucible" made of relocated materials.

During the reflooding phase, all codes predict a rapid increase of the water level in the core and the core bypass with approximately the same velocity, and a stable final water level in the core. The average predicted pressure increase of about 50 bars is in agreement with the TMI-2 scenario. But the codes do not agree on the final coolability of the core.

For almost all the participants, there is no significant increase of the degradation or core melting during reflooding, and the hydrogen production is quite limited: this does not agree with experimental findings and the TMI-2 observation where half of the total hydrogen release was supposedly produced during reflooding (see Chapter 2). This may be a consequence of an inaccurate modeling of melt formation and relocation, as well as of oxidation of metal-rich corium during its relocation.

Overall, the results of this exercise benchmark are quite encouraging with respect to the situation some years ago: first on the numerical robustness of the codes, and second on the consistency of the results. The general agreement between codes can be considered as good for the initial thermal-hydraulic phase and the degradation phase, up to reflooding. For the reflooding phase, not all code results are able to reproduce the hydrogen production as shown by experiments and by the TMI-2 assumed evolution. Although the abilities and robustness of codes have been considerably improved, more efforts are needed on modeling and assessment for an accurate prediction of the reflooding phase. An extension of this benchmark is starting in the SARNET frame, still in collaboration with OECD/NEA/CSNI.

8.4.4. Overall Status of the Validation of Integral Codes

For the main codes, namely, MAAP, MELCOR and ASTEC, their assessment is performed within their respective users groups: MCAP (MELCOR Code Assessment Program) for MELCOR assessment by SNL and international partners, administered by USNRC; ASTEC Users Club in the frame of SARNET; and the MAAP4 Users Group. The main conclusions of the current status of validation are as follows:

- Regarding the in-vessel stage of the accidents, the predictions are good for RCS thermalhydraulics, heat-up of core structures and cladding oxidation. Larger differences can be observed in the in-vessel late phase, from time of melt generation and relocation to the vessel lower head, but they can be judged as acceptable. However, quenching phenomena are still not well modeled, in particular related to the debris bed and melt-pool coolability in the late in-vessel phase. These deficiencies are being addressed by ongoing experimental and theoretical efforts.
- For the ex-vessel stage, MCCI phenomena still need complementary knowledge through experimental programs to reduce the uncertainties, especially

concerning the 2D heat distribution along the corium–concrete interface and the debris coolability. DCH models, although they can be fitted to the different existing cavity geometries thanks to several experimental programs, are not accurate enough for plant applications.

- For FP behavior, the models of release from the core can be considered as validated in most cases. The main uncertainty remains the gaseous iodine behavior in RCS, in particular due to the importance of the chemistry speciation in gas phase. Progress was made in the last years for iodine and ruthenium behavior in containment but further work is still necessary.
- For containment behavior, the thermal-hydraulic and aerosol models can be considered as globally acceptable.

8.5. SOME PERSPECTIVES FOR INTEGRAL CODES

Since the beginning of their development, the main goal of these codes was to provide an integrated description of the major safety-related areas—core degradation, fission product release, ex-vessel phenomena, and containment response—which could support scoping estimations of the source term and of SA consequences, for example, containment integrity, for risk evaluations in plant PSAs. Because of their inherently fast execution time and their integrated feature, these codes are appropriate for sensitivity studies and for analyses of uncertainties. Recently, improvements have enabled them to be used in a much wider range of situations, to meet a larger set of needs, and to achieve more stringent demands as regards accuracy, completeness, run-time practicality and user image. Most of these codes include progressively more mechanistic models.

In complement, the mechanistic codes continue to be employed to refine the estimated consequences that are calculated by the integral codes. In some specific cases, the CFD codes should be used for detailed analysis of specific phenomena.

Nowadays, the integral codes are viewed as state-of-the-art tools for source term calculations, and they serve as a repository of knowledge of SA phenomenology. Their flexibility makes them applicable to other accidents such as in spent fuel pools and in future NPP designs. The adaptations to Gen.III reactors, like EPR or AP-1000, are already under way. Adaptations are also being planned, in the case of MELCOR and ASTEC, to severe accidents in Gen.IV reactor types (respectively for VHTR and SFR). Another trend is the wider use of these codes or their modules in simulators for operator training and for emergency response tools.

REFERENCES

[1] H.-J. Allelein et al., Validation strategies for severe accident codes (VASA), in: G. Van Goethem (Ed.), EU Co-sponsored research on Containment Integrity, EUR 19952 EN, Brussels, pp. 295–324.

[2] H. Karwat, CSNI: International Standard Problems (ISP), brief descriptions (1975–1999), NEA/CSNI/R(2000)5, February 2000.

[3] J. Baccou, E. Chojnacki, Contribution of the mathematical modelling of knowledge to the evaluation of uncertainty margins of a LBLOCA transient (LOFT-L2-5), Nucl. Eng. and Design 237 (19) (2007) 2064–2074.
[4] B.E. Boyack, V.K. Dhir, J.A. Gieseke, T.J. Haste, M.A. Kenton, M. Khatib-Rahbar, M.T. Leonard, R. Viskanta, MELCOR Peer Review, Los Alamos Lab. Report LA-12240, 1992.
[5] J.A. Gieseke, P. Cybulskis, H. Jordan, K.W. Lee, P.M. Schumacher, L.A. Curtis, R.O. Wooton, S.F. Quayle, V. Kogan, Source term Code Package: A User's Guide (Mod1), G Report NUREG/CR-4587, BMI-2138, Battellle Colombus Division, Ohio, USA, July 21, 1986.
[6] A. Hidaka, T. Kudo, J. Ishikawa, T. Fuketa, Radionuclide Release from Mixed-Oxide Fuel under High Temperature at Elevated Pressure and Influence on Source Terms, Journal of Nuclear Science and Technology 42 (5) (2005) 451–461.
[7] B. Linet, A. Maillat, ESCADRE code development and validation—An overview, Proceedings of the International Topical meeting Severe Accident Risk and management (SARM'97), Piestany, Slovakia, June 16–18, 1997.
[8] W.P. Baek, J.E. Yang, J.J. Ha, Safety assessment of Korean nuclear facilities: current status and future, Nuclear Engineering and Technology 41 (No. 4) (May 2009) 391–402.
[9] J.-P. Van Dorsselaere, C. Seropian, P. Chatelard, F. Jacq, J. Fleurot, P. Giordano, N. Reinke, B. Schwinges, H.-J. Allelein, W. Luther, The ASTEC integral code for severe accident simulation, Nuclear Technology 165 (No. 3) (2009) 293–307.
[10] J.-P. Van Dorsselaere, J.-C. Micaelli, H.-J. Allelein, ASTEC and SARNET—Integrating severe accident research in Europe, ICAPP'05, Seoul, Korea, May 15–19, 2005.
[11] H.-J. Allelein, et al., European validation of the integral code ASTEC (EVITA), Nucl. Eng. and Design 221 (2003) 95–118.
[12] P. Chatelard, J. Fleurot, O. Marchand, P. Drai, Assessment of ICARE/CATHARE V1 severe accident code, ICONE-14, Miami, USA, July 17–20, 2006.
[13] F. Rahn, Technical Foundation of Reactor Safety—Knowledge Base for Resolving Severe Accident Issues—Revision 1, EPRI Report No. 1022186, October 2010.
[14] R.O. Gauntt, J.E Cash, R.K. Cole, C.M. Erickson, L.L. Humphries, S.B. Rodriguez, M.F. Young, MELCOR Computer Code Manuals, NUREG/CR-6119, Rev. 3 (2 vols.), September 2005.
[15] T.J. Haste, J. Birchley, J. Vitazkova, E. Cazzoli, MELCOR/MACCS simulation of the TMI-2 severe accident and initial recovery phases, off-site fission product release and consequences, Nucl. Eng. and Design 236 (2006) 1099–1112.
[16] T.J. Haste, J. Birchley, M. Richner, Accident management following loss-of-coolant accidents during cooldown in a Westinghouse two-loop PWR, Nucl. Eng. and Design 240 (2010) 1599–1605.
[17] J. Birchley, T.J. Haste, M. Richner, Accident management following loss of residual heat removal during mid-loop operation in a Westinghouse two-loop PWR, Nucl. Eng. and Design 238 (2008) 2173–2181.
[18] A.A. Zaitsev, A.A. Lukyanov, T.V. Popova, A.Y. Kiselev, V.F. Strizhov, A.L. Fokin, Development of the integral code for simulation of beyond design basis accidents at NPP with VVER, Proceedings of the International Conference "Provisions to safety of NPP with VVER," Podolsk, Russia, 2005.
[19] L.A. Bolshov, V.F. Strizhov, SOCRAT—The System of Codes for Realistic Analysis of Severe Accidents, ICAPP '06, Reno, NV (USA), June 4–8, 2006.
[20] H. Ujita, T. Ikeda, M. Naitoh, Severe accident analysis code SAMPSON improvements for IMPACT project, JSME International Journal, Series B 45 (No. 3) (2001) 607–614.

[21] M. Naitoh, S. Hosoda, C.M. Allison, Assessment of water injection as severe accident management using SAMPSON code, ICONE-13, Beijing (China), May 16–20, 2005.

[22] K. Trambauer, T.J. Haste, B. Adroguer, Z. Hózer, D. Magallon, A. Zurita, In-vessel core degradation code validation matrix. Update 1996–1999, NEA/CSNI/R(2000)21.

[23] H. Glaeser et al., CSNI Integral test facility validation matrix for the assessment of thermal hydraulic codes for LWR LOCA and transients, NEA/CSNI/R(96)17.

[24] B. Clement, T.J. Haste, Comparison report on International Standard Problem ISP46 (Phébus FPT1), report SAM–THENPHEBISP–D005, THEmatic Network for a PHEBus FPT-1 International Standard Problem, EC Contract No. FIKS-CT-2001-20151, July 2003.

[25] J.-P. Van Dorsselaere, A. Auvinen, D. Beraha, P. Chatelard, C. Journeau, I. Kljenak, A. Miassoedov, S. Paci, Th.W. Tromm, R. Zeyen, Some outcomes of the SARNET network on severe accidents at mid-term of the FP7 project, Proceedings of ICAPP 2011, Nice (France), May 2–5, 2011.

[26] S. Weber, H. Austregesilo, F. Fichot, O. Marchand, G. Bandini, M. Barnák, P. Matejovič, S. Paci, K.Y. Suh, M. Buck, L. Humphries, A Benchmark Exercise on an Alternative TMI-2 Accident Scenario, 13th International Topical Meeting on Nuclear Reactor Thermal Hydraulics (NURETH-13), Kanazawa City (Japan), September 27–October 2, 2009.

Appendix 1

Corium Thermodynamics and Thermophysics

Outline

1. CORIUM THERMODYNAMICS

(M. Barrachin)

For prevention, mitigation, and management of severe accidents, several phenomena related to the core melt have to be understood: fuel degradation, melting and relocation, convection in the core melt(s) (named corium), coolability and retention of the corium, fission product release, hydrogen production, behavior of the materials of the protective layers, and ex-vessel spreading of the core melt(s). To determine the impact of these phenomena, it is necessary to call on several disciplines, including thermodynamics, fluid dynamics, mechanics, and others. Among these fields, thermodynamics plays a very particular role. It provides knowledge on the state of order of a material (in other words, its phases in an equilibrium state); the manner in which this state is modified depending on parameters, such as composition and temperature; and subsequently, the conditions in which a transformation may occur in a given direction. Of course, it does not say anything about the transformation mechanisms themselves, or their duration, and thus it says nothing about the kinetics of the equilibrium approach. But it is important to make a distinction between those reactions that are possible, and those that are not, which is what thermodynamics is capable of achieving with certainty.

The specific difficulties of treating the thermodynamics of corium and fission products arise from the fact that the chemical nature of materials can change depending on the temperature and the oxygen potential, which can vary during a severe accident transient. Difficulties also derive from the fact that numerous chemical elements have to be considered in order to have a reliable representation of the thermodynamic behavior of the core melt. For example, a thermodynamic study of fuel degradation must include not only the chemical reactions between fuel and fission products, but also any possible interaction between the degraded fuel and its cladding (and some structural materials, as necessary), for various gaseous atmospheres representative of accident scenarios (oxidizing or reducing conditions), and for temperatures that may range up to the fuel melting point. It quickly becomes apparent that this represents an extremely complex chemical system, especially in terms of scope (i.e., the number of chemical elements to be considered), and the possible associations that may be formed between the various chemical elements (through the formation of compounds, for example). It is therefore necessary to work with a thermodynamic model that covers a sufficient number of chemical elements to accurately describe the thermodynamic behavior of each one.

1.1. Different Approaches

Conventionally, thermodynamic knowledge of a material is ascertained by establishing a phase diagram—that is, a graphic representation of the state of order of the material, which generally depends on its composition and temperature. The phase diagram is determined experimentally by measuring different properties (phase-change temperature, phase composition after quenching, etc.). Compendiums list these diagrams, established experimentally for simple materials, binary systems (i.e., compounds formed by two chemical elements, as in [1], and some ternary systems (three elements). For corium, the task is much more complex, given the large number of chemical elements to be considered and the wide temperature range to be covered. It is easy to understand that an experimental approach alone cannot meet this challenge, even if it remains essential, as it will be seen.

Another approach consists of building a phase diagram through calculation. The most fundamental approach consists of writing the thermodynamic partition function (Z) representative of the microscopic properties of the material, and studying the singularities of Z to determine the phase boundaries. Z provides access to all the thermodynamic magnitudes (for example, the Gibbs energy is written $G = -kTLnZ$ at the thermodynamic limit). Calculation (even approximate) of a partition function is certainly the most difficult problem to solve in thermodynamics. It requires having the appropriate model to express the internal energy of the system (generally based on its electron structure) and an approximation to calculate entropy (mean field, Monte Carlo simulation, etc.) [2] [3]. This methodology, already quite complex for a binary system, quickly becomes unworkable when considering materials consisting of more than three chemical elements.

For complex materials, the selected approach consists of determining the macroscopic thermodynamic behavior of each phase that may appear in the phase diagram. For a constant-pressure phase diagram, this behavior may be described using the Gibbs energy, G. The mathematical form of G for a given phase is generally a function of composition and temperature. To write the function explicitly, it is usually necessary to call on models designed to describe the microscopic reality of the material as closely as possible, while being careful to express G in a relatively simple manner. The coefficients used in these models are determined from experimental data, or based on assumptions when data are not available. Given the Gibbs energies for the different phases, it is possible to determine the Gibbs energy function for all the composition and temperature domains, and by minimizing this functional, to obtain the phase stability domains—that is, the phases that may occur for a given composition and temperature. This method, called the CALPHAD method (CALculation of PhAse Diagram), described for the first time in [4] at the end of the 1960s, can be used to merge the phase equilibrium thermodynamics of Gibbs and numerical calculation. This mixed approach, based on both calculation and experimentation and widely used to describe complex materials, was usually employed to describe corium thermodynamics.

Contrary to the "fundamental" approach described previously, the CALPHAD method is only partially predictive, since the construction of the Gibbs energies for the various phases is based on adjusting models to reproduce experimental results. It nonetheless offers the advantage of being able to predict the thermodynamics of a complex material based on modeling of the binary and ternary systems alone, making it a very powerful method. It can also be used to integrate experimental data deduced from phase diagrams (mainly stability domain boundaries and transition temperatures) and data obtained by measuring thermodynamic magnitudes (enthalpy of formation, activity, chemical potential, etc.) in a single function (G), thereby ensuring consistency between these various sources of information. It is nonetheless important to remember that these thermodynamic magnitudes unequivocally determine the energy level in the Gibbs energies of the various phases, which cannot be achieved using only the topology data from the phase diagram. This is a very important point to be considered when attempting to model the thermodynamics of a complex material. From this point of view, it is the abundance of experimental data, especially measured thermodynamic quantities, that determines the significance level assigned to the thermodynamic model of the material in question.

1.2. The CALPHAD Approach

To set the CALPHAD method to work, it is first necessary to describe how Gibbs energy, G, in the different phases depends on temperature and composition. This dependence is a function of the type of phase: a distinction is made between pure substances (U, O, etc.), stoichiometric compounds (U_3O_8, ZrO_2, etc.), and solutions (solids, for example, $UO_{2\pm x}$, or liquids). For elements and stoichiometric compounds, the fundamental thermodynamic quantities are the standard enthalpy of formation $\Delta H^0_{f,298.15K}$, entropy at ambient temperature ($S^0_{f,298.15K}$) and the change in specific heat as a function of temperature at constant pressure $(C^0_p(T))$. Based on these data, G can be calculated using the following classical equations:

$$\Delta G = \Delta G(T) = \Delta H(T) - T\Delta S(T) \tag{1}$$

where:

$$\begin{cases} \Delta S(T) = S^0_{f,298.15K} + \int_{298.15K}^{T} \frac{C^0_p(T)}{T} dT \\ \Delta H(T) = \Delta H^0_{f,298.15K} + \int_{298.15K}^{T} C^0_p(T)\, dT \end{cases} \tag{2}$$

The analytical form used to express specific heat at constant pressure $(C^0_p(T))$ leads to a general equation for $\Delta G(T)$ expressed as:

$$\Delta G(T) = a_0 + a_1 T + b_0 TLnT + a_2 T^2 + a_3 T^3 + a_{-1} T^{-1} + \ldots \tag{3}$$

For solutions that may be stable over a wide concentration domain, G is developed as a function of temperature T and composition $c = (c_i)_{i=1,N}$ (N being the number of constituents in the solution). The double dependence of ΔG is generally taken into account by calculating the sum of two terms:

$$\Delta G(c,T) = \Delta G^{ref}(c,T) + \Delta G^{mix}(c,T) \tag{4}$$

The first term, $\Delta G^{ref}(c,T)$, is the weighted sum of the Gibbs energies of the solution constituents.

$$\Delta G^{ref}(c,T) = \sum_{i=1}^{N} c_i \Delta G_i(T) \tag{5}$$

The second term, $\Delta G^{mix}(c,T)$, corresponds to the solution's Gibbs energy of mixing. The difficulty lies in estimating this mixing term, for which there are different models; each new model is developed to improve the reliability of the extrapolation of ΔG to temperature and composition domains for which no experimental data are available. In general, the Gibbs energy of mixing is expressed as the sum of two terms, the first corresponding to the Gibbs energy of ideal mixing (mixing "without excess interaction" between the different constituents in the solution, i.e., the perfect gas hypothesis), and the second expressing the gap between the Gibbs energy of mixing and this hypothetical ideal (referred to as the "excess Gibbs energy of mixing"):

$$\Delta G^{mix}(c,T) = \Delta G^{mix}_{id}(c,T) + \Delta G^{mix}_{ex}(c,T) \tag{6}$$

where:

$$\Delta G^{mix}_{id}(c,T) = RT\sum_{i}^{N} c_i \ln c_i \tag{7}$$

The excess Gibbs energy of mixing is usually approached using polynomials given by [5]:

$$\Delta G^{mix}_{ex}(c,T) = \frac{1}{2}\sum_{i=1}^{N}\sum_{j=1}^{N} c_i c_j \sum_{m} L^{m}_{ij}(T)(c_i - c_j)^m \tag{8}$$

The above expressions are very general, making it possible to separate the different contributions (reference, ideal, and excess). Nonetheless, each of these contributions can take a noticeably different form, depending on the model.

To describe the thermodynamics of oxide fuels, it is particularly important to carefully choose the models relative to the solution phases present in the O-U binary system (the principal system to be modeled). Uranium dioxide, UO_2, crystallizes in the fluorite-type cubic structure (CaF_2). Atom ordering in the nonstoichiometric solid solution $UO_{2\pm x}$ can be described using a three-sublattice model with defects (see [6]), the defects making it possible to cover the entire solution composition domain using the same model. The first sublattice contains the uranium atoms, the second contains the oxygen atoms and vacancy-type defects used to describe the substoichiometric composition domain (UO_{2-x}), and the last sublattice contains interstitial oxygen atoms for UO_{2+x}. This representation does not correspond strictly to microscopic reality as described by [7], but it makes possible to represent the experimental data obtained to establish the phase diagram and the measured thermodynamic quantities, which are quite abundant for the U-O system. For the liquid solution, the Gibbs energy can be calculated using the associated model, developed by Prigogine [8], which assumes non-ideal mixing between the uranium and oxygen atoms and the uranium dioxide molecules. This description covers the entire composition domain, from the liquid solution of the metal (pure uranium) to the overstoichiometric oxide (UO_{2+x}).

The usual strategy for assessment of multicomponent systems (as the corium) is then the following. First, a critical analysis of all the available experimental data has to be done to select a consistent set of experimental points. The thermodynamic descriptions of the constituent binary systems are derived; that is, the coefficients of the Gibbs energy functions of the different phases for each binary system are evaluated. This work is performed by using an optimization procedure to take into account simultaneously all the selected experimental information, equilibrium-phase diagram, and thermodynamic properties. Thermodynamic extrapolation methods are then used to extend the thermodynamic functions of the binaries into ternary and higher order systems. The results of such extrapolations can then be used to design critical experiments. In view of the experimental results, the thermodynamic description of the higher order system may be adjusted, if necessary, by added interaction terms (or excess terms) to the Gibbs energy. In principle, this strategy is followed until all 2, 3, ..., n constituent systems of an n-component system have been assessed. However, experience has shown that, in most cases, no or very minor corrections are necessary for reasonable prediction of quaternary or higher order systems. That is, assessment of most ternary constituent systems is often sufficient to describe an n-component system.

The Gibbs energy coefficients of the different phases are usually stored under a database format. The types of nuclear thermodynamic databases can be broadly separated into two groups. First, there are pure substance databases that are typically general in application, contain data for thousands of species, and can be used in calculations for a wide temperature range, but do not consider the solution phases. A lot of databases of this type exist. They do not allow really treating the thermodynamics of corium. Second, databases are available that can be used to model solution phases. These databases contain data for stoichiometric substances as well as non-ideal interaction terms for different components of the solution phases, which enable complex-phase equilibrium calculations to be performed, including the calculation of phase diagrams. The solution databases are very specific in their applications and are usually focused on a relatively small number of elements.

1.3. Sources of Data

As mentioned previously, the quality of the CALPHAD approach depends primarily on the experimental measurement data available to determine the Gibbs energy coefficients. Most metal binary systems have been studied experimentally, but data on oxide binary systems, particularly those containing uranium and plutonium oxides, are not as extensive. The amount of experimental data on ternary and quaternary systems is even less.

For more than 10 years, a vast experimental effort has been undertaken through various national and international programs to fill in these gaps. These different programs include CIT [9] [10] (European Commission or EC project), COLOSS (EC), RASPLAV (OECD/CSNI project), MASCA (OECD/CSNI) [11], ENTHALPY (EC) [12], MASCA2 (OECD/CSNI) [13], and CORPHAD (ISTC project) [14]. Particular focus has been given to the U-O-Zr-Fe quaternary system [15] [16] [17], a key system for describing fuel rod degradation in accident conditions. This effort is being pursued today in the ISTC PRECOS program.

The thermodynamic databases are gradually incorporating new experimental data obtained through the above programs, as well as information published in open technical literature, by performing regular updates on coefficients used to express Gibbs energy, thus contributing to a constant improvement of the significance level. In this aspect, these databases can be considered as a unique tool for building knowledge on nuclear fuel thermodynamics. In spite of these continuous efforts, there are composition and temperature domains for which there are no experimental data. Phase-equilibrium thermodynamics being a relatively old discipline (dating back to the nineteenth century), whatever has not been accomplished so far is not easy to achieve. Calling on modeling hypotheses to build Gibbs energy data for certain phases can open a new road to this information. In this context, it is important to confront these hypotheses with other alternatives. This confrontation can take place by comparing thermodynamic databases devoted to corium thermodynamics. In 2003, as part of the ENTHALPY program [12], the NUCLEA database developed in the framework of a IRSN-THERMODATA-INPG-CNRS (all are French organizations) collaboration was compared with the THMO database established by AEA-Technology (United Kingdom), with the aim of reaching a consensus on modeling hypotheses [18].

Another source of information consists of the results obtained from calculating electron structures, which involves resolving quantum mechanics equations. Without entering a detailed theoretical discussion, it should be noted that this approach can be applied to obtain values of the enthalpy of formation for stoichiometric compounds with a satisfactory degree of accuracy. These values can contribute to the determination of coefficients used to express Gibbs energy (equations (1) and (2)).

1.4. Existing Nuclear Thermodynamics Databases for Corium

Regarding the thermodynamics of the corium, until recently, the number of databases able to describe the thermodynamics of corium from its formation in the reactor vessel until its interaction with concrete (molten-corium-concrete-interaction or MCCI) is rather reduced. Two databases existed, one developed by AEA-T (United Kingdom) and one developed by THERMODATA (France). With the support of the European Commission (fifth Framework Program), these two databases were fused in order to constitute a common database, the European Nuclear Thermochemical Database (NUCLEA) dedicated to the corium thermochemistry. It contains the Gibbs energies of all the phases, including one or several of the following elements U-O-Zr-Fe-Ni-Cr-B-C-Ag-In-La-Ru-Sr-Ba-Mg-Al-Si-Ca (Ar-H being treated in the gas phase). The modeling and the critical assessment of all the binary and the most important higher-order subsystems (metallic, oxide, and metal-oxide/oxygen) of this complex system were reviewed.

Today, the NUCLEA database is used by a large number of institutes, industrial partners, and universities, including EDF, CEA, AREVA (France),

VTT (Finland), AEA-T (United Kingdom), Alexandrov RIT (Russia), KAERI (South Korea), AECL (Canada), Risø National Laboratory (Denmark), Boise State University (USA), and others.

2. CORIUM THERMOPHYSICS

(P. Piluso)

Safety studies are required under accident and severe accident conditions for current and future water-cooled reactors. In a hypothetical severe accident, very high temperatures of around 3300 K could be reached. The materials of the nuclear reactor-nuclear oxide fuel and fission products, cladding, metallic alloys, moderator, absorbers, structural materials, coolants, concrete, and so on, could melt to form complex, multiphases, and aggressive mixtures known by the general term *corium*. In the framework of severe accident studies, accurate data for the thermophysical properties [19] [20] are necessary to model the corium behavior (thermal-hydraulics, physico-chemistry, etc.) at different steps during the various stages of severe accident progression (steam explosion, in-vessel interaction, corium concrete interaction, corium spreading) and for use in severe accident codes (Figure 1). For experimental interpretation, modeling, or code calculations of severe accident progression in a reactor, it is necessary to estimate the corium physical properties as a function of composition and temperature. Analysis of therma-hydraulic and physico-chemical processes is based on the data on the thermophysical properties of corium that determine the material, and energy balance, dissipation of energy and momentum, radiation, and so on. High temperatures and the lack of an adequate practice of studying make the task of corium data compilation and evaluation

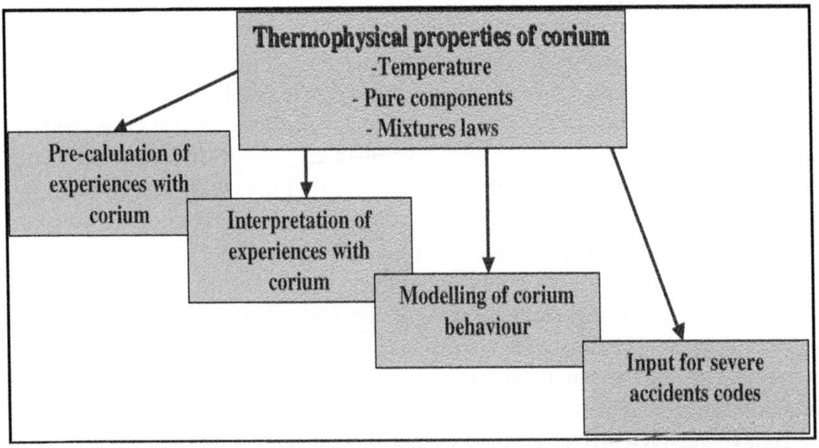

FIGURE 1 Use of thermophysical properties for severe accident analysis.

rather problematic. While concluding the research program on thermophysical properties of materials for water-cooled reactors, the IAEA [21] has stressed the lack of valid and reviewed data on melts. At present a number of severe accident-related national and international research programs have been conducted, for example, OECD projects such as RASPLAV/MASCA [22] and MACE [23]. Nevertheless, it must be stressed that there are no general thermophysical laws applicable for complex mixtures of uranium oxide and zirconium, the two main components of corium.

2.1. Modeling of the Thermophysics of Complex Mixtures

Main Thermophysical Properties for Severe Accident Applications

Five groups of thermophysical properties are of major importance for severe accident analyses: (1) density, linear and volumetric expansion; (2) thermal conductivity and thermal diffusivity; (3) emissivity (total emittance); (4) viscosity; and (5) surface tension. A strong coupling with thermodynamic properties is necessary to estimate the thermophysical properties of corium according to the scenario [24].

Calculation of Thermophysical Properties

The calculation of corium thermophysical properties is based on both a review of individual constituents and the development of mixing laws, as well as the use of validity criteria for various applications. The following approach in two stages is generally used to calculate corium thermophysical properties [20]:

- Estimation of the physical property of all the phases present in a given corium (these phases can be either pure substance or solutions; in the latter case, the property for the solution is estimated by taking into account the contributions of the solution constituents);
- Estimation of the corium apparent property, by using mixing laws applicable to the given topological configuration of the phases.

In the next sections, this approach is applied on three main thermophysical properties: density, viscosity, and thermal conductivity.

Density

Three properties are linked by the following equations:

- The molar volume V_{molar} is the volume occupied by one mole of the species (note that it depends on the definition of the species, e.g., one mole of $FeO_{1.5}$ occupies half the volume of 1 mole of Fe_2O_3).

$$V_{molar} = \frac{M}{\rho}; \quad \boldsymbol{M}: \ \textbf{\textit{molar mass}} \quad \boldsymbol{\rho}: \textbf{\textit{density}} \tag{9}$$

- The coefficient of volume expansion is defined as the ratio of the temperature derivative of the molar volume by the molar volume.

$$\alpha_V = \frac{1}{V_{molar}} \cdot \left(\frac{\partial V_{molar}}{\partial T}\right); \; T \; temperature \tag{10}$$

Volume expansion is the driving force for natural convection. Therefore a good knowledge of this derivative is necessary.
- The linear thermal expansion α in the x, y, z directions is related to the coefficient of volume expansion by:

$$\left(1 + \int \alpha_x \, dT\right)\left(1 + \int \alpha_y \, dT\right)\left(1 + \int \alpha_z \, dT\right) = 1 + \int \alpha_V \cdot dT \tag{11}$$

For solutions (either solid or liquid), the independence of partial molar volumes of each constituent from bulk composition and ideal mixing is assumed. Nelson and Carmichael [25] have verified, on silicate liquids, that the above assumption could lead to errors of the order of 1% or less. For solid-state compound oxides, Kamigaito [26] showed that assuming ideal mixing of the single oxides also gives acceptable results. This also has been promoted for metallic alloys with errors of less than 4% [28]. Nevertheless, some metallic alloys are non-ideal [29], such as Na-Pb, Na-Bi, Na-In, and Fe-Si. The maximum non-ideal behavior is observed at compositions generally corresponding to the solid state to intermetallic compounds. It must be noted that Crawley [29] proved a lack of correlation between the excess volumes (and even the sign of excess volume) and excess enthalpies or entropies, where the excess parameter (X^{ex}) is defined as the difference between the actual value of magnitude and the value for an ideal system:

$$X^{ex} = X(\text{actual}) - X(\text{ideal}) \tag{12}$$

Nevertheless, in the absence of pertinent data, the excess volumes are assumed to be negligible. Neglecting excess volumes, the volume of one mole of solution is given by:

$$V = \sum_i y_i \cdot V_i \tag{13}$$

where y_i is the molar fraction of species i (having a "partial" molar volume V_i and a molar mass M_i) in the solution. The site fractions are used as estimators of the species molar fraction y_i, and the site fractions are calculated by a thermodynamic code (Gibbs energy minimizer) such as GEMINI2 software using the thermodynamic database (e.g., the TDBCR or NUCLEA databases [30]). In this modeling, for the solid and liquid solution phases, a general multi-sublattice model has been used:

$$G^{ref} = \sum_i P_i(Y) \cdot {}^0G_i \tag{14}$$

where ^0G_r represents the Gibbs Energy of all reference substances, $P_i(Y)$ is the corresponding product of site fractions from the matrix: $Y = y_i^{sl}$, atomic fraction of the component I (pure or associate species) on the sublattice sl.

The density of the solution is then given by:

$$\rho = \frac{\sum_i y_i \cdot M_i}{\sum_i y_i \cdot V_i} \tag{15}$$

The values for pure species can be calculated adding partial molar volumes to the data on pure species molar volumes and densities. The value of density or molar volume, which must be taken for liquid components in mixtures that are liquid at temperatures below the melting temperature of the pure component, is an important factor. For the density calculations, we recommend that the expansion coefficient be taken as a constant below the melting point.

For a mixture of different condensed phases (liquid(s), solid solutions), the following mixing law is used. The volume of one mole of a mixture made of phases having molar fractions x_j and molar volumes V_j is given by:

$$V = \sum_j x_j \cdot V_j \tag{16}$$

The above approach is, for instance, proposed for multiphase solids such as concretes [31].

In case of gaseous inclusions, the major effect is an increase in the global molar volume. If the volume fraction of gases is P and the density of the condensed phases is ρ, then the global density taking into account of porosity is:

$$\rho_{global} = (1 - P)\rho_{condensed} \tag{17}$$

Viscosity

Viscosity of Nonsilicate Liquid Phases For corium with less than $5\%_{mol}$ of silica, it is recommended [32] to use the Andrade [33] relationship:

$$\eta = \boldsymbol{K} \cdot \frac{(\boldsymbol{M} \cdot \boldsymbol{T_m})^{1/2}}{V^{2/3}} \cdot \exp\left[\frac{\boldsymbol{Q}}{\boldsymbol{R}} \cdot \left(\frac{1}{\boldsymbol{T}} - \frac{1}{\boldsymbol{T_m}}\right)\right] \tag{18}$$

where the coefficient K recommended for corium is the value proposed by Nazaré [34]. Here, T_m refers to the melting point, Q the activation energy, and V the molar volume. The activation energies [32] are 35 kJ/mol for UO_2 and 247 kJ/mol for ZrO_2. For the other materials, the following empirical relationship is recommended (although it was derived for metals) [34]:

$$Q \sim 1.8\, T_m^{1.348} \tag{19}$$

This approach was validated [32] against experimental data obtained from the RASPLAV project.

Viscosity of Silicate Liquids For silicate melts, the presence of silicate chains significantly increases the viscosity. Andrade's [33] model is no longer valid, and we recommend the approach of Urbain [36], which has been extended to corium [37].

The viscosity is described by the Weymann relationship:

$$\eta = 0.1\, \boldsymbol{AT} \exp(1000\, \boldsymbol{B}/\boldsymbol{T}) \tag{20}$$

where A and B are linked by the empirical relationship:

$$-\ln \mathrm{A} = 0.29\,\mathrm{B} + 11.57 \tag{21}$$

The molten silicate melt constituents are divided into three families: glass formers (SiO_2 and complex silicate molecules); modifiers (CaO, MgO, FeO, $U_{1/2}O$, $Zr_{1/2}O$); and amphoterics (Al_2O_3, Fe_2O_3). The parameter B in equation (21) is obtained from the molar fractions of glass formers and modifiers. Ramacciotti [37] validated this approach against experimental data both with and without uranium dioxide.

Viscosity of Suspensions Viscosity is strongly affected by the presence of solid particles. This property is important during some solidification processes as those of spreading or rapid cooling. For a semi-solid configuration, Ramacciotti [37] recommends the following equation to estimate the mixture viscosity from the actual-liquid viscosity and the solid volume fraction (estimated from thermodynamics) as follows:

$$\eta = \eta_{liquid}\, e^{2.5C\phi} \tag{22}$$

where ϕ refers to solid volume fraction and the constant C is approximately 6.

This relationship was validated using data from corium viscosity measurements and data from corium spreading calculations of prototypic material experiments [38].

Viscosity of Emulsions Another phenomenon to take into account in corium behavior is the emulsion of a liquid phase in another liquid. In this case, the Taylor [39] model can be used:

$$\eta_r = \frac{\eta}{\eta_c} = 1 + \frac{5\,\boldsymbol{K} + 2}{2(\boldsymbol{K} + 1)}\phi \tag{23}$$

Here η_c refers to the viscosity of the continuous fluid. It must be noted that for the solidification of corium with a miscibility gap, the viscosity calculation leads to different results depending on whether one considers an emulsion of metal in a semi-solid oxide or a suspension of oxide crystals in an oxide-metal emulsion. The choice of the pertinent configuration is thus crucial. For spreading, experiments [38] show that the first assumption is true.

Viscosity of Bubbly Flows Another phenomenon of interest is an emulsion of gas bubbles in corium. In this case, we recommend the following Llewellin [40] relationship established for stationary flows:

$$\eta = \eta_0(1 + 9\phi) \tag{24}$$

Thermal Conductivity

Thermal Conductivity at Solid State At a macroscopic scale, the thermal conductivity is described by the following equation:

$$\lambda = \alpha\rho\, C_P \tag{25}$$

The thermal conductivity is described by the following equation on a microscopic scale with three contributions (lattice vibration, electronic, and radiation):

$$\lambda = \lambda_\mathrm{L} + \lambda\mathrm{e} + \lambda_\mathrm{r} \tag{26}$$

Below 1800 K, the thermal conductivity is essentially governed by the lattice vibration (phonons). Thus, the following general law can describe the thermal conductivity due to the lattice vibration [41]:

$$\lambda_\mathrm{L} = (\mathrm{a} + \mathrm{bT})^{-1} \tag{27}$$

Above 1800 K, the radiation component becomes important. Therefore, the radiation contribution is as follows (knowing that this phenomenon starts being appreciable at 1000 K):

$$\lambda_r = \frac{16}{3}\sigma n^2 T^3 l_r \tag{28}$$

For metals such as U, Zr, or Fe at high temperatures, the thermal conductivity is governed by the electronic contribution [42]:

$$\lambda_e = \frac{2m^{*1/2}}{3^{1/2}E_g e}(kT)^{3/2}\mu^* \frac{\left(2 + \frac{E_g}{kT}\right)^2}{\frac{3}{2} + \frac{E_g}{kT}} \tag{29}$$

For oxides at very high temperatures, another contribution has been recently identified: the polaron contribution (ion-electron contribution). This component contributes to the increase of thermal conductivity at temperatures above 2000 K. Macroscopically, it means that the thermal diffusivity remains constant, whereas the heat capacity increases [43]. Applying this approach to oxides present in the corium, it is possible to propose a general law for the macroscopic evolution of the thermal conductivity, applicable to oxides such as zirconia or iron oxide [44].

$$\lambda = \frac{1}{a+bT} + c(n)T^3 + \frac{d(\sigma)}{T} \tag{30}$$

$d(\sigma)$ is a function of the electrical conductivity.

Thermal Conductivity at Liquid State For very high temperatures, especially for oxides (except for UO_2 [45][46][47]), only very few experimental or calculated data are available for the constituents of corium at liquid state. Usually, severe accident researchers use, for the thermal conductivity in the liquid state, the value recommended for the solid state, which is an extrapolation without any scientific justification.

A different empirical approach can be proposed to calculate the thermal conductivity in the liquid state using an analogy of the relationship between the thermal conductivity in the solid and the liquid states at the melting point.

$$\lambda_{liquid} = \alpha_{solid} \cdot \rho_{liquid} \cdot C_{v\ liquid} \tag{31}$$

We can make the following two hypotheses:

- Continuity of the thermal diffusivity at the melting point ($\alpha_{liquid} = \alpha_{solid}$)
- And, for liquid state, we must consider C_v as heat capacity. But there are no data available for C_v of liquid oxides.

For UO_2, Ronchi [43] proposes that $\rho \cdot C_V$ at solid state is about 40% lower than the liquid state. With this assumption for the liquids and extending the relationship to all the oxides, we get the following relationship:

$$\rho_{liquid}(T) \cdot C_{v\ liquid} = 0.4\rho_{solid} \cdot Cp_{solid} \tag{32}$$

Thermal Conductivity of Two Phases: Solid and Gas For mixtures including a solid phase and a gas phase, the general thermal conductivity laws will depend on their volume fractions. We recommend the following general physical laws, depending on the void fraction (gas phase) and neglecting the radiation contribution:

- Void fraction <5%, Loeb's modeling [48]:

$$\lambda_t = (1 - V)\lambda_{dense} \tag{33}$$

V refers to the volume of the gas phase

- 5% < Void fraction < 25%, Maxwell-Eucken's model [49]:

$$\lambda_t = \frac{2\lambda_d + \lambda g + 2V(\lambda_g - \lambda_d)}{2\lambda_d + \lambda g - V(\lambda_g - \lambda_d)}\lambda_d \tag{34}$$

λ_d is thermal conductivity of dense phase, λ_g thermal conductivity of gaseous phase. The model is not valid when the porosity is higher than 35%

$\lambda_g = 0{,}024\ W \cdot m^{-1} \cdot K^{-1}$ at 20°C [50]

- 25% < Void fraction < 35%, Percolation model:

$$\lambda_t = \frac{1}{4}\left(\lambda_g(3V - 1) + \lambda_d(2 - 3V) + \left[\left(\lambda_g(3V - 1) + \lambda_d(2 - 3V)\right)^2\right]^{1/2}\right) \quad (35)$$

2.2. Survey of the Existing Thermophysics Databases

Different databases have been developed for the chemical and physical properties of the compounds, including nuclear materials. In the field of severe accidents, we can mainly indicate:

- MATPRO (http://www.insc.anl.gov/matprop/#comp),
- AIEA/THERSYST (http://www.iaea.org/inis/inisdb.htm (International Nuclear Information System),
- CORPRO (CORium PROPerties data base) [24].

REFERENCES

[1] M. Hansen, K. Anderko, Constitution of binary alloys, Genium Publishing Corporation, New York, 1991.

[2] A. Finel (1987), Contribution à l'Etude des Effets d'Ordre dans le Cadre du Modèle d'Ising: Etats de Base et Diagrammes de Phase, Doctorat d'Etat de l'Université Pierre et Marie Curie.

[3] M. Barrachin (1993), Contribution à l'Etude de l'Ordre Local dans le Système Nickel-Vanadium par Diffusion Diffuse de Neutrons, Doctorat de l'Université Paris XI-Orsay.

[4] L. Kaufman, H. Bernstein, Computer Calculation of Phase Diagrams with Special Reference to Refractory Metals, Academic Press, New York, 1970.

[5] O. Redlich, A.T. Kister, Algebraic Representation of Thermodynamic Properties and the Classification of Solutions, Ind. Eng. Chem. 40 (1948) 345.

[6] B. Sundman, J. Agren, A Regular Solution Model for Phases with Several Components and Sublattices Suitable for Computer Applications, J. Phys. Chem. Solids 42 (1981) 297.

[7] B.T.M. Willis, Structures of UO_2, UO_{2+x} and U_4O_9 by Neutron Diffraction, J. Phys. France 25 (1964) 431.

[8] I. Prigogine, Thermodynamique Chimique, R. Defay (Eds.), Dunod (1950).

[9] B. Adroguer et al., FISA 1997, EURATOM, EUR 18258, (1998) p.103–112.

[10] B. Adroguer et al., FISA 1999, EURATOM, EUR 19532, (2000) p. 202–210.

[11] M. Barrachin, F. Defoort, Thermophysical Properties of In-Vessel Corium: MASCA Programme Related Results, Proceedings of the MASCA Seminar 2004, Vol. 1, (2004) pp. 331–370, 10–11 June 2004, Aix-en Provence (France)

[12] A. De Bremaecker et al., FISA 2003, EURATOM, EUR 21026, (2004) 202–210.

[13] B. Cheynet, E. Fischer, P.-Y. Chevalier, M. Barrachin, Progress in nuclear thermodynamic databanking for MCCI applications, Proceedings of the OECD MCCI Seminar, October 10–11, 2007, Cadarache (France) (2007).

[14] S.V. Bechta, V.S. Granovsky, V.B. Khabensky, V.V. Gusarov, V.I. Almiashev, L.P. Mezentseva, E.V. Krushinov, S.Yu. Kotova, R.A. Kosarevsky, M. Barrachin, D. Bottomley, F. Fichot, M. Fischer, Corium phase equilibria based on MASCA, METCOR and CORPHAD results, Nuclear Engineering Design, 238, (2008) 276–277.

[15] P.-Y. Chevalier, B. Cheynet, E. Fischer, Progress in the thermodynamic modeling of the U-O-Zr ternary system, Calphad 28, (2004) 15.
[16] S.V. Bechta, E.V. Krushinov, V.I. Almjashev, S.A. Vitol, L.P. Mezentseva, Yu.B. Petrov, D.B. Lopukh, V.B. Khabensky, M. Barrachin, S. Hellmann, K. Froment, M. Fischer, W. Tromm, D. Bottomley, F. Defoort, V.V. Gusarov, Phase Diagram of the ZrO_2–FeO System, J. Nucl. Mater. 348 (2006) 114.
[17] S.V. Bechta, E.V. Krushinov, V.I. Almjashev, S.A. Vitol, L.P. Mezentseva, Yu.B. Petrov, D.B. Lopukh, V.B. Khabensky, M. Barrachin, S. Hellmann, K. Froment, M. Fischer, W. Tromm, D. Bottomley, F. Defoort, V.V. Gusarov, Phase Diagram of the UO_2–FeO_{1+x} System, J. Nucl. Mater. 362 (2007) 46.
[18] B. Cheynet, P. Chaud, P.-Y. Chevalier, E. Fisher, P. Masson, M. Mignanelli, NUCLEA Propriétés Dynamiques et Equilibres de Phases dans les Systèmes d'Intérêt Nucléaire, Journal de Physique IV 113 (2004) 61.
[19] P. Piluso, J. Monerris, C. Journeau, G. Cognet, Viscosity measurements of ceramic oxides by aerodynamic levitation, International Journal of Thermophysics, 2002.
[20] C. Journeau, P. Piluso, K.N. Frolov, Corium Physical Properties for Severe Accident R&D, Intern. Cong. on Advanc. Nuclear Power Plant (2004).
[21] IAEA TECDOC Series No. 1496 (2006).
[22] V.G. Asmolov, S.S. Abalin, Yu.G. Degaltsev, O.Ya. Shakh, E.K. Dyakov, V.F. Strizhov, Behavior of the molten pool in the reactor bottom head (RASPLAV Project), Atomic Energy (in Russian), 84 (1998) 303–318.
[23] M.T. Farmer, S. Lomperski, S. Basu, Status of the Melt Coolability and Concrete Interaction (MCCI) Program at Argonne National Laboratory, Proceedings ICAPP '05, Paper 5644, Seoul, Korea, May 15–19, (2005).
[24] V.F. Strizhov, R.G. Galimov, V.D. Ozrin, V.Yu. Zitserman, G.A. Kobzev, L.R. Fokin, P. Piluso, H. Chalaye, Thermo-physical properties of corium: Development of an Assessed Data Base for Severe Accident Applications, ICAPP 2007, Nice, France, May 13–18, (2007).
[25] S.A. Nelson, I.S.E. Carmichael, Partial molar volume of oxide components in silicate liquid, Contrib. Mineral. Petr. 71 (1979) 117–124.
[26] O. Kamigaito, Density of compound oxides, J. Ceram. Soc. Jap. 108 (2000) 944–947.
[27] T.Z. Harmathy, Thermal properties of concrete at elevated temperatures, J. Mater. 5 (1) (1970) 47–74.
[28] F.C. Hull, Estimating alloy densities, Metal Progress, November 1969, pp. 139–140 (1969).
[29] A.F. Crawley, Densities of liquid Metals and Alloys, Int. Metal. Rev. 19 (1974) 32–48.
[30] A. De Bremaecker, et al., European Nuclear Thermodynamic Database validated and applicable in Severe Accidents Codes, Proc. FISA-2003, 460–465, European Commission, Luxembourg (2003).
[31] A.S. Gandhi, A. Saravan, V. Jayaram, Materials Science and Engineering, A221, 68 (1996).
[32] F. Sudreau, G. Cognet, Corium viscosity modelling above liquidus temperature, Nucl. Eng. Design 178 (1997) 269–277.
[33] E.N.D.C. Andrade, A theory of the viscosity of liquids, Phil. Mag. S17 (1934) 497–511 and 698–732.
[34] S. Nazaré, G. Ondracek, B. Schultz, Stoffwerte von LWR-Coreschmelzen, Wissen. Ber. KFKA 2217 (1976) 381–393.
[35] S.S. Abalin, et al., Corium kinematic viscosity measurement, Nucl. Eng. Des. 200 (2000) 107–115.
[36] G. Urbain, Viscosity estimation of slags, Steel res. 58 (1987) 11–116.

[37] M. Ramacciotti, C. Journeau, F. Sudreau, G. Cognet, Viscosity models for corium melts, Nucl. Eng. Design 204 (2001) 377–389.

[38] C. Journeau, et al., Ex-vessel corium spreading: results from the VULCANO spreading tests, Nucl. Des. Eng. 223 (2003) 75–102.

[39] G.I. Taylor, The viscosity of a fluid containing small drops of another fluid, Proc. Royal Soc. 138A (1932) 41–48.

[40] E.W. Llewelin, et al., The rheology of a bubbly liquid, Proc. R. Soc. Lond. A458 (2002) 987–1016.

[41] C. Kittel, Physique de l'état solide, Editions DUNOD, 103–149 (1983).

[42] P. Gardie, Contribution à l'étude thermodynamique des alliages U-Fe et U-Ga par spectrométrie de masse à haute température, et de la mouillabilité de l'oxyde d'yttrium par l'uranium, Institut National Polytechnique, Grenoble, 1992. PhD Thesis.

[43] C. Ronchi, M. Sheindlin, M. Musella, G.J. Hyland, Thermal conductivity of uranium up to 2900K from simultaneous measurement of the heat capacity and thermal diffusivity, Journal of Applied Physics 85 (2) (1999) 776–789.

[44] S. Fayette, P. Piluso, D. Smith, Conductivité thermique des oxydes à moyennes et haute températures, Congrès MFHT–1, session propriétés physiques, Aix-en-Provence, France (2002).

[45] C. Otter, D. Damien, Mesure de la diffusivité thermique de UO2 fondu, High Temperatures-High Pressures, 16 (1984) 1–6.

[46] H.A. Tasman, D. Pel, J. Richter, H.-E. Schmidt, Measurement of the thermal conductivity of liquid UO_2, High Temperatures–High Pressures, 15 (1983) 419–431.

[47] C.S. Kim, et al., Measurement of thermal diffusivity of Molten UO_2, Proceedings 7th Symposium on Thermophysical Properties, ASME, New York, 1977. Gaithersburg, MD, 338–343.

[48] L. Loeb, Thermal conductivity: A theory of thermal conductivity of porous materials, J. Am. Ceram. Soc. 37 (2) (1954) 96–99.

[49] J.C. Maxwell, A treatese on electricity and magnetism, Ed. Clarendon, 440 (1892).

[50] Y. Litovski, M. Shapiro, Gas pressure and temperature dependences of thermal conductivity of porous ceramic materials: part 1, refractories and ceramics with porosity below 30, J. Am. Ceram. Soc. 75 (12) (1992) 334–349.

Appendix 2

Severe Accidents in PHWR Reactors

Outline

(P. M. Mathew)

This Appendix briefly outlines some of the pressurized heavy water reactor (PHWR) plant features, with special reference to its severe accident evolution. Distinctive and inherent safety-related characteristics of PHWRs are as follows:

- On-power refueling helps maintain low excess reactivity.
- There are two independent fast-acting shutdown systems of diverse nature.
- High-pressure and high-temperature coolant is separated from the low-pressure and low-temperature moderator. Reactivity and shutdown mechanisms in the moderator are unaffected by the disturbance in the coolant loop.
- A large and subcooled inventory of moderator can act as an ultimate heat sink, thus preventing severe core degradation under accident conditions. Water surrounding the calandria vessel in the calandria vault/shield tank system can hold the core debris and corium within the calandria vessel during severe core damage accidents.

  Thus, a large inventory of water and multiple barriers has inherent capability to prevent corium release to the containment. PHWR accidents that result in damage to the reactor core fall naturally into two classes [1]:

- Limited Core Damage Accidents (LCDA), where the core geometry is preserved,
- Severe Core Damage Accidents (SCDA), where the core geometry is lost.

1. LIMITED CORE DAMAGE ACCIDENTS

The LCDAs can involve single channels or the entire core. For example, a feeder break can result in overheating of the fuel in the affected channel. Or a LOCA with loss of Emergency Core Cooling (ECC) could lead to widespread fuel damage. In both cases, however, the presence of the moderator as a secondary heat sink prevents failure of the fuel channels and core degradation. In this sense, the LCDAs are distinct from accident sequences for a LWR, as they can involve fuel damage without core relocation.

The distributed nature of the core of a PHWR, with a network of feeder pipes circulating coolant between large-diameter headers and individual fuel channels, requires consideration of accident sequences for which the coolant flow to an individual channel is disrupted. These are commonly referred to as single-channel events. These accidents can be initiated by partial or complete blockage of the flow, or a break in a feeder or a fuel channel (Figure 1).

Since only a limited number of channels are instrumented, the event can continue at full power until a reactor trip point is reached (containment pressure, for example). The outcome of the event depends on the degree to which cooling of the affected channel is reduced. In the extreme case of complete blockage or flow stagnation, the fuel in the affected channel can be damaged. The consequences of a single-channel event are determined by the extent of fuel damage in the affected channel. Assessments are performed to ensure there are no phenomena that can lead to propagation to other fuel channels.

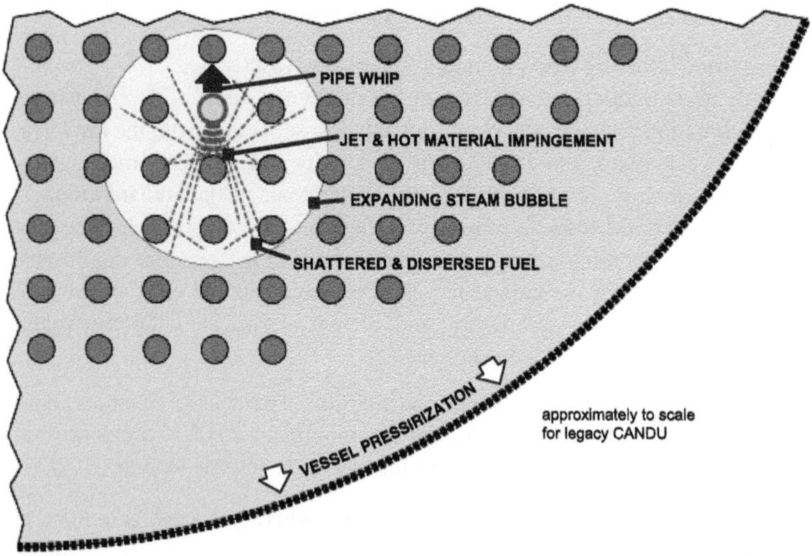

FIGURE 1 Core behavior phenomena in single-channel events (the red circle represents the failed single channel).

For the fuel in more than one fuel channel to be damaged, the primary cooling flow must be interrupted and emergency core cooling must be impaired. Interruption of the primary coolant flow to more than one channel requires an initiating event such as a LOCA that will trip the reactor. Thereafter, these events proceed at decay power. The presence of a secondary heat sink, in the form of the moderator around every fuel channel, limits the consequences of these accidents. Under extreme conditions, such as large LOCA with loss of ECC, the fuel in many fuel channels can undergo a high-temperature transient. Heat loss from the fuel to the moderator, via the pressure and calandria tubes, prevents gross melting of the fuel and preserves the channel core geometry of the reactor.

1.1. Single-channel Events

When the coolant flow to an individual channel is reduced, the remaining flow will be increasingly converted to steam as the mass flow decreases. The reduced flow is still capable of removing heat from the fuel, and high reductions in mass flow (equivalent to blockages greater than 90% of the flow area) are required before fuel temperatures increase significantly. Eventually, the fuel channel will fail, allowing for reintroduction of coolant from the unbroken or unblocked side. This cools the fuel and trips the reactor as for a small-break LOCA.

The phenomena of interest are those that lead to fuel bundle deformation and to failure of the fuel cladding. These include fuel element bowing, zircaloy oxidation, and beryllium-braze penetration. The high-coolant pressure tends to prevent clad ballooning. If the affected pressure tube does not fail due to circumferential temperature gradients, there can be clad melting and relocation onto the pressure tube, which is predicted to cause failure. If delayed failure is assumed, there can be fuel melting.

As heat is transferred to the pressure tube, its temperature will rise and the tube will start to deform [2] under the influence of the full-system pressure as shown in Figure 2.

Stratification of the coolant in the channel will keep fuel temperatures lower in the bottom of the fuel channel and lead to a top to bottom circumferential gradient. Under these conditions, the top of the pressure tube will deform more quickly and fail [3] [4] [5]. If the temperature around the tube is more uniform (perhaps due to rapid expulsion of the coolant with little stratification), the pressure tube will expand uniformly (balloon) up to the contact with the calandria tube.

Contact of the pressure tube with the surrounding calandria tube provides support and cools the pressure tube. The fuel will continue to heat up, and either molten zircaloy or a combination of molten zircaloy and UO_2 will form and relocate to the bottom of the fuel channel. The combined pressure and calandria tubes will then heat up locally and fail. In principle, localized "hot spots" on the pressure tube wall could also be caused by contact of fuel

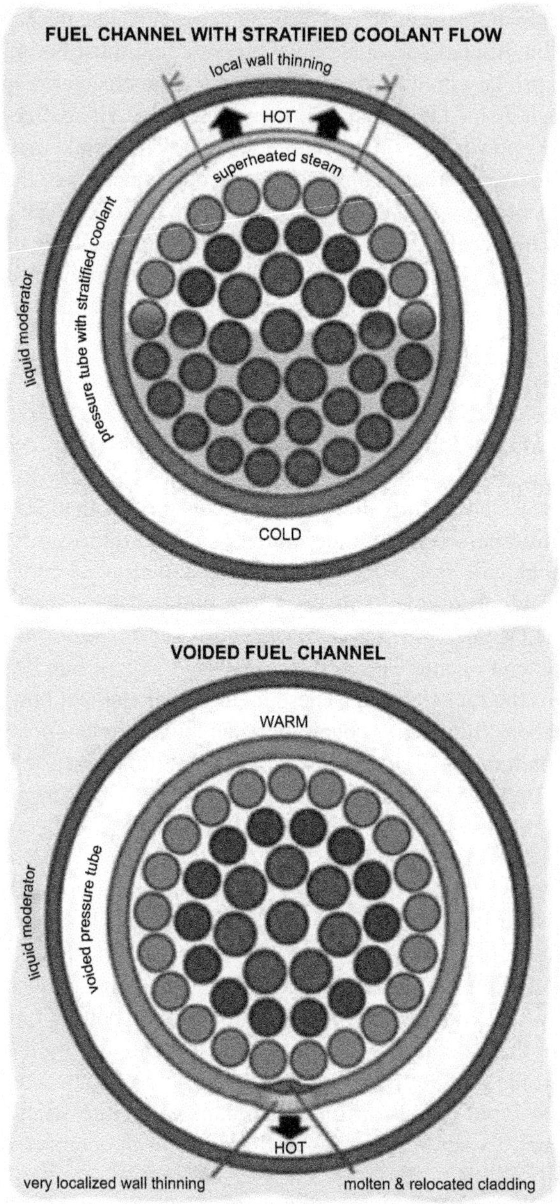

FIGURE 2 Examples of single-channel events (above) and pressure tube failure in single-channel events (below).

appendages or of bowed fuel elements. These hot spots could only lead to earlier failure of the pressure tube, thereby avoiding the formation of molten clad or fuel.

1.2. Full-Core Events

Extreme scenarios for this family of accidents are LOCAs with loss of ECC. Deformation of the fuel channel creates heat transfer paths from the fuel to the moderator, thereby limiting the consequences.

The early stages of a full-core LCDA are dominated by the blowdown behavior. The pulse in the reactor power is quickly terminated by the reactor shutdown. A characteristic of PHWRs is the figure-of-eight configuration for the Primary Heat Transport System (PHTS). In a CANDU 6 reactor with two loops, the pass with the break will start depressurizing first in a LOCA, followed by the unbroken pass. Once the PHTS has depressurized, the remaining liquid coolant boils off, and the resulting steam flows out the break. The flows in the primary system can be complex, with buoyancy forces and steam-induced pressure gradients producing interchannel flows through the connections provided by the headers. When steaming ceases, the remaining vapor in the PHTS, consisting of steam and hydrogen, continues to flow, driven by buoyancy forces. Not much hydrogen is produced because there is no fresh steam available to maintain the zircaloy oxidation.

During the early stages of a LOCA with loss of ECC, changing thermal-hydraulic conditions can result in flow reversals that forcibly translate the fuel bundles within the fuel channel. Since fuel is still relatively cool, break-up of the fuel bundles is not expected. As the accident proceeds, the fuel will heat up and axially expand. If there is not enough free axial gap, the expansion will be constrained, leading to deformation and possible break-up of fuel bundles.

Peak fuel rod temperatures are reached around the time the fuel bundle starts slumping. The values of peak temperature are accident-scenario specific. The interior of high-power fuel bundles could reach the zircaloy melting point if the channel were to void rapidly and remain voided thereafter. The additional energy deposition into the fuel caused by the positive void reactivity feedback before the reactor is tripped contributes to high fuel rod temperatures. In the long term, the temperatures decrease slowly with decreasing decay power.

Hot fuel cladding will balloon due to the difference between the internal gas pressure and the coolant pressure. The free volume in a CANDU fuel rod is relatively small, and therefore ballooning in a localized region with the highest cladding temperature will balance the internal and external pressures. The fuel cladding may fail during ballooning by various mechanisms, including over-strain, oxide cracking, oxygen embrittlement, and beryllium-braze crack penetration. The ballooning of fuel cladding causes some flow area obstruction. Following ballooning, the fuel pellets relocate to the bottom of ballooned fuel

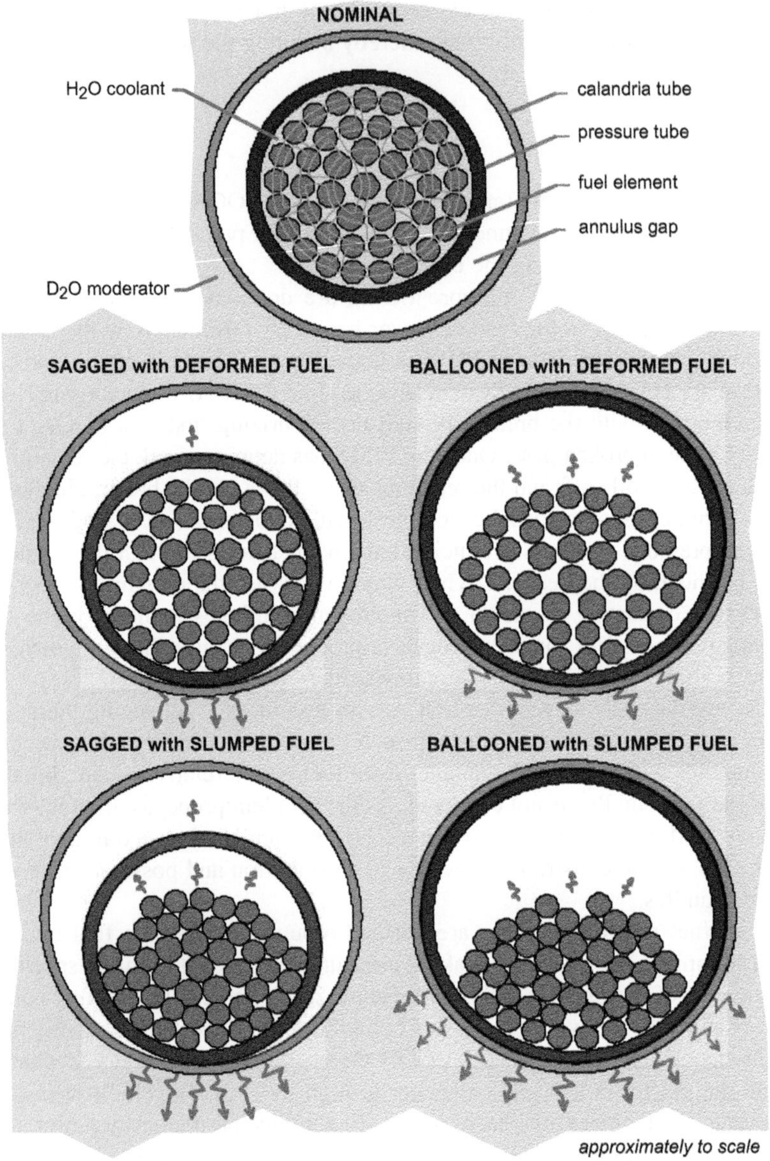

FIGURE 3 Channel deformation models in a LOCA.

cladding, and cracking during the temperature excursion may lead to further relocation of fuel fragments.

Prolonged exposure to high temperatures causes sagging of the fuel rods and deformation of the softened end plates. While end plates may maintain some spacing between the ends of fuel rods, within a short distance, the fuel rods sag into contact and fuse with each other [6] [7] [8]. For very high-temperature excursions, the bundle becomes a coarse debris pile at the bottom of the pressure tube. The fuel pellets are retained in perforated and oxidized cladding shells. All distorted bundle geometries impede access of steam to the interior subchannels, which reduces the energy contribution from zircaloy oxidation. Fuel geometries may be further compacted by thermal fragmentation of embrittled fuel rods during water refilling of the fuel channels when emergency cooling is eventually introduced.

The fuel debris bed at the bottom of the pressure tube will be quite shallow (a few centimetres deep), so there is no issue of fuel coolability after the liquid water enters the channel.

Gross fuel bundle deformation is accompanied by thermal-chemical interactions of UO_2 and zircaloy. Limited amounts of molten material may be formed due to interaction between UO_2 and zircaloy below their melting points, depending on the interface temperature and contact pressure. UO_2/zircaloy interaction also leads to reduction of the fuel as uranium oxide fuel dissolves in molten zircaloy. Some molten material (largely from the end caps and end plates) may relocate onto the pressure tube, causing intense localized hot spots on the pressure tube.

For LOCA transients that lead to pressure tube heat-up while the internal pressures remain relatively high, the dominant mode of pressure tube deformation is ballooning [3] [5]. If the channels void more gradually, the pressure tubes will not heat up until the residual pressure is low. Under these conditions, the pressure tube will sag into contact with the calandria tube. Figure 3 illustrates the modes of channel deformation for a LOCA.

Upon contact between the pressure and calandria tubes, the stored energy in the hot pressure tube augments the heat transfer from radiation and conduction. If there is insufficient subcooling margin during ballooning contact, there can be extensive film boiling on the outside of the calandria tube. Under these conditions, the temperatures of the combined calandria and pressure tubes can escalate to the point of failure. Such failures are not a concern for sagging contact because any film boiling is localized and does not lead to temperature escalation.

Extreme temperature excursions to the zircaloy melting point result in the relocation of molten metal onto the composite wall of the submerged tube. An unstable dry-out patch arises at the point of molten material contact that does not produce any noticeable deformation of the composite tube.

2. SEVERE CORE DAMAGE ACCIDENTS

For an accident to proceed to severe core damage, there must be multiple fuel channel failures leading to significant core degradation. Such a situation is only possible if the moderator heat sink is lost [9]. Typically, SCDA initiates as a LCDA, as described in the previous section, but there is no moderator cooling to remove heat and there is no moderator make-up. In this case, the moderator boils off, allowing uncovered fuel channels to heat up and fail. Thus the early stages of a SCDA for a PHWR are different from those of a LWR since the presence of the heavy water moderator slows down the progression of core disassembly. Once the core has become fully disassembled and has collected on the bottom of the calandria vessel, subsequent behavior is similar to that of a LWR with the core relocated into the lower vessel head. However, there are differences in corium composition (i.e., proportions of UO_2, Zr, and other materials) and corium geometry (i.e., a surface-to-volume ratio, which is given by the shape of the vessel or the containment compartment).

As long as the fuel channels remain surrounded by the moderator, the core geometry will be maintained. If moderator cooling and make-up are unavailable, the water level in the calandria will start to drop and uncover the upper fuel channels. The moderator level may also drop suddenly for an in-core break that leads to discharge of the moderator.

Uncovered fuel channels heat up and deform by sagging. Eventually, channel segments break off and form a coarse debris bed, which rests on still-intact lower channels [10] [11] [12] [13] [14]. This suspended debris bed imposes a load on the channels below and alters the steam flow patterns in the calandria vessel. Along time, the suspended debris also include materials from uncovered in-core devices.

When the weight of the suspended debris exceeds the load-bearing capacity of the fuel channel plane below the water level, most of the core collapses into the water pool at the bottom of the calandria vessel.

Perforations of calandria and pressure tube walls allow steam access into the annulus gap, which contains zirconium-alloy surfaces that are not protected by appreciable ZrO_2 layers. Steam also gains access to fuel surfaces within the pressure tube, which may have been steam-starved before the channel broke apart. Such conditions are favorable to the Zr-steam reaction. The main impediment to a "runaway" reaction is the absence of pressure gradients across the length of channel debris segments to drive the steam into the interior of tubular debris.

There are two modes of suspended debris relocation to the bottom of the calandria vessel:

- Intermittent, small pours of liquid zirconium alloy and/or a dropping of a small mass of fragmented solid debris; and
- Sudden drop of a large mass of hot, solid material.

The pours of liquid zirconium alloy are relatively small amounts (kilograms to tens of kilograms) of melt. Upon contact with liquid water, the melt reacts with it and partially oxidizes. There are no particular safety concerns with these small, intermittent pours of reactive, molten metal into water at saturation temperature.

A large mass of solid materials relocates rapidly (tens of seconds) when enough suspended debris accumulate above the water level to exceed the load-bearing capacity of the first plane of submerged channels. The load is carried mainly by the calandria tubes, which are relatively cool. The thicker pressure tubes within the calandria tubes are hot and thus weaker. Once the first submerged plane of channels cannot support the weight of the suspended debris, the planes at lower elevations invariably cannot support the load. A "cascading" process occurs in which the load is transmitted to channels below. The whole core, except the channel stubs left behind during the formation of the suspended debris bed, collapses on a rather short time scale (minutes) into the residual liquid pool (Figure 4).

Once the core has collapsed, all core materials are under water. The terminal debris consists of channel segments and fragments at temperatures well below the melting point of zircaloy and the solid channel debris below the melting point of UO_2. The embrittled debris are likely to fragment during the core collapse. The channels that were submerged and failed by pull-out maintain their tubular geometry.

The quenching of long channel segments is complex because their exterior is not hot and their interior can only be accessed via a limited flow area at the ends of the segments. The quenched debris are a coarse mixture of ceramic and metallic materials at low temperatures. As water evaporates, this mixture is gradually uncovered. Eventually, the dry terminal debris reheats in a nonoxidizing environment and compacts. A "crucible", a frozen solidified layer of corium, is formed where materials adjacent to the externally cooled calandria vessel wall and materials at the top of the debris pile are solid. The interior of the crucible may contain molten or liquefied materials.

Corium can be retained in the calandria vessel or the shield tank (Figure 5). The surface-to-volume ratio of the debris bed is large, resulting in low heat fluxes as well as short heat conduction distances within the corium bed. Low heat fluxes avoid possible problems with heat removal on the waterside. Water-cooled walls completely surround the top debris surface to ensure effective heat transfer by thermal radiation.

The corium configuration within the calandria vessel is stable. Two alternate heat sinks are available (i.e., active heat removal by heat exchangers and passive heat removal by boiling and make-up). Corium can be contained within the calandria vessel as long as it is cooled on the outside to remove the decay heat. For the corium bed to relocate to the shield tank/calandria vault, the water level around the calandria vessel must decrease to approximately the elevation of the corium bed surface (Figure 5). The radiation heat sink will have deteriorated during voiding of the shield water, so that the corium crust at the top surface has become thin. Eventually, the corium thermally attacks the calandria vessel wall just above the water surface. Liquid corium flows out of the hole, ablating the crust as well as the wall. Interactions of molten corium with liquid water will cause steam surges and associated pressure loads on the walls of both vessels.

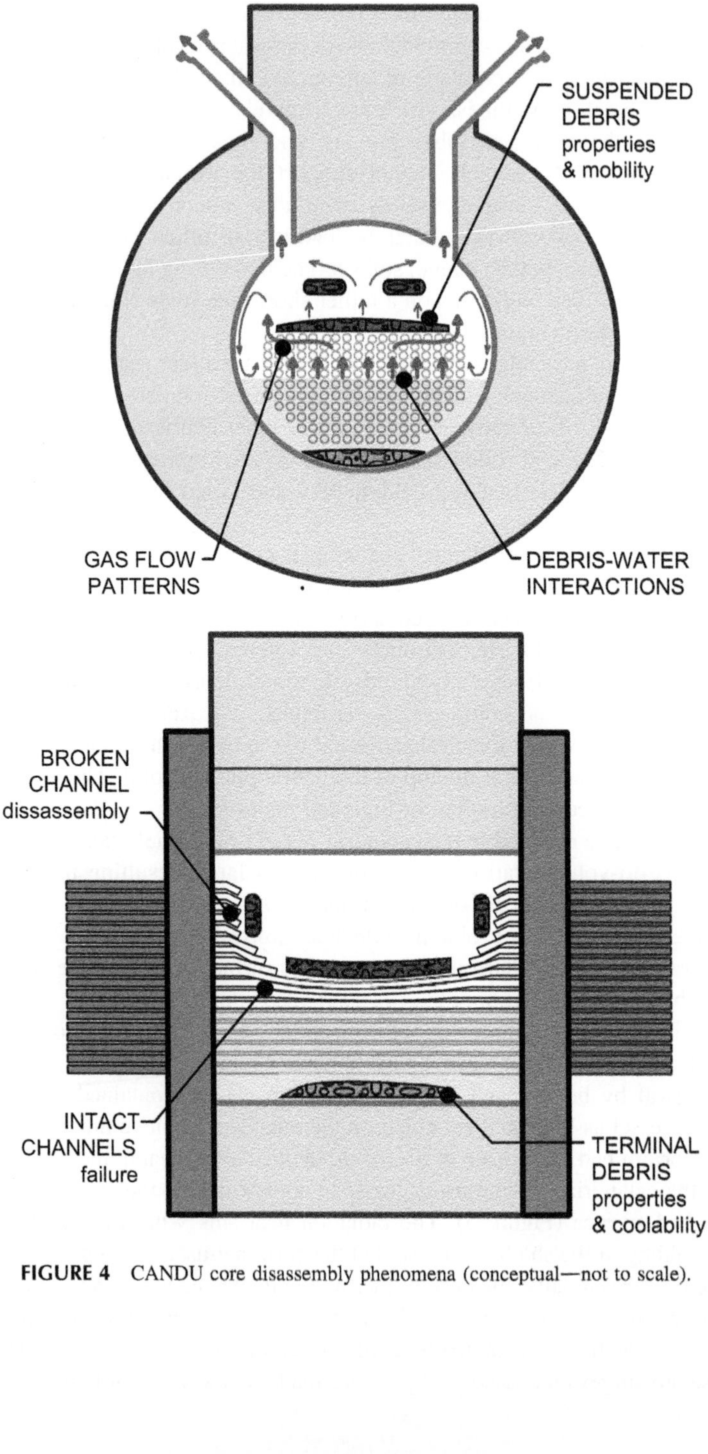

FIGURE 4 CANDU core disassembly phenomena (conceptual—not to scale).

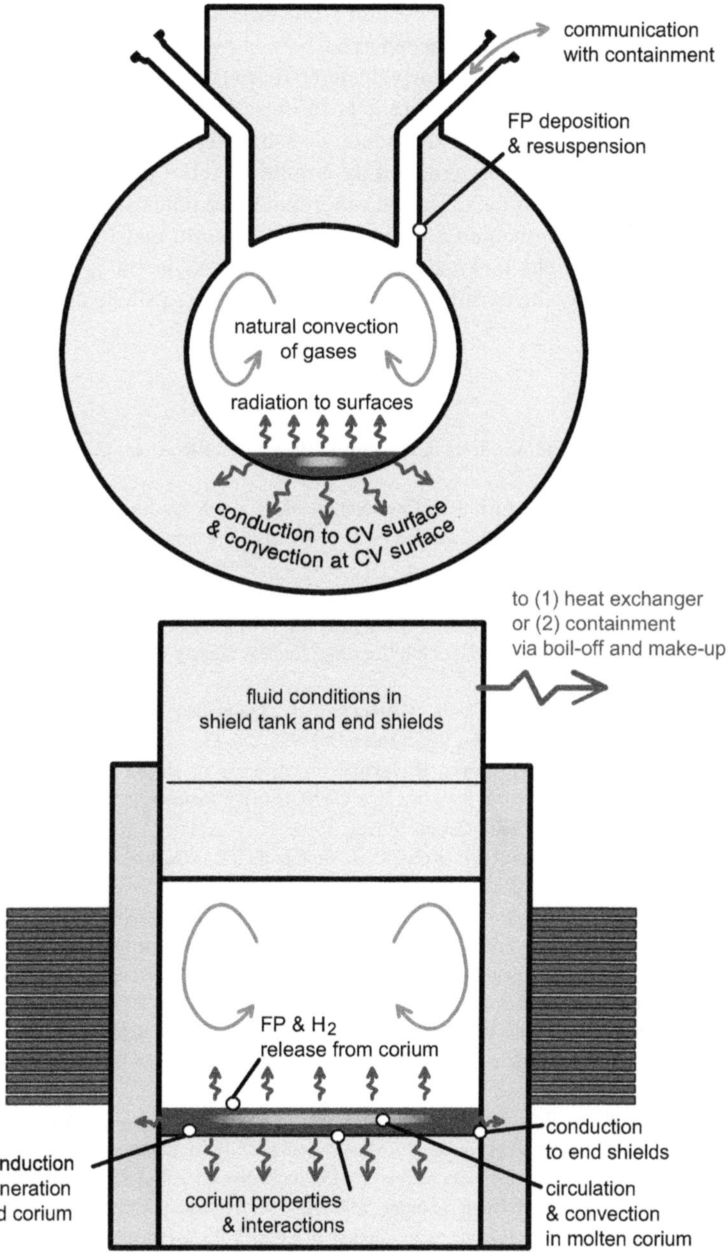

FIGURE 5 Corium retention phenomena in calandria vessel (conceptual—not to scale).

The relocation of molten corium from the calandria vessel to the shield tank/calandria vault will proceed relatively slowly. The melt pool near the corium surface is not particularly deep (so there is no significant head to drive the melt flow), and the pressure acts in opposition to the melt flow. These conditions can be expected to produce slow transfer rates and moderate steam surges that can be accommodated by combined relief paths of the calandria vessel and the shield tank. On the other hand, the potential for steam explosions arises as fragmented corium droplets can form quasi-stable slurries in water. If the shield tank/calandria vault survives the integrity challenges brought about by the corium interactions with water, a stable configuration is again reached.

REFERENCES

[1] Analysis of Severe Accidents in Pressurized Heavy Water Reactors, IAEA-TECDOC-1594, June 2008.

[2] P.S. Kundurpi, G.H. Archinoff, Development of Failure Maps for Integrity Assessment of Pressure Tubes, Proc. 7th Canadian Nuclear Society Conference, Toronto, Ontario, Canada, 1986.

[3] G.E. Gillespie, R.G. Moyer, G.I. Hadaller, J.G. Hildebrandt, An Experimental Investigation into the Development of Pressure Tube/Calandria Tube Contact and Associated Heat Transfer under LOCA Conditions, Proc. 6th Canadian Nuclear Society Conference, Ottawa, Ontario, Canada, 1985.

[4] R.S.W. Shewfelt, D.P. Godin, Ballooning of Thin-walled Tubes with Circumferential Temperature Variations, Res. Mechanica 18 (1986) 21–23.

[5] D.B. Sanderson, R.G. Moyer, R. Dutton, Effectiveness of the Moderator as a Heat Sink during a Loss-of-Coolant Accident in a CANDU-PHW Reactor, Proc. Int. Centre for Heat and Mass Transfer Conf., Cesme, Turkey, 1995.

[6] G.I. Hadaller, R. Sawala, E. Kohn, G.H. Archinoff, S.L. Wadsworth, Experiments Investigating the Thermal-Mechanical Behavior of CANDU Fuel under Severely Degraded Cooling, Proc. 5th Int. Meeting on Thermal Nuclear Reactor safety, Karlsruhe, Germany, 1984.

[7] S.L. Wadsworth, G.I. Hadaller, R.M. Sawala, E. Kohn, Experimental Investigation of CANDU Fuel Deformation during Severely Degraded Cooling, Proc. Int. ANS/ENS Topical Mtg on Thermal Reactor Safety, San Diego, California, USA, 1986.

[8] P.J. Mills, D.B. Sanderson, K.A. Haugen, G.G. Hacke, Twenty-eight-element Fuel-Channel Thermal-Chemical Experiments, Proc. 17th Annual Canadian Nuclear Society Conference, Ottawa, Ontario, Canada, 1996.

[9] P.M. Mathew, T. Nitheanandan, S.J. Bushby, Severe Core Damage Progression within a CANDU 6 Calandria Vessel, ERMSAR Seminar, Third European Review Meeting on Severe Accident research, September 23–25, 2008, Nesseber, Bulgaria.

[10] P.M. Mathew, WC.H. Kupferschmidt, V.G. Snell, M. Bonechi, CANDU-specific Severe Core Damage Accident Experiments in Support of Level 2 PSA, Proceedings of the 16th International Conference on Structural Mechanics in Reactor Technology, SMiRT-16 Conference, Washington, DC, August 12–17, 2001.

[11] P.M. Mathew, A.J. White, V.G. Snell, M. Bonechi, Severe Core Damage Experiments and Analysis for CANDU Applications, published in Structural Mechanics in Reactor Technology (SMiRT 17) Proceedings, Prague, August 17–22, 2003.

[12] P.M. Mathew, Phenomenology during the Progression of a Severe Core Damage Accident within a CANDU 6 Calandria Vessel, Proceedings of the AECL/IAEA/CNSC Technical Meeting on Severe Accident Analysis, Accident Management and PSA Applications for Pressurized Heavy Water Reactors, Mississauga, ON, November 10–13, 2008.
[13] J.T. Rogers, A Study of the Failure of the Moderator Cooling System in a Severe Accident Sequence in a CANDU Reactor, Proc. Fifth Int. Meeting on Thermal Nuclear reactor safety, KfK 3880/1 (1984).
[14] J.T. Rogers, Thermal and Hydraulic behavior of CANDU Cores under Severe Accident Conditions- Final report, vol. 1 & 2, Atomic Energy Control Board, 1984. INFO-0136.

Index

Note: Page numbers followed by *f* indicate figures and *t* indicate tables.

B

C

D

F

G

H

I

Q

R

S

T

U

Printed and bound by CPI Group (UK) Ltd, Croydon, CR0 4YY
11/04/2025
01843905-0003